A Bridge Between Conceptual Frameworks

HISTORY OF MECHANISM AND MACHINE SCIENCE
Volume 27

Series Editor
Marco Ceccarelli
Univ Cassino & South Latium Lab of Robotics & Mechatronics, Cassino, Italy

Aims and Scope of the Series

This book series aims to establish a well defined forum for Monographs and Proceedings on the History of Mechanism and Machine Science (MMS). The series publishes works that give an overview of the historical developments, from the earliest times up to and including the recent past, of MMS in all its technical aspects.

This technical approach is an essential characteristic of the series. By discussing technical details and formulations and even reformulating those in terms of modern formalisms the possibility is created not only to track the historical technical developments but also to use past experiences in technical teaching and research today. In order to do so, the emphasis must be on technical aspects rather than a purely historical focus, although the latter has its place too.

Furthermore, the series will consider the republication of out-of-print older works with English translation and comments.

The book series is intended to collect technical views on historical developments of the broad field of MMS in a unique frame that can be seen in its totality as an Encyclopaedia of the History of MMS but with the additional purpose of archiving and teaching the History of MMS. Therefore the book series is intended not only for researchers of the History of Engineering but also for professionals and students who are interested in obtaining a clear perspective of the past for their future technical works. The books will be written in general by engineers but not only for engineers.

Prospective authors and editors can contact the series editor, Professor M. Ceccarelli, about future publications within the series at:

LARM: Laboratory of Robotics and Mechatronics
DiMSAT–University of Cassino
Via Di Biasio 43, 03043 Cassino (Fr)
Italy
E-mail: ceccarelli@unicas.it

More information about this series at http://www.springer.com/series/7481

Raffaele Pisano
Editor

A Bridge Between Conceptual Frameworks

Sciences, Society and Technology Studies

Springer

ress, then it must be possible to write intellectually decent histories of it. It may be very important to do this. Establishing the *a priori* dogma that *any* such history must be Whiggish—that is, based on the assumption that progress is inevitable or, worse, mere propaganda on behalf of the endeavour—just ensures that no intellectually decent history of any progress-achieving endeavour will be written, an appalling impoverishment of what history should be.

Science is one of those rare human endeavours that does make progress across generations and centuries. It is vitally important that good, intellectually responsible histories of this progress-achieving endeavour of science are written. How is this to be done so as to avoid Whiggishness? There are a few very obvious points to make.

1. Do not assume progress is inevitable.
2. Do not write propaganda on behalf of science and scientific progress. Praise where praise is deserved, and criticize where criticisms need to be made. Do not conceal deplorable incidents—faking of results, plagiarism, petty disputes about priorities, immoral or criminal behaviour of scientists. Explore controversial issues about science and politics, science and war, science and the arms industry, science funding,
3. Do not just write about scientific success. In order to understand how and why scientific progress occurs it is absolutely essential to take into account the blind alleys, the research projects that led nowhere, the false leads, the ideas that turned out to be unproductive.
4. Do not hesitate to make judgements about how good or bad a piece of scientific research was. Do not assume, however, that scientific work is good if it turns out to be true, successful, or productive, and bad if it turns out to be false, unsuccessful or unproductive. Do not judge the intellectual merit of scientific work purely in terms of the contribution it ultimately makes to scientific progress. Brilliant scientific work may lead nowhere, and contributions that turn out subsequently to be important may come out of shoddy work, even out of mistakes.
5. In writing about past scientific episodes, try to see things from the actors' points of view so as to understand their problems, aims, ideas, theories, prejudices, standards, methods, as they saw them and experienced them. Seek to assess scientific work and contributions in terms of the standard prevalent at the time. But do not shrink from assessing the merit and significance of past work from the standpoint of the best standards and ideals available to us today, in an attempt to assess the significance of past contributions to overall scientific progress—where it is relevant to do this. Do not shrink from criticizing past work from the standpoint of our best current intellectual standards, should it be relevant to do this.
6. Keep in mind that what constitutes *progress* depends on what *aim* is presupposed. There are a range of aims that may be assigned to science, all more or less problematic (see below). Whether science as a whole, or a particular science, makes progress or not during a specific period may depend crucially on what *aim* for science is presupposed. Consider, for example, the aim of science of "improving human knowledge". This may be interpreted as (1) improving knowledge of scientific experts, or (2) improving knowledge of humanity as a whole. A science might make splendid progress given aim (1), but very little progress or none at all given aim (2).

Raffaele Pisano
Editor

A Bridge Between Conceptual Frameworks

Sciences, Society and Technology Studies

Springer

Editor
Raffaele Pisano
Department of Physics
Lille 1 University Science and Technology
Villeneuve d'Ascq
France

ISSN 1875-3442 ISSN 1875-3426 (electronic)
History of Mechanism and Machine Science
ISBN 978-94-017-9644-6 ISBN 978-94-017-9645-3 (eBook)
DOI 10.1007/978-94-017-9645-3

Library of Congress Control Number: 2015935742

Springer Dordrecht Heidelberg New York London

Printed on acid-free paper

Springer is part of Springer Science+Business Media (www.springer.com)

Foreword

Science Streams and Engineering Practices

This book deals with important issues linking Science concepts to Engineering practices. Therefore, it can be considered meaningful to both theoreticians and practitioners, as well as to scholars and students of the History of MMS (Mechanism and Machine Science) to which this book series is addressed.

The history of how the fundamentals of Science and its developments have been implemented to devise procedures in applications to problems with engineering solutions is an interesting part of the historical evolution of engineering expertise. Even more interesting is how those transmissions of knowledge were accepted and indeed used in fields of engineering related with machinery.

I have found the approach of this book to be very stimulating in its discovery of insights and implications that usually we engineers are not accustomed to considering.

I congratulate the editor Dr Pisano for his efforts and time required to gather so many contributors from so many different disciplines. He has attained a valuable result in a comprehensive survey of historical achievements in the bridging of Science, Society and Technology.

I strongly believe that a reader will get satisfaction and indeed inspiration for her/his work both on historical matters and/or technical practices, while reading and absorbing the contents of this book.

Enjoy the stimulating reading!

The Editor–in–Chief for History of MMS — Marco Ceccarelli

Prologue

What's Wrong with Science and Technology Studies? What Needs to Be Done to Put it Right?

Nicholas Maxwell
Science and Technology Studies, University College London, U.K.
E-Mail: nicholas.maxwell@ucl.ac.uk

High Aspirations of History and Philosophy of Science in the 1960s

I came to *Science and Technological Studies* (STS) by means of a rather circuitous route, via a passionate, childhood desire to understand the nature of the universe which, after reading Eddington, transformed into an obsession with mathematics which in turn, when adolescence struck, transformed into a desire to understand people via the novel—all of which I failed at dismally.[1] I then took up the study of philosophy in the early 60s at Manchester University. As a part of the undergraduate course, I was introduced to Oxford philosophy, which appalled me. It struck me as a species of anti-philosophy. I concentrated on philosophy of science. Philosophy might not matter, but clearly science does. Then, in the Summer of 1961 I had a revelation: philosophy ought to be, not about the meaning of words, but about how to live! The profound mystery is not even "What is the ultimate nature of the universe?" but rather "What is ultimately of value in life and how is it to be realized?." The problem with academic philosophy is that it is produced by academic philosophers who have already decided how to live, and have thereby lost all interest in real philosophy, which concerns what to do with our agonizingly brief time alive.

---

[1] See my (2012a, pp. 673–679).

I decided to do an MA at Manchester, say what needed to be said, and then escape from the madhouse of academic philosophy.[2]

And then I discovered the works of Karl Popper, and I became an occasional student at the LSE. Attending Popper's seminars, I was both immensely impressed and somewhat alarmed.[3] Here at last was a philosopher passionately concerned with profound, real problems of the real world which he tackled with fierce intellectual integrity and great originality. There was first his transformation of science—or at least his transformation of our conception of science. Laws and theories cannot be verified in science, but they can be empirically falsified, and that is how science makes progress. As a result of subjecting theories to fierce sustained attempted empirical refutation, we eventually discover where they go wrong, and are thus provoked into thinking up theories which do even better, until they are in turn refuted. Scientific knowledge is simply made up of our best, boldest imaginative guesses that have survived all our most ruthless attempts at empirical refutation.[4]

Then there was his generalization of this falsificationist conception of science to form a radically new conception of rationality. To be rational is to be critical. Just as science makes progress through subjecting our best conjectures to fierce attempted falsification, so more generally, in all areas of human life, we can best hope to make progress by subjecting our best attempts at solving our problems to fierce criticism. Empirical testing in science is just an especially severe form of criticism.[5]

The entire tradition of western philosophy had got it wrong. Scepticism is not the enemy to be vanquished—or to be indulged until it can go no further, thus revealing a bedrock of certainty, as with Descartes, and many empiricists. Quite the contrary, scepticism is our friend, the very soul of reason. It is by means of imagination subjected to sustained, ferocious scepticism that we can learn, and make progress. Science is institutionalized scepticism.

What impressed me most, however, was the application of these ideas to the profound problem of creating civilization or, as Popper called it, "the open society". Rationality is the critical attitude. But this is only really possible in an "open" society, a society, that is, which tolerates a diversity of views, values and ways of life. In a "closed" society, in which there is just one view of things, one set of values, one way of life, there can be no possibility of criticism, since to criticize A we need, at least as a possibility, some alternative view B. Thus the rational society is the open society—not a society enslaved to some monolithic, dictatorial notion of "reason", but simply a liberal society that tolerates and sustains diversity of views, values and

[2] See my (2012a, pp. 679–688). See also my (2009a).

[3] Popper could be ferociously critical in his seminars. Rarely did the speaker get past the title before Popper's attack began. Once he reduced a young visiting speaker—now a well known philosopher of science—to tears.

[4] See Popper (1959, 1963).

[5] "inter-subjective *testing* is merely a very important aspect of inter-subjective *criticism*, or in other words, of the idea of mutual rational control by critical discussion." Popper (1959, p. 44, note 1*). Popper refers the reader to his (1969, Chaps. 23 and 24)—first published in 1945.

ways of life, and can, as a result, learn, make progress, and even create and pursue science.[6]

But the move from the closed to the open society has a severe penalty associated with it. We move from certainty to doubt. Living in the open society requires that we shoulder the adult responsibility of living in a state of uncertainty, of doubt. Everything we believe, everything we hold most dear, and value—the very meaning and value of our whole way of life—may be wrong or misconceived. Doubt is the price we pay for civilization, for reason, for humanity, and for science. In his masterpiece *The Open Society and Its Enemies*, Popper calls this essential doubt "the strain of civilization", and he points out that all too many people cannot bear it, and seek to return to the false certainties of the closed society. Even some of our greatest thinkers have sought to do this, and they are the enemies of the open society—above all, for Popper, Plato and Marx.[7]

I breathed a great sigh of relief. Popper had, it seemed, solved the problems that had so tormented me. The anguish of the twentieth century—the nightmare of not knowing how to live with only a few measly decades available to try to find out—had been explicated as being due to our new exposure to global society and to history: exposure to a multitude of contradictory beliefs, values and ways of life which, inevitably, had the effect of throwing into doubt the validity of one's own entire way of life and set of values.

Popper demonstrated, it seemed to me, that it was possible to be an academic philosopher and yet retain one's intellectual integrity.[8] I moved down to London and got a job as lecturer in philosophy of science in the Department of History of Philosophy of Science at University College London. Larry Laudan and Paul Feyerabend were among my departmental colleagues.

It was an exciting time and place to be doing history and philosophy of science (HPS). London felt like the HPS capital of the world. HPS seemed to be a fledgling academic discipline, having associated with it all the excitement, freshness, high aspirations and optimism of a new discipline. There was the idea that each wing needed the other: history of science would be blind without philosophy of science, which in turn would be empty without history of science. Natural science seemed to be the one great human endeavour that undeniably made progress across generations and centuries. Aside from mathematics, in no other sphere of human endeavour did this happen—not in art, music, literature, politics, or morality. There was technological progress, certainly, and economic progress too, but these were closely linked to, and dependent on, scientific progress. It was the great task of HPS to work out how science did make progress, and what might be learned from scientific progress about how to make progress in other areas of human life: art, literature, law, education, politics, economics, international relations, personal flourishing and fulfilment. Popper had shown the way. But he could hardly be the last word on the subject. Popper's philosophy needed to be applied to itself, and

[6] See Popper (1969).

[7] As in note 6.

[8] See my (2012a, pp. 688–699).

subjected to sustained critical scrutiny in an attempt to improve on it. And there were plenty of contending ideas around. There was Thomas Kuhn's *The Structure of Scientific Revolutions*, which in part agreed with Popper in stressing the existence and likelihood of scientific revolutions, but also violently disagreed with Popper in holding that the dogmatic puzzle solving of normal science was an essential and desirable aspect of science as well.[9] Popper, outraged, called normal science "a danger to science and, indeed, to our civilization"[10] (which makes perfect sense, of course, given his viewpoint). Then there was Imre Lakatos's attempted resolution of Popper and Kuhn in his "Methodology of Scientific Research Programmes" which acknowledged that research programmes have a "hard core" (Kuhn's "paradigm" under another name), and legitimately get pursued with a degree of dogmatism.[11] And there was Paul Feyerabend, who went one further than Popper, and argued, in effect, that the plurality of views of the open society would need to be imported into science itself. Severe testing—essential, according to Popper, for empirical scrutiny of theories—requires at least the germ of an alternative theoretical idea. We need actively to develop alternative theories simply to be in a position to test severely the reigning, accepted theory—almost exactly the opposite of what goes on, according to Kuhn, during a period of normal science.[12]

Beginnings of the Decline of HPS

I am now going to tell the tale of the sad decline of HPS into confusion, irrationality and irrelevance. But before I do so, I want to stress that good work has been done and continues to be done in both history and philosophy of science despite the fashionable stupidities of both disciplines.[13] My complaint is that those who study science and technology—philosophers, historians, sociologists and others—could have done so much better during the period under consideration, the mid 1960s up to 2013. Much energy has been expended on idiotic disputes and urgent and fundamental problems, of great importance for science, and for humanity, have been ignored. HPS lost its way.

There are, on the one hand, those sociologists and historians of science—and a few philosophers—who stress the importance of attending to the social dimension of science but, disastrously, abandon such ideas as that science makes progress, acquires authentic knowledge about the world, improves knowledge of fact and truth, and embodies rationality, and puts progress-achieving methods into scientific practice. On the other hand there are some scientists, and some philosophers and historians of science who defend orthodox conceptions of science against these

[9] Kuhn (1962).

[10] Popper (1970, p. 53).

[11] Lakatos (1970).

[12] Feyerabend (1965).

[13] To cite just one recent book in the field that I find very impressive: Harper (2011).

sociological, anti-rationalist attacks. I must make it very clear, at the outset, that I am critical of both wings of this dispute. The dispute itself—the "science wars" as the dispute came to be called—is the wrong argument to engage in. It is a symptom of the decline in the high aspirations of HPS in the 1960s. It is a distraction from what really needs to be done: to get the scientific community to acknowledge the real, and highly problematic aims of science which have, inherent in them, highly problematic assumptions concerning metaphysics, values and politics. It is here that really dramatic and enormously fruitful developments are to be made—as I shall try to indicate towards the end of this essay. If those who study science had combined with sympathetic scientists to create greater honesty about the problematic aims of science among the scientific community, we might have today a different kind of science, more intellectually rigorous and of greater human value. We might even have a different kind of academic inquiry, rationally devoted to helping humanity create a wiser world. We might even have a different, wiser world—as I will try to explain in what follows. But first I must tell the sad story of decline.

Somewhat arbitrarily, we may begin with a dreadful blunder made by Feyerabend. On Popper's behalf, he assailed the logical empiricists, Hempel, Carnap and Nagel, for holding that meaning had to be transported up from evidence to theory.[14] No, Feyerabend argued, that was not possible, for observational terms are "theory laden", so that conflicting theories would have conflicting, or at any rate different, observation terms, conflicting or different accounts of observational phenomena. There can be no such thing, Feyerabend argued, as a stable observational language independent of theory (an argument to be found in Kuhn as well). But logical empiricism depends utterly on there being just such a theory-independent observation language. The whole position takes it for granted. Its non-existence destroys logical empiricism completely. Its foundations do not exist! So far, so so good.[15] But then Feyerabend made an idiotic mistake. If meaning cannot be transported up, from observation to theory (because a theory independent observation language does not

[14] Logical positivism held that the meaning of a proposition, or theory, is the method of its verification. The idea was to render scientific theories meaningful, but metaphysics, which cannot be verified empirically, meaningless. This failed for the simple reason that scientific theories cannot be verified. So logical positivism morphed into the very much weaker doctrine of logical empiricism which held that theoretical terms acquire their meaning as a result of being linked to observational terms by means of bridge statements. It was this doctrine that Feyerabend set out to demolish.

[15] Not really very good, of course, for even if observational terms are theory-laden, nevertheless given any two conflicting theories ostensibly about the same, or overlapping, ranges of phenomena, one can always concoct observational terms that are such that the theory presupposed by them is neutral between the two theories: see my (2014a). That this can always be done means that empirical predictions of conflicting theories about overlapping phenomena can always be assessed in terms of these phenomena described by means of terms that presuppose low-level theory that is neutral between the conflicting theories in question. Feyerabend's argument for incommensurability, methodological anarchy and dadaism collapses completely. I did my best to point this out to Feyerabend in person, but he was having none of it. And nor was Kuhn when I tried to point out that his argument for incommensurability rested on the same fallacy. The problem was solved long ago by Michael Faraday in scientific practice in connection with his work on electrolysis. How extraordinary that, over a century later, two leading philosophers of science could not grasp what Faraday had understood long ago: see my (2014a).

exist), then meaning must be transported down, from theory to observation terms. But this means in turn, Feyerabend argued, that conflicting theories, with different theoretical terms, must have different observational terms as well, which in turn means that the predictions of the conflicting theories cannot be compared. And so the very basis for Popper's philosophy of science—his falsificationism—collapses.[16] Not just logical empiricism, but falsificationism too must be thrown on the rubbish dump of history. Scientists should follow their instincts, Feyerabend concludes. Anything goes. Methodological anarchy reigns supreme. There is no such thing as the rationality of science. It is irrational. And it is damaged when it attempts to conform to some misguided idea of rationality dreamed up by a philosopher of science.[17]

Feyerabend had an absolutely disastrous influence. He became a sort of approved intellectual court jester. All those who deplored what they perhaps saw as the illegitimate mighty authority of science were entranced by Feyerabend's annihilation of science's claim to be rational and methodological, upon which its mighty authority rested. The emperor had no clothes. Feyerabend had stripped science bare. Or so it seemed to all too many.

HPS began to take an absolutely disastrous turn for the worse. The initial great ambitions and optimism of the fledgling discipline were lost sight of. HPS began to tear itself to pieces in an orgy of stupidity, like a political party thrown out of power, or a political movement with no hope of ever gaining power. It came in wave after wave of idiocy.

At about the same time as Feyerabend began to drum up support for relativism and unreason, a very different kind of disastrous stupidity was being incubated in Edinburgh. It was called "the strong programme", and its authors were Barry Barnes and David Bloor.[18] They argued that science is social in character, and therefore needs to be studied by sociologists. This means, they held, that there is no such thing as scientific truth, knowledge, rationality or progress. There is just change of scientific belief, as science goes on its way. Traditionally it has been held that science is rational, its theories being established by evidence, science being entitled to claim it acquires genuine knowledge of factual truth, science thus progressively increasing and improving our knowledge and understanding of the universe. But all this has been shown to be untenable—by Kuhn, Feyerabend and others. Those philosophers of science who do, absurdly, still claim that science makes progress, is rational, and acquires genuine knowledge of factual truth, are unable to say how this is done. The problem of induction remains unsolved. Even Popper, who almost alone does claim to have solved the problem, has not really solved it. So science must be treated as social in character, purely *social* factors determining what is accepted and rejected in science—namely observational and experimental results, laws and theories. It is the sociologist of science, not the philosopher of

[16] Feyerabend (1970).

[17] Feyerabend (1975, 1978, 1987).

[18] Harry Collins, John Henry and others were, and still are (at the time of writing) associated with the movement.

science, who can improve knowledge about science, how it proceeds, and modifies its beliefs, its "scientific myths" one might say. Truth, fact, knowledge, scientific progress, method and reason all fly out of the window. These are fantasy ideas of old fashioned philosophy of science, illusory notions that have nothing to do with science as it really is, an integral part of society, social through and through.[19]

At about the time "the strong programme" was being launched on the world, *The British Society for the Philosophy of Science* held its annual conference in Edinburgh, and naturally the Edinburgh school was given its chance to air its ideas. I remember thinking at the time that ideas as foolish as these would never get anywhere. How wrong I was. I also remember wondering why proponents of "the strong programme" had not bothered to read Popper, for in *The Open Society and Its Enemies* Popper anticipated and decisively dealt with and obviated the need for this sociological programme.[20]

Popper makes the point that rationality—critical rationality, that is—is essentially social in character, in that criticism requires diversity of views (as we have seen) and so many people in communication to hold and discuss these diverse views. Furthermore, science is fundamentally social in character too, and owes its rationality, its scientific character, to its social character. Far from the social character of science somehow cancelling the scientific character of science, as proponents of "the strong programme" seemed to believe, it is all the other way round: the scientific character of science actually requires science to be inherently social.

Furthermore, what *methods* are implemented in scientific/social practice may well, quite obviously have an immense impact on whether science meets with success in improving knowledge about the world. Compare M_1: "accept theories that are empirically refuted, and reject theories that are empirically confirmed" with M_2: "accept the best explanatory theories that are empirically confirmed, and reject theories that are decisively refuted". We would all agree that a community of scientists that puts M_2 into social/scientific practice is more likely to meet with success and improve knowledge than one that puts M_1 into practice. It is, in short, utterly trivially obvious that what methods are implemented in social/scientific practice may well make a profound difference to the intellectual success or failure of science—its success in acquiring knowledge about the world.

What *methods* science puts into practice is a vital part of the whole social structure of science which the sociological study of science cannot possibly ignore if it is to be remotely adequate. Both Popper and Kuhn are very good, in their different ways, in pointing out that what matters are the methods that are implicit in scientific practice.[21]

Construing science to be a social endeavour thus does not obviate the intellectual or rational character of science, and certainly does not do away with crucial questions about what methods science does, and ought to, adopt and implement. It

19 Bloor (1976); Barnes (1977, 1982, 1985); Barnes, Bloor and Henry (1996).

20 Popper (1969, vol. 2, Chaps. 23 and 24).

21 See previous note, and Popper (1974, section 32, 'the institutional theory of progress', pp. 152–159). See also Kuhn on normal science: Kuhn (1962, Chaps. III–V).

does not mean that science does not acquire genuine factual knowledge, and make progress.

Furthermore, science in particular, and our social world more generally, is imbued with values, whether intellectual, moral, legal, or aesthetic, some better than others. It certainly ought to be a part of the professional job of academics to try to discriminate between good and bad intellectual values, and promote the former. Sociologists of science, like scientists themselves, philosophers of science and all other academics, ought to do what they can, in their professional work, to promote good intellectual values—ones having to do with rationality, validity, the successful pursuit of knowledge of fact and truth—at the very least.

"The Strong Programme" is a kind of acid which eats all these things away, and leaves science as a value-denuded, knowledge-denuded, truth and reason denuded, empty social practice. But all this arises from elementary and appalling misunderstandings about the nature of our social world in general, and that bit of it that is science, in particular—a refusal at the outset to see that values and standards, whether intellectual or humanitarian, are essential features of our social world. To exclude all values from the social world *a priori*, as it were, is to adopt something close to a psychopath's vision of things. Ironically, it probably comes from the unconscious adoption of a very crude philosophy science which says values have no place in science, and hence no place in sociology, or the sociology of science either. (I say "ironically" because, according to the proponents of "the strong programme", philosophy of science is a sort of irrelevant fantasy.)

It is as if proponents of "the strong programme" had convinced themselves of the correctness of the following argument.

1. Reason, validity, valid scientific methods, truth, fact, knowledge, scientific progress are all inherently purely *intellectual*.
2. The *intellectual* is not *social* (and no part of the *social* is *intellectual*).
3. But science is wholly and purely *social*.
4. Hence science is wholly free of the *intellectual*. It has nothing to do with reason, validity, valid scientific methods, truth, fact, knowledge, scientific progress.

The argument may be valid, but step 2 is false. The intellectual is wholly social in character. That makes step 4 false as well. As I have said, one cannot begin to do justice to the character of our social world if one refuses, at the outset, to acknowledge that the social is quite essentially imbued with values of all kinds, intellectual, moral, legal, aesthetic—imbued not just with values but with *what is of value*.[22]

Social Constructivism and Anti-Whiggism

I have so far concentrated on the damage done to HPS by Feyerabend's methodological anarchism and the blunders of "the strong programme". But damage came from another source as well: French philosophy, Foucault, Derrida and others. The

[22] For value realism see my (1984 or 2007a, Chap. 10; 1999, 2001, Chap. 2).

upshot was a whole new way of construing science, which may be called "social constructivism". This is the view indicated above that I have attributed to "the strong programme". Scientific knowledge is merely a social construct, having nothing to do with knowledge, truth and falsity, or reason. In studying science and its history, we must entirely forego the idea that science makes progress, and we must refrain from making intellectual or scientific judgements about one theory being "better", "truer", or "more firmly established" than another. In the main sociologists and historians took to social constructivism, while philosophers of science looked on in amazement and horror, at the idiocy of it. As a result, HPS broke asunder. The integrated enterprise, bringing together history and philosophy of science, which had started out with such high hopes and aspirations, and which was still alive and kicking when I began my academic career around 1965, was no more.

An even more devastating consequence, perhaps, of the widespread adoption of social constructivism among historians of science was that it annihilated the fundamental problem of the discipline. As I stressed at the beginning of this essay, science is almost unique among human endeavours in that it makes genuine progress. We know and understand vastly more about the universe, and ourselves as a part of the universe, than was known to Darwin, to Faraday, to Newton, or to Aristotle. The fundamental problem of HPS is: How has scientific progress come about? And for philosophy of science in particular: How is scientific progress possible? What methods have brought it about? What methods give the best hope of progress?

Social constructivism annihilates these fundamental problems. What ought to be the central problem of the history of science just disappears from view. This is perhaps the strongest indication of the intellectual poverty and destructive character of social constructivism.

Where did this idea that science does not make progress come from? In addition to the intellectual blunders that I have already indicated, it came from a blunder about history. The historian Herbert Butterfield wrote a little book against what he called "Whiggish history".[23] This is history that takes for granted that progress, the spread of enlightenment, democracy and justice are inevitable, and it is the job of the historian to describe this process. An even cruder kind of Whiggish history would have built into it dogmatic assumptions about what does constitute progress, history being written as propaganda to help the process along, or fool the reader into believing that progress in this sense has occurred and is occurring when nothing of the kind is the case.

Whiggish history in these senses is intellectually disreputable. It is, however, utterly absurd to think that this means historians can't ever write histories of any human endeavour whatsoever that does in fact make progress towards some goal, or seeks to make progress towards some goal. That is, clearly, an absurd position to adopt. If there is a human endeavour that makes progress, or seeks to make prog-

[23] Butterfield (1951). Butterfield seems to believe that ideally history would be about everything. He says at one point "The value of history lies in the richness of its recovery of the concrete life of the past" (1951, p. 68). He ignores that history is always about something specific—power, the black death, the potato, or whatever—and may quite legitimately be about something that seeks, and even perhaps achieves, progress.

ress, then it must be possible to write intellectually decent histories of it. It may be very important to do this. Establishing the *a priori* dogma that *any* such history must be Whiggish—that is, based on the assumption that progress is inevitable or, worse, mere propaganda on behalf of the endeavour—just ensures that no intellectually decent history of any progress-achieving endeavour will be written, an appalling impoverishment of what history should be.

Science is one of those rare human endeavours that does make progress across generations and centuries. It is vitally important that good, intellectually responsible histories of this progress-achieving endeavour of science are written. How is this to be done so as to avoid Whiggishness? There are a few very obvious points to make.

1. Do not assume progress is inevitable.
2. Do not write propaganda on behalf of science and scientific progress. Praise where praise is deserved, and criticize where criticisms need to be made. Do not conceal deplorable incidents—faking of results, plagiarism, petty disputes about priorities, immoral or criminal behaviour of scientists. Explore controversial issues about science and politics, science and war, science and the arms industry, science funding,
3. Do not just write about scientific success. In order to understand how and why scientific progress occurs it is absolutely essential to take into account the blind alleys, the research projects that led nowhere, the false leads, the ideas that turned out to be unproductive.
4. Do not hesitate to make judgements about how good or bad a piece of scientific research was. Do not assume, however, that scientific work is good if it turns out to be true, successful, or productive, and bad if it turns out to be false, unsuccessful or unproductive. Do not judge the intellectual merit of scientific work purely in terms of the contribution it ultimately makes to scientific progress. Brilliant scientific work may lead nowhere, and contributions that turn out subsequently to be important may come out of shoddy work, even out of mistakes.
5. In writing about past scientific episodes, try to see things from the actors' points of view so as to understand their problems, aims, ideas, theories, prejudices, standards, methods, as they saw them and experienced them. Seek to assess scientific work and contributions in terms of the standard prevalent at the time. But do not shrink from assessing the merit and significance of past work from the standpoint of the best standards and ideals available to us today, in an attempt to assess the significance of past contributions to overall scientific progress—where it is relevant to do this. Do not shrink from criticizing past work from the standpoint of our best current intellectual standards, should it be relevant to do this.
6. Keep in mind that what constitutes *progress* depends on what *aim* is presupposed. There are a range of aims that may be assigned to science, all more or less problematic (see below). Whether science as a whole, or a particular science, makes progress or not during a specific period may depend crucially on what *aim* for science is presupposed. Consider, for example, the aim of science of "improving human knowledge". This may be interpreted as (1) improving knowledge of scientific experts, or (2) improving knowledge of humanity as a whole. A science might make splendid progress given aim (1), but very little progress or none at all given aim (2).

7. Take into account that, in so far as scientific knowledge is conjectural in character, judgements about scientific progress will be conjectural too. Thus the historian's judgements as to whether scientific progress has taken place, what it consists in, and how it was achieved, will be conjectural as well, and may be falsified or at least modified as current scientific knowledge is modified. This is of course more likely to happen to history of recent scientific developments than it is to history of scientific developments a century or so ago.
8. Far from it being assumed at the outset by a history of a progress-seeking endeavour, whether scientific or not, that progress occurs (let alone is inevitable), such a history should be open-minded about the matter. Whether progress has been made, of what type, towards what goal, and of what mixture of advance and regression, are all questions open for historical research to discover. It might indeed emerge that no progress has been made, or that the opposite has happened, and the endeavour has regressed. (Perhaps this is the case as far as HPS itself is concerned.)
9. Make no *a priori* judgements about whether *intellectual* or (non-intellectual) *social* factors influenced some specific piece of scientific work.[24] Much that scientists do is probably influenced by a complicated mixture of these factors. Thus the decision to work on a specific scientific problem may be influenced by (1) curiosity, (2) availability of funds, (3) the guess that the problem will be easily solved, and will thus enhance career prospects, (4) the hunch that it will turn out to be important to solve from the standpoint of social applications (medical, industrial, etc.), (5) a request from the scientist in charge of a scientific team, (6) the presence in the laboratory of relevant equipment. Are any of these considerations wholly "intellectual" or wholly "social"?
10. Science is a human endeavour different for the historian from other, non-intellectual endeavours—even endeavours that also make progress. In the case of science, what the historian studies, and the discipline of history itself, have some common goals: to improve human knowledge and understanding. This means that in the particular case of science, it may well be legitimate for the historian to write history which seeks to help promote the very thing he is writing about. The historian of science may quite legitimately seek to highlight neglected work from the past that may, if better known, have important implications for the future of the science in question.[25] This kind of science-promoting history can be done in a thoroughly intellectually responsible way even though, if done about other kinds of endeavour, it might well amount to no more than a kind of propaganda for the endeavour in question. Serious history of science of this kind should not, however, degenerate into the simplified,

[24] Intellectual factors at work in science are social in character. We can thus distinguish two kinds of social factors influencing science: the intellectual, and the non-intellectual.

[25] I have attempted something along these lines in my (2010, Chap. 10, especially pp. 276–289). Following Alister Hardy's lead in his (1965), I call upon some neglected and misrepresented history of evolutionary thought to provide support for an interpretation of Darwinian evolution which gives an increasingly important role to purposive action in evolution, and which holds that the mechanisms of evolution themselves evolve.

distorted, potted history that scientists tell their students for pedagogic purposes. The all-important point, furthermore, is that history of science does not have to be science-promoting, in this way, as the above points, 1–9, indicate.

As long as these and similar strictures are kept in mind and observed, there is no reason whatsoever why histories of science that depict science as making progress should not be done that meet the highest standards of intellectual excellence, there being not the faintest whiff of Whiggishness in any of the bad senses.

It is quite extraordinary that so many historians of science have been unable for decades to draw the distinction between "Whiggish history" in the bad senses, and "history of some endeavour that makes progress" that is intellectually responsible and excellent. It is all the more extraordinary, when one considers that the failure to draw this obvious distinction has meant that, for these historians, the fundamental problem of the history of science, "How and why has scientific progress come about?" has died, and disappeared entirely from view. It is as if cosmologists managed to reach the conviction that there is no such thing as the cosmos, or biologists convinced themselves that there is no such thing as life on earth. Intellectual history is turned into mere social gossip.

I encountered the consequences of these elementary intellectual blunders in my professional life as a lecturer in philosophy of science in the Department of History and Philosophy of Science at University College London. We taught a joint MSc Programme with the Wellcome Institute, with Bill Bynum, Chris Lawrence and Mike Neve—these latter all firmly committed to social constructivism in the history of medicine. Students were baffled. At the Wellcome Institute they learnt there is no such thing as scientific progress, rationality, truth or knowledge. In my lectures they heard that there is a fundamental problem concerning the rationality of science—a big, serious, unsolved problem about how it is that science acquires knowledge of truth and makes progress. How to choose between holding that truth, knowledge, progress and reason are of fundamental importance, and holding that there are no such things at all? In the end most shut their eyes and made a Kierkegaardian leap of faith into one or other position. I pleaded with Bynum, Lawrence and Neve to hold a seminar with me and the students to discuss these issues. They refused. One year I did persuade one historian, Rob Iliffe, to take part in such a discussion of the issues, but only if it was informal, after hours as it were, and with beer to drink. He pointed out how bad it is just to assume dogmatically that science makes progress when there is much to criticize in modern science. I replied that if rationality is abandoned, the very possibility of being critical of modern science is abandoned too, for criticism presupposes and requires rationality. Iliffe had no answer. In the end he was reduced to arguing that he had to go along with social constructivism in order to get an academic job as a historian of science.

One bizarre feature of social constructivism is that its proponents are often left wing and highly critical of aspects of modern science. But of course as a result of abandoning rationality, the very possibility of criticism disappears. My attempts to point this out to proponents of "anti-Whiggism" over the years invariably fell upon deaf ears.

Sometime in the 1970 and 1980's a new branch of HPS emerged which came to be called *Science and Technology Studies* (STS). This emerged out of the sociology of science, out of a concern to give far great emphasis to technology and the technological sciences, and out of a concern to tackle issues associated with science and society—the impact of science on society, and *vice versa*. From the outset, much of the potential inherent in STS has however been subverted by the influence of ideas stemming from "the strong programme", social constructivism, "anti-Whiggism", and anti-rationalism.

There has been a tendency too for Philosophy of Science to degenerate into a kind of scholasticism in that it has splintered into a multitude of specialized disciplines: philosophies of the specialized sciences—physics, chemistry, neuroscience, astronomy, botany, and so on. As a result, Philosophy of Science has rather lost sight of the magnificent endeavour of natural science as a whole, and has come to ignore the great, fundamental problems that were, initially, the whole raison d'être for its existence: the problem of induction, the problem of the rationality of science, the problem of how, by what means, science makes progress.

Alan Sokal's Hoax and The Science Wars

In 1996 the worst excesses of the social constructivists and anti-rationalists were brilliantly satirized by a spoof article by Alan Sokal called "Transgressing the Boundaries; Toward a Transformative Hermeneutics of Quantum Gravity".[26] This was published in an American academic journal called *Social Texts*, the editors of which took the paper to be a serious academic contribution. Actually, it was a tissue of hilarious nonsense decked out with liberal quotations from the constructivists—although Sokal admitted subsequently that, despite considerable effort, he did not always succeed in the article in attaining the dense obscurity of what he satirized. One of the editors was interviewed on the BBC, on the Today Programme, and made the dreadful mistake of protesting at the immorality of Sokal's hoax, instead of laughing and admitting that they had been had.

Around this time, and partly in response to Sokal's hoax, the "science wars" exploded onto the scene, some scientists and philosophers of science springing to the defence of science against the corrosive acid of social constructivism, anti-rationalism and postmodernism. Paul Gross and Norman Levitt wrote a book assailing the worst excesses of postmodernist writing about science, and subsequently edited a book that continued the argument.[27] Alan Sokal and Jean Bricmont outraged French intellectuals with devastating criticisms of French philosophers' writings about science: Jacques Lacan, Luce Irigaray, Bruno Latour, Gilles Deleuze and

[26] Sokal (1998). See also Sokal (2008) for an annotated version of the hoax article, and essays on related matters.

[27] Gross and Levitt (1994); Gross, Levitt and Lewis (1996).

others.[28] Noretta Koertge edited *A House Built on Sand: Exposing Postmodernist Myths About Science*.[29] Others joined the affray. Social constructivists protested that distinctions were being ignored, contexts overlooked.

Did this counter attack on behalf of orthodox conceptions of science win the day, and rid STS of anti-rationalist views? No. They continued to be influential, but in perhaps a slightly muted way. Here is just one fairly recent example of this influence, and how damaging it can be.

In 2009 a young practitioner of STS, Sergio Sismondo, gave a good lecture on the scandal of medical "ghost writing" in my very Department of STS at UCL. "Ghost writing" is the process whereby a drug company writes a paper specifically designed to be published in a particular medical journal, in terms of such spurious features as layout, references, etc. The paper praises a new drug the firm has produced, and then gets an academic who is an acknowledged authority in the field to author the paper, even though he or she has not seen data from trials, in particular data about harmful side effects. The paper is duly published, and what is essentially an advertisement is treated by GPs and other researchers as if it is a genuine contribution to scientific knowledge.

When the talk was over, I made the point, dressed up as a question, that such a contribution could not be regarded as an authentic contribution to knowledge. The deception might well lead to deaths—as happened in connection with Vioxx. No, Sismondo responded, such a paper did constitute a contribution to scientific knowledge because it had satisfied all the criteria for publication of the journal in question—and there could be no question about some of these criteria being epistemologically irrelevant. Social constructivist habits of thought had rendered Sismondo incapable of acknowledging the full extent of the scandal, even the criminality, his talk was about.[30]

Social constructivists and their sympathizers are absolutely right to stress the vital importance of taking social aspects of science and technological research into account. The way this has been done, however, has been an intellectual disaster. It has helped sabotage urgently needed developments in thinking about science which would have brought together scientific and social thinking in sensible, rational and fruitful ways, as I shall try to show in a moment. The point that there is no agreed solution to the problem of induction, the problem of the rationality of science, is absolutely correct. The solution is, however, waiting in the wings to be taken note of. And this solution leads on to a profound transformation in the way we think about the aims of science, science itself, and academic inquiry as a whole, more generally.

---

[28] Sokal and Bricmont (1998).

[29] Koertge (1998).

[30] See Sismondo (2009a). For a criticism see McHenry (2009b); and for a reply see Sismondo (2009b).

Metaphysics, Values and Politics Inherent in the Aims of Science

Most scientists and philosophers of science take for granted one or other version of a view of science that I have called *standard empiricism* (SE). This holds that the basic intellectual aim of science is factual truth (nothing being presupposed about the truth), the basic method being to assess claims to knowledge impartially with respect to evidence. Considerations such as the simplicity, unity or explanatory character of a theory may influence what theory is accepted, but not in such a way that the universe or the phenomena are permanently assumed to be simple, unified or comprehensible. According to SE, what theory is accepted may even be influenced for a time in science by some paradigm or metaphysical "hard core" in the kind of way depicted by Kuhn and Lakatos[31] as long as, in the end, empirical success and failure are the decisive factors in determining what theories are accepted and rejected. The decisive tenet of SE is that *no substantial thesis about the nature of the universe can be accepted as a permanent part of scientific knowledge independently of empirical considerations* (let alone in violation of empirical considerations).

Even those who—like Feyerabend, social constructivists and postmodernists—reject the whole idea that science is rational, delivers authentic knowledge, and makes progress, nevertheless tend, in a way, to uphold some version of SE as the only possible rationalist conception of science. No rational account of science is possible, they hold in effect, because the only candidate, SE, is untenable (as shown by the failure of SE to solve the problem of induction).

Despite being almost universally taken for granted by scientists, SE is nevertheless untenable. SE very seriously misrepresents the aims of science. The intellectual aim of science is not to improve knowledge of factual truth, nothing being presupposed about the truth. On the contrary, science cannot proceed without making a very substantial and highly problematic *metaphysical* hypothesis about the nature of the universe: it is such that some kind of unified pattern of physical law governs all natural phenomena. Science seeks, not truth per se, but rather *explanatory* truth—truth presupposed to be explanatory. More generally, science seeks *valuable* truth—truth that is of intrinsic interest in some way or useful. This aim is, if anything, even more problematic. And science seeks knowledge of valuable truth so that it can be used in social life, ideally so as to enhance the quality of human life. There are, in other words, problematic *humanitarian* or *political* assumptions inherent in the aims of science. In holding that the basic intellectual aim of science is *truth per se*, the orthodox position of SE misrepresents the real and highly problematic aims of science.

The vital task that needs to be done to develop STS in fruitful directions—a task not performed because of the influential absurdities of "the strong programme", social constructivism and the science wars debate—is to give absolute priority to two fundamental questions: What are the real aims of science? What ought they to

[31] Kuhn (1962); Lakatos (1970).

be? Ever since around 1970, when I began to consider these questions, those associated with HPS and STS ought to have put these two questions at the heart of science studies. If this had been done, science studies, in conjunction with sympathetic scientists, science journalists and others, might have helped develop a conception of science, and even a kind of science, both more rigorous and of greater human value than what we have today. Indeed, a new kind of academic inquiry might have emerged that is rationally devoted to helping humanity make social progress towards as good a world as possible. We might even have begun to see the beginning of a new kind world capable of tackling its immense global problems in increasingly effective and cooperatively rational ways. None of this has come about because the academic disciplines most directly responsible for helping to initiate these developments, HPS and STS, have been distracted by intellectual stupidities.

The key step that needs to be taken to permit these urgently needed intellectual, institutional and humanitarian developments to unfold is the widespread recognition that standard empiricism (SE) is indeed untenable, and needs to be replaced by something better. So, let us see why SE is untenable.

As it happens, reasons for rejecting SE have been spelled out in the literature again and again, ever since 1974.[32] But these refutations of SE have been ignored. In outline, the refutation goes like this.

Theoretical physics persistently only ever accepts *unified* theories—theories that attribute the same dynamical laws to the phenomena to which the theory applies. Given any such accepted theory—Newtonian theory, classical electrodynamics, quantum theory, general relativity, quantum electrodynamics, or the standard model—endlessly many disunified rivals can be easily concocted to fit the available phenomena even better that the accepted unified theory.[33] These disunified

[32] See my (1974, 1993, 1998, 2000b, 2002, 2004, 2005, 2007a, Chaps. 9 and 14; 2009b, 2011, 2014a).

[33] Here is a demonstration of this point. Let T be any accepted fundamental physical theory. There are, to begin with, infinitely many disunified rivals to T that are *just as empirically successful* as T. In order to concoct such a rival, T_1 say, all we need to do is modify T in an entirely *ad hoc* way for phenomena that occur after some future date. Thus, if T is Newtonian theory (NT), NT_1 might assert: everything occurs as NT predicts until the first moment of 2050 (GMT) when an inverse cube law of gravitation comes into operation: $F = Gm_1m_2/d^3$. Infinitely many such disunified rivals can be concocted by choosing infinitely many different future times for an abrupt, arbitrary change of law. These theories will no doubt be refuted as each date falls due, but infinitely many will remain unrefuted. We can also concoct endlessly many disunified rivals to T by modifying the predictions of T for just one kind of system that we have never observed. Thus, if T is, as before, NT, then NT_2 might assert: everything occurs as NT predicts except for any system of pure gold spheres, each of mass greater than 1000 tons, moving in a vacuum, centres no more than 1000 miles apart, when Newton's law becomes $F = Gm_1m_2/d^4$. Yet again, we may concoct further endlessly many equally empirically successful disunified rivals to T by taking any standard experiment that corroborates T and modifying it in some trivial, irrelevant fashion—painting the apparatus purple, for example, or sprinkling diamond dust in a circle around the apparatus. We then modify T in an *ad hoc* way so that the modified theory, T_3 say, agrees with T for all phenomena except for the trivially modified experiment. For this experiment, not yet performed, T_3 predicts—whatever we choose. We may choose endlessly many different outcomes, thus creating endlessly many different modifications of T associated with this one trivially modified experiment. On top of that, we can, of course, trivially

rivals that postulate different laws for different phenomena in a "patchwork quilt" fashion, are (quite properly) never taken seriously for a moment despite being empirically more successful. This persistent acceptance of unified theories in physics even though endlessly many empirically more successful, patchwork quilt rivals can readily be formulated means that physics makes a persistent assumption about the universe: it is such that all seriously disunified theories are false. The universe is such that some kind of underlying unified pattern of physical law runs through all phenomena.

If physicists only ever accepted theories that postulate atoms even though empirically more successful rival theories are available that postulate other entities such as fields, it would surely be quite clear: physicists implicitly assume that the universe is such that all theories that postulate entities other than atoms are false. Just the same holds in connection with unified theories. That physicists only ever accept unified theories even though endlessly many empirically more successful, disunified rival theories are available means that physics implicitly assumes that the universe is such that all such disunified theories are false.

In accepting the unified theories that it does accept—Newtonian theory, classical electrodynamics and the rest—physics thereby adopts a big, highly problematic metaphysical hypothesis, H, about the nature of the universe: it is such that all rival, grossly disunified, "patchwork quilt" but empirically more successful theories are false. H, though a metaphysical hypothesis, is nevertheless a permanent, even if generally unacknowledged, item of theoretical knowledge. Theories that clash with it, even though empirically more successful than accepted physical theories, are rejected—or rather, are not even considered for acceptance. Whenever a fundamental physical theory is accepted, endlessly many empirically more successful rivals, easily formulated, are not even considered just because, in effect, they clash with H. Thus H is a permanent item of theoretical knowledge in physics, more securely established in scientific practice indeed than any physical theory. Physical theories

modify endlessly many further experiments, each of which generates endlessly many further disunified rivals to T. Each of these equally empirically successful, disunified rivals to T—T_1, T_2, … T_∞—can now be modified further, so that each becomes *empirically more successful* than T. Any accepted fundamental physical theory is almost bound to face some empirical difficulties, and is thus, on the face of it, refuted—by phenomena A. There will be phenomena, B, which come within the scope of the theory but which cannot be predicted because the equations of the theory cannot (as yet) be solved. And there will be other phenomena C that fall outside the scope of the theory altogether. We can now take any one of the disunified rivals to T, T_1 say, and modify it further so that the new theory, T_1*, differs further from T in predicting, in an entirely *ad hoc* way, that phenomena A, B and C occur in accordance with empirically established laws L_A, L_B and L_C. T_1* successfully predicts all that T has successfully predicted; T_1* successfully predicts phenomena A that ostensibly refute T; and T_1* successfully predicts phenomena B and C that T fails to predict. On empirical grounds alone, T_1* is clearly more successful and better corroborated, than T. And all this can be repeated as far as all the other disunified rivals of T are concerned, to generate infinitely many empirically more successful disunified rivals to T: T_1*, T_2*, … T_∞*.

tend eventually to be shown to be false, but H persists through theoretical revolutions in physics.[34]

Nevertheless, H is a hypothesis, a pure conjecture. How can we make sense of the idea that science is rational and delivers authentic knowledge if the whole enterprise depends crucially on accepting such an unsupported hypothesis as a secure item of scientific knowledge—a hypothesis that exercises a major influence over what theories are accepted and rejected in physics?

Aim-Oriented Empiricism

In order to answer this question, we need to adopt a conception of science that I have called *aim-oriented empiricism* (AOE). Precisely because H is a substantial assertion about the nature of the universe, an assertion that, though purely conjectural in character, nevertheless exercises a major influence over what theories are accepted and rejected, even to the extent of over-riding empirical considerations, it needs to be made explicit within physics so that it can be critically assessed, rival hypotheses if possible being developed and assessed, in the hope that H can be improved on. We need a new conception of science which represents the metaphysical hypotheses of physics in the form of a hierarchy of hypotheses, as one goes up the hierarchy hypotheses becoming less and less substantial, and more nearly such that their truth is required for science, or the pursuit of knowledge, to be possible at all. In this way we create a relatively unproblematic framework of hypotheses, and associated methodological rules, high up in the hierarchy, within which much more substantial and problematic hypotheses, and associated methodological rules, low down in the hierarchy, can be critically assessed and, we may hope, improved, in the light of the empirical success they lead to, and other considerations: see Fig. 1.

All this can be reformulated in terms of aims and methods. The aim of science is not truth per se, as SE holds. It is rather truth presupposed to be explanatory—or at least knowable. Precisely because this aim of science presupposes a problematic metaphysical hypothesis, the aim (or the hypothesis presupposed by the aim) needs to be represented in the form of a hierarchy of aims (or hypotheses) as indicated in Fig. 1, so that attempts to improve aims (or hypotheses) may receive the best possible help. As our scientific knowledge and understanding improve, so aims and methods improve as well. There is something like positive feedback between improving scientific knowledge and improving aims and methods—improving knowledge about how to improve knowledge. Science adapts itself to what it finds out about the universe. It is this positive feedback, this interaction between improving scientific knowledge on the one hand, and improving aims and methods (improving assumptions and methods) on the other, that helps explain the explosive growth of modern science.

[34] For expositions of this argument see Maxwell (1974, part 1; 1993, part 1; 1998, Chap. 2; 2000b, 2002, 2004, Chap. 1; 2005, 2011, 2014a).

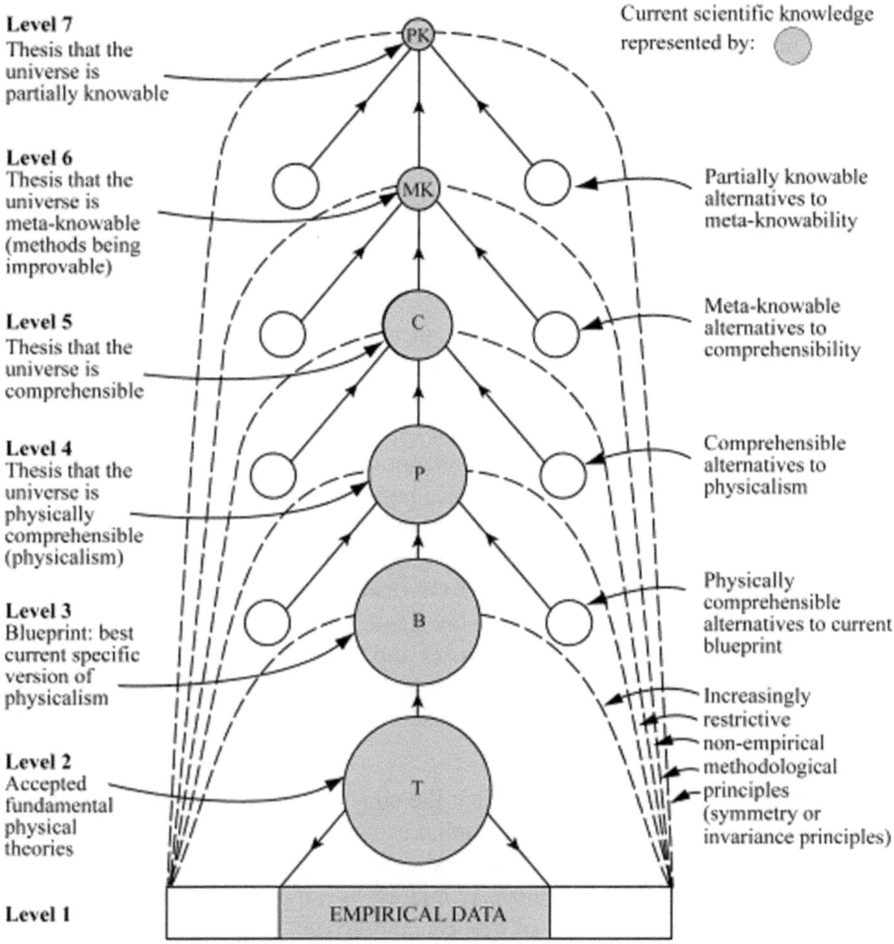

Fig. 1 Aim-Oriented Empiricism (AOE)

For all this has gone on in scientific practice despite scientists paying lip service to SE. Allegiance to SE has been sufficiently hypocritical to permit aim-oriented empiricism (AOE) to be put into scientific practice, to some extent at least. Allegiance to SE has nevertheless obstructed full implementation of AOE, and has had damaging consequences for science as a result.[35]

There are now three key points to note about AOE.

1. It is not just theoretical physics that has a problematic aim because of problematic hypotheses inherent in the aim. This is true of most—perhaps all—scientific disciplines. Thus most, or perhaps all, scientific disciplines need to be understood in terms of diverse versions of the hierarchical, meta-methodological structure of AOE depicted in Fig. 1. The aims and methods of science change as

[35] For expositions of, and arguments for AOE see works referred to in note 32.

we move from one science to another, and as we move within any given science from one time to another. The common factors are (a) something like the hierarchical, interacting structure depicted in Fig. 1; (b) the common endeavour to improve knowledge and understanding of the universe, and ourselves and other living things as a part of it. AOE provides a general solution to the problem of the nature of the progress-achieving methods of science.[36]
2. AOE solves fundamental problems in the philosophy of science: in particular, the problem of induction (the problem of the rationality of science); the problem of verisimilitude; and the problem of what it means to say of a physical theory that it is unified.[37]
3. AOE transforms the nature of science, the nature of philosophy of science, and the nature of the relationship between the two. And all this impacts on the nature of the history of science, the sociology of science, and STS. Traditionally, philosophy of science has been conceived of, and practised, as a meta discipline, studying science in the same way as astronomers study the moon or distant galaxies. This might make sense if science had a fixed aim and fixed methods, as SE holds science does. But AOE asserts that, because the basic aims of science are profoundly problematic, they evolve as scientific knowledge evolves, and change from one science to another. AOE demands that there is a two-way interaction between science itself, on the one hand, and its aims-and-methods, or philosophy, on the other hand. Metaphysics and the philosophy of science become vital ingredients of science itself, concerned to help science make progress. The nature of science, the philosophy of science, and the relationship between the two, all change dramatically.[38]

Exploring probing questions about what the aims of science are, and ought to be, goes much further. For science seeks truth presupposed to be explanatory—explanatory truth as one might say—as a special case of the much more general aim of *valuable truth*—truth that is of intrinsic interest in some way, or of use. A science which increased our knowledge of irredeemably trivial, useless, utterly uninteresting truth would not be said to be making progress. Science both does, and ought to, seek truth that is of use or of value. Merely in order to be accepted for publication, a scientific paper must report a finding that meets some threshold of potential interest. Counting leaves on trees or pebbles on beaches does not, in itself, contribute to scientific knowledge even if the information is new and true.

But the aim of valuable truth is almost more problematic than that of explanatory truth. Of value to whom? And in what way? Is what science seeks to discover always of value to humanity, to those whose needs are the greatest? What of the links that science funding has with the military, corporations of one kind or another, and governments? Do the aims of science always respond to the curiosity and wonder of scientists, or sometimes to their career ambitions and vanity? Given that modern

[36] Maxwell (2004, pp. 39–47).

[37] Maxwell (1998, Chaps. 3–6; 2004, Chaps. 1, 2, and appendix; 2007a, Chap. 14; 2014a).

[38] See works referred to in note 32.

science is expensive, is there not always going to be an inherent conflict between the interests of those who pay for science—the wealthy and powerful—and those whose needs are the greatest—the poor and powerless?

If science is to pursue the problematic aim of valuable truth rationally, and in such a way that justice is done to the best interests of humanity, it is vital that science is pursued within the framework of a generalized version of AOE—humane AOE I have called it—so that three domains of discussion are recognized: (1) evidence; (2) theory; and (3) aims. The third domain of discussion, aims, is as important as the first two. At present it is "repressed"; it goes on in fund giving committees, and in private between scientists, but not openly in journals and conferences along with (1) and (2). Sustained exploration of the problematic aim of valuable truth needs to attempt to articulate (a) what we conjecture to be scientifically discoverable, and (b) what we conjecture it would be of value to discover, so that we may try to determine the all-important region of overlap between the two. The scientific community may have expertise when it comes to (a), but cannot have any exclusive expertise when it comes to (b). If science is to come to serve the best interests of humanity, it is vital that scientists and non-scientists alike cooperate in engaging in sustained imaginative and critical exploration of what it would be of most value for science to attempt to discover—what ought to be the aims and priorities of scientific and technological research. The institutional/ intellectual structure of science needs to be changed to facilitate such aim-exploration. Journals and conferences need to be set up. Science journalism needs to contribute. SE, in misrepresenting the aim of science to be truth per se, in effect "represses" the real, problematic aim of valuable truth, and thus damages science by inhibiting the kind of sustained, cooperative exploration of actual and possible valuable aims science does, and might, pursue.[39]

It is important to appreciate that all this comes within the province of philosophy of science which is centrally concerned with problems about the aims and methods of science. Philosophy of science, in order to be done properly, must concern itself with moral, social, value questions about science. It must seek to call into question the less praiseworthy human aspirations science may seek to fulfil—the greed of corporations, the military might of some governments, the self-interests of some scientists. And it must explore neglected avenues of research that might lead to discoveries and technological developments of great potential value to humanity.

It does not stop here. For of course science seeks knowledge of valuable truth so that it may be used by people in life—ideally, so as to enhance and enrich the quality of human life. Science is to be used by people, either culturally, to aid the quest to know, to understand, or practically, as a means to the realization of other goals of value—health, security, travel, communications, entertainment, and so on. Science aims to contribute to the social world. There is a political dimension to the aims of science—once again, profoundly problematic. Everything said above about the value dimensions of the aims of science applies here too to the social, humanitarian or political dimensions. And this, too, comes within the province of philosophy of science, properly conceived. The orthodox distinction between "internal" factors

[39] See my (1976, 1984, 2001, 2004, 2007a, 2010, 2014b).

(purely intellectual) and "external" (social, political, economic, evaluative) is a nonsense. At least, the way this distinction is usually drawn is a nonsense.[40]

Damaging Irrationality of Knowledge-Inquiry

We come now to the really substantial step in this exploration of problematic aims. We need to look, not just at the aims of natural science and technological science, as we have done so far. In addition, we need to look at the aims of social science too—social science and the humanities, and indeed, the aims of academic inquiry as a whole. The upshot of such an examination of aims is dramatic. We urgently need to bring about a revolution in academic inquiry so that the basic aim becomes wisdom, and not just knowledge.[41]

The official, overall aim of academia, it can generally be agreed, is to help promote human welfare by intellectual and educational means—help people realize what is of value to them in life, help humanity make progress towards as good a world as possible. From the past we have inherited the view that the best way academic inquiry can do this is, in the first instance, to acquire knowledge. First, knowledge has to be acquired; then, secondarily, it can be applied to help solve social problems. Academia organized in this way may be called *knowledge-inquiry*.

Knowledge-inquiry has, associated with it a severe censorship system. Only that may enter the intellectual domain of inquiry relevant to the pursuit of knowledge: observational and experimental results, factual claims to knowledge, valid arguments, theories, and so on. Everything else must be ruthlessly excluded: values, feelings and desires, politics, political ideas, policies, cries of distress, problems of living and proposals for their solution, philosophies of life—although knowledge about these things can of course be included.

Not everything that goes on in universities today conforms precisely to the edicts of knowledge-inquiry. It is, nevertheless, the dominant view, and exercises a profound influence over what goes on in universities. Knowledge-inquiry is nevertheless profoundly and damagingly irrational in a wholesale, structural way. The irrationality of knowledge-inquiry is so damaging that it is in part responsible for our current incapacity to learn how to tackle effectively our current global problems.

Rationality, as I use the term—and this is the notion that is relevant to the issues we are considering—assumes that there is some probably rather ill-defined set of methods, strategies or rules which, if put into practice, give us our best chances, other things being equal, of solving our problems or realizing our aims. The rules of reason don't tell us precisely what to do (they tell us what to attempt), and they

[40] See previous note.

[41] The "from knowledge to wisdom" argument I am about to sketch was first expounded in my (1976). It was spelled out in much greater detail in my (1984); see also my (2007a). See also my (1998, 2001, 2004, 2010, 2014b). For summaries that expound different aspects of the argument see Maxwell (1980, 1992, 2000a, b, 2008, 2011, 2012a, b).

don't guarantee success. They assume that there is much that we can already do, and they tell us how to marshal these already solved problems in order best to tackle new problems.[42]

There are four elementary rules of reason any problem-solving endeavour must implement if it is to be rational, and stand the best chances of meeting with success.

1. Articulate, and try to improve the articulation of, the basic problem to be solved.
2. Propose and critically assess possible solutions.
3. If the basic problem we are trying to solve proves to be especially difficult to solve, specialize. Break the problem up into subordinate problems. Tackle analogous, easier to solve problems in an attempt to work gradually towards the solution to the basic problem.
4. But if we do specialize in this way, make sure specialized and basic problem-solving keep in touch with one another, so that each influences the other.

Any problem-solving endeavour that persistently violates just one of these rules will be seriously irrational, and will suffer as a result. Knowledge-inquiry violates *three* of these rules. It is as bad as that.

Knowledge-inquiry puts rule (3) into practice magnificently, especially as exemplified in universities around the world. Endless specialization, disciplines being endlessly subdivided into ever more specialized disciplines, is a striking feature of academia as it exists today. But rules (1), (2) and (4) are all violated.

If we take seriously that academia has as its basic task to help promote human welfare—help people realize what is of value to them in life—then the basic problems academia needs to help solve are *problems of living*, problems of action in the real world, and not, fundamentally, problems of knowledge. It is what we do—or refrain from doing—that enables us to achieve what is of value in life, and not what we know. Even where new knowledge or technology is relevant, as it is in medicine, for example, or agriculture, it is always what this knowledge or technology enables us to do that enables us to achieve what is of value in life, not the knowledge as such (except when knowledge is itself of value).

So, in order to put rules (1) and (2) into practice, academia needs to give absolute intellectual priority to the tasks of (1) articulating our problems of living, including our global problems, and (2) proposing and critically assessing possible solutions—that is, possible *actions, policies, political programmes, strategies, new institutions, new social endeavours, new social arrangements, new ways of living, philosophies of life.* But the censorship system of knowledge-inquiry excludes all this from the intellectual domain of inquiry because it does not constitute contributions to knowledge. Just that which academia most needs to do in order help people, humanity, solve problems of living in increasingly cooperatively rational ways is not done within knowledge-inquiry because it does not contribute to the pursuit of knowledge. And in practice in universities today, thinking about problems of living and policy issues is pushed to the periphery of academia, and does not proceed at

[42] For more about rationality see my (1984, pp. 69–71 and Chap. 5; or 2007a, pp. 82–84 and Chap. 5).

the heart of the academic enterprise, as the most fundamental intellectual activity. It is in part because universities today fail to do what most needs to be done to help us make progress towards as good a world as possible, that we are in the mess that we are in.

Having violated rules (1) and (2), knowledge-inquiry also violates rule (4). If you fail to engage in thinking about fundamental problems, you cannot interconnect specialized and fundamental problem-solving, as rule (4) requires. As a result, specialized research is likely to become unrelated to our most urgent needs which, one may well argue, is what has happened in our universities today.

Wisdom-Inquiry: Problem-Solving Version

We need urgently to transform academic inquiry so that all four basic rules of reason are put into practice in a structural way. The outcome is what I have called *wisdom-inquiry*. Wisdom-inquiry is what emerges when knowledge-inquiry is modified just sufficiently to correct its severe rationality defects. At the heart of wisdom-inquiry there are the absolutely fundamental intellectual tasks of (1) articulating and improving the articulation of problems of living, including global problems, and (2) proposing and critically assessing possible solutions—possible actions, policies, political programmes, ways of life and so on. More specialized problem-solving, and in particular scientific and technological research, emerge out of this and feed back into it, in accordance with rules (3) and (4). Thinking about our problems of living and what to do about them influences the aims and priorities of scientific and technological research, and the results of scientific and technological research of course influence thinking about problems of living: see Fig. 2.

Almost every branch and aspect of academia is modified as we move from knowledge-inquiry to wisdom-inquiry. Within knowledge-inquiry, social inquiry is primarily social science. The social sciences and humanities have, as their basic task, to improve our knowledge and understanding of social phenomena, the human world. Within wisdom-inquiry, by contrast, the diverse branches of social inquiry have, as their basic task, to articulate problems of living and propose and assess possible solutions. The basic task is to help people, humanity, tackle conflicts and problems of living in the real world in increasingly cooperatively rational ways so that humanity may make progress towards a genuinely good, wise world—or at least as good a world as possible. Social inquiry, so conceived, within wisdom-inquiry, is intellectually more fundamental than natural science.

As we move from knowledge-inquiry to wisdom-inquiry the relationship between academia as a whole and the rest of the social world is transformed. Knowledge-inquiry seeks to shield itself from the social world to preserve the objectivity and integrity of the pursuit of knowledge.

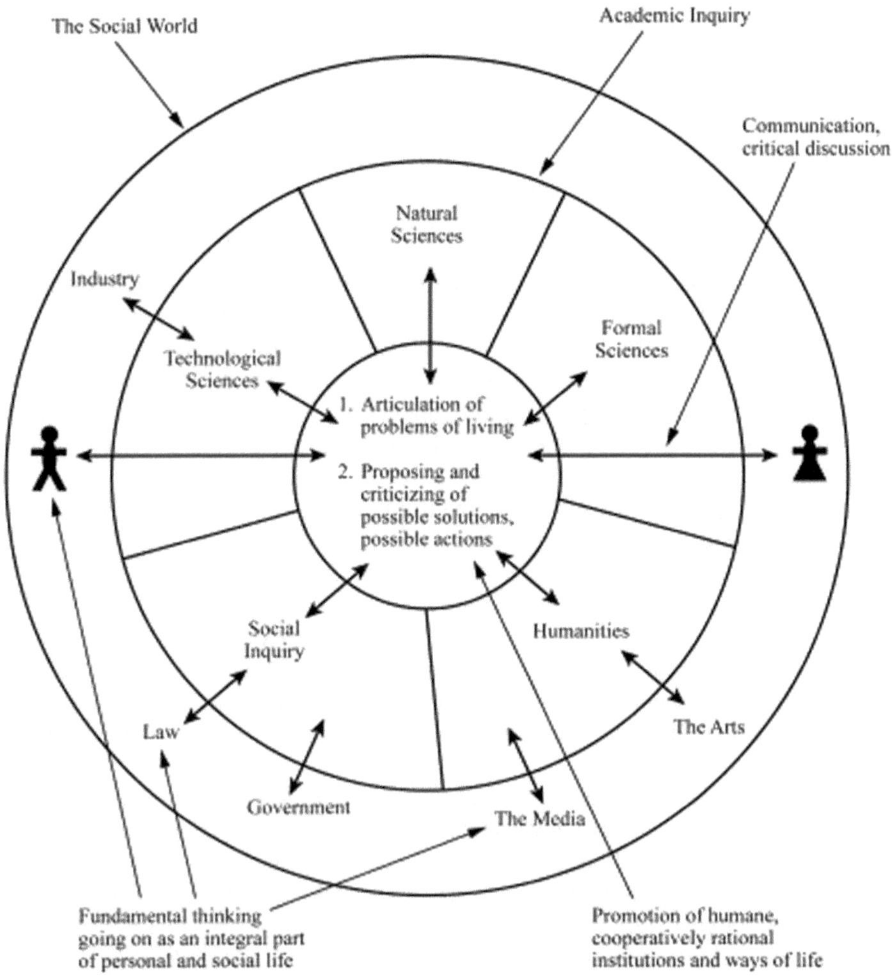

Fig. 2 Wisdom-inquiry implementing problem-solving rationality

Wisdom-inquiry, by contrast, seeks to interact with the social world, ideas, experiences and arguments going in both directions, so that academia may help humanity learn how to tackle our immense global problems more effectively. Wisdom-inquiry might be regarded as a kind of civil service for humanity. What actual civil services are supposed to do in secret for governments, wisdom-inquiry academia does openly for the public.

Knowledge-inquiry has two quite distinct fundamental aims: the intellectual aim of knowledge, and the social or humanitarian aim of helping to promote human welfare. There is a sense in which wisdom-inquiry fuses these together in the one basic aim of seeking and promoting *wisdom*—wisdom being the capacity, and perhaps the active desire, to realize what is of value in life, for oneself and others, wisdom thus including knowledge and technological know-how but much else besides.

Wisdom-Inquiry: Aim-Pursuing Version

Granted that the argument of the previous section is correct, and universities today, dominated as they are by knowledge-inquiry, are damagingly irrational in a structural way, an obvious question to ask is: When and how did this come about?

It all goes back to the eighteenth century Enlightenment, especially the French Enlightenment. The *Philosophes* of the Enlightenment, Voltaire, Diderot, Condorcet and the rest, had the magnificent idea that it might be possible to learn from scientific progress towards greater knowledge how to make social progress towards an enlightened world. Unfortunately, in developing and implementing this magnificent idea, they blundered. They botched the job. They thought the task was to develop the social sciences alongside the natural sciences. This got developed throughout the nineteenth century, and got built into universities in the early twentieth century with the creation of Departments of social science. The outcome is what we have, by and large, today: knowledge-inquiry.

But all this represents a series of dreadful blunders. In order to implement the profound, basic idea of the Enlightenment properly, there are three crucial steps it is essential to get right. The *Philosophes* got all three steps wrong.

First, it is essential to get clear about what the progress-achieving methods of science are, what methods, precisely, make scientific progress possible.

Second, these methods need to be correctly generalized so that they become potentially fruitfully applicable to any worthwhile, problematic human endeavour, whatever the aims may be, and not just applicable to the scientific endeavour of improving knowledge.

Third, These correctly generalized progress-achieving methods then need to be got into the social world, into government, industry, agriculture, education, the media, the law, international relations, and so on, so that they may be exploited correctly in the great human endeavour of trying to make social progress towards an enlightened, wise world.

From the eighteenth century down to today, scientists and philosophers of science have accepted one or other version of standard empiricism (SE) which, as we saw above very seriously misrepresents the aims and methods of science. In order to get the *first* step right we need to adopt aim-oriented empiricism (AOE).

In order to get the *second* step right, we need to generalize AOE so that it becomes potentially fruitful to any problematic worthwhile human endeavour, and not just science, in this way creating a conception of rationality that helps us improve aims when they are problematic. I have called this aim-pursuing conception of rationality *aim-oriented rationality* (AOR). The vital point to appreciate is that it is not just the aims of science that are problematic; this is true in life as well, in all sorts of personal, social and institutional contexts. Aims conflict. They have unforeseen, undesirable consequences. They are not as desirable as we suppose, or not as realizable, or both. We may misrepresent our aims. The more "rationally"—that is, effectively—we pursue a bad aim, the worse off we will be. We need to try to improve our aims as we act.

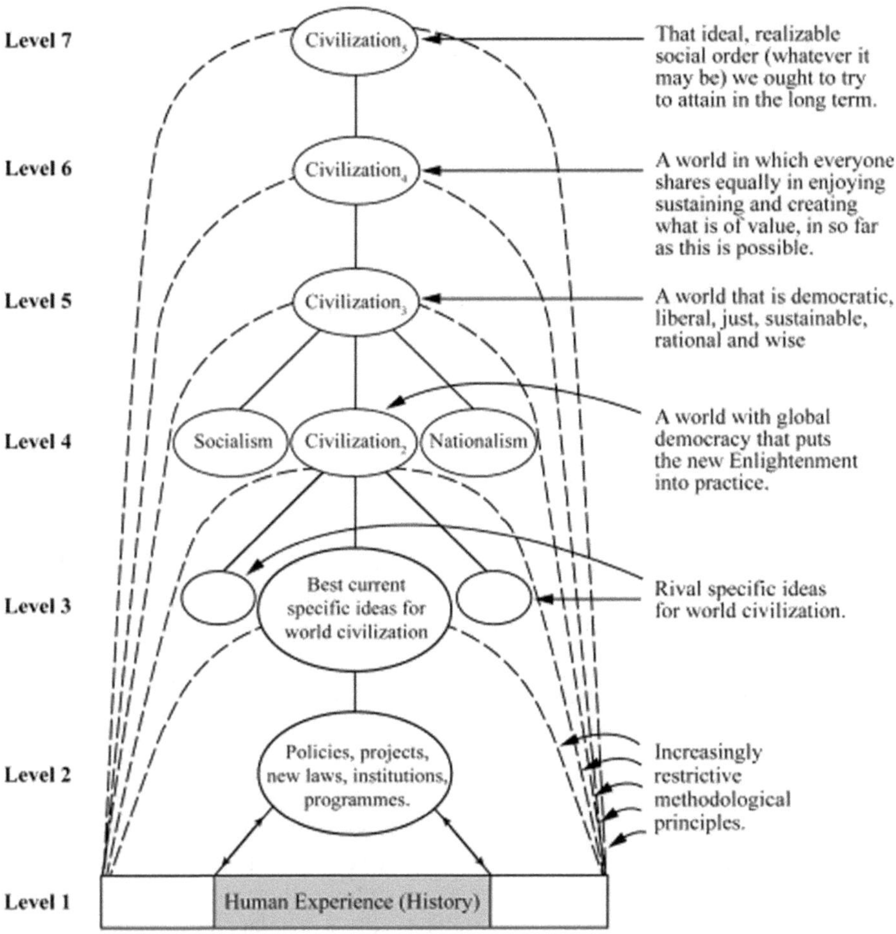

Fig. 3 Hierarchical social methodology generalized from science

Quite generally, whenever we pursue problematic aims, we need to represent them in the form of a hierarchy, along the lines depicted in Fig. 1, thus giving ourselves the best chances of improving our aims and methods as we act.

In order to get the *third* step right, we need to try to get AOR, arrived at by generalizing AOE, the progress-achieving methods of science, into all our other worthwhile, problematic endeavours besides science—into government, industry, finance, agriculture, education, the media, the law, international relations, and so on. Above all, we need to get AOR into the endeavour to make progress towards the profoundly problematic aim of creating an enlightened world: see Fig. 3. The *philosophes* made the disastrous mistake of applying a misconceived conception of scientific method, SE, to the task of improving *knowledge* of social phenomena, thus creating social science, when what they ought to have done is apply AOR to *social life itself* so that humanity may make progress towards an enlightened world. According to this second version of wisdom-inquiry (building on the first version), social inquiry is not social *science*, but rather social *methodology* or social *philosophy*. What ought to be the relationship between philosophy of science and science, within the framework of AOE, so too that ought to be the relationship between social inquiry and society. Sociology thus emerges as social *methodology*, and the sociology of science, in particular, emerges as *scientific methodology*, or in other words, *philosophy of science*. At present, philosophy of science and sociology of science are at loggerheads with one another—partly because of social constructivist disagreements. Within this second version of wisdom-inquiry, however, philosophy of science and sociology of science emerge as one and the same discipline, both concerned with what ought to be the intellectual and social aims and methods of science.

The Future of Science and Technology Studies

When these arguments for AOE and wisdom-inquiry, just summarized, were spelled out in detail in my book *From Knowledge to Wisdom* in the Orwellian year of 1984, Christopher Longuet-Higgins in a glowing review in *Nature* said:

> Maxwell is advocating nothing less than a revolution (based on reason, not on religious or Marxist doctrine) in our intellectual goals and methods of inquiry [...]. There are altogether too many symptoms of malaise in our science-based society for Nicholas Maxwell's diagnosis to be ignored.[43]

But this is just what has happened. By and large, my diagnosis has been ignored—especially by those who should be most concerned professionally, those engaged in HPS, STS and philosophy. Instead of bringing to scientists, to fellow academics and to the public the message that universities, in so far as they put knowledge-inquiry into practice, betray both reason and humanity, these scholars have, rather, devoted themselves to wrangles about social constructivism, and the traditional fare of science studies. The fundamental question *What kind of inquiry can best help humanity learn how to make progress towards as good a world as possible?* continues, scandalously, to be ignored.

I can only hope that this essay will provoke some STS folk to take up intellectual cudgels on behalf of reason and humanity.

[43] Longuet-Higgins (1984).

Conclusion

In order to create a better, wiser world, we need to learn how to do it. That in turn requires that our institutions of learning, our schools and universities, are well-designed, rationally designed and devoted for the task. At present they are not. It is this that is in part responsible for our global problems and our current incapacity to tackle them effectively. We urgently need to bring about a revolution in universities around the world so that they become devoted to seeking and promoting wisdom—helping humanity create a better world. As far as the long term interests of humanity are concerned, there is probably no more important thing that we need to do. Is this academic revolution really needed? What would it imply? What are its advantages and disadvantages? How ought universities to develop? If the revolution is required, what can be done to help bring it about? These are some of the questions STS ought to tackle.

References

Barnes B (1977) Interests and the growth of knowledge. Routledge & Kegan Paul, London
Barnes B (1982) T. S. Kuhn and social science. Macmillan, London
Barnes B (1985) About science. Blackwell, Oxford
Barnes B, Bloor D, Henry J (1996) Scientific knowledge: a sociological analysis. University of Chicago Press, Chicago
Bloor D (1976) Knowledge and social imagery. Routledge and Kegan Paul, London
Butterfield H (1951) The Whig interpretation of history. Bell and Sons, London. (First published in 1931)
Feyerabend P (1965) Problems of empiricism. In: Colodny RG (ed) Beyond the edge of certainty. Prentice-Hall, New York, pp 145–260
Feyerabend P (1970) Problems of empiricism, part II. In: Colodny RG (ed) The nature and function of scientific theories. University of Pittsburgh Press, Pittsburgh, pp 275–354
Feyerabend P (1975) Against method: outline of an anarchistic theory of knowledge. New Left Books, London
Feyerabend P (1978) Science in a free society. New Left Books, London
Feyerabend P (1987) Farewell to reason. Verso, London
Gross P, Levitt, N (1994) Higher superstition: the academic left and its quarrels with science. John Hopkins University Press, Baltimore
Gross P, Levitt N, Lewis M (eds) (1996) The flight from science and reason. John Hopkins University Press, Baltimore
Hardy A (1968) The living stream. Collins, London
Harper WL (2011) Isaac Newton's scientific method. Oxford University Press, Oxford
Koertge N (1998) A house built on sand: exposing postmodernist myths about science. Oxford University Press, Oxford
Kuhn TS (1962) The structure of scientific revolutions. Chicago University Press, Chicago
Lakatos I (1970) Falsification and the methodology of scientific research programmes. In: Lakatos, Musgrave, pp 91–195
Lakatos I, Musgrave A (eds) (1970) Criticism and the growth of knowledge. Cambridge University Press, London
Longuet-Higgins C (1984) For goodness sake nature, vol. 312, p. 204. http://www.ucl.ac.uk/from-knowledge-to-wisdom/reviews/#goodness

Maxwell N (1974) The rationality of scientific discovery. Philosophy of Science 41:123–153 and 247–295
Maxwell N (1976) What's wrong with science? Towards a people's rational science of delight and compassion. Bran's Head Books, Hayes (2nd edn, 2007)
Maxwell N (1980) Science, reason, knowledge and wisdom: a critique of specialism. Inquiry 23:19–81
Maxwell N (1984) From knowledge to wisdom: a revolution in the aims and methods of science. Blackwell, Oxford
Maxwell N (1992) What kind of inquiry can best help us create a good world? Science Technology & Human Values 17:205–227
Maxwell N (1993) Induction and scientific realism: Einstein versus van Fraassen. British Journal for the Philosophy of Science 44:61–79, 81–101 and 275–305
Maxwell N (1998) The comprehensibility of the universe: a new conception of science. Clarendon Press, Oxford
Maxwell N (1999) Are there objective values? The Dalhousie Review 79(3):301–317
Maxwell N (2000a) Can humanity learn to become civilized? The crisis of science without civilization. Journal of Applied Philosophy 17:29–44
Maxwell N (2000b) A new conception of science. Physics World. 13(8):17–18
Maxwell N (2001) The human world in the physical universe: consciousness, free will and evolution. Rowman and Littlefield, Lanham
Maxwell N (2002) The need for a revolution in the philosophy of science. Journal for General Philosophy of Science 33:381–408
Maxwell N (2004) Is science neurotic? Imperial College Press, London
Maxwell N (2005) Popper, Kuhn, Lakatos and aim-oriented empiricism. Philosophia 32(1–4):181–239
Maxwell N (2007a) From knowledge to wisdom: a revolution for science and the humanities. Pentire Press, London. (2nd edn. of my 1984, revised and extended)
Maxwell N (2007b) From knowledge to wisdom: the need for an academic revolution. London Review of Education 5(2):97–115
Maxwell N (2008) Do we need a scientific revolution? Journal for Biological Physics and Chemistry 8(3):95–105
Maxwell N (2009a) How can life of value best flourish in the real world? In: McHenry, pp 1–56
Maxwell N (2009b) The metaphysics of science: an account of modern science in terms of principles, laws and theories (review of book by Craig Dilworth). International Studies in the Philosophy of Science 23(2):228–232
Maxwell N (2010) Cutting God in half—and putting the pieces together again: a new approach to philosophy. Pentire Press, London. http://discovery.ucl.ac.uk/view/people/ANMAX22.date.html
Maxwell N (2011) A priori conjectural knowledge in physics. In: Shaffer M, Veber M (eds) What place for the a priori? Open Court, Chicago, pp 211–240
Maxwell N (2012a) Arguing for wisdom in the university: an intellectual autobiography. Philosophia 40:663–704
Maxwell N (2012b) How universities can help humanity learn how to resolve the crises of our times—from knowledge to wisdom: the University College London experience. In: Heam G, Katlelle T, Rooney D (eds) Handbook on the knowledge economy, vol. 2. Edward Elgar, Cheltenham, pp 158–179
Maxwell N (2014a) Unification and revolution: a paradigm for paradigms. Journal for General Philosophy of Science 45(1):133–149. http://philpapers.org/rec/MAXUAR
Maxwell N (2014b) How universities can help create a wiser world: the urgent need for an academic revolution. Imprint Academic, Exeter
McHenry L (ed) (2009a) Science and the pursuit of wisdom: studies in the philosophy of Nicholas Maxwell. Ontos Verlag, Frankfurt a. M.
McHenry L (2009b) Ghosts in the machine: comment of Sismondo. Social Studies of Science 39:943–947

Popper K (1959) The logic of scientific discovery. Hutchinson, London. (First published in German in 1934)
Popper K (1963) Conjectures and refutations. Routledge and Kegan Paul, London
Popper, K (1969) The open society and its enemies. Routledge and Kegan Paul, London. (First published in 1945)
Popper K (1970) Normal science and its dangers. In: Lakatos, Musgrave. pp 51–58
Popper, K (1974) The poverty of historicism. Routledge and Kegan Paul, London
Sismondo S (2009a) Ghosts in the machine: publication planning in the medical sciences. Social Studies of Science 39:171–198
Sokal A (1998) Transgressing the boundaries; toward a transformative hermeneutics of quantum gravity. In: Sokal, Bricmont. pp 199–240
Sokal A (2008) Beyond the hoax: science, philosophy and culture. Oxford University Press, New York
Sokal A, Bricmont B (1998) Intellectual impostures. Profile Books, London

Introduction

History of Science, Society and Technology Studies

Raffaele Pisano
Department of Physics, Lille 1 University Science and Technology, France
Villeneuve d'Ascq Cedex
E-Mail: pisanoraffaele@iol.it

The study of science and its history has long been central to the shared scholarly efforts of both the Eastern and Western Worlds. Most theoretical and practical results depended mainly on their individual scientific and disciplinary ambitions and their unique technological innovations. For example, scientific traditions were established and individual contributions made by such scholars as Nicolaus Copernicus (1473–1543), Galileo Galilei (1564–1642), Johannes Kepler (1571–1630) and René Descartes (1596–1650), resulting in a thoroughly scientific framework within which to interpret celestial and terrestrial phenomena.

Societies, from ancient times (and generally speaking) to nowadays, have usually constructed a perception according with the idea that science is synonymous with *progress & modernity*, especially during periods of technological materializations. We historically know that anomalies, inversions and controversies also belong to periods of so-called (erroneously) *progress & modernity*. It would be interesting to investigate the literature author by author as a means of understanding precisely how science and society work in tandem i.e., the concept of civilization. For example, Newtonian science certainly produced a strong impact on humanity, particularly on Western civilization both with respect to the *scientific* and *supernatural* background of the laws of nature, including mathematical interpretations of phenomena like non–physical laws, sometimes outside the context of theory (i.e., providence, religion etc.). Here we have insufficient room to engage in a *dialogue* between specialists (advanced and applied researches) and non–specialists (versus a scientific civilization) as a debatable matter, in history as well as in the present. A formulation of the question could be "*How is it possible to pass from science to technique and to technologies? And, who is really able to be mediator in any other context of society?*"

The social and political environment in which scientists live has profound influences on how their scientific results and methods are framed. This is specifically true for the seventeenth century, the epoch of the scientific revolution and a century of deep social and political transformations. Nevertheless, we think influence of the social–political situation on the work of a scientist has to be deduced directly from the analysis of his or her scientific works. In other terms: an analysis of a society in a certain period can be useful to understand the general direction taken by science in that period. But, in itself, it is not enough to understand the specific work and results of an individual scientist. This kind of general analysis risks becoming a sort of *a priori passepartout* through which scientific work risks to be seriously misunderstood by the way in which individual scientists have presented the results of their researches. It is always necessary to begin a historical investigation—including an investigation into the relation between science and society in some determined period—from the existing theoretical and technical work of the scientific community as a whole. If, within analysis of the complete work of a scientist, a historian of science reveals some lack of clarity, or internal inconsistencies or a lack of coherence between the methods used by other scientists in different works and, if the questions raised cannot be explained either for technical reasons (for example the lack or the misunderstanding of certain mathematical methods) or with the general methodological and epistemological convictions of the scientist himself, then it is necessary to look into the general structure of the society in that period. In this way, technical analysis of the results and methods used by the scientist can a priori be considered and evaluated within the relevant civilization.

Traditionally, a discussion of the history of science and technique/technology has been presented as a tool within the history of science to understand the relationship between science and the developments of art and crafts produced by non–recognized scientists during a certain historical time period. The relationship between science and society and consequent civilizing by science is centred on the possibility that society could effectively develop a fundamental organization with capacity to *absorb* science and produce technologies (i.e., water and electrical supply, transportation systems etc.); this capability was lacking in the past. Thus, *a development of civilization was necessary parallel to the development of science within society? Did that happen? Did Descartesian and Newtonian physical works develop as a response to the needs of society?*

Alexandre Koyré (1892–1964) emphatically linked the history of science and the role played by mathematics between Descartes and Newton in the history of scientific thought. Through his intuition that the fundamentals of scientific theories contain two basic choices, Koyré's intellectual matrix has been clarified.

> The new science, we are told sometimes, is the science of craftsman and engineer, of the working, enterprising and calculating tradesman, in fact, the science of rising bourgeois classes of modern society. There is certainly some truth in this descriptions and explanations [...]. I do not see what the *scientia activa* has ever had to do with the development of the calculus, nor the rise of the bourgeoisie with that of the Copernican, or Keplerian, astronomy theories. [...] I am convinced that the rise and the growth of experimental science is not the source but, on the contrary, the result of the new *theoretical*, that is, the new *metaphysical* approach to nature that forms the content of the scientific revolution of

> the seventeenth century, a content which we have to understand before we can attempt an explanation (whatever this may be) of its historical occurrence.[1] [...] I shall therefore characterize this revolution [the birth of the modern science] by two closely connected and even complementary features: *(a)* the destruction of the cosmos and therefore the disappearance from science—at least in principle, if not always in fact—of all considerations based on this concept, and *(b)* the geometrization of space, that is, the substitution of the homogeneous and abstract—however now considered as real—dimension space of the Euclidean geometry for the concrete and differentiated place–continuum of pre–Galilean Physics and Astronomy.[2]

According to the distinguished Russian historian[3] we can consider that: (1) the history of scientific thought has never been entirely separated by philosophical thought. (2) the most important scientific revolutions have always been determined by a replacement of philosophical speculations. Thus, i.e., the history of scientific thought (i.e., for physical Descartesian and Newtonian sciences) was not developed in a *vacuum*, but it moves within a set of ideas, foundational principles, or axiomatic evidence.

The general purpose of the book you are now reading is to analyse historical problems related to the use of physics, mathematics and geometry in applied sciences, to be covered by a series of invited speakers. A main question is: *When and why did the tension between mathematics, physics, chemistry, engineering and astronomy, give rise to a new scientific discipline, as well as to modern technologies?* Thus, the book begins with the realities of science and continues to investigate historical development of sciences and technology. Each argument is developed from historical and epistemological standpoints. It emphasizes the need to establish the role played by both theoretical and conceptual frameworks by means of the bridge between applied science & technologies. Scholars from different traditions discuss "emergency style" thinking in methodology and in theoretical perspectives. It then proceeds to the fundamental question: *What is the conceptual bridge between science research and the organization of technological research in the development of industry?* Moreover, the book also deals with sciences individually and their relationships with technologies in history. Some papers point out two main branches of applied mechanics, one that examines mechanism and machines (today known as industrial engineering), and another that examines the structures of civil and military constructions (today known as civil and military engineering). Reference to and emphasis on the technologies, material requirements, and production methods are added to the historical and scientific discourse. An interdisciplinary topic between sciences and engineering is presented, *A Bridge between Conceptual Frameworks,*

---

[1] Koyré A (1965) *Newtonian Studies*. The Harvard University Press, Cambridge–MA, pp. 5–6. 50th from his death is celebrate in 2014. With Joseph Agassi et al. an edited book I proposed.

[2] *Ivi*, pp. 6–7. The following explanation of Alexandre Koyré's choice for the history of science, "*The destruction of the cosmos*", is a replacement of the finite world, as it had been hierarchically classified by Aristotle, with an infinite universe. "*The geometrization of space*" is a replacement of Aristotle's physical (concrete) space with the abstract space of Euclidean geometry.

[3] A conference (1954, Boston) of *American Association for the Advancement of Science* (*The scientific Monthly*, 1955; Koyré A (1971) *Études d'Histoire de la pensée philosophique*. Gallimard, Paris).

Sciences, Society and Technology Studies concerning the interaction between historical–epistemological methods of investigations and science and technologies including their integration in foundations of science (pure and applied) and society in the history of sciences. Therefore, it is a pleasure to offer outstanding studies given by specialists from various parts of the world. The editor and authors[4] herewith bring the history of science and technology to a wider audience, synthesizing *discoveries*, *controversies*, *events* and *facts* of scholarship in the history of sciences, with a narration and an analysis of scientific activities of major periods in World history and civilization. All abstracts and papers submitted were accepted only after having been peer–reviewed both for style and contents: consequently, a high and recognized level of content is offered for individual and collective research projects appropriate for a wide international audience.

Nicholas Maxwell's (United Kingdom) prologue deals with the impact that social constructivism had on history and philosophy of science. He deems such an impact as extremely negative and thinks that the field of science and technology is still affected by this tendency. To overcome the situation, science and history and philosophy of science have to collaborate till creating a unique discipline. The author advances some proposals in this direction.

Chris Bissell (United Kingdom) deals with modelling in electronics and information engineering. According to the author, a conspicuous amount of literature exists on modelling in natural sciences and economics, but, with regard to the engineering of electronics, telecommunications, signal processing and control, material is not abundant. The author tries hence to open a relatively new field of research.

Assunta Bonanno, Michele Camarca, Peppino Sapia (Italy) face a classical subject: the scientific method. The crisis of this method, due to recent experimental results, jeopardizes the bases of knowledge. The authors analyze the crisis through a historical survey of the scientific method. They try to identify the cross hybridization between science and technology in the last three centuries as a possible factor to overcome the crisis.

Vito Bonifàcio (Portugal) highlights that, before the invention of *cinematographe* in 1895, Jules Janssen ideated a photographic revolver that permitted one to track the contact instant of 1874 Venus' transit with precision. One could expect this kind of apparatuses, to require a great fortune, which was not the case. In this paper, the author explains the reasons of the early misfortunes of movie cameras in observational astronomy.

Luca Borghi (Italy) stresses that the success of medicine between the late nineteenth century and the first half of the twentieth century was also due to the use of technology and engineering inside medicine. Especially between 1930 and 1980, scientific couples existed composed—in the words of the author—"by a surgeon and an above–average–skilled engineer". The author analyzes the contributions made by these couples to medicine in the mentioned period.

[4] All authors are reported in alphabetical order.

Mario Calamia, Giorgio Franceschetti and Alessandro Mori (Italy) deal with the discovery of electromagnetic waves and on their impact on daily life. The authors start by commenting on Maxwell's equations and proceed by underlining the applications of electromagnetic waves, finally reaching preliminary experiments of wireless power transmission and outlining a possible impact of Maxwell's discovery on society.

Yulia Petrovna Chukova (Russian Federation) also starts from Maxwell. In particular, the huge use of electromagnetic waves has also given origin to new illness: "The radio wave illness". This caused a wave of new studies in biology and medicine. The author deals with the nonthermal effects of MM radiation, underlining the importance of thermodynamics for systems under electromagnetic radiation. This synergy has allowed determination of a correlation between Western–science and Chinese–medicine languages.

Vincenzo Cioci (Italy) stresses that construction of the atomic bomb represents one of the most significant cases of relations between science, technology and society. The author traces the most relevant scientific discoveries and technical applications that led to construction of the bomb. He deals with the behaviour of the involved scientists, too, and also comments some writings by them, a part of which is still unpublished.

Javier Echávarri, Eduardo de la Guerra and Enrique Chancón (Spain) address the history of tribology, the science and engineering of rubbing surfaces. The authors recall the observations, experiences and experiments concerning the process of friction that led to tribology as a science. Afterwards they analyse the impact of tribology on economy and ecology.

Fernando Bandeira Figueiredo (Portugal) traces the history of construction of the astronomical observatory of Coimbra from the project's groundbreaking (1772) until its final construction (1799). He examines the changes that occurred on the initial plane, mainly because of financial problems, the phase in which construction was stopped and, finally, the positive solution.

Rémi Franckowiak (France), within the history of chemistry and technology studies, presents a revaluation of Jean Hellot's status in relation to the French *Académie des Sciences*. The amazing history deals with chemistry, science, and his relationship with the public authorities and alchemy speculations, as well. The aim is a revaluation of the history of *Académie of Sciences* as unique scientific paradigm in the development of the economy of France in the middle of the eighteenth century.

Elena Alexandrovna Gavrilina (Russian Federation) deals with the problem of creativity inside engineering. She claims that creativity enters into several phases of an engineering–process. The author develops an investigation into peculiar forms and characteristics, connoting nowadays this kind of creativity inside society itself.

Vitaly Gorokhov (Russian Federation) takes into account a particular aspect of Galilei's activity: according to the author, the Pisan scientist did not limit himself to creation of an experimental model of science; he also developed scientific knowledge for practical and technical aims. Because of this, it is appropriate to speak of "technoscience" when referring to the work of Galilei. The author justifies this opinion, also comparing Galilei's approach to the relations of art–science–technology with those of other famous men of his epoch.

Ladislav Kvasz (Czech Republic) identifies two different ways of carrying out experiments inside physics: one in Galilean style, the other in Newtonian style. The difference between the two relies upon the relation between experiment and mathematics. The author analyses the reason that connects Galilei's conception of experiment to Newton's. The role played by Descartes' physics in this context is highlighted.

Roberto Lalli (Germany) examines the figure of Karl Darrow. His work was fundamental to the spread of quantum physics in an industrial context, primarily the Bell Telephone Laboratories where Darrow spent a great part of his career. The author points out Darrow's approach to the spreading use of quantum mechanics, but he also deals with Darrow's epistemological convictions, according to which physics is an evolutionary process.

Annibale Mottana and Augusto Marcelli (Italy) have, since the second half of the twentieth century, tackled the problem of X–ray absorption fine spectroscopy (XAFS). They speak of the role played by synchrotrons in this context. The authors also stress the contribution given by XAFS to the technical refinement and theoretical improvement of material science, despite a still missing unified theory of X–ray fine absorption.

Peeter Müürsepp (Estonia) exposes the development that research in mathematics and astronomy has enjoyed at the University of Tartu since 1802, when this institution was reopened. Many important personalities worked in Tartu. The most famous was the astronomer Wilhelm Struve (1793–1864), born in Altona, who but spent his scientific career in Tartu, at the observatory (of which he was the director since 1820) and at the university, where he became full professor.

Moon–hyon Nam and Il–Seong Nha (Korea) narrate the story of how and why King Sejong of the Joseon dynasty created the Royal Astronomical Observatory *Gunai–dae* in the period 1432–1438. Through this article one learns the scientific, religious, social and economical reasons that induced the King to construct such a facility. The instruments created for the observatory are analyzed as well.

Moon–hyon Nam (Korea) reconsiders some aspects of the preceding paper. He describes the astronomical instruments installed in the observatory, revealing that in the period 1432–1439 fifteen kinds of astronomical equipment and timekeeping devices were installed. He further explains the nature of these instruments.

Agamenon Rodrigues Eufrásio Oliveira (Brazil) inspects Lazare Carnot's theory as far as Carnot applied his theory to machines. The author highlights the importance of the concept of *work* to include Carnot's theory inside rational mechanics. This concept is a fundamental intermediation between theoretical and applied physics. The way in which Carnot used D'Alembert's principle is also remarked.

Dominique Pécaud (France), in this paper, deals with the concept of work, not starting from the machines or from the influence science had on technology; rather he analyses—as he himself says—the "human factor". The manner in which human work has changed has a history and an engineering parallel to the case for science and technology. The author explains the nature of human factor–history and the aims of human–factor engineering.

Raffaele Pisano (France) and Paolo Bussotti (Germany) identify a fundamental moment starting from which the construction of machines became a technological activity, well supported by a scientific theory. This moment is thermodynamics and heat machines theory by Sadi Carnot. The authors, also providing their basic theoretical elements, show how profoundly these two disciplines – as early technosciences – changed the way of conceiving and constructing machineries, both in civil and military contexts, that is within a technological and social context of science.

Birutė Railienė (Lithuania) supplies a picture of open access sources, with particular reference to history of technology. The author analyses the role of traditional academic libraries that preceded this new deal. Furthermore, he presents various types of open access providers, including journals, institutional repositories and several databases. The possible social and educative perspectives connected to the open access sources are examined.

Pedro Miguel Pinto Raposo (Portugal) tells the story of the naval officer and hydrographic Portuguese engineer Hugo de Lacerda (1860–1944). He was appointed to the chair of hydrography at the Lisbon Naval School. The author explains how de Lacerda overcame long–period polemics on hydrographic activity in Portugal, putting together a number of elements derived from pedagogical strategies, personal experience, military values and a patriotic rhetoric.

Lisa Rougetet (France) presents some machines developed for mathematical games: these are cases in which mathematics and technology interact. This is particularly true for combinatorial games, starting from the simplest one, the Nim game, and a machine constructed to play Nim.

Maria Sirago (Italy) provides a historical contribution concerning steamships and the history of shipping industry in the Two Sicilies kingdom. The author shows the most important phases of this story: from the development of new steamships (1816) until the edification of Pietrarsa factory in 1840 and the construction of Ettore Fieramosca frigate in 1851.

Bratislav Stojiljković and Svetislav Lj. Marković (Serbia) deal with a project by Tesla: in 1917 he ideated five different kinds of table fountains. The authors underline that Tesla developed the project of one of these fountains to the point that he supplied four different versions. To write this paper, the authors explored original archival documents to clarify how research can proceed in a field that is not as famous as others to which Tesla contributed.

Emanuele Zamperini (Italy) faces a particular aspect of technique in which—as in many other cases—an empirical approach was progressively replaced by a structural analysis: the design of timber trusses in Italy. The author traces this history and identifies the middle of the nineteenth century as the turning point from one approach to another. In the course of the paper, numerous historical and technical details are introduced to justify this idea.

A point of view is presented by Bertrand Bocquet (France). His review concerns the relationship between scientific research as large international activities and individual researchers both inhabiting an economic-technological World. His standpoint deals with *participative research* as a new possible paradigm of the *knowledge society* promoted by the European Community.

Finally, scientific problems within history of physics, engineering, chemistry, astronomy and medicine correlated with technological applications in the social context are analyzed. The authors explain the various ways in which the sciences allowed advanced modelling on the one hand, and the development of new technological ideas on the other hand. An emphasis on the role played by mechanisms, productions and instruments highlights the benefits to historical and scientific discourse: they have produced theories, institutions, universities, schools for engineers, and other social implications as well.

Scholars from different traditions have discussed in this book the traditional emergency style thinking in methodology and in theoretical perspective, aiming to gather and re–evaluate the current thinking on this subject. The book brings together contributions from leading experts in the field, and gives much-needed insight in the subject from a historical point of view. It should prove to be absorbing reading for historians, philosophers and scientists.

The Editor
Lille, France, January 2015

Acknowledgments

The genesis of such a lengthy book as this collection of articles needed to be found in deep roots. The book has been a long time in the making and has fulfilled its promise. Therefore, I heartily express my appreciation to all contributing Authors for their efforts to produce papers of interest and of elevated quality worthy of this Springer volume. The result is excellent.

I express my warm and pleasant gratitude to all our distinguished Authors. In particular I thank Paolo Bussotti for his friendly and professional collaboration during book composition.

And finally yet importantly, my acknowledgments one more time are addressed to Marco Ceccarelli (here author of a Foreword, as well), Nathalie Jacobs, Anneke Pot, respectively Springer book Series Editor, Springer Publishing Editor–in–Chief, and Springer Editorial Assistant, for their good job and positive reception of our project in the *Science, Society and Technology Studies*.

The Editor
Lille, France, January, 2015

Remarks for the Reader

All of the papers in this volume have been independently refereed. However, the Editor has respected different individual ideas, historical and epistemological accounts from each of the eminent Authors. The Authors' contributions appear in alphabetical order. A final section provides a summary in the form of an independent point of view. Each of the Authors is responsible for his or her own opinions, which should be regarded as personal scientific and experienced background.

Contents

Contributors

Chris Bissell Open University, Milton Keynes, UK

Bertrand Bocquet Sciences, Societies, Cultures in their evolutions, Lille 1 University Science and Technology, Villeneuve d'Ascq Cedex, France

Assunta Bonanno Physics Education Research Group, Physics Department, University of Calabria, Arcavacata di Rende, Cosenza, Italy

Vitor Bonifácio Physics Department, Research Centre Didactics and Technology in Education of Trainers, University of Aveiro, Aveiro, Portugal

Luca Borghi FAST—Istituto di Filosofia dell'Agire Scientifico e Tecnologico, Università Campus Bio-Medico di Roma, Roma, Italy

Paolo Bussotti The Alexander von Humboldt Foundation, Berlin, Germany

Mario Calamia University of Florence, Florence, Italy

Michele Camarca Physics Education Research Group, Physics Department, University of Calabria, Arcavacata di Rende, Cosenza, Italy

Enrique Chacón Grupo de Investigación en Ingeniería de Máquinas, Universidad Politécnica de Madrid, Madrid, Spain

Yulia Petrovna Chukova The Moscow Society of researchers of nature, Krasnopresnenskiy Ecological Fund, Moscow, Russian Federation

Vincenzo Cioci Research Centre for the Theory and History of Science, University of West Bohemia, Pilsen, Czech Republic/Research Group on Physics Education and History of Physics, University of Calabria, Cosenza, Italy

Eduardo de la Guerra Grupo de Investigación en Ingeniería de Máquinas, Universidad Politécnica de Madrid, Madrid, Spain

Javier Echávarri Grupo de Investigación en Ingeniería de Máquinas, Universidad Politécnica de Madrid, Madrid, Spain

Fernando B. Figueiredo Department of Mathematics/Astronomical Observatory and Institute of Geophysics, University of Coimbra, Coimbra, Portugal/Centre François Viète, University of Nantes, Nantes, France

Giorgio Franceschetti University of Naples "Federico II", Naples, Italy

Rémi Franckowiak Department of Physics, Lille 1 University Science and Technology, Villeneuve d'Ascq, France

Elena Alexandrovna Gavrilina Bauman Moscow State Technical University, Moscow, Russian Federation

Vitaly Gorokhov Research Center for Philosophy of Technology and Engineering Ethics of the Institute of Philosophy, Russian Academy of Sciences, Russian Federation, Moscow, Russia

Ladislav Kvasz Institute of Philosophy, Academy of Sciences of the Czech Republic, Prague, Czech Republic

Roberto Lalli Max Planck Institute for the History of Science, Berlin, Germany

Augusto Marcelli Istituto Nazionale di Fisica Nucleare, Laboratori Nazionali di Frascati, Frascati, Italy

Svetislav Lj. Marković Technical College, Čačak, Serbia

Alessandro Mori Ingegneria dei Sistemi S.p.A, Pisa, Italy

Annibale Mottana Department of Geosciences, University of Roma Tre, Roma, Italy

Peeter Müürsepp Tallinn University of Technology, Tallinn, Estonia

Moon-hyon Nam Konkuk University and Jagyeongnu Research Institute, Seoul, South Korea

Il–seong Nha Yonsei University and The Nha Museum of Astronomy, Yecheon, South Korea

Agamenon Rodrigues Eufrásio Oliveira Polytechnic School, Federal University of Rio de Janeiro, Rio de Janeiro, Brazil

Dominique Pécaud Centre François-Viète, University of Nantes, Nantes, France

Pedro Miguel Pinto Raposo Inter-University Centre for the History of Science and Technology, University of Lisbon, Lisbon, Portugal

Raffaele Pisano Department of Physics, Lille 1 University Science and Technology, Villeneuve d'Ascq, France/Centre François Viète, University of Nantes, Nantes, France

Birutė Railienė Head of Scholarly Information Department, The Wroblewski Library of the Lithuanian Academy of Sciences, Vilnius, Lithuania

Lisa Rougetet Laboratoire d'Informatique Fondamentale de Lille, Lille 1 University Science and Technology, Villeneuve d'Ascq, France

Peppino Sapia Physics Education Research Group, Physics Department, University of Calabria, Arcavacata di Rende, CS, Italy

Maria Sirago Institute Jacopo Sannazaro, Naples, Italy

Bratislav Stojiljković Nikola Tesla Museum, Belgrade, Serbia

Emanuele Zamperini Department of Civil Engineering and Architecture, University of Pavia, Pavia, Italy

Electronics and Information Engineering: A New Approach to Modelling 1880–1950

Chris Bissell

Abstract Most of the literature on mathematical modelling has been devoted to the natural sciences and economics; comparatively little has been written on the specific characteristics of modelling for engineering and technology. This chapter will briefly examine some such specifics in the context of information engineering, by which is meant here the engineering disciplines of electronics, telecommunications, signal processing and control. It is claimed that there are some very significant differences between modelling for engineering—at least for information engineering—and modelling in the natural sciences.

Keywords Modelling · Information engineering · History · Linear systems

1 Introduction

A number of the major scientific figures of the nineteenth century considered deeply the question of modelling: what, precisely, a model is, and how it relates to reality. These figures (Maxwell, Boltzman, Hertz, Lord Kelvin, for example) have been discussed in depth elsewhere, and nothing more will be said on this. (For an interesting overview, however, see Monk 2012). Towards the end of the nineteenth century, electrical and telegraph engineers turned their attention both to modelling the phenomena they observed in the new technologies and to designing systems that would behave in the desired way. Although the mathematical techniques they used were mostly well established, such engineers developed novel ways of developing and employing them. Oliver Heaviside, for example, as well as simplifying and re-casting Maxwell's equations in the vector form that subsequently became universal, developed his operational calculus (effectively equivalent to Laplace transforms) in order to model the behaviour of electrical circuits. By the early decades of the twentieth century the whole array of Fourier techniques was being developed, including convolution in the time domain as an equivalent to multiplication in the frequency domain.

C. Bissell (✉)
Open University, Milton Keynes MK7 6AA, UK
e-mail: c.c.bissell@open.ac.uk

R. Pisano (ed.), *A Bridge between Conceptual Frameworks,* History of Mechanism and Machine Science 27, DOI 10.1007/978-94-017-9645-3_1

Feedback circuits presented a particular challenge, as they were known to become easily unstable. Researchers at Bell Labs studied this problem in detail, resulting in the Nyquist stability criterion of 1932 and Bode's monumental work on circuit design some years later. Karl Küpfmüller in Germany carried out similar, but rather less well-known work.

Some of the most impressive, and still under-estimated, techniques involved graphical tools for design. The Nichols Chart removed the need for difficult computation of closed-loop behaviour based on open-loop modelling or experimental recording, while the Smith Chart, although it will not be discussed here, provided a similar resource for engineers concerned with transmission lines (Bissell 2012). Such charts now form an integral part of computer tools—not, now, to replace calculation, but because they are still unsurpassed as ways of presenting information to the skilled engineer.

2 Oliver Heaviside: Changing the Paradigm

Until comparatively recently, Heaviside was largely neglected in the history of technology, but in the last 25 years or so several significant books and a number of articles have appeared (Nahin 1988, 2002; Mahon 2009). Heaviside was a strange character, largely self-taught, and he engaged in sometimes vituperative exchanges with those with whom he disagreed—particularly with Sir William Preece, chief engineer of the British Post Office.

Born in 1850, Heaviside's major contribution to the development of modelling in information engineering was his operational calculus, mostly published in the 1880s and essentially an application of the Laplace transform. Heaviside used the operator p^n to represent the *n*th derivative in a differential equation, thus transforming an *n*th order differential equation into an *n*th order algebraic one—exactly as the *D*-operator is sometimes taught today. Using this technique, as well as some other quite advanced mathematical methods, he revisited a result derived by William Thomson (Lord Kelvin) on modelling the transmission of telegraph signals. Kelvin had neglected the self-inductance of the transmission line (which was valid, as signalling speeds at that time were sufficiently low for inductance to be negligible), but as speeds increased, the inductance of the cable played a major role, significantly distorting the signals, much to the puzzlement of practising engineers. Exploiting his mathematical expertise, which most electrical engineers at that time were unable to follow, Heaviside came to the counter-intuitive conclusion that loading the line periodically with additional inductance would greatly reduce the problem. Devices to do this were subsequently introduced in practice by M I Pupin and G A Campbell, and became vital to long-distance telecommunication cables.

Heaviside's operational methods received a lot of criticism at the time, and were slow to be accepted. This is hardly surprising, since although the methods were based implicitly on Fourier and Laplace transforms—and some even earlier results—Heaviside was quite happy, for example, to expand his *p*-operator as an infinite series, multiply several infinite series together and even—in a famous paper correcting Kelvin's analysis of the cooling of the earth—use the square root of p! Most mathematicians at the time were baffled that apparently correct results could be obtained in this way and Heaviside himself remarked that the notion of the square root of p is "unintelligible by ordinary notions of differentiation" (although in fact it can be justified with recourse to the gamma function). Among Heaviside's other achievements were to popularize vector calculus and hence to cast Maxwell's equations in their now familiar form. But because of his often cavalier approach to mathematical rigour, he ultimately became alienated from the scientific establishment; and even though he had been elected a Fellow of the Royal Society, one of his papers was famously rejected by them in 1894 as being insufficiently rigorous. He died in poverty and obscurity in 1925.

The claim that Heaviside 'changed the paradigm' is based on two partially conflicting observations. First, although having left school at 16 with only an elementary knowledge of mathematics, he was happy to develop advanced techniques for solving practical problems, many of which were quite beyond the competences of most electrical or telegraph engineers. Note, however, that although he did not pursue formal education beyond the age of 16, he was in the top 1 % of the candidates for the College of Preceptors school-leaving examinations (Mahon 2009). So, in a sense he was crucial to the mathematicisation of communications engineering. Second, he believed that rigorous proofs could be left to others: if his techniques worked, then engineers could use them. This approach to mathematics has coloured information engineering ever since, and the tension between the two observations is still to be found in the teaching of engineering mathematics today, when students and professional engineers often query the usefulness of the formal mathematics taught at university. Finally, although Heaviside did use his operational methods for certain analytical problems (for example, the cooling of the earth problem mentioned above), his methods were also clearly oriented towards synthesis and design, as in the inductive loading of cables. This latter trend marked much subsequent development of modelling for information engineering, and will now be explored in more detail in the following sections.

3 The Development of Linear Systems Theory

Heaviside's operational calculus was given a rigorous foundation by 1920, particularly through the work of T J I'A Bromwich and J R Carson (Bennett 1979). Bromwich related Heaviside's work explicitly to Fourier analysis and contour integration, thus justifying his techniques to the satisfaction of mathematicians. Carson also linked the frequency domain and time domain approaches. Figure 1 illustrates this in a modern form.

A time-invariant linear system (that is, one that obeys the principle of superposition) can be modelled in the frequency domain by its frequency response or transfer function.

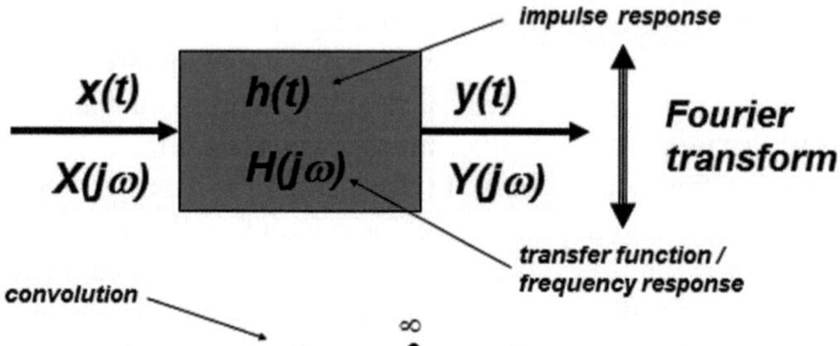

$$y(t) = x(t) * h(t) = \int_{-\infty}^{\infty} x(t).h(t-\tau)d\tau$$

$$Y(j\omega) = X(j\omega) \times H(j\omega)$$

Fig. 1 Linear system input-output relations in a modern form

Multiplying the input spectrum by the system frequency response gives the output spectrum. Alternatively, as Carson showed, the system can be modelled by the convolution integral. Convolving the input signal as a function of time with the impulse response (the ideal response of the system to a delta function) gives the output signal. In fact, Carson wrote the convolution integral in terms of the step response of the system, which he called the indicial admittance, but this is essentially identical.

By the 1920s these approaches were being used for various electrical and telecommunications problems, and electrical and telecommunications engineers were becoming highly adept in moving between the time- and frequency domains as necessary both to understand system behaviour and to design devices such as filters. One important advance was made by the German Karl Küpfmüller, although again it met with some resistance at the time (Bissell 1986, 2006). Küpfmüller realised that important conclusions could be drawn about system behaviour without any knowledge about its component parts. In particular, a model of a perfect so-called 'brick wall' filter (a perfectly rectangular frequency response, with constant gain or attenuation in the pass band, and complete rejection of all other, out-of-band, frequencies) allowed important general conclusions about the limiting behaviour of filters in general, whatever their implementation. Although his name is little known outside Germany, in his native country he is considered to be one of the great founding fathers of information engineering.

4 Filters

By the 1920s there was a need to be able to synthesise filters to a particular specification in order to separate out the various channels in frequency division multiplexing systems. Some of the most important advances were made by

Campbell, Foster and Zobel (USA) and Cauer and Wagner (Germany). One of the most accessible introductions to their ideas is still that of Guillemin (1935), where full bibliographic references can be found. Guillemin had studied with Arnold Sommerfeld in Germany, and was a major conduit of German work on filter design to the English-speaking world.

The emphasis of modelling by now was increasingly on design, rather than analysis. Such wave filters were modelled, like transmission lines, as sequences of lumped passive elements (particularly capacitors and inductors, but sometimes resistors and transformers), but the key was to develop various so-called canonical forms, which could be used to synthesise a particular required filter characteristic. The first significant attempt to do this was by the American R M Foster, who used partial fraction expansions as the basis of the mathematical model, but far more wide-reaching was the work a few years later by the German Wilhelm Cauer, who used continued fractions, and was the first to put circuit synthesis on a sound mathematical basis (Cauer et al. 2000). Figure 2 shows some basic topologies of filter sections which could be linked directly to the corresponding mathematical model. A series of transformations enabled lowpass topologies to be converted to bandpass or highpass topologies simply by manipulating the diagrammatic structure, which was directly isomorphic with the mathematical model. For further information on the history of circuit design, and the mathematical techniques involved, see Belevitch (1962) and Darlington (1999).

A few words should be said here about realisability. The ideal 'brick wall' filter is non-realizable: it is simply impossible in the real world to have an infinitely steep cut-off after a perfectly flat passband. By the 1930s various realisable approximations had been discovered, again exploiting complex mathematical ideas for synthesis and design, and then simplifying the design approach so as to avoid the need for the electronics engineers to carry out—or even fully understand—the underlying mathematics.

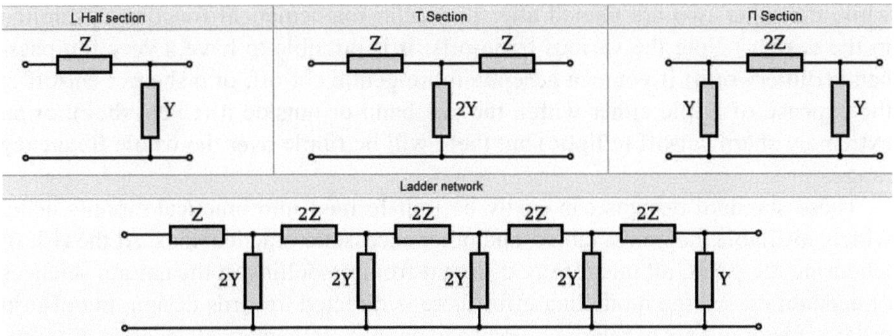

Fig. 2 Some capacitor/inductor topologies for filter design. (Source: Wikipedia, Electronic Filter Topology. Available under the Creative Commons Attribution-ShareAlike License. Accessed 10 May 2013)

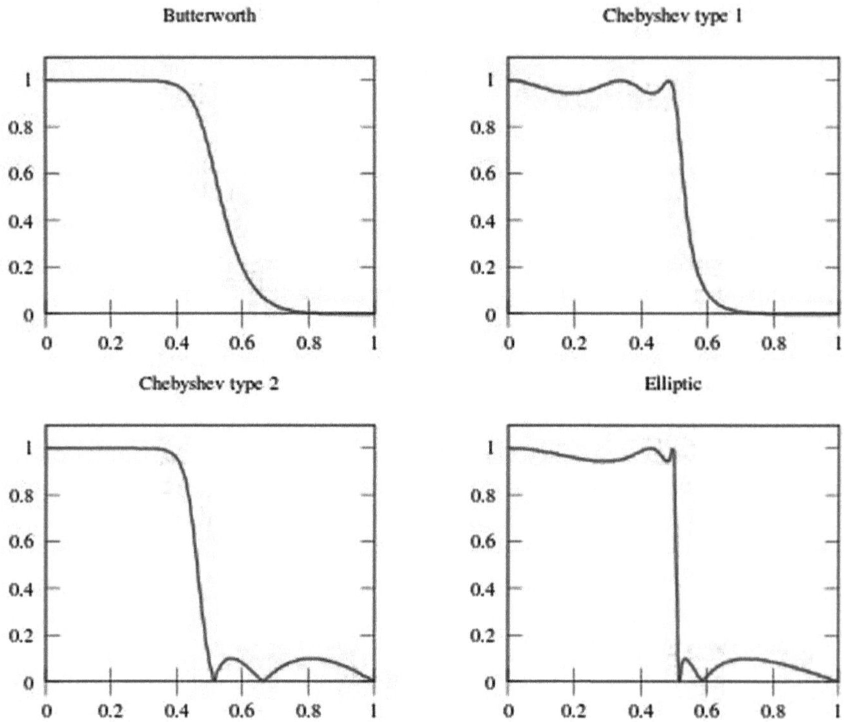

Fig. 3 Some realisable filter approximations. (Source: Wikipedia, Butterworth Filter. Available under the Creative Commons Attribution-ShareAlike License. Accessed 11 May 2013)

Figure 3 shows some normalised frequency response plots for various filter synthesis techniques. (The Butterworth filter is named after its inventor in 1930, while the other two are named after particular mathematical functions exploited in the design.) Note the various trade-offs: it is possible to have a very flat pass-band (Butterworth) if you can accept a more gentle cut-off; or a sharper cut-off at the expense of ripple either within the passband or outside it (Chebyshev); or an extremely sharp cut-off (elliptic) but there will be ripple over the whole frequency range.

These standard designs can easily be transformed into practical circuits using widely available diagrams, tables, and other accessible practical aids. At the risk of labouring the point, all this is very different from modelling in the natural sciences or economics. All the modelling effort here is directed towards design, but also in order to result in a set of relatively easy and barely 'mathematical' techniques (compared with the mathematical effort in deriving and proving results such as those in Fig. 3) that can be applied in practice by the circuit designer.

5 Electronic Feedback Circuits

Feedback circuits had been used in electronics since the early part of the twentieth century, and it was well known that they could become unstable. Sometimes this was a desirable quality—for the design of an oscillator circuit for radio transmission, for example, but in other cases it was very troublesome. In 1927 Harold Black (1898–1983) realised that by means of negative feedback, an amplifier with very low distortion could be realised, something required for transcontinental telephony. He famously sketched his idea on that day's New York Times (Fig. 4).

Black discovered that although such a design could become unstable, its behaviour was not in accordance with the naïve model of the time, a version of the Barkhausen criterion developed for modelling oscillators. The assumption was that instability would occur when the overall loop gain was > 1 and the input and feedback signals were in-phase, thus reinforcing the signal indefinitely each time around the loop. The problem was resolved in 1932 by Black's colleague at Bell Labs, Harry Nyquist (1889–1976). Nyquist realised that you had to take account of the frequency response of the system. If you plotted this on polar coordinates of amplitude and phase, stability was determined when the frequency response curve encircled a certain critical point—in Nyquist's original model (1, 0) but later revised to (− 1, 0) because of a slight generalisation of the model, so that the curves were plotted in a different way. This is illustrated in Fig. 5.

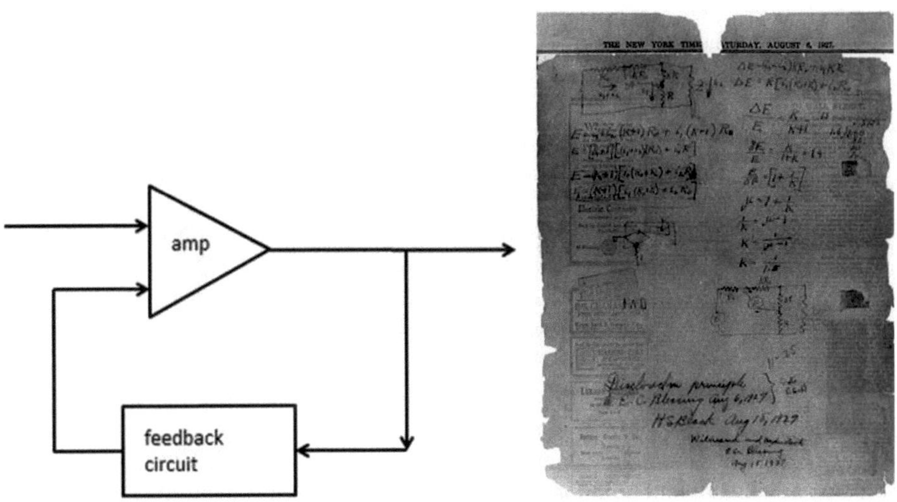

Fig. 4 Black's feedback amplifier. Schematic diagram (**a**) and Black's original sketch (**b**) (Reprinted with permission of Alcatel-Lucent USA Inc)

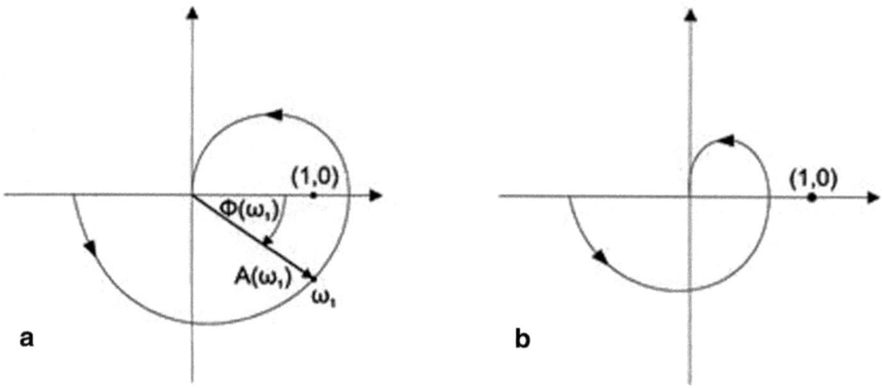

Fig. 5 Unstable (**a**) and stable (**b**) Nyquist plots. (Source: Bissell and Dillon (2012), p 56)

The distance of the curve from the critical point is a semi-quantitative measure of the closeness to (in)stability, as will be discussed further in the following section: note, for example, that if it is possible to decrease the gain in (a) sufficiently, the curve will shrink, ultimately into the stable region.

6 Feedback Control Systems

At the same time as electronics and communications engineers were developing an understanding of linear systems, feedback loops and stability, engineers dealing with control systems were running up against similar problems. Although invented in the latter half of the nineteenth century, servomechanisms underwent a huge development during the early part of the twentieth century for such widely differing applications as ship steering and differential analysers, and the same issues of stability arose. Towards the end of the 1930s, and even more during WW2, there was a coming together of telecommunications and control engineers, as well as mathematicians, particularly in such US wartime centres of R&D such as Bell Labs and MIT. Other important work was carried out in the UK, Germany and the USSR, but it was the American results which determined to a large extent the later development of what became known as classical control theory, so this brief account will be restricted to the main players in the USA.

The history of automatic control has been well documented: see Bissell (2009) for a short account and references to other sources. Basically, a number of researchers realised that the feedback model of Fig. 6, essentially that used by Nyquist in his analysis, could be applied to any other linear feedback system, even to control systems in which the variables might be flow rate, temperature, position, velocity and so on, rather than electrical waveforms.

The problem was, given a knowledge of the open loop transfer function H, how could one determine the closed loop transfer function in order to apply the Nyquist criterion and, if necessary, make changes to the system to ensure desired performance.

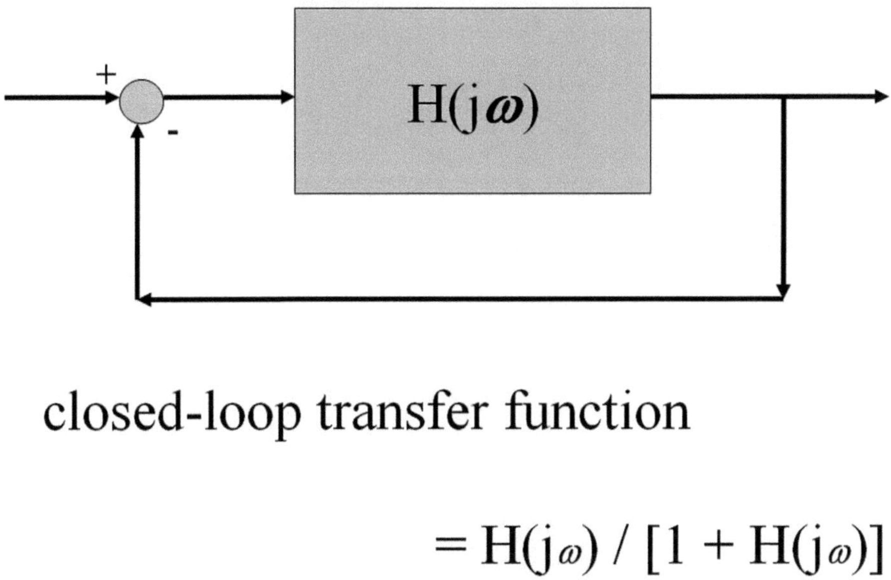

Fig. 6 A feedback system showing the closed-loop transfer function

Yet again, an ingenious mathematical analysis was turned into a straightforward design tool. Nathaniel Nichols (1914–1997) realised during his WW2 work on gun control servos that it was possible to derive closed-loop amplitude and phase loci for any open-loop function. Plotting these loci in amplitude (decibels) and phase (degrees) form resulted in the famous chart shown in Fig. 7. For each open loop point defined on the rectangular grid, a corresponding close-loop frequency response point could be read from the curved lines as amplitude and phase. And that was not all. The closeness of approach of the open-loop locus to the critical point (halfway up the vertical axis in the figure) gave a direct measure of the transient behaviour of the system and the degree of (in)stability. A compensating network or controller could then be designed to be inserted into the feedback loop and shift the locus to a region with a desirable closed-loop dynamic response. The uncompensated closed loop frequency response, very close to the critical point, and the compensated one much further to the right of the critical point, are included in Fig. 7.

The approach to control system design just outlined was developed just before and during WW2, mainly in the US, but also in the UK and to a lesser extent elsewhere. It reached the public domain immediately after the war, but many control engineers from mechanical or process engineering backgrounds found it difficult to understand or accept. The idea that you could talk about the frequency response of mechanical systems or even chemical process plant—where the frequencies involved might be fractions of a hertz—took considerable time to assimilate. The immediate post-war control engineering literature is littered with reports of meeting discussions where participants were still bewildered, or with arguments about the precise best way to plot graphs or specify gain and frequency.

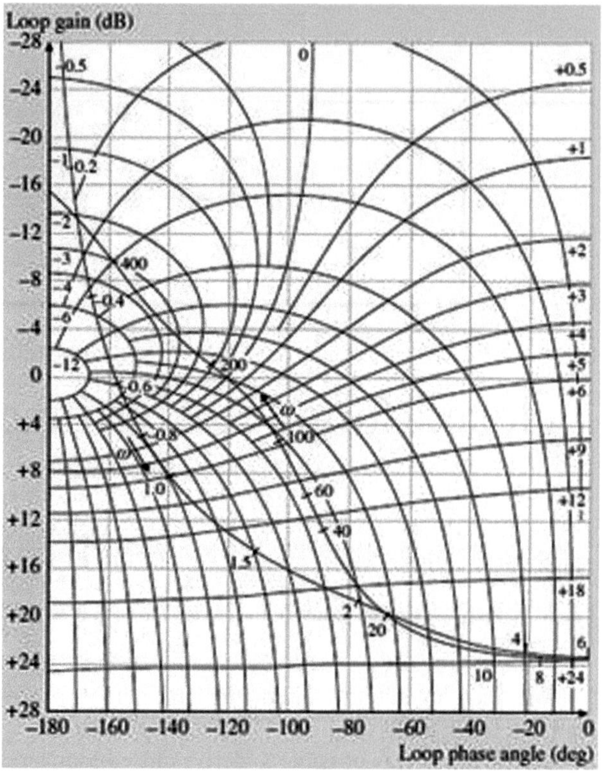

Fig. 7 Redrawing of the original published form (1947) of the Nichols Chart. (Source: Bissell (2009). A history of automatic control. In: Nof, Shimon Y. ed. Springer handbook of automation. Springer handbook series (LXXVI). Heidelberg, Germany: Springer Verlag, pp. 53–69)

7 System Identification

Our final example of the special nature of modelling in information engineering is what is known as system identification. Assuming, as throughout this chapter, that a system can be modelled sufficiently closely as a linear system, how might we obtain a suitable model in order to design a controller? The obvious approach, as is common in the natural sciences, is to make models of each element—such as an electric motor, a valve, a pump, a hydraulic cylinder, and so on—and then combine them into an overall model. While this is sometimes done, more often the system to be controlled may not consist of easily modelled components. Furthermore, strict analysis using Newton's or Kirchhoff's laws can result in an overly complicated or high-order model. When this is the case, direct input-output testing can often identify an appropriate model.

For example, it may be possible to subject the system to an input step change in variable, and from the response directly deduce a model of an appropriate order. A second possibility is to subject the system to a frequency response test—that is, apply an input sinusoid, wait until any transient has died away, and record the output sinusoid. Repeating this over the appropriate range of frequencies gives a direct measure of the frequency response. A third possibility is to apply white noise to the input and then correlate the output with the input: this also results in a knowledge of the system transfer function. All these techniques are valid because the input contains, in principle, all frequencies, and so can identify the complete system frequency response. In the early days of system identification such testing required significant manual input, but it is now highly automated, with much more complicated techniques using computer algorithms to match a model of desired order to the test results. Discussion here has been limited to simple linear cases, but ultimately techniques for non-linear systems were also developed.

Note that all of these methods use a 'black box' approach: no knowledge of the internal constituent parts of the system is necessary to obtain a model. And very often it is possible to obtain an adequate lower-order model of a system which, if analysed in terms of the physical behaviour of its individual components, would result in an unwieldy higher-order model. Furthermore, neither the Nyquist nor the Nichols plots discussed above require an analytical model of the overall system; a model derived by system identification suffices for the ultimate design.

8 The Mathematical Education of Information Engineers

Over the period under discussion the mathematical training of information engineers changed radically, reflecting the increasing use of the approaches outlined in the previous sections. Clearly, any detailed historical account of the development of the teaching of engineering mathematics would be impossible here. However, it is worth briefly mentioning one of the pioneers of the new curricula from the 1930s onwards: Ernst Guillemin (1898–1970) at MIT, who published six seminal works in three decades (Bissell 2008). His first, two-volume work, Communication Networks was considered by many at the time to be too 'advanced' for a supposedly introductory text, introducing transient and steady-state response; network theory; the Heaviside approach; and Fourier analysis—in other words the very material needed to understand the developments of the previous few decades outlined in this chapter. In his preface to the first volume he is unapologetic, and comments: "Methods are frequently designated as advanced merely because they are not in current use. To the student the entire field is new; the advanced methods are no exception. If they afford better understanding of the situation involved, then it is good pedagogy to introduce them into an elementary discussion. It is well for the teacher to bear in mind that the methods which are very familiar to him are not necessarily the easiest for the student to grasp." In Volume 2 (1935) he was also one of the first to stress synthesis, "which has been an important motivating influence in the enlargement of our view and process with regard to network theory, not only had its inception in the field of communications but owes its development almost wholly to workers in that field. Nevertheless, these ideas and principles are too general in nature to remain confined to one field of application [...]" (Fig. 8).

Fig. 8 The most important volumes of Guillemin's pedagogical output

Over a period of several decades Guillemin's pedagogical approach was highly influential in the USA and elsewhere, breaking much new ground on what to teach and how to teach it, although not uncontroversial. His teaching philosophy is clearly presented in Guillemin (1962). While not shirking from presenting students with advanced mathematics, he was also a great proponent of heuristic arguments in order to firmly ground the theory in engineering practice. But, as noted in Sect. 2 of this chapter, tensions of this nature still remain in the teaching of mathematics to information engineers.

9 Conclusion

This short chapter has aimed to present some of the major special characteristics of the way models are used in information engineering, in contrast to much of the literature on modelling in the natural sciences or economics. These characteristics include:

1. The primary aim of the modelling is for system synthesis or design, rather than analysis or explanation.
2. Many of the models are based on quite complicated mathematics, such as complex analysis and Fourier and Laplace transforms, and thus were not immediately accepted by practising engineers when they were introduced.
3. Practising engineers had to cope with considerable changes over the period outlined in this chapter, accepting increasingly more sophisticated models of electrical, electronic or control systems than they had been used to, and learning new languages with which to discuss their design processes.

4. The new models were converted into much simpler form for the use of engineers, particularly graphs and charts which, often isomorphic with the mathematical foundations of the techniques, were able to hide the complexities of the underlying models from practitioners.

The history of modelling in information engineering is thus a complicated story of both mathematicisation and demathematicisation. With the advent of the digital computer, it became possible to carry out very complex engineering calculations automatically. Yet the graphical techniques presented above—as well as many others not mentioned here—remain an essential part of the user interface owing to the succinct and accessible way in which they present ideas.

Acknowledgment Parts of this chapter draw on material presented in much greater detail in Chapters 3 and 4 of Bissell and Dillon (2012).

References

Belevitch V (1962) Summary of the history of circuit theory. Proceedings of the IRE, May, pp 848–855

Bennett S (1979) A history of control engineering 1800–1930. Peter Peregrinus, Stevenage

Bissell (1986) Karl Küpfmüller: a German contributor to the early development of linear systems theory. International Journal of Control 44(4):977–989

Bissell CC (2006) Karl Küpfmüller, 1928: an early time-domain, closed-loop, stability criterion. IEEE Control Systems Magazine 26(3):115–116, 126

Bissell CC (2008) "He was the father of us all." Ernie Guillemin and the teaching of modern network theory. In: History of Telecommunications Conference (HISTELCON '08), Paris, 11–12 Sept 2008

Bissell C (2009). A history of automatic control. In: Nof SY (ed) Springer handbook of automation. (Springer Handbook Series (LXXVI)). Springer, Heidelberg, pp 53–69

Bissell C (2012) Metatools for information engineering design. In: Bissell C, Dillon, C (eds) Ways of thinking, ways of seeing: mathematical and other modelling in engineering and technology. Automation, collaboration and e-services, vol 1. Springer, Berlin, pp 71–94

Bissell C, Dillon C (eds) (2012) Ways of thinking, ways of seeing: mathematical and other modelling in engineering and technology. Automation, collaboration and e-services, vol 1. Springer, Berlin

Cauer E et al (2000) Life and work of Wilhelm Cauer (1900–1945). In: Proceedings of the Fourteenth International Symposium of Mathematical Theory of Networks and Systems (MTNS2000), Perpignan, France, June 19–23, 2000

Darlington S (1999) A history of network synthesis and filter theory for circuits composed of resistors, inductors, and capacitors. IEEE Trans Circuits Syst I: Fundam Theory Appl 46(1):4–13

Guillemin E (1931, 1935) Communication networks, Vols I and II. Wiley, New York

Guillemin EA (1962) Teaching of circuit theory and its impact on other disciplines. Proceedings of the IRE, May 50(5), pp 872–878

Mahon B (2009) Oliver Heaviside: maverick mastermind of electricity. The Institution of Engineering and Technology, London

Monk J (2012) Creating reality In: Bissell C, Dillon C (eds) (2012) Ways of thinking, ways of seeing: mathematical and other modelling in engineering and technology. Automation, collaboration and e-Services, vol 1. Springer, Berlin, pp 1–28

Nahin PJ (1988) Oliver Heaviside, sage in solitude. IEEE Press, New York
Nahin PJ (2002) Oliver Heaviside: the life, work, and times of an electrical genius of the Victorian age. Johns Hopkins University Press, Baltimore (Reprint of the 1988 book with a new, extensive, introduction)

Science-Technology Cross-Hybridization and its Role in the Crisis of the Scientific Method: An Historical Perspective

Assunta Bonanno, Michele Camarca and Peppino Sapia

Abstract Nowadays, in the science-related community, an urgent need has emerged to clarify a crisis of the methodological paradigm known as "Scientific Method". Such a crisis, arising from some recent striking experimental results achieved in experimental sciences, undermines the very foundations of knowledge, with potential serious consequences to the development of future technological applications. In this work, the crisis is analysed within an historical survey on the evolution of the Scientific Method. Furthermore, the role played by cross-hybridization between sciences and technological development is highlighted, throughout the last three centuries, as a possible factor in overcoming that crisis.

Keywords Scientific method · Science · Technology · Scientific knowledge · Scientific method crisis · Science-technology relation

1 Introduction

All those involved in physics, and more generally in experimental sciences, as well as people working in many applicative disciplines related to them, warn today of an urgent need to clarify a crisis of the methodological paradigm known as "Scientific Method" (SM), arising especially from some recent striking experimental results achieved in these sciences. Just to give some examples, we mention here experimental results related to quantum teleportation (Bouwmeester et al. 1997; Ma et al. 2012), a topic challenging common sense and the usual way SM is intended; or the experimental enterprise that lead to the discovery of the Higgs boson (Riordan et al.

A. Bonanno (✉) · M. Camarca · P. Sapia
Physics Education Research Group, Physics Department, University of Calabria, 87036 Arcavacata di Rende, Cosenza, Italy
e-mail: assunta.bonanno@fis.unical.it

M. Camarca
e-mail: michele.camarca@fis.unical.it

P. Sapia
e-mail: peppino.sapia@fis.unical.it

R. Pisano (ed.), *A Bridge between Conceptual Frameworks,* History of Mechanism and Machine Science 27, DOI 10.1007/978-94-017-9645-3_2

2012), a context in which the experimental "evidence" is far beyond the possibility of direct understanding and control of a single individual, or even a small group of experimenters. Conclusions can be drawn from these (and many more) results that undermine the very foundations of knowledge and create a state of permanent confusion, with potential serious consequences on the development of future technological applications. This conceptual framework finds, in recent decades, its philosophical correspondence in the fact that theories such as the so-called "weak thought", or crypto-idealistic, gained ground, leading to a dangerous atmosphere of uncertainty and doubt (Vattimo and Rovatti 2012). In order to highlight the way in which the recent evolution of SM can affect the rate of development of technological implementation of scientific discoveries, it is necessary to define—in a contemporary perspective—what the SM is, and how it has been developed over time.

In the present work we accomplish this task, highlighting in particular the role played by cross-hybridization between sciences, on one side, and technological development, on the other side, throughout the last three centuries. We show by an historical analysis that (i) the achievement of cognitive goals and the formulation of more and more general principles, and (ii) the ability of this knowledge to change (via technological applications) the world and the same structures of thought, are cross-linked in a bi-directional causal way. This way of thinking has been challenged by many recent scientific discoveries.

2 An Overview on Scientific Method

Let us start speaking about the SM in an historical perspective, focusing on how it has been put together over the course of time, through achievement of the fundamental goals of knowledge and formulation of more general principles. The first period of development of scientific thought had its roots in ancient Greece, based its early stage on the experience and doctrines of older Ionian physicists (Britannica 2013), and reached its highest point in the period of the reorganization and consolidation of knowledge coinciding with the age of Aristotle (384 BC–322 BC). This is the context, in fact, in which the concept of "cause" (which constitutes one of the foundations of science) appears and consolidates. Moreover, it is still in ancient Greece that uninterrupted flow of thought going under the name of "realism" is manifested.

We will consider here only three stages in the development of the SM: (i) its birth with the physics of Aristotle (384 BC–322 BC), (ii) the change of perspective with Galileo Galilei (1564–1642) and Isaac Newton (1642–1727) and (iii) its crisis in modern physics (both in astrophysics and in microphysics) mainly due to phenomena which are no longer explainable in intuitive terms and which violate laws for a long time considered valid (at least within the solar system).

We will try to illustrate, first of all, that there actually was a sudden transition from one "unscientific" thought (which lasted almost 2000 years) to a "scientific" one (which was born and developed suddenly). Furthermore, we will show that the

Aristotelian doctrine of "Potency" and "Act" continued to possess unchanged liveliness and explanatory power throughout centuries. Even in the twentieth century, in fact, it will be invoked by Werner Heisenberg (1901–1976), one of the founders of modern physics, to explain the meaning and interpretation of one of the crucial points of quantum mechanics.

The sharp contrast between the Aristotelian philosophy and the Galilean one has produced on the former the layering of an incredible amount of negative criticisms, eventually ruling out its exclusion from the history of scientific thought. From the educational point of view, Aristotelian physics is nearly ignored today and it would almost seem that, not only scientific thought, but also the thought *tout court*, was born with the Copernican revolution. This negative judgment also results from the contribution of authors of primary importance, which led to the persuasion that the Greek civilization was only the "world of approximation". For example, Alexandre Koyré (1892–1964) argued that the Greeks had no real technology, no real physics in our sense (Koyré 1961). However, in spite of this current of thought (which eventually relegated the Aristotelian philosophy to the role of an unfruitful theory, far from any empirical aspect and having a decidedly metaphysical and speculative character), in spite of this we believe that Aristotelian physics has been a rigorous, flexible and sophisticated architecture of thought. In this regard, some illuminating reflections on the subject by Mary B. Hesse (1924–) comfort us. She recognizes that:

> Comparing the arguments by which Aristotle reaches his primary qualities with those by which the atomists reached theirs, it is remarkable that Aristotle relies on common experience of actual properties of bodies, however superficially he may interpret this, while the atomists on the other hand were influenced by the most sophisticated metaphysical speculations. This example and others like it make it somewhat ironical that in the seventeenth century it is atomism which is regarded as progressive and empirical, while the Aristotelian tradition carries the stigma of non-empirical speculation.[1]

We shall not debate here whether Aristotelianism had been a school of thought that gave course to practical applications, because surely they were poor or non-existent. However, we think that Aristotelianism is to be placed in a stream of thought having some degree of continuity with Galilean thought, and we believe that this fact should be properly emphasized. Continuity does not mean identity between the natural philosophies of the two paradigms, but rather the recognition that both have in common the spirit and the desire to be in front of nature with the aim of explaining phenomena, using terms and concepts sometimes similar. The main difference between the methods adopted, however, is to be found in the purposes that they were intended to achieve and in different social references.

Aristotelian physics has a profoundly observational character and looks for "causes" of phenomena (*cognitio certa per causas*), even if it runs out in purely qualitative descriptions. It establishes the rules for correct reasoning because the exercise of reason was then a practice widely appreciated by a gentry who abhorred any manual activity. The Galilean physics, on the contrary, looks for the "laws" of

[1] Hesse (2005).

phenomena (*cognitio certa per leges*); it begins setting up a quantitative representation of phenomena, using mathematical tools and employing a new experimental method. Lucio Russo (1944–), in a fundamental book on the history of science (Russo 2004), shows that in ancient Greece, as well as in the late Hellenistic period, there were a large number of technical applications, and this leads him to believe that ancient societies had created and exploited technological knowledge. However, cited examples suggest that, in the long historical period under consideration, only isolated individuals (such as Archimedes, c. 287 BC–212 BC, just to give an example) were individually able to turn into practical applications the evidence hitherto accumulated in the various fields of knowledge (hydraulics, optics, etc.). Anyway, science needed to become a *social enterprise*, in order to produce true technological applications: and this was done only in the seventeenth century. Indeed, it is not a coincidence that in this period arose "academies", which, unlike "universities", were not expected to simply transmit "knowledge" but to strive to acquire it in an experimental and direct way. In addition, "academies" were required to make public their results so that they could be socially exploited and be useful to the community (Bernal 1971).

As we said, there was not a clean break between a "before" Galileo and an "after" him; there was not a non-scientific thought previous to Galileo and a sparkling scientific thought after him. Indeed, ancient Greece had already reached a high degree of formal mathematical perfection. Just think, to give some examples, of: Euclidean geometry, the development of conics theory by Apollonius (262 BC–190 BC), the work of Diophantus (c. fourth Century BC) on equations or the invention of the method of exhaustion by Eudoxus of Cnidus (408 BC–355 BC), later expanded and used by Archimedes to find the area of a circle. It is likely that the latter method, had it been properly studied, might have led to the birth of calculus with millennia of advance. Despite having developed such a sophisticated mathematics, however, the ancient world produced little progress in the *application*s of mathematics to physical phenomena, since it considered impossible that phenomena of the "sublunary world" could have aspects correlated with each other in a precise and quantitative way. The ancient world, in fact, had a predilection for the "proof" of phenomena over the "discovery" of them, and considered deductive logic superior to inductive logic, which was based on observation and experiment. Unavoidably, therefore, such a conception definitely hindered the development of technology, and consequently delayed the satisfaction of human needs. In this context, it makes sense to ask what the reasons were for the lack of scientific and technical production, so extensive and long lasting for civilizations such as the Greek, the Roman and later the medieval. Some historians of science believe that societies based on slavery (and the bondman was just a slave with a new name), had neither stimuli nor interest to promote the birth of scientific enterprise and technological progress. Such societies, indeed, found the workforce necessary to produce needed goods at practically zero cost in slavery. These researchers admit that there may have been other reasons, however they believe that what we have said above constitutes the basic reason of scientific poverty that lasted two millennia. In this regard, the following passage from Benjamin Farrington (1891–1974) is enlightening:

> The failure of ancient science was in the use that was made of it. It failed in its social function. Even when the acquisition of slaves became more and more difficult the ancients still did not turn to a systematic application of science to production. It is not claimed that such applications never occurred.... But the general truth remains that ancient society had set in a mould which precluded the possibility of an effective search for power other than the muscles of slaves. The dependence of society on the slave is everywhere reflected in the consciousness of the age.[2]

In the period from the mid-sixteenth century to the end of the seventeenth century (coinciding with the birth of commercial enterprises, industries, stock exchange, newspapers and academic reports) a great ferment of ideas appears together with an eagerness to communicate them by all means. Still relying on old institutions such as the monarchy, new ways of goods production and transportation have been growing and this reduced the strict division, until then existing, between the free man and the servants bound to the earth. For goods production, the nascent industry needed a new workforce and so a growing necessity for technical applications of science arises.

The spirit of the time in which Galileo worked (revitalizing the "practical" vocation of the Ionian physicists) was characterized by an underground ferment, by a complex network of subterranean currents of thought which finally resulted in the Enlightenment of the eighteenth century. Galileo, in addition to being an intellectual and a scientist, became an instrument maker, turned into a "vil meccanico" (ancient Italian for "mere technician"). The instruments that came into use in scientific practice were not neutral entities. In their construction, in fact, are already involved a number of ideas and assumptions about aspects of the phenomena that they are aimed to investigate. Moreover, they begin to represent—for the first time in the history of knowledge—a not-neutral channel through which information are acquired, or also a filter between the experimenter and the world under observation. Newly-invented scientific instruments amplified what otherwise could have not been perceived directly by senses and so allowed the measurement of important properties related to the study of phenomena. In this period there is the remarkable circumstance that technological tools, constructed from existing scientific knowledge, in turn led to the acquisition of new scientific knowledge. This is a first significant example of a kind of "feedback" of technology in science, which will become more and more relevant, as we shall see. In the light of this, we can certainly say that one of the peculiar characteristics of the development of Galilean science was constituted by the overcoming of the conception according to which intellectual and manual provisions would be opposite: on the contrary, they can coexist in the same individual. In other words, in that period:

> [...] has gained ground a new consideration of the manual work and of the cultural function of mechanical arts. Moreover, was established the idea of knowledge as a progressive construction, since it consists of a series of results that are placed, one after the other, to an always increasing level of complexity or perfection.[3]

---

[2] Farrington (2000).

[3] Rossi (2001).

The main reason (though certainly not the only one) of the paradigm shift is to be found in the changed social conditions, which favored the release of new forces, and this gave rise to a new way of thinking and being. Nature was seen as an entity that needed to be questioned, forcing her to give answers, and no longer a mother who just needed to be contemplated.

3 From Aristotle to Modern Physics Through Galileo and Newton

In this section, we expose a quick overview of the evolution of SM from its origins to the crisis that arose in modern times, in particular with relativity and quantum physics. The analysis of specific aspects, here and there, provides an opportunity to highlight the claimed science-technology cross-hybridization. A more extensive analysis of this topic will be given in Sect. 4.

3.1 Aristotelian Physics

We said that Aristotelian physics is ultimately a qualitative physics. It observes phenomena in their wholeness and, limited to those occurring in the "sublunary sphere", renounces the possibility of developing a mathematical quantitative analysis of them. On the contrary, Aristotelian physics, instead, leaves this opportunity to the "extralunary world": the motion of stars and planets. The physics of Aristotle conceives of nature as a living body, as a whole: dissecting it would not have resulted in a greater understanding, but, on the contrary, would have meant to kill it. The overall vision of such a physics is therefore entirely holistic: it investigates the substantial unity of the cosmos and the order of the cosmos rests on a "telos" (from the Greek τέλος for "end", "purpose", or "goal"): the search for purpose is the fundamental premise of philosophical inquiry of Aristotle. However, the unity of the universe does not mean its uniformity: the cosmos is finite and is divided into a sub-lunar sphere (where bodies are changeable and alterable in quality and where generation and corruption operate as a part of Being) and an extra-lunar sphere (which is the realm of the eternal and unalterable). Planets (including the Sun) and stars moving around the Earth are formed by a solid crystalline, transparent, weightless and incorruptible substance: the cosmic ether or *quinta essentia*. These planets are embedded in the ethereal homocentric spheres, which rotate in a circular motion around the centre of the universe (which coincides with the centre of the Earth). The perfection of uniform circular motion, without beginning or end, will give rise to a kind of axiom and will become an element of continuity with either later Ptolemaic astronomy or with Galilean physics (and then with Keplerian astronomy). As regards the ultimate composition of the world, Aristotle believed that matter is infinitely divisible (as a process, he meant: an infinite in "potency" not in "act") and that the atomic theory of Leucippus (fifth Century BC) and Democritus (460 BC–370 BC) (who hypothesized an ultimate reality made up of indivisible atoms) is incorrect. In fact, atoms are objects lacking in quality, indivisible and unchangeable, and then, in the Aristotelian view, they are necessarily motionless. Atoms cannot move for two reasons: firstly because the movement would have resulted in a change in them, and secondly because atoms would have had to move in a vacuum, and this is not allowed in Aristotelian physics since this fact would have caused their motion with infinite speed. In the sub-lunar sphere, there are two kinds of local motion: *natural motion* or *violent motion*. Natural motions are those of bodies tending to reach their natural places, respectively, the centre of the Earth and the lunar sphere, and their trajectories are straight. Violent motions, instead, are those caused by an external motor agent. In the latter case motion is determined by contact between the agent and the body, and motion ceases as soon as the contact finishes (*cessante causa, cessat effectus*, in Latin, i.e., once the cause ceases, the effect ceases too). Without going into depth about this delicate subject, we will say that it constituted a lacerating contradiction. In fact, if a body, to move, needs something to move it (*omne quod movetur ab alio movetur*, in Latin, i.e., everything that moves is moved by something else), then you cannot understand what could be the cause of the motion after the body has detached from the motor agent. This view had a real weakness and provoked a multitude of criticism both inside and outside of Aristotelianism. Aristotle, in fact, was forced to define the medium in which the motion takes place as, simultaneously, a resistant and a boosting agent. Aware of this contradiction, he proposed as a solution the following mechanism (*antiperistasis*) to explain the motion of an object thrown into the air. The object in motion leaves behind a void that the surrounding air rushes in to fill, so impressing a boost to the object, which will continue to move forward in a straight line. However, due to the resistance of the medium, the so-generated propulsive thrust will gradually diminish and eventually the object will stop (*Nullum violentum potest esse perpetuum*, in Latin, i.e., no violent motion can be forever). This solution, albeit ingenious, was not convincing, because the medium played two opposite roles. An acceptable solution was proposed, some seventeenth centuries later, by Johannes Buridanus (Jean Buridan, 1295–1361), who introduced the concept of "impetus", a notion looking something like the modern concept of momentum. Before detaching, the thrust agent provides the bullet a quality (the *impetus*), which makes it move forward in a straight line. This quality is progressively consumed during the motion, becoming eventually zero. From now on the motion will cease to be a "violent motion" becoming a "natural motion" and the projectile will fall. Such a description is pictorially shown in Fig. 1 by a sixteenth century drawing depicting the Buridan theory of impetus (AB portion of trajectory), its progressive consumption (BC portion) and then the "natural" falling motion (CD portion). This aspect of Aristotelian physics can be a useful matter for reflection because it allows us to illustrate in an exemplary manner the concept of *common sense representation* of phenomena. This is of great interest because similar problems of representation are widely present in modern quantum physics. We mean here by "common sense" that kind of knowledge that comes from the popular dissemination of concepts and representations

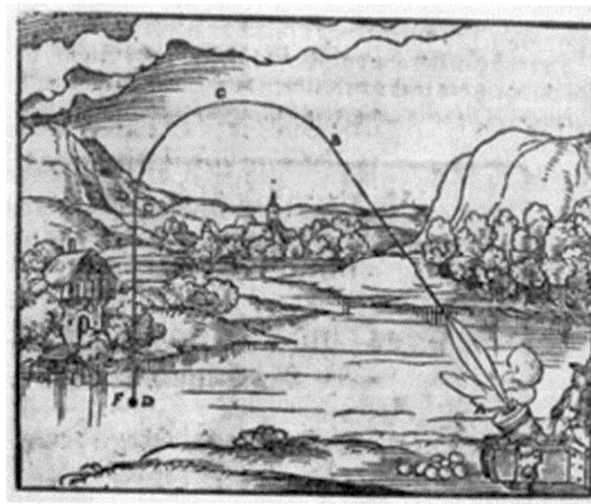

Fig. 1 A sixteenth century drawing by Walther H Ryff (1582) depicting the Buridan theory of *impetus*. (http://www.yorku.ca/lbianchi/nats1800/lecture16a.html)

and which constitutes the opinions, judgments, beliefs and anything else of ordinary people even on specialized topics that go beyond those of daily life. Without writing a treatise on the subject, we note that several authors refer to this concept, considering the theme also worthy of experimentation in physics education (Halloun and Hestenes 1985; Whitaker 1983).

3.2 Galilean-Newtonian Physics

The SM resting on Galilean "sensate esperienze" (i.e. "experiences made through senses"), sometimes considered revolutionary, is only one of the two instruments that Galileo used. The other conceptual tool was the mathematical method (what Galileo called "necessary demonstrations") which was used to formulate hypotheses and express scientific explanations, deducing the consequences that the assumptions implied. In this way, given the state of a mechanical system, one could predict not only its evolution in time, but also new aspects of the phenomena that could then be controlled experimentally, thereby closing a circular chain that, in this way, was always in progress. The failure in verifying predicted values or effects resulting from the hypothesis called for a revision, a settlement or a complete rejection of the hypothesis, and then for the re-formulation of new hypotheses. In Galilean physics, the aspects of phenomena were isolated in a manner appropriate to the possibility of describing them in quantitative and mathematical terms. This physics, moreover, began using the word "experiment" in the modern meaning of the term: man was no longer a passive subject who only watches the phenomenal aspects which nature consents to show him; on the contrary, he interrogates nature, setting the conditions in the belief that nature can adequately respond.

Let us consider, for example, the problem of falling bodies. Galileo realized that, for bodies in free fall, it was not possible to perform reasonable measurements due to their excessive speed. He noted, however, that making them move along an inclined plane could be a flexible way by which to slow down bodies to a desired speed. He conceived then a process that progressively eliminates the roughness of the parchment lining the groove in which he ran the bronze balls. Using pumice to make the parchment more and more smooth, Galileo performed space-time measurements in each state of smoothness. Measuring travelled space posed no particular problems, because the procedure for this was already well known from geometry; many problems, instead, arose from measuring time intervals (note that he could certainly not evaluate them with an hourglass!). Galileo built then a tool that would allow him to execute sufficiently precise time measurements: the *water clock*. He realized that, if you let water gush out of a small hole drilled in the wall of a large container, the amount of water gushing out in a given time interval should be proportional to the time elapsed. In this way, it was possible to make the quantitative comparison (i.e., the "measurement") of time intervals. Notice that, for his purposes, the mathematics of ratios was sufficient, and therefore he did not consider important to measure the absolute elapsed time, but only relations between time intervals (or the amount of water gushed out, collected in a bowl and then weighed). Of course, this "stopwatch" had provided proper relationships only in the case that the flow of water was uniform: Galileo had not proved this, but he sensed it, believed it, and then went on as if this had been proved. This was an approach to the knowledge of the physical world very different from any other previous experience, although it was a mode of proceeding that in many ways resembles the aforementioned "world of approximation", the uncertain, the unproved.[4] Indeed, this is the paradigmatic way forward of physics. The experiment contains errors, unexpected features, some incorrect or not fully tested assumptions, but it is still carried out until you can decide if the results are or not consistent with the assumptions you have made. Yet, you will have the courage to apply mathematics to this mass of inaccurate data. There will be time later to go back to correct, to give account of assumptions made, to repeat the experiments by making the experimental apparatus less uncertain. In this regard, it is remarkable to note that, when successively the mechanics was strictly founded in its principles, the Galilean intuition on the proportionality relationship between the amount of water and the time interval was confirmed true with a high degree of accuracy, depending on the ratio between the hole area and the area of the large container. However, Galileo had stated this already dozens of years before!

The evolution of the late seventeenth century, and then throughout the entire eighteenth century, did not change the overall picture of SM that had started with Galileo. Its subsequent development was both experimental and theoretical. The

[4] Koyré (1961) thinks that the "world of approximation" is that of the Aristotelian philosophy, but on a closer inspection, we find that this is not true. Galilean physics does not reject the "world of approximation", but instead uses it, because it receives suggestions for a glimpse of the "world of precision".

paradigm shift about the nature of motion came with Newton, whose working was characterised, on one side, by the achievement of an exemplary clarity in the enunciation of principles of the new physics of motion, and, on the other side, by alchemical studies of which we will not talk here. He established that bodies in rectilinear and uniform motion do not need any cause to do this, while causes (efficient, in the Aristotelian sense) are required to change the state of motion of bodies, and these causes are called *forces*. He stated that any force acting on a body, causes a variation in the amount of motion (the product of mass by velocity), and also affirmed that pairs of bodies mutually exchange interactions equal in intensity, in contrary to and directed along the line joining them. Starting then from Johannes Kepler's (1571–1630) laws, Newton clearly stated the universal gravitation law, which unified the laws of physics of the earth with those of the heavens, so overcoming the Aristotelian distinction between the "sublunary sphere" and the "extralunary world". Newton stated, moreover, that there was no need for any animism, or for ad hoc assumptions ("*hypoteses non fingo*"). In order to accomplish this enormous task of systematization he had the need to introduce absolute space and time to describe the motion. In the Newtonian picture, these two concepts represent the frame, we could say the passive "container", of physical phenomena (Jammer 2007, 2012).

In the eighteenth century and the first half of the nineteenth century, areas such as analytical mechanics and mathematical physics born and developed that deduced all the consequences contained in the Newtonian and Galilean postulates and at the same time applied SM to the study of physical phenomena in many different fields of mechanics, from statics to hydraulics, to atmospheric phenomena. Other areas not yet included within mechanical phenomena (such as optics, thermal phenomena, electricity and magnetism) began to seek their experimental and theoretical systematization. The success of SM, between the beginning of the eighteenth century and much of the nineteenth century, led to the unification of different phenomenal fields seemingly unrelated to each other. In this period, under the impulse of manufacturing industries, ironworks and the nascent railways, the theory of heat (thermodynamics) and electromagnetism found a fertile ground for development. The need arose for a solid theory of thermal machine's efficiency so as to take the most out of them, consuming the minimum amount of coal needed to produce a given mechanical energy. In this context, it is not surprising that the formulation of the second principle of thermodynamics (which deals with the efficiency of thermal machines) saw the light before the first principle, even before a clear definition about the nature of heat had been achieved! The method is always the same: if you go steadfastly ahead, successively someone will systematize the results. In the mid-nineteenth century, Joule showed that mechanical work and heat were both forms of the same physical quantity, energy, convertible into one another (although not symmetrically) and this laid the foundation for the formulation of a new principle of conservation of energy including heat.

Electrical and magnetic phenomena throughout the seventeenth and eighteenth centuries had been known only in their static form and therefore had appeared as different phenomena that did not influence each other. However, already in the early nineteenth century, Hans Christian Ørsted (1777–1851) showed that dynamic forms of electricity, i.e. electric currents, affect the orientation of magnetic needles, and after a few years Michael Faraday (1791–1867) conversely demonstrated that magnets in relative motion with respect to an electric circuit cause the appearance of electric currents. So the conclusion was reached that electricity and magnetism were just different aspects of a single entity, i.e., the electromagnetic field. A unified mathematical formulation of electromagnetic phenomenology (made shortly thereafter by James Clerk Maxwell (1831–1879), with strong influences of Faraday's approach) provides important insights about the dialectical relationship between physics and mathematics that has characterized the development of these disciplines in the nineteenth century (Barbin and Pisano 2013). In particular, some author traces back to this context the emergence of a new discipline: *physical mathematics*.[5] The concept of a field, which had begun to enter into physicists' reasoning with Faraday, almost in low-key, had become pervasive, dominating the whole of physics to the present day. The emergence of the concept of a field, in a sense, came to revive the old dichotomy of "natural motion—violent motion" in the new form of "action-at-a-distance—contact-action".

3.3 Modern Science and the Crisis of SM

The evolution of the scientific method to the beginning of the twentieth century not only attained specific theoretical and applicable achievements, but also the reflection on its foundations brought to light some of the cornerstones on which rested the whole epistemological framework. The image of the world that science has been building rests on three assumptions that must be accepted without being able to give any proof. The first is *realism*, namely the view that regularities observed in physical phenomena are caused by the existence of a physical reality that is independent of the observer. The second hypothesis concerns the free applicability of *inductive inferences*, a kind of reasoning allowing us to draw valid conclusions, starting from consistent observations. The third hypothesis, called *Einstein separability* or *locality*, states that no action (or influence) of any kind can propagate faster than the speed of light in a vacuum. These three assumptions, accepted as postulates, are the basis of so-called *local realistic theories* (Nagata 2008; Hardy 1992). Experiments performed in particle physics have achieved results that pose a dilemma: predictions obtained by applying quantum mechanics rules are different from those obtained by imposing the validity of the above-mentioned premises. It follows that either local realistic theories are incorrect or quantum mechanics is wrong. Let us proceed in an orderly manner. In the last quarter of the nineteenth century, numerous experimental results accumulated which did not find any explanation, neither within mechanics nor within electromagnetism. The processing of these experimental results and their consequences led, after a long journey

[5] Pisano (2013). See in particular the discussion on the peculiarity of *physical mathematics* with respect to *mathematical physics*.

from which we cannot digress here, to give up one of the fundamental characteristics owned by the Aristotelian-Galilean-Newtonian time, i.e., the uniqueness of the temporal ordering of events. In other words, if an event A temporally precedes another event B, when these are observed by a given reference system, it is possible that B precedes A, if observing by another reference system. This characteristic is the basis of one of the most important epistemological problems posed by the special theory of relativity: the possibility of a *causal inversion* of events. It is in fact a peculiarity of the principle of cause and effect that the cause precedes in time the effect. If then in a reference system, A was the cause of B, it could happen that in a second system in motion with respect to the first, B was the cause of A. Indeed, this apparent paradox is resolved if we consider the peculiar causal structure of special-relativistic spacetime, the so-called Minkowski spacetime (see Fig. 2). This profound modification of the causality structure of space-time ordering of events deeply challenged the traditional structure of the SM. The results of the crisis further worsened (as regards the SM applicability) by the peculiarities introduced by the other big scientific revolution of the twentieth century (quantum mechanics) which gave a leading role to the observer, besides the observed system. In fact, the Aristotelian-Galilean-Newtonian SM implicitly postulates that an observer exists, independent (*absolutus*, in Latin) on the observed world and, in turns, that the properties of the observed world are independent of the way the observer detects them. In other words, there is a conceptual separability between the two entities.

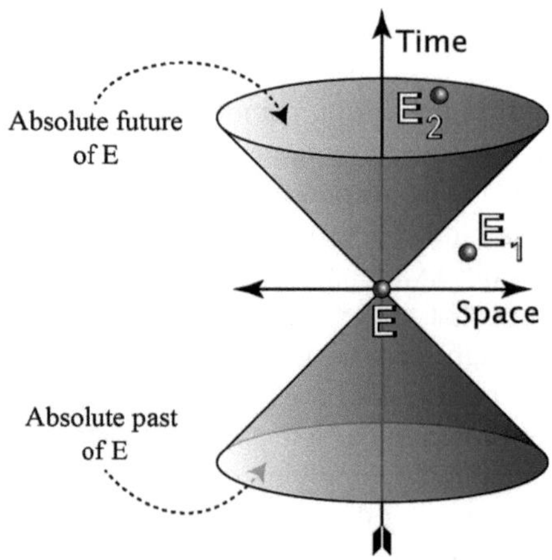

Fig. 2 The existence of an upper limit to the propagation speed of any signal induces a peculiar causal structure on the spacetime of special relativity. Considering a given event *E* (the present) spacetime is divided into three zones: the two inside the cone are causally connected with the event *E* and represent respectively the absolute past and the absolute future, because between the events of these regions and the event *E*, it is not possible to reverse the temporal ordering. The region of spacetime external to the cone, instead, constitutes the elsewhere of E: points in this region cannot be causally connected with *E*, because this would require the transmission of a signal at a speed exceeding that of light (In the figure, *E2* belongs to the absolute future of *E*, while *E1* belongs elsewhere with respect to *E*)

The classical notion of SM assumes also that the portion of the observed world under study (which we usually call "the system") is unambiguously defined. Special Relativity, moreover, adds to these characteristics the further requirement that the interactions between subparts of the system underlie the peculiar causal structure described above, which excludes in particular that among such parts, causal connections can exist propagating at a speed greater than that of the light in a vacuum. Such an idea of SM is strongly challenged by a class of phenomena expected by quantum mechanics, characterized by a counterintuitive property called "entanglement" (Gilder 2008). Scientific research on these properties has already led to implementing, on a subatomic scale, a phenomenon known as "quantum teleportation" (Bouwmeester et al. 1997; Ma et al. 2012). Such a phenomenon, whose name brings to mind Star Trek, far from having anything sci-fi, can be an important step in the realization of the next generation of computers: the quantum computer.

4 Science-Technology Cross-Hybridization

As we have seen so far, the history of the SM shows more or less clearly that the relationship between scientific inquiry and technological applications is by no means simple and unidirectional. However, if you try asking the ordinary person (the men-in-the-street, but also—in-the-school, or—in-the-university!) about the relationship between science and technology, almost certainly you will get in response a statement tending to describe the latter as an applicative consequence of the former; in other words, you probably get the opinion that science precedes technology, determining and driving its evolution. To see a proof, just take a trip through some of the forums on the internet to run into statements like:

> Science discovers fundamental information about how the universe works. Technology is the practical application of that information, or knowledge. A computer is an example of technology; in order to invent one, it is necessary to know a lot of fundamental science. Science sets the stage for technology, which produces useful devices. There would be no laptops without the fundamental discoveries of science.[6]

However, a reflection on the evolution of the Scientific Method, which we have outlined so far, clearly suggests that such a view is extremely simplistic. The supposed relationship of subordination of technology in science has seemed simplistic since the days of Galileo, had configured itself as increasingly reductive by the end of the nineteenth century, to reach a crescendo in the present day. A positive consequence of the growth of this awareness at the academic level is the fact that now, since the nineties of the last century, a growing recognition is establishing of the challenges that the real nature of the relationship between technology and science raises, in connection (also, but not only) to the education of new generations (Layton 1993;

[6] Answer to the question "What is the relationship between science and technology?", retrieved online on April 2013 at the URL: http://wiki.answers.com/Q/What_is_the_relationship_between_science_and_technology.

Barnes 1982; Hamburg 1984; Lederman 1987; Meyer 2000; Subramanian and Soh 2010). Moreover, the bidirectional interdependence science-technology is progressively extended to include engineering too, in a kind of triangle whose sides represent the multidimensional interdependence among scientific inquiry, engineering design and technological development. This view has recently been authoritatively acknowledged and reaffirmed in some official documents aimed at outlining the developments in scientific and technological education for the next decades:

> The fields of science and engineering are mutually supportive, and scientists and engineers often work together in teams, especially in fields at the borders of science and engineering. Advances in science offer new capabilities, new materials, or new understanding of processes that can be applied through engineering to produce advances in technology. Advances in technology, in turn, provide scientists with new capabilities to probe the natural world at larger or smaller scales; to record, manage, and analyze data; and to model ever more complex systems with greater precision. In addition, engineers' efforts to develop or improve technologies often raise new questions for scientists' investigations.[7]

The enlargement of the traditional science-technology connection framework, to include engineering, allows us to re-interpret, in a more complete way, the last centuries' progress. In fact, the interdependence of scientific inquiry (with consequent discovery of laws and principles) and engineering implementations (with its resulting technologies) comprises a number of ideas about how the fields of science and engineering interconnect. First of all there is the suggestion that scientific discoveries enable engineers to do their work. For example, in the last decades of the nineteenth century there have been a number of scientific developments regarding the theoretical existence of electromagnetic waves (Maxwell), and then the confirmation of their actual existence (Hertz, Hallwachs). At the turn of the nineteenth and twentieth century, engineers fully developed the technology allowing them to produce and detect electromagnetic waves of different wavelengths. These technological and engineering advances grew up exponentially over the following decades, driven mainly by military needs, culminating during the World War II with the full development of radar technology. On the other hand, the maturity of microwave technology led a few decades later (1964) Arno Penzias and Robert Wilson (who worked in developing a satellite radio system) to accidentally discover the cosmic microwave background radiation (Chown 1988). In turns, this scientific discovery—a direct (and accidental) consequence of a technological development—opened a new horizon in science, even driving the birth of a new science: experimental cosmology (Brush 1992). Furthermore, radiofrequency and microwave technologies gave impulse to other branches of science, besides cosmology: to cite an example, they allowed develop of nuclear and electronic magnetic resonance systems, which in turns have contributed so much to new scientific discoveries in the field of the structure of matter. Moreover, magnetic resonance technology gave a burst to modern neurological sciences, allowing researchers to see the brain "in action" by means of functional magnetic resonance imaging (fMRI) and so leading

[7] NRC (2012).

to new scientific modeling of the brain-mind relation and then opening new routes in neuropsychological sciences (Greene et al. 2001; deCharms 2008; Miller 2012; Smith 2012; Underwood 2013). This interplay between research and applications is pictorially illustrated in Fig. 3.

We mentioned above the circumstance that the development of radar technology was accelerated by the demands of wars. This is not an isolated case: regardless of any opinion you may have about it, it is a matter of fact that war has often been a key driver of progress either technological (directly) or scientific (as a side effect). An historically sounding example is constituted by the well-known correlation between the development of military technology during the Napoleonic Wars and the birth of modern thermodynamics. In fact, the French physicist Léonard Sadi Carnot (1796–1832) worked intensively on those ideas that would become the foundations of thermodynamics, believing that the improvement of engine efficiency was essential to help France win the Napoleonic Wars. Similarly, Benjamin Thomson (count Rumford, 1753–1814) gave fundamental contributions in understanding the nature of heat (exceeding the then-current "caloric" theory) by reflecting on the problem of great heating affecting the boring procedure (drilling) of war cannon at the Munich arsenal.

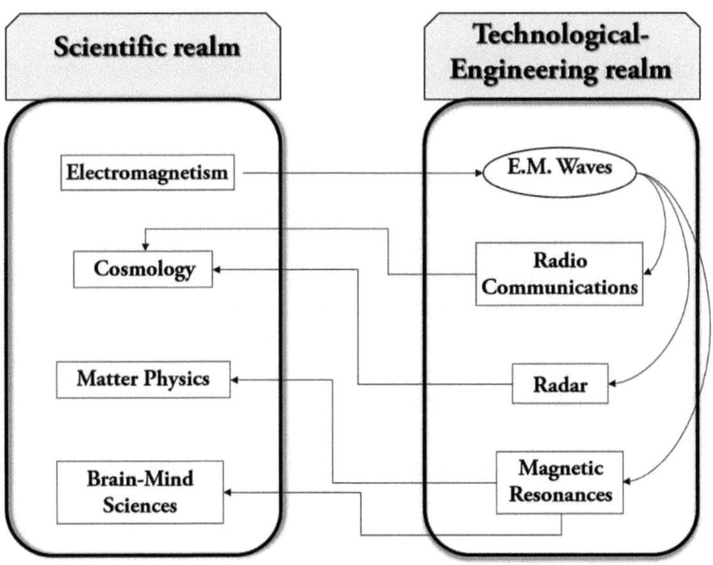

Fig. 3 An example of bidirectionality of the relationship between scientific inquiry and engineering technological implementations. Scientific discoveries trigger technological applications, which in turns may lead to new scientific discoveries: this is what we called "Science-technology cross-hybridization"

Many other illustrations could be given on the route linking war needs to techno-scientific advances. A very timely example of the correlation often existing between wartime necessities and scientific and technological development (in that order) is a consequence of the 2001 Afghanistan War succeeding the September 11 terrorist attacks. Following this war, the growing problems related to limb losses (due to landmines and heavy gunnery) led the US Defense Advanced Research Projects Agency (DARPA) to launch in 2006 the "Revolutionizing Prosthetics" program,[8] aiming to create a neurally controlled artificial limb in the attempt to restore near-natural motor and sensory capability to upper-extremity amputee patients. The technological efforts prefunded in this program recently led to interesting progress in neurosciences, regarding nerve tissue growth (Kullen and Smith 2013): a cutting-edge fully scientific topic, directly derived from a technological/engineering effort (in this case biomechanical engineering), in turns promoted by a war context.

Let us close this section by emphasizing once again that science and technology continuously interact and move each other forward, by quoting another passage from the National Research Council Framework for Science Education:

> New insights from science often catalyze the emergence of new technologies and their applications, which are developed using engineering design. In turn, new technologies open opportunities for new scientific investigations.[9]

5 Conclusion

In this work we have addressed two issues concerning the technical/scientific evolution, which are in our opinion widely correlated: (i) the creeping crisis of the classically understood Scientific Method, and (ii) the mutual-inspiration relationship between science and technology, which has been growing parallel to the above mentioned crisis. To illustrate the nature of this crisis (mainly resulting in the radical change of the role of the observer in modern science), we have first accomplished an overview of the birth and evolution of the Scientific Method, from Aristotle through Galileo and Newton, to reach the century just elapsed. In particular, we have pointed out that Aristotelian physics has a profoundly observational character and looks for "causes" of phenomena, elaborating purely qualitative descriptions of them in a teleological conceptual frame, establishing the rules for correct reasoning. Galilean physics, on the contrary, looking for the "laws" of phenomena (avoiding any finalistic connotation) begins setting up a quantitative representation of an observed realm, using mathematical tools and employing a new experimental method. This method, which is what we commonly call the Scientific Method, continues essentially unchanged from almost the past four centuries. However, recent decade's development of both theoretical and experimental physics have severely

[8] http://www.darpa.mil/Our_Work/DSO/Programs/Revolutionizing_Prosthetics.aspx.

[9] NRC (2012).

challenged SM by questioning some of its formal prerequisites, such as the observer-observed conceptual separation and the causal structure of space-time events.

Moreover, the historical overview on the SM we have performed, allowed us to introduce the issue of the bi-directionality of the science-technology relationship, through the illustration of various examples in different epochs. In particular, we have given an interpretation of nearly 2000 years of stagnation in scientific progress (since the time of Aristotle to that of Galileo) by tracing it back to the lack of effort of technological innovation in this long period. Such a stagnation was in turn determined by the wide availability of a workforce at low cost (or no cost at all) because of various (and gradually changing through eras) forms of slavery. Finally, we have given some historical examples suggesting the role of wartime needs as thruster of technological (and, consequently, scientific) advancement and as moments in which the science-technology interplay was particularly evident. As regards this thesis, last, we want to emphasize that we state it as a matter of fact, apart from any ethical evaluation of the role played by the wars in the wider social context.

References

Barbin E, Pisano R (eds) (2013) The dialectic relation between physics and mathematics in the XIXth century. Springer, Dordrecht

Barnes B (1982) The science-technology relationship: a model and a query. Social Studies of Science 12:166–172

Bernal JD (1971) Science in history, volume 1: the emergence of science. MIT Press, Cambridge

Bouwmeester D, Pan J-W, Mattle K, Eibl M, Weinfurter H, Zeilinger A (1997) Experimental quantum teleportation. Nature 390:575–579

Britannica (2013) "Ionian school": entry in the Encyclopaedia Britannica and resources therein. http://www.britannica.com/EBchecked/topic/292908/Ionian-school. Accessed April 2013

Brush SG (1992) How cosmology became a science. Scientific American 267:34–40

Chown M (1988) A cosmic relic in three degrees. New Scientist 119:51–55

deCharms C (2008) Applications of real-time fMRI. Nature Reviews Neuroscience 9:720–729

Farrington B (2000) Greek science: its meaning for us. Spokesman Books, Nottingham

Gilder L (2008) The age of entanglement: when quantum physics was reborn. Knopf, New York

Greene JD, Sommerville RB, Nystrom LE, Darley JM, Cohen JD (2001) An fMRI investigation of emotional engagement in moral judgement. Science 293:2105–2108

Halloun IA, Hestenes D (1985) Common sense concepts about motion. American Journal of Physics 53:1056–1064

Hamburg DA (1984) Science and technology in a world transformed. Science 224:943–946

Hardy L (1992) Quantum mechanics, local realistic theories, and Lorentz-invariant realistic theories. Physical Review Letters 68:2981–2984

Hesse MB (2005) Forces and fields: the concept of action at a distance in the history of physics. Dover Publications, Mineola

Jammer M (2007) Concepts of time in physics: a synopsis. Physics in Perspective 9:266–280

Jammer M (2012) Concepts of space: the history of theories of space in physics. Dover Publications, Mineola

Koyré A (1961) Du monde de l'Ã peu près Ã l'univers de la précision. Critique 4, 28:806–823. In Etudes d'histoire de la pensée philosophique, pp 311–329, Paris

Kullen DK, Smith DH (2013) How artificial arms could connect to the nervous system. Scientific American 308:52–57

Layton D (1993) Technology's challenge to science education. Open University Press, Bristol
Lederman L (1987) Science and technology policies and priorities: a comparative analysis. Science 237:1125–1133
Ma XS, Herbst T, Scheidl T, Wang D, Kropatschek S, Naylor W, Wittmann B, Mech A, Kofler J, Anisimova E, Makarov V, Jennewein T, Ursin R, Zeilinger A (2012) Quantum teleportation over 143 kilometres using active feed-forward. Nature 489:269–273
Meyer M (2000) Does science push technology? Patents citing scientific literature. Research Policy 29:409–434
Miller G (2012) Why are you and your brain unique? Science 338:35–36
Nagata K (2008) Classification of local realistic theories. Journal of Physics A: Mathematical and Theoretical 41:155308–155318
NRC (2012). National Research Council. A framework for K-12 science education: practices, crosscutting concepts, and core Ideas. Committee on a conceptual framework for New K-12 science education standards. Board on Science Education, Division of Behavioral and Social Sciences and Education. The National Academies Press, Washington, DC
Pisano R (2013) Historical reflections on the physics mathematics relationship in electromagnetic theory. In: Barbin E, Pisano R (eds) The dialectic relation between physics and mathematics in the XIXth century. Springer, Dordrecht, pp 31–58
Riordan M, Tonelli G, Wu SL (2012) The Higgs at last. Scientific American 307:66–73
Rossi P (2001) The birth of the modern science. Blackwell, Oxford
Russo L (2004) The forgotten revolution: how science was born in 300 BC and why it had to be reborn. Springer, Berlin
Smith K (2012) Brain imaging: fMRI 2.0. Nature 484:24–26
Subramanian AM, Soh P-H (2010) An empirical examination of the science–technology relationship in the biotechnology industry. Journal of Engineering and Technology Management 27:160–171
Underwood E (2013) How to build a dream-reading machine. Science 340:21–27.
Vattimo G, Rovatti PA (eds) (2012) Weak thought. SUNY Press, Albany
Whitaker RJ (1983) Aristotle is not dead: student understanding of trajectory motion. American Journal of Physics 51:352–356

How the Movie Camera Failed to Become Part of the Standard Astronomical Observational Toolkit (1895–1914)

Vitor Bonifácio

Abstract A series of technological developments driven both by scientific pursuits, particularly Étienne-Jules Marey's motion studies, and commercial reasons led to the birth of Lumières' 1895 'cinematographe'. Its ability to automatically record a sequence of photographic images had previously been attained by Jules Janssen's photographic revolver, an instrument developed to time with high precision the contact instants of the 1874 transit of Venus. While with this pedigree one might expect a rich use of movie cameras in astronomical observations after 1895, current historical accounts of the development of both cinema and astronomy usually cite none. Is this due to historiographical reasons and/or the new technology failed to become part of the astronomers' observational toolkit? Analysing all astronomical movies attempted or shot between 1895 and 1914, we concluded that the low usage of movie cameras in this time period was a consequence of a lack of suitable observable subjects and the small film frames used. While new technological apparatus may open unexpected lines of scientific enquiry, they must also struggle to find a place and function against already established ones. It was precisely this inability to stand out that led to the astronomical moving pictures' fate as a rarely used and indeed seldom useful technique.

Keywords Science movies · Scientific films · Early cinema · Cinema development · Astronomy · Solar eclipses

1 Introduction

Despite its pedigree, earlier astronomical films have been largely ignored by contemporaneous accounts of early-cinema history. Recent research into the role played by scientific pursuits in the development of what would later be called cinema only alludes to Jules Janssen's (1824–1907) photographic revolver and the 1874 transit of Venus observation (Tosi 2007). The DVD, "La vera nascita del cinema. Le origini

V. Bonifácio (✉)
Physics Department, Research Centre Didactics and Technology in Education of Trainers, University of Aveiro, Aveiro, Portugal
e-mail: vitor.bonifacio@ua.pt

R. Pisano (ed.), *A Bridge between Conceptual Frameworks,* History of Mechanism and Machine Science 27, DOI 10.1007/978-94-017-9645-3_3

del cinema scientifico" (The true birth of cinema. Origin of scientific cinema) has chapters describing early cinematographical applications in the fields of Botany; Biology and Physiology; Medicine and Surgery; Technical Sciences (ballistic studies); Mathematics and Ethnology but not Astronomy (Tosi 2005). Likewise the "Encyclopedia of Early Cinema" does not cite a single astronomical application despite containing two entries entitled "Scientific films: Europe" and "Scientific films: USA" (Curtis 2010; Lefebvre 2010). Furthermore a recent and otherwise excellent book misunderstands, in our opinion, the reason behind the lack of applicability of astronomical cinematography to high precision time measurements since a time stamp could be simultaneously recorded on film by a convenient choice of apparatus. (André 1912; Carvallo and Vlès 1912; Vlès 1914; Canales 2010, p. 151). On the other hand we found that the few previous works tackling the history of astronomical cinematography although knowledgeably written are incomplete (Vlès 1914; Korff 1933; Bourgeois and Cox 1933; Atkinson 1953; Leclerc 1956; Bianchi 1994).

In this paper we endeavoured to bring to light all astronomical cinematographic attempts made with film cameras for scientific purposes between 1895 and 1914 (Sect. 2). To the extent that astronomers split the continuous flow of time to extract data from different stills, the coincidence of recorded and viewed images per second ratios was usually irrelevant. In fact slow motion and speed up techniques were considered useful research tools earlier on in several scientific areas (Vlès 1914; Chaperon 1995). Any possible pedagogical and commercial value of astronomical moving pictures will not be discussed here.

Possible astronomical applications of movie cameras are analysed in Sect. 3. Finally, in Sect. 4 we present our conclusions.

2 Moving Pictures: 1898–1914

On 10 February 1873 Janssen presented at the Paris *Académie des Sciences* his plan to construct a new instrument "that would enable one to obtain a series of photographs at very short regular intervals" (Janssen 1873). This would, in principle, allow timing with high precision the instants of contact between Venus and the Sun at the 1874 transit. At least nine photographic revolvers of either Janssen or Warren De La Rue (1815–1889) design were used in 1874. A few plates still survive today although not Janssen's expedition original obtained by the Brazilian Francisco António d'Almeida (?–?) at Nagasaki (Japan) (Launay and Hingley 2005; Mourão 2005).

Current film history recognizes Janssen's photographic revolver as the earliest of all cinema precursors (Sicard 1998; Launay and Hingley 2005; Tosi 2005, 2007). In the following years Janssen's 'idea' was developed by, amongst others, Eadweard Muybridge (1830–1904), Etienne-Jules Marey (1830–1904), Georges Demenÿ (1850–1917) and Thomas Edison (1847–1931). A process that led to Auguste (1862–1954) and Louis (1864–1948) Lumière public presentation of the 'cinématographe' on 22 March 1895 in Paris (Tosi 2007)

2.1 *William Edward Wilson Sunspot*

Less than 2 years later, in early 1897, the Irish amateur astronomer William Edward Wilson (1851–1908) ordered, for his personal observatory, a "special form of Cinematograph […] in order to try whether it would be possible to show visually the changes in the forms of Sun-spots" (Wilson 1898). The instrument was only delivered at the end of the year when the Sun was, according to Wilson, too low in the sky to use it. The first 'film' was shot on August 9th 1898 while 4 days later "400 photographs of a sunspot between 10:45 am and 2:30 pm" were obtained (Wilson 1899; Mc Connell n.d.).

Wilson's film experiments were probably short lived since they are neither mentioned in later observatory reports presented to the Royal Astronomical Society nor in the 1900 book "Astronomical and physical researches made at Mr. Wilson's observatory, Daramona, Westmeath" (Wilson n.d., 1900, 1901, 1902, 1903, 1904a, 1905a). The local weather apparently played an important role in this outcome. "The intervals of sunshine that we get are too short to make it [the film camera] of any value" wrote Wilson in April 1906 to George Ellery Hale (1868–1938) (Wilson 1906).

We are unsure when Wilson and Hale got in touch. From the extant correspondence it seems likely that the first contact occurred in 1904 as a consequence of Hale's plan to establish what would later become the International Union for Cooperation in Solar Research (Wilson 1904b). Wilson was not present at the St. Louis 1904 meeting but attended, as well as Hale, the 1905 Oxford conference (Anonymous 1905a, 1906). On October 11, less than 2 weeks after the conference ended Wilson sent Hale the "little cinematograph which I hope you will be able to try on a Sun Spot soon" (Wilson 1905b). On 1906 January 12 Hale acknowledged the instrument safe arrival and commented this "is exactly what I wanted, not only for photographing the spots directly but also with the spectroheliograph, to which I think it can be adapted without much trouble" (Hale 1906). Hale's attempts were likely unsuccessful since a few years later he complained to his brother that the "good seeing does not last long enough to get a full set of pictures" (Wright 1994, p. 280). Later, in the 1930s, Hale revisited the idea of using a moving-picture attachment in solar observation but apparently was unable to try it out (Wright 1994, p. 426).

2.2 *Total Solar Eclipse 1898 January 22*

The first known use of film cameras to record an astronomical phenomenon occurred in 1898 in the course of the January 22 total solar eclipse. Joseph Norman Lockyer (1836–1920) led the South Kensington Observatory expedition to Viziadrug (Vijayadurg, India). The expedition had two film cameras, one to register the eclipse and the other the shadow bands. No results were obtained since the films "were too badly fogged to serve any useful purpose" (Lockyer 1898).

Fig. 1 Bacon at Buxar, India. (Bacon 1907, in front of page 208)

The main eclipse party of the British Astronomical Association (BAA) expedition, led by John Mackenzie Bacon (1846–1904), stood approximately 1500 km away, at Buxar. Bacon planned to study possible coronal variations during the eclipse brief minutes of totality. Following his friend John Nevil Maskelyne's (1863–1924) advice that the "newly invented animatograph might settle the matter" (Bacon 1907, p. 204). Bacon took an "animatograph telescope, specially designed [by Maskelyne] for the expedition" to India (Bacon 1899). During the eclipse Bacon was in charge of the instrument, which worked flawlessly (Fig. 1). That night the 'precious' film was removed from the machine and carefully stowed away to be developed in England (Bacon 1907, p. 210).

Upon receiving the packing-case in London Maskelyne realised that the film box "was empty and the film had disappeared!" (Bacon 1907, p. 214). According to Bacon's daughter "Many theories were promulgated in the press and elsewhere, nor were there wanting ill-natured folk who declared the whole thing a hoax—that there was no film, nor ever had been!" (Bacon 1907, p. 214). An advertisement offering a reward in exchange for information appeared in several journals to no avail (Anonymous 1898).

2.3 Total Solar Eclipse 1900 May 28

For his next eclipse expedition, in 1900, Bacon accompanied by Maskelyne travelled to the United States of America.

Fig. 2 Photographed at Wadesborough, USA, by Mr. J. N. Maskelyne, with his 3.5 in. kinematograph it shows the 1900 May 18 total solar eclipse second contact. (Maunder and Maunder 1901, in front of page 128)

The expedition started on the wrong foot when it was realised that the "kinematograph telescope" optical part had been left in London (Bacon 1901). Maskelyne being a skilled mechanic managed to improvise a solution and the fine weather on the day of the eclipse at the expedition location, Wadesborough (North Carolina, USA) allowed them to film for

> about 5¾ min, commencing some 25 s before totality, and running for nearly 4 min after totality was ended. In all 1187 exposures were made, 87 before totality, 299 during totality, and 801 after. The corona is seen very definitely on the first exposure, and can be traced right away to number 841, that is to say, to number 455 after the return of sunlight.[1]

A film frame, the earliest from any astronomical film known today, was printed in the BAA 1900 eclipse report (Fig. 2).

Annie (1868–1947) and Edward Maunder (1851–1928) made a positive assessment of the film's "special interest by the way in which it enables us to trace the gradual fading of the corona in the face of the increasing sunlight" (Maunder and Maunder 1901). The film was exhibited both at BAA and Royal Astronomical Society meetings (Anonymous 1900, 1901a). At this last venue Maskelyne pointed out the film's shortcomings due to "unsteadiness of the cinematograph" (Anonymous 1900). Still Edward Ball Knobel (1841–1930), the society president congratulated him

> on the singularity ingenious and interesting exhibition [...] It is the first time we have seen anything of the sort, and we are much interested. We shall look forward to the time when Mr. Maskelyne has perfected his instrument, and we may see an eclipse of the Sun without the expense and annoyance of taking such journeys as we have to at present.[2]

The film quality may be better inferred by David Peck Todd's (1855–1939) statement that the first successful eclipse movie was only shot in 1914 despite his awareness of Maskelyne's earlier effort (Todd 1900, 1922).

[1] Maunder and Maunder 1901, p. 143.

[2] Anonymous 1900, p. 435.

On 28 May 1900 another quite different attempt to record the eclipse was tried by Henry Deslandres (1853–1948).

In the late nineteenth and early twentieth centuries the two 'hottest' solar research topics were the corona and the 'flash spectra' both only observed during a total solar eclipse. The 'flash spectrum' in particular provided information about the vertical structure of the inner solar atmosphere, the chromosphere. Typically the solar spectrum exhibits a series of absorption lines superimposed upon a continuum. Near a total solar eclipse 2nd and 3rd contacts as the Moon covers and uncovers the solar surface, respectively, for a few seconds one may observe the solar atmosphere in the absence of the solar surface and detect the bright chromospheric emission lines. This 'flash spectrum', as it was then known, was first observed by Charles Augustus Young (1834–1908) at the solar eclipse of 22 December 1870 and photographed by William Shackleton (1871–1921) in 1896 (Langley 1871; Anonymous 1911a, 1922; Meadows 1970). Observing from Argamasilla (Spain) the *Bureau des Longitudes* mission led by Deslandres used a mobile chronophotograph lent by Marey to record fast variations in the ultraviolet spectra and in this way complement the longer exposure photographs. The chronophotograph was a late addition to the expedition equipment being brought to Spain by Fallot (?–?), an amateur astronomer, only 4 days prior to the eclipse (Fig. 3).

Four crown prisms placed in front of the chronophotograph allowed the study of the 3500–3800 Å spectral range. The second and third contacts were recorded at six to ten images per second.

Fig. 3 Marey's chronophotographe used by Deslandres mission. (Leclerc 1956)

The images obtained were in general jumbled or made out of double spectra due to vibrations provoked by the rotation of the motion handle and gears. The experience proved nevertheless, according to Deslandres, the possibility of obtaining a flash spectrum with short exposures and the practicability of recording the spectral changes in a way more complete than had been previously done (Deslandres 1900a, b, 1905).

2.4 *Total Solar Eclipses of 1901 May 18 and 1905 August 30*

The total eclipse on 18 May 1901 was particularly compelling due to its long duration of approximately six and a half minutes. Maskelyne lent his film camera to Edward Maunder to be used in the BAA's eclipse expedition to Mauritius but unfortunately the "kinematograph gave no result, the film tearing across before totality was reached" (Anonymous 1901b; Maunder 1902).

The next successful film observation was accomplished by the Spanish astronomer, director of the Fabra observatory, Josep Comas Solà (1868–1937) at the 1905 August 30 solar eclipse (Ruiz-Castell 2008, p. 201). At the May 1900 eclipse Comas Solà had obtained two chromospheric spectra photographs (Solà 1900). In 1905, planning to further his studies, he placed a Mailhat prism in front of a Gaumont film camera. In an October communication to the Paris *Académie des Sciences* Comas Solà made a brief reference to the film results—they confirmed the photographic and visual observations. Nevertheless in the same article Comas Solà pointed out that "the spectro-cinematograph process" was a powerful ally of other spectroscopic observations (Solà 1905). One should point out that this article was partially reprinted in the influential journal of the *Société Astronomique de France*, *L'Astronomie*, with all references to the movie observation edited out (Anonymous 1905b).

2.5 *Movie Cameras Galore—The Hybrid Solar Eclipse of 1912*

A series of eclipses observable either from the ocean or locations of difficult access may explain why the next astronomical motion pictures were only shot during the hybrid solar eclipse of 1912 April 17. The eclipse started as annular in Venezuela crossed Portugal and Spain as total before becoming annular again over the golf of Biscay. It ended in Russia, after crossing France, Belgium, Germany, Latvia and Estonia. According to recent predictions the eclipse totality lasted at best only 2 s (Espenak n.d. a) and consequently against contemporaneous practice the main scientific interest of its observation was astrometrical rather than astrophysical (Lobo 1912a). In 1912 the slightly different eclipse elements used by different calculators led to conflicting predictions. The eclipse could, within the uncertainties, either be annular or hybrid. Mutually exclusive shadow paths upon the Earth's surface were also predicted (Bonifácio et al. 2010 and references therein).

In the day of the eclipse at least ten movie cameras were placed in observing stations from Portugal to Germany. This bounty was a probable consequence of the eclipse characteristics, a favourable shadow path and the films commercial potential.

In 1912 two types of movie observations were performed—visual and spectroscopic. Comas Solà took a "spectro-cinematograph" to Barco de Valdeorras, Galicia (Spain) whereas all the other films were visual. Francisco Miranda da Costa Lobo (1864–1945) placed one movie camera at his main observing station in Ovar (Portugal; Lobo 1912b). Fred Vlès (1885–1944) and Jacques Carvallo (?–?) took two cameras to Cacabelos (Spain) (Carvallo and Vlès 1912). In France an unknown number of movie cameras from Gaumont's film company were located between Trappes and Neauphle in the Paris Polytechnic School observing line set-up by Emmanuel Carvallo (1856–1945) while Aymar de La Baume-Pluvinel (1860–1938) shot the eclipse from Saint-Germain-en-Laye (France) (Carvallo 1912; Baume-Pluvinel 1912b). Father Fernand Willaert (1877–1953) at Namur (Belgium) and the Hamburg observatory expedition positioned at Hagenow (Germany) recorded an annular eclipse (Schorr 1912; Lucas and Willaert 1912). Finally at Lyon Observatory the partially eclipsed Sun was projected onto a screen beside which a chronometer was placed. A film camera recorded them simultaneously at approximately ten images per second (Andrè 1912). Good weather and almost faultless instruments allowed for the successful recording of several films.

As usual in the following months several eclipse observation reports were published. Fred Vlès, Jacques Carvallo and Richard Schorr (1867–1951) used their films to estimate the camera location relatively to the eclipse central line. De la Baume-Pluvinel determined the time of the middle eclipse with a 0.2 s precision from the Baily's Beads assuming equal lunar valley depths on both the East and West sides of the Moon (Baume-Pluvinel 1912b; Carvallo and Vlès 1912; Schorr 1912). Using Lyon's film, Charles Andrè (1842–1912), timed the eclipse first and second contacts with an uncertainty of approximately one second, an improvement upon equivalent visual observations where uncertainties of a few seconds were common (Márquez 1861; Andrè 1912). In his May 30 paper, written in Spanish but published in the *Astronomische Nachrichten* journal, Comas Solà described the spectra in 16 instants in the vicinity of the local eclipse maximum, T. He concluded that the spectral evolution was asymmetrical around T and that the complete spectral inversion—from absorption to emission—of the Calcium H and K lines occurred at $T+2$ s. This indicated, in his opinion, a non-uniform gaseous distribution of the lower solar chromosphere (Solà 1912). One should nevertheless point out that due to the special characteristics of the 1912 April 17 eclipse any observations are highly dependent on the observer's line of sight and Comas Solà was, by his own reckoning, a few kilometres outside the eclipse's narrow shadow path.

From the above summary one quickly realises that, with the possible exception of Comas Solà, no new information was extracted from the eclipse movies. By contrast in a paper read at the Paris *Académie des Sciences* in May 20th Costa Lobo proposed an unforeseen result based solely upon a film analysis. Having realised that the Baily's Beads were not uniformly distributed around the lunar limb (Fig. 4)

Fig. 4 Consecutive frames from Costa Lobo's Ovar film. (Lobo 1912b)

and assuming the observed asymmetry arose from a lunar polar flatness Costa Lobo, proceeded to estimate it in two limiting situations. Initially he proposed a polar flatness in the range [1/1800; 1/600], a value he later revised to [1/1136; 1/380] (Lobo 1912b, c; Bonifácio et al. 2010). This was unexpected since, at the time, the scientific community believed that the Moon was either a sphere or a prolate spheroid with major axis in the Earth-Moon direction. A small article "The Moon is not round: Moving Pictures of the Eclipse Accepted as Proof of This" even appeared on the New York Times newspaper (Anonymous 1912b). In the following weeks several authors supported Costa Lobo's conclusion. Camille Flammarion (1842–1925), for instance, thought that Léon Gaumont's (1864–1946) film, shot in Grand-Croix (France) equally showed "a bigger Moon in the orientation of its movement than in the perpendicular direction" (Flammarion 1912a).

Father Fernand Willaert reported that his annular eclipse film displayed a solar ring thicker at the poles than at the equator. Assuming a circular Sun this implied, in his opinion, a lunar disc slightly flattened at the poles albeit by a lower value than Costa Lobo's, 1/2050 (Lucas and Willaert 1912).

Possibly induced by Costa Lobo's paper Fred Vlès studied the effect produced by different conveniently scaled geometric figures moving in front of each other (two circles and a circle and an ellipse). In mid-September he communicated to the Paris *Académie des Sciences* that the Moon and Sun could not both have circular projections on the sky and that an elliptical Sun provided a better fit to his results. More damaging, we believe, was Flammarion's change of mind. Following his analysis of the eclipse reports sent to him and received by the *Société Astronomique de France,* Flammarion proclaimed that the observed Baily's Beads asymmetry was due to the irregularities of the lunar profile (Flammarion 1912b).

In fact both effects were of the same order of magnitude and no final decision could be made in the absence of new observations (Bonifácio et al. 2010). Costa Lobo himself carefully remarked, "It is evident that other observations are necessary in order to establish definitive values". Unfortunately he also knew that one would be unable to repeat a similar observation before 1927 due to the 1912 eclipse's particular characteristics (Lobo 1912b).

To make matters worse, cinematographic observations although not new had not been discussed in this manner before and the scientific community apparently did not attribute great weight to them. Their perceived value may be judged by how quickly the movie results were forgotten (Bonifácio et al. 2010). The 1913 Royal Astronomical Society report on "Solar Research in 1912" simply mentions that

“Kinematograph records were obtained by some of the French observers” (Anonymous 1913). While at the 1913 Fifth International Union for Co-operation in Solar Research conference the committee for the organization of eclipse observations completely ignored the 1912 films (Anonymous 1914). An unexpected oversight since De la Baume-Pluvinel, the committee secretary, shot one of them and prior to the event believed that “to follow all the eclipse details and the rapid appearance and disappearance of the Baily’s Beads we cannot do better than cinematograph the Sun during its maximum phase” (Baume-Pluvinel 1912a).

2.6 *After 1912*

The fact that several early practitioners, Deslandres, De la Baume-Pluvinel, Lockyer and Maunder did not persevere in their cinema pursuits may hint to the medium inadequacy for astronomical research. Still at least four 1912 April 17 eclipse observers undertook astronomical movies in the following years.

As early as 1911, Nicolae Donici (1874–1956) (also known as Nicolae Donitch) planned to “undertake kinematographic observations of the flocculi and prominences” at his Dubasarii Vechi (Moldova) private observatory (Anonymous 1911b). Curiously Donici’s 1912 eclipse station was located at Ovar (Portugal) in the vicinity of Costa Lobo’s one. At the 1922 International Astronomical Union (IAU) first general assembly held in Rome, Donici stated that he “was recording changes in the forms, of granules, faculae, and prominences by means of a cinematograph” (Fowler 1922, p. 161).

Fred Vlès had been involved with scientific cinema prior to the 1912 eclipse. In 1909 he co-authored a paper about the kinematics of the segmentation and growth of an urchin’s egg observed by micro-cinematography (Chevroton and Vlès 1909). Following his 1912 observation Vlès made various unsuccessful astronomical movies and published the first thematic book on the topic (Vlès 1914). In particular, his plan to cinematograph the 1914 March 12 partial lunar eclipse failed due to cloudy skies. Vlès also tried to use a film camera as a transit instrument attachment, hoping to increase the timing precision of solar meridian passages by simultaneously recording a time stamp on the film via a two-hand chronometer. One chronometer hand ticked each 0.2 s while the other had a continuous motion. In this manner one could measure time fractions of 0.2 s without interpolations. According to Vlès this would “probably greatly improve the precision attained by visual observations, at least those done with a fixed reticule” (Vlès 1914). Despite this positive assessment from 1914 onwards Vlès seems to have abandoned all astronomical pursuits and focused instead on his life sciences research interests.

To observe the next total solar eclipse visible from Europe on 1914 August 21 Costa Lobo devised a new camera with an equatorial mount (Fig. 5). Fully aware that he could not repeat the 1912 observation, Costa Lobo planned to use the movie camera to study any Baily’s Beads brightness variability due to the presence of a tenuous lunar atmosphere in the deepest lunar valleys, the likelihood of which, he thought, was shown in his 1912 eclipse film.

Fig. 5 Costa Lobo's 1914 film camera. (Lobo 1914)

Travelling inland to Feodosia, Crimea (Ukraine, then part of Russia) the Portuguese expedition members were in Berlin in August 1, 1914, the day the German Ambassador to St. Petersburg presented Germany's Declaration of War to Russia. The expedition was cancelled and its members returned home via Switzerland. Costa Lobo observed a partial eclipse at Coimbra University Astronomical Observatory (Lobo 1914). A few years later, in 1927, he had yet another chance to film a solar eclipse. Unfortunately adverse weather conditions at Stonyhurst (Great Britain) impeded the observation of the June 29 total solar eclipse (Anonymous 1927).

The outbreak of the First World War (WWI) thwarted several 1914 eclipse expeditions and disrupted scientific research and international co-operation in the years to come (Todd 1915; Anonymous 1918). In particular the failed Hamburg's (Bergedorf) observatory 1914 expedition to Feodosia is worth mention since movie observations were planned (Anonymous 1916). A cinematographic record was, however, obtained by the Swedish amateur astronomer Nils Viktor Nordenmark (1867–1962) with the aim of determine the eclipse contact times (Rodès 1914). The film shot at Solleftea captured a few hundred "quite perfect pictures of the corona, the coronal ring being clearly caught for several seconds of the partial phase" (Todd 1915).

3 A Limited Choice of Movie Subjects

At this point the reader surely has already realised that the majority of astronomical moving pictures attempts were directed at the Sun (Table 1).

Table 1 Scientific moving picture made between 1898 and 1914

Date	Event	Responsible	Nr	Obs. type	S	Images/s
1898 Jan 22	Total solar eclipse	J. M. Bacon	1	V	?	5–6
		J. N. Lockyer	2	V	Films fogged	
1898 Aug	Sunspot	W. E. Wilson	1	V	×	
1900 May 28	Total solar eclipse	J. N. Maskelyne	1	V	×	~3.4
		H. Deslandres	1	S	*p*	6–10
1901 May 18	Total solar eclipse	E. Maunder	1	V	Film tore up	
1905 Aug 30	Total solar eclipse	J. Comas Solà	1	S	*p*	
1912 Apr 17	Hybrid solar eclipse	J. Comas Solà	1	S	×	
		F. Costa Lobo	1	V	×	560 frames per minute
		F. Vlès and J. Carvallo	1	V	×	15–20
			1	V	p	15
		Léon Gaumont	?	V	×	
			1	V colour	×	
		A. Baume-Pluvinel	1	V	×	13–14
		C. André	1	V	×	
		F. Willaert	1	V	×	14
		R. Schorr	1	V	×	9 frames in 1.2 s
1914?	Solar transit	Fred Vlès	1	V	×	
1914 Mar 12	Lunar total eclipse	Fred Vlès	1	V	Cloudy	
1914 Aug 21	Total solar eclipse	F. da Costa Lobo	1	V	WWI	
		R. Schorr	1	?	WWI	
		N. V. Nordenmark	1	V	×	~6

Meaning of abbreviations: *Nr* number of movie cameras; observation type is either visual, *V*, or spectroscopic, *S*; *S* successful observations are indicated by × while partial ones by *p*, in case of failure reason is presented if known.

The reason behind this ‘bias’ stems from the lack of suitable objects, as we will show below.

In 1914 a few days prior to the March 12 lunar eclipse, Vlès used a 0.72 m focal length and ten F-number apparatus to register, in an Eastman film, the Moon in less than 0.1 s (Vlès 1914).

Table 2 Exposure times using Vlès 1914 apparatus calculated for two limiting situations: the observation day occurred at first quarter, t_L and full Moon, t_U

Celestial object	t_L (s)	t_U (s)
Sun	4.0×10^{-14}	2.5×10^{-13}
Proeminences and inner corona	0.067	0.37
Moon (quarter)	0.10	0.56
Full Moon	0.018	0.10
Light total lunar eclipse	64	360
Mercury	0.046	0.26
Venus	0.0032	0.018
Mars	0.032	0.18
Jupiter	0.11	0.60
Saturn	0.40	2.25
Brightest star (other than the Sun)	0.00042	0.0023

In Table 2 we used the brightness values provided by Convigton (1999, Appendix A) to estimate the exposure times required to film various celestial objects with Vlès apparatus. Two different situations were considered since the Moon's brightness varies considerably throughout the lunar cycle (Espenak n.d. b). As it is unlikely that Vlès would made a test as early in the lunar cycle as the first quarter, one may regard the exposure times, t_L and t_U, presented in Table 2 as lower and upper limits, respectively (IMCCE n.d.). The brightness of planets is highly dependent on their relative position to both the Sun and the Earth. Values were computed around maximum planetary brightness. One should nevertheless point out that these results do not take into account, for example, atmospheric absorption, film wavelength response and reciprocity failure. As such they may be considered only as a crude estimation.

Not surprisingly one concludes that the Sun, Moon, Baily's Beads, chromosphere, proeminences and, at least, inner corona, are all phenomena bright enough to be captured by Vlès's apparatus. It is clear from Table 2 that early movie cameras could record other solar system planets and the brightest stars at a few frames per second, although no attempt to do so was found in the time period under consideration.

Absent in Table 1 are Venus and Mercury transits movies. This cannot be attributed to any exposure difficulty since in a transit the planet is seen, in silhouette, against the Sun. Following the disappointing 1874 transit of Venus photographic results, visual observations were in 1882 once more preferred by many, namely the British and the French expeditions, i.e., those who had previously used the 'photographic revolver'. In 1882, Janssen himself opted to perform astrophysical rather than astrometric observations (Launay 2008, p. 118). The next transit of Venus occurred only in 2004 well outside the time period under study.

Mercury, on the other hand, transits the Sun more often. Transits occurred in 1878, 1881, 1891, 1894, 1907 and 1914 (Espenak n.d. c). The lack of observations is, in our opinion, related with their perceived unimportance since Mercury transits were unsuitable for astronomical unit determinations.

4 Conclusion

The low film speed was a handicap clearly perceived at the time. For instance, Frederico Oom (1864–1930) sub-director of the Lisbon Astronomical Observatory wrote, in 1900, that

> A great future is undoubtedly foreseen for this new species of solar eclipse photographs as soon as one manages to solve the film sensitivity difficulties.[3]

While 11 years later Colin Bennett wrote in "The Handbook of Kinematography" that

> There would seem to be a considerable field for the application of the motion picture camera to the telescope, especially to the astronomical telescope. [...] For this purpose undoubtedly, some system of gearing down the rate of taking to compensate for want of light in the bodies themselves, as also for reasons of economy of film length exposed, would, however, be necessary.[4]

Notwithstanding the small number and limited focus of moving picture attempts cannot be, as shown, simply explained by the low brightness of celestial objects. These are, we believe, a consequence of two other impediments: film frame size and the timescales of many celestial phenomena.

The small solar diameter on film, approximately 6.7 mm with Vlès apparatus, limited the amount of detail that could be extracted from individual frames. Especially taking into account that by the end of the nineteenth century daily solar photographical images were already at least ten times larger (Bonifácio et al. 2007). In another example of the problems created by the small film frames, when Paul Bourgeois (1898–1974) and Jacques Cox (1898–1972) tried to record Mercury's 1927 November 10 transit they realised that the planet's image was too small for its position to be obtained (Stroobant 1927). Mercury's apparent angular diameter at inferior conjunction varies between 10″ and 12″. That is, the Sun to Mercury angular diameter ratio falls in the range 158–196 (Bigourdan 1907). At the 1927 transit Mercury's angular diameter was near its lower value (IMCEE n.d.) and consequently if the Sun's image in the movie had a diameter of 10 mm, then Mercury would appear as a 0.051 mm circle.

On the other hand the long time scales of several celestial phenomena implied that if time-lapse images were required it would be preferable to use an already standard piece of equipment, the photographic camera, with its larger plates despite the difficulties experienced by contemporary astronomical sequential photographers and the slim results obtained (Bonifácio 2011)

It seems that the lack of convenient subjects explains the small number of attempts made and the almost non-existent in-depth analysis of the few movies obtained. An exception occurred in 1912 when at least ten film cameras recorded the April 17 solar eclipse and the first astronomical hypothesis solely based on an astronomical movie was put forward. A scientific discussion of the different film

[3] Oom 1900, p. 71.

[4] Bennett 1911, p. 243.

results ensued in several international journals but despite this visibility they were quickly ignored by the astronomical community. As Table 1 shows, April 1912 also marks the usage peak of movie cameras in a single astronomical observation in the time period studied. The eclipse characteristics and favourable shadow path played a part in the high number of movie cameras used, likely helped by the interest in exploiting the eclipse film's commercial potential. The 1912 interaction between film companies and astronomers still needs to be analysed. In the Portuguese case the company *União Cinematographica Limitada* provided, at least, the manpower and equipment necessary to shoot the Ovar film. In May the eclipse film was exhibited in cinemas in Porto and Lisbon (Anonymous 1912a; Ribeiro 1978). A 1912 eclipse film also became part of Gaumont's educational series known as *L'encyclopédie Gaumont* (Delmeulle 2001).

The fact that astronomical cinema results were forgotten almost as soon as they were obtained clearly reflects their lack of relevance to the scientific community. The 'gap' between early practitioners' claims concerning the possibilities of astronomical cinema and their failure to pursue them also leads us to infer that the technology was not yet ripe for use in an astronomical context.

In a nutshell, in our opinion the lack of suitable movie subjects constituted the Achilles' heel of astronomical cinematography. With the notable exception of solar eclipses there simply weren't many bright fast celestial events of interest.

Technological developments may improve experimental data and/or determine new ideas and open-up new lines of research. Still, as it is known, every new medium 'struggles' to find its place and function against already established ones. While the details of the implementation process are defined by a broad set of conditions, a new scientific tool needs to be more efficient than its predecessors in, at least, a particular useful situation (Pingree and Gitelman 2003). It was precisely this inability to stand out that led to the astronomical moving pictures ultimate fate as a rarely used and indeed seldom useful technique.

Acknowledgements We are grateful to professor Bob Morris who brought to our attention the 1912 New York Times article referring to Costa Lobo's eclipse movie results.

References

André C (1912) Observation de l'éclipse du 17 avril 1912 à l'Observatoire de Lyon. Comptes Rendus de l'Académie des Sciences de Paris 154:1056–1058

Anonymous (1898) Notes. The Observatory 21:245–256

Anonymous (1900) 1900 November 9 meeting of the Royal Astronomical Society. The Observatory 23:429–437

Anonymous (1901a) 1900 November 28 meeting of the British Astronomical Association. The Observatory 24:46–49

Anonymous (1901b) Scientific societies. British Astronomical Association. English Mechanic 72:353–355

Anonymous (1905a) International Union for Co-operation in Solar Research: meeting at Oxford, 1905 September 27–29. The Observatory 28:382–387

Anonymous (1905b) Observations de MM. José Comas Sola, directeur de l'Observatoire Fabra et Salvador Raurich, à Vinaroz (Espagne). Bulletin de la Socié Astronomique de France et Revue Mensuelle d'Astronomie Populaire, de Météorologie et de Physique du Globe 19:495

Anonymous (1906) Papers communicated to the second conference, held at Oxford, Sept 1905. In: International Union for Cooperation in Solar Research (ed). Transactions of the International Union for Cooperation in Solar Research, University Press, Manchester, vol 1. pp 178–200

Anonymous (1911a) Part III—Proceedings of the conference, second session. In: International Union for Cooperation in Solar Research (ed). Transactions of the International Union for Cooperation in Solar Research, University Press, Manchester, vol 3. pp 44–79

Anonymous (1911b) Part III—Proceedings of the conference, third session. In: International Union for Cooperation in Solar Research (ed). Transactions of the International Union for Cooperation in Solar Research, University Press, Manchester, vol 3. pp 79–105, 93

Anonymous (18 April 1912a) O Eclipse do Sol. Comércio do Porto, 1

Anonymous (9 June 1912b) The Moon not round: moving pictures of the eclipse accepted as proof of this. The New York Times

Anonymous (1913) Council note on solar research in 1912. Monthly Notices of the Royal Astronomical Society 73:273–278

Anonymous (1914) Part III—Proceedings—fifth conference. In: International Union for Cooperation in Solar Research (ed). Transactions of the International Union for Cooperation in Solar Research, University Press, Manchester, vol 4. pp 151–153

Anonymous (1916) Observatories. The Observatory 39:64–69

Anonymous (1918) General notes. Popular Astronomy 26:658–661

Anonymous (1922) Obituary notices: fellows—Shackleton, William. Monthly Notices of the Royal Astronomical Society 82:255–256

Anonymous (25 July 1927) O Estudo do Sol. O Século, 1–2

Atkinson R (1953) Cinematography of partial solar eclipses: I. General. Monthly Notices of the Royal Astronomical Society 113:18–33

Bacon JM (1899) The expedition to Buxar. In: Maunder EW (ed). The Indian Eclipse 1898. Hazell, Watson, and Viney, Ld., London, pp 31–46

Bacon JM (1901) Wadesborough, North Carolina. In: Maunder EW (ed) The total solar eclipse 1900. "Knowledge" Office, London, pp 6–17

Bacon G (1907) The record of an aeronaut being the life of John M. Bacon. J. Long., London

Baume-Pluvinel A de la (1912a) L'éclipse de soleil du 17 avril 1912. Bulletin de la Société Astronomique de France et Revue Mensuelle d'Astronomie Populaire, de Météorologie et de Physique du Globe, 26:161–178

Baume-Pluvinel A de la (1912b) Sur l'observation de l'éclipse de Soleil du 17 avril 1912. Comptes Rendus de l'Académie des Sciences de Paris 154:1139–1140

Bennett CN (1911) The handbook of kinematography, the history, theory, and practice of motion photography and projection. Kinematograph Weekly, London, p 243

Bianchi B (1994) Le cinéma et l'astronomie. In: Martinet A (ed). Le cinéma et la science. CNRS Éditions, Paris, pp 120–136

Bigourdan G (1907) Sur les passages de Mercure devant le Soleil et, en particulier, sur celui du 14 novembre 1907. Comptes Rendus de l'Académie des Sciences de Paris 145:647–654

Bonifácio V (2011) Early astronomical sequential photography, 1873–1923. Journal of Astronomical History and Heritage 14:203–210

Bonifácio V, Malaquias M, Fernandes J (2007) Solar photography in the nineteenth century: the case of the Infante D. Luiz Observatory in Lisbon (1871–1880). Journal of Astronomical History and Heritage 10:101–113

Bonifácio V, Malaquias M, Fernandes J (2010) The first astronomical hypothesis based upon cinematographical observations: Costa Lobo's 1912 evidence for polar flattening of the Moon. Journal of Astronomical History and Heritage 13:159–168

Bourgeois P, Cox JF (1933) Eclipse cinematography. Review of Scientific Instruments 4:340–341

Canales J (2010) A tenth of a second. A history. University of Chicago Press, Chicago

Carvallo E (1912) Observation de l'éclipse de Soleil du 17 avril par l'École Polytechnique. Comptes Rendus de l'Académie des Sciences de Paris 154:1072–1075

Carvallo J, Vlès F (1912) Enregistrement de l'éclipse de Soleil du 17 avril 1912 sur la portion espagnole de sa trajectoire. Comptes Rendus de l'Académie des Sciences de Paris 154:1142–1144

Chaperon D (1995) Le cinématographe astronomique. Camille flammarion: un parcours de 1864 à 1898. 1895 18:53–64

Chevroton L Vlès F (1909) La cinématique de la segmentation dé l'oeuf et la chronophotographie du développement de L'Oursin. Comptes Rendus de l'Académie des Sciences de Paris 149:806–809

Covington MA (1999) Astrophotography for the amateur. Cambridge University Press, Cambridge

Curtis S (2010) Scientific films: USA. In: Richard A (ed) Encyclopedia of early cinema. Routledge, London, pp 569–572

Delmeulle F (2001) Contribution à l'histoire documentaire en France: le cas de l'Encyclopédie Gaumont, 1909–1929. Atelier National de Repruduction de Thèses, Villeneuve d'Ascq

Deslandres H (1900a) Observations de l'éclipse totale de Soleil du 28 mai 1900, aàrgamasilla (Espagne). Comptes Rendus de l'Académie des Sciences de Paris 130:1691–1695

Deslandres H (1900b) Observations of the total solar eclipse of May 28, 1900, at Argamasilla, Spain. Astrophys J 12:287–290

Deslandres H (1905) Spectres ultra-violets de la couche renversante pendant l'éclipse totale de Soleil du 28 mai 1900. Comptes Rendus de l'Académie des Sciences de Paris 141:409–414

Espenak E (n.d. a) Six Millennium Catalog of Solar Eclipses. http://www.eclipsewise.com/solar/SEcatalog/SEcatalog.html. Accessed 23 Dec 2014

Espenak E (n.d. b) How to photograph a solar eclipse. (2008). http://www.mreclipse.com/SEphoto/SEphoto.html. Accessed 23 Dec 2014

Espenak E (n.d. c) Seven century catalog of Mercury transits: 1601 CE to 2300 CE. http://eclipse.gsfc.nasa.gov/transit/catalog/MercuryCatalog.html. Accessed 23 Dec 2014

Flammarion C (1912a) Forme de la Lune déduites des observations cinématographiques. Bulletin de la Société Astronomique de France et Revue Mensuelle d'Astronomie Populaire, de Météorologie et de Physique du Globe 26:321–324

Flammarion C (1912b) Notices Scientifiques—L'éclipse de Soleil du 17 avril 1912. Annuaire Astronomique et Météorologique pour 1913 49:247–283

Fowler A (ed) (1922) Transactions of the International Astronomical Union—First General Assembly held at Rome. Imperial College Bookstall, London

Hale GE (1906) Letter to Wilson, January 12, California Institute of Technology Archives. George Ellery Hale Papers. I. Correspondence with individuals. Box 44

IMCCE (n.d.) Institut de mécanique céleste et de calcul des éphémérides, Ephémérides générales de position des corps du système solaire. http://www.imcce.fr/fr/ephemerides/

Janssen J (1873) Passage de vénus; méthode pour obtenir photographiquement l'instant des contacts, avec les circonstances physiques qu'ils présentent. Comptes Rendus de l'Académie des Sciences de Paris 76:677–679 (Our translation)

Korff S (1933) Eclipse cinematography. Review of Scientific Instruments 4:86–87

Langley SP (1871) The American eclipse expedition. Nature 3:228–229

Launay F (2008) Un Globe-Trotter de la Physique Céleste. Vuibert, Observatoire de Paris, Paris

Launay F, Hingley PD (2005) Jules Janssen's 'Revolver photographique' and its British derivative, 'The Janssen slide'. Journal for the History of Astronomy 36:57–79

Leclerc J (1956) Le Cinéma astronomique. L'Astronomie 70:423–432

Lefebvre T (2010) Scientific films: Europe. In: Richard A (ed) Encyclopedia of early cinema. Routledge, London, pp 566–569

Lobo FMC (1912a) Eclipse central do sol. Revista da Universidade de Coimbra 1:179–190

Lobo FMC (1912b) Enregistrement cinématographique de l'éclipse du 17 avril, et forme un peu allongée du contour lunaire. Comptes Rendus de l'Acad emie des Sciences de Paris 154:1396–1399

Lobo FMC (1912c) L'éclipse de soleil du 17 avril 1912. Revista da Universidade de Coimbra 1:548–585, 553

Lobo FMC (1914) O eclipse de 21 de Agosto de 1914 (Coimbra, 1914)

Lockyer JN (1898) Total eclipse of the Sun, January 22, 1898: expedition to Viziadrug. Monthly Notices of the Royal Astronomical Society 58:A27–A42, A31 (Appendix to Volume 58)

Lucas D, Willaert F (1912) L'éclipse de soleil du 17 avril 1912 observée au Laboratoire du Collège N.-D. de la Paix Ã Namur. Revue des questions scientifiques 72:187–225

Márquez F (1861) Memoria sobre el eclipse de sol de 18 de Julio de 1860. Imprenta de Fortanet, Madrid, pp 11–14

Maunder EW (1902) Preliminary note on observations of the total solar eclipse of 1901 May 18, made at Pamplemousses, Mauritius. Monthly Notices of the Royal Astronomical Society 62:A57–A62 (Appendix to Volume 62)
Maunder EW, Maunder ASD (1901) Photographs of the partial phase In: Maunder EW (ed) The total solar eclipse 1900. "Knowledge" Office, London, pp 142–146
Mc Connell JC (n.d.) William Edward Wilson and the story of Daramona. http://www.eaas.co.uk/news/daramona.html. Accessed 23 Dec 2014
Meadows AJ (1970) Early solar physics. Pergamon Press, Oxford
Mourão RRF (2005) The Brazilian contribution to the observation of the transit of Venus In: Kurtz DW (ed) IAU Colloq. 196: transits of Venus: new views of the solar system and galaxy. Cambridge University Press, Cambridge, pp 154–160
Oom F (1900) O Eclipse de Sol de 1900 Maio 28 em Portugal. Imprensa Nacional, Lisboa (Our translation)
Pingree GB, Gitelman L (2003) Chapter introduction: what's new about new media? In: Gitelman L, Pingree GB (eds) New media, 1740–1915. MIT Press, Cambridge, pp xi–xxii
Ribeiro MF (1978) Os mais antigos Cinemas de Lisboa. Cinemateca Portuguesa, Lisboa
Rodès L (1914) Viaje científico—Hacia el eclipse. Iberica 41:233–237
Ruiz-Castell P (2008) Astronomy and Astrophysics in Spain (1850–1914). Cambridge Scholars Publishing, Newcastle
Schorr R (1912) Die Beobachtung der ringförmigen Sonnenfinsternis 1912 April 17 durch die Hamburger Sternwarte. Astronomische Nachrichten 191:429–438 (Our translation)
Sicard M (1998) Passage de Vénus. Études photographiques 4 [Online], put online 04 February 2005. http://etudesphotographiques.revues.org/157. Accessed 23 Dec 2014
Solá JC (1900) El eclipse de Sol del 28 mayo 1900. Astronomische Nachrichten 153:85–87
Solá JC (1905) Observation de l'éclipse totale du Soleil du 30 aot 1905. Comptes Rendus de l'Académie des Sciences de Paris 141:616–617
Solá JC (1912) Sonnenfinsternis 1912 April 17. Astronomische Nachrichten 192:205–207
Stroobant P (1927) Observations du passage de Mercure sur le Soleil du 10 novembre 1927, faites à l'Institut d'Astronomie de l'Université de Bruxelles. Bulletin de la Classe des sciences. Académie royale de Belgique 13:755
Todd DP (1900) On the adaptation of the principle of the wedge photometer to the biograph camera in photographing total eclipses In: British Association for the Advancement of Science (ed) Report of the seventieth meeting of the British Association for the Advancement of Science held at Bradford in Sept 1900, 70. John Murray, London, pp 674–675
Todd DP (1915) The Amherst eclipse expedition to Russia. The Observatory 38:53–56
Todd DP (1922) Astronomy the science of the heavenly bodies. Harper & brothers, New York
Tosi V (2005) La vera nascita del cinema. le origini del cinema scientifico. DVD produced by the Institut für den wissenschaftlichen film (Göttingen) the Centre National de la Recherche Scientifique (Paris) and the Istituto Luce S.p.A. (Roma)
Tosi V (2007) Il cinema prima del cinema. Editrice Il Castoro, Milano
Vlès F (1914) La Photographie Astronomique. Charles-Mendel, Paris
Wilson WE (1898) Report of his observatory. Monthly Notices of the Royal Astronomical Society 58:189
Wilson WE (1899) Report of his observatory. Monthly Notices of the Royal Astronomical Society 59:265
Wilson WE (1900) Mr. W. E. Wilson's observatory, Daramona, Streete, co. Westmeath. Monthly Notices of the Royal Astronomical Society 60:360
Wilson WE (1901) Proceedings of observatories: Daramona observatory (Mr. W. E. Wilson's). Mon Not R Astron Soc 61:234–235
Wilson WE (1902) Proceedings of observatories: Daramona observatory (Mr. W. E. Wilson's). Mon Not R Astron Soc 62:280
Wilson WE (1903) Report of his observatory. Monthly Notices of the Royal Astronomical Society 63:233
Wilson WE (1904a) Report of his observatory. Monthly Notices of the Royal Astronomical Society 64:332

Wilson WE (1904b) Letter to George Ellery Hale, March 7. California Institute of Technology Archives. George Ellery Hale Papers. I. Correspondence with individuals. Box 44

Wilson WE (1905a) Report of his observatory. Monthly Notices of the Royal Astronomical Society 65:374

Wilson WE (1905b) Letter to George Ellery Hale, Oct 11. California Institute of Technology Archives. George Ellery Hale Papers. I. Correspondence with Individuals. Box 44

Wilson WE (1906) Letter to George Ellery Hale, April 26. California Institute of Technology Archives. George Ellery Hale Papers. I. Correspondence with Individuals. Box 44

Wilson WE (n.d.) Astronomical and physical researches made at Mr. Wilson's observatory, Daramona, Westmeath. [Printed for private circulation]

Wright H (1994) Explorer of the universe: a biography of George Ellery Hale. American Institute of Physics, New York, p 280

Heart Matters. The Collaboration Between Surgeons and Engineers in the Rise of Cardiac Surgery

Luca Borghi

Abstract Triumphant medicine in the late nineteenth and in the first half of the twentieth centuries involved increasing amounts of technology and engineering skills. The present contribution gives an example of such a momentous and irreversible convergence between two specific fields: surgery and mechanical engineering. In less than 50 years (*c.* 1930–1980) several "scientific couples" (usually a visionary surgeon and an above-average-skilled engineer) had made a reality of a long dreamed "impossibility" of modern medicine such as open-heart surgery. This contribution will focus not only on technical details but also on the "human factor" which was the hallmark of the protagonists of this revolution.

Keywords Medicine and technology · Cardiac surgery · Human factor

1 Introduction. René Laennec, the Stethoscope and the Snowball Effect

When, in 1816, French physician René Laennec (1781–1826) rolled up a piece of paper for the auscultation of the chest of a young female patient and, by doing so, invented the stethoscope, medicine finally met technology (Reiser 1978). Moreover, it was love at first sight.

The stethoscope not only turned out to be a crucial diagnostic tool, especially for heart and lung diseases, but it was also a source of inspiration for many other physicians and scientists. They were eager to exploit the newly discovered laws of physics, optics and mechanics in order to improve medical ability to explore (and possibly to cure) hidden aspects of the sick human body. Otoscope (1820s), cystoscope (1826), spirometer (1846) and ophthalmoscope (1850) were only the first objects of a very long series of instruments that quickly developed in a sort of technological snowball effect (Borghi 2012, pp. 91–93).

L. Borghi (✉)
FAST—Istituto di Filosofia dell'Agire Scientifico e Tecnologico,
Università Campus Bio-Medico di Roma, Via Alvaro del Portillo, 21 - 00128 Roma, Italy
e-mail: l.borghi@unicampus.it

R. Pisano (ed.), *A Bridge between Conceptual Frameworks,* History of Mechanism and Machine Science 27, DOI 10.1007/978-94-017-9645-3_4

Triumphant medicine in the late nineteenth and in the first half of twentieth centuries involved an increasing amount of technology and engineering skills. Surgery was not an exception. On the contrary, surgeons of that time—already much more self confident thanks to earlier introduction of anaesthesia (1846) and antisepsis (*c.* 1870)—constantly looked for technological improvements (such as the Röntgen x-ray apparatus or the Sauerbruch pressure chamber) in order to perform new and more challenging operations (Gedeon 2006).

There was only one notable exception: the heart. The great Austrian surgeon Theodor Billroth (1829–1893) was not so opposed in principle to cardiac surgery as it has been suggested by someone. Anyway, he wrote in 1882 that due to its risks even a *minor* operation on the heart as the paracentesis of the hydropic pericardium seemed a sort of "prostitution of the surgical art" or, at best, "surgical frivolity"(Absolon 1983).

In the following pages, I will try to account for that momentous and irreversible convergence between two specific fields, i.e. surgery and mechanical engineering, that would at last have made a reality of a long dreamed "impossibility" of modern medicine such as open-heart surgery[1].

We will see how, in less than 50 years (*c.* 1930–1980), some "scientific couples" (usually a visionary surgeon and an above-average-skilled engineer) would accomplish such a revolution, not only on the basis of their scientific and technological knowledge, but also well grounded on the "human factor" of their personalities, histories, moods, feelings and dreams.

What material and intangible conditions made it possible for Nobel Prize winner Alexis Carrel and *solo* aviator Charles Lindbergh to envisage a prototypic "artificial heart" in the 1930s? What 'love affair' enabled Philadelphia surgeon John H. Gibbon and IBM staff to actually build the first heart-lung machine? What children-striking tragedy spurred pioneer cardiac surgeon Walton Lillehei and 'Medtronic' founder Earl Bakken to make the pacemaker a portable life-saving tool? Why was salmon fishing so relevant for the development of the Starr-Edwards artificial heart valve? What social and ethical concerns led dialysis father Willem Kolff and medical engineer Robert Jarvik to devise and implant a real artificial heart into a human being?

To begin to answer these questions, let us proceed in a proper order.

2 Lindbergh, Carrel and Pump

Charles A. Lindberg (1902–1974) was not, strictly speaking, an engineer. After a very troubled school career, he enrolled in Mechanical Engineering at the University of Wisconsin, but he dropped out of the university in his sophomore year and went to the Nebraska Aircraft Corporation in 1922. At that time, a private company was probably much more advanced "in every aspect of aircraft building, maintenance,

[1] I've already briefly outlined this history in a chapter of Borghi 2012, pp. 275–286 (in Italian).

and flying" (Berg 1998, p. 63) than any possible university, and the Nebraska experience gave Lindbergh all the technical knowledge and skills he needed. From there Lindbergh literally took flight for the enterprise that would make him famous all over the world: the first trans-Atlantic *solo* flight on the *Spirit of St Louis* in May 1927 (Lindbergh 1927)[2].

Lindbergh became interested in biomedical problems at the end of the 1920s, when his sister-in-law, Elizabeth Morrow, developed a severe and untreatable heart disease (Malinin 1979, p. 127). In his travels from one doctor to another, Lindbergh often asked why the heart could not be cured surgically. Doctors generally answered that it "could not be stopped long enough for the surgery to be performed" (Berg 1998, p. 221), but he remained unsatisfied due to a certain *engineering* naivety:

> Knowing nothing about the surgical problems involved, it seemed to me it would be quite simple to design a mechanical pump capable of circulating blood through a body during the short period required for an operation. (Lindbergh, quoted in Berg 1998, p. 221)

We can easily imagine the undisguised perplexity of many of his listeners and we are not surprised that one of them would finally refer him to a surgeon who was well known for his eccentricity and visionary ideas.

Alexis Carrel (1873–1944) had already earned the 1912 Medicine Nobel Prize for his revolutionary work "on vascular suture and the transplantation of blood vessels and organs". Marginalized by the French medical community for his public support of miraculous healings at Lourdes, Carrel emigrated in the United States at the beginning of the Twentieth Century and, in 1906, joined the brand new Rockefeller Institute for Medical Research in New York, where he organized and directed the Division of Experimental Surgery for many years (Malinin 1979, pp. 30 and ff.).

It was at the Rockefeller Institute that Lindbergh met Carrel for the first time in November 1928. Despite the difference of age, an immediate sympathy and harmony of views rose between the two men. After carefully listening to his young and famous visitor, Carrel showed Lindbergh the laboratories where he was trying to keep alive and healthy various organs isolated from the body in view of possible transplantation. It was not an easy task to artificially perfuse isolated organs and it seemed virtually impossible to protect them from bacterial infections (Malinin 1979, p. 126).

When Carrel showed him, on the laboratory shelves, some apparatus designed for the artificial perfusion of organs which had not proven satisfactory, Lindbergh immediately "felt he could design better ones" (Malinin 1979, p. 128). An agreement was made immediately: Lindbergh offered himself to work as a volunteer for the design of a new kind of pump for the perfusion of organs and Carrel placed at his disposal all the laboratory facilities (Malinin 1979, pp. 128–129).

And so, *the legend of aviation* started a daily 2-h journey from his house in New Jersey to the Institute in Manhattan to help *the legend of surgery* to devise the "pump":

---

[2] From an academic point of view, Lindbergh afterwards refused many offers of honorary degrees, but made an exception in 1930 for a honorary Master of Science from Princeton University (Berg 1998, p. 224).

> Not one of the 22 great medical scientists—as *Time* magazine will notice in September 1935—who are members of the Rockefeller Institute for Medical research in Manhattan has a reputation with the man-in-the-street equal to that of a minor volunteer worker at the Institute named Charles Augustus Lindbergh. (Carrel's Man 1935)

Lindbergh had to undergo an intense training about "the problems of infection, the sensitivity of blood, the complicated character of living tissue, the hereditary qualities in every cell" (Lindbergh, quoted by Berg 1998, p. 224), but then he could unleash his creative power. Carrel, deeply impressed by Lindbergh industry, was ready to test every new apparatus designed by him:

> No matter how often infection developed or a mechanical breakdown occurred, he was ready to schedule another operation. (Lindbergh, quoted in Berg 1998, p. 225)

Clearly, mutual admiration between the two men played a relevant role for the success of their enterprise.

Lindbergh worked hard for several years overcoming technical problems that presented themselves one after the other, problems related to the low pressure of the perfusion fluid, to the need of correctly simulating the pulsating rhythm of the natural heart and to the causes of bacterial contamination of the organs. At last, he succeeded in developing "an all-glass apparatus which produced sterile pulsating circulation" (Malinin 1979, p. 125). The survival of isolated organs now depended only on the biological conditions of the tissues and on the quality of the nutrient fluid (Malinin 1979, p. 130).

In June 1935 an article in *Science*, "The culture of whole organs", made the new pump aware to the scientific world (Carrel and Lindbergh 1935). A few weeks later popular magazines started to spread the news and the already quoted *Time* magazine referred to the pump designed by the "Bio-mechanic Lindbergh" as "an artificial heart" (Carrel's Man 1935). This suggestive expression, which was completely absent in the scientific paper of *Science*[3], seemed to open to the public an entirely new world of dreams.

But the two scientists, while fully aware of the revolutionary perspectives of their work, probably envisaged a development much more in the line of the future heart-lung machine. In fact, Lindbergh wrote to Carrel in 1937:

> might we not be able to construct an apparatus to maintain artificial circulation during operations on the heart? You may remember that I did some preliminary work on such an apparatus at the Institute two or three years ago. (Malinin 1979, p. 143)

The very idea of an implantable *artificial heart* for the moment was too crazy also for two visionaries like Alexis Carrel and Charles Lindbergh.

In the meanwhile somebody else, in Boston, was already focused on the idea of artificially maintaining blood circulation during surgical operations.

---

[3] I haven't found it even in the monograph on the same subject published by Carrel and Lindbergh in 1938 (see References).

3 Gibbon, His Wife and the Heart-Lung Machine

Many have had the experience that a long-searched-for-solution to some difficult problem may occur suddenly during a sleepless night. This was, more or less, what happened to John Heysham Gibbon, Jr. (1903–1973), the father of the heart-lung machine.

As a junior surgeon at Massachusetts General Hospital, in Boston, on the afternoon of October 3, 1930, Gibbon was charged with the task of monitoring every 15 min the vital signs of a middle-aged woman suffering for a massive pulmonary embolus and who was designed to undergo a very risky operation on the following morning (Romaine-Davis 1991, pp. 21–22).

In 1970, Gibbon himself gave an account of his reflections during that sleepless night:

> During that long night, watching helplessly the patient struggle for life as her blood became darker and her veins more distended, the idea naturally occurred to me that if it were possible to remove some of the blue blood from the patient's swollen veins, put oxygen into that blood and allow carbon dioxide to escape from it, and then inject continuously the now red-blood back into the patient's arteries, we might have saved her life. We would have bypassed the obstructing embolus and performed part of the work of the patient's heart and lungs outside of the body. (Shumacker 1999, p. 74)

As it can be seen easily, while Lindbergh was mainly concerned with the artificial pumping of the blood, Gibbon from the very beginning focused his attention on the extracorporeal blood oxygenation and this will be his primary technical objective. The problem of artificial pumping appeared to him as a relatively minor issue (Shumacker 1999, p. 78).

In any case, since that night of October 1930 Gibbon began to work on a project that, with many ups and downs, would take more than 20 years to reach its full realization, clearly indicating that tenacity was not a secondary feature of its protagonist.

Gibbon became immediately aware that he needed help from the technical point of view, but an early attempt of collaboration with a professor of steam engineering at MIT was not satisfactory (Shumacker 1999, p. 122). The first real help came to him by a young and skilled female laboratory technician at Massachusetts General, named Mary "Maly" Hopkinson (1905–1986). John Gibbon and Maly Hopkinson started to design "a machine that could take over the functions of the heart and lungs during short critical periods, just long enough for emergency surgery to correct a life-threatening problem or a cardiac defect" (Romaine-Davis 1991, p. 22).

But the hearts of the sick were not the only ones involved and the two young scientists soon fell in love, became engaged and married on March 1931, so inaugurating a long-life prolific partnership, both on the professional and the family side (Shumacker 1999, p. 80).

At the time, the main obstacle was the large number of professional commitments that prevented Gibbon from concentrating on his project and so he showed great satisfaction when, in 1934, he could obtain under the auspices of Edward Churchill, professor of surgery at Harvard, a 2-year grant to devote himself full-time to the construction of *the machine*:

> Certainly I would have asked for nothing more than you so gently supplied me with: a fellowship, a laboratory and supplies, and a wife (mine) to help me. (Shumacker 1999, p. 103)

The problem of extracorporeal oxygenation of the blood was initially dealt with a revolving cylinder. The venous blood flowed downward along the sides of the cylinder and the centrifugal force of the revolutions turned it into a thin film which could easily pick up the oxygen flowing into the cylinder and give off carbon dioxide (Romaine-Davis 1991, p. 29). In order to pump the blood through the extracorporeal circuit—which included also blood reservoirs and devices for controlling and maintaining the temperature of the blood—Gibbon, in 1934–35, used a pump that required internal valves, but in subsequent experiments (1938) adopted the valveless roller-type pump developed by Michael DeBakey (Romaine-Davis 1991, p. 35).

Thanks to an extraordinary professional consonance and a good amount of hard work, towards the end of the '30s, John and Maly were able to perform a temporary substitution of the heart and lungs functions of little animals (mainly cats) with an entirely mechanical apparatus, followed by prolonged survival of some animals (Romaine-Davis 1991, p. 61).

But then the Second World War broke out, John Gibbon was enlisted as a Medical Officer in 1942 and assigned to the South Pacific front, just when his and Maly's four children ranged in age from 1 to 10. Shortly after his return from the war, at the end of 1945, John was offered the position of Professor of Surgery and Director of Surgical Research at Jefferson Medical College, in Philadelphia (Romaine-Davis 1991, pp. 66–68). It was time to resume work on the apparatus, after an interruption of almost 5 years.

Gibbon was aware that, in order to build a machine that could work with larger animals and, in perspective, with humans, he needed more sophisticated engineering skills and increased financial support. Both came to him in a very serendipitous way, through a first year student who was a relative to none other than the legendary chairman and CEO of IBM, Thomas J. Watson (1874–1956). But let's hear Gibbon himself:

> Through a freshman medical student at Jefferson I learned that the research department of the International Business Machines Corporation might be interested in a project such as ours. There followed a very fruitful association for seven years with IBM and Mr. Thomas J. Watson, then chairman of the board. I shall never forget the first time I met Mr. Watson at his office in New York City. He came in the anteroom where I sat, carrying reprints of my publications. He shook his head and sat down beside me. He said that the idea was interesting and asked what he could do to help. (…) I then explained that what I needed was engineering help in the design and construction of a heart-lung machine large and efficient enough to be used on human patients. "Certainly. You name the time and place and I shall have engineers there to discuss the matter with you." From that time on we not only had engineering help always available, but IBM paid the entire cost of construction of the various machines with which we carried on the work for the next seven years. (Shumacker 1999, pp. 153–154)

Thanks to the enthusiastic collaboration of three innovative IBM engineers, led by Gustav Malmos, Gibbon could overcome one by one all the intervening obstacles: selection and testing of better materials for the components of the machine, correction of mechanical factors producing gaseous emboli or blood cells rupture, etc. (Shumacker 1999, pp. 154–155).

However, the critical factor remained the oxygenating capacity of the apparatus (for adult humans, a four or fivefold capacity was needed) and a new machine, of a very different scheme, was finally designed in 1951. It was called "Model II" and

> it consisted of stationary screens vertically suspended in parallel inside a cabinet with appropriate mechanisms for distributing the venous blood at the top and for collecting the oxygenated blood at the bottom. (Shumacker 1999, pp. 156–157)

With the new model a series of successful experiments of cardiopulmonary bypass and open-heart surgery was performed, between autumn 1951 and spring 1952, on large dogs. As Gibbon said, in May 1952, at a meeting of the American Association for Thoracic Surgery the time of testing the machine on human patients was approaching (Shumacker 1999, pp. 157–158).

After a tragic false start with a 15-month-old infant, when the ill-fated outcome was due to error in diagnosis and not to malfunction of the apparatus, the first successful open-cardiac procedure using complete cardiopulmonary by-pass finally took place on May 6, 1953. A serious atrial septal defect of an 18-year-old girl, Cecelia Bavolek, was perfectly repaired by John Gibbon and his assistants. The heart and lungs of the young woman had been completely substituted by the machine for 26 very long minutes (Romaine-Davis 1991, pp. 118–122).

But now, after a more than 20-year-long struggle, John Gibbon was tired. A couple more of unfortunate procedures and he decided to step down in favour of his younger collaborators, turning his interests towards general thoracic surgery and lung cancer (Shumacker 1999, pp. 188–189).

It was in fact John Kirklin (1917–2004), a surgeon at the Mayo Clinic in Rochester, who, with full support by Gibbon and the IBM engineers, ameliorated little by little the apparatus and its use in the following years, finally establishing the heart-lung machine as an indispensable and routine tool in heart surgery (Shumacker 1999, pp. 193–197).

Gibbon's sleepless night dream had become a reality.

4 Lillehei, Bakken and Pacemaker

Open-heart surgery was, at last, a real option, but pioneers of this field very soon had to face a new serious obstacle: cardiac arrest. The shock caused by cardiac surgical procedure resulted in the often fatal complication known as "complete heart block" in 10–20 % of the patients (Goor 2007, pp. 40–41; Lillehei and Bakken 1960, p. 76).

Since the beginning of the 1930s, pioneering and controversial experiences of a New York physician, Albert Hyman (1893–1972), demonstrated that a cardiac arrest could be reversed by electrical impulses directed into the patient's right atrium through a bipolar needle electrode (Bakken 1999, p. 48). But for that time, those attempts of *resuscitation* seemed far too defiant and unnatural.

Research on electro-stimulation of the heart languished until the early 1950s, when Boston cardiologist Paul Zoll (1911–1999) developed and used for the first time on a patient a "portable" AC-driven pacemaker (Cooper 2010, p. 114). But the machine was bulky, limited in movement by the electric cable and painful, especially for pediatric patients, since electric stimuli were provided by plate electrodes placed on the skin in front of the heart (Bakken 1999, p. 43; Gedeon 2006, p. 473).

Walton Lillehei (1918–1999) was one of those pioneer surgeons who, at the University of Minnesota Medical Center, were trying to make the most of the heart-lung machine in advancing the open-heart surgery. He was especially interested in pediatric surgery and it was with the aim of relieving the sufferings of his baby patients that he made a first substantial improvement in the Zoll-type pacemaker at the beginning of 1957: he tried to connect the electrical wires directly to the heart muscle and it worked. A very low voltage could produce a single contraction in a completely non-painful way, the pace of stimulation being determined by the doctor (Goor 2007, pp. 40–43).

But later, on the same year, an unexpected tragedy accelerated history. On October 31, 1957, a 3 h blackout affected Minneapolis. Lillehei's hospital had emergency power generators for operating and intensive care units but not for the ordinary rooms and one of his little patients, who was connected to an AC-driven pacemaker, died (Bakken 1999, p. 49).

A quite distressed Lillehei immediately turned to a young electrical engineer who was already providing occasional technical assistance with the hospital's medical devices. Earl Bakken (1924) was the founder of a little company, Medtronic, based in a dilapidated garage and still "in search of a mission", where—in his own words—

> we were constantly building custom devices at the request of our growing number of friends and acquaintances within the Upper Midwest medical profession. If we weren't building new devices from scratch, we were modifying existing equipment or adapting it for new applications. (Bakken 1999, p. 45)

Lillehei showed Bakken the state-of-the-art pacemakers and asked if he could do better than that, especially as to their portability and power supply. Bakken took up the challenge and here is his own account:

> Back at the garage, I dug out a back issue of *Popular Electronics* magazine in which I recalled seeing a circuit for an electronic transistorized metronome. The circuit transmitted clicks through a loudspeaker; the rate of the clicks could be adjusted to fit the music. I simply modified that circuit and placed it, without the loudspeaker, in a four-inch-square, inch-and-half-thick metal box with terminals and switches on the outside—and that, as they say, was that. What we had was a small, self-contained, transistorized battery-powered pacemaker that could be taped to the patient's chest or bed free of any cords and AC connections. The wires that carried the pulse to the heart could be passed through the patient's chest wall. When pacing was no longer needed, the wires could be carefully withdrawn without having to re-open the chest. Without any grandiose expectations for the device, I was moderately optimistic about what it might eventually do for Lillehei's patients. I drove

> the device over to the university's animal lab where it could be tested on a dog. Of course it worked. (Bakken 1999, p. 50)

That was that! Quite an understatement for a medical device that was going to save millions of lives in the years to come. But, for the moment, Bakken was quite surprised to see, on the very following day, that his prototype was already worn by a little girl (Bakken 1999, p. 50). As soon as Lillehei became aware that the battery-powered pulse generator assembled by Bakken actually worked, he didn't want to deprive his baby patients of that life-saving device: he sutured an end of the insulated wire electrodes to the heart of the girl, trailed the other ends through her skin and connected them to Bakken's little box. Lillehei could select whatever rate of heart contractions he wanted, simply by regulating the control knob (Goor 2007, p. 43).

Further refinements and miniaturization of the equipment were achieved in the following years (Lillehei and Bakken 1960), but a major breakthrough was obtained when Medtronic, in October 1960, managed to secure the collaboration of another electrical engineer called Wilson Greatbatch (1919–2011) who had devised the first implantable pacemaker powered by a primary battery (Bakken 1999, p. 53).

Walton Lillehei died in 1999, after a life of ups and downs, including a conviction for income tax evasion. Denton Cooley (1920), another giant of cardiac surgery history who was his friend, remembered:

> Because he tended to challenge established beliefs and was his "own man," he was sometimes a target for critics. However, he was also warm, sociable, and friendly, with enormous energy and concern for his patients. Indeed, he was remarkably generous to many patients who could not afford his services. Moreover, he successfully blended the academic and industrial aspects of surgery in his work with Medtronic and later St. Jude Medical. (Cooley 1999, p. 1365)

Today Medtronic is one of the biggest medical device makers in the world, its implantable pacemakers have been sold by millions and saved lives by the same rate, while his founder, Earl Bakken, is completing his own "full life" in Hawaii:

> Success in almost any field, I assume, will give a person some confidence. Success in the field of medical technology gave me, in addition, whole new worlds of opportunity and ideas. It allowed me to work with and learn from some of the great minds of contemporary science and medicine. It allowed me to travel to all parts of the globe, finally leading me to a place that may be as close to paradise as I could ever expect to find while still on earth. (Bakken 1999, pp. vii–viii)

5 Starr, Edwards and Valve

Now you have a machine that can take over the functions of the heart and lungs during surgery; now you have a device that can help to solve (or even to prevent) a cardiac arrest. Now, what's next?

In the mid '50s Albert Starr (1926) was one of the young surgeons eager to give an answer to this question and make the best of the new field of open-heart surgery. The surprising thing was that, after having his medical studies and his surgical training between New York (Columbia University) and Baltimore (Johns Hopkins), he decided to look for an answer in Portland, Oregon, at the time not exactly in the mainstream of cutting-edge science (Cooper 2010, pp. 261–263).

Legend says that he longed to "go West" and to go fishing salmons, but more probably he accepted the offer by the Oregon Heart Association to finance brand new clinical and research facilities for an open-heart surgical division in the local University hospital (Tesler 2012, p. 178).

At the time, the major focus of cardiac surgery was on treatment of congenital heart disease and most of procedures were still done without opening it. Then the heart-lung machine had already showed new perspectives and, in the fall of 1957, Starr was already working in his new animal laboratory on surgical issues such as atrial septal defect and pulmonary hypertension (Starr 2007, p. 1160).

In the spring of 1958 he treated his first heart patient in the operating room and then he received an unexpected visit:

> At that time I had a visitor, a retired engineer named M. Lowell Edwards, who asked if I would collaborate with him in the development of an artificial heart. He was serious. I told him it was too soon, that we did not even have satisfactory artificial valves, and that simple vascular grafts were not yet fully satisfactory. The quality of valve surgery itself was still relatively crude for both closed- and open-heart techniques. But we made a deal. We would start the project by developing one valve at a time, taking the mitral first and later doing the heart. We shook hands. In the West that was it. (Starr 2007, p. 1160)

Miles Lowell Edwards (1898–1982) was an electrical engineer whose early interests turned towards mechanics and hydraulics (Mullins 2013). When he met Starr for the first time he had more than 60 patents to his credit, his major technical contribution being a fuel-injection system for rapidly climbing aircraft that had been used on 85% of US military airplanes during the World War II. The royalties allowed him to open his own research lab, *Edwards Laboratories*, where his interest finally focused on human "hydraulics"(Starr 2007, p. 1160; Tesler 2012, p. 179).

The morbid anatomy of rheumatic mitral disease—the chronic mitral valve damage that can occur after a person has had an episode of acute rheumatic fever—is such that in many instances nothing short of excision and replacement could allow adequate relief of the hemodynamic abnormality (Starr and Edwards 1961, p. 726).

Starr and Edwards were neither the first nor the only ones working on artificial heart valves. Initially following a biomimetics approach (aiming to *imitate nature*), they tried to build a two-leaflets plastic valve similar to the natural mitral valve, but the animal experiments were unsatisfactory: within hours or days clots formed, compromising the functionality of the valve, and the animals died from thrombotic occlusion (Starr 2007, p. 1161; Cooper 2010, p. 265). *Biomimesis* had to be abandoned:

> We considered the possibility of a ball design for the valve, hoping that thrombus formation would stop at the margin of the valve orifice and not interfere with valvular function. Such devices had a history of use both industrially and medically. In the years before open-heart surgery, [Charles] Hufnagel developed such a device to palliate aortic regurgitation in the form of an acrylic tube implanted quickly into the descending aorta using a nonsuture technique. If a ball design could help us achieve longer follow-up than a few days, we might

> learn more about the long-term suitability with regards to ventricular function and late valve-related complications. (Starr 2007, p. 1161)

After a deep reengineering of Hufnagel's valve by Edwards, they started to implant it in dogs in the fall of 1958 and after a year or so they could show visitors many healthy dogs with long-term survival after substitution of their mitral valve with the artificial one. In the early summer of 1960, Starr received in the experimental lab a visit of Herbert Griswold, then Professor and Chief of Cardiology at Portland University Hospital. Griswold was so impressed by the vitality of the dogs with the artificial mitral valve that, thinking of his patients helplessly dying of heart failure with mitral valve disease, asked Starr to anticipate the programs of testing on humans. And they were on it (Starr 2007, p. 1162)!

Starr and Edwards not only had to face many unprecedented technical, legal and commercial concerns:

> There also were ethical issues. As this was to be the first implantable life-support device of any kind in people, did we have additional responsibilities beyond the usual associated with open-heart surgery? From the beginning we committed ourselves to the lifetime follow-up of the patients, and to objective and scientific reporting of long-term results. The experiment would continue, but this time in humans. Patients would be selected only when there was no alternative therapy and a limited life expectancy—of weeks or months in the absence of treatment. (Starr 2007, p. 1162)

In March 1961, Albert Starr and Lowell Edwards presented their first results during a meeting of the American Surgical Association in Boca Raton, Florida. The mitral ball valve prosthesis had been considered available for initial trial in humans in July 1960, and by February 1961, eight patients had undergone resection and replacement therapy for rheumatic mitral disease. Six out of eight had survived the operation and were in good general conditions (Starr and Edwards 1961, p. 728). Their paper on "Mitral Replacement" became a classic of cardiac surgery literature.

Starr and Edwards turned very quickly to devise artificial substitutions for aortic and tricuspid valves and by 1963 Starr was already doing triple valve replacements: replacing the aortic, mitral and tricuspid valves in the patient's heart during the same surgical procedure (Cooper 2010, pp. 266–267).

Amanda Rao, a 28-year-old university student of Chinese origin with pure mitral insufficiency, underwent mitral replacement on October 27, 1960. The procedure was completely successful and the young lady was soon at work again (Starr and Edwards 1961, pp. 728 and 732). Her long-term and healthy survival was made popular by a long article by Thelma Wilson published on the June 1964 issue of *McCall's*, the famous American women's magazine: "A Miracle for Amanda" became just the paradigm of a long and uninterrupted series of *miracles* made possible by the invention of artificial heart valves (Tesler 2012, p. 182).

By the mid '60s, heart valve replacement had become surgical routine and operative mortality had fallen from 50% to zero (Starr 2007, p. 1163). Once again, visionary and confident cooperation between surgeon and engineer had proven to be effective!

6 Kolff, Jarvik and Artificial Heart

As we have already noted, the "artificial heart" had been a recurrent dream along the early history of cardiac surgery. But the real starting point of this last adventure came along with the very apex of *traditional* surgery.

Shortly after Christiaan Barnard (1922–2001) performed successfully the first human heart transplantation on Louis Washkansky on December 3, 1967 (Barnard 1969), a flood of people with severe and life-threatening heart diseases started to besiege the few hospitals in the world where the new *magic* procedure was already attempted.

But the waiting lists soon became too long for the short life-expectancy of the majority of those sick people. A temporary solution was urgently needed: was it time for a *real* artificial heart?

Dutch physician Willem Johan Kolff (1911–2009) was already a legend in the field of artificial organs when, in May 1967, he arrived with his family at the University of Utah, in Salt Lake City, the Mormons' *Land of the Pure in Heart* (Broers 2006, p. 158). Since 1943 his artificial kidney had turned haemodialysis into a worldwide life-saving procedure (Broers 2006, pp. 62ss.). Then, in the mid '50s he had also given substantial contributions in the refinement of the heart-lung machine (Broers 2006, pp. 135ss.).

When, by the end of 1967, first rumours about Barnard's epoch-making surgical procedure reached Salt Lake City, Kolff was one of the first in grasping the idea that, as the artificial kidney had been a good bridge solution for people waiting for a kidney transplant, an artificial heart could answer well for the problem of heart transplant's waiting lists.

As a matter of fact, Kolff had been working on prototypic artificial hearts at least since 1957 when, with the help of Japanese doctor Tetsuzu Akutsu, he managed to substitute a dog's heart with an artificial one and to keep the animal alive for 90 min (Broers 2006, pp. 147–149).

But now the challenge was far more complex and Kolff was aware of the need for highly qualified engineering skills. Kolff had very good experience and clear ideas about engineering-medicine collaboration:

> I bring people together, let them work together. Enough engineering will bounce off the engineers and enough medicine will bounce off the doctors and they will understand each other. As soon as they do, I give them enough rope to do well, or hang themselves. (Broers 2006, p. 164)

But, on April 4 1969, there was a false start by Denton Cooley, an already famous heart surgeon, who implanted in Houston an artificial heart on a patient. The device performed well for 65 h and then the patient underwent a human heart transplantation, dying at last after a couple of days for complications. Even if the unfortunate outcome probably was not due to malfunction of the artificial heart, Cooley's experiment raised a storm of criticism. Neither the National Institute of Health (NIH) nor the Food and Drug Administration (FDA) had given permission for the experiment and now a new procedural obstacle was placed in the way of the artificial heart (Broers 2006, p. 163).

At the time, Kolff did not enter the heated debate; on the contrary, he publicly praised Cooley's attempt as "a great step forward in medical history":

> The implantation of an artificial heart in a human being by Cooley and his team proves that an artificial heart can indeed replace a natural one in man. In the very near future an artificial heart will be available for two thousand dollars to people who need one. (Broers 2006, p. 163)

His words revealed an indomitable optimism combined, probably, to a good degree of recklessness. But Kolff was also very able in gathering a number of talented young people round him. One of those people was Robert Jarvik.

Robert K. Jarvik (1946) was a creative young man, very talented for art and handicraft, but his academic results were not so brilliant for admission to a medical degree programme at a good American University so, after a 2 years "exile" in Italy, he entered a training programme in biomedical engineering at New York University (Broers 2006, pp. 166–167).

Barely at the end of his studies, in 1971 Jarvik was hired in Kolff's team with an initial salary of 100 $ a week. Despite Jarvik's troublesome temper, Kolff immediately shared with his young collaborator the view of hands-on work and entrusted him the task of improving the design of the artificial heart previously devised by Clifford Kwan-Gett and Thomas Kessler (Broers 2006, pp. 167–168).

From the beginning, Kolff's team was focused on the *total* artificial heart and didn't pay much attention to the supporters of the "more modest solution": a sort of pacemaker-like motor that could help a weak heart, with only partial substitution of the *natural machine* (Broers 2006, p. 161).

The Seventies were a decade of hard work for Kolff's team in Salt Lake City, but every time Bob Jarvik produced a new model of the heart (Jarvik-3... Jarvik-5... Jarvik-7...) the calves in which they were implanted survived longer and in better conditions (Broers 2006, pp. 172–173). But, with a broad brush, what was the artificial heart and how was it designed to work?

> A two-chambered structure that replaced the patient's ventricles (pumping chambers) and was sutured in place through connections with the two atria (collecting chambers) and the aorta and pulmonary artery. Within each artificial "ventricle" was a polyurethane diaphragm, which was moved by air pressure pumped in and out from an external pump through drivelines. When the membrane relaxed, the ventricles filled with blood. When air pushed the membrane upward, the blood was ejected. Four metal mechanical valves ensured that the blood travelled only in the correct direction. The drivelines could be connected to a portable power source for periods of time to allow the patient some degree of mobility. (Cooper 2010, p. 357)

When a calf named *Alfred Lord Tennyson*, in 1981, survived for 268 days with a Jarvik-heart, it became clear to everyone that the time for human implant was rapidly approaching (Wakely 1980; Broers 2006, p. 174). On that same year, Kolff submitted to the Food and Drug Administration (FDA) a formal request for permission to implant an artificial heart into a human being. Even if with much hesitation and many conditions, the permission was granted (DeVries et al. 1984, p. 274; Broers 2006, p. 177). Immediately the search for a suitable patient started.

On late November 1982, the lot fell on 61-year-old Barney Clark, a heavily built heart patient whose "death appeared imminent within hours to days in the opinion of consulting physicians" (DeVries et al. 1984, p. 274). William Castle DeVries was the brilliant young surgeon to lead the surgical team:

> With the consent of the patient and the institutional review committee, a decision was made to proceed with implantation of artificial heart. Anesthesia was induced with fentanyl and metocurine. The chest was opened through a midsternal incision. The patient was placed on cardiopulmonary bypass, and the heart excised at the level of the right and left atria near the atrioventricular valves. Connections to the great vessels were made with vascular grafts. The left and right artificial ventricles were installed and connected to the atrial cuffs and great-vessel grafts. The internal drive lines were tunneled subcutaneously to exit from the left flank and were connected to the drive lines of the pneumatic heart driver. (…) [T]he patient was taken off bypass readily, with stable hemodynamics, and the chest was closed. The total pump time was four hours and nine minutes, and the operative time was seven hours. (…) The patient opened his eyes and moved his extremities on command approximately three hours after operation. He was extubated successfully on the second postoperative day. The patient was conscious and communicated with his family and physicians. (DeVries et al. 1984, pp. 274–275)

Despite some critical moments, Barney Clark survived 112 days with his artificial heart and his death was not directly due to malfunction of the machine but to progressive renal failure and circulatory shock. At post-mortem examination "the artificial heart system was intact, well anchored by connective tissue and uninvolved by thrombotic and infectious processes" (DeVries et al. 1984, pp. 274–275).

Was it a success or a *fiasco*? Not easy to decide. As a matter of fact, after a few more operations over the next 2 years with not fully convincing outcomes, institutional and public consensus started to come down and money to shorten (Cooper 2010, pp. 390–392). A sad side effect of that critical phase was the breaking off between Jarvik and Kolff: Kolff described the crisis to DeVries as "the son rebelling against the father"(Cooper 2010, p. 392). This too is a "human factor"…

Anyway, for the time being, the artificial heart was doomed to remain the dream—or the nightmare—of cardiac surgery. Or rather, the *total* artificial heart. In fact, insofar, it is the "more modest solution" of the so-called Left Ventricular Assist Device (LVAD) to come off best: with such a sort of *partial* artificial heart the failing native heart is not removed but only helped (Cooper 2010, pp. 365–366). It was a device of this kind that recently kept former US Vice-President Dick Cheney's blood flowing for 20 months while he waited for a heart transplant (Moisse 2012).

In the meantime, Willem Kolff died in 2009 at 98. Robert Jarvik is still at work. While more directly interested in Left Ventricular Assist Devices (he himself a *convert*), he still keeps a close eye on an evolution of his total artificial heart called CardioWest, approved by FDA in 2004 as a bridge solution towards the human heart transplant and now in use in more than 30 centres in the US and Europe (Tesler 2012, pp. 421–422).

7 Conclusion

This could possibly be a "conclusion" but certainly not an "end". The history I have briefly sketched hasn't yet come to an end. The question about the future, or the *futures*, of cardiac surgery is under discussion. For sure, the state of siege by interventional cardiologists makes uncertain the future physiognomy of heart surgeons (Tesler 2012, pp. 435–437). What is beyond all dispute is the need for surgeons to carry on their collaboration with biomedical engineers, if they want to remain competitive and on the cutting-edge of medical science.

The highly interconnected flux of ideas, facts and feelings which characterized the rise of cardiac surgery in the twentieth Century clearly shows how manifold is always the history of scientific and technological progress.

Albert Starr said, in a memorial article on the artificial heart valve:

> So much in life is determined by being in the right place at the right time, and by being prepared and bold enough to seize opportunities as they present themselves. (Starr 2007, p. 1160)

No less important is being able to collaborate with the right people, learning how to appreciate and value their specific technical and professional skills.

This is how a handful of enlightened surgeons and visionary engineers could cooperate in one of the most thrilling and momentous scientific and technical adventure of the last century.

"*No man is an island*", John Donne wrote in 1624, and this is true also from the professional point of view.

References

Absolon KB (1983) Theodor Billroth and cardiac surgery. Editorial. The Journal of Thoracic and Cardiovascular Surgery 86:451–452

Bakken EE (1999) One man's full life. Medtronic, Inc., Minneapolis

Barnard C (1969) Una vita. Una biografia scritta da Christiaan Barnard e Curtis Bill Pepper. Mondadori, Verona

Berg AS (1998) Lindbergh. Putnam's Sons, New York

Borghi L (2012) Umori. Il fattore umano nella storia delle discipline biomediche. Società Editrice Universo, Roma

Broers H (2006) Inventor for *life*. The story of W.J. Kolff, Father of artificial organs. B & Vmedia Publishers, Kampen

Carrel A, Lindbergh C (1935) The culture of whole organs. Science 81:621–623

Carrel A, Lindbergh C (1938) The culture of organs. Hoeber, New York

Carrel's Man (1935) The Time Magazine, Monday, Sept. 16

Cooley DA (1999) C. Walton Lillehei, the "Father of Open Heart Surgery". Circulation 100:1364–1365

Cooper DKC (2010) Open heart. The radical surgeons who revolutionized medicine. Kaplan, New York

DeVries WC, Jarvik RK, Kolff WJ et al (1984) Clinical use of the total artificial heart. New England Journal of Medicine 310–315:273–278

Gedeon A (2006) Science and technology in medicine. Springer, New York
Goor DA (2007) The genius of C. Walton Lillehei and the true history of open heart surgery. Vantage Press, New York
Lillehei CW, Bakken EE et al. (1960) Transistor pacemaker for treatment of complete atrioventricular dissociation. JAMA 172:2006–10
Lindbergh CA (1927) We—pilot & plane. Putnam's Sons, London
Malinin TI (1979) Surgery and life. The extraordinary career of Alexis Carrel. Harcourt Brace Jovanovich, New York
Moisse K (2012) Heart Failure Patients Seek 'Cheney Pump'. ABC World News, March 26. http://abcnews.go.com/Health/HeartFailureTreatment/cheney-heart-pump-high-demand/story?id=16003183#.UWV4c8j-6Kd. Accessed 15 April 2013
Mullins R (2013) Miles Lowell Edwards (1898–1982). The Oregon Encyclopedia. http://www.oregonencyclopedia.org/entry/view/edwards_lowell_1898_1982_/. Accessed 15 April 2013
Reiser SJ (1978) Medicine and the reign of technology. Cambridge University Press, Cambridge
Romaine-Davis A (1991) John Gibbon and His Heart-Lung Machine. University of Pennsylvania Press, Philadelphia
Shumacker HB (1999) A dream of the heart. The life of John H. Gibbon, Jr., Father of the heart-lung machine. Fithian Press, Santa Barbara
Starr A (2007) The artificial heart valve. Nature Medicine 13(10):1160–1164
Starr A, Edwards ML (1961) Mitral replacement: clinical experience with a ball-valve prosthesis. Annals of Surgery 154(4):726–740
Tesler UF (2012) Viaggio nel cuore. Storia e storie della cardiochirurgia. UTET, Torino
Wakely R (1980) Artificial heart successful in calf—humans are next. Nashua Telegraph 112(255):12

Discovery of Electromagnetic Waves and Their Impact on Our Life-Style

Mario Calamia, Giorgio Franceschetti and Alessandro Mori

Abstract The history of discovery of electromagnetic waves and its impact on our life style is presented in the framework of its historical context. Discovery of the relevant equations by James Clerk Maxwell are first commented, and the subsequent applications discussed: wired transmission of electric power, cable transmission of messages via telegraph and telephone, up to wireless propagation over the ground and via satellites. More recent preliminary experiments of wireless power transmission are also noted. We conclude with a conjecture about possible additional impacts of Maxwell's discovery on the organization of our society.

Keywords Electromagnetic theory · Maxwell equations · Cable communications · Wireless communications · Satellite communications · Power lines · Wireless power transmission

1 Introduction

As stated in the title, this chapter deals with the impact of an outstanding scientific discovery in our lifetime. In other words, we show that the history of our planet has drastically changed, and is still changing, as a result of newly acquired knowledge. Science is one of the key elements in the development of history, but also history may play a relevant role in scientific advancement. Accordingly, we introduce the development of the *electromagnetic theory* in a wider historical context.

In the history of our evolution, the need to communicate with others, and to reduce the time needed to cover the distances that separate each other, has always

M. Calamia (✉)
University of Florence, Florence, Italy
e-mail: mario.calamia@unifi.it

G. Franceschetti
University of Naples "Federico II", Naples, Italy
e-mail: gfrance@unina.it

A. Mori
Ingegneria dei Sistemi S.p.A, Pisa, Italy
e-mail: a.mori@idscorporation.com

R. Pisano (ed.), *A Bridge between Conceptual Frameworks,* History of Mechanism and Machine Science 27, DOI 10.1007/978-94-017-9645-3_5

been evident. We do not need to look too far to consider, and reflect on two scientific events that radically modified, or offered outstanding alternatives to solution of that communication need.

The first event is credited mainly to Isaac Newton (1642–1727) and his important theories about gravitation and motion. We focus on Newton's results concerning the physical relationship—as hypothesized by Galileo Galilei (1564–1642)—between strength and acceleration, and not between strength and speed, as was believed by Aristotle. Newton's equation was the starting point of mechanical science.

There is no doubt that the development of this science, initiated in France and Great Britain on the basis of Newton's equation, was the starting point of the industrial revolution. However, a second revolutionary event started in the eighteenth century, and has not yet been concluded. A very condensed history of this second revolution, from its starting set of equations up to the latest influence on our lifestyle, is reported hereafter. This second revolution has deep roots in the past, but reached its pinnacle with James Clerk Maxwell (1831–1879), and his famous equations. We retrace its footsteps in the following section.

2 Electric, Magnetic and Electromagnetic Phenomena

Magnetic phenomena first became known in the ancient populations of China and Greece. Examples include the magnetic compass and the (at that time mysterious) attraction of small particles by a rubbed amber stone. However, the effort to model these two phenomena, and formalise them in a set of mathematical equations, essentially started only in the nineteenth century. The experiments of Hans Christian Ørsted (1777–1851)—Copenhagen, 1820—, the subsequent analysis of Francois Arago (1786–1853), André-Marie Ampère (1775–1836), Jean-Baptiste Biot (1774–1862) and Félix Savart (1791–1841)—Academy of Science in Paris, 1820—, the additional experiments of Michel Faraday (1791–1867)—Royal Academy in London, 1831—led to the following set of equations, whose essential meaning can be understood even without specific mathematical knowledge:

$$\left[\begin{array}{l} \nabla \times e = \mu \dfrac{\partial h}{\partial t} \\ \nabla \times h = j \end{array}\right. . \tag{1}$$

In Eq. 1 $e(r, t)$ and $h(r, t)$, functions of space, r, and time, t, are the electric and magnetic fields, respectively; $j(r, t)$ is the electric (density) current that generates the fields; and μ is a parameter (permeability) that accounts for the magnetic properties of the medium, where the field is present. Reading the meaning of Eq. 1 is simple. The bottom equation shows that the source current generates a magnetic field, which mimics the time variation of the current, and curls around it (Oersted experiment), due to the presence of $\nabla\times$, the curl operator. The top equation shows that this magnetic field generates in turn an electric field, spatially confined to its magnetic counterpart, and curling around it (Faraday experiment).

However, James Clerk Maxwell did not like the equations: why? was it the lack of symmetry? their odd appearance? another reason? We do not know. While at King's College of London—1864—, he *invented* another term, without any experimental evidence, and inserted it in the equations, which now read as follows:

$$\begin{cases} \nabla \times e = -\mu \dfrac{\partial h}{\partial t} \\ \nabla \times h = \varepsilon \dfrac{\partial e}{\partial t} + j \end{cases}, \tag{2}$$

where ε is another parameter (permittivity) that accounts for the electric properties of the medium (Maxwell 1873). These equations are universally known as *Maxwell's equations*. This new term seems to be a small addition, but it drastically changes the reading of the set of equations. The source currents generate a magnetic field (bottom equation), which in turn generates an electric field (top equation); but now the electric field appears also in the bottom equation, so that it generates a magnetic field, which curls around the electric field, and not around the source current. This process iterates, with mutual successive concatenations of the two fields, the electric and the magnetic ones. The conclusion: an *electromagnetic wave* is generated that propagates in space, with a velocity related to the parameters of the medium (in free-space, the speed is equal to 300,000 km/s), and electric and magnetic fields are not independent physical quantities (unless they are static, i.e., do not change in time), being physically related. The existence of these waves (*invented*, as stated before, by Maxwell) was experimentally verified by Heinrich Rudolph Hertz (1857–1894)—Karlsruhe Polytechnic, 1886—and the electromagnetic theory was born. Maxwell himself condensed all of these occurrences in a witty way, composing this nice sonnet (Maxwell 1860):

> "O tell me, when along the line
> From my full hearth the message flows,
> What currents are induced in thine?
> One click from thee will end my woes"
> Through many an Ohm the Weber flew,
> And clicked the answer back to me, -
> "I am thy Farad, staunch and true,
> Charged to a Volt with love for thee"

The electromagnetic theory has been the basis for a large number of applications, as presented in the following. When the frequency of the electromagnetic waves is very small, Maxwell's equations simplify. On the basis of this simplified approach, propagation along wires has been formalised (circuit theory), for the transmission of electrical energy, with the design of *electrical transmission lines*, together with *electrical machines* and *transformers*. The electromagnetic waves were empirically used, even before the development of Maxwell's equations, as carriers of telegraph and telephone signals, propagating along wires and cables. Connections between continents required the use of submarine cables: first the telegraphic one across the Atlantic (1853), and then the telephonic cable (1957): in both cases, very important engineering accomplishments were achieved. Nowadays submarine optical fibres, downloaded at the sea-bottom, are also used for signal transmission. However, this wired connection at very large distances has not been the final goal: additional expectations have materialised, as described in the following.

Wireless transmission was pioneered by Guglielmo Marconi (1874–1937), who established a preliminary experiment (1895), and then permanent radio connections: the first transoceanic transmission (1901); Europe-America bilateral link (1902, open to the market 1907); and radio connection with Australia (1930). The connection of points at large distances was implemented by using *radio-bridge* networks, i.e., a chain of ground-located transmitting-receiving antennas, so that the signal may cover large distances by means of successive jumps. From an engineering viewpoint, the development of innovative antennas (parabolic reflectors, steerable arrays, etc.) has been accomplished. The era of wireless transmission of information, *sound and images*, was born, with great expectations.

Use of satellites for long distance communications began in 1960. These satellites developed from being just *passive mirrors* (reflecting the transmitted signal down to the Earth), up to being an *electronic platform*, where the transmitted signal is received, amplified, and radiated back to the Earth at the receiver point. Starting from 1970, satellites have been upgraded to generate also microwave images of the Earth, and other planets, by synthesising very long antennas (SAR →Synthetic Aperture Radar) in space, with sub-meter attainable spatial resolution. Satellites have also recently been organised in constellations (Cosmo SkyMed and TerraSAR, 2007), with potential additional capabilities (Calamia et al. 2012).

Along a different line, transmission of power, instead of information, was implemented. This has been initially (from the end of nineteenth century) accomplished by using transmission lines and cables; but nowadays wireless power, for both acquisition and transmission of energy, is being studied, and initial experiments have taken place. Possible acquisition of power is postulated from solar cells in space, and then downloaded to the Earth via a microwave beam, thus providing renewable, non-polluting energy. Wireless Power Transmission (WPT) is intended to refuel the batteries of remote ground stations, and unmanned aerial platforms (UAV).

A brief history of these studies and implementations is presented in the following sections, in the frame of information as well as power transmission.

3 The Connection Between Remote Points, with and Without a Physical Support for the Information Transfer

The history of remote message transfer (today we refer to this as distance connection) has a particularly fascinating flavour. Just remember the Roman messengers, who were able to cover huge distances by the combined usage of horses and ships (when necessary): the transfer time, and related accomplishments, continue to amaze us even today. However, nowadays the message may travel by itself, without the help of man, making use of the discovered electromagnetic theory. The message can be graphical, vocal, and visual as well.

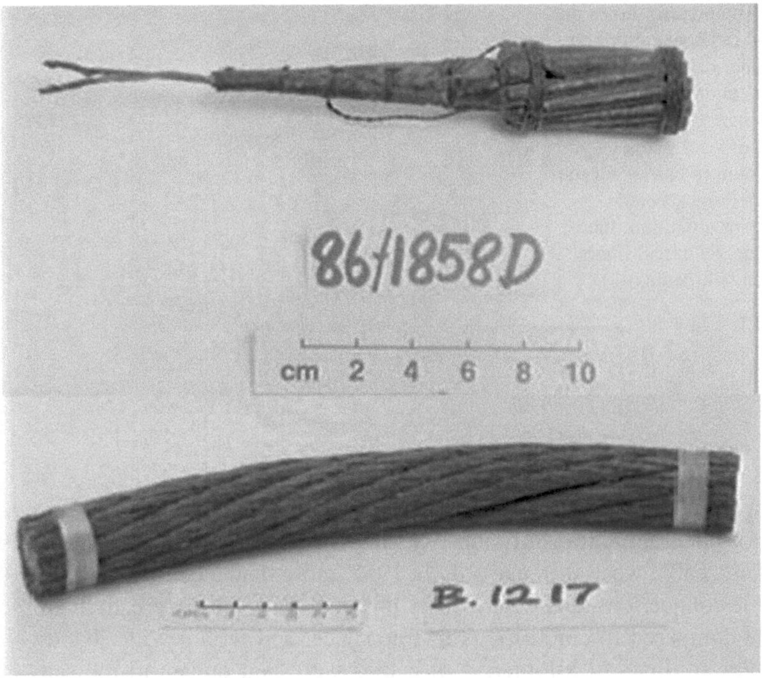

Fig. 1 Sections of an underwater cable from 1858. (Powerhouse Museum, http://from.ph/61309, http://from.ph/207179, Australia)

The first possibility is to code the message as an electric signal, transmitted by using cables. When connections between places with territorial continuity were in order, transmission at large distances was implemented by inserting electronic amplifiers along the cable, in order to compensate for the inevitable attenuation of the signal with the increase of the distance. To connect the continents Europe and America (where there was no such territorial continuity), it was necessary to use the ocean floor, awaiting 1853, when the first underwater telegraph service became operational. Note the relevance, and significance of this 1853 date: it was finally possible to send and receive news in real time through telegrams between continents! A picture of two sections of the underwater cable, used in 1853, is depicted in Fig. 1.

Every important event, happy or sad, can be transmitted, and consigned at destination, by the telegram. Its arrival was (and in part is still, though the technology is becoming obsolete, and alternative procedures have been implemented) always an important event. The telegraph system needs a limited bandwidth, thus being appropriate to dispatch the message via cable, whose bandwidth is also limited. It is important to note that the bandwidth provides an assessment of the quantity of information that is transmitted. For example, a telephonic transmission corresponds to a band of about 5 kHz. Later, developments in technology allowed for improvements in the quality of the cables, so that these could be used even for real-time telephone conversations. Reduction of cable weight was implemented, so that larger cables with larger bandwidth were used, the most recent result being the use of

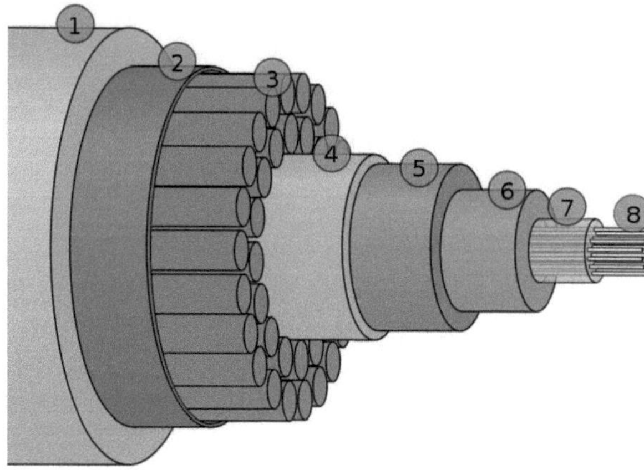

Fig. 2 The picture shows the diagonal cross section of an underwater cable used for modern communication. *1* chemistry, *2* mylar tape, *3* twists of steel wires, *4* aluminum (a barrier for the water), *5* Polycarbonate, *6* copper or aluminum tubes, *7* vaseline, *8* Optical fibers. (http://en.wikipedia.org)

fiber optic cables, see Fig. 2. It is important to note that, while the first underwater cable for telegraphic transmissions was implemented in 1853, the first underwater telephonic cable was set in use only in 1957, more than a century later.

Cable connections, characterized by physical continuity, were, at that time, more reliable compared to connections in free space, and with a wider bandwidth. Accordingly, millions of kilometers of cables were laid out in all the seas, to form a huge web of incredible dimensions, see Fig. 3 (http://www.atlantic-cable.com/Maps/index.htm 2013). The laying of underwater cables was a massive engineering feat. The ships "Citta di Milano" and "Giulio Verne", see Fig. 4, that dug the ocean to put down those cables at the bottom of the ocean, are still remembered, and their logs quoted. The first cables exhibiting large dimensions were made of twisted insulated wires and contained a section that was able to renew the signal.

In the meantime, wireless connections were being developed following Marconi's footsteps. His advancements in this area were made between 1895 and 1930, until the eventual takeover of satellite communication techniques. Those advancements, all implemented in the wireless scenario. are hereafter summarized (Corazza 1998):

1895	First wireless radio connection experiment
1901	First transoceanic experiment, between Newfoundland (Canada) and Cornwall (Great Britain)
1902	Bilateral communication between Europe and America
1907	Commercial communication between Europe and America
1923	Radio telephonic connection
1930	Connection with Australia

It is important to note that the development of radio connections did not interrupt the installation of underwater cables that, with simple technological advances, were becoming lighter, more reliable and broader, as already noted. Today, a large number of fibre optic cables are installed on the ocean floor, see Fig. 3.

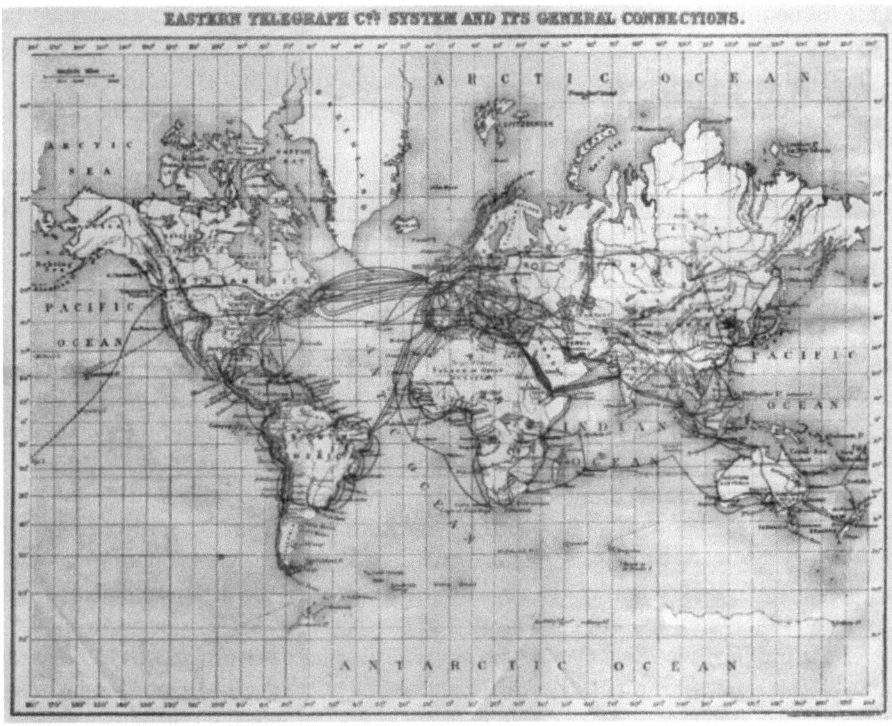

Fig. 3 Map of the laid out cables (Eastern Telegraphy Company, 1901). (http://en.wikipedia.org)

Fig. 4 The cable ship "Città di Milano" (on the *left*) and "Giulio Verne" (on the *right*). (http://en.wikipedia.org, http://hudsonproject.com/gallery/)

In 1960 the satellite era was born. Just looking only at the wireless communications between Europe and America in 1960, it is easy to recognize that the satellite signified a complete revolution of the intercontinental communication. Let us consider the United States, and imagine that we have two ideal vertical planes: one from north to south, thus dividing the States in two halves; the other in the Atlantic Ocean, in front of the Oriental coast. In 1960, telephonic and telegraphic lines, suitable for complex transmissions of about 100 MHz of bandwidth, were crossing the first plane; this is at variance with the second plane, where the potential crossing bandwidth was limited to only 1 MHz. We already noted, and is recalled here, that the bandwidth provides an assessment of the quantity of information that is transmitted. We can therefore conclude as follows.

The 100 MHz bandwidth was transmitted via the already developed radio and microwave links, requiring the physical support of antennas separated by some tens of kilometers, appropriate over solid land, but not implementable over the ocean. On the other side, the 1 MHz bandwidth was transmitted by using short wave radio links (HF) via ionosphere reflection (developed following Marconi's 1901 experiments), and by means of underwater telegraphic and telephonic connections. It is therefore easy to understand the enthusiasm surrounding the success of the TELSTAR satellite, put in orbit in 1962, see Fig. 5, that, with its 5 MHz bandwidth in one single shot, was able to increase the power of transoceanic communication by five times. The satellite was recognized to be the right physical support needed to use microwaves in intercontinental communication.

There have been many attempts to try to insure the transmission of information over great distances by using structures positioned in space. We cannot mention all of them, so we limit ourselves to shedding some light on the different philosophies used for attacking the problem. The first idea (derived from similar arrangements over the ground) was to locate in the sky a reflecting device that would function as a mirror: the transmitter over the Earth radiates a signal toward the satellite, where it is automatically reflected to the receiver point over the Earth.

Fig. 5 An image of the Telstar (AT&T, 1962) Satellite. (NASA image, http://en.wikipedia.org/wiki/Telstar)

Fig. 6 Photo of the Echo satellite (1960). (NASA image, http://dayton.hq.nasa.gov/IMAGES/LARGE/GPN-2002-000122.jpg)

The first realization along this line was implemented in 1960, launching the satellite ECHO, see Fig. 6, into space.

Later in 1963, following another line of thought, the West Ford Project was presented, aimed to construct a sort of artificial ionosphere, see Fig. 7, by disseminating a large number of needles (very thin wires) in orbit around the Earth. However, the Astronomers opposed the project, and it was not approved.

The real successful alternative was the use of active artificial satellites that did not function only as passive mirrors. They were instead equipped with an electronic board to receive signals from Earth, amplify and retransmit them to Earth using a different frequency. There was no opposition to the placement of these satellites in space, and they are still there in orbit today.

The SYNCOM, satellite, see Fig. 8, is also of important significance. The satellite was launched in the same year as the failed West Ford project (1963), and was the first *synchronized* satellite, set at the 36,000 km orbit, and therefore rotating at the same angular speed as the Earth. This far-away location was not considered acceptable, because of the excessive time delay between the transmission and the reception of the signal. As a matter of fact, it was believed that the average user would not tolerate this time delay. Such a belief was proven wrong, as users rapidly adapted to this (however annoying) delay.

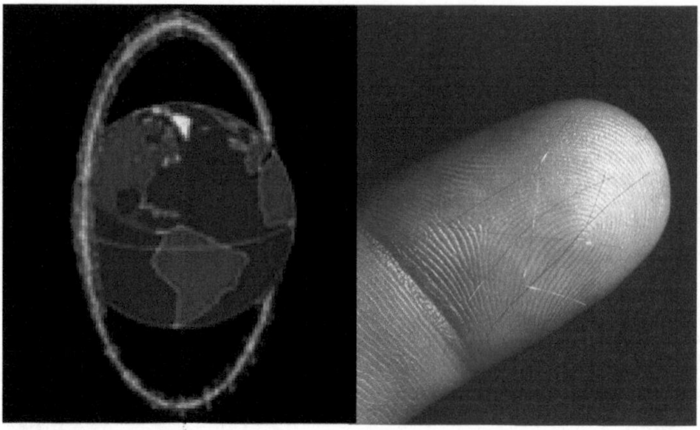

Fig. 7 An image of the orbit of needles' reflectors *artificial ionosphere* (on the *left*), suggested in the West Ford Project in 1963. Note the picture on the *right*, where the dimensions of the needles can be appreciated with comparison to the finger of a human hand. (NASA image, http://history.nasa.gov)

Fig. 8 An image of the active Syncom satellite (1963). Note the presence of the antennas at the border, typical of an active satellite. (NASA image, http://nssdc.gsfc.nasa.gov/nmc/spacecraftDisplay.do?id=1963-031A)

Today satellite communications still use only synchronized satellites.

The SYNCOM satellite allowed for the live transmission of a major sporting event for the first time: the Tokyo Olympic Games.

4 Remote Transfer of Large Amounts of Electrical Power

In the realm of development of communication at large distances (low power application), the development of transfer techniques for large amounts of electrical power has been studied, and implemented. This is a very important issue, being an essential tool for both industrial applications and domestic use. The basic instrument for electrical power transmission is the power line.

The history of the power lines is also useful, in order to understand the interconnection between engineering and electromagnetics. The necessity to have continuous access to large quantities of energy, available far from the production zone (for instance, large hydroelectric plants, or large thermal centres, close to the sea) required choices, due to the two factors appearing in the electrical power, i.e., voltage and current. Unable to increase the current, which would have required bigger, and therefore heavier cables, it was decided to operate on the voltage, which was increased from 20 to 40 kV, and so on up to 380 kV, by using new infrastructure, see Fig. 9. In fact, the convenience of using these suspended wires (combined with their ever-increasing voltage) for carrying the electric power, lead to the design of power lines that generated relevant electric and magnetic fields around them of up to tens of meters. The problems linked to these types of transmission, and their control, became rather serious, when it was recognised that electric and magnetic fields might cause health problems to human beings.

Fig. 9 High voltage (380 kV) overhead power line. (authors' image)

Evaluation of the effect of both the electric and magnetic fields, at industrial frequency, on the human body, was therefore a problem to be considered. The values allowed, in populated areas, were set corresponding to maximum acceptable value of the current (and associate heating) they would induce in the human body. Accordingly, prescribed exposure limits (which still vary nowadays from country to country) were officially established. However, the debate is still open, because it turned out that the effects of electromagnetic exposure to electromagnetic fields are not only thermal, but also non-thermal (effects on the biological structure of the human tissue) (WHO 2007). The presence of these additional effects have been finally accepted, at least by the scientific community, and are now widely studied and experimented. This problem is still being studied.

Control of the allowed limits, of the electric and magnetic fields generated by power lines, has brought about various consequences, in particular the necessity for evaluation and measurement of the fields around them. Both mathematical and numerical models have been developed, in order to carry out the calculation. These models account for the power line geometry, its operating voltage and current, and the shape of the surrounding environment. They range from more simple models up to tri-dimensional numerical computational procedures, which are more accurate, see Fig. 10.

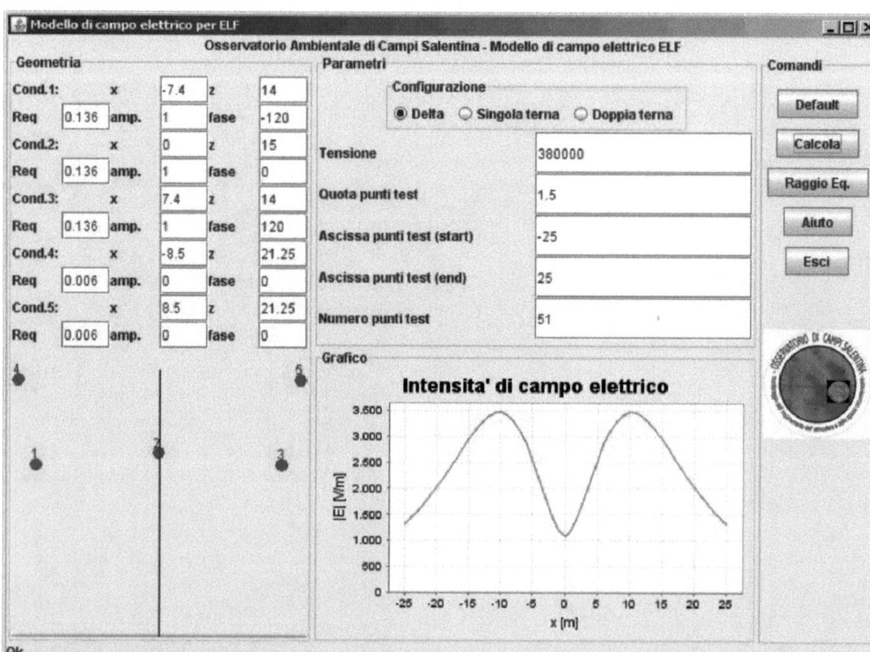

Fig. 10 An example of a software tool able to evaluate, by means of numerical models, the electric and magnetic fields generated in the surrounding environment by a power line. (authors' image)

Fig. 11 Measurement station, devoted to continuous control of the electromagnetic fields in the environment. (authors' image)

Specifically constructed instruments have also been designed for accurate measurement of the electric and magnetic fields, the operating frequency and its several harmonics, being further equipped with control units for continuous monitoring, see Fig. 11.

The necessity to limit the electromagnetic field around the power lines led to development of mitigation techniques. These consist of properly designed geometry and spatial configurations of the electro-conductors, such that the fields around the power line would be lower, compared to the conventional classical geometrical compositions of the conductors (Conti et al. 2003).

Previous development of operating procedures, models, and computations can be derived by using specific equations, e.g., the Biot-Savart one, but the full scenario is regulated by Maxwell's equations. The adopted procedures are just convenient approximations, simply because the coupling between the electric and magnetic fields is pretty low, and they are slightly concatenated, due to the use of low frequencies (50 or 60 Hz) for electric power transmission.

5 Wireless Transmission of Energy

The most recent application of electromagnetic fields is Wireless Power Transmission (WPT). The necessity to find new forms of renewable and ecologically acceptable energy requires the development of a new science, where power (i.e., the Poynting-Umov vector), and not the individual electric and magnetic fields, plays a significant role: new scientific methods must be developed but, in any case, they need to rely on, and use Maxwell's equations. With this in mind, the new WPT science should be generated.

Collecting large amounts of energy in space through solar cells, and transmitting the collected power down to the Earth via microwave beams, is not a new idea. In fact, Peter Edward Glaser (born 1923), who proposed the Solar Power Satellite program, developed this real possibility about 50 years ago (Glaser 1973). Following this innovative idea, there have been various proposals (mostly in the US and Japan) in the industrial world, trying to tackle this problem. The key challenges to be faced are the following ones.

- How to place large and complex equipment in the right orbit around the Earth with acceptable costs.
- How to securely transmit the collected power, down to the Earth, in the format of electromagnetic waves (microwave and infrared) by using the best and most efficient techniques.

Currently, no experiment has been carried out regarding the above-mentioned queries. However, the usual increases in the cost of petrol, largely used for electric energy production, and the spread of environmental concerns, have inspired a first series of preliminary WPT experiments. The efficiency of realised systems and procedures is yet to be evaluated: it is therefore important to implement a more complete, reliable experimentation.

At the same time, suitable technologies for various applications of WPT have been widely tested. These include high power microwave connection between points on the Earth's surface, or in its vicinity, for ground applications. Significant examples of this area are the SHARP and HALE-UAV projects.

The SHARP (Stationary High Altitude Relay Platform) project, proposed in 1985 (Maryniak 1996), was aimed to permanently maintain an unmanned aerial platform, UAV, in space. This was achieved by refuelling the UAV, operating with batteries, by means of a microwave beam (2.45 GHz) from the Earth: the beam was rectified on board, and the UAV batteries were recharged. The same experiment, on a reduced scale, was done in Canada in 1987, with positive results.

The HALE-UAV (High Altitude Long Endurance—Unmanned Aerial Vehicle) project is essentially similar to the previous one, but tailored to a specific application, namely to assure wireless communication between non-visible points over the Earth. The proposed procedure was to radiate a microwave beam from the Earth station, and illuminate an UAV in the space: the incoming signal is reflected by the UAV toward the receiving point over the Earth, the platform being stationary, or almost stationary, in space, with reference to predetermined co-ordinates on Earth.

Also in this case, the permanence of the UAV in the sky is guaranteed by the refuelling procedure of batteries, recharged via a microwave beam from the Earth.

There are therefore others applications of great interest, in addition to the main one, namely to gather renewable and clean energy incoming from outer space.

All of these applications require new scientific findings, especially in the electromagnetics area. However, the key players are no longer the electric and magnetic fields, but rather, as already stated, the Poynting-Umov vector, which describes the power density associate with the field, and propagates with the combined presence of the electric and magnetic fields. All this new science needs to be generated and implemented in the Engineering Schools. Therefore, future projects will be devoted to the design and testing of new antennas and new propagation channels. Detailed consideration of the mechanical effects induced by electromagnetic waves on physical objects also needs to be studied in depth, and fully understood.

6 Conclusions

Without needing to look back at the prehistoric era, there is no doubt that the development of human beings has long been related to relocation, in the constant search for a better life. These relocations have been, for a long time, migrations, documented by historians and anthropologists. To facilitate these relocations, man has invented various methods of transport. At first our ancestors used animals, then roads were constructed, and ships were made; later, flying systems were developed, continuing increasing safety and sophistication, always with the objective of facilitating movement, satisfying a seemingly innate need for humans. Reflecting on this need, we find that until 1600/1700, man improved roads to increase speed, and made ships that were safer and faster.

It was with Newton that a new era began. Newton's equations are the basis of mechanical theory, and he is recognised as the father of the industrial revolution. Machines were invented to lighten the load of both man and animals. The invention of the motor, and the subsequent improvement of roads and related infrastructures, dramatically increased the speed of movement. However, the price to pay has been significant.

Use of polluting and not renewable fuels, development of traffic mess, generating unacceptable transfer times, significant increase of costs, are not only deteriorating the quality of our life, but at same time making the physical interaction among the population difficult and less convenient. The old small αγορά (*agorà*) has been very widely enlarged, but the village, i.e., the previously increased interaction between people, is now rather decreasing.

In the second half of the nineteenth century, Maxwell arrived. With his two equations, he was able to unify the electrical and magnetic fields. This understanding of electromagnetism has paved the way for the transmission of information over very large distances, essentially at any distance. To communicate, we no longer need to move physically: by using the right equipment it is sufficient that the information travels for us. Then, Maxwell's equations play their role, and a new revolution has begun: information transfer, in its increasingly more advanced forms, may substitute for physical transfer. This is already possible, as far as voice, images and, pretty soon, virtual reality is concerned; even feelings and emotions might be able to be transferred, when their biological counterpart is totally assessed. The conclusion is that both the *agorà* and the village are now becoming really global, and the quality of our life will improve again: this is the upgraded life-style that the (Maxwell) counter-revolution is offering us.

References

Calamia M, Franceschetti G, Mori A (2012) Elettromagnetismo e Ingegneria. Uno stretto connubio. In: Storia dell'ingegneria, Atti del 4° Convegno Nazionale, Cuzzolin, Napoli, 2012, pp 7–16

Conti R, Giorgi A, Rendina R, Sartore L, Sena EA (2003) Technical solutions to reduce 50 Hz magnetic fields from power lines. In: Power tech conference proceedings, IEEE, 2003, vol 2, pp 23–26

Corazza GC (1998) Marconi's history. Proceedings of the IEEE 86(7):1307–1311

Glaser PE (1973) Method and apparatus for converting solar radiation to electrical power. United States Patent 3,781, 647, 1973

Maryniak GE (1996) Status of international experimentation in wireless power transmission. Solar Energy 56(1):87–91 (Pergamon Press, Oxford)

Maxwell JC (1860) Valentine by a telegraph clerk to a telegraph clerk. In: Campbell M, Garnett W The life of James Clerk Maxwell (1882). Macmillan and Company, London, pp 320–321

Maxwell JC (1873) A treatise on electricity and magnetism, Chap. IX, Vol 2. Macmillan and Company, London

Submarine Cable Route. History of the Atlantic Cable & Undersea Communications, Maps. http://www.atlantic-cable.com/Maps/index.htm. Accessed May 2013

WHO—World Health Organization (2007) Electromagnetic fields and public health. In: Fact sheet No. 322, 2007. http://www.who.int/peh-emf/publications/facts/fs322/en/index.html

Interactions of Science, Technology and Medicine: Electromagnetic Radiation During the Twentieth Century

Yulia Petrovna Chukova

Abstract Maxwell's original prediction for existence of electromagnetic waves was quickly confirmed by experiment. And at once the basic sphere of use of electromagnetic radiation for communication was defined. Development of more and more powerful radar devices led to high morbidity of attendants and personnel. The term "radio wave illness" appeared, stimulating development of a new scientific direction in biology and medicine and demonstrating the need for standards of hygienic safety.

Detection of nonthermal effects of MM radiation radically increased the social importance of scientific medical researches, due to the proven existence of their positive medical effects. Presence of positive medical effect has changed orientation of technical workings out for radar devices. The orientation on creation of portable radar devices appeared. This allowed an expansion in the sphere of medical use, including about 80 diagnoses of illnesses.

Research into nonthermal effects had shown a poor reproducibility of results. Use of the thermodynamic theory of systems under electromagnetic radiation has allowed us to establish the reason for poor reproducibility of isothermal processes in the radio-frequency region. Moreover, thermodynamic theory has allowed us to establish a new approach to hygienic standardization of harmful actions of electromagnetic radiation, to explain a wide generality of applicability of the Weber-Fechner law, to establish a correlation between terminology of Western science and Chinese medicine and to estimate prospects of their integration.

Keywords Thermodynamics · Efficiency · Endergonic processes · The Weber-Fechner law

Y. P. Chukova (✉)
The Moscow Society of researchers of nature, Krasnopresnenskiy Ecological Fund, Moscow, Russian Federation
e-mail: y.chukova@mtu-net.ru

R. Pisano (ed.), *A Bridge between Conceptual Frameworks,* History of Mechanism and Machine Science 27, DOI 10.1007/978-94-017-9645-3_6

1 A Short Historical Review

The English theoretical physicist James K. Maxwell predicted in 1865 the existence of electromagnetic waves. In 1888 this existence was demonstrated experimentally by the German physicist H. Hertz. In the same year the Russian physicist A.S. Popov repeated Hertz's experiment, and in 1889 he was the first to demonstrate the use of electromagnetic waves for signaling over a distance. Popov designed the generator and the receiver (detector) of signals and showed them to the Russian physical- and -chemical society for the first time on May, 7th 1895. He showed a signal transmission over a distance of 250 m to the same society on March, 24th, 1896. It was the first-ever radiogram and consisted of two words: "Henry Hertz".

In 1897 Italian physicist G. Markoni received a patent for use of electromagnetic waves for wireless communication (in England). The scheme of Marconi's receiver was the same as Popov's receiver. Thanks to an abundance of material resources and energy, Marconi achieved wide practical application of this new method of communication. In 1901 Marconi sent a radio communication through the Atlantic ocean. His activity played a large role in the science of radio communication. In 1909 he was awarded the Nobel Prize in Physics.

As a result, in the twentieth century electromagnetic radiation became a highway of technical progress of our civilization. The second half of the twentieth century was marked by an unprecedented growth of quantity of the diversified man-made sources of electric and magnetic fields used in personal, industrial and commercial aims. All of these devices have made our life more comfortable. Communication possibilities between widely separate people has considerably extended. The work of ambulance drivers and police responders has been facilitated. Use of radar has improved safety of air flights and enhanced control over speed of movement on roads, etc.

But there is also a medal back.

2 Radars and Radio Wave Illness

The whole region of electromagnetic waves has 22 orders of frequency and consequently is usually given in logarithmic scale. Division of the whole region into ranges has been made by a number of experimenters, using a variety of means for generation and reception of radiation. Now there are seven divisions (Fig. 1):

- Extremely low frequencies
- Radio wave radiation
- Infra-red radiation
- Visible radiation
- Ultra-violet radiation
- X-ray radiation
- Gamma radiation.

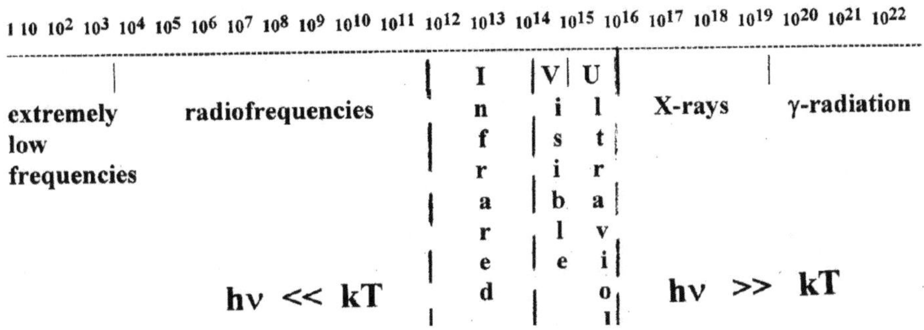

Fig. 1 Partition of the whole region of electromagnetic radiation by experimentalists

Experimenters have faced the fact of overlapping of some ranges:

- Radio wave and infra-red radiation
- Ultra-violet and x-ray radiation
- X-ray and gamma radiations.

A more exact division is a result of deliberations of the International organizations that are responsible for standardization, consequently it can be considered as a social decision.

The works of Hertz, Popov, Marconi and their followers encompass a wide area of radio frequencies among which use of microwave radiation ($3\ 10^8$–$3\ 10^{12}$ Hz) developed most quickly. They were widely used in military technology. This area will be the main object of our consideration, because scientists began to study biological and medical effects of other radiation (in particular visible light) much earlier.

Harmful action of radiofrequency radiation came to light already in the first half of the twentieth century and was studied by hygienists in connection with fast growth of radar power. In postwar years, power of radar devices increased 10–30 times over a decade. The staff of people working with radar devices became the first object of research on harmful actions of microwave radiation.

The response of an organism after exposure was so clearly expressed that a new term “radio wave illnesses” arose among researchers. However it later appeared that the majority of scientists did not have a definitive opinion on necessity in separate nosological[1] unit. Hygienists and physicians fixed (detected) and studied diverse known diseases that arose from the influence of radiofrequency radiation. There were neuro-physiological diseases (Baranski and Edelwejn 1974), diseases of the endocrine system (Micolajczyk 1974), the blood system (Pazderova et al. 1974), diseases of the eyes (mainly, cataracts, Ferri 1974; Tengroth and Aurell 1974), teratogenic diseases (Rugh et al. 1974) and others (Yagi et al. 1974; Miro et al. 1974; Baillie 1974).

An International Symposium on the biological effects and health hazards of microwave radiation ($3\ 10^8$–$3\ 10^{11}$ Hz) was held during the 4-day period October 15–18, 1973, at a conference center in Jadwisin, near Warsaw, under the joint spon-

[1] Nosology is a separate part of medicine.

sorship of the Government of Poland and the United States of America and of the World Health Organization.

Sixty participants from the following countries and from WHO attended: Canada, Czechoslovakia, Denmark, the Federal Republic of Germany, France, the German Democratic Republic, Japan, Poland, Sweden, the Union of Soviet Socialist Republics, the United Kingdom and the United States of America. English and Russian were the official languages of the Symposium.

The participants were scientists and program directors from various institutions, universities, agencies and laboratories concerned with the physical, biomedical and behavioral sciences.

Thirty-nine scientific papers were presented in six sessions:

- General effects of microwave radiation
- Influence of microwave radiation on the nervous system and behavior
- Effects of microwave radiation on the cellular and molecular level (Schwan 1974; Baranski and Edelwejn 1974)
- Measurements of microwave radiation
- Occupational exposure and public health aspects of microwave radiation
- Future research needs, conclusions and recommendations

Differences in approaches and findings, principally between countries in western Europe and North America and those in eastern Europe, have led to considerable variance in microwave exposure criteria and standards (Michaelson 1974; Gordon et al. 1974; Siekierzynski 1974).

3 Standards of Hygienic Safety

Standardization of harmful action is in many respects a social process. At the heart of a system of standards of safety always lies the scientific-and-practical work of experimenters. Figure 2 schematically explains the process of a birth (occurrence) of a standard of safety.

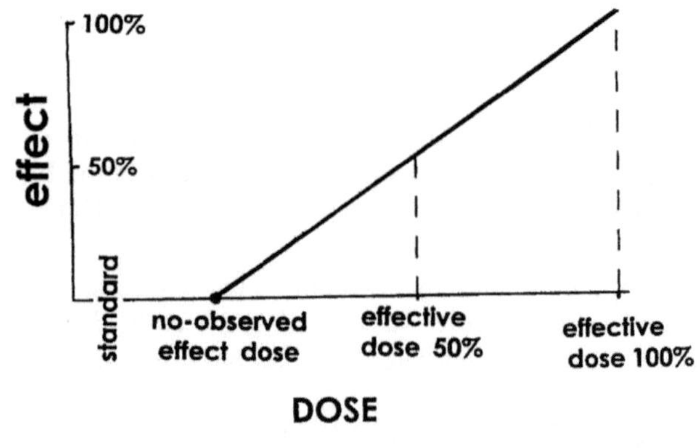

Fig. 2 Dependence effect—dose. (Chukova 2006)

Hygienists, being afraid of artifacts, do not love measurements under weak influences. Therefore they tend to begin measurements under strong influences (the big doses) when reaction of an organism (death, illness and other) is quite distinctly expressed; they then reduce the influence and discontinue it when, in a studied ensemble of individuals, any individual has not shown reaction under that influence. The point "no-observed effect dose" is the last stage of work of experimenters. The line on Fig. 2 is a natural-scientific basis (substantiation) for acceptance of the standard of safety.

The safety standard is the social decision of a commission of experts confirmed by the government or other supervising organization. The unique distinct requirement for a standard is to be smaller than the "no-observed effect" dose. This means that the safety standard should be more rigid than the "no-observed effect" dose. But the distance between a standard and a "no-observed effect" dose is defined independently in each State (or other organization). Therefore safety standards in the different States can differ in allowed percentage. American hygienists Johnson and Guy (1972) have written: "We hear strong arguments between the microwave oven industry, the military, and the Public Health Service, and also between reputable scientists, on where realistic safety level for microwave exposure should lie".

The USA and the USSR, leaders of the arms race in the middle of the twentieth century, were the first to have accepted standards of safety for microwaves. All works in this area were confidential. Frequencies of work of transmitters, their powers and configurations were confidential, but despite that, the first standards of safety appeared to be identical in the USA and the USSR. This standard of safety was equal to 10 mW/cm^2.

The reason for such coincidence is extremely simple. Powerful microwave radiation causes heating. Conservation of constant temperature of a live organism is the first and main requirement of preservation of a life. Therefore the first standards of safety answered the requirement of absence of appreciable heating.

The beginning of specifications of microwave irradiation for the standard of the USA was necessary still in 1942. Even in the time of its establishment (1960s) among scientists there was no unanimity. Some private companies (for example, Bell Labs and General Electric) adopted more rigid specifications, specifically, – 1 and 0.1 мВт/см^2. The definitive statement of the standard of the USA occurred in 1966. However it was only the beginning of further discussions of a problem.

At once there was a question about the appropriate standard of microwave irradiation for nonprofessionals. Here scientists were more unanimous. They were assured that specifications for the population would be more rigid than specifications for professionals. At the same time the standard of the USA accepted for professionals was, for a long time, considered as admissible for the population at large.

At the first stage of standardization of microwave radiation, only the thermal effect in a living system under absorbed microwave radiation was taken into consideration. It is possible to tell that the first stage of standardization of electromagnetic radiation was a stage of maximum consent between hygienists, but it did not proceed along this path for long.

Soviet hygienists working under the guidance of Gordon (1974) quickly understood that microwaves of high intensity erase features of biological effect and initial shifts in characteristics of live objects, giving a total thermal effect. Weak intensity allowed the Soviet hygienists to see a weaker influence on nervous and cardiovascular systems an albuminous and carbohydrate exchange, activity of some enzymes, etc (Tolgskaya and Gordon 1973). They fixed morphological changes of living tissues under very low levels of influence (10 μW/cm^2). This value as a result was accepted for the Soviet standard of hygienic safety. So there appeared two standards of safety for long-term human exposure: 10 mW/cm^2 in the United States and 10 μW/cm^2 in the USSR.

These two standards of safety represented essentially different approaches to hygienic rationing. The American standard has a base of rise in temperature of a body of a person at 1°, which is observed under 10 mW/cm^2. If we take into consideration that the interval of existence of warm-blooded beings is very narrow (for a person about 10°), it becomes obvious that the influence which is taken into consideration by American hygienists is strong. Soviet hygienists have researched thermal and nonthermal effects of microwave radiation, but in a basis of the standard of safety have put a weak nonthermal influence.

4 Discovery of Nonthermal Effects of MM Radiation

The history of science attests to the extreme events connected with aceptance of the law of conservation of energy. The fact is that this law was formulated the first time not by physicists, but by physician Julius Robert Mayer (1841)—though James Joule (1843) and Herman Helmholtz (1847) had been considered as pioneers for a long time. And the second case of participation of physicians in the great discovery of physicists is connected with electromagnetic radiation The rigid Soviet standard of hygienic safety for microwave radiation (10 μW/cm^2) forced Soviet physicists to carry out researches under weak influences on various objects, therefore hastening the discovery of nonthermal resonant bioeffects of MM radiation.

On Jan. 17–18, 1973, there was a scientific session of the Division of General Physics and Astronomy of the USSR Academy of Science in Moscow. At this session, Academician Devyatkov (1974) reported for the first time frequency-dependent microwave bioeffects in the frequency range from 39 to 60 GHz. These effects were observed at the molecular, cellular, organic, and organismic levels of organization. Results were arrived by different experimentalists in laboratories of different institutes under the direction of Acad. Devyatkov. that the following facts were discovered for almost all of the biological objects studied:

- the existence of irradiation effects which depend on frequency, sometimes in a resonant manner,
- the existence of a threshold intensity necessary for induction of such effects, and that over an intensity range of several orders of magnitude above the threshold, the induced effects do not vary with intensity,
- the effects are observed to depend significantly on duration of irradiation.

Fig. 3 Influence of the power density of radiation on induction of colicin synthesis. (Smolyanskaya and Vilenskaya 1974)

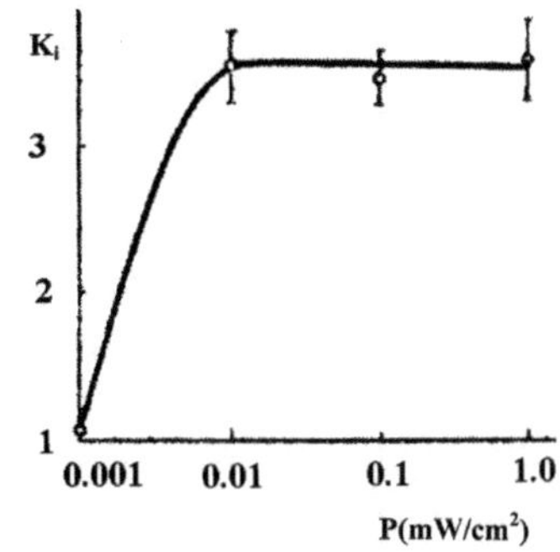

The results obtained are of great scientific and practical interest. For example, it was established that a vital activity of microorganisms is affected by millimeter wave radiation. The effect may be a both positive and negative one, depending on the particular conditions of irradiation and system characteristics. Different systems and many tests were studied, especially cell division under MM radiation. Some experiments, for example an experiment of Smolyanskaya and Vilenskaya (1974), now are considered as classical references (Fig. 3).

Smolyanskaya and Vilenskaya reported that under MM wave radiation, intracellular systems are responsible for lethal synthesis in bacteria, i.e., the synthesis of substances that results in the death of the cell. The so-called colicinogenic (col) factor of *E. coli* was chosen as the test paradigm. The col-factor is an extrachromosomic genetic element. The functional activity of this element is normally repressed, but a number of physical and chemical treatments can induce its activity. Under MM radiation the col-factor results in synthesis of a special protein substance known as colicin. The cell then perishes. The colicin that it produces has an antibacterial action with respect to other bacteria of the same and similar species.

The activity of the colicin synthesis was determined by the method of lacunas, in which the number of individual colicin-synthesizing bacteria are counted. The effect was evaluated by the so-called induction coefficient, which is determined by the ration of the lacuna formation frequencies in the experiment and the control.

It was found that the number of colicin-synthesizing cells increased sharply upon irradiation of the colicinogenic strain with millimeter waves of certain wavelengths. Thus, the number of cells that synthesized colicin increased by an average of 300 % upon irradiation at wavelengths of 5.8; 6.5 and 7.1 mm. At the same time neighbouring wavelengths (6.15 and 6.57 mm) showed no such effect. Thus the effect of *E. coli* to produce colicin was induced only at particular frequencies. The results obtained were reproduced with high regularity.

Smolyanskaya and Vilenskaya also studied the variation of the influence of power density of radiation on induction of colicin synthesis. They found that a variation by a factor of 100, from 0.01 to 1.00 mW/cm^2, had no influence on the induction coefficient, and only a further reduction of power density down to 0.001 mW/cm^2 (1 μW/cm^2) resulted in a sharp decrease in the biological effect.

That the effect does not depend on power was a strong argument in favor of the nonthermal effects of millimeter waves, since all thermal effects depend primarily on intensity of radiation.

As another example we cite the results by Sevast'yanova and Vilenskaya (1974). The investigations deal with damage to bone marrow cells of mice arising from exposure to X-rays and the protecting effect of microwave irradiation. When prior irradiation with microwaves has taken place, then at certain frequencies and intensities the damage is vastly reduced. Experimentalists investigated this protection effect of microwave fields at various power densities from 1 to 75 mW/cm^2. The X-ray dose was 700 rad.

A considerable number of experiments have been connected with irradiation of yeast cultures. A resonant effect of millimeter-wave irradiation on the division intensity of cells was observed. Thus, for example, irradiation of a culture of Rhodotorula rubra for 15 h at wavelengths of 7.16; 7.17, 7.18 and 7.19 mm showed a sharp frequency dependence. Irradiation of a Candida culture caused a marked change in the nature of cell division as compared with the control (Devyatkov 1974).

Academician Devjatkov concluded a report at a session of the Academy of Sciences with the words: "An explanation of the mechanism of the resonant effect of irradiation and some of its other properties would be of enormous interest from the scientific standpoint. As yet, we have no rigorous scientific explanations for the effects of millimeter-band electromagnetic waves. There have been only a few attempts to develop approximate hypotheses to account for the resonant effect, and they require further experimental and theoretical confirmation…."

"In addition to the scientific groundwork, a more active search should be made for fields of practical application of the effects of millimeter-wave irradiation"

The results of a scientific session of the USSR Academy of Science were more than simply pilot results. Around 1965, a number of organizations in the USSR began, under the direction of N. D. Devyatkov, systematic researches on the effects of low-power millimeter waves on biological objects. This scientific session of the USSR Academy of Science was a culmination of several years of research and scientific discussions at the highest level. At this scientific session, specialists working in the fields of biophysics, microbiology, biochemistry, and medicine, were invited. The brief contents of some papers were published in the Soviet journal "Uspekhi physicheskikh nauk". An English translation entitled "Soviet Physics—USPEKHI" was published in 1974. Therefore scientists of other countries learned of the Soviet results and tried to replicate some of them.

Further events developed in two directions. We will consider them below.

5 New Orientation of Development of Technical Devices

Before the Devyatkov report, a refinement of a radar device with constantly increasing power was the basic direction of development of radio engineering in the USSR. After the report, a new opposite direction appeared—designing of compact generators of MM with low power (less than 6–8 mW/cm^2). This problem was addressed by a big collective of scientists under the guidance of N.D. Devjatkov. Leading developers of these portable devices won the State award of the Russian Federation for science and technology in the year 2000.

As a result the new noninvasive method of MM-therapy was developed. It won all kinds of high level approbation and certification (Betskii and Lebedeva 2001).

The method was approved in more than 90 Russian clinics, including many large medical centers.

High efficiency of MM-therapy was confirmed in such areas as:

- Gastroenterology (a stomach ulcer and a duodenal gut, hepatitises, a cholecystitis, a pancreatitis)
- Neurology (a painful syndrome, neuritis, a radiculitis, an osteochondrosis)
- Cardiology (ischemic illness of heart, a stenocardia)
- Urology (a pyelonephritis, an impotence, prostatitis)
- Pulmonology (a tuberculosis, sarcoidosis of lungs)
- Skin diseases (psoriasis, neurodermite)
- Gynecology (erosion of a neck of a uterus, a fibroma)
- Surgery (acceleration of processes of regeneration)
- Oncology (protection of blood formation systems, elimination of by-effects in chemotherapy).
- Preventive maintenance of an aggravation of chronic diseases, cold and virus infections.

Experience of clinical use of this method allows us to speak about absence of remote consequences. Besides, the method combines well with other methods of treatment (medicinal, physiotherapeutic etc.). It has no absolute contra-indications.

MM radiation can be used as monotherapy.

Now in Russia there exist two authoritative journals devoted to these problems:

- "Millimeter waves in biology and medicine" (from 1992),
- "Biomedical technologies and radio electronics" (from 1998).

6 Difficulties of Experimental Measurements

After the publication of the Devyatkov report in English (1974), many foreign scientists tried to repeat the results of the Soviet scientists on different living systems.

Grundler et al. (1977, 1988), Grundler and Keilmann (1978) studied the growth behavior of yeast cultures under coherent microwave irradiation. To avoid thermal effects of irradiation he chose to use a yeast cell culture in stirred aqueous suspension, as this assures an efficient thermal exchange between the cells and the surrounding medium. The temperature of the suspension (32± 1°C) was continuously monitored.

Similar to a yeast experiment described by Devyatkov (1974), frequencies were chosen in a range near 42 GHz. When the cultures were irradiated by CW microwave fields of 1–2 mW/cm^2, the growth rate was considerably enhanced or reduced depending on the frequencies varying by no more than a few megahertz in the 42 GHz range. A spectral fine structure with a width of the order of 10 MHz was observed. Careful temperature monitoring excluded a trivial thermal original of this effect.

Webb (1979) published a paper in which the induction of lambda prophages by of *E. coli* under irradiation by millimeter microwaves was chosen as the test object. Lambda prophage is a particular virus that is parasitic within a bacterium: it is normally dormant, but when "woken up" by, for example, weak external microwave radiation of the correct frequency, it multiplies and kills the bacterium. The effect was distinctly frequency-dependent and has dependence on power density that looks like a step process.

At a symposium in Helsinki in 1978 and in Seattle in 1979, reports were presented by Dandanoni et al. (1978), who studied the action of MM waves on a culture Candida Ablicans. The first such research was described by Devyatkov (1974). Irradiation of a Candida culture caused a marked change in the nature of cell division as compared with the control. The difference between control and experiment at various stages during irradiation was illustrated. Dandanoni et al established that irradiation by MM waves on a frequency of 72 GHz for 90 min with power density level at a few mW/cm^2, and with pulse modulation on 1 kHz, considerably reduces cell growth at room temperature, whereas irradiation in the CW mode under the same conditions has no effect, or a slight growth.

It was quickly found that researchers in this discipline were having great difficulty in replicating results obtained by others. Moreover, the same scientist was sometimes able to observe the effect and sometimes not.

The strangest history was with colicin induction by *E. coli*. As noted above this effect was first reported by Smolyanskaya and Vilenskaya in 1973 (1974). It was confirmed by Swicord et al. (1978) at the 19th General assembly of URSI in Helsinki. It was reported that when *E.coli* W3110 (col E1) was exposed to MM waves on low power levels of less than 1.0 mW/cm^2 in the frequency range of 45.5–46.2 GHz, an effect was observed on induction of colicin synthesis. On the end points of this frequency band, the coefficient of induction was equal to unity. Maximum response was observed on a frequency of 45.9 GHz. The coefficient of induction was 8.

However, at the Bioelectromagnetics Symposium in Seattle, Athey (1979) reported: "We devoted about a year to the colicin induction experiment. In the pilot study using a temporary experimental system, we seemed to be getting some positive results; but after refining our experimental system we never again saw any increased colicin induction".

In contrast to Athey, Motzkin and others (1979) described facilities for irradiating objects in the 26.5–75 GHz band. Colicin production was doubled on a frequency of 51.7 GHz at power density 0.5 mW/cm^2. This effect was observed with irradiation only on a few definite frequencies at temperatures of 25 and 37 °C. At the Seattle Symposium S.M. Motzkin reported results in support of the Smolyanskaya and Vilenskaya original observation.

At this symposium Partlow (1980) remembered in addition, that Hill et al. (1978) were unable to replicate the frequency-specific effects previously reported by Berteaud et al. (1975) on the rate of growth of *E. coli.*

So, it is no surprise that at the Bioelectromagnetics Symposium in Seattle, Partlow (1980) declared: "All that is needed is a confirmation that the reported effect is real". These words relate to many bioeffects of MM radiation.

The Bioelectromagnetics Symposium in Seattle on June 22, 1979 was the most serious discussion of the problem after the scientific session of the USSR Academy of Science in Moscow in Jan. 1973.

In 1991 in Moscow there was an International symposium "Millimeter waves of nonthermal intensity in medicine". Motzkin (1991) presented an abstract with the title "Low power continuous wave millimeter irradiation faills to produce biological effects in lipid vesicles, mammalian muscle cells and *E. coli*".

It was clear that the real confusion can only be resolved by further millimeter wave bioeffects study. The following researches did not change the general situation. Brilliant works have been performed on studying of resonant bioeffects of MM of radiation, for example Belyaev et al. (1996); but simultaneously there still appeared works denying existence of effects.

It was clear that the science of electromagnetic radiation had faced the next great problem. In the history of electromagnetic radiation, there had been previous such problems. Such expressions as "ultra-violet catastrophe" and "thermal death of the Universe" remind us of them. All these problems were solved by a thermodynamic method. Therefore it was expedient to involve this method for the decision of a problem of bioresonant effects of MM radiation.

7 Thermodynamic Theory as a Key to Understanding of Laws of Energy Conversion

Consideration will be given to the basis of the thermodynamic method that was the leader in research into many problems of electromagnetic radiation. Stages of development of thermodynamics of electromagnetic radiation are distinctly connected with names of the scientists of different countries. Their names in chronological order of the published works are presented at first for equilibrium radiation, and then—for nonequilibrium radiation.

7.1 Equilibrium Radiation

W. Wien (1864–1928) generalized concepts of absolute temperature and entropy for thermal radiation and formulated two laws for the short-wave wing of equilibrium electromagnetic radiation (The Wien region of EMR). Nobel Prize 1911.

Lord Rayleigh (1842–1919) described the long-wave wing of thermal radiation (The Rayleigh-Jeans region).

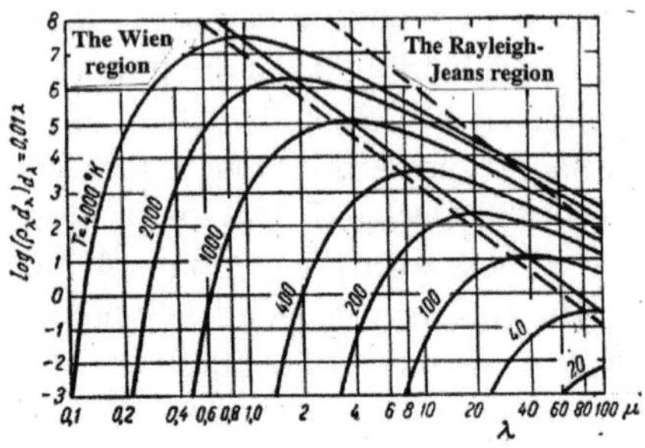

Fig. 4 Equilibrium thermal radiation for different temperatures. (Levich 1954)

Max Planck (1858–1947) (1914) introduced the concept of "action quantum", solved the problem of the theoretical description of full function of thermal radiation (the universal function of Kirchhoff $\varepsilon_{\nu,\ T}$) and deduced the formula for entropy of thermal radiation. Nobel prize 1918.

Albert Einstein (1879–1955) introduced representation about quantum structure of light and explained a photoeffect. Nobel prize 1921.

S. N. Bose (1894–1955) (1924) elaborated quantum statistics for particles with integer spin and deduced the Planck law for thermal radiation on this base.

The result of their works (Fig. 4) is well known. It has completely exhausted a problem of thermal electromagnetic radiation and has moved (transferred) Planck's law from independent autonomous position into a strictly designated place in the general structure of exact knowledge about Nature.

7.2 A Nonequilibrium Radiation

The first step from equilibrium radiation to nonequilibrium radiation was made by L.D. Landau.

Lev Landau (1908–1968) generalized the Planck formula for entropy of thermal radiation for nonequilibrium radiation (1946).

M.A. Weinstein (1960) deduced the formula for the efficiency limit of electroluminescence, taking into consideration its entropy.

Yu.P. Chukova (from 1969) formulated the general law of efficiency of electromagnetic radiation energy conversion into other kinds of energy in irreversible isothermal processes.

M.A. Leontovich (1903–1981) has calculated efficiency limit for direct conversion of sunlight into electric energy.

P.T. Landsberg (1923–2010) (1980) published the review of works on thermodynamics of nonequilibrium electromagnetic radiation.

8 The General Law of Efficiency of Electromagnetic Radiation Energy Conversion into Other Kinds of Energy

In Chukova's works (1969, 1985, 1999, 2001, 2004, 2009) a simple scheme of conversion of energy fluxes has been created. It is shown in Fig. 5, where the following designations are used:

$\dot{W}_a$ is the rate of electromagnetic energy absorption,

$\dot{S}_a$ is the flux of entropy of absorbed energy,

$\dot{W}_L$ is the flux of irradiated electromagnetic energy,

$\dot{S}_L$ is the flux of entropy of irradiated energy,

$\dot{U}_r$ is the flux of the internal energy of reactants,

$\dot{S}_r$ is the flux of entropy of internal energy of reactants,

$\dot{U}_P$ is the flux of the internal energy of products,

$\dot{S}_P$ is the flux of entropy of internal energy of products,

$\dot{S}_i$ is entropy generation rate due to irreversible processes inside the system,

$\dot{Q}$ is the flux of thermal energy,

T is the temperature of the system,

$\dot{Q}/T$ is the rate at which the entropy of thermal flux carries from the system.

The scheme in Fig. 5 indicates three ways of energy conversion:

- conversion of energy of electromagnetic radiation into other kinds of electromagnetic radiation. It is more often various kinds of luminescence. The efficiency is designated η_L.

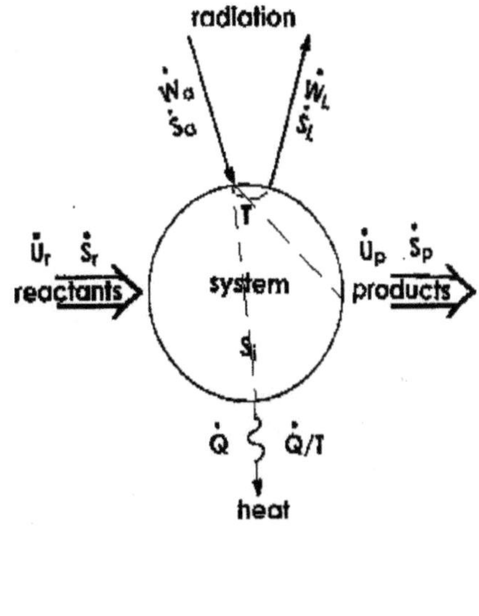

Fig. 5 Fluxes of energy and entropy in a thermodynamic open system. (Chukova 2001)

- conversion of energy of electromagnetic radiation into energy of chemical bonds and into electric energy. This efficiency is designated η and is introduced by the formula

$$\eta = \Delta F/\Delta W_a \tag{1}$$

where F is the Helmholtz free energy and is introduced by the formula $F = U - ST$,

$$\eta = [(\dot{U}_p - \dot{S}_p T) - (\dot{U}_r - \dot{S}_r T)]/\dot{W}_a \tag{1a}$$

- conversion of energy of electromagnetic radiation into thermal flux $\dot{Q}$. This conversion is considered as energy losses.

On the basis of the law of conservation of energy and the law of balance of entropy in equilibrium conditions, Yu. P. Chukova (2001) received the following generalized formula:

$$\eta_L + \eta = 1 + T(\dot{S}_L - \dot{S}_a - \dot{S}_i)/\dot{W}_a. \tag{2}$$

Within the limits of this article, the case of luminescence does not represent interest, therefore $\eta_L = 0$ and $\dot{S}_L = 0$, then the formula (2) becomes more simply

$$\eta = 1 - T(\dot{S}_a + \dot{S}_i)/\dot{W}_a. \tag{3}$$

According to rules of thermodynamics, calculations begin from a thermodynamic limit ($\dot{S}_i = 0$). The efficiency for a thermodynamic limit is designated η*

$$\eta^* = 1 - T\dot{S}_a/\dot{W}_a. \tag{4}$$

For calculation of the upper limit of efficiency, it is necessary to calculate the absorbed energy and its entropy according to formulas (5)–(7),

$$\dot{W}_a = \int E_\nu d\nu \tag{5}$$

where ν is the frequency of absorbed electromagnetic radiation and E_ν is the spectral intensity of absorbed power.

$$\dot{S}_a = 2\pi k c^{-2} \int \nu^2 [(1+\rho)\ln(1+\rho) - \rho \ln \rho] d\nu, \tag{6}$$

$$\rho = c^2 E_\nu / 2\pi h \nu^3 \tag{7}$$

where c is the light velocity, h is Planck's constant, k is Boltzmann's constant, ρ is distribution function.

The result of calculation η* is presented in Fig. 6 for a full region of frequencies of electromagnetic radiation.

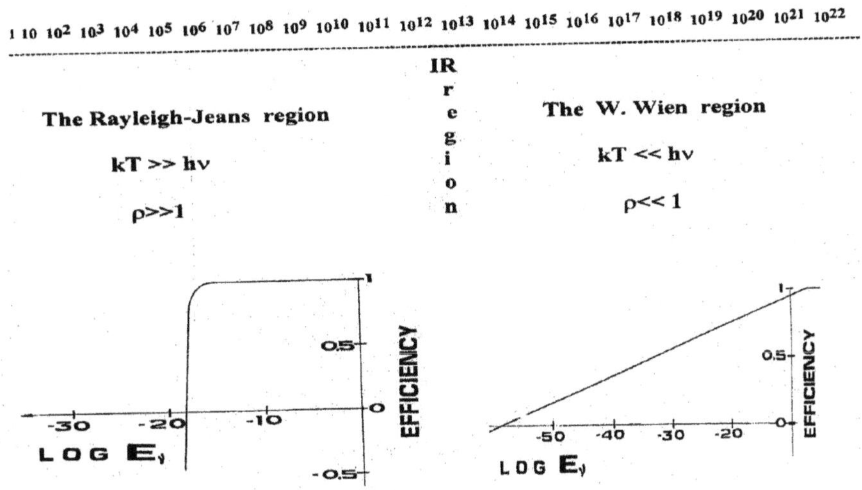

Fig. 6 Partition of the whole region of electromagnetic radiation by theoretical physicists (Chukova 2009)

Figure 4 shows that, according to fundamental laws of nature, a full region of electromagnetic waves breaks up on two areas: the Wilhelm Wien's region and the Rayleigh-Jeans region. Infra-red radiation is the boundary between them. Laws of energy conversion efficiency of electromagnetic radiation differ radically in these two areas. Functional dependence of conversion efficiency η^* (for which the linear scale is used) from spectral density of absorbed power $E\nu$ (for which the logarithmic scale is used) is presented at the bottom of Fig. 6.

Thus, we have a result in a semi-logarithmic scale. The straight line in the Wien region indicates (means), that this dependence is logarithmic. It is known to experimenters of different specializations (physiologists, biologists, light engineering, chemists and others) as the Weber-Fechner law which was formulated 150 years ago.

In the Rayleigh-Jeans region, dependence looks like a step process. Such results were obtained for in experiments for the first time by Russian scientists and were published in 1973 (the Devyatkov law).

Dependence on the absorbed power in the Rayleigh-Jeans region (the Devyatkov law), is rather unusual and is not joint with the main principle of an exact experimental (Chukova 1995). This principle demands averaging of results on a series of measurements. Such averaging is significant when the effect cannot change a sign. Figure 6 shows that, in the Rayleigh-Jeans region, endergonic processes (with increase of the Helmholtz free energy) which are located above an axis of absciss, and exergonic processes (with reduction of the Helmholtz free energy) which are located under an axis of abscisses are possible. At averaging of such results the probability of a total zero result is rather great, which we have seen in results of some authors mentioned above. When researchers experiment with an ensemble of microorganisms, the ensemble itself makes an averaging.

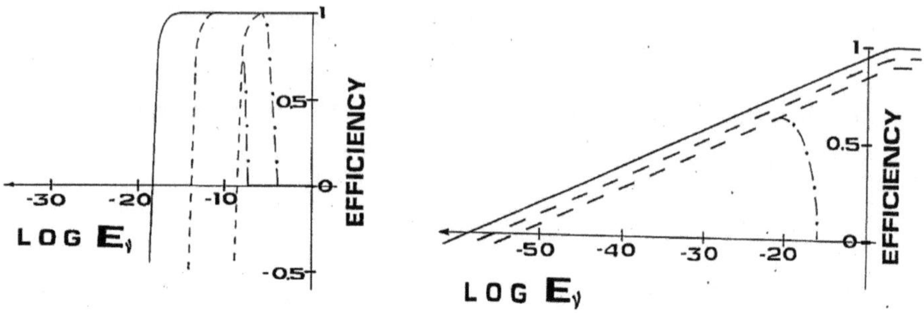

Fig. 7 Allowing for irreversibility processes in the Rayleigh-Jeans region (*left*) and in the Wien region (*right*). (Chukova 1999)

Exclusive skill of the experimenter in this case is required to avoid averaging and to fix an effect. The physician who works with an individual patient is always relieved of such averaging. Therefore medical researches with use of radio-frequency radiation have much higher repeatability than microbiological ones.

But that is not yet all we can say. In Fig. 6, dependences on spectral density of absorbed power E_ν is given for a thermodynamic limit of efficiency which are fair only for reversible processes. Reversible processes do not exist in nature. All natural processes are irreversible. To receive dependences, which is fair for them, it is necessary to consider speed of generation of entropy in the system under electromagnetic radiation. For this purpose it is necessary to calculate according to the formula (3).

The result of such calculation is presented in Fig. 7 for the Rayleigh-Jeans region and for the Wien region (Chukova 2001, 2002, 2009). A continuous line indicates a thermodynamic limit; a shaped line indicates efficiency of the account of linear irreversibility, and a dot-dash line indicates efficiency with the account's super-linear irreversibility. Results show that the account of linear irreversibility does not change the curve form: it remains the same, as in the case of a thermodynamic limit. But in the Rayleigh-Jeans region, it shifts (moves) along an axis of the absorbed energy, and this shift is very great. As calculations on the basis of experimental data have shown, the curve is displaced on 4–8 orders of absorbed power. As a result the experimenter in the Rayleigh-Jeans region has at once two inconveniences in comparison with the Wien region: very narrow interval of change η from 0 to 1 and the strongest change of position of this interval on an axis of the absorbed energy. The sum of these two factors can lead to "loss" of effect by the experimenter. This phenomenon is called the Cheshire cat effect (Chukova 2008). And if an experimentalist adheres more diligently to the standard rules in processing of results of experiments which are developed for visible light (the Wien region), the fixation of zero result ("loss" of effect) becomes more reliable (Chukova 2004).

It means that experimentalists who prepared for work with visible radiation or other radiation of the Wien region will have many extreme difficulties in successfully investigating effects under the Rayleigh-Jeans radiation. Knowledge of thermodynamics will allow one to avoid some errors in experiments and to give clarity to scientific discussion.

9 The Weber-Fechner Law

Now we will pass to consideration of the Wien region which includes (contains) visible light. It is most well investigated and is extremely important for all processes of life on the Earth, but first of all for processes of photosynthesis of plants and human vision. Human vision places first among other sense organs, providing a person with 80–90 % of the information available in today's world.

Research on activity of sense organs is of great importance not only for medicine and physiology, but also for philosophy. It is not an exaggeration to say that now a path of the most active struggle of idealism and materialism lies exactly here. Throughout the last two centuries scientists have moved steadfastly ahead in this area. There is a set of proofs of that, the Nobel Prize in 2004 being a bright demonstration. From this point of view the history of the Weber-Fechner law is interesting and certainly instructive.

9.1 Successes of the Nineteenth Century

In 1834 the German anatomist and psychologist E. H. Weber (1795–1878) presented outcomes of his research into the activity of sense organs. He studied a differential threshold of sensation by sense organs, i.e., he was interested in knowing what minimum change of a stimulus can be detected by a human observer. For example, he found that the addition of one candle to 60 burning candles allows a person to detect an increase of intensity of stimulus, and the addition of one candle to 120 burning candles appears poor for fixation of increasing of stimulus brightness. In the presence of 120 burning candles, for detection of an increase of intensity of stimulus, it was required to add, as minimum, 2 candles, and in the presence of 300 candles—5 candles. Weber concluded that two signals can be distinguished from each other, if the difference between them is proportional to their value. So, Weber's rule is

$$\Delta I/I = \text{const} \tag{8}$$

where I is stimulus intensity, ΔI is a value of the differential threshold, which being added to stimulus intensity I will call a hardly discernible difference in sensations and the numerical value of a constant (const) depends on the sensory system investigated. Weber investigated hearing, vision and touch.

The mathematical rule formulated by Weber did not break the norms of medical habitualness (i.e., did not upset the paradigm that had been adopted in medicine and in physiology) and has not become a source of stinging debate. Weber's experiment was simple and clear, and his outcome became a base for all following researches in sensory physiology.

The situation changed when the physicist, philosopher and physiologist Gustav Theodore Fechner (1801–1887), a contemporary of Weber, published a book under the title "Elemente der Psychophysik" (Fechner 1860). The essence of Fechner's ideas is extremely simple from the mathematical point of view: he integrated the ratio obtained by Weber and deduced the formula

$$\psi = \chi \log(I / I_0) \tag{9}$$

for a stimulus in conditions at the absolute threshold of sensation (Fechner 1860). Formula (9) is called the Weber-Fechner law. It became the fundamental law of psycho-physics and one of the fundamental laws of general sensory physiology. From a world point of view, formula (9) represents for physiologists and doctors something grandiose and continues to be an object of debate in the present day.

The intense discussion of this formula began at once after the publication of Fechner's book. It was started by some researchers who were certain that sensations are not subject to any measurements. Formula (9) represented a mathematical method for ordinary research of mental processes!!! It could not be accepted by those who stood on idealistic stands, contending that mental processes can not be a subject of measurement at all. But even for those who thought differently, the Fechner formula appeared unusual because of its use of a logarithmic function.

Perhaps it was not so important even then whether Fechner was or was not by nature a formidably argumentative person (controversialist, eristikos), or if he had been steeled in the heat of the passions that had boiled around his book; suffice it to say that in discussion, where he was always a victor, a phrase was born and recorded in the history of science: "The Tower of Babel was never finished because the workers could not reach an understanding on how they should build it; my psychophysical edifice will stand because the workers will never agree on how to tear it down".

Fechner had not taken into account that the scientific idea had appeared in a single head, that of his contemporary, Frenchman M.H. Plateau, to be exact. Plateau's objections had essentially been addressed to Weber, but not to Fechner. However Fechner accepted the impact upon himself.

As distinct from Weber, Plateau was sure that in sensory processes the ratio of sensations is a constant, if the ratio of the acting stimuli is constant. In this case, sensation is determined by a power function of stimulus (Plateau 1872).

The debate in the nineteenth century was finished by Fechner's victory, and the Weber-Fechner law became the basic law for five sense organs.

9.2 Events of the Middle of the Twentieth Century

Plateau's published article attracted attention only occasionally. In the middle of the 20th century Stevens S.S. has attempted to give it an experimental verification.

In September 1960 in Chicago, Fechner was honored at a centennial symposium sponsored by the American Psychological Association and Psychometric Society. The main report was made by S.S. Stevens, director of the Psychological Laboratory of Harvard University. In 1961, his report in the adapted form was published in the journal "Science" (Stevens 1961) with a tendentious title "To honor Fechner and repeal his law" with the subheading "A power function, not a log function, describes the operating characteristics of a sensory system".

In the beginning of the paper, Stevens honored Fechner and the unique traits of his character, and then he presented a problem, which was bound up with the formula (2). He noted that no "opponents" of Fechner could adduce essential experimental arguments against his formula. Stevens himself adduced such arguments on the basis of his own experiences 100 years after appearance of the Fechner formula.

Stevens studied nine stimulus: electric shock (~60 Hz), warmth, lifted weights, pressure on the palm, cold, white noise, tone (~1000 Hz), and white light. He used the log-log scale for his experimental outcomes. The intensity of stimuli was given in arbitrary units on the abscissa. The sensation intensity, which was also investigated, was given on the ordinate axis. Intensity of sensation as a function of stimulus intensity was a straight line. The fan of straight lines in an accompanying figure showed that all these processes are quite satisfactorily described by a power function

$$\psi = m(I - I_0)^n, \qquad (10)$$

where the above-mentioned symbols are used, m and n are constants.

For electric shock (~60 Hz), warmth, lifted weights, pressure on the palm and cold, the range of stimulus was equal to one order of magnitude, i.e., the stimulus had changed not more than ten times during the experiment. For vibration (60~), the range was equal to a 1.5 order of magnitude, for white noise and 1000~tone if it was two orders. And only for white light was this value equal to a 3.5 order. The dotted line was in the figure, which corresponds to $n=1$, i.e., it is a linear function.

As evident from the figure, among nine effects studied by Stevens, eight have power $n<1$, and only for electric shock $n>1$.

For a tenfold change of a parameter, the desire of an experimenter to present outcomes of experience in linear coordinates would be quite normal. It was made by Chukova for power functions with $n=0.33$; 0.5 and 1.2 (all three values correspond to Stevens experiment). The result was as follows: at tenfold change of stimulus, the power functions with the exponent $0.33<n<1.2$, can be quite satisfactorily described by a linear function with different coefficient of proportionality. The problem of legitimacy of linear approximating of outcomes can arise only for the smallest values of stimulus. But such measurements in physiology are specially difficult and are not often made. Usually a physiologist works at any average values of stimulus, when it is possible to expect the best repeatability of outcome and to claim large accuracy of outcome.

Thus, it is impossible to distinguish between linear dependence and a power function, if the exponent is less than 1.2. These functions are quite reliably distinct, if the power n is much higher than unity (2, 3 etc.) So for tenfold changes of parameter, mathematics does not allow us to consider Stevens' power functions as a convincing advantage before a linear function for approximation of experimental data.

The range of change of stimulus in Stevens's experiments is too small (an eye works in a range of change of stimulus on 10–16 orders, and an ear works on 6–9 orders) and it is insufficient for repeal of the Weber-Fechner law. Stevens did not manage to cancel the logarithmic law.

What was possible? Fechner had been separated from Weber. So the separate Fechner law had appeared and was called in publications a dubious, shady, notorious and so-called law. The word "law" appeared inside inverted commas. It was possible to put at the same level the Weber-Fechner law, the power function of Stevens and linear dependence. Formula (3) can be found in any modern textbook on psychophysics and sensory physiology alongside the Weber-Fechner formula. Thus this branch of science had become a field for voluntarism: each experimenter could approximate his or her outcomes by any function.

I will give two examples to show how the relation to the Weber-Fechner law changed after the Chicago anniversary.

The first example:

In the 1930's, teachers in Moscow schools speak to pupils about the Weber-Fechner law in lessons on biology.

The second example:

In the beginning of the twenty-first century a reviewer of the Russian journal "Successes of physiological sciences" declares that "the mental act "sensation" is not a subject for measurements and gradations".

9.3 *Rigid Dictatorship of Thermodynamics*

But the situation has changed considerably once again. In 2010 at the Lomonosov State University in Moscow, scientists with bunches of flowers celebrated the 150th anniversary of the Fechner book "Elemente der Psychophysik". Some organizations (including the German Embassy in Moscow) prepared this jubilee. Alexander Slobodin's video film "Anniversary" and the Chukova book "The Weber-Fechner law" in English (2010a) is devoted to this event.

After development of the thermodynamic theory of nonequilibrium electromagnetic radiation and the result presented in Fig. 6, there is no possibility to speak about cancellation of the Weber-Fechner law because logarithmic dependence follows from the law of conservation of energy, and it is not connected with any private restrictions or assumptions.

It is necessary to pay attention to the very big interval of change of illumination in the operating characteristics of human vision.

An eye sees badly at night, when the brightness of the night sky without stars is 10^{-6} cd/m^2 (cd is abbreviation of candelas). If in the night sky there are stars but there is no moon, the brightness of such a sky is 10^{-3} cd/ m^2, and at the full moon it is 10^{-1} cd/m^2. Metallic filament of an incandescent lamp has brightness more than 10^6 cd/m^2. At a cloudless summer solar noon, the sky brightness reaches 10^7 cd/m^2. Thus the change of illumination, where a human eye works normally and permanently, has more than 12 orders of magnitude, i.e., the human visual system during evolution has achieved such perfection that it was made possible to fix a change of stimulus in huge limits (more than 10 orders, i.e., more than ten billion times!!!). Brightness of the Sun is about 10^9 cd/m^2. But to look at the Sun on a bright summer day is bad for one's health. In this case, the working range of an eye is 15 orders, and the hygienists recommend wearing sunglasses in summer. The brightness of a pulse stroboscopic lamp is near 10^{11} cd/m^2, but it is already an adverse condition, which requires a special reliable protection for the eye. So, a human eye can work in a range of change of sunlight intensity on 17 orders of magnitude.

This raises a question: "Is there a mathematical function that is capable of making the eye work?"

Yes, it is the logarithmic function. The decimal logarithm is varied on 17 units, if the parameter is varied on 17 orders of magnitude, and if its parameter is varied on 15 orders of magnitude, the logarithm is varied on 15 units. So, the mathematical formula within a statute of the scientific law can describe mathematically the full operating characteristic of the human eye. The brightness of a pulse stroboscopic lamp is near 10^{11} cd/m^2, but it is already an adverse condition, which requires a special reliable protection for the eye.

Thus, from the point of view of mathematics everything is all right. What is it possible to say about physiology? Nature could not create a receptor that would be efficient in such a wide range. The human eye has two receptors of light: rods and cones. They have different spectral sensitivity and different efficiency of conversion of the absorbed radiation. The situation is described in detail by Chukova (2002a, b, 2010a). Now it is important for us to know the following: at transition from day sight (cones) to night sight (rods) on a straight line (in semi-logarithmic scale of Fig. 6), there is a change. And it is normal! Only in this range, Stevens's power function has some sense. Transition from cones to rods occupies about three orders of intensity. In this range it is possible to use the Stevens power function for approximation of experimental data. But we must never confuse mathematical operation of approximation and the physical law. These are essentially different things. The physical law gives logarithmic dependence!!! (Chukova 2010)

Change of spectral sensitivity of the human eye at transition from day illumination to evening illumination has been known from the beginning of the nineteenth century and is called the Purkyne effect. The Purkyne effect is well studied, but till now nobody could answer distinctly the question of why it exists. The exhaustive answer to this question was given by the thermodynamic theory of systems under electromagnetic radiation.

Thus, cancellation of the logarithmic law is impossible, because it is impossible to repeal the law of conservation of energy. It means, that there are no problems with human vision. The foregoing thermodynamic theory is quite right for it.

The thermodynamic theory stated above is capable of explaining not only the basic characteristics of human vision, but also many other photobiological phenomena with logarithmic dependence, such as photosynthesis and phototropism of green plants, all kinds of photomovement (Nultsch 1961a, b, 1962; Diehn and Tollin 1966) of the elementary organisms (phototaxis, phobotaxis and photokinesis), and also many other phenomena.

Apparently, the logarithmic base was understood by those who stood near the source of quantitative research of human hearing, whereas in this area the logarithmic scale has already been used for measurement of sound pressure level for a long time.

It is evident according to hygienic research, that a change of sound pressure from 0 to 80 decibels is a normal condition for operating of the human ear.

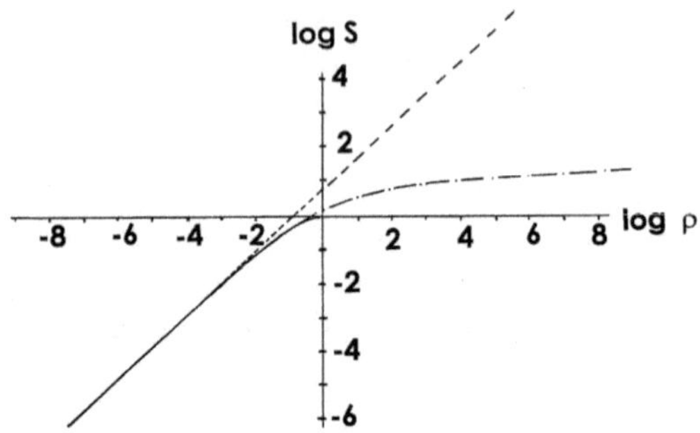

Fig. 8 Dependence of entropy from ρ for Fermi-Dirac distribution (*a continuous line*), Bose-Einstein distribution (a *continuous line* and a *dot-dash line*) and Maxwell-Boltzmann distribution (a *continuous line* and a *dotted line*). (Chukova 2010b)

Sound pressure from 80 to 110 decibels is considered as marginal conditions, and pressure of sound from 110 to 170 is a very bad condition for the work of the human ear, which can be a reason for damage and originating of disease. So it follows from hygienists studies that the human ear in normal conditions works in the range of 4 orders of stimulus magnitude. The general range of human audibility is 5.5–8.5 orders of magnitude.

But how is it be with other sense organs?

The system of Fig. 5 is right for electromagnetic radiation and for other influences. The same is possible to tell about formulas (1)–(4). Specificity of electromagnetic radiation is concluded in formulas (5)–(7), and the specificity of the Wien region is caused by value of distribution function ρ, which is less than unity. In the Rayleigh-Jeans region it is more than unity.

Dependence of entropy flux from ρ is presented in Fig. 8 for three distributions most often used by physicists: Fermi-Dirac distribution (a continuous line), Bose-Einstein distribution (a continuous line and a dot-dash line) and Maxwell-Boltzmann distribution (a continuous line and a dotted line). For the particles submitting to Fermi-Dirac statistics, values ρ can be only below unity. And for the particles submitting to two other statistics, they can be both lower and above unity. But now only the interval ρ below unity, which defines occurrence of logarithmic dependence, is of interest for us. Its presence in all three statistics says that logarithmic dependence is the general dependence for a huge circle of absolutely different phenomena which are now considered not only as independent, but absolutely alien to each other. Therefore, the presence of logarithmic dependence at quantitative studying of other sense organs of a person is absolutely natural.

So, the thermodynamic theory has answered all basic questions connected with the Weber–Fechner law.

10 Western and Chinese Medicine on an Integration Way

In the history of our civilization there is one big riddle connected with independent existence of Oriental (Chinese) and Western (European) medicines during two millenniums. This independence arose in a time when China was artificially divided from the world around.

Now in the period of globalization all artificial boundaries have fallen, and independence between European medicine and Chinese medicine becomes a subject of discussion. At the 13th International Congress of Logic, Methodology and Philosophy of Science (2007, Beijing, China) there appeared for the first time a Special Symposium "Chinese Traditional Medicine vs. Western Medicine". At this Symposium, scientists of China and France attempted to compare these independent sciences. But this comparison was like that of fire and ice because it contained only enumeration of characteristics of one and the other, which demonstrated clearly their deep difference.

Now we have a method that allows us to build a bridge from Western science to traditional Chinese medicine (TCM). This bridge was given by a new branch of thermodynamics, namely thermodynamics of irreversible processes in systems under electromagnetic radiation. This new branch of thermodynamics has been described above and can be used now with success.

Thermodynamics of irreversible processes in systems under electromagnetic radiation (the irreversible thermodynamics of physiological processes of electromagnetic radiation) can be a dictionary for translation of terminology of Chinese medicine into the language of Western science.

The key concept of Chinese medicine (energy Chi) corresponds to Helmholtz's free energy which is a basis of the foregoing theory. From this point of view, it is easy to understand why before it was not possible to find any conformity between Chinese medicine and western medicine. These are different sciences: western medicine has concentrated on morphology of patient, and Chinese medicine is a science about energy. More importantly, Chinese medicine is a science about weak influences while the western medicine studied until recently strong influences. Only in recent years has western medicine paid attention to weak influences of electromagnetic radiation. It has given us a chance to construct the bridge, connecting both sciences. As a result, there is now appearing a clear contour of new world medicine of the middle and the second part of the twenty-first century. But it is a theme for that calls for extensive research.

11 Conclusion

The face of our twentieth century civilization was defined by successes of space researches, cybernetics and semiconductors. Theoretically based works advanced experiments in these three areas. Theoretical science lit a lantern in areas where experimenters can work productively. In the middle of the twentieth century Sergey Ivanovich Vavilov, President of the Academy of Science of the USSR, repeatedly underlined the importance of continuing development of theory in science. All the above discourse shows his correctness: thermodynamic theory, for example, has allowed us to understand the most complicated contradictions of experiments with MM radiation, having explained the features of biological and medical effects which can be both beneficial and harmful for living organisms (Chukova 2001, 2011).

Even more to the point, this theory has shown the importance and generality of those problems which were solved in the twentieth century for radiation MM, for all regions of radio-frequency radiation and extremely low frequencies, including terahertz radiation, which only recently began to be used in biological and medical experiments.

The thermodynamic theory can be crucial in resolving a question on hygienic standards of safety on a strictly scientific basis. It is of the utmost importance that the WHO not ignore this opportunity (Chukova 2006, 2008, 2011). Chukova's Memorandum on this problem was addressed to Margaret Chen, the Director General of WHO. It was sent on July 30, 2012, arrived at WHO headquarters on August 17, 2012 and was fully expected to be taken up for consideration by qualified experts (Chukova 2013).

The thermodynamic theory has allowed us to thoroughly examine and assess the proposal made by Stevens to abandon the Weber-Fechner law and has proved that it cannot be discarded without also discarding the law of conservation of energy. Moreover, it has shown that the Weber-Fechner law has a much wider generality than physiological science only. It is one of the basic general laws of nature.

Such generality of the law of energy conversion of electromagnetic radiation allows us to use the general power basis not only for all phenomena studied by photobiology, but also for many other physical phenomena. All of this argument implies the need to write and publish new textbooks that will allow us to further our progress in knowledge about natural phenomena and to facilitate the process of acceptance of reasonable social decisions which do not contradict nature's laws.

A profound example is that thermodynamic theory has found a key (approach) for an explanation of distinctions between Western and Chinese medicine. Now it is possible to expect that this direction will open a way for treatment of many chronic diseases which now have no treatments within the limits of the regulations accepted by Western medicine

References

Athey TW (1980) Report of a panel discussion. Proceedings of IEEE 68(1):116–117

Baillie HD (1974) Rapid warming from hypothermia by microwaves. In: Proceedings of an international symposium biological effects and health hazards of microwave radiation. Polish Medical Publishers, Warsaw, pp 15–21

Baranski S, Edelwejn Z (1974) Pharmacologic analysis of microwaves effects on the central nervous system in experimental animals. In: Proceedings of an international symposium biological effects and health hazards of microwave radiation. Polish Medical Publishers, Warsaw, pp 119–127

Baranski S, Szmigielski S, Moneta J (1974) Effects of microwave irradiation in vitro on cell membrane permeability. In: Proceedings of an international symposium biological effects and health hazards of microwave radiation. Polish Medical Publishers, Warsaw, pp 173–177

Belyaev IY, Shcheglov VS, Alipov YD, Polunin VA (1996) Resonance effect of millimeter waves in the power range from 10–19 to 3.10–3 W/cm^2 on *Escherichia coli* cells at different concentrations. Bioelectromagnetics 17:312–321

Berteaud J, Dardalhon M, Rebeyrotte N, Averbeck D (1975) Action d'un rayonnement electromagnetique a longueur d'onde millimetrique sur la croissance bacterienne. Comptes Rendus de l'Académie des Sciences Paris 281:843–846

Betskii OV, Lebedeva NN (2001) История становления КВЧ-терапии и десятилетние итоги работы Медико-технической ассоциации КВЧ. Миллиметровые волны в биологии и медицине. No 4(24), c. 5–12

Bose SN (1924) Plancks Gesetz and Lichtquantenhypothese. Zeitschrift für Physik 26:178

Carpenter RL, Ferri ES, Hagan GJ (1974) Assessing microwaves as a hazard to the eye—progress and problems. In: Proceedings of an international symposium biological effects and health hazards of microwave radiation. Polish Medical Publishers, Warsaw, pp 178–185

Chukova YuP (1969) Thermodynamic limit of the luminescence efficiency. JETP Letters 10:294–296

Chukova YuP (1985) Thermodynamic limitation of nonthermal processes of the millimeter waves conversion, obtained by Landau-Weinstein method. In: The application of millimeter waves of low intensity in biology and medicine. Academy of science USSR, Moscow, pp 147–156. [in Russian]

Chukova YuP (1988) Spectral sensitivity of the human eye. Docl. biological science. Proceedings of the Academy of Sciences of the USSR 300:306–308

Chukova YuP (1990) Biological response to radiofrequency electromagnetic radiation Docl. Biological Science. Proceedings of the Academy of Sciences of the USSR 311:237–239

Chukova YuP (1995) The radiofrequency bioeffects and the crisis of paradigm of the precision biological measurement. 10th International Congress of logic, methodology and philosophy of science. Italy, Florence, p 509

Chukova YuP (1999) Reasons of poor replicability of nonthermal bioefects by millimeter waves. Bioelectrochemistry and Bioenergetics 48:349–353

Chukova YuP (2001) Advances in nonequilibrium thermodynamics of the systems under electromagnetic radiation, Moscow, Khrizostom, 2001, ISBN 5-7508-0285-X

Chukova YuP (2002a) The general laws of biological effects of optical electromagnetic fields—proceedings of the 2nd international workshop "Biological effects of EVFs". Rhodes, Greece, pp 318–326

Chukova YuP (2002b) Low-level influence effects. Thermodynamic, experimental (medical and biological), social, legislative, international and philosophical aspects of the problem. Moscow, ISBN: 5-89047-013-2 [in Russian]

Chukova YuP (2004) The difficulties in experimental study of isothermal bioeffects under radiation of radio-frequency region. In Kostarakis P (ed) Proceedings of the 3rd international workshop "Biological effects of EVFs", vol 1. Demokritos Publishers, Kos, pp 399–407

Chukova YuP (2006) Thermodynamic law and the problem of standardization of radiofrequency electromagnetic radiation. In: Kostarakis P (ed) Biological effects of EMFs 4th International Workshop Proceedings, vol 1. Crete, pp 383–391

Chukova YuP (2008) The Cheshire cat phenomenon in bioprocesses of millimeter waves. 5th international workshop on biological effects of electromagnetic fields. Palermo-Terrasini. CD-R, ISBN: 978-88-95813-03-5

Chukova YuP (2009) Bose condensation and non thermal processes in living systems under millimeter (MM) radiation. Electromagnetic Biology and Medicine 28(1):41–45

Chukova YuP (2010a) The Weber-Fechner law. To the 150th anniversary of the publication of the book "Elemente der Psychophysik" by GT Fechner. Moscow. ZAO MP "Gigiena". ISBN: 978-5-904207-07-6

Chukova YuP (2010b) Nonthermal bioeffects of MM radiation in the light of max planck's papers. Proceedings 6th Workshop on biological effects of electromagnetic fields. 2010. Bodrum. http://www.istanbul.edu.tr/6internatwshopbioeffemf

Chukova YuP (2011) Doubts about nonthermal effects of MM radiation have no scientific foundations. Journal of Physics: conference series 324 2011012032. http://iopscience.iop.org/1742-6596/329/1/012032/pdf/1742-96_329_1_012032.pdf

Chukova YuP (2013) Thermal vs. nonthermal effects. The Devyatkov law and the health safety standard. In: Youth in a postindustrial society. Dusseldorf-Saratov, ANO Press-Lizei, pp 62–74. http://hsjournal.org/youth001

Dandanoni L, Torrengossa MV, Zanforlin L, Spalla M (1978) XIX general assembly abstracts. Biological effects of EM waves. URSI, p 38

Devyatkov ND (1974) Influence of millimeter-band electromagnetic radiation on biological objects. Soviet Physics – USPEKHI 4:568–569
Diehn B, Tollin G (1966) Phototaxis in euglena.11. Physical factors determining the rate of phototactic response. Photochemistry and Photobiology 5:523–532
Fechner GT (1860) Elemente der Psychophysik. Leipzig
Frölich H (1975) Evidence for bose condensation-like excitation of coherent modes in biological systems. Physics Letters 51A(1):21
Gordon ZV, Roscin AV, Byckov MS (1974) Main directions and results of research conducted in the USSR on the biological effects of microwaves. In: Proceedings of an international symposium biological effects and health hazards of microwave radiation. Polish Medical Publishers, Warsaw, pp 22–35
Grundler W, Keilmann F (1978), Nonthermal effects of millimetre microwaves on yeast growth. Zeitschrift für Naturforschung, 33 c, 15–22
Grundler W, Jentzsch U, Keilmann F, Putterlik V (1988) Resonant cellular effects of low intensity microwaves. In: Herbert F (ed) Biological coherence and response to external stimuli. Springer, Berlin, pp 65–85
Hill DW, Hagman MJ, Riazi A, Gandhi OP, Partlow LM, Stensaas LJ (1978) Effect of millimetre waves on bacteria and viruses. Abstracts of scientific papers open symposium on the biological effects of electromagnetic waves 19 General assembly of URSI, Helsinki, Finland, p 35
Jonson CC, Guy W (1972) Nonionizing electromagnetic wave effects in biological materials and systems. Proceedings of the IEEE 60(6):692–718
Landau LD (1946) On the thermodynamics of photoluminescence. Journal of Physics-USSR 10:503–506
Landsberg PT, Tonge G (1980) Thermodynamic energy conversion efficiencies. Journal of Applied Physics 51(7):R1–R20
Levich WG (1954) Левич В.Г. Введение в статистическую физику. Учебник для университетов. ГИТТЛ, Москва. С.492
Michaelson SM (1974) Thermal effects of single and repeated exposures to microwaves. In: Proceeding of an international symposium Biological effects and health hazards of microwave radiation. Polish Medical Publishers, Warsaw, pp 1–14
Micolajczyk HJ (1974) Microwave irradiation and endocrine functions. In: Proceedings of an international symposium biological effects and health hazards of microwave radiation. Polish Medical Publishers, Warsaw, pp 46–51
Miro L, Loubiere R, Pfister A (1974) Effects of microwaves on the cell metabolism of the reticulohistocytic systems. In: Proceedings of an international symposium biological effects and health hazards of microwave radiation. Polish Medical Publishers, Warsaw, pp 89–97
Motzkin SM (1980) Report of a panel discussion. Proceedings of the IEEE 68(1)
Motzkin SM (1991) Low power continuous wave millimetre irradiation falls to produce biological effects in lipid vesicles, mammalian muscle cells and E. coli, Digest of papers International symposium "Millimetre waves of non-thermal intensity in medicine", Vol 2, Moscow, pp 367–368
Motzkin S, Birenbaum L, Rosenthal S, Rybenstein C, Davidow S, Remily R, Melnic R (1979) Millimeter wave CW cells irradiation studies I. Microwave aspects: colicin induction in E. coli. National radio science meeting bioelectromagnetics symposium, Seatlle, pp 460–461
Nultsch W (1961a) Der einfluss der lichtes auf die bewegung der cyanophyceen. 1.Phototopotaxis von phormidium autumnale. Planta 56:622–647
Nultsch W (1961b) Der einfluss der lichtes auf die bewegung der yanophyceen.11.Photokinesis bei phormidium autumnale. Planta 57:613–623
Nultsch W (1962) Der einfluss der lichtes auf die bewegung der cyanophyceen.111. Photophobotaxis von phormidium uncinatum. Planta 58:647–663
Partlow LM (1980) Report of a panel discussion. Proceedings of the IEEE 68(1):117–118
Pazderova J, Pickova J, Bryndova V (1974) Blood proteins in personnel of television and radio transmitting stations. In: Proceedings of an international symposium biological effects and health hazards of microwave radiation. Polish Medical Publishers, Warsaw, pp 281–288

Planck M (1914) The theory of heat radiation—1959
Plateau MH (1872) Sur la mesure des sensations physique, et sur la loi qui lie l'intensite de ces sensations a l'intensite de la cause excitante. Bulletins de l'Academie Royale des Sciences de Belgique 33:376–388
Rugh R, Ginns EI, Ho HS, Leach WM (1974) Are microwaves teratogenic? In: Proceeding of an international symposium biological effects and health hazards of microwave radiation. Polish Medical Publishers, Warsaw, pp 98–108
Schwan HP (1974) Principles of interaction of microwave fields at the cellular and molecular level. In: Proceedings of an international symposium biological effects and health hazards of microwave radiation. Polish Medical Publishers, Warsaw, pp 152–159
Sevast'yanova LA, Vilenskaya RL (1974) A study of the effects of millimetre-band microwaves on the bone marrow of mice. Soviet Physics USPEKHI, pp 570–571
Siekierzynski M (1974) A study of the health status of microwave workers. In: Proceedings of an international symposium biological effects and health hazards of microwave radiation. Polish Medical Publishers, Warsaw, pp 273–280
Smolyanskaya AZ, Vilenskaya RL (1974) Effects of millimetre-band electromagnetic radiation on the functional activity of certain genetic elements of bacterial cells. Soviet Physics USPEKHI, p 571
Stevens SS (1961) To honor Fechner and repeal his law. A power function, not a log function, describes operating characteristic of a sensory system. Science 133:80–86
Swicord ML, Athey TW, Buchta FL, Krop BA (1978) Colicin induction by exposure to millimetre-wave radiation. Abstracts of scientific paper open symposium on the biological effects of electromagnetic waves, 10 general assembly of URSI, Helsinki, p 35
Tengroth B, Aurell E (1974) Retinal changes in microwave workers. In: Proceedings of an International symposium biological effects and health hazards of microwave radiation. Polish Medical Publishers, Warsaw, pp 302–305
Tolgskaya MS, Gordon ZV (1973) Pathological effects of radio waves. Consultants Bureau, London, p 146
Webb J (1979) Factors affecting the induction of lambda prophages by millimeter microwaves. Physics Letters 73A(2):145–148
Weinstein MA (1960) Thermodynamic limitation on the conversion of heat into light. Journal of the Optical Society of America 50(6):597–602
Yagi K, Ueyama R, Kurohane S, Hiramine N, Ito H, Umehara S (1974) In: Proceedings of an international symposium biological effects and health hazards of microwave radiation. Polish Medical Publishers, Warsaw, pp 75–88

On the History and Technology of the Atomic Bomb. The Commitment of the Scientists

Vincenzo Cioci

Abstract The development of the atomic bomb in the first half of the twentieth century marked a turning point in the history of nuclear science because it revealed the close relationships that exist among science, technology and society. In this paper the main discoveries that led to the scientific and technological development of the atomic bomb are presented together with the commitment of scientists who tried to avoid possible harmful uses of the results of their researches.

In this work a deep examination of the writings of some of these scientists is introduced, some of which are still unpublished, although already quoted by several authors, with the purpose to highlight the relevance of their position against the bomb to the present day. In this sense the content of the paper may appear as a novelty within the history of science and technology; even if it cannot be a unique story.

Keywords Atomic bomb · Soddy · Szilard · Rasetti · Oppenheimer · Commitment of the scientists

1 Introduction

It is generally believed among non-scientists that the invention of the atomic bomb took place in the United States of America during the Second World War and that it was made possible by the discovery of uranium fission, by Otto Hahn (1879–1968) and Fritz Strassmann (1902–1980), in Germany, in December of 1938. This isn't completely correct. The process that led to invention of the bomb was actually much longer, lasting about half a century, and was characterized by a series of discoveries of which the first and certainly the most important was that of the enormous amount of energy associated with natural radioactive phenomena made at the beginning

V. Cioci (✉)
Research Centre for the Theory and History of Science, University of West Bohemia, Pilsen, Czech Republic
e-mail: vincenzocioci@gmail.com

Research Group on Physics Education and History of Physics, University of Calabria, Cosenza, Italy

R. Pisano (ed.), *A Bridge between Conceptual Frameworks,* History of Mechanism and Machine Science 27, DOI 10.1007/978-94-017-9645-3_7

of the twentieth century by Ernest Rutherford (1871–1937) and Frederick Soddy (1877–1956). Soddy was the first to pose the problem of the social significance of this discovery, even before Einstein formulated his famous law of energy.

As we shall see in the next paragraphs, in the 1930s there was a succession of discoveries: existence of the neutron, the radioactivity produced by alpha-particle bombardment, the neutron induced radioactivity, the properties of slow neutrons, nuclear fission, the nuclear chain reaction, until the invention of the atomic bomb.

Physicists immediately realized the great potentialities and possible applications of nuclear physics. But before the certainty that a bomb of unprecedented destructive power could actually be built, there were some scientists who undertook to avoid potentially harmful uses of science.

In this paper I want to emphasize the commitment of these individuals who tried by all means in their power to involve other scientists, scientific institutions and societies in their struggle.

Frederick Soddy, based on his personal experience, said that perhaps as they were constituted, scientific organizations were not suitable to deal with ethical issues (Soddy 1945, p. 9). Yet this work aims to demonstrate that the most reasonable way to address these issues is just internal debate within the scientific community.

2 The Discovery of Atomic Energy by Rutherford and Soddy

At the beginning of the twentieth century, at McGill University in Montreal, Rutherford and Soddy made important contributions to the study of radioactivity.

In 1901, Soddy brought his experience as a chemist to the study of gaseous emanation of radioactive thorium which had been observed by Rutherford a few years earlier. Soddy realized that thorium was transformed spontaneously in an inert gas of the argon family. This was the first, clear, experimental evidence of the direct formation of a chemical element known from another one.[1]

Their collaboration led quickly to the complete interpretation of radioactive phenomena as natural processes of spontaneous sub-atomic disintegration. Between 1902 and 1903, the two scientists published a series of articles describing the importance of the general theory of radioactivity with the laws of radioactive decay (Rutherford and Soddy 1902a, 1902b, 1903).

[1] [1]Rutherford won the Nobel Prize for Chemistry in 1908 "for his investigations into the disintegration of the elements and the chemistry of radioactive substances", although the chemist of the team was Soddy. He gained due recognition only with the Nobel Prize in 1921 "for his contributions to our knowledge of the chemistry of radioactive substances and his investigations into the origin and nature of isotopes". In 1913, in fact, he had found that certain elements exist in two or more forms with different atomic weights but they are chemically indistinguishable. Between 1911 and 1913, he had also formulated the Law of radioactive displacement, which refers to the fact that the emission of an α particle by an element moves it back two places in the periodic table, while the emission of a β particle situates it in one position forward. His naming to the Royal Swedish Academy of Sciences was proposed by Rutherford (and supported by Thomson), as if to repay the debt contract with Soddy on the occasion of the Nobel Prize in 1908.

As for the purposes of this study, the most relevant result obtained by the two scientists concerns their discovery of the huge amount of energy associated with radioactive phenomena. They, in fact, in 1903, estimated the total energy released during the radioactive decay of 1 g of radium, evaluating the kinetic energy of α particles emitted. They discovered that for a given mass, much more energy was emitted (up to a million times) than was produced in any known chemical reaction (Rutherford and Soddy 1903).[2]

In 1904 Ramsay and Soddy arrived at a more accurate estimate of this energy by multiplying the heat generated (per unit of time elapses) by the radioactive emanation from radium in a unit of time by the average life of emanation, which they measured. They found that the relation between the energy emitted by the radium emanation during its disintegration and the one released in the association of hydrogen and oxygen for the formation of water is, for the same weight, about 216,000 (Ramsay and Soddy 1904, p. 357).

The discovery of the immense amount of energy associated with atomic disintegration preceded, a few years, the famous formula $E = m \cdot c^2$ of special relativity. Einstein himself, in 1905, had suggested that his theory could perhaps be confirmed by "using bodies whose energy content is variable to a high degree (e.g. salts of radium)" (Einstein 1905, p. 174, line 19). Then he returned to this subject to express his doubts about the effective possibility of testing, in an experimental way, his theory because of the limits dictated by the technology of the time, unless other radioactive phenomena were discovered in which there was a greater mass fraction that would be transformed into energy (Einstein 1907, p. 288; 1910, p. 144). See also Pais (1982) pp. 148–149.

3 Soddy in the Face of the Risks and the Expectations of Atomic Energy

In 1903, Soddy returned to England and became adviser for and a commentator on the recent discoveries made together with Rutherford on radioactivity, putting particular emphasis on the inexhaustible energy associated with this phenomenon.

At the Bodleian Library, Oxford University, a document has been preserved that gives an account and the interpretative key of his speeches and articles of that period. It is a letter addressed to Harold Hartley, dated 22 May 1953, in which Soddy says:

> As I have indicated in these years one could not possibly have discussed radioactivity at all without reference to the hitherto completely unsuspected colossal store of energy latent in the atom and only knowable when the atom disintegrates.[3]

[2] In March 1903, Pierre Curie (1859–1906) and Albert Laborde (1878–1968) had measured the heat developed in a Bunsen ice calorimeter from a known amount of radio in a defined time. They found enormous values of the order of 100 calories per hour for a single gram of radium (Curie and Laborde 1903). These authors, however, did not consider it possible that this heat was developed at the expense of the internal energy of the radium but they thought that it came from a source outside the atom, of unknown nature (on this topic see Soddy 1904b, Chap. XI entitled "The energy of radio-active change", pp. 165–170).

[3] Soddy (1953, f 282, 2, line 6).

In Contemporary Review, particularly in May of 1903, he used for the first time the term *atomic energy* in referring to the inexhaustible amount of energy stored in matter, and he asserted that radioactivity would have led to "alter our attitude towards inanimate matter", that it had to be considered as a vast reservoir of energy (Soddy 1903a, p. 720, line 10).

In his writings of that period he pointed out that the internal energy of the elements had to be really great because only a minimal fraction of it was released: the difference between the energy stored in the atom before and after its radioactive transformation was small (Soddy 1906). He pointed out, moreover, that, if 1 day scientists were able to accelerate the speed of radioactive transformations, it would be possible to solve problems related to the depletion of energy resources (Soddy 1903b). But also that human beings could create a bomb capable of destroying the whole world, if only they want to (Soddy 1904a).

The fundamental popular work, written in non-technical language by Soddy, was *The interpretation of radium*, 1909. It exposes the contents of six experimental lessons of popular character held at the University of Glasgow in 1908. The book had a wide circulation, including being translated also into Russian and to having two subsequent editions, in 1912 and in 1920, revised and updated with the advance of scientific knowledge.

In the first edition, in particular, Soddy underlines the possibility of catastrophes, consequences of the irresponsible use of atomic energy. On the whole, however, the volume expressed great hope in the potential of science and humanity. Thanks to atomic energy, scientists could explore distant worlds, make the desert habitable, and transform the whole planet into "one smiling Garden of Eden", freeing man from his daily needs and changing his relationship with nature (Soddy 1909, p. 244, line 17). Reading the 11th chapter of this book inspired Herbert George Wells (1866–1946) to write his novel *The world set free*[4] that in its turn would influence the choices of Leo Szilard (1898–1964) and other scientists against the atomic bomb.[5] In 1926, Soddy expressed his appreciation to Wells for "his customary brilliance and insight" that he had shown in analyzing the possible consequences of the discovery of atomic energy (Soddy 1926, p. 28, line 18).

The reading of the novel and even more the disasters of the First World War (with the transformation of many technological processes in devices of war) convinced Soddy to spend all his energies (as did several nuclear physicists many years after him) to warn humanity because

> the social effect of recent advances in physical science promises to be annihilating, unless, before it is too late, these arises an equal and compensating advance, of which there is at the present no sign, in the moral and spiritual forces of society[6]

---

[4] In this science fiction novel, H. G. Wells predicts, about 20 years in advance, the discovery of artificial radioactivity, the industrial use of atomic energy and a global conflict resulting from the use of "atomic bombs", with the devastation of the main cities of the planet (Wells 1914).

[5] See the 6th paragraph and Cioci (2008).

[6] Soddy (1915, p. 13, line 19).

Soddy tried to reform the Royal Society, transforming it into an organization engaged in functions similar to those that the BMA (British Medical Association) performed for doctors, who were required to utter the Hippocratic oath before they began to exercise their profession.

Soddy undertook to increase the democratic participation of Fellows in the life of the Society, proposing a series of measures including the possibility to elect the Council or new Fellows by postal vote (Soddy 1934). Moreover, he considered "that there should be a new system to make known to the public the achievements of scientists at the earliest possible moment"[7] in order to prevent, before it is too late, possible harmful uses of science.

Even later, he returned again to the crucial role to be played by the Society of specialists in making scientists responsible for the use of their discoveries and inventions (Soddy 1945, p. 9). Soddy advocated the establishment of a strong international authority, linking scientific institutions around the world and forcing scientists

> to obey a code of ethics drawn up for their protection and guidance, and requires from them an oath that they will not be a party to assisting in war work before allowing them to engage in scientific work, having adequate power to withhold the means for their doing so.[8]

4 The Discovery of Neutron-Induced Radioactivity

A study by Spencer Weart, former director of the Center for History of Physics, American Institute of Physics, showed that during the first decades of the 1900s, radium took almost the same proportion of space in printing as nuclear energy did in the 1960s (Weart 1982).

After the First World War and publication of The World Set Free, numerous short stories and novels were published in which was described the general destruction caused by new scientific weapons and in some cases also atomic ones.

In 1919, Rutherford performed the first artificial transmutation of the atom according to the nuclear reaction

$$^{14}N + \alpha \rightarrow {}^{17}O + p. \tag{1}$$

An alpha particle collides with a nitrogen atom knocking out a hydrogen nucleus—which Rutherford dubbed the proton in 1920—and changing the nitrogen in oxygen into the form of an oxygen isotope with mass number 17.

However, about 30 years had passed since discovery of the huge amount of energy associated with radioactive decay without having seen any considerable advances towards using macroscopic atomic energy. But in 1932 John Cockcroft (1897–1967) and Ernest Walton (1903–1995) split lithium in alpha particles, bom-

[7] The Sydney Morning Herald (1935, f 124, line 43).

[8] Soddy (1949, p. 128, column 1, line 52). For the significant contribution made by Soddy to economics, linked to his commitment to the prevention of war, see Cioci (2009a).

barding it with high energy protons, and James Chadwick (1891–1974) discovered the neutron in that same year.

At the beginning of 1934, Irène Curie (1897–1956) and her husband Frédéric Joliot (1900–1958) discovered artificial radioactivity induced by alpha-particle bombardment: bombing, with alpha particles, boron or aluminum, they obtained respectively new isotopes of nitrogen and phosphorus that emitted positrons.

After the discovery of Curie and Joliot, Enrico Fermi (1901–1954) had the idea to use neutrons to induce radioactivity: neutrons, being neutral, would not have been repelled by the positive charge of the atomic nucleus and therefore they would have been more effective in producing nuclear reactions. In March of 1934, Fermi began to irradiate the known elements with neutrons using a radon-beryllium neutron source constituted by a glass tube containing beryllium powder and radon.

A few weeks later, also Franco Rasetti (1901–2001), Edoardo Amaldi (1908–1989), Emilio Segre (1905–1989) and the chemist Oscar D'Agostino (1901–1975) gave a valuable contribution, together with Fermi, to the systematic study and classification of the different mechanisms by which neutron-induced radioactivity, for different elements, took place (Fermi et al. 1934). In May of 1934, that interpretation of the case of uranium created many difficulties. The reaction products were elements whose atomic number was not included between that of lead and that of the same uranium. It was hypothesized that the irradiation of uranium with low-energy neutrons was to produce one or more elements in the periodic table that occupied successive positions and therefore called transuranic (Fermi 1934). Actually, the members of the Via Panisperna group (so called from the address of the Institute of physics, at University of Rome "La Sapienza", in Via Panisperna) had produced uranium fission without being aware of it.

In October 1934, after several months of work, "the Via Panisperna boys", who in the meantime had also seen the entrance of Bruno Pontecorvo (1913–1993), observed, interposing some hydrogenated means such as paraffin between the neutron source and a silver target, an amplification of the intensity of the activation. Contrary to what was assumed, this occurred as a consequence of the slowing down of neutrons caused by collisions with nuclei of hydrogen, since a slow neutron has a higher probability of being absorbed by a silver nucleus than a fast neutron is expected to do. Rome had become "the capital of the nuclear world". Aware of the possibility of industrial applications of the new discovery, Orso Maria Corbino (1876–1937) convinced the Roman physicists to apply for a patent in relation to their method of producing radioactive substances by bombardment with slow neutrons. Soon, however, Via Panisperna group dispersed.

5 Rasetti's Refusal to Participate in Researches for Military Use of Atomic Energy

In 1938, Fermi received the Nobel Prize for his "demonstrations of the existence of new radioactive elements produced by neutron irradiation, and for his related discovery of nuclear reactions brought about by slow neutrons". Straight from

Stockholm, Fermi left for the United States of America. The following year also Rasetti left Rome and moved to Canada and took over the Department of Physics of the nascent Faculty of Science and Engineering at the Catholic University of Laval in Quebec. Different choices had been made by the two physicists with respect to the creation of the atomic bomb. While Fermi gave an essential contribution to the Manhattan Project, Rasetti, in January 1943, was approached by Hans von Halban (1908–1964) and George Placzek (1905–1955) who proposed that he become a part of the group of French and English specialists who had moved from England to Montreal for security reasons (because of the constant bombing) and who wanted to build a nuclear pile, for military purposes, using heavy water as a moderator.[9]

Rasetti refused on grounds of morality. Not even the presence in Montreal of Bruno Pontecorvo, his old colleague in the group of Via Panisperna, convinced him to change his mind. He never regretted this choice.

In his biographical notes, Rasetti explained clearly his "opinion on the atom bomb question", since the renunciation of participating in the project of nuclear energy for military purposes marked his whole future scientific work:

> I was convinced that no good could ever come from new and more means of destruction [...][10]

His position was of great moral integrity, starting from consideration of the tragedies that were characterizing the performance of the Second World War and expressed his refusal to submit science to any degeneration:

> Evil as the Axis powers were, it was apparent that the other side was sinking to a similar moral (or rather immoral) level in the conduct of the war, witness the massacre of 200000 Japanese civilians at Hiroshima and Nagasaki.[11]

In this precious document is also reported a "stern judgment" formulated by Rasetti against those scientists "including Fermi" who made a different choice from his about the making of the atomic bomb (Rasetti 1958–1968, p. 11, line 10). This judgment has already been analyzed by several scholars.[12] In this work, I try to give an interpretation in terms of a confrontation inside the scientific community on ethical issues in relation to the discovery of nuclear energy and its use for military purposes.

---

[9] Heavy water was considered to be slow motion par excellence because deuterium has a light nucleus, suitable to subtract kinetic energy from neutrons, and being already formed by a neutron and a proton it was believed to be not "inclined" to absorb neutrons. The research started eventually in Montreal after the war with construction of the atomic pile ZEEP, the origin of the CANDU reactors.

[10] F. Rasetti (1958–1968, p. 11, line 2). Information about this document can be found in Amaldi (1990, p. 175, footnote 15). The document was widely quoted by Battimelli and De Maria in the "Preface" of Amaldi (1997) and by Maltese (2003).

[11] Ivi, line 4.

[12] Battimelli (2002), Maltese (2003). These authors reported Edoardo Amaldi's assessments; he considered the decision to work in the Manhattan Project as a necessary assumption of responsibility by scientists to prevent the world being conquered by the Nazis using nuclear weapons: "If I had found myself there (in front of this dramatic dilemma), after deep and painful considerations on which it was my moral duty of man asked to decide whether cooperate in the defense of democracies ... or lock myself in my private life, doing nothing to fight the dictatorship, I would have eventually opted for the first solution" (Amaldi 1997, p. 98, line 32).

Rasetti's statements about his colleagues, particularly in regard to Enrico Fermi, should not be considered, in my opinion, as personal judgments of conviction: they were expressed mainly within the community of Italian physicists, so that if only Rasetti had wanted to, he could have expressed them with more visibility at the death of Fermi. Both in the article published in Science in 1955 and in that written in 1968 for the Celebration of the Accademia dei Lincei in honour of the great scientist who died prematurely, however, he praised his "ability to reach the summits of creative thought", his greatness both as a theoretical physicist and in the perfect "combination of a theorician and an experimenter", his carelessness of the "personal advantages" (Rasetti 1955, pp. 449–450), "his sense of duty, his unyielding spirit honesty" as well as the merit of having established "the tradition of integrity, scrupulous scientific seriousness, high-level research that reigns today in Italian physics." (Rasetti 1968, pp. 17–18). The esteem that had always bound him to his friend Fermi testifies that Rasetti did not intend to judge anyone but that he wanted only to emphasize his ethical choices (Cioci 2007, p. 213).

Among the various positions taken by the physicists of the Via Panisperna Group against the atomic bomb, we should emphasize the one assumed by Eoardo Amaldi. He was not as critical about Fermi as Rasetti, but he made important choices to avoid any nefarious use of the results of his research (Cioci 2009b, p. 56).

In the winter of 1940–1941, he was working, together with his collaborators, on the measurement of the cross section of fast neutrons of various energies against the nuclei of different atomic number and to the study of dependence of the cross section for fission on uranium energy of the neutrons incident. He become aware of the possible military applications of these studies. Then, "after extensive discussion," Amaldi and other Roman physicists decided to abandon the problem of fission and to engage in a general theme of research as far as possible from the previous one:

> we feared that being active and recognized experts on this topic could expose us to the invitation or coercion to work for the Axis powers to the development of military applications of nuclear fission.[13]

6 Leo Szilard and the Chain Reaction

Soon after the Joliot-Curie discovery of artificial radioactivity, Rutherford, on September 11, 1933, declared before the British Association for the Advancement of Science, in Leicester:

> The energy produced by the breaking down of the atom is a very poor kind of thing. Anyone who expects a source of power from the transformation of these atoms is talking moonshine.[14]

[13] Amaldi (1979, p. 199, column 2, line 13).

[14] Associated Press (1933, p. 1, line 18). A summary of the presentation at the British Association is published in the Times of September 12, 1933, p. 7 and of Nature, no. 132, pp. 432–433 (16 September 1933).

The reading of the report of Rutherford's speech in The Times gave birth to Szilard's interest and then to the idea of a practical method for using nuclear energy: an element that, "split by neutrons", emits two of them after having absorbed one, could sustain a nuclear chain reaction. On March 12, 1934, he presented a patent application (British patent application no. 7840) which included both the generation of radioactive elements by means of neutrons and the concept of a nuclear chain reaction (Szilard 1972, pp. 605–621).

In his memoirs, Szilard said he had read, in 1932, Wells's The World Set Free and that it made him realize the consequences that could have been derived from practical applications of nuclear energy. Thus he divided the patent into two parts, ensuring that the second (British Patent 630.726), relative to the chain reaction, would not become public domain when yielding the patent to the British Admiralty. This device allowed delay in publication of that part of the patent until 1949 (Szilard 1972, pp. 639–651).

In view of the possible significant applications, in 1936 he wrote to Fermi, Segre and Rutherford to ask them to participate in the formation of a sort of association in order to exercise a form of control on possible developments of their research through managing the patents granted relatively to it.[15] His attempts did not have the desired effect. According to Edoardo Amaldi

> Nothing came out of this proposal, mainly (I guess) because Fermi thought (in 1936) that the applications of our discovery were too remote.[16]

In December of 1938, the nuclear chemists Otto Hahn and Fritz Strassmann proved unequivocally the production of barium in the disintegration of uranium bombarded by neutrons (Hahn and Strassmann 1939). Lise Meitner (1878–1968) and her nephew Otto Frisch (1904–1979) interpreted the result of the experiment in terms of the atomic nucleus fission of uranium. In January 1939, they published two articles in Nature. In the first, the two scientists calculated that about 1/5 of the mass of a proton would be transformed into energy. For Einstein's law $E=m\ c^2$ this was equivalent to about 200 MeV, an energy much higher than that associated with radioactive phenomena known at the time (Meitner and Frisch 1939). In the second article, Frisch described the physical evidence obtained by him of the uranium fission into two fragments of nearly equal size, with high kinetic energy and electrical charge, which were revealed in an ionization chamber (Frisch 1939).

A complete theoretical description of the process was prepared by Bohr (Bohr 1939; Bohr and Wheeler 1939) who had already built a few years back a model of the nuclear compound system that described the capture of a neutron by the nucleus and the resulting nuclear transmutations (Bohr 1936; Bohr and Kalckar 1937).

The only question that remained open was whether neutrons were formed in the process of fission of uranium.

Szilard knew at once that this would happen. He, concerned about this forecast, tried to convince the researchers who were working on the problem—the two

---

[15] Szilard (1972, pp. 729–732; 1978, pp. 45–46).

[16] E. Amaldi (1984, p. 160, line 27).

groups at Columbia University in New York, the one formed by the same Szilard with Walter Henry Zinn (1906–2000) and that directed by Fermi, and the group in Paris that consisted of Joliot, von Halban and Kowarski—not to publish anything about it. Fermi decided that, if the majority had been opposed to the publication, even he would have abstained. In the end, researchers at Columbia University sent two items to the Physical Review, demanding that their publications be delayed until they had decided whether to keep these results secret or make them known (Zinn and Szilard 1939, dated April 15, received March 16; Anderson, Fermi and Hanstein 1939, dated April 15, received March 16). There was no way, however, to convince Joliot who published the results of his research in Nature (von Halban et al. 1939, dated March 18).

Only after each attempt to prevent the realization of monstrous means of destruction did Szilard propose to Einstein, in August 1939, the famous letter to the President of the United States of America, Franklin Delano Roosevelt (1882–1945), with which he recommended financial support and acceleration of atomic research: to ensure that Germany was not the only state in possession of the bomb.

Indeed, after the defeat of Hitler, Szilard tried in all possible ways to prevent nuclear weapons being used on Japan.

7 Oppenheimer and the Making of the Bomb

Einstein's letter to Roosevelt determined the allocation of only $ 6000 for research on uranium. The decision of the United States for a large-scale effort was made only after the news from England. Otto Frisch and Rudolph Pierls (1907–1995) had calculated that, to build a fission bomb based on a uranium isotope of mass 235 initiated by an impact with fast neutrons, a critical mass of only 5 kg[17] would have been sufficient to generate a self-sustaining chain reaction (compared to several tons necessary for uranium 238). This meant that an atomic bomb could be made by the end of the war. The two scientists prepared two memoranda (Frisch and Peierls 1940a, b)—the second of which was more technical—in March 1940 for the British government. These documents also hinted at the mechanism to detonate the bomb:

> a sphere should be made in two (or more) parts which are brought together first when the explosion is wanted. Once assembled, the bomb would explode within a second or less, since one [cosmic] neutron is sufficient to start the reaction[18]

In the summer of 1942, the U.S. government assigned to Colonel (later General) Leslie Groves (1896–1970) the task of realizing a project to create the first atomic bomb, known under the code name "Manhattan Project". Oppenheimer suggested to Groves that the development of the bomb was concentrated in a single laboratory

[17] The exact critical mass for uranium 235 is 52 kg. The critical mass for plutonium 239 is 10 Kg.

[18] Frisch and Peierls (1940b, p. 86, line 15).

> where people could talk freely with each other, where theoretical ideas and experimental findings could affect each other, where the waste and frustration and error of the many compartmentalized experimental studies could be eliminated, where we could begin to come to grips with chemical, metallurgical, engineering, and ordnance problems that had so far received no consideration[19]

Groves followed Oppenheimer's advice in the creation and location (at Los Alamos, New Mexico) of the laboratory and chose Oppenheimer himself as its director.

Meanwhile, December 2, 1942, the first controlled nuclear chain reaction took place under the stands of Stagg Field Stadium by the group of researchers from the University of Chicago under the direction of Fermi. While they had demonstrated that a chain reaction of uranium could be generated, they had also made manifest the impossibility (already known) of using natural uranium with a moderator, as occurs in an atomic pile, for the production of bombs of unprecedented power. As a matter of fact, the size would be too large for an explosive device (the equatorial axis of the pile was almost 4 m long) and "the thermal neutrons take so long (so many micro-seconds) to act that only a feeble explosion would result" (Smyth 1945, p. 209, line 23). Another (not secondary) effect of the reactions that occur in a stack is that part of ^{238}U absorbs a neutron in ^{239}U that with subsequent β decays can change into plutonium, extremely fissile material even if bombarded with fast neutrons.

The Manhattan project's success owed much to the leadership of Oppenheimer. Since March 1943, he attracted a first-class team of scientists and was able to delegate responsibilities and to trust his collaborators. Despite the problems of security and secrecy, Oppenheimer managed to keep free the flow of information and the in-depth discussions among scientists involved in the project.

During Oppenheimer's memorial session of the American Physical Society meeting held in Washington, D.C., in April 1967, Victor Weisskopf (1908–2002), speaking of the spirit of collaboration among scientists from different nations inspired by Oppenheimer, said that "he was to create at Los Alamos a new form of scientific life, … the new ways of big science, in nuclear physics and particle physics, have been inspired by the Los Alamos venture".[20]

For purposes of producing atomic explosive, however, one must consider the work done by several laboratories including the Metallurgical Laboratory at Chicago, the Clinton Laboratory at Oak Ridge, the Radiation Laboratory at Berkley, the Hanford Engineer Works at Richland, the Argonne Laboratory, the Jersey City laboratories for the separation of uranium isotopes U-235 from U-238 by electromagnetic and gaseous diffusion methods and for production and chemical separation of plutonium.

The number of people employed to build atomic bombs (nearly 130,000 including construction workers and military personnel) and the cost of the project ($ 2 billion in 1945) was the largest technological enterprise in the history of mankind (Jones 1985, p. 344).

---

[19] United States Atomic Energy Commission (1971, p. 12, line 4).

[20] Weisskopf (1967, p. 40, column 1, line 19; 42, column 1, line 11).

On 16 July 1945 the first atomic bomb was tested at Alamogordo. It demonstrated the power of the new weapon. The official (censored) report on the development of the atomic bomb, written by Henry De Wolf Smyth (1898–1986) shortly after the war, reported that

> No man-made phenomenon of such tremendous power had ever occurred before. The lighting effects beggared description. The whole country was lighted by a searing light with the intensity many times that of the midday sun.[21]

Oppenheimer said, later, that when the bomb detonated, he became aware of the verse of the Bhagavad Gita "I am become Death, the destroyer of worlds" (Peierls 1974, p. 216, column 1, line 14).

8 The Frank Report and the Decision to Drop the Bomb on Japan

Once it became clear that Germany would not have been able to acquire the bomb by the end of the war, doubts about the meaning of their work began to spread among the scientists involved in the project.

Joseph Rotblat (1908–2005), a Polish-born physicist who had worked in England to research on the atomic bomb, had moved to Los Alamos in early 1944 where he got to work on the experimental study, using a cyclotron, of secondary effects of fast neutron irradiation with the fission products. He also participated in the restricted meetings with the coordinators of the project (Brown 2012, p. 47).

After learning from General Groves that by then "the real purpose in making the bomb was to subdue the Soviet" and "when it became evident, toward the end of 1944, that the Germans had abandoned their bomb project", the whole reason of his "being in Los Alamos ceased to be", and he got the "permission to leave and return to Britain".[22]

Leo Szilard, however, tried to stop the military use of the atomic bomb by the United States of America.

In March of 1945 he wrote a memorandum to President Roosevelt. Einstein enclosed a letter of introduction; this letter was dated 25 March 1945 (Szilard 1978, pp. 205–207). Szilard's Memorandum drew the attention of the President to the consequences that the use of the bombs over Japan would engender, such as the arms race that would follow and the dangers that the United States would have to face due to a possible nuclear war. A copy of the letter was sent to Mrs. Roosevelt asking for an appointment with her husband, but this never took place because the President died April 12, 1945.

Subsequently Szilard had the opportunity to present, without success, his memorandum (rewritten with greater care for the occasion) to James Byrnes, who had the

---

[21] Smyth (1945, p. 254, line 16).

[22] Rotblat (1985, p. 18, line 22).

confidence of President Truman and who would become Secretary of State (Szilard 1978, pp. 196–204).

He promoted, among the scientists who participated in the Manhattan Project, and on moral grounds alone, a petition against the use of the bomb against Japan (Szilard 1978, p. 211).

Szilard also made a contribution to the proposal made by James Franck (1882–1964). A Nobel Prize-winner, together with Gustav Ludwig Hertz (1887–1975) "for their discovery of the laws governing the impact of an electron upon an atom", and senior physicist in the Metallurgy Laboratory of Chicago, he had been made aware of the social responsibility of scientists after participating in the First World War in a German program for development of chemical weapons. He agreed to participate in the atomic bomb project in 1942, with the promise by Arthur Holly Compton (1892–1962) that, when the time had come for a decision on the use of the bomb, he would have had the opportunity to present his views to high-level politicians.

Franck was chairman of the Committee of Compton's Metallurgical Laboratory in Chicago on Social and Political Implications of the atomic bomb that included also Donald J. Hughes (1915–1960), James Joseph Nickson (1915–1985), Eugene Rabinowitch (1901–1973), Glenn Theodore Seaborg (1912–1999) and Leo Szilard. In June 1945 they prepared a memorandum known as the Franck Report for Secretary of War Stimson which proposed the use of the bomb on an uninhabited island before the representatives of all nations.

The motivations of the Chicago group, to limit the military use of the bomb, were based on the impact it would have had on the international and post-war situation, providing as fundamental objective "an international agreement on the prevention of nuclear warfare". From this point of view, the use of atomic weapons could easily destroy all the future possibilities of reaching an agreement, because it would be extremely difficult to persuade the world that a nation which had used such a weapon of mass destruction, could then be trusted in its proclaimed desire to abolish these weapons, by means of an international agreement.[23]

The memorandum was submitted to the Interim Commission of the War Department composed of Arthur H. Compton, Ernest O. Lawrence (1901–1958), Enrico Fermi and J. Robert Oppenheimer. They argued that a demonstration on an uninhabited island would not have been effective, and that the only way in which the atomic bomb could be used to end the war was to use it on a military objective in a densely populated area.

The four scientists were told that it was impossible to cancel or postpone the planned invasion of Japan, certainly very costly in terms of human lives, if Japan would not surrender in advance as a consequence of being told about the bomb. They knew nothing of attempts by the Japanese government to enter into negotiations for peace, which could have led to a diplomatic solution of the conflict.

On 6 August 1945, the uranium bomb was dropped on Hiroshima: 140,000 of its citizens were killed in a year and 200,000 in 5 years. Three days later the plutonium bomb was dropped on Nagasaki, 70,000 people died before the year was out and more than 70,000 died in the next 5 years as a result of radiation (Rhodes 1986, p. 734, 740–742).

---

[23] Committee on Social and Political Implications (1946, p. 3, column 3, line 34).

9 The Commitment to Peace by Robert Oppenheimer

Philip Morrison (1915–2005) and Robert Serber (1909–1997) went to Hiroshima at the beginning of September 1945 to study the effects of nuclear weapons. They reported to the Los Alamos scientists the terrible suffering endured by the civilian population. As the days passed, revulsion grew for what had been done, even by those who believed that the successful end of the war was the justification for the bombing.

On April 16, 1954, interviewed by Robert Robb, when asked if he had scruples about the fact that 70,000 civilians were killed or injured by dropping the bomb on Japan, Oppenheimer said: "Terrible ones" (United States Atomic Energy Commission 1971, p. 235, line 61).

Oppenheimer, in 1947, during a conference at the Massachusetts Institute of Technology (MIT) in Cambridge entitled "Physics in the Contemporary World", declared that "the physicists felt a peculiarly intimate responsibility for suggesting, for supporting, and in the end, in large measure, for achieving the realization of atomic weapons", so they "have known sin, and this is a knowledge which they cannot lose".[24]

Much is written about this opinion expressed by Oppenheimer and his repentance after Hiroshima. According to Alice Kimball Smith, who edited the account of the commitment to peace of atomic scientists between 1945 and 1947[25] and published a remarkable collection of letters and memories of Oppenheimer,[26] it expresses more "an intensely personal experience of the reality of evil [...] and not a feeling of guilt in the ordinary sense" (Smith 1971, p. 77, line 25).

Many years later, in the face of representations, even theatrical ones (Kipphardt 1964), which gave Oppenheimer as a broken man for what he had done, the scientist wrote that

> My principle remaining disgust with Kipphardt's text is the long and totally improvised final speech I am supposed to have made [...] My own feelings about responsibility and guilt have always had to do with the present. and so far in this life that has been more than enough to occupy me.[27]

The awareness of the committed evil generated in Oppenheimer a new attitude. Even while waiting for news of the capitulation of Japan after the bombing of Nagasaki, Oppenheimer worked on the final report on post-war planning that the Interim Committee's Scientific Panel was preparing for the Secretary of War. In the report that Oppenheimer brought to Washington to submit to Secretary Stimson, it is pos-

[24] Oppenheimer (1948, p. 66, line 45).

[25] The account of A. K. Smith is entitled "A peril and a Hope" by a famous expression of Oppenheimer according to which nuclear weapons would constitute a danger and a hope for humanity because given the power of these terrible means of destruction humankind would have to give up war to settle international disputes and would have to create a united world under the law and humanity.

[26] Smith and Weiner (1980).

[27] Oppenheimer (1966, line 20).

sible to grasp what would be his next commitment to peace, so that "all steps be taken, all necessary international arrangements be made, to this one end"(Smith and Weiner 1980, p. 294, line 31).

Oppenheimer then gave up the direction of Los Alamos to devote himself to teaching, to the social implications of atomic energy and to the project for its international control.

In 1946 he was the only atomic scientist who took part in drafting the "Acheson Lilienthal Report", developed under the auspices of the State Department after the first resolution of the General Assembly of the United Nations, held in London in January 1946, which advocated the elimination from national arsenals of nuclear weapons and of all weapons of mass destruction. Central to the proposal was the recommendation to set up an International Atomic Development Authority, to assist the United Nations,

> with exclusive jurisdiction to conduct all intrinsically dangerous operations in the field.... The international agency would also maintain inspection facilities to assure that illicit operations were not occurring[28]

Oppenheimer foreshadowed thus the birth of the International Atomic Energy Agency (IAEA).[29]

Oppenheimers greatest contribution to peace and to disarmament was undoubtedly his opposition to the hydrogen bomb.

After the explosion of the first Russian atomic bomb, in August 1949, the Commission for Atomic Energy of the United States of America convened a special session of the General Advisory Committee chaired by Oppenheimer to discuss what should be the U.S. response to the "aggressive" policy of the Soviet Union and in particular in order to express its opinion about the realization of a super bomb, a nuclear bomb (based on the fusion of hydrogen) about a 1000 times more powerful than the bombs of Hiroshima and Nagasaki.

The Committee members observed that

> once the problem of initiation has been solved, there is no limit to the explosive power of the bomb itself except that imposed by requirements of delivery. This is because one can continue to add deuterium—an essentially cheap material—to make larger and larger explosions[30]

They then were unanimous in recommending that the development of the bomb must somehow be avoided, since its use "carries much further than the atomic bomb itself the policy of exterminating civilian populations".[31]

The majority of the Committee, with Oppenheimer, believed that "this should be an unqualified commitment", while the minority, by Fermi and Isidor Isaac Rabi

---

[28] Lilienthal et al. (1946, p. 24, line 15).

[29] For further arguments in favour of this conclusion see Cioci (2004).

[30] General Advisory Committee (1949, p. 155, line 39).

[31] Ivi, line 21.

(1898–1988), felt that this commitment "should be made conditional on the response of the Soviet government to a proposal to renounce such development".[32]

After the decision taken by President Harry S. Truman (1884–1972), dated January 31, 1950, to begin a program of development of the hydrogen bomb, Oppenheimer, invited by Teller, refused to move to Los Alamos to work on the project for the construction of the super bomb (United States Atomic Energy Commission 1971, p. 232)

From April to May of 1954, Oppenheimer had to stand trial because of some contacts he had had at the beginning of the 1930s and 1940s with the Communists, but also because in 1949 he had opposed the development program of the hydrogen bomb. Oppenheimer was accused of having slowed down, with his influence on American scientists, the effort that was to lead to development of the bomb (United States Atomic Energy Commission 1971, p. 1011). The trial ended with suspension of his security clearance, the authorization for access to secret information.

Oppenheimer was fully rehabilitated in 1963, when President Lyndon Jonson gave him the "Enrico Fermi Award", the greatest honor that the U.S. government can bestow for outstanding service in the field of nuclear energy. The proposal was in fact approved by John F. Kennedy shortly before his assassination, recognizing that a great injustice had been done against Oppenehimer, who died a few years later, on February 18, 1967.

10 Conclusion. The Moral Responsibility of the Scientist

The experiences of scientists presented in this work indicate an example to follow. They have questioned the possible consequences of their findings, have discussed and have compared their views with those of their colleagues, expressing their concerns within the scientific community. They have tried to delay the publication of research results or even to change their field of study for not carrying out, in a particular historical moment next to the war, terrible means of destruction. This has provided a model of behavior also for other branches of science.[33]

In some topical cases, the scientists have expressed their fears to political institutions and to the public. It is what Einstein and Szilard did when they wrote the letter to the President of the United States of America. Roosevelt then recommended financial support and the acceleration of atomic research for fear that the research on uranium fission could lead to creation of a Nazi atomic bomb.

Later, Einstein promoted shortly before his death, together with Bertrand A. W. Russell (1872–1970), the publication in London in 1955, of an Appeal for Abolition of War (known as the Russell-Einstein Manifesto), almost a spiritual testament, in which he informed the authorities in the world, and through them the scientists and

---

[32] Ibidem, 156, line 6.

[33] See Capuozzo and Cioci (2010).

the public, that nuclear "weapons threaten the continued existence of mankind" (Russell and Einstein 1955, p. 25, column 5, line 57). It was then hoped that the "scientists should assemble in conference to appraise the perils that have arisen as a result of the development of weapons of mass destruction, and to discuss a resolution in the spirit of the appended draft" (Russell and Einstein 1955, p. 25, column 1, line 62). See also Butcher (2005).

The Manifesto was signed shortly by thousands of scientists around the world and was the basis for the emergence in 1957 of the Pugwash Conferences on Science and World Affairs, which took place at the Canadian village of Pugwash, where scientists met from around the world to respond to the call made by Einstein and Russell.

Unfortunately, while on one hand, scientists informed civil society, on the other they continued to do their "duty" in military laboratories.[34]

This position, in which the man of science can often be found, refers to the distinction, which Einstein also does, of the "two roles of the intellectual worker: his role as scientist or man of letters, and his role as citizen" (Hinshaw 1949, p. 652, line 27). Only on the basis of this differentiation could a scientist preserve his or her own methodological objectivity in their scientific activity, which requires no showing of concern for value judgments that are of a very subjective nature and often deal with social issues.

According to the philosopher Hans Jonas (1903–1993), who had been engaged in the construction, in a pluralistic society, of an ethic of responsibility in favor of future generations (Jonas 1984), it is important to reconsider the question of so-called "freedom from values" of science, because leaving out of consideration the value of the object of knowledge allows a degree of freedom in treating and manipulating it without any limits and respect. Jonas, in his essay "On Technology, Medicine and Ethics" about the practice of the responsibility principle, cites precisely the experience of Oppenheimer as a starting point for a renewal of scientific practice (Jonas 1997, p. 55).

Oppenheimer in fact solves the problem in an original way, synthesizing the task of the scientist and that of the citizen, by opposing, in institutional settings, construction of the hydrogen bomb and refusing to move to Los Alamos to work on it.

There is a substantial difference between informing people about the risks connected with the results of scientific research and simply objecting to them. In the second action the scientist assigns values to the science in which he participates and he becomes a sign of hope for humanity.

The position of Oppenheimer, the most advanced, has been isolated. Following disapproval by the Commission for Atomic Energy of the United States of America, scientists preferred the attitude of Einstein, certainly less uncomfortable but perhaps the time is ripe for it to resume the position of Oppenheimer (Cioci 2004, p. 143).

---

[34] E.g. see: Teller (1950).

Acknowledgments I thank Professor Giovanni Battimelli, Referent for Documents and Sources for the History of Physics, Department of Physics, University "La Sapienza" of Rome, who helped me in the consultation of the Edoardo Amaldi Archives.

My special thanks is addressed to Colin Harris, Superintendent of the Special Collections Reading Rooms, for his help in finding the unpublished writings of Frederick Soddy, to Dr. Chris Fletcher, Keeper of Special Collections, and to the Bodleian Library, Oxford University, which owns them, for permission given on the basis of 'non-exclusive rights' to quote and to paraphrase them.

I am grateful to Bonnie Coles for helping me in the examination and retrieval of the writings of Robert Oppenheimer at the Manuscript Division of the Library of Congress (USA).

References

Amaldi E (1979) Gli anni della ricostruzione (The years of reconstruction). Giornale di Fisica 20(3):186–225

Amaldi E (1984) From the discovery of the neutron to the discovery of nuclear fission. Physics Reports 111(Sept):1–332

Amaldi E (1990) The case of physics (Il caso della fisica. Le conseguenze culturali delle leggi razziali. Atti dei convegni lincei, Roma 84: 107–133). In: Battimelli G, Paoloni G (eds) 20th century physics: essays and recollections. A selection of historical writings by Edoardo Amaldi (1998). World Scientific, Singapore

Amaldi E (1997) Da via Panisperna all'America (From via Panisperna to America). Editori Riuniti, Roma

Anderson HL, Fermi E, Hanstein HB (1939) Production of neutrons in uranium bombarded by neutrons. Physical Review 55:797–798.

Associated Press (September 1933) Atom-powered world absurd, scientists told. New York Herald Tribune, pp 1, 37

Battimelli G (2002) Era un ragazzo di via Panisperna (He was a boy from via Panisperna). Lettera Matematica 42:15–21

Bohr N (1936) Neutron capture and nuclear constitution. Nature 137:344–348, 351

Bohr N (1939) Disintegration of heavy Nuclei. Nature 143:330.

Bohr N, Kalckar F (1937) On the transmutation of atomic nuclei by impact of material particles. I. General theoretical remarks. Mat-Fys Medd Dan Vidensk Selsk 14(10):1–40

Bohr N, Wheeler JA (1939) The mechanism of nuclear fission. Physical Review 56:426–450

Brown A (2012) Keeper of the nuclear conscience. The life and works of Joseph Rotblat. Oxford University Press, New York

Butcher SI (2005) The origins of the Russell-Einstein Manifesto. Pugwash History Series 1:3–35

Capuozzo L, Cioci V (2010) La Fisica, le origini della biologia molecolare e dell'Ingegneria genetica: il caso della conferenza di asilomar (Physics, the origins of molecular biology and genetic engineering: the case of the Asilomar Conference). In: Giannetto E, Giannini G, Toscano M (eds) Relatività, quanti, chaos e altre rivoluzioni della fisica (Proceedings of the XXVII SISFA Congress). Guaraldi, Rimini, pp 357–364

Cioci V (2004) Una rivisitazione del caso Oppenheimer (A review of Oppenheimer case). In: Tucci P, Garuccio A, Nigro M (eds) Proceedings of the XXIII SISFA Congress. Progedit, Bari, pp 131–144

Cioci V (2007) Storia e confronto di tre diverse posizioni di fisici contro la bomba atomica (History and comparison of three different positions of physicists against the atomic bomb). In: Leone M, Preziosi B, Robotti N (eds) L'eredità di Fermi Majorana e altri temi. Atti del XXIV Congresso della Società Italiana degli Storici della Fisica e dell'Astronomia. (Proceedings of the XXIV SISFA Congress). Bibliopolis, Naples, pp 211–215

Cioci V (2008) Science fiction e realtà: La liberazione del mondo di H. G. Wells e il suo influsso sugli scienziati atomici (Science fiction and reality: The world set free by HG Wells and his influence on the atomic scientists). In: Pitrelli N, Ramani D, Sturloni G (eds) Proceedings of the VI Italian national conference on science communication. Polimetrica, Monza, pp 97–102

Cioci V (2009a) Frederick Soddy, un chimico alle frontiere della conoscenza, fra Fisica, Economia, Matematica ed Ecologia (Frederick Soddy, a chemist at the frontiers of knowledge, among physics, economics, mathematics and ecology). In: Calascibetta F (ed) Proceedings of the XIII National Congress of History and Fundamentals of Chemistry, Excerpt from vol 127, Memoirs of Physical and Natural Sciences, Proceedings of the Italian National Academy of Sciences called Academy of Forty, ser. 5, vol 33, part II, vol II, pp 319–330

Cioci V (2009b) Edoardo Amaldi e il rifiuto delle applicazioni militari della fisica nucleare (Edoardo Amaldi and the refusal of the military applications of nuclear physics). Proceedings of the XLVII AIF Congress, Supplement to LNFS 42(3): 50–58

Committee on Social and Political Implications (1946) A report to the secretary of war (Frank report). Bulletin of Atomic Scientists 1(10):2–4, 16

Curie P, Laborde A (1903) Sur la chaleur dégagée spontanément par les sels de radium. Comptes Rendus 136(sem.1):673–675

Einstein A (1905) Does the inertia of a body depend upon its energy content? (Ist die Trägheit eines Körpers von seimen Eneirgieinhalt abhängig? Annalen der Physik, ser. 4, 18:639–641). In: Einstein A (1989) Collected papers of Albert Einstein, vol 2. Princeton University Press, Princeton, pp 172–174

Einstein A (1907) On the relativity principle and the conclusion drawn from it (Relativitätsprinzip und die aus demselben gezogenen Folgerungen, Jahrbuch der Radioaktivität 4:411–62; 5:98–99). In: Einstein A (1989) Collected papers of Albert Einstein, vol 2. Princeton University Press, Princeton, pp 252–311

Einstein A (1910) Principe de relativité et ses conséquences dans la physique moderne. Archives des sciences physiques et naturelles (Series 4), 29:5–28,125–244

Einstein A (1989) Collected Papers of Albert Einstein, vol 2. The Swiss years: writings, 1900–1909. Princeton University Press, Princeton. (English translation supplement. Beck A (tr), Havas P (con), Buchwald DK (g ed))

Fermi E (1934) Possible production of elements of atomic number higher than 92. Nature 133:898–899

Fermi E, Amaldi E, D'Agostino O, Rasetti F, Segré E (1934) Artificial radioactivity produced by neutron bombardment—I. Proceedings of the Royal Society of London A 146(857):483–500

Frisch O (1939) Physical evidence for the division of heavy nuclei under neutron bombardment. Nature 143:276

Frisch O, Peierls R (1940a) Memorandum on the properties of a radioactive "superbomb". In: Serber R (ed) The Los Alamos primer. University of California Press, Berkeley, pp 80–83

Frisch O, Peierls R (1940b) On the construction of a "super-bomb"; based on a nuclear chain reaction in uranium. In: Serber R (ed) The Los Alamos primer. University of California Press, Berkeley, pp 83–88

General Advisory Committee (1949) Report of GAC to the U. S. atomic energy commission, October 30. In: York HF (ed) The advisors. Oppheneimer, teller and the superbomb (1976). Freeman and Company, San Francisco, pp 150–159

Hahn O, Strassmann F (1939) Concerning the existence of alkaline earth metals resulting from neutron irradiation of uranium (Über den Nachweis und das Verhalten der bei der Bestrahlung des Urans mittels Neutronen entstehenden Erdalkalimetalle. Die Naturwissenschaften 27:11–15). Am J Phys 32:9–15 (English Translation)

Hinshaw VGJ (1949) Einstein's social philosophy. In: Schilpp AP (ed) Albert Einstein: philosopher-scientist. Volume VII in the Library of Living Philosophers, MJF Books, New York, pp 649–661

Jonas H (1984) The imperative of responsibility: in search of an ethics for the technological age (Das Prinzip Verantwortung, Insel Verlag, Frankfurt am Main, 1979) (trans: H Jonas, D Herr). University of Chicago Press, Chicago

Jonas H (1997) Tecnica medicina ed etica. Prassi del principio responsabilità. (Technik, Medizin und Ethik: Zur Praxis des Prinzips Verantwortung. Insel Verlag, Frankfurt am Main, 1985) (trans: P Becchi, A Benussi). Einaudi, Torino. (Not translated into English yet)

Jones VC (1985) Manhattan: the army and the atomic bomb. United States Army, Washington (Center of military history)

Kipphardt H (1964) In the matter of J. Robert Oppenheimer; a play freely adapted on the basis of the documents. Hill & Wang, New York

Lilienthal DE (Chairman), Barnard CI, Oppenheimer JR, Thomas CA, Winne HA (1946) A report on the international control of atomic energy prepared for the secretary of state's committee on atomic energy (Acheson-Lilienthal report). U. S. Government Printing Office, Washington

Maltese G (2003) Enrico Fermi in America. Una biografia scientifica: 1938–1954 (Enrico Fermi in America. A scientific biography: 1938–1954). Zanichelli, Bologna

Meitner L, Frisch OR (1939) Disintegration of uranium by neutrons: a new type of nuclear reaction. Nature 143:239–40

Oppenheimer JR (1948) Physics in the contemporary world. Bulletin of Atomic Scientists 4(3):65–68, 85–86

Oppenheimer JR (1966) Draft letter to David Bohm, 2 December Library of congress. Manuscript division, Papers of J. R. Oppenheimer, box 20

Pais A (1982) Subtle is the lord. The science and the life of Albert Einstein. Oxford University Press, New York

Peierls R (1974) Oppenheimer J. Robert. In: Gillispie CC (ed) Dictionary of scientific biography, vol X. Charles Scribner's Sons, New York, pp 213–218

Ramsay W, Soddy F, (1904) Further experiments on the production of helium from radium. Proceedings of the Royal Society of London 73:346–358

Rasetti F (1955) Fermi. Science 121(1344):449–451

Rasetti F (1958–1968) Biographical notes and scientific work of Franco Rasetti, unpublished typescript preserved in the Amaldi Archives, Department of Physics, Rome University "La Sapienza", sc. 1E, fasc. 22

Rasetti F (1968) Enrico Fermi e la fisica italiana (Enrico Fermi and the Italian physics). In: Celebrazioni Lincee in ricordo di Enrico Fermi. Accademia Nazionale dei Lincei, Roma, pp 11–19

Rhodes R (1986) The making of atomic bomb. Simon & Schuster, New York

Rotblat J (1985) Leaving the bomb project. The Bulletin of the Atomic Scientists 41(7):16–19

Russell BAW, Einstein A (1955) Texts of scientists' appeal for abolition of war. The New York Times, 10 July, p 25. (The Russell-Einstein Manifesto). Issued in London, 9 July 1955, and signed also by Born M, Bridgman PW, Infeld AL, Joliot-Curie F, Muller HJ, Pauling L, Powell CF, Rotblat J, Yukawa H. Reprinted in: Bone AG (ed) The Collected Papers of Bertrand Russell, vol 28. Man's Peril, 1954–1955. Routledge, London and New York, pp 259–262

Rutherford E, Soddy F (1902a) The cause and nature of radioactivity, parts I and II. Philosophical Magazine (Series 6) 4:370–396, 569–585

Rutherford E, Soddy F (1902b) The radioactivity of thorium compounds, parts I and II. Transactions of Journal of the Chemical Society 81:837–860, 837–851

Rutherford E, Soddy F (1903) Radioactive change. Philosophical Magazine (Series 6) 5:576–591

Smith AK (1971) A peril and a hope. The scientists' movement in America 1945–47. The MIT Press, Cambridge

Smith AK, Weiner C (eds) (1980) Robert Oppenheimer. Letters and recollections. Harvard University Press, Cambridge

Smyth HD (1945) Atomic energy for military purposes. The official report on the development of the atomic bomb under the auspices of the United States Government, 1940–1945. Princeton University Press, Princeton. (Reprinted by Maple Press, York, Pennsylvania)

Soddy F (1903a) Some recent advances in radioactivity. Contemporary Review 83(May):708–720

Soddy F (1903b) Possible future applications of radium. Times Literature Supplement 17:225–226

Soddy F (1904a) Radium. Professional Papers of the Royal Corps of Engineers 29:237–251

Soddy F (1904b) Radio-activity: an elementary treatise from the standpoint of the disintegration theory. The electrician, London

Soddy F (1906) The internal energy of elements. Journal of the Institution of Electrical Engineers 37:140–147
Soddy F (1909) The interpretation of radium. G. P. Putnam's Sons, New York
Soddy F (1915) The social effect of recent advances in physical sciences. In: Professor Frederick Soddy Papers, Bodleian Library, Oxford, MS Eng Misc b179, ff. 542–556
Soddy F (1926) Wealth, virtual wealth and debt. The solution of the economic paradox. George Allen & Unwin Ltd, London
Soddy F (1934) Suggestion to the ordinary fellows of the royal society for a reform group to uphold their rights. In: Professor Frederick Soddy Papers, Bodleian Library, Oxford, MS. Eng. misc. b170, f 147
Soddy F (1945) The moving finger writes [...] Cavalcade, August 18, pp 8–9
Soddy F (1949) The story of atomic energy. Nova Atlantis, London
Soddy F (1953) Correspondence with Sir Harold Hartley. In: Professor Frederick Soddy Papers, Bodleian Library, Oxford, MS Eng Misc b189, ff 281–285
Szilard L (1972) The collected works of Leo Szilard: scientific papers. Edited by Feld BT, Weiss Szilard G. The MIT Press, Cambridge
Szilard L (1978) His Version of the Facts. Selected Recollections and Correspondence. Edited by Weart SR, Weiss Szilard G. The Mit Press, Cambridge
Szilard L, Zinn WH (1939) Instantaneous emission of fast neutrons in the interaction of slow neutrons with uranium. Physical Review 55:799–800
Teller E (1950) Back to the laboratories. Bulletin of Atomic Scientists 6(3):71
The Sydney Morning Herald (1935) Royal society. Demand for Reform., December 23, N.S.W., Australia. In: Professor Frederick Soddy Papers, Bodleian Library, Oxford, MS. Eng. misc. b170, f 124
United States Atomic Energy Commission (1971) In the matter of J. Robert Oppenheimer. Transcript of hearing before personnel security board and texts of principal documents and letters. Foreword by Stern PM. The MIT Press, Cambridge
von Halban H, Joliot F, Kowarski L (1939) Liberation of neutrons in the nuclear explosion of uranium. Nature 143:470–471
Weart S (1982) The road to Los Alamos. Journal De Physique (Colloque C8, suppl au n 12) 43:301–313
Weisskopf V (1967) The Los Alamos years. In: A memorial to Oppenheimer. Physics Today 20(10):39–42
Wells HG (1914) The world set free. A story of mankind. Macmillan and Company, London

Tribology: A Historical Overview of the Relation Between Theory and Application

Javier Echávarri, Eduardo de la Guerra and Enrique Chacón

Abstract Tribology is the science and engineering of rubbing surfaces. This article presents a historical overview of tribology evolution from prehistory to the highly advanced stage reached nowadays. Phenomenological analyses based on observation and experimentation gave rise to a progressive understanding of the processes of friction, wear and lubrication, with application to many industrial problems. Multiple industrial reports published during the last few decades have shown the huge impact of tribology on economy and ecology, such as the classical Jost Report in the United Kingdom (1966), that has been followed up to the present by similar investigations worldwide.

Keywords Tribology · Friction · Wear · Lubrication · Bearing · Hydrodynamics

1 Introduction to Tribology

The word "tribology" was introduced in the 1960s, as the science and engineering of rubbing surfaces. It includes the study and application of the principles of friction, wear and lubrication.

Therefore, tribology is multidisciplinary in nature, and closely related to physics, chemistry and mechanics of materials among other disciplines. In recent years, research in tribology has reached a highly advanced stage from a theoretical and experimental standpoint, with application to every sector of industry, e.g., automotive industry: each automobile presents approximately 2000 tribological contacts.

Learning more about tribological phenomena has led to improvements in mechanical systems, namely reduction of wear and the likelihood of failure, increase in energy efficiency, reduction of maintenance and repair costs, savings in raw materials and the reduction of noise and vibration.

Some organizations historically produce outstanding activities in tribology, e.g. the Institution of Mechanical Engineers (IMechE, founded in London in 1847), the

J. Echávarri (✉) · E. de la Guerra · E. Chacón
Grupo de Investigación en Ingeniería de Máquinas, Universidad Politécnica de Madrid, Madrid, Spain
e-mail: jechavarri@etsii.upm.es

R. Pisano (ed.), *A Bridge between Conceptual Frameworks,* History of Mechanism and Machine Science 27, DOI 10.1007/978-94-017-9645-3_8

American Society of Mechanical Engineers (ASME, founded in 1880), the American Society for Testing Materials (ASTM, founded in 1898), the American Gear Manufacturers Association (AGMA, founded in 1916), the Society of Tribologists and Lubrication Engineers (STLE, founded in 1944 by ASLE, the American Society of Lubrication Engineers) and the Japanese Society of Tribologist (JAST, founded in 1956). Several of these organizations belong to the International Tribology Council.

These associations and many others promote meetings and conferences on tribology worldwide, such as the "Leeds-Lyon Symposium on Tribology", held every year since 1973, and the "World Congress on Tribology", held every 4 years since 1997. They also publish journals of tribological interest, e.g. "Journal of Engineering Tribology" (Proceedings of the IMechE), "Journal of Tribology" (Transactions of ASME), "Tribology International", "Wear" and "Tribology Letters".

2 On the Significance of Tribology

Several authors (Reti 1967; Dowson 1998) have drawn attention to a sixteenth century indication of the economical importance of tribology in archives kept in Simancas, Spain. These archives refer to a large, complex machine, constructed by Juanelo Turriano (1501–1585), for raising considerable amounts of water under atmospheric pressure by means of a set of towers. Each tower included pivoted pipes and buckets with backward and forward motion which raised the water in stages (Fig. 1). The size of its moving parts and the dynamic forces received during operation made the tribological effects of this machine very significant in relation to the continuous wear of the machine and the ensuing need for costly reparations.

This early attention to the economic significance of tribology is in line with Renaissance authors like Leonardo da Vinci (1452–1519), who recognized the interest of studying friction forces and performed experiments with remarkable conclusions.

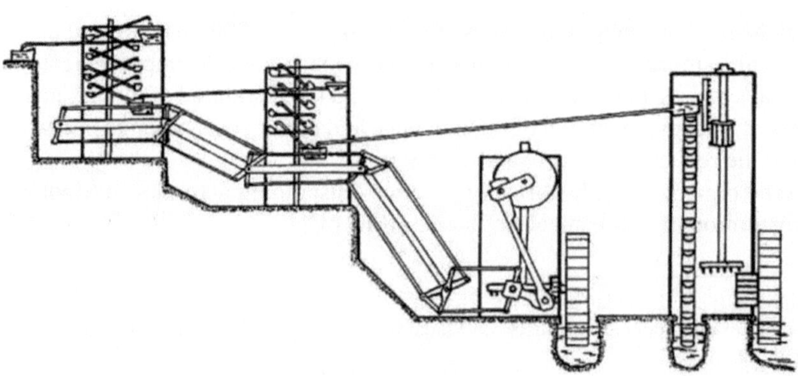

Fig. 1 Reconstruction of Juanelo's machine. (Reti 1967)

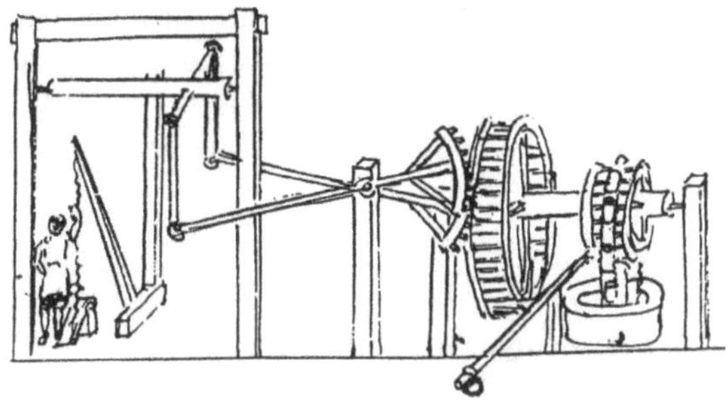

Fig. 2 Jerónimo de Ayanz's machine for raising water. (García Tapia 2001)

Furthermore, the interest in studying energy loss due to tribological phenomena was depicted by Jerónimo de Ayanz (1553–1613). Figure 2 shows a lever with a counterweight in a machine for raising water (García Tapia 1990, 2001). Thus, a comparative measurement of the efficiency of the different machines may be possible by the position of a weight on a rocker arm.

Later on, influenced by Leonardo da Vinci's work, Guillaume Amontons (1663–1705) wrote in 1699 "…among all those who have written on the subject of moving forces, probably not a single one has given sufficient attention to the effect of friction in machines…".

Despite the evolution in tribology during the Renaissance and even more so during the Industrial Revolution, it was not until the last few decades that several published reports clearly quantified the massive impact of tribology on economy and ecology. Peter Jost's classical Report for the Department of Education and Science of the United Kingdom (1966) first used the term "tribology", defined as "the science and technology of interacting surfaces in relative motion and related subjects practices". This report quantified possible savings by means of improved tribological practices and gave an idea about the tribological needs of industry. Figure 3 presents the distribution of potential savings according to Jost's findings.

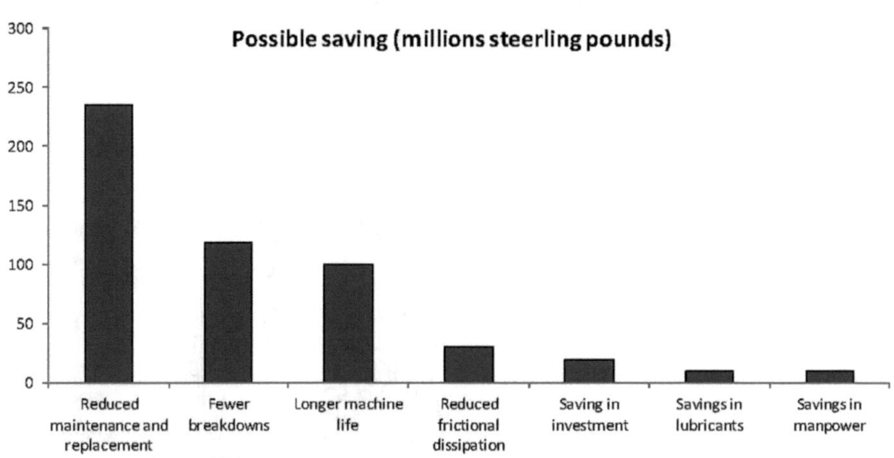

Fig. 3 Savings indicated by the Jost report in the United Kingdom—1966 levels. (Based on Jost 1966)

This report was followed by similar investigations worldwide (Pinkus and Wilcock 1977; Jost 2001). Each one established that the application of tribology could save between 1 and 2 % of an industrial country's gross national product (GNP). In other words, up to 30 % of consumed energy was exhausted in trying to overcome friction.

3 Tribology in the Past

By looking at anthropological remains used to generate fire by friction in primitive cultures, it may be assumed that the history of tribology covers developments from prehistory. In fact, the use of flintstones or dry wood for making fire suggests that early human beings had some knowledge about temperature and frictional heating.

Figure 4 represents different fire-generating techniques used during prehistoric times (Mauss 1967). Some of them involve certain complexity, e.g. the bow-based device requires converting linear into rotary motion, generating initial tension in a rope-pulley system, multiplying the speed, using primitive bone or stone bearings, etc.

As early human's basic life needs of food and heat were secured, he was able to devote more time to crafts and construction. This led to the development of devices which frequently required the use of plain bearings. Figure 5a shows an evolution of the bow system designed to generate fire, used in the Stone Age for drilling (Strandh 1979). Further evidence of the use of plain bearings are door-sockets, which could have been made with these sorts of drilling machines. Figure 5b shows a stone door-socket that includes an inscription, found in Mesopotamia and dated 2500 BC (Singer 1954).

The discontinuous motion of these previous devices gave rise to the first evidence of continuous rotary motion in the potter's wheel. Later on, we can find other examples in wheeled vehicles used in early civilizations, e.g. the south-pointing chariot built in China between 2600 and 1100 BC. It was used for pointing the way home by using a figure moved by an early differential gear. Figure 6 shows two drawings of the south-pointing chariot with its bearings.

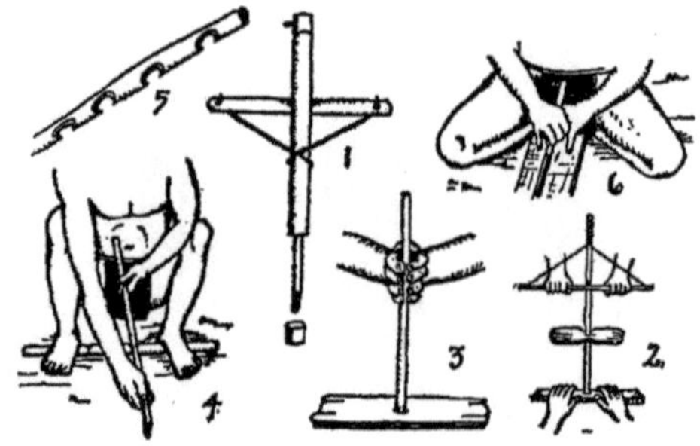

Fig. 4 Fire-making techniques: *1* Yukaghir bow drill, *2* pump movement drill, *3* hand drill, *4* saw for making fire, *5* instrument from Queensland, *6* Melanesian fire plow method. (Mauss 1967)

a **b**

Fig. 5 **a** Bow drilling machine (Strandh 1979); **b** stone door-socket, Mesopotamia. (Singer 1954)

Fig. 6 Drawings of the south-pointing chariot. (Needham 1975)

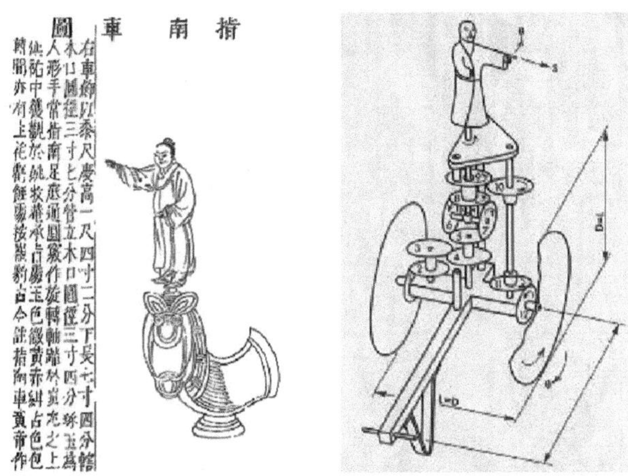

Other remarkable examples are Greek tethrippons and Roman quadrigas, together with Dejbjerg celtic carts dated first century BC. It seems reasonable to think about the possibility of the existence of lubrication for the contacts in carts and chariots, taking into account the previous evidence of the use of lubricants for moving heavy statues and building blocks. By way of example, Fig. 7 shows how a statue was transported during the 12th dynasty of ancient Egypt. Just in front of the statue, there was a slave pouring lubricant on the ground in order to reduce friction, wear and heating (Dowson 1998).

Very significant technological achievements in tribology were made during the Greek and Roman periods, including lathes, wheeled transport, pulleys, gears, cranes, mills and other mechanical systems that utilized rotary motion.

We can find a remarkable example in Hero of Alexandria's system for opening doors automatically, a forerunner of the steam engine (Fig. 8). In his design, described in his work "Pneumatics" (written in the first century, translated by

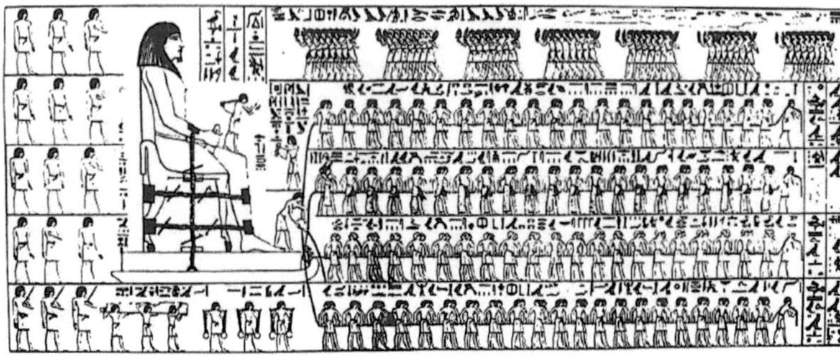

Fig. 7 Moving an ancient Egyptian statue. (Bautista et al. 2010)

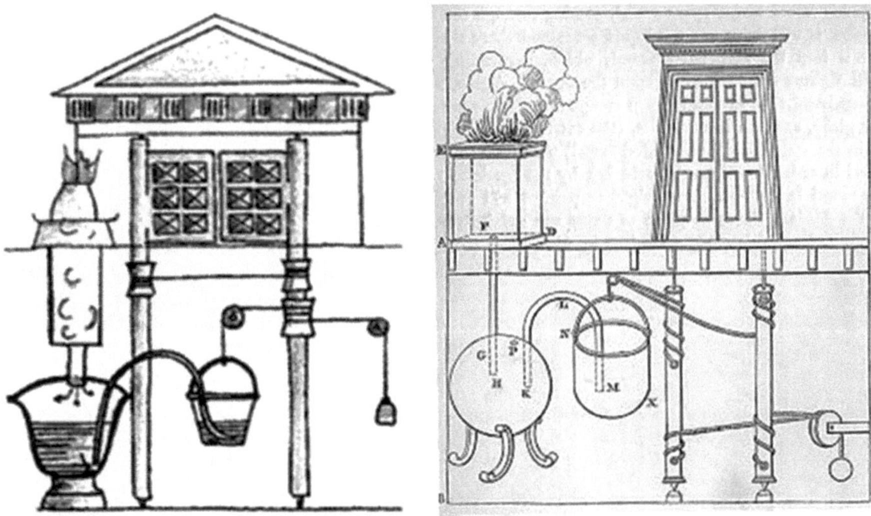

Fig. 8 Reconstruction of automatically opening doors. (Woodcroft 1851)

Woodcroft 1851), we can find pivot bearings for the door spindles. He wrote, "Let the hinges of the doors be extended downwards and turn freely on pivots in the base".

Marcus Vitruvius provided further information on the state of tribology in Greek-Roman times in his books "De Architectura" (ca. 15 BC, reedited in 1511). He digressed to explain the ingenious contrivance of Chersiphron used during the construction of the great Artemision of Ephesus for moving the columns. "When he removed from the quarry the shafts of the columns, … not thinking it prudent to trust them on carriages, lest their weight should sink the wheels in the soft roads over which they would have to pass, he devised the following scheme. He made a frame of four pieces of timber, two of which were equal in length to the shafts of the columns, and were held together by the two transverse. In each end of the shaft he inserted iron pivots, whose ends were dovetailed thereinto, and run with lead. The pivots worked in gudgeons fastened to the timber frame, whereto were attached oaken shafts.

Fig. 9 Astronomical clock tower. (Lu 2000)

The pivots having a free revolution in the gudgeons, when the oxen were attached and drew the frame, the shafts rolled round, and might have been conveyed to any distance".

Tribological progress led to the development of early forms of cylindrical, taper roller and ball bearings (Williams 1994). Evidence of this was found, namely rolling elements which were trunnion mounted, in the remains of two large ships in Lake Nemi, near Rome and dating from about 50 AD.

In the Middle Ages the evolution of general machinery was modest and therefore there was no need for major evolution of existing materials, lubricants and bearings. However, the use of wooden and stone bearings gave rise to the use of metal bearings, like those of iron-on-iron included in Fig. 9, which represents an astronomical clock tower (Su Song 1089).

Indeed, main Medieval developments in tribology were found in mechanical clocks, transportation such as harvest carts and wheelbarrows, mechanical power generation such as water and wind driven machines, and in the use of hard stone inserts to protect the mouldboards of ploughs and the axles of carts. In addition, there was a steady rate of development in the use of lubricants, namely vegetable oils and animal fats. In most cases, there was a lack of theoretical knowledge and therefore the progress was related to phenomenological analyses. Thus, most advances were gained through observation and experimentation but without sufficient understanding of processes involved.

Unlike what happened in Medieval times, during the Renaissance there was a period of recovery and revitalisation in Western Europe, where classical Greek and Roman works were re-examined, improved and illustrated. Moreover, the use of the printing press disseminated the existing knowledge on a large scale. Therefore, two separate activities may be recognised at that time: professional practice of a traditional experimental nature and new theoretical studies of a scientific nature.

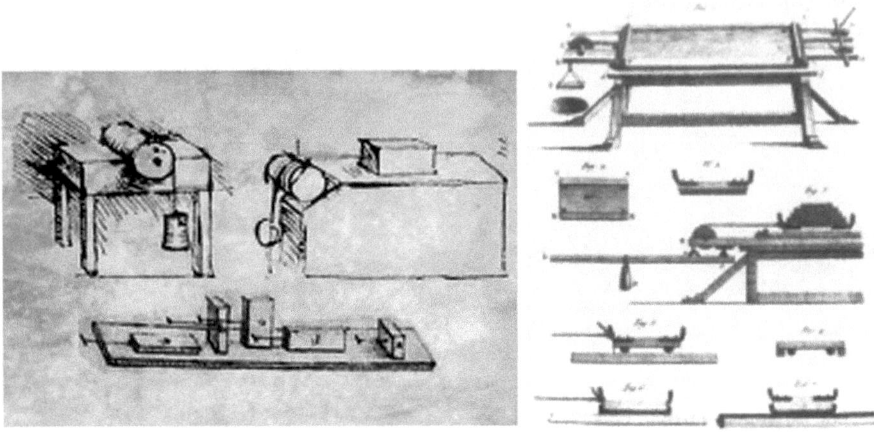

a b

Fig. 10 **a** Sketches from Leonardo da Vinci notebooks. (Da Vinci 15th century); **b** devices used by Coulomb (Coulomb 1785)

Both lines reached important progress and gradually converged until they came together in the seventeenth century.

During the Renaissance we can find first indications of in-depth studies on the force of friction, which had been recognized by Aristotle approximately twenty centuries earlier. Leonardo da Vinci (ca. 1480) presented the first quantitative studies on friction by measuring the force of friction between objects on both horizontal and inclined surfaces. Figure 10a shows the use of cords attached to the objects to be moved, which passed over fixed rollers to weights, giving a measure of the friction force. In the same way, he measured the torque on a roller placed in a semicircular section. He observed that without lubrication "every frictional body has a resistance of friction equal to one-quarter of its weight", this observation is incorrect but quite realistic for the materials commonly used in bearings at that time (Szeri 2005).

In 1495, Leonardo da Vinci formulated the first two basic laws of sliding friction: "Friction is independent of contact area, and friction is proportional to load". These laws were rediscovered by Amontons in 1699, who used springs and was able to measure static and kinetic friction forces, and they were further developed (Fig. 10b) by Charles-Augustin de Coulomb (1736–1806).

Another law of friction is due to Isaac Newton (1642–1727): "Moving friction is not dependent on speed or velocity". These observations were physically supported by the work of Frank Philip Bowden and David Tabor (Bowden and Tabor 1950 and 1964), when they analyzed micro-scale friction taking into account the interaction between surface asperities of the contacting bodies.

Figure 11 includes Leonardo da Vinci's analysis on the difference between rolling and sliding friction. The potential of true rolling motion for low-friction supports was recognized by him, who also wrote in the Codex Madrid I: "…I do not see any difference between balls and rollers save the fact that balls have universal motion while rollers can move in one direction alone. But if balls or rollers touch each other in their motion, they will make the movement more difficult than if there

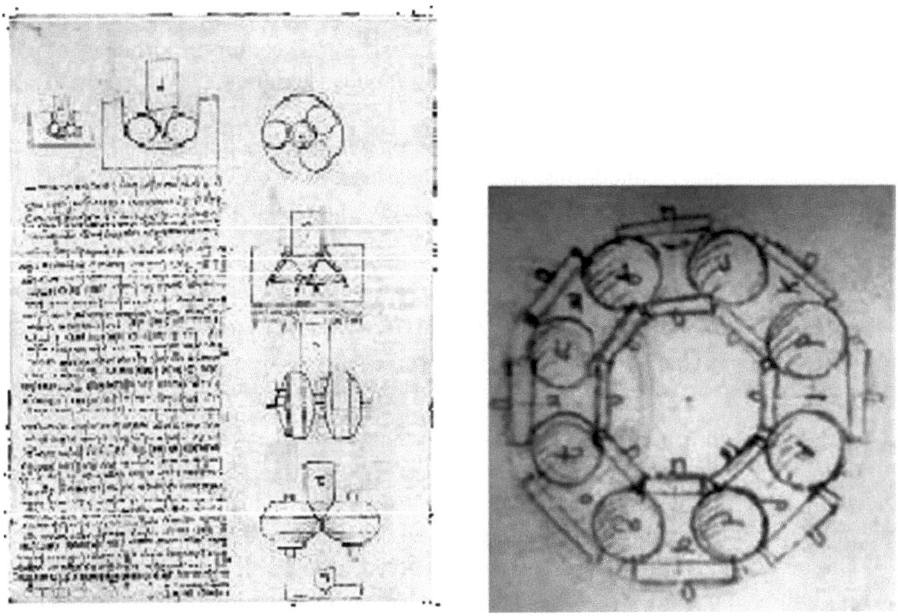

Fig. 11 Leonardo da Vinci's sketches on rolling and sliding friction. (Da Vinci 15th century)

were no contact between them, because their touching is by contrary motions and this friction causes contrariwise movements".

Subsequently, there were attempts to apply knowledge in friction to develop low-friction bearings. A first approach presented in Leonardo da Vinci's Codex Madrid I, is a bearing alloy consisting of three parts of copper and seven parts of tin melted together.

By combining their understanding of the theory and their practical genius, Renaissance engineers reached remarkable tribological developments, leading to new proposals of different types of bearings for machines. Figure 12 presents illustrations

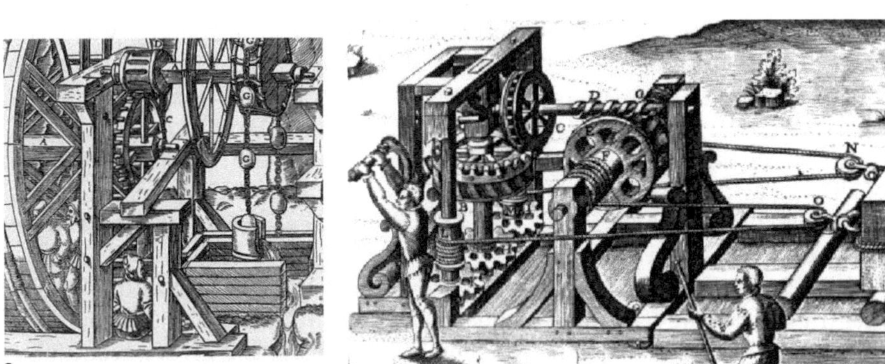

Fig. 12 **a** Details of plain bearings in a rag-and-chain pump (Agricola 1556); **b** plain bearings and rollers in a hand-operated worm gear (Ramelli 1588)

of Georgius Agricola (1556) and Agostino Ramelli (1588) which show the common use of both plain bearings and rollers in the sixteenth century. Further examples can be found in the treatises of machines by Jaques Besson (1578), Vittorio Zonca (1607) and many others.

Finally, it is worthwhile to note the experimental studies performed by Vittorio Zonca (1568–1603) concerning wear in bearings made of dissimilar materials, where steel and bronze contacts were analyzed.

4 From the Renaissance to the Industrial Revolution: The Beginnings of a New Science

During the seventeenth century there were changes in social and production structures, namely population growth, first factories (Royal Manufactures) and transport developments. These modifications were the seed for the Industrial Revolution of the second half of the eighteenth century, when all the fields of technology grew rapidly.

On the one hand, population growth involved changes in food supply, together with clothes and other services that promoted an increase of commercial activities. On the other hand, changes in production structures led to a rise in the power demand due to the change from hand-made workshops to early factories that were created to cover the increasing demand of products in the textile, metallurgy, mining and agriculture sectors.

In this context, different studies related to friction and wear were carried out in order to improve the production and performance of the machines installed in the factories. These machines were initially driven by water or animal traction, until the generalized use of the steam engine during the Industrial Revolution (Bautista et al. 2010).

4.1 Friction and Wear in Bearings

In line with the Renaissance engineers, friction and wear in bearings was identified as essential for the correct operation of machines. Therefore, the concept of Leonardo da Vinci's low-friction bearing was further developed by Hooke (1684) and Babbitt.

Robert Hooke (1635–1703) focused on friction and wear reduction in bearings of chariots by means of experimental analyses. His improvements led him to present a sailing chariot in the Royal Society in London, i.e. a chariot movable by air-power due to its low friction.

Isaac Babbitt (1799–1862) greatly improved the friction bearing design with the first friction bearing using low-friction "babbitt metal": an alloy of tin, antimony and copper. This formulation presents low shear strength and gives a reduction of 25 % in dynamic friction coefficient, if compared to previous steel-steel contacts in bearings (Hellemans and Bunch 1988).

In the same way, Leonardo da Vinci's early studies of wear in bearings finally led to Archard's wear law in 1953, based in the theory of asperity contact. It states that the volume of wear is proportional to the distance slid and to the applied load, and inversely proportional to the hardness of the softer material (Archard and Hirst 1956).

4.2 Hydrodynamic Lubrication

Despite the increasing advancement of bearings, it was not until the nineteenth century when fluid film lubrication was introduced and thus minimized friction. Fluid film lubrication, or hydrodynamic lubrication, is essential in machines that operate at high velocity, where frictional heating and wear can produce negative effects.

The mechanisms that allow the reduction of friction under fluid film lubrication were initially unknown (Hamrock 1994). The first theories of Leupold considered that the fluid film reduced the roughness or asperities of the surfaces in contact (Stachowiak et al. 2004). Then, Leslie (1804) exposed that the lubricant filled the gap between asperities diminishing the friction. Later on, Rennie (1829) proposed that the lubricant separated the surfaces in contact.

More studies on hydrodynamic lubrication were carried out by Von Pauli (1849), Barrans (1850), Hirn (1854) and Thurston (1879). Figure 13 shows a scheme of the test device used by Gustave Adolphe Hirn (1815–1890) in order to analyze the effect of diverse loads and sliding velocities.

Nikolai Pavlovich Petrov (1836–1920) and Beauchamp Tower (1845–1904) in Russia and England respectively, further developed these studies with almost simultaneous developments related to railway applications, in which the availability of mineral oils during the first half of nineteenth century played an important role. First distillation plants were installed in Prague (1810) and France (1834) and by 1850 James Young founded an oil refinery (Dowson 1998).

Mineral oils substituted previous lubricants of animal or vegetable origin (sperm whale, olive, rapeseed, etc.) because of their better behavior under severe operating conditions, typical of machines developed during the Industrial Revolution. Nowadays, mineral oils are the most widely used lubricants worldwide, though they are being substituted by synthetic oils in several specific applications, e.g. when working temperature is very high.

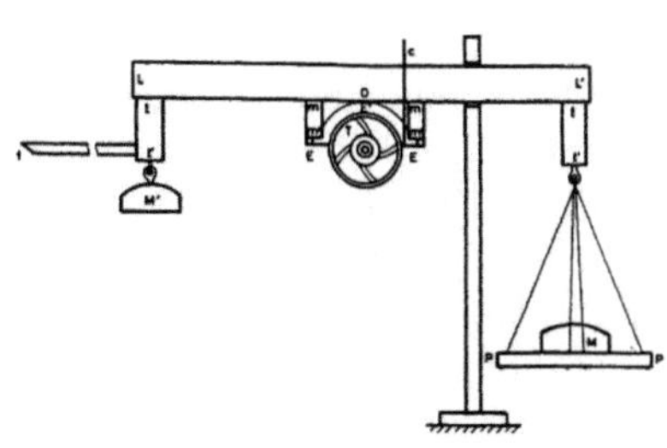

Fig. 13 Test device used by Hirn. (Hirn 1854)

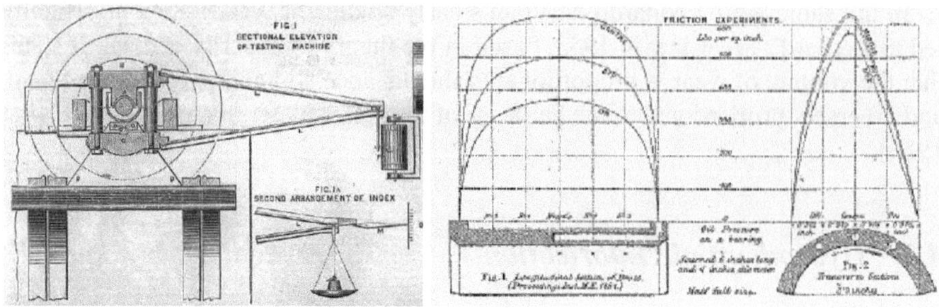

Fig. 14 Experiments performed by Tower. (Tower 1883)

Petrov (1883) analyzed the specific characteristics of Russian mineral oils, whose high viscosity caused high losses by friction in diverse applications in railway transportation. His experimentation led to a first theoretical law that gave the friction coefficient in a journal bearing as a function of the geometry, the lubricant viscosity and the operating conditions.

In 1883, Tower experimentally studied the evolution of the lubricated friction coefficient in journal bearings at high sliding velocities, under operating conditions which simulated railway axle boxes. This study was requested by the Institution of Mechanical Engineers in order to improve lubrication of journal bearings used in trains.

Tower's first tests were performed with bath lubrication and he found very low friction under high loads which suggested the existence of a fluid film. In his second series of tests, Tower made a hole in the bearing in order to analyze forced lubrication. However, he accidentally discovered that lubricant escaped through this hole at very high pressure levels, and in this way he proved the existence of a hydrodynamic wedge that carried the load. Figure 14 shows the layout of the experiment and the measurements of the pressure, obtained by placing a gauge in the hole.

Osborne Reynolds (1842–1912) derived and published (Reynolds 1886) the differential Eq. (1), which supported the results of Tower. This equation provided a physical explanation of fluid film lubrication and related the variables involved, namely pressure (p), viscosity (η), average velocity between surfaces (u), time (t) and lubricant film thickness (h).

$$\frac{d}{dx}\left[h^3\frac{dp}{dx}\right]+\frac{d}{dy}\left[h^3\frac{dp}{dy}\right]=6\eta\left[u\frac{dh}{dx}+\frac{dh}{dt}\right] \tag{1}$$

In 1904, Arnold Sommerfeld (1868–1951) solved Reynolds' equation and presented a formal theory of hydrodynamic lubrication with an analytical solution for journal bearings. The solution was based on dimensionless parameters and was highly useful for design purposes. Soon after, Michell (1905) applied the Reynolds theory to thrust plain bearings.

In this way, lubrication of bearings constitutes an example of the evolution of our understanding, beginning with observations and concluding with models of the phenomena involved.

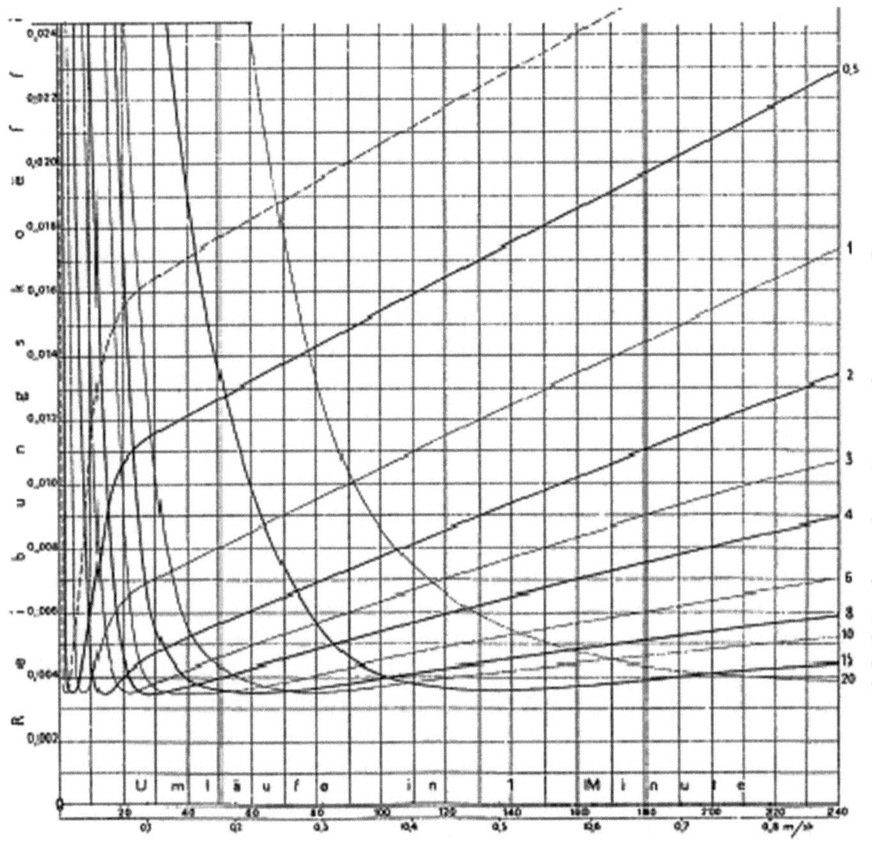

Fig. 15 Detail of Richard Stribeck's curves. (Stribeck 1902)

It is important to note the distinction between fluid film lubrication and other lubrication regimes, according to Richard Stribeck's work, published in 1902. He performed a set of systematic tests for a range of velocities (N) and loads, and represented the friction coefficient as a function of the Gumbel number ($\eta N/p$). Figure 15 shows a detail of Stribeck's results, where very different behaviors can be observed for the friction coefficient.

5 Tribology as a True Science

Before the twentieth century, tribology was primarily a technological endeavor, with specific matters of different sciences (Halling 1975). This consideration began to change with the development of hydrodynamic theory and its later generalization to elastohydrodynamics.

5.1 Elastohydrodynamic Lubrication

Martin (1916) applied hydrodynamic theory to two rigid discs under a high load in conditions representative of the line contact in gears. This model gave an insignificant value for the lubricant film thickness, insufficient for forming a fluid film when compared with the roughness of the contacting surfaces. Therefore, a high value was expected for the friction coefficient. In contrast, experimental results of Martin showed a low value of friction. Subsequent modification of hydrodynamic theory led to elastohydrodynamic lubrication, which took into account the elastic deformation of the contacting surfaces and the lubricant viscosity dependence on pressure.

The first remarkable approaches to the solution of the elastohydrodynamic problem were proposed in the middle of the twentieth century by Grubin (1949) and Petrusevich (1951). In 1959, Duncan Dowson and Gordon Robert Higginson (Dowson and Higginson 1959) completely developed the elastohydrodynamic theory predicting the existence of a pressure peak and a reduction of the film thickness at the outlet of the contact. These characteristics of elastohydrodynamic contacts were experimentally confirmed by Crook (1961) using an oscilloscope to measure the electrical resistance between two lubricated bodies in contact, as shown in Fig. 16.

Solving this complex problem required the development of specific models and techniques, together with advanced computing facilities. The initial solutions were improved by Bernard J. Hamrock and Duncan (1976, 1977), among others. These results considered Newtonian behavior of the lubricant and approximate isothermal work regime, where the thermal effects due to sliding contacts were neglected. However, these hypotheses were only valid under particular operating conditions.

In fact, under the severe operating conditions inherent to elastohydrodynamics, frequently common lubricants cease to behave as Newtonian fluids and exhibit pseudoplastic behavior (Bair 2007). These fluids present a rapid rise in shear stress with the shear rate, but the increase gradually drops as the shear rate rises. Various models have been developed to explain the rheological behaviour of lubricants (Jacobson 1991; Höglund 1999), such as Carreau's equation (Carreau 1972).

In general, heat generated by friction can cause a very considerable local increase in temperature and the corresponding reduction of viscosity (Gohar 1988).

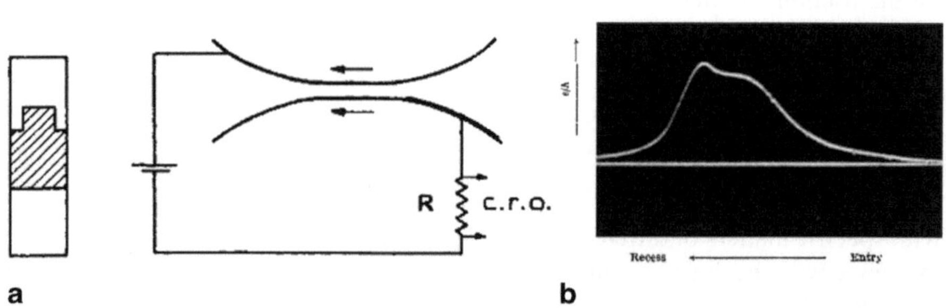

Fig. 16 Crook's experiments. **a** Layout; **b** pressure results. (Crook 1961)

This requires a thermal elastohydrodynamic study to examine how the viscosity characteristics of the lubricant change with temperature, and the subsequent effects on its behavior (Stachowiak et al. 2004). Further research (Olver and Spikes 1998) opened up the way to the so-called thermal-elastohydrodynamic lubrication, which presents multiple applications in mechanical systems that work under very high pressure, such as gears, rolling bearings and cams (Habchi et al. 2008).

5.2 Non-Fluid Lubrication

Despite the wide application field of fluid film theories, conventional liquid lubricants are not suitable for machinery working under particularly severe operating conditions, e.g., extreme temperature, very high pressure, vacuum conditions or ambient dust concentrations.

By the way of example, Kingsbury developed air-lubricated bearings (Kingsbury 1897). They presented a very low friction coefficient and worked within high temperatures but their application was limited by their low load capacity.

In other cases, solid lubricants or self-lubricating materials are used that exhibit lubricity by themselves and can be applied as coatings (Carnes 2005). They are also used as additives dispersed in liquid and grease media. The introduction of graphite as a self-lubricating material for applications such as air compressors or open gears dates back to 1906, whereas the molybdenum disulfide (patent issued in 1939) was used in oxidizing media with extreme temperatures and guaranteed a high stability with vacuum in space vehicles. In 1938 teflon was discovered, one of the most famous self-lubricating polymer materials. It is very slippery and relatively inert and can be used for biotribology applications (Harris 1991), e.g. in medical prostheses. Recently, it has also been extended the use of ceramic materials (Ziebig and Luber 1983), which are used in high-speed ball bearings and femoral heads and are very hard, inert and produce low wear rate (Fig. 17).

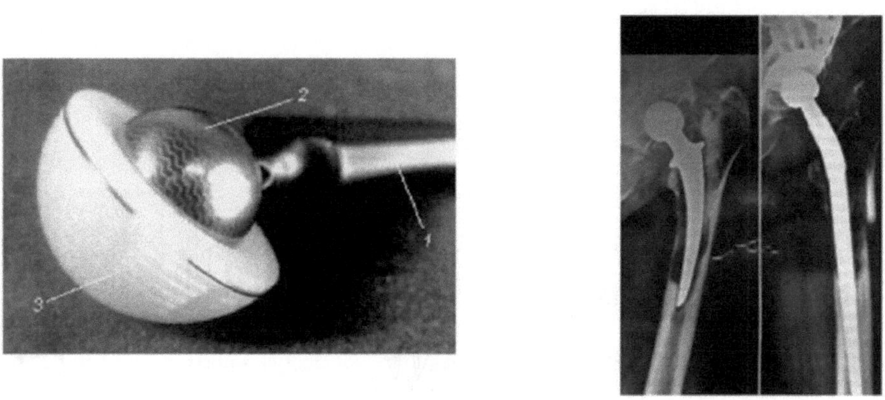

Fig. 17 Ceramic hip prosthesis. (Levitin 1997; Pinchuk et al. 2006)

5.3 Testing in Tribology

The difficulty of predicting the in-service behavior of the lubricant in new machinery increased due to the large number of parameters with influence: operating parameters (load, sliding velocity, temperature…), parameters of the material (hardness, elasticity, thermal properties…), surface geometry, roughness and lubricant parameters (viscosity, density, thermal properties…). We must bear in mind that there are a wide variety of mineral and synthetic lubricant bases (Spikes 1994). Moreover, its behavior is highly influenced by additives added to improve lubricant performance. This is one of the reasons why theoretical predictions frequently need to be complemented or verified through experimentation.

Testing in tribology became widespread since 1927, with the Pin and Vee Block tester shown in Fig. 18, a first commercial tester from Falex. It was designed to quantify the anti-wear and extreme pressure properties of fluid and solid lubricants measuring the wear with a ratchet wheel. Load was applied to the vee-blocks against a rotating device.

This tribometer was followed by other classical equipment (Pin-on-Disc apparatus, Four-ball tester, Timken apparatus, etc.) designed by different companies, such as Falex, Cameron-Plint and PCS-Instruments. Some of this equipment, shown in Fig. 19, constitutes a bridge between theory and industrial applications, as they are designed to determine the influence of each parameter of interest in the phenomena analyzed, providing fast and reliable results under controlled conditions.

The generalized use of certain test machines led to the development of standard tests performed in them, as the case of the back-to-back FZG gear test rig (Fig. 20). The test gears are connected to slave gears by two shafts and a static torque can be applied to the system by means of a torsion bar and a clutch. In addition, there are other complementary testers used in tribology, e.g., equipment for the characterization of properties of lubricants or test benches used for studying mechanical failures. All this testing equipment supports the theoretical progress of tribology.

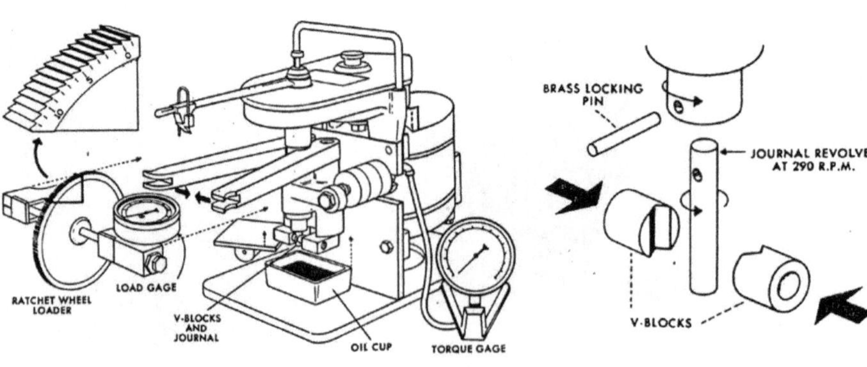

Fig. 18 Pin and vee block tester. (Falex catalog, http://www.falex.com)

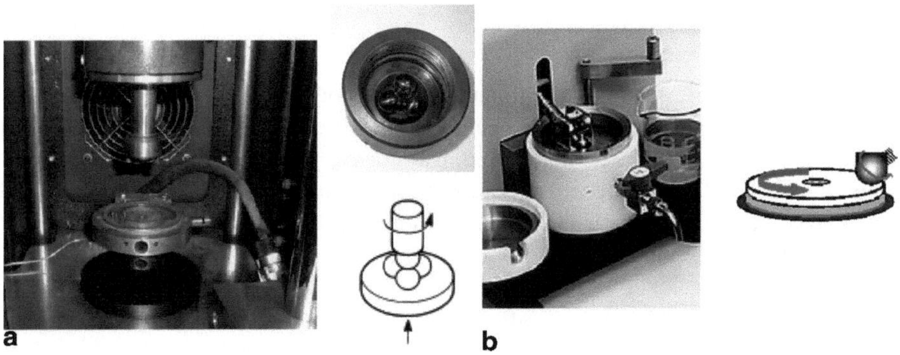

Fig. 19 **a** Four ball tester; **b** mini traction machine, with details of the contacts. (Courtesy of the company Repsol)

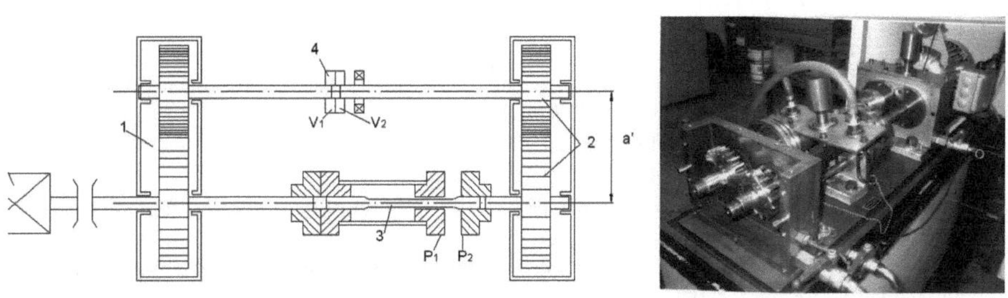

Fig. 20 FZG gear test rig: *1* Slave gear, *2* Test gear, *3* Torsion bar, *4* Clutch. (Lafont et al. 2009, and courtesy of Repsol)

5.4 Present and Future of Tribology

Tribology has become deeply embedded in science, engineering, design, manufacture and life cycle management across a huge range of application areas. As such we can now regard tribology as a mature scientific and engineering discipline.

Nowadays, the design of machines (and mechanical systems in general) can take advantage of extensive existing knowledge about tribology in order to optimize their operation and life. In fact, understanding the complex interrelationship between materials, lubricants and the working conditions and environments of machines constitutes the key to reduce friction and avoid failures caused by tribological phenomena. Even small benefits of one machine element can produce huge global savings due to the size of the marketplace.

Friction phenomenon has also been studied at the nanometer scale (nanotribology), where the atomic forces affect the behavior of the system (Harrison et al. 1998). Other challenges in tribology are the development of new generation lubricants and additives, the improvement of contacting surfaces (microgeometry, coatings, etc.), further development of tribometry and biotribology applications for tissue growth and prostheses (Spikes 2001; Carnes 2005).

6 Conclusion

A historical overview of the significance of tribology during its long history has been presented, with focus on the theoretical developments and their application to technical problems. As demonstrated, tribology is related to the performance of every mechanical device because wear, friction and lubrication present a key influence on critical aspects such as efficiency, in-service life, breakdowns and need for maintenance and repair.

References

Agricola G (1556) De Re Metallica. Reprinted in 1950 by Dover publishing

Amontons G (1699) De la Résistance causée dans les Machines. Mémoires de l'Académie Royale A, pp 257–282, 1699

Archard JF (1953) Contact and rubbing of flat surface. Journal of Applied Physics 24(8):981–988

Archard JF, Hirst W (1956) The wear of materials under unlubricated conditions. Proceedings of the Royal Society of London A-236:397–410

Bair S (2007) High pressure rheology for quantitative elastohydrodynamics. Tribology and interface engineering series, No. 54. Elsevier, Amsterdam

Barrans J (1850) On an improved axle box for railway engines and carriages. Proceedings of the Institution of Mechanical Engineers 2, pp. 3–8

Bautista E, Ceccarelli M, Echávarri J, Muñoz JL (2010) A brief illustrated history of machines and mechanisms. Springer, Berlin

Besson J (1578) Theatrum Instrumentum et Machinarum, Lugduni Vincent

Bowden FP, Tabor D (1950 and 1964) The friction and lubrication of solids. Oxford University Press, Part I (1950); Part II (1964)

Carnes K (2005) The ten greatest events in tribology history: as chosen by the readers of TLT magazine. Tribology & Lubrication Technology 61(6):38–47

Carreau PJ (1972) Rheological equations from molecular network theories. Transactions of The Society of Rheology 16(1):99–127

Coulomb CA (1785) Theories des machines simples, en ayant égard au frottement de leurs parties, et a la roider des cordages. Mém Math Phys X:161–332

Crook AW (1961) Elastohydrodynamic lubrication of rollers. Nature 190(4782):1182–1183

Da Vinci L (15th century) Codex Madrid I and II, Codex Atlanticus and Codex Arundel

Dowson D (1998) History of tribology, 2nd edn. Longman, London

Dowson D, Higginson GR (1959) A numerical solution to the elastohydrodynamic problem. The Journal of Mechanical Engineering Science 1(1):6–15

García Tapia N (1990) Patentes de Invención Españolas en el Siglo de Oro. Industrial Property Register, D.L

García Tapia N (2001) Un Inventor Navarro. Jerónimo de Ayanz y Beaumont. Pamplona Department of Education and Culture

Gohar R (1988) Elastohydrodynamics. Ellys Horwood Ltd., London

Grubin AN (1949) Fundamentals of the hydrodynamic theory of lubrication of heavily loaded cylindrical surfaces. In: Ketova KF (ed) Proceedings of symposium on investigation of the contact of machine components, Book No. 30, pp. 115–166. Moscow, Russia

Habchi W, Eyheramendy D, Bair S, Vergne P, Morales-Espejel G (2008) Thermal elastohydrodynamic lubrication of point contacts using a Newtonian/generalized Newtonian lubricant. Tribology Letters 30:41–52

Halling J (1975) Principles of tribology. Edition, illustrated. MacMillan Press, London

Hamrock BJ (1994) Fundamentals of fluid film lubrication. McGraw-Hill, New York
Hamrock BJ, Dowson D (1976) Isothermal elastohydrodynamic lubrication of point contacts, Part I-theoretical formulation. ASME Journal of Lubrication Technology 98:223–229
Hamrock BJ, Dowson D (1977) Isothermal elastohydrodynamic lubrication of point contacts, Part III-fully flooded results. ASME Journal of Lubrication Technology 99:264–275
Harris TA (1991) Rolling bearings analysis, 3rd edn. Wiley, New York
Harrison JA, Stuart SJ, Brenner DW (1998) Atomic-scale simulation of tribological and related phenomena. In: Bhushan B (ed) Handbook of micro/nanotribology. CRC, New York, pp 525–594
Hellemans A, Bunch B (1988) The timetables of science. Simon & Schuster, New York
Hirn GA (1854) Etudes sur les principaux phénomènes que présentent les frottements médiats et sur les diverses manières de déterminer la valeur mécanique des matières employées au graissage des machines. Bulletin de la Société Industrielle de Mulhouse, 26, No. 129, pp 188–277
Höglund E (1999) Influence of lubricant properties on elastohydrodynamic lubrication. Wear 232:176–184
Hooke R (1684) The posthumous works of Robert Hooke. Waller and Seer, London
Jacobson BO (1991) Rheology and elastohydrodynamic lubrication. Tribology series, vol 19. Elsevier, Amsterdam, pp 1–381
Jost P (1966) Lubrication (tribology), education and research. A Report on the present position and industry's needs, published as an English government report on 9 March 1966
Jost P (2001) The tasks of tribology societies on a changing world. Opening address, Second World Tribology Congress
Kingsbury A (1897) Experiments with an air lubricated bearing. Journal of the American Society for Naval Engineers 9, pp 267–292
Lafont P, Díaz A, Echávarri J (2009) Diseño y Cálculo de Transmisiones por Engranajes. Publicaciones de la ETS de Ingenieros Industriales, Madrid
Leslie J (1804) An experimental inquiry into the nature and propogation of heat. T. Gillet Printer, Salisbury Square
Levitin M (1997) Friction reduction in hip implants treated with regular microrelief. Regmi-Tech, Haifa, p 12
Lu Z (2000) The development of water-powered machines in China in 10–14th Century. In: Proceedings of HMM 2000. Kluwer Academic Publishers, Dordrecht
Martin HM (1916) Lubrication of gear teeth. Engineering, vol 102. London, pp 199–221
Mauss M (1967) Manuel D'Ethnographie. Payot, Rivages (eds), Paris, France
Michell AGM (1905) The lubrication of plane surfaces. Zeitschrift für Mathematik und physik 52(Pt. 2):123–137
Needham J (1975) Science and civilisation in China. Cambridge University Press, Cambridge
Olver AV, Spikes HA (1998) Prediction of traction in elastohydrodynamic lubrication. Proceedings of the Institution of Mechanical Engineers Part J-Journal of Engineering Tribology 212:321–332
Petrov NP (1883) Friction in machines and the effect of the lubricant. Inzh Zh St-Peterb 1:71–140; (2:227–279; 3:377–436; 4:435–464)
Petrusevich AI (1951) Fundamental conclusions from the contact-hydrodynamic theory of lubrication. Izvestiya Akademii Nauk SSSR SSSR (OTN) 2:209–223
Pinchuk LS, Nikolaev VI, Tsvetkova EA, Goldade VA (2006) Tribology and biophysics of artificial joints. Elsevier, Amsterdam
Pinkus O, Wilcock DF (1977) Strategy for energy conservation through tribology. Tribology in energy technology workshop, Washington: Tribology in Energy Technology Workshop 1977, ASME
Ramelli A (1588) Le Diverse et Artificiose Macchine del Capitano Agostino Ramelli dal Ponte della Tresia, Paris, France
Rennie (1829) Experiments on the friction and abrasion of the surfaces of solids. Philosophical Transactions of the Royal Society of London 119:143–170
Reti L (1967) El artificio de Juanelo en Toledo. Magazine of Toledo provincial council, No. 60, 1967, pp. 3–46

Reynolds O (1886) On the theory of lubrication and its application to Mr. Beauchamp Tower's experiments, including an experimental determination of the viscosity of olive oil. Philosophical Transactions of the Royal Society of London 177:157–234
Singer C (1954) A history on technology. Clarendon Press, Oxford
Sommerfeld A (1904) Zur hydrodynamischen theorie der Schmiermittelreibung. Zeitschrift für angewandte Mathematik und Physik 50:97–155
Spikes HA (1994) The behaviour of lubricants in contacts; current understanding and future possibilities. Proceedings of the Institution of Mechanical Engineers, Part J: Journal of Engineering Tribology 208(1): 3–15
Spikes HA (2001) Tribology research in the twenty-first century. Tribology International 34:789–799
Stachowiak GW, Batchelor AW, Stachowiak GB (2004) Experimental methods in tribology. Tribology series, 44. Elsevier, Amsterdam
Strandh S (1979) Machines. An illustrated history. Littlehampton Book Services, Worthing
Stribeck R (1902) Die Wesentlichen Eigenshaften der Gleit und Rollen lager. Z Ver dt Ing 46(38): 1341–1348
Su Song (1089) Xin Yi Xiang Fa Yao. (Horological treatise)
Szeri AZ (2005) Fluid film lubrication. Theory and design. Cambridge University Press, Cambridge
Thurston RH (1879) Determination of the laws and coefficients of friction by new methods and with new apparatus. The Railroad Gazette, New York
Tower B (1883) First report on friction experiments (friction of lubricated bearings). Proceedings of the Institution of Mechanical Engineers Nov., pp 632–659
Vitruvius M (1511) De Architectura. Edited by Fra Giocondo, Verona, (reprinted in 1513, 1522 and 1523).
Von Pauli FA (1849) Uberden Widerstand der zapfenreibung. Kunst and Gewerbeblatt des polytechnischen Verein des Konigreich Bayern, vol 8/9, pp 452–469
Williams JA (1994) Engineering tribology. Oxford University Press, Oxford
Woodcroft B (1851) The Pneumatics of Hero von Alexandria. Walton & Maberly, London
Ziebig A, Luber H (1983) Bioceramic hip joint components-industrial production and testing procedures to ensure high functional reliability. In: Vincenzini P (ed) Ceramics in surgery. Elsevier, New York
Zonca V (1607) Novo Teatro Di MACHINE ET EDIFICII Per uarie et Sicure operationi Co[n] le loro figure tagliate in Rame é la dichiaratione, e dimostratione di ciascuna. Opera necesaria ad Architetti, e a quelli, ch[i] di tale studio si dilettano. Di VITTORIO ZONCA. Architetto della Magnifica Communita di Padoua. In Padoua Appresto Pietro Bertelli.

From Paper to Erected Walls: The Astronomical Observatory of Coimbra: 1772–1799

Fernando B. Figueiredo

Abstract The idea to establish an astronomical observatory at the University of Coimbra (OAUC) was inspired by a major reform of that university in 1772. At that time the observatory was planned for the site where the Castle of the city was built. However, mainly due to financial difficulties, its construction stopped after 3 years. Meanwhile a small building was erected (1775–1777) to serve astronomical lessons. The problem of lack of a real and effective observatory to serve true scientific research required a solution that began to be formulated around the years 1785–1790. In this article we explore the problems related with the foundation of the OAUC and discus the importance of astronomical instruments in implementing its primary scientific mission—elaboration of the astronomical ephemeris (1803).

1 Introduction

On October 7, 1781 José Monteiro da Rocha (1734–1819), professor of Physics and Applied Mathematics at the Faculty of Mathematics of the University of Coimbra, in a letter to the Secretary of the *Academia das Ciências de Lisboa* (Academy of Sciences of Lisbon, ACL)[1], Luis António Furtado (1754–1830), shows no excitement at the idea of ACL organizing and publishing 'some kind' of astronomical/nautical almanac[2] (ACL Ms. Azul 1944). Monteiro da Rocha, besides finding that there was no such need, noted that the few astronomers in Portugal had easy access

[1] The ACL was created on December 24, 1779. José Monteiro da Rocha was one of its first members (elected in January 16, 1780).

[2] This letter of Monteiro da Rocha to the ACL's Secretary is, as much as we can understand, his answer to another one (unknown) of Luis António Furtado asking about his opinion on that ACL's project.

F. B. Figueiredo (✉)
Department of Mathematics/Astronomical Observatory and Institute of Geophysics, University of Coimbra, Coimbra, Portugal
e-mail: bandeira@mat.uc.pt

Centre François Viète, University of Nantes, Nantes, France

R. Pisano (ed.), *A Bridge between Conceptual Frameworks,* History of Mechanism and Machine Science 27, DOI 10.1007/978-94-017-9645-3_9

to the French *Connaissance des Temps* (CDT) or the English Nautical Almanac (NA); neither does he think that such publication was possible at that time in Portugal. As Monteiro da Rocha believed, there was neither technical nor scientific national capacity to carry out such a project. However, he thought that it would be rather easy to produce a recalculated copy almanac from CDT or NA for the meridian of Lisbon. This hypothesis required a relatively small computing effort when compared with a direct calculation of ephemeris from astronomical tables, because to do so it would be enough to take into account the difference in longitude between the meridians of Greenwich and Lisbon[3]. What would be desirable, but impossible he argues, is to calculate the lunar distances directly from other astronomical tables "than those of Mayer's, in which the calculations of the English Nautical Almanac were based, as also its copy of *Connaissance des Temps*, such as Clairaut and Euler's tables". So he adds,

> An almanac of this kind would be interesting throughout all maritime European countries, and it would be glorious to the Court of Portugal, as it is the English Nautical Almanac founded in Mayer's Astronomical Tables.

The problem of determining the terrestrial longitude was a central issue of the astronomical and nautical science of the eighteenth century (Andrewes 1993; Boistel 2001), and the conception, calculation and elaboration of astronomical ephemeris were one of the main objectives of the scientific activity of the large European astronomical observatories, until about 1820–1830 (Wolf 1902; Lovell 1994). The Greenwich Observatory was founded by Charles II (1630–1685), in 1675, with the specific purpose of 'rectifying the tables of the motions of the heavens, and the places of the fixed stars, so as to find out the so much desired longitude of places for the perfecting of the art of navigation'[4]. This same problem of determining the longitude and cartographing the Earth, a matter of so great importance for the maritime and economic ambitions of Louis XIV (1638–1715), was also behind the creation of the Paris Observatory and the publication of *Connaissance des Temps* (1678).

At the time José Monteiro da Rocha was already recognized as one of the most prominent personalities of science in Portugal. He had been one of the main designers of the new curricula for mathematics and astronomy within the framework of Pombal's Reform of the University and he will also play a central role in all subsequent teaching, scientific and administrative activities of University life[5]. Monteiro

[3] Instead of the meridian of Lisbon, Monteiro da Rocha suggested the meridian of the island of Ferro, commonly used at the time by national and foreign mariners.

[4] The need to stimulate a satisfactory solution to the longitude problem would prompt the British government to create in 1714 a prize of £ 20,000—the famous Longitude Act.

[5] Very little is known about the first years of Monteiro da Rocha's life. It is known that he joined the Jesuits in his youth (1752) and left Portugal to go to Brazil where he studied at the Jesuit school of Salvador da Bahia (1752–1759). Following the expulsion of the Jesuits from Portugal in 1759, Monteiro da Rocha left the Society of Jesus and later returned to Portugal (1766). In 1771, he was called by the marquis of Pombal to participate in the educational Reform of the University of Coimbra. Henceforth he will be the lecturer in charge of the courses of Physics and Applied Mathematics (1772–1783) and Astronomy (1783–1804). In 1795 he was appointed Dean and Permanent Director of the Faculty of Mathematics and Director of the Royal Astronomical

da Rocha will also be very important for the future scientific activity of the Astronomical Observatory of the University of Coimbra (OAUC), namely publication of the *'Ephemerides Astronomicas, calculadas para o meridiano do Real Observatório Astronómico da Universidade de Coimbra'* (Astronomical Ephemerides calculated for the Meridian of the Royal Observatory of the University of Coimbra, EAOAUC). That is why Monteiro da Rocha was certainly one of the best people in Portugal to advise ACL on the astronomical/nautical ephemeris project.

Actually the plan suggested by Monteiro da Rocha where astronomical ephemeris "will not be deduced or copied from the Greenwich Nautical Almanac, nor from any other, but instead determined directly from the astronomical tables", was carried out by himself about 20 years later in the OAUC, with publication in 1803 of the EAOAUC, as the Pombaline Statutes of the University of 1772 had settled for about 30 years before.

As we will see later on in this article the ACL, along with the *Academia Real da Marinha* (Royal Navy Academy, ARM), will end up publishing in Portuguese some nautical ephemeris copied from the English Nautical Almanac. This fact will be one of the main reasons that motivate the discussion and the future solution for the Astronomical Observatory of the University of Coimbra around 1785–1790.

2 The Reform of the University Of Coimbra (1772) and the Institutionalization of Mathematical Science in Portugal in the Late Eighteenth Century

The reform of the University of Coimbra performed between 1770 and 1772 and known as the Pombaline Reform is the realization of a master plan initiated by a group of men who, under the auspices and control of King José I (1714–1777) and his prime minister Sebastião José de Carvalho e Melo, 1st Marquis of Pombal, (1699–1782), aimed at synchronizing Portugal with the ideas of Enlightened Europe[6]. All of the regime's education policy was based upon Enlightenment ideas,

Observatory of the University of Coimbra. He was also vice-principal of the University from 1786 to 1804. In 1800 Monteiro da Rocha became a member of the royal council of Prince Regent João VI (1767–1826) and in 1804 became tutor to Prince Pedro (1798–1834) (future Emperor of Brazil and King of Portugal) and moved to Lisbon where he died on 11 December 1819. His scientific work covered quite separate mathematical and astronomical domains. The astronomical work of Monteiro da Rocha spans from theoretical to practical astronomy, the most significant elements being: a work on the determination of comet's orbits, several papers on calculation of eclipses, on longitudes; astronomical tables of the sun, moon and planets and charts of Jupiter satellites, on the use of the rhomboidal reticle and on the use and calibration of the transit instrument. About Monteiro da Rocha's scientific work see (Figueiredo 2005, 2011).

[6] When speaking of the eighteenth century European Enlightenment, we can't speak of a single and uniform movement that had taken root all over Europe. The Enlightenment is a plural motion of idiosyncrasies and paradoxes which manifested itself in various forms and ways. However these various expressions emphasize a common denominator: the use of the laws of reason as the way of knowledge of the world and of man himself. It had presented science as an ideal, a liberating force,

whose protagonists wanted to remove from the hands of the Church, specifically from the Jesuits and later from the Oratorians, the right to teach and to center it in the state's hands.

The University, as an instrument of the State, would be an irreplaceable and forceful tool for the development of a modern society 'enlightened' by science and technological progress. With that purpose new Faculties of Mathematics, Natural Philosophy and Medicine were created and their scientific courses organized in modern syllabi and curricula. In fact, in this reformist program science is seen as the only power capable of generating a real change in society (Maxwell 1995, pp. 87–110, 131–148). Pombal wanted the University to be become not only an educational center but also a center for the production of knowledge, as an answer to the technical and scientific demands of a country that urged modernization.

In this reformist context, mathematical sciences and in particular astronomy will play a special role[7]. Mathematics is seen not only as a fundamental theoretic subject (Geometry is required for all university students), but also as a very important practical one.

In the report that university chancellor D. Francisco de Lemos (1735–1822) wrote to the new Queen Maria I (1734–1816) after the death of King José in 1777 defending the continuity of the Pombaline University project, he underlines with respect to mathematics that,

> beyond its private excellence, which it enjoys by the lights of the purest evidence and the most accurate proceeding of its own demonstrations, directing virtually all the human understanding,

the mathematical sciences contain,

> a great amount of subjects of the utmost importance, to: regulate and measure the time, the measurement of the lands and the boundaries of the countries and determination of the geographical places, the military maneuvers and the campaign practices of the navy; the naval construction; the civil and military architecture; the industry and factories; as all kind of devices and artefacts to help the weakness of men, as well a plethora of other benefits to advantageously promote a large number of useful Arts, necessary to the State.[8]

In the same report Francisco de Lemos, concerned about the low attendance rate of mathematics courses, proposes that the Government legislates so that some professions like engineers and cosmographers would only be carried out by qualified mathematicians graduated from the University[9]. For example, in 1777, José

---

a demonstrably successful method of interpreting the natural world, which exemplified human progress (Bektas and Crosland 1992). About the Portuguese Enlightenment see (Calafate 1990).

[7] The entire ideology that underlies the different science courses program, particularly those relating to the structure of the mathematics course syllabus, strongly materialized the scientific matrix corpus of the French Enlightenment, reflecting the ideas of d'Alembert, as well as other French authors (such as the authors of the textbooks that were adopted, Bezout, Bossut, Marie, Lacaille, Lalande). On the influence of the French Enlightenment in the Reform of the University of Coimbra see (Carvalho 2008; Figueiredo 2011, pp. 45–91).

[8] Lemos 1980, p. 81.

[9] Some years later, in June 9, 1801, was expressly created legislation establishing in each administrative and judicial district of the country a mathematician, whose functions would be drafting

Francisco de Lacerda de Almeida (1750–1798) and António Pires da Silva Pontes (1750–1805), who had got their PhD from the Faculty of Mathematics, were sent by the Government to a Geodetic Commission Survey of the southern borders of Brazil, as a result of treaty limits signed with Spain in that same year.

At the beginning of the new century, in 1801, the Faculty of Mathematics, facing new challenges, restructured the mathematics course, creating two new disciplines: Hydraulics and Practical Astronomy.

The construction of the bar of Aveiro (1781–1808) and the channeling of the Mondego River (1788–1808) and other hydraulic engineering public works undertaken by the government and for which the University had been asked technical advice, demanded new scientifically updated answers. The creation of the discipline of Practical Astronomy was closely related to the future activity of the OAUC inaugurated in the meantime (1799), which involves working,

> assiduously in more accurate observations, to contribute, verify and rectify the Astronomical Tables [...] and to cooperate with more accredited European Observatories[10].

Manuel Pedro de Melo (1765–1833), a former student of the University and PhD in mathematics, at the time professor of the Royal Navy Academy, would be sent to Europe to organize the discipline of Hydraulics, for which he had been appointed professor. In France, in the 1800s, he will work with Jean B. Delambre (1749–1822) at the Observatory of Paris. In fact, due to this connection between Pedro de Melo and Delambre the EAOAUC got very favorable book reviews in CDT (1806, 1807 and 1808). This voyage of Pedro Manuel de Melo was in line with the new chart in law of OAUC (December 4, 1799), which established the need to initiate some scientific voyages to astronomical observatories and other foreign scientific institutions in order to improve and exchange knowledge and scientific practice on a regular basis (from 10 to 10 years).

With the Pombaline Reform of Coimbra's University studies the process of institutionalization of science in Portugal, namely of mathematics and astronomy, begins. For the latter this occurred with the inauguration of the OAUC in 1799.

2.1 The Mathematics and Astronomy Syllabuses Required by the University's Pombaline Statutes

The mathematics course was organized based on seven disciplines (four from the Faculty of Mathematics and the others from Faculty of Natural Philosophy) spread over 4 years. In the 1st and 2nd years the disciplines of pure mathematics were taught and in the last two the mixed or applied mathematics: 1st year, Geometry (+ Moral and Rational Philosophy + Natural History at the Faculty Natural of Philosophy), 2nd year Algebra (+ Experimental Physics at the Faculty of Natural

a topographic map of that region according to the rules established by the Kingdom Geographic Charter of 1790.

[10] OAU char in law December 4, 1799 (EAOAUC 1803, p. v).

Philosophy); 3rd year Physics and Applied Mathematics, and in the 4th year, Astronomy. There was an extra discipline of Drawing and Architecture to be attended in the 3rd or 4th year.

Geometry and Algebra consisted of arithmetic, geometry, trigonometry, algebra, differential and integral calculus. The 3rd year discipline consisted of the study of kinematics and dynamics, hydrodynamics, acoustics and optics (which included the study of optical instruments). Astronomy, "applied to the movement of the celestial bodies", although considered a branch of applied mathematical physics (studied in 3rd year) was instead studied separately in the 4th year. This was justified by the vastness of its subject and of its own importance within the branch of mathematical sciences. Astronomy includes: history of astronomy; spherical trigonometry (spherical astronomy); the study of the so called 'physical astronomy', the planetary movements, the three body problem and theory of the Moon, comets movements, Sun and Moon eclipses, transits of Venus and Mercury; and practical astronomy to be undertaken at the astronomical observatory. Students were expected to acquire skills in the use of the observational instruments and knowledge in astronomical calculations—"throughout this course the theory and the practice should always be together", reinforces the Statutes.

In the discipline of Drawing and Architecture students were given the basics of drawing and perspective, as well as civil and military architecture and representation of maps and topographic charts.

Regarding the adopted textbooks, they fall in the French tradition of 'livres élémentaires' (elementary textbooks), designed to teach the fundamentals of sciences (Schubring 1997). With the only exception imposed by Statutes of Euclid's Elements for the teaching of geometry, the others are all of French authors. Out of ten books used for teaching the different matters, seven were translated into Portuguese and Monteiro da Rocha was responsible for translating six of them. For pure mathematics the volumes relating to arithmetic, plane trigonometry, algebra and differential and integral calculus of the Cours de Mathématiques de la Marine (Paris, 1764–1769) of Étienne Bézout (1730–1783) were translated. For Physics-Applied Mathematics the textbooks of Marie's (1738–1801) Traité de Méchanique (1774), Bossut's (1730–1814) Traité Élémentaire d'Hydrodynamique (1771) and the Leçons Élémentaires d'Optique of Lacaille (1713–1762)[11] were chosen. For Astronomy the Lacaille's Leçons Élémentaires Physique et d'Astronomie Geometrique (1746), and also Lalande's Astronomie (1764) was adopted.

[11] Although the adopted authors have formal approaches similar to Bézout, who devotes in his Cours two volumes to the 'Principes generaux de la Mechanique', the reasons for not adopting the latter is to us related with Marie's and Bossut's books up-to-date. Those books, published respectively in 1774 and 1771, incorporated the latest scientific developments of its fields. The fact that each book was devoted to one branch of mechanics was also of some importance (Marie's compendium deals with the rigid bodies and Bossut's compendium with the fluids). An additional advantage for the Marie's compendium adoption was a single chapter devoted to the study of the central forces, a very important subject for the discipline of Astronomy to be taught in the following year, but which was not present in Bézout's book.

In 1801, about 16 months after the OAUC came into service, the primitive 4th year Astronomy was split into two autonomous disciplines: Theoretical Astronomy and Practical Astronomy, each with their own teacher. The justification for this is presented in the preamble of the document of its creation,

> Because of its vastness, Astronomy cannot be taught and understood with the breadth and depth that it should be, in only one discipline[12].

Therefore other textbooks were introduced. For celestial mechanics and the "*last discoveries of secular inequalities*", that would be studied in Theoretical Astronomy, the Mécanique Céleste (1799–1825) of Laplace (1749–1827) was adopted. In Practical Astronomy, which studied the theory and use of astronomical instruments as well as the methods of calculation and reduction of observations, especially the "calculation of Astronomical tables in all its parts", the book of Biot (1774–1862), Traité Élémentaire d'Astronomie Physique (1805) was adopted.

It is in this context of the Pombaline Reform of the University studies that the Statutes proclaim the construction of an astronomical observatory. That observatory was not only for students to carry out their astronomy practicals but also for teachers to investigate. Its main function was not that of a typical teaching activity facility, another dimension is required. In fact what the Statutes want is a University-based astronomical observatory, although with aspects of a National one. The role and practice that was required fulfilled the dual role of a university teaching/research facility with specific competences usually attributed to national observatories[13]—an observatory to carry out regular observations in order to verify the fundamental astronomical measurements and theories in order to determine the geographical longitude. This branch of national observatory is reinforced in the chart-in-law of December 4, 1799, that will establish the OAUC organization and mission, tuning it absolutely with the astronomical program of the great European national observatories of that time. This law's seventh paragraph establishes unambiguously the calculus of the astronomical ephemeris as its main scientific purpose,

> The Astronomical Ephemeris should be calculated for the Meridian of the Observatory, for its own use (a common practice of the most famous Observatories of Europe at this time), and for the use of the Portuguese mariners; the Ephemeris should not be reduced or copied from the English Nautical Almanac, or from any other, but calculated immediately from the Astronomical Tables.[14]

In fact, after the OAUC came into full operation in 1799 all its activity focuses on the development of astronomical ephemeris. The entire teaching activity is completely minimized to not interfere "with the daily astronomical observations and practices of the Observatory".

---

[12] Chart in law of April 1, 1801 (AUC IV-1ªE-8-3-4).

[13] Regarding this concept of national versus university observatories see (Hutchins 1999, pp. 4–22).

[14] OAUC chart in law December 4, 1799 (EAOAUC 1803, p. viii).

2.2 *The Astronomy of Precision in the Eighteenth Century and the Astronomical Practice for the Observatory of the University of Coimbra*

The astronomical practice of an observatory is obviously linked to its instrumental collection[15], or, to be more precise, we should say that it is the instrumental collection that commands its possible observational program, that is, its real and effective astronomical practice.

The program of the major national astronomical observatories of eighteenth and nineteenth centuries moves around celestial mechanics (Loewy 1881, p. i). It's characterized by a constant request for accurate positions of the solar system bodies and stars, which may contribute to the improvement of the Newtonian theory and the mathematical tools involved in it,

> L'Astronomie, considérée de la manière la plus générale, est un grande problème de Mécanique, dont les éléments des mouvements célestes sont les arbitraires; sa solution dépend à la fois de l'exactitude des observations et de la perfection de l'analyse, et il importe extrêmement d'en bannir tout empirisme et de la réduire à n'emprunter de l'observation que les données indispensables.[16]

In this ongoing process (development of instrumental methods of observation, observational data reduction and refinement of the theory)[17] the astronomy practice takes place mainly around the angular measurement of the right ascensions and declinations of the celestial bodies passing through the meridian of the observatories—which is referred to as an international meridian program consensus by Jim Bennett,

> Thus programmes of meridian measurement came to be pursued in all the active observatories of Europe [...] they [the observational data] were accumulated by the activity that became the sine qua non of an astronomical observatory[18].

Practical astronomy became the main routine of any observatory, a routine that is renewed continuously in search of greater observational accuracy and new methods of instrumentation and observation. This astrometric program was the basis of a triumphant progress in astronomical science, and promotes the development of real astronomical instruments industry, dominated by the English manufacturers, from the 1720s on (Daumas 1972, pp. 121–135). This instrumental industry makes the dissemination of this astronomical program through the European observatories possible according to J. Bennett (Bennett 1987, p. 113–129, 1992, p. 13).

The instrumental core of a typical observatory of that time was then anchored in a set of half a dozen instruments. In the center of this group is the mural quadrant,

[15] For instance, the use of instruments is conditioned by the nature of the astronomical phenomenon; such is the case, for example, of the transit of Mercury and Venus over the solar disk.

[16] Laplace 1878–1882, vol. 1, p. i.

[17] "Le seul moyen de connaître la nature, est de l'interroger par l'observation et le calcul" (Laplace 1835, p. 207).

[18] Bennett 1992.

or the quarter-circle[19], which becomes the quintessential of the late eighteenth and early nineteenthcentury's observatory (Turner 2002). The quadrant, which already occupies a prominent role in the great medieval Arabic observatories, and later at the Tycho Brahe's (1546–1601) observatory and afterwards at Greenwich with John Flamsteed (1646–1719), assumes in the late eighteenth century a primacy, becoming the first of a new class of very precise instruments[20].

Along with the quadrant other instruments comprise that essential core of very precise instruments. The entry 'Observatoire', which Lalande writes for Methodique Encyclopedie, the following instruments are specified as essential:

> Un quart de cercle mobile [...], une lunette méridienne [...], un mural [...], une bonne lunette achromatique de 3 à 4 piés, montée sur un pied parallactique [...], pendule & le compteur.[21]

Those are also the instruments that Lalande devotes two chapters to in his book Astronomie: "*des instruments d'astronomie*" (Lalande 1771–1781, vol. 2, pp. 722–830) e *"de l'usage des instrumens & de la pratique des Observations"* (Lalande 1771–1781, vol. 3, pp. 1–82). The French astronomer Antoine Darquier de Pellepoix (1718–1802) also indicates those same main instruments as being necessary for an effective study of the heavens,

> [Avec les instruments ci-dessus détaillés, un observateur exercé & laborieux pourra faire beaucoup d'observations utiles]: 1° un quart de cercle de cuivre [...]; 2° un bon instrument de passages de deux pieds [...]; 3° une bonne pendule à secondes, à verge simple ou composée [...]; 4° un compteur, vous savez que c'est un mouvement de pendule simple qui marque les minutes, se sonne les secondes, 5° une lunette ordinaire de deux pieds [...]; 6° un petit quart de cercle de 18 à 20 pouces de rayon [...]; 7° une lunette de 7 à 8 pieds, ou un télescope à réflexion de 18 pouces au moins.[22]

The quadrant with a telescope with a micrometer or a rhomboid reticle will eventually be the most versatile and widely used instrument in the observatories of the late eighteenth century[23], supplanting the mural quadrant that was difficult to build and install and above all too expensive and unaffordable for the budget of most observatories (Turner 2002; Brooks 1991).

In the eighteenth century these are the fundamental instruments that allow any astronomical observatory of that time to establish an effective astrometric program. And it is also this type of instruments that the Coimbra's Statutes of 1772 point out as the "collection of good instruments" that the University's observatory should have,

---

[19] A quadrant is a quarter of a circle, and the term refers to several different types of instruments covering an arc of that size (Bennett 1998).

[20] It is from the 8-foot mural quadrant made by George Graham in 1725 to be used by Edmund Halley (1656–1742) at the Greenwich Observatory that the model evolves, after becoming almost ubiquitous in all astronomical observatories (Learner 1981, pp. 52–72).

[21] Encyclopédie Méthodique (mat.) 1784–1789, t. II p. 481.

[22] Darquier 1786, pp. 5–7.

[23] "Le quart-de-cercle mobile est de tous les instruments d'Astronomie, celui dont l'usage est le plus ancien, le plus général, le plus indispensable, le plus commode" (Lalande 1771–1781, vol. 2, p. 743). (The corresponding Fig. (Fig. 149) of the portable quarter-circle is on p. 768).

> A mural quadrant, made it by some of the best European instrument makers, a good assortment of Quadrants; Sextants of different magnitudes, of Micrometers; transit instruments; parallactic machines; telescopes; levels; horologes and pendulum clocks [...] and everything else needed for an observatory work in behalf of the progress of astronomy.[24]

And there are also those instruments which will actually be acquired in the 1780s. The OAUC didn't actually have a mural quadrant; instead it had several portable quadrants, the most important of which is the Troughton portable quadrant (c.1787–1792).

3 From Paper to Erected Walls: The Construction of the Royal Astronomical Observatory of the University of Coimbra

3.1 The Initial Plan for the Observatory of the Castle (Not Concluded)

The site of the Castle of the city of Coimbra located on the highest hill of the city "unshaded by all parties" and not far from the University, corresponded to one of the main requirements that an observatory demands (Estatutos 1772, vol. 3, p. 214)[25].

Guilherme Elsden (?-1779), an English engineer living in Portugal since the 1760s, will be the architect in charge of the works of Reform of the University and he will develop two versions for the building project of the Observatory of the Castle[26]. In the first version he makes use of the two towers of the castle, one having a square form and the other pentagonal, those frame both sides of a three floors building (Fig. 1).

However, due to the bad structural conditions of the pentagonal tower, it had to be demolished, so Elsden advances to other version where the square tower is the central element—the tower of the observatory—, standing in middle of the building and not at the top side. This second version, the most monumental of both, will be approved by the government in the last quarter of 1773 (Fig. 2).

In 1775 (September) when the bulk of the ground floor was built "at a height not less than 8 m" the construction works stopped and would never be resumed.

---

[24] Estatutos 1772, vol. 3, p. 214.

[25] "The astronomical observatory must be unencumbered by all parties, it should dominate the horizon freely to able the observation of all phenomena which succeed in upper hemisphere. Furthermore, it should be spacious and comfortable, so several astronomers can simultaneously make their own observations." (Estatutos 1772, vol. 3, p. 214). Darquier concerning the most suitable place to install an astronomical observatory writes: "La position le plus avantageuse, pour un observatoire, seroit sans contredit d'être situe au rez-de-chaussée, isole de toute parts, & ayant un ciel découvert de tous les côtés jusqu'à l'horizon [lettre de 10 Juillet 1777]" (Darquier 1786, p. 4).

[26] In fact throughout the year 1773 there were huge uncertainties regarding the final draft of the observatory building, with several and successive plans drawn and discussed with all teachers, especially with Miguel Ciera (1725?–1782) who was at that time the professor of Astronomy.

Fig. 1 Elevation view of the observatory of the Castle (c.1773) (MNMC Inv. 2945/DA 23)

Fig. 2 Architectural plan of the main front of the Castle's observatory (September 1773) (Franco 1983)

The high cost that the construction works achieved in about two and a half years when a significant part of the building was to be built, was the main cause.[27]

Meanwhile, a small interim observatory was built in the University courtyard for teaching classes, between the years 1775 and 1777. Distant from most of the city's accesses, at a more secluded place, the University courtyard had a wide open space in the north-south direction, but some constraints in the east-west horizon because of the Joanine Library and St. Peter College. Between 1777 and 1790 this small building would serve exclusively for teaching. The construction of the Observatory of the Castle would never be resumed.

3.2 The problem of the Inexistence of a Genuine Astronomical Observatory Call for Scientific Premises

The lack of a real and effective astronomical observatory at the University to carry out a genuine scientific research requires a solution that began to be drawn around the 1780s. There are several reasons why this problem has been put on the table. On the one side external constraints: tensions between the University and the ACL, and the government final support to University's requests. On the other side reasons

[27] The Book of Expenses related to the works of the Observatory of the Castle closed the month of September 1775 with a total amount of 18879$582reis (AUC liv. R/D 1772–1775). This amount represents about 15 % of the total cost of all construction works of the University until that year.

related to scientific aspects: astronomical instruments and proper physical conditions to perform observational and theoretical work.

With the death of king José I (1777) and the consequent dismissal of the marquis of Pombal, due to the change of government carried out by Queen Maria I (1734–1816), the action of some more conservative agents, who until then had been controlled by the king, reappear. In this context the Pombaline project for the University of Coimbra was seriously threatened, because it hadn't had the time required for its solid implementation. On the other side, in the first years of Queen Maria's reign, the teaching reform shifted to a more technical education system. In 1779 the *Academia Real da Marinha* (ARM) and in 1782 the *Academia Real dos Guardas-Marinhas* (Royal Academy of Midshipmen) were created. Teaching naval science and nautical astronomy was crucial for the development of navigation and maritime trade with overseas (mainly with Brazil and Far East). The ARM was created as a theoretical teaching establishment which set out to prepare officers for the Military Navy, Merchant Navy and Army Engineers; the latter to train officers for the Portuguese Royal Navy. The study program of those academies consisted, among other matters, in theoretical and practical mathematics, navigation and nautical astronomy. In 1779, in the city of Oporto, the discipline of Sketching and Drawing related to the nautical course that was offered in this city was created. Later other higher education schools and technical and scientific institutions appeared. For example, the *Academia de Fortificação, Artilharia e Desenho* (Academy of Fortification, Artillery and Design) in 1790, the *Real Corpo de Engenheiros* (Royal Corps of Engineers) in 1790–1793, the *Sociedade Real Marítima, Militar e Geográfica* (Royal Society Maritime, Military and Geographic) in 1798, the *Arquivo Militar* (Military Archive) in 1802–1805 and the *Academia de Comércio da Cidade do Porto* (Academy of Marine and Trade of Porto city) in 1803.

It was also during the reign of Maria I, in 1779, as we referred, that the ACL was created. The founders, the most important of which is the Second Duke of Lafões, D. João Carlos de Bragança (1719–1806), influenced by the Enlightenment values, wanted the country to develop economically and socially thanks to science and technology[28]. Because of this context the University sees its unique role seriously threatened. The opening of the ACL caused multiple and well documented reactions of some professors. During the chancellorship of José Francisco de Mendonça (1725–1808), which lasted almost 7 years (1779–1786), the University faced serious problems. All the construction works that were being done at the University stopped. The construction of the Castle Observatory remains unfinished. After the dismissal of Mendonça, the University was 'delivered' (1786–1799) to Francisco Antonio Rafael de Castro (1750–1816), with José Monteiro da Rocha as vice-chancellor at his side. Under the leadership of these two men a set of attempts were made to redirect and strengthen the role and purposes of the University Reform of 1772. At this time, some of Pombal's unfinished projects are finally resumed, which was the case of the astronomical observatory.

[28] The royal denomination was only established in May 13, 1783 with the government recognition of its public utility status.

One of the driving forces for the construction of the definitive astronomical observatory of the University of Coimbra (in the middle of 1780s) was the already mentioned ACL project of 1781 for astronomical/nautical ephemeris to be published.

As we saw before, Monteiro da Rocha caused this project not to be carried out. In reality ACL's project is only suspended. About 6 years later (1787) it will be discussed again, but this time formally. In the academic session of December 5, 1787, the ACL decided to publish it. A scientific commission composed by professors of the ARM, who had graduated from the University of Coimbra, Custódio Gomes Villas-Boas (1741–1808), Francisco Antonio Ciera (1763–1814) and Francisco Garção Stockler (1759–1829), was formally chosen to elaborate a nautical almanac to be published under the auspices of ACL[29]. In 1788 the first volume of the '*Ephemerides Nauticas, ou Diario Astronomico*' (Nautical Ephemeris or Astronomical Diary, ENDA), calculated to the meridian of Lisbon for the year of 1789, was published. The ENDA were not calculated directly from the astronomical tables, but copied from the English NA with the data recalculated to the meridian of the ACL astronomical observatory as Monteiro da Rocha had proposed in the past. The ACL's observatory had been constructed in the Castle of S. Jorge, in Lisbon. The construction of this building began in 1785 and was inaugurated on January 9, 1787. Its first director was precisely Custódio Gomes Villas-Boas. In 1797 the scientific responsibility of the ENDA will be assumed by the astronomical observatory of the Royal Academy of Navy (but the ACL will remain the institution responsible for its publication)[30].

As we can see ACL's project collides with the scientific one outlined by the University Statutes for the astronomical observatory of the University, the elaboration of some astronomical ephemeris "for use of the Portuguese Navigation". This purpose was somehow threatened by the advance of the ACL's own ephemeris project.

In fact the interim observatory of the University has neither physical nor material conditions to allow an effective astronomical scientific investigation, to make observations and to do the theoretical work needed to elaborate the astronomical ephemeris. It was only used to teach practical astronomy. It doesn't have the necessary instruments to start any kind of scientific activity, and mostly it doesn't have the necessary physical conditions to accommodate and install the future instruments. Because of this Monteiro da Rocha begins a campaign to pressure the government to find a solution for the astronomical observatory at the University (around 1785–1787). This campaign implies not only a pressure to build the OAUC, but also to purchase the necessary astronomical instruments. It seems that the Royal Notice from October 1, 1787, which states the definitive construction of the OAUC,

[29] All the scientific training of these men had been made at the University of Coimbra, where they were students of Monteiro da Rocha.

[30] The construction of the astronomical observatory of the Royal Academy of Navy, "to teach students in matters of nautical astronomy", was proposed by Francisco Antonio Ciera, teacher of Navigation, in 1791. The Royal Observatory Marine was only built in 1798 (its regulation date from July 23 of that year).

is a direct consequence of the successive interpellations of José Monteiro da Rocha to the government[31].

The government formed (in 1777) after the rise to power of Queen Maria I did not contribute to give new impetus to the construction works of the University. Rather, the minister who would replace Pombal, Tomás Xavier de Lima (1727–1800), lacked commitment and persistence to make a lot of already planned public construction works. This government (1777–1788) is known for hesitating and procrastinating. Only in 1788 when another government led by José de Seabra da Silva (1732–1813), a former secretary of the marquis of Pombal, was formed a new impetus would be given to the public plans of the construction works.

It is then (in 1788) that the project to build the astronomical observatory at the University is given credit. The instruments had already been ordered from several London instrument makers (Sect. 4). Now, it is necessary to construct a proper building to install them and be able to provide working conditions for astronomers. Several projects are then conceived under the considerations of Monteiro da Rocha.

3.3 After Such a Long Wait Finally a Building: The Royal Astronomical Observatory of the University of Coimbra

The final location of the OAUC was thought to be in the southern part of the University courtyard (the interim observatory was then demolished). Guilherme Elsden's primitive design for the Observatory of the Castle is therefore definitely abandoned.

The construction plan of the OAUC was highly debated. There are several proposals of the architect Manuel Alves Macomboa (?-1815), in line with Monteiro da Rocha's considerations. In a 4 year period (1788–1792) five architectural projects are conceived (three of them in less than half a year): a first version in 1788, a second version in September 1790, a third version in November of 1790, a fourth version in February 1791 and the final one in September of 1792.

The project of the future OAUC is a very rich process of drawings, all of which have common points: the program, the location and the lasting nature of the building; what changes is its shape and the volumetric disposal.

On February 5, 1791 the University Council approves the architectural plan of the OAUC, and in 1799 the building is completed. It consists of a horizontal body with a flat roof and a central tower with three floors (a front of 41 m; side of 11 m and a height of 24 m), with specific rooms for the storage of instruments and observation, rooms for the ephemeris calculators, rest rooms for astronomers, a common living room, a library, as well as class rooms for the students (the chart-in-law of OAUC is published in December 4, 1799)[32].

31 "Her Majesty the Queen considers that this construction work will be done as you present it [...] but nothing should be done until further notice" (Almeida 1937–1979 vol. 2, pp. 177–78).

32 EAOAUC 1803, pp. iv-xii.

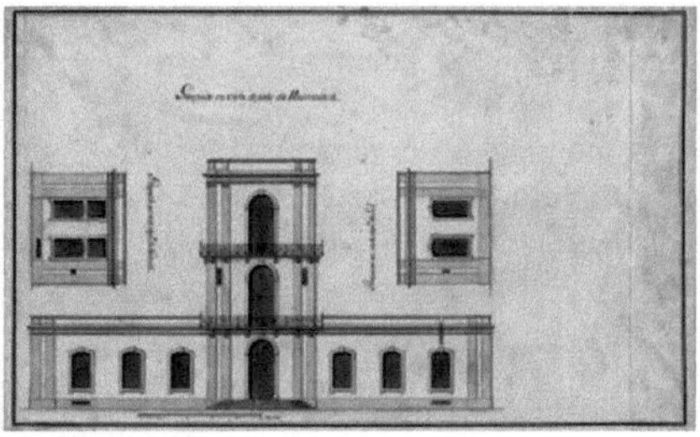

Fig. 3 View from the university's courtyard (BGUC Ms. 3377–44)

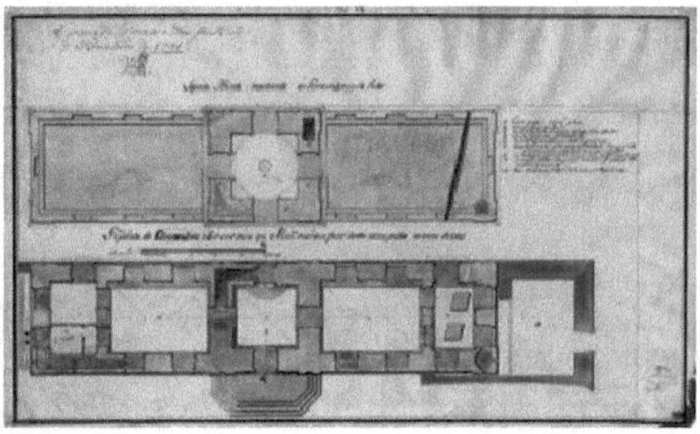

Fig. 4 "Architectural plan of the astronomical observatory that the university had to make in its courtyard in the year 1791" (BGUC Ms. 3377–44)

This final project of the OAUC is a good example of the gap between the initial ambitions (1772) of the Pombaline Reform and the new reality of the 1780s and 1790s. The astronomical observatory for the site of the Castle was thought as a building detached from all the other scientific buildings of the University. But in the project of 1792 the OAUC will end up being one of the most modest architectural designs of all. The final plan gave up the symbolic function and the urban design of the Castle's Observatory and focused its attention on a simple and functional astronomical building (Figs. 3, 4, 5 and 6).

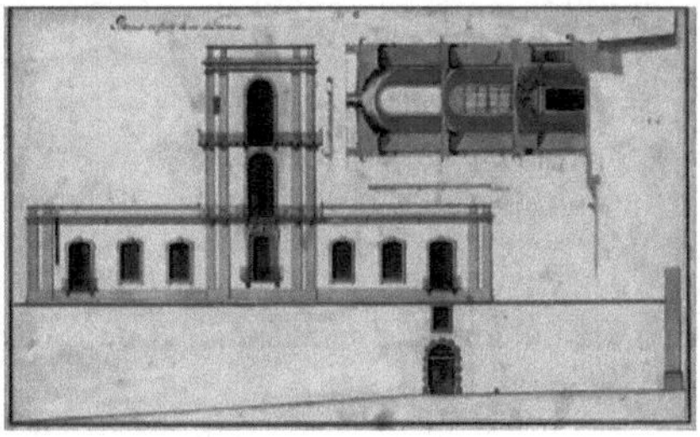

Fig. 5 View from Trindade's street (BGUC Ms. 3377–45)

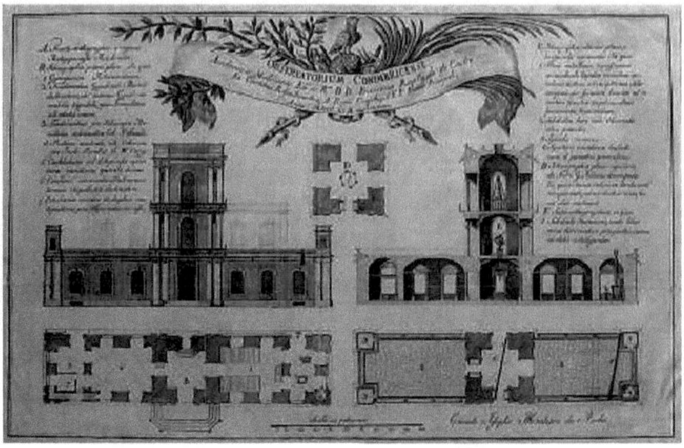

Fig. 6 "Observatorium Conimbricense Academian [...], Anno MDCCXCII" (Astronomical observatory of the Coimbra University [...], year 1792) (OAUC G-006)

4 The Acquisition of Instruments for the OAUC

The instruments commissioned by Monteiro da Rocha for OAUC are clearly marked on the architectural plan of 1792 that will serve for its construction (Fig. 7).

They are: a Troughton portable quadrant, a Dollond transit instrument, a parallactic machine from Cary and an Adams' sector.[33]

This set of instrumental levels the OAUC to the good European observatories of the time. In 1808, the Italian statistician and geographer, Adrien Balbi (1782–1848), in a visit to the OAUC says it is well built and situated, and "il était aussi-trés bien fourni d'instrumens" (Balbi 1822, vol. 2, p. 95). Lalande also says something about the OAUC and its Troughton quadrant,

> Nous avons reçu encore une description de l'Observatoire de Coimbra, par laquelle on voit qu'il y a des instruments considérables; un secteur de dix pieds, une lunette méridienne de cinq pieds, un quart-de-cercle de trois pieds et demi, divisé à Londres par Troughton.[34]

Those instruments were purchased in London through João Jacinto de Magalhães (1722–1790), a Portuguese scientist member of the Royal Society living in London[35].

[33] On the architectural plant they are labeled as: "Fundamentum Quadranti Mutrali destinatum ubi interim Quadraras mobilis tripedalis, opus Troughtoni absolutissimu; Fuandamentum pró Telescópio Meridiano acromático Cel. Dollondi; Podium australe, ubi columna Inst. Parallat. De W. Cary; Ichnographia plani superioris, ubi Sector G. Adams decempedalis, quem ternae columnae limbo ortu respiciente, ad occidentem verso, ternae aliae sustinente". They are also represented three pendulum clocks and small telesopes ("speculae minors").

[34] Lalande 1803, pp. 871–872.

[35] In English he is known as John Hyacinth de Magellan. He was born in the city of Aveiro. He joined the Order of Canons Regular of the Holy Cross with 11 years-old and it was during this time that Magellan became familiar with science, particularly astronomy. From 1758 to 1762 (he left the Order in 1758) he travels around Europe and he lives in Paris. During these years he began an important contact network with several European scientists and philosophers. In 1763 he travelled to London, where he would reside until his death. Magellan's work and notoriety earned him membership in Académie Royale des Sciences de Paris (1771) and in the Royal Society of London (1774) among others. For his life and work see (Malaquias and Thomaz 1991; Malaquias 1994).

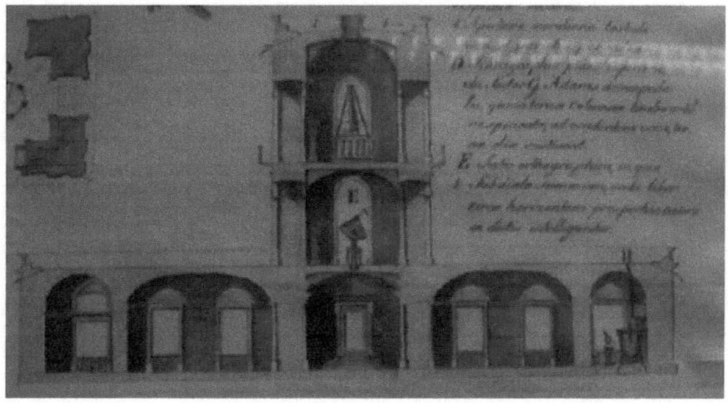

Fig. 7 Detail of the architectural plan of 1792 (Fig. 6), where we can see the instruments and its disposal on the OAUC building

Hyacinth de Magellan was the obvious choice for carrying out this service, due to his privileged personal contacts network among the leading London instrument makers. In the past (1778–1783) Magellan had already acquired instruments for the Portuguese government[36].

When Magellan settles in London the British industry of scientific instruments is at its peak. English precision mathematical instrument makers enjoyed universal reputation, "precision mechanics is an activity typically British" (Bennett 1993, 1997). All over Europe scientific establishment, observatories and experimental physics cabinets included had scientific instruments manufactured by Graham, Sisson, Dollond, Adams, Bird, Ramsden and Troughton (Bennett 1987 pp. 88–92, 113–129).

As these kinds of instruments were not immediately available for sale, they had to be ordered and then built under certain specifications defined by the customer. In a letter from March 26th 1787 Monteiro da Rocha instructs Magellan about his requirements on the stability, stiffness and division of the limbo scale of the quadrant[37].

Monteiro da Rocha wants the instrument basis to have a triangular shape instead of a quadrilateral one, "you should prevent that notwithstanding the reasons that seem contrary, the pedestal must rest on three points only, and no way in four". Monteiro da Rocha also worries about the stability (the rigidity of the axis structure of the instrument) and shape of the perfect body of the quadrant, so he demands that it be taken into account, "you for your great ability and wide experience must try that an instrument as capital like this one has all the perfection possible". Monteiro da Rocha didn't need to worry about that because Magellan was well versed in that kind of instruments, since he had already written a small essay on the quadrant, in 1779[38]. Some years later, in 1788, he sent a communication on astronomical

---

[36] Magellan collaborated with several personalities and governments (Portugal, Spain, France and Prussia) in the acquisition of instruments (Turner 1974). To Portugal, he sent several scientific instruments for the demarcation of the borders of Brazil and to an offer that queen Maria I made to the Chinese emperor Che-K'ien Long (1736–1796), and also to several Portuguese institutions (Carvalho 1990–1991).

[37] The manuscript is located in the Oxford Bodleian Library, MS. Rigaud 38, fols. 151–153verso.

[38] "Lorsque instruments ces sont bien faits, tous les & garnis avantages, dont ils ont été par les fournis Astronomes & Artistes Anglais, ils sont les plus comodes, les plus utiles, & les estimables

instruments to the Academy of St. Petersburg which was read at the meeting of April 7, 1788 by the Secretary of the Academy.[39]

Regarding the instrument's limbo, Monteiro da Rocha requires that the quadrant has two scale divisions, one in degrees and the other divided in 96 parts:

> I must remember you that it must have two divisions, one in 90° and the other in 96 parts, and must have its azimuth circle at least one foot in diameter, with its Nonius showing the minutes.

A parallactic machine "with a small telescope mounted, and a rhomboidal lattice" and a transit instrument were also commissioned. With respect to the latter nothing special was required[40].

The quadrant was ordered to the firm that belonged to the brothers John (1739–1807) and Edward Troughton (1753–1835). This firm was already one of London's most prestigious firms for the making mathematical instruments, in the mid-1780s[41].

In 1788 J. D. Cassini (1748–1825) considers the Troughton brothers the best instrument makers after Jesse Ramsden (1735–1800) (Cassini 1810, p. 30). In 1801, after the death of Ramsden, Lalande says about Edward "il est actuellement le plus célèbre artiste d'Angleterre" (Lalande 1803, p. 861)[42]. The construction of the quadrant for the future OAUC was then delivered to one of the best craftsmen of the time[43].

The instruments' arrival at Coimbra is uncertain. Between the commissioning and the instrument delivery many months or years might have gone by. Most likely, they didn't arrive all at the same time, once the order had been made to different craftsmen. Regarding the parallactic machine Monteiro da Rocha saw no problem in its arriving before the other instruments. Besides that, John Hyacinth de Magellan died in February 1790, putting an end to a privileged relationship between Coimbra and the British instrument makers.

The instruments probably arrived at Coimbra around 1797 (at the moment there is no conclusive primary font that can provide an answer to that). In 1798 the first

---

plus tous les instrumens" (Magalhães 1779, p. 25).

[39] "Le 7 Avril. Le Secretaire a lu une lettre de M. de Magellan, datée de Londres le 17 Mars, qui communique une nouvelle construction pour le Quart-de-cercle & les autres instrumens astronomiques, inventée & exécutée par M. Troughton." (Nova Acta Academiae 1790, pp. 12–13). This memory (not founded today in the Academy's archives) is possibly related with the order of the instruments for the Observatory of Coimbra.

[40] The parallactic machine which was acquired is described in the inventory of 1810 as "Parallactic Machine, construction of W. Carry. London"; the transit instrument is described in the same inventory as, "Transit Instrument, 42-inch focus, aperture 2.5, and 40 shaft. Construction of Dollond. London" (Anonymous 1810, 1824).

[41] About the Troughton's firm see (Skempton and Brown 1973).

[42] Also Marc Pictet (1752–1825), professor of physics in Geneva, refers the Troughton brothers among the best instrument makers, "Troughton me paroit jouir ici de l'une des premières réputations en ce genre." (Cited in Turner 1976, p. 4).

[43] In Allan Chapman's opinion the work of E. Troughton took the élite of craftsmanship to impressive levels of excellence (Chapman 1993, p. 418).

observations were made with the quadrant for the determination of the precise latitude and longitude of the OAUC.

5 The Astronomical Activity of the OAUC: The Astronomical Ephemeris

After its inauguration in 1799 the scientific activity of the OAUC was entirely focused on the calculation and publication of the EAOAUC[44]. The work of calculating the ephemeris demanded an intense observational activity linked to a huge theoretical work. Monteiro da Rocha as the OAUC's director was the person in charge to guide all that work. In fact all future astronomical and scientific activity of the OAUC was under his responsibility. He was not only responsible for the design and instrumentation of the OAUC, but he was also the scientific mentor behind the applied mathematical and astronomical methods, algorithms and tables that allowed the OAUC to establish and publish its most important and significant scientific production, the EAOAUC.

The activities involving the observations to establish the OAUC's geographical coordinates were made in the early 1798. The latitude was determined using 20 observations (between January 19 and February 2) of the Polar heights in its upper and lower passages through the meridian. The value obtained was 40° 12' 29.6"(40.208°N) (EAOAUC 1803, p. 240). There is a perfect match between these observations because they differ not less than 6".[45]

The determination of the longitude was achieved by comparing the observations of the solar eclipse of August 17, 1803 in Coimbra (OAUC) and Paris (at the College de France by Messier and Lalande). Monteiro da Rocha determines that there is a difference of 43 min of time (10.75°) between both meridians. Through other observations (eclipses of the Sun, stars and by satellites of Jupiter) Monteiro da Rocha had already determined that the difference in longitude between Paris and Coimbra lies between 42 m 55 s and 43 m 6 s of time (so he fixed 43 m as the average value). This value was confirmed by observations of the eclipse of August 17, 1803[46].

The first volume of the EAOAUC was published in 1803 (by the University's press) with the astronomical data for the year 1804 and providing all the conventional outputs (12 pages for each month with the position of Sun, Moon, planets and

---

[44] The chart-in-law of the OAUC clearly expresses that all activity should start with the essential tasks for the preparation of the astronomical ephemerides for the year 1804 (EAOAUC 1803, p. viii).

[45] In the 1850s and 1860s Rodrigo Ribeiro de Sousa Pinto (1811–1893), director of the OAUC between 1858 and 1866, makes with a Troughton & Simms meridian circle, purchased in 1851 and installed on OAUC in 1855, new observations and he determines the new latitude of OAUC as to be 40° 12' 25.75"N—a difference of 3.8" compared with Monteiro da Rocha's value.

[46] In 1867, using the culminating stars method Sousa Pinto determines the longitude of the OAUC as to be 33 m 34.51 s west of Greenwich, i.e. 42 m 56.51 s west of Paris (taking as the difference between Paris and Greenwich the value of 9 m 22 s) (Pinto 1887, pp. 24–35).

lunar distances). Since the first volume the EAOAUC adopted some particularities: they were calculated in reference to the mean Sun instead of the true Sun (like CDT or NA did). The EAOAUC used the 360° measure and not the widely used sign unit; they provide the lunar distances not only for the Sun and the stars but also for Mars, Jupiter and Saturn. Unlike the foreign ephemeris where the positions of the Moon were calculated for both noon and midnight directly from the astronomical tables, at the EAOAUC only the noon position was directly calculated from those tables, being the position for midnight calculated using a particular interpolation method proposed by Monteiro da Rocha (Rocha 1808, pp. 121–180). This method that used finite differences up to eighth order, also served to calculate the lunar distances to other instants (in the EAOAUC the lunar distances were tabulated every 12 h).

Similarly to the CDT, the EAOAUC also published scientific articles most of them written by Monteiro da Rocha. Some of these were translated and published in France by Manuel Pedro de Melo, *Mémoires sur l'Astronomie Pratique (Paris, 1808)*.

In 1813 Monteiro da Rocha published the *Taboas Astronomicas ordenadas a facilitar o calculo das Ephemerides da Universidade de Coimbra* (Astronomical Tables to facilitate the calculation of the Ephemerides of the University of Coimbra) (Coimbra 1813), that would be the basis for calculating the EAOAUC until the mid-nineteenth century[47].

6 Conclusion

The creation of the OAUC was fundamental for the institutionalization of astronomical science in Portugal in the second half of the eighteenth and early nineteenth centuries, a period in which astronomy, supported by major theoretical advances of celestial mechanics and of applied mathematics, tries to finally solve the major issues that it had been facing since Newton's time. These issues related to the problems of navigation, geodesy and cartography, determination of planets and comets' orbits, measurements of time, which were part of the work plan of any national observatory at that time, are also the basis for the creation and planning of the Astronomical Observatory of the University of Coimbra. Through it Portugal would tune in with the scientific Europe of the time—"All over the world Astronomy has deserved the attention of all the Kings, who built magnificent Observatories for its progress" (Estatutos 1772 vol. 3, p. 213).

The construction of the observatory began in 1773 at the site of the Castle, but due to the lack of financial means the works quickly stopped. And so the project

[47] The first volumes of the EAOAUC were calculated using the astronomical tables published by Lalande in the 3rd edition of his Astronomie (1792) (except the positions of the planet Mars that were calculated using some tables composed by Monteiro da Rocha in 1802). The positions of the Sun and Moon, published between 1807 and 1813 in the EAOAUC, were calculated upon the tables of Burg and Delambre, published by the Bureau des Longitudes in 1806.

to construct an astronomical observatory in the University was delayed for several years. In the 1780s, the problem of the non-existence of a true scientific astronomical observatory at the University began to be questioned.

There are several reasons why this problem was only then put on the table. One is related to astronomy itself: astronomical instruments and proper conditions to do observational reductions and theoretical work. There are also external constraints, namely the tensions that develop between the University and the ACL.

Monteiro da Rocha knew that only with good tools and good facilities would it be possible to undertake the astronomical project designed for University of Coimbra, of which he was the main mentor and future responsible. He wanted the project to be a national one, focused on the training of astronomers for the research and development of astronomical ephemeris, to serve the interests of the nation.

So, when the ACL threatened his project, Monteiro da Rocha turned to the government demanding that the OAUC be built. It is in this context that he decides to commission the instruments to several important manufacturers in London. The fact that he ordered the instruments before the existence of a building to shelter and install them, gave Monteiro da Rocha a strong argument for the construction of the OAUC—he has virtually all the necessary astronomical instruments but no place for them. At the end of the 1780s began the discussion around the architectural plan for the future OAUC, which will be designed taking into account those instruments. In 1790 the construction work began and in 1799 the OAUC was finished and ready for the astronomers to work in.

From its inception and throughout its history the OAUC tried to follow and contribute to contemporaneous astronomical research trends and developments. While at the beginning celestial mechanics and its applications were the institution's main research topic, in the nineteenth and twenty centuries it pursued the new avenues opened up by the development of astrophysics and in particular of solar studies.

Acknowledgments This work was carried out under a post-doc scholarship financed by the Portuguese Government through the FCT—Foundation for Science and Technology (SFRHBPD/76075/2011).

References

[ACL Ms. Azul] (1944) Arquivo da Biblioteca da ACL. "Correspondência da Academia Real das Ciências de Lisboa desde 1780 até 1790", Ms. Azul 1944

Almeida ML (1937–1979) Documentos da Reforma Pombalina (2 vols). Imprensa da Universidade, Coimbra

Andrewes WJH (1993) The quest for longitude: the proceedings of the longitude symposium, Harvard University, Cambridge, Massachusetts, November 4–6, 1993. Collection of historical scientific instruments, Harvard University, Harvard

Anonymous (1810) Catálogo i enventario no obseruatorio da universidade [manuscript], Observatório Astronómico da Universidade de Coimbra

Anonymous (1824) Inventario dos instrument. livros e moveis do Observator. R. da Universidade de Coimbra em 1824 [manuscript], Observatório Astronómico da Universidade de Coimbra

[AUC liv. R/D] (1772–1775) Arquivo da Universidade de Coimbra. Livro de Receita e Despesa da Universidade: 1772–1775

[AUC IV-1ªE-8-3-4] Arquivo da Universidade de Coimbra. Legislação Académica, AUC IV-1ªE-8-3-4

Balbi A (1822) Essai Statistique sur le Royaume de Portugal et d'Algarve (2 vols). Rey & Gravier, Paris

Beales D (2005) Enlightenment and reform in eighteenth-century Europe. I.B. Tauris & Co. Ltd, New York

Bektas MY, Crosland M (1992) The Copley Medal: the establishment of reward system in the Royal Society, 1731–1839. Notes and Records of the Royal Society of London 46:43–76

Bennett JA (1987) The divided circle, a history of instruments for astronomy, navigation and surveying. Phaidon-Christie's, Oxford

Bennett JA (1992) The English quadrant in Europe: instruments and the growth of consensus in practical astronomy. Journal for History of Astronomy 23(1):1–14

[BGUC Ms. 3377] Biblioteca Geral da Universidade de Coimbra, Secção de Reservados. "Planta do Castelo e Casas a ele contíguas em a Universidade de Coimbra [Guilherme Elsden, 1773]", Ms. 3377

Boistel G (2001) L'Astronomie Nautique au XVIIIème siècle en France: tables de la Lune et Longitudes en Mer. vol 2. [Ph.D. thesis, University of Nantes], Nantes

Brooks RC (1991) The development of micrometers in the seventeenth, eighteenth and nineteenth centuries. Journal for History of Astronomy 22(68):127–173

Burns WE (2003) Science in the enlightenment, an encyclopedia. ABC-Clio Inc, Santa Barbara

Calafate P (1990) Iluminismo em Portugal. In: Logos: Enciclopédia Luso-Brasileira de Filosofia vol 2, pp 1316–1326, Verbo. Lisboa

Carolino LM (2012) Measuring the heavens to rule the territory: filipe folque and the teaching of astronomy at the Lisbon polytechnic school and the modernization of the state apparatus in nineteenth century Portugal. Science & Education 21:109–133

Carvalho R (1985) A astronomia em Portugal no séc. XVIII, Biblioteca Breve. Lisboa

Carvalho R (1990–1991) As requisições de "instrumentos matemáticos" dirigidas de Lisboa a João jacinto de Magalhães. Memória da Academia das Ciências de Lisboa, Classe das Ciências 31:129–173

Carvalho FR (2008) Um Iluminismo Português? A reforma da Universidade de Coimbra (1772). Annablume, São Paulo

Cassini JD (1810) Mémoires pour servir a l'Histoire des Sciences et e celle de l'Observatoire Royal de Paris [...]. Bleuet, Paris

Chapman A (1993) The Victorian amateur astronomer: William Lassell, John Leach, and their worlds. In: Moore P (ed) 1994 yearbook of astronomy. Sidgwick & Jackson, London, pp 159–177

Darquier A (1786) Lettres sur l'Astronomie Pratique. Didot, Paris

Daumas M (1972) Scientific instruments of the 17th & 18th centuries and their makers. B. T. Batsford, London

Encyclopédie Méthodique (1784–1789) Mathématiques, par D'Alembert, Bossut, Lalande, Condorcet, et al. (3 vols). Panckoucke, Paris

Estatutos da Universidade de Coimbra (1772), 3 vols. Imprensa da Universidade, Coimbra

Figueiredo FB (2005) A contribuição de José Monteiro da Rocha para o cálculo da órbita de cometas [MSc. Thesis, New University of Lisbon] (unpublished)

Figueiredo FB (2011) José Monteiro da Rocha e a actividade científica da 'Faculdade de Mathematica' e do 'Real Observatório da Universidade de Coimbra': 1772–1820. Ph.D. thesis, University of Coimbra (unpublished)

Franco MS (1983) Riscos das obras da Universidade de Coimbra: o valioso álbum da Reforma Pombalina. Museu Nacional de Machado de Castro, Coimbra

Freire FC (1872) Memoria Histórica da Faculdade de Mathematica nos cem annos decorridos desde a Reforma da Universidade em 1772 até o presente. Imprensa da Universidade, Coimbra

Hankins TL (1999) Science and the enlightenment. University Press, Cambridge

Hellman CD (1932) John Bird (1709–1776) mathematical instrument-maker in the strand. Isis 17(1):127–153

Hutchins R (1999) British University Observatories c.1820–1939: Ideals and resources (2 vols). Ph.D. thesis, University of Oxford, Oxford, British Thesis Service

Lalande J (1771–1781) Astronomie, par M. de La Lande (4 vols). Veuve Desaint, Paris

Lalande J (1803) Bibliographie Astronomique, avec l'Histoire de l'Astronomie depuis 1781 jusqu'à 1802. L'Imprimerie de la Republique, Paris

Laplace PS (1835) Exposition du système du Monde. Courcier, Paris

Laplace PS (1878–1882) Mécanique Céleste, Oeuvres complétes de Lapalace, publiées sous les auspices de L'Académie des Sciences. Gauthier-Villars, Paris

Learner R (1981) Astronomy through the telescope. van Nostrand Reinhold Company, New York

Lemos Fde (1980) Relação Geral do Estado da Universidade [facsimile of the 1777 manuscript]. Imprensa da Universidade, Coimbra

Livingstone DN (2000) Making space for science. Erdkunde (Archive for Scientific Geography) 54(4):285–296

Loewy M (1881) Observatoires astronomiques de province, année 1880. Impimerie Nacionale, Paris

Longitude Act (1714) Anno Regni Anna Regina Magrna Britannia, Francia, & Hibernia, Duodecimo. At the paliament summonned to be Helda t Westminster, the Twelfth Day of November, Anno Dom. 1713. [...]. John Baskett, London

Lovell B (1994) The Royal Society, the Royal Greenwich Observatory and the Astronomer Royal. Notes and Records of the Royal Society of London 48(2):283–297

Magalhães JJ (1779) Description et usages des instrumens d'astronomie et de physique, faits Ã Londres, par ordre de la cour de Portugal en 1778 et 1779. P. Elmesley et W. Brown, London

Malaquias I (1994) A obra de João Jacinto de Magalhães no contexto da ciência do séc. XVIII. Ph.D. thesis, University of Aveiro (unpublished)

Malaquias I, Thomaz F (1991) Scientific communication in the XVIIIth century: the case of Jean-Hyacinthe Magellan. Physics 31:817–834

Maxwell K (1995) Pombal, paradox f the enlightenment. University Press, Cambridge

Misa TJ (1988) How machines make history, and how historians (and others) help them to do so. Science, Technology, & Human Values 13(3/4):308–331

[MNMC Inv. 2945/DA 23] Museu Nacional Machado de Castro. "Por Ordem do Ill.mo e Ex.mo Senhor Marques de Pombal, Ministro e Secretario de Estado Visitador da Universidade de Coimbra, Plenipotenciário, e Lugar Tenente de Sua Majestade na nova fundação da Universidade de Coimbra Projecto Principal do Observatório Astronómico da mesma Universidade [Elsden, c.1773]", MNMC Inv. 2945/DA 23

Monteiro NG (2007) The patterns of Portuguese politics in the 18th century or the shadow of Pombal. A reply to António Manuel Hespanha. e-journal of Portuguese History 5(2):1–6

Nova Acta Academiae Scientiarum Imperialis Petropolitanae (1790), tomus VI praecedit historia eiusdem Academiae ad annum MDCCLXXXVIII, Petropoli Typis Academiae Scientiatum 4:12–13

[OAUC G-006] Arquivo da Biblioteca do OAUC. "Observatorium Conimbricense Academian Moderante Ex.mo ac Rmo D. D. Francisco Raphaele de Castro Ex Comitibus Resendiensibus, A Regiis Consiliis, S. E. P. Lisbon principali, Anno M.DCC.XCII exstructum [1792]", G–006

Pereira M (2008) Um manuscrito de cerca de 1767, do P. José Monteiro da Rocha, S.J. com uma solução matemática para a obtenção da longitude pelas distâncias lunares. Cuadernos de Estudios Borjanos 50–51:339–394

Pinto RRS (1887) Estudos instrumentaes no observatorio astronomico da universidade de Coimbra. Imprensa da Universidade, Coimbra

Pond J (1806) On the declinations of some of the principal fixed stars; with a description of an astronomical circle, and some remarks on the construction of circular instruments. Philosophical Transactions Royal Society of London 96:420–454

Rocha JM (1808) Mémoires sur l'Astronomie Pratique par M. J. Monteiro da Rocha. Courcier, Paris

Rocha JM (1813) Taboas Astronomicas ordenadas a facilitar o cálculo das Ephemerides da Universidade de Coimbra. Imprensa da Universidade, Coimbra

Rocha JM (n.d.) Methodo De achar a Longitude Geografica no mar e na terra Pelas observaçoens e calculos da Lua Para o uso da navegação Portugueza pelo P. Jozé Monteiro da Rocha [manuscript], Colecção Pombalina Ms. 511, Biblioteca Nacional de Portugal
S.A. [Hyp.: José Monteiro da Rocha] (1803) Ephemerides Astronomicas do Real Observatório da Universidade de Coimbra. Imprensa da Universidade, Coimbra
Schubring G (1997) Analysis of Historical textbooks in Mathematicas. PUC, Rio de Janeiro
Skempton AW, Brown J (1973) John and Edward Troughton, mathematical instrument makers. Notes and Records of the Royal Society of London 27(2):233–249
Turner GL'E (1976) The London trade in scientific instrument-making in the eighteenth century. Vistas in Astronomy 20:173–82
Turner AJ (2002) The observatory and the quadrant in eighteenth-century Europe. Journal for History of Astronomy 33(4.113):373–385
Wolf C (1902) Histoire de l'Observatoire de Paris de sa fondation à 1793. Gauthier-Villards, Imprimeur-Libraire, Paris

Jean Hellot and 18th Century Chemistry at the Service of the State

Rémi Franckowiak

Abstract The paper proposes a re-examination of Jean Hellot's status in relation to the *Académie des Sciences* but also in relation to the public authorities and to the matter of alchemy. The result will be a re-evaluation of the role of the *Académie* as a unique highly qualified scientific authority in the development of the economy of France in the middle of the eighteenth century, as well as a re-evaluation of alchemical practices perceived from a strictly economic standpoint.

Keywords Jean Hellot · Science academy · Chemistry · Alchemy · State · Economic development · Social scientist status · Eighteenth century

Jean Hellot (1685–1766) is a somewhat forgotten character of the history of science[1]. Since the end of the 1960's some studies were published highlighting his major role in the development of the dyeing, metallurgical and porcelain work in France, through his essential contribution to the progress of what was called chemical technology[2]. Indeed since 1752 Hellot was general inspector of the French royal

[1] This paper is a work in progress.

[2] The expression 'chemical technology' was proposed by Todericiu (1975) *Chimie théorique et appliquée au milieu du XVIIIe siècle. Œuvre et vie de Jean Hellot (1685–1766)*, PhD thesis, University of Paris-Sorbonne [handwritten 522 p. and annexes]. It was taken up in Wisniak (2009) Jean Hellot. A pioneer of chemical technology, *Revista CENIC Ciencias Químicas* 40/2:111–121. About Jean Hellot, see also Thuillier and Birembaut (1966) Une source inédite: les cahiers du chimiste Jean Hellot (1685–1766), *Annales. Economies, Sociétés, Civilisations*, 21/2:357–364; and the other articles of Todericiu (1976) La bibliothèque d'un savant chimiste et technologue parisien du XVIIIe siècle. Livres et manuscrits de Jean Hellot (1685–1766), *Physis. Rivista internazionale di Storia della Scienza Firenze*, 18/2:198–216; the same (1976) Correspondance de Jean Hellot, *Revue de Synthèse* 97/81–82:129; the same (1977) Le traité de chimie inachevé de Jean Hellot (1685–1766), *Physis. Rivista internazionale di Storia della Scienza Firenze*, 19/1–4:355–375; the same (1983) Les mines du pays de Liège dans les papiers du Savant Français Jean Hellot (1685–1766), *Technologia*, Histoire des techniques, archéologie industrielle 6/2:61–68; and d'Albis A. (1983) Procédés de fabrication de la porcelaine tendre de Vincennes, d'après les livres de Hellot, *Faenza* 69/3–4: 202–216.

R. Franckowiak (✉)
Department of Physics, Lille 1 University Science and Technology, Villeneuve d'Ascq, France
e-mail: remi.franckowiak@univ-lille1.fr

R. Pisano (ed.), *A Bridge between Conceptual Frameworks,* History of Mechanism and Machine Science 27, DOI 10.1007/978-94-017-9645-3_10

porcelain factory. In 1750 he was the author of a book about the diffusion of the best scientific and technical knowledge of dyeing, *L'art de la teinture des laines et des étoffes de laine en grand et petit teint, avec une instruction sur les déboüillis*. In 1750 and 1753 he was also the author of an important two-volume book, based on the reshaped translation of Schlutter's book *De la Fonte des mines, des fonderies, &c.*, in order to train the workers of French mines, to encourage exploitation of the kingdom's mines and to prepare future directors of mines to exploit most effectively and economically their mines. So Hellot's image established by the historians of science is an image of a chemist who strongly condemned alchemy and was fully involved in the development of craft and factory production during the eighteenth century through his participation as an expert in a lot of committees set up by the *Académie Royale des Sciences* at the public authorities' request but also through his appointment in a personal capacity as an adviser of the Controller-General of Finances.

What is proposed in this paper is a re-evaluation of Hellot's status in relation to the *Académie des Sciences* that he integrated rather late inlife—when he was 50 years old—but also in relation to the public authorities and to the matter of alchemy. The result will be a re-evaluation of the role of the *Académie des Sciences* as a unique highly qualified scientific authority in the development of the economy of France in the middle of the century, as well as a re-evaluation of the alchemical practices perceived from a strictly economic standpoint. Not only did Hellot advise ministers independently of the ordinary procedures instituted within the Science Academy—not without some tensions with its perpetual secretary—but his alchemical competence could serve the public authorities. Hellot indeed seemed to have fully decided on and accepted a role as a public servant, charged by the State to study all proposals concerning mine exploitation—for which good alchemical knowledge could be sometimes necessary—as well as those concerning the production of several arts and crafts sectors. Thus, thanks to Hellot, the State systematized the use of scientific expertise, standardizing expertise procedures and tending to make official the function of science advisor. The State actually substituted the expertise of Hellot for that of members of the *Académie des Sciences* with respect to functions which were until then exclusively its own.

It should be noted that Hellot was not so different from the other Academy chemists he knew: Geoffroy, Grosse, Duhamel and Du Fay; however his involvement was deeper. What joined these men together, apart from an obvious friendship, was the combination of the highest level recognition of their scientific activities with alchemical practices and a full submission to the interests of the State.

The permanence of alchemical topics at that time is not surprising. This even allowed, especially since the bankruptcy in 1720 of the economist, then Minister of Finance, John Law's system, the development of deceit that the State wished from now on to reduce in order to re-orientate the financial investments of the kingdom's subjects to the profit of, for example, the exploitation of mining resources technically and legislatively well defined. In a general way, the State had to base a part of its efforts for the development of the economy of France on the competences of the *Académie Royale des Sciences*' members and, in the case of metallurgy, on that of its chemists who also had knowledge in the field of alchemy. In this chemists' circle, plenty of time was devoted to these activities; this was especially the case of Jean Hellot. They worked sometimes together, they helped one another, they exchanged manuscripts and quoted one another willingly but they obviously had limited contacts with those who were not so much devoted to the State and the Public: Lemery, Rouelle and Venel among others.

1 Hellot and the *Académie des Sciences*

Jean Hellot was short, plump with lively eyes. He was sometimes roguish in his words, but never ironic. He was destined for an ecclesiastic career, but he discovered chemistry quite early through papers and library of his grandfather, whose name was also Jean Hellot and who had translated in 1651 the 1633 Latin book of William Davisson with the title *Éléments de la philosophie de l'art du feu ou chimie*; so, his interest in chemistry had been already excited by quite old chemistry texts. The perpetual secretary of the *Académie Royale des Sciences*, Grandjean de Fouchy, in the funeral eulogy that he composed for Hellot's death, remarked, on the 9th April 1766, that Hellot devoted himself to chemistry, "unreservedly", during his youth; this brought him quite early close to Claude-Joseph Geoffroy, who married, in his second marriage, Hellot's niece in 1729. Hellot left afterwards for England in order to meet the scholars of the Royal Society.

Hellot became also editor in the *Gazette de France* from 1718. As he was severely struck by the bankruptcy of John Law's financial system, Hellot found himself constrained to continue this job until 1732. Then, according to Grandjean de Fouchy, freed from this constraint as he was,

> his friends thought he would claim a position in this Academy & encouraged him to put himself in for it: we say 'his friends' because he was not one of those who, with false modesty, have confidence in their own merit, & he, more than anybody else, needed someone to have designs & ambition for him […].[3]

According to the minutes of the *Académie*'s sessions of 16 February 1735, it was the pensionnaires and the associated chemists of the *Compagnie* who proposed Hellot's candidature,[4] namely Geoffroy, Du Fay, Boulduc and Lemery; the election of Hellot was validated on the 2nd of March 1735.[5] Almost 4 years later he became directly supernumerary pensionnaire of the *Académie*—in which he became twice the

---

[3] "ses amis le crurent en état de prétendre à une place dans cette Académie, & l'engagèrent à s'y présenter: nous disons ses amis, car il n'étoit pas de ceux qui, sous une fausse modestie, savent se faire confidence à eux-mêmes de leur propre mérite, & il avoit besoin plus que personne qu'on se chargeât d'avoir des vues & de l'ambition pour lui […]", Grandjean de Fouchy (1976) Éloge de M. Hellot, *Histoire et Mémoires de l'Académie Royale des Sciences*, 169.

[4] To substitute La Condamine nominated associate. Two other people were actually proposed too: Haber and de la Rivière.

[5] See Procès-Verbaux de séances de l'Académie Royale des Sciences (1735), t. 54, ff. 35r et 41r.

assistant director (1750 and 1763) and twice also director (1751 and 1764)—and then Fellow of the Royal Society (1740).[6]

However, we have to point out that his election had been well-prepared by his friends. In fact we may consider that Hellot's first paper at the *Académie* had been given before his election. In 1734, Duhamel and Grosse had devoted one third of their dissertation on the liqueur of Frobenius to an extract of a letter that Hellot had recently sent to Duhamel.[7] In this letter, we learn that Hellot had worked on this liqueur at least since the year 1731, that was just a little time after Frobenius had sent several little flasks of his liqueur to Etienne-François Geoffroy, as the dissertation revealed anyway.[8] The dissertation refers also to the works of Claude-Joseph Geoffroy, his great friend, concerning this liqueur since 1731.

Jean Hellot was already closely related for quite a long time to Claude-Joseph Geoffroy,[9] Du Fay, Duhamel du Monceau and Johann Grosse, when he integrated the *Académie des Sciences* in 1735. These chemists, all of them, shared an obvious interest in alchemical works on metal transmutation. And yet, all these chemists took part, in a privileged manner, in the economic development effort of France initiated in particular by the Regent, Philippe II Duke of Orleans. Philippe, very interested in alchemy, placed directly under his protection, by 1715, the *Académie Royale des Sciences* which he wanted to make an essential instrument of his projects of reform and re-establishment of the kingdom, especially through the inquiry that he launched from 1716 to 1718 in order to create a national economy to replace the local economies.[10] So the Duke of Orleans appears as a key character of the evolution of eighteenth century chemistry: he named academicians people of his entourage related to alchemy, who in their turn proposed candidates at the *Académie* when a place in the class of chemistry was released. The result was the constitution of a core of alchemical chemists in the *Académie* who all conformed to the requirement of utility and subordination to the interests of investors looking for

[6] On 23 October 1740, see The record of the Royal Society of London for the promotion of nature knowledge (1940), London, 1940, 406.

[7] Duhamel and Grosse (1734) Recherche chimique sur la composition d'une liqueur tres-volatile, connuë sous le nom d'Éther, *Histoire et Mémoires de l'Académie Royale des Sciences*, 41–54.

[8] Grosse, as for him, received from the German chemist Godfrey Hanckwitz, living for long time in London, who left in the *Philosophical Transactions* (May 1730) a dissertation on the subject.

[9] Here is what was written by Monnet A. G., *Démonstration de la fausseté des principes des nouveaux chimistes, pour servir de supplément au "traité de la dissolution des métaux*, an 6, 360–361: "Hellot: simple homme de lettres d'abord, et rédacteur de la *Gazette de France*. En fréquentant les Geoffroy, ses parents, et en les voyant opérer dans leur laboratoire, qui étoit alors le rendez-vous de tous ce qu'il y avoit à Paris des savans en chymie, il prit du goût pour cette science, suivit leurs travaux, s'attacha principalement à ceux qui se faisoient sur les minéraux. Il devint essayeur des mines de France, sous les contrôleurs généraux des finances Orry et Trudaine le père, et entra à l'académie des sciences. Il composa un traité des teintures par ordre du gouvernement, en rassemblant tous les procédés en usage chez les ouvriers, et mit en françois le *Traité de la fonte des mines*, par Schlutter, qu'avoit traduit de l'allemand [Koenig], directeur des mines de Poulluen en Bretagne, et obtint une pension considérable de l'état."

[10] See Demeulenaere-Douyère and Sturdy (2008) *L'enquête du Régent 1716–1718. Sciences, techniques et politique dans la France pré-industrielle*, Éditions Brepols, Turnhout.

scientific and technical competences provided by the *Académie des Sciences*. The most representative character here was Jean Hellot, practising alchemist and collector of alchemical texts.[11]

By the way, the link between science and development of the economy was also clear from the opposite direction, from politics towards science. It should be indeed recalled that the Chancellors of the Exchequer or Intendants of Finances Fagon, Trudaine, Orry, Le Pelletier, Marchault d'Arnouville, Moreau de Séchelles, Bertin, Rouillé were all of them honorary members of the *Académie des Sciences*, even sometimes vice-presidents and presidents of the *Académie.*[12]

After having recalled the State's interest in the *Académie*, we should go back to the first years of Hellot's academic career. Du Fay, general inspector of dyeing, intendant of the *Jardin des Plantes* and close colleague of Hellot, died on the 16th of July 1739. Hellot was designated his legatee. In spite of his regular attendance, he was absent during the *Académie* sessions of the 22 and 29 of July. On the 19th of August, he was proposed, as along with two other academicians, for the available post of pensionnaire after the death of Du Fay. Hellot was absent again two more times, on the 2nd and the 5th of September. Only 3 months later, on the 18th of November 1739,[13] Hellot was nominated by the king as supernumerary pensionnaire. The exceptionally long delay for the nomination (usually from 4 to 15 days), Hellot's repetitive absences and at the same time the exceptional status of a supernumerary, without having passed through the intermediate grade of the associate, allow us to suppose that during this lapse of time Hellot had been contacted by the State, he had accepted to take over Du Fay's responsibilities and to commit himself entirely to the service of the Controller-General of Finances and of the Intendant of Finances; and during this lapse of time the negotiations with the *Académie* must have taken place so that he may be proposed a compensation by this exceptional status offering him a quite comfortable position.

Hellot's particular situation, inside and outside the *Compagnie*, caused tension at least on two occasions. The *Académie* was very often asked by a Minister or the State services to study several dissertations, books, machines, substances and natural or artificial matters (hundreds of times from 1730 to 1760). A committee was then designated by the director of the *Compagnie.*[14] For instance, on the 7th of January Hellot and Geoffroy were designated commissioners to study Sieur Kemerlin's pewter, as it had been demanded by the minister Maurepas; but contrary to the usual procedure, Hellot refused to hand his report over to the secretary of the *Académie*, who recorded the event; Hellot wanted to communicate it directly to Maurepas. He finally handed his report over to the secretary Dortous de Mairan

[11] See Franckowiak R., forthcoming in 2014 in *Almagest.*

[12] But not Herault, the State Counsellor who gave the authorisation of 1738 in favour of Chéron to produce artificial realistic tin, as we will see further in this paper.

[13] See respectively the Registres Procès-Verbaux de séances de l'Académie Royale des Sciences (1739), t. 58, ff. 143r, 145r, 213r et 221r.

[14] See Hahn (1994) *L'Anatomie d'une institution scientifique: l'Académie des sciences de Paris, 1666–1803*, (1st English edition in 1969), Paris, Yverdon, Éditions des archives contemporaines.

a week later.[15] It should be noted that a first study for the same pewter had taken place 2 years earlier but in very different conditions, as the commissioners reported. A judgment of the court of the Mint of 23 September 1739 had ordered Kemerlin to reproduce his experiment of purification of the pewter before his lawyer, the general judge of the Mint named Bichon, a clerk of the court, the king's prosecutor and a potter worker known as Charles Julien de Quevanne, appointed as an "expert" by the court; the experiment took place 2 weeks later in Kemerlin's laboratory, Baffroy Street, Faubourg St Antoine. A report was made and completed the next day after some more analyses by Quevanne in his workshop during the night. This report concluded in a few lines only that the examined pewter was pure, so it did not contain other metals nor harmful substances.[16] Hellot and Geoffroy were surprised by the rapidity of the test and by the absence of precise information concerning the chemical analysis; they, in contrast, made their test during a period of 3 months before concluding with a very long, thoroughly detailed report. The difference between these two procedures is astonishing: on the first case, it was based on testimony, in a private laboratory, by sworn officers but ignorant of chemistry, with just the opinion of a worker, and on the second one, the procedure took place in the *Académie des Sciences*, by order of the minister, with specialists in chemistry who took the time to ask, in their turn, the opinion of a worker and to make all the necessary experiments themselves, so as to produce an indisputable report.[17]

Another tension with the *Compagnie* broke out in May 1749: by the end of September of the previous year, Hellot had recommended the academician Macquer to reveal nothing to the *Académie* concerning his discoveries on Prussian blue, thinking that they would interest the State; he also advised him to report them first to the Controller-General of Finances in Fontainebleau. Hellot made on the 26th of March 1749 a public confession, during a session, concerning his initiative, mostly in order to certify the paternity of Macquer on those discoveries, and not really by remorse regarding his colleagues (a controversy started then with the abbot François

[15] See the Registre des Procès-Verbaux de séances de l'Académie Royale des Sciences (1741), t. 60, 6–8, about this affair which took place between the 6th and the 14th January.

[16] See Jean Hellot's manuscript, *Notes et observations des mines de France*, Bibliothèque Mazarine, Ms. 2756, pp. 133–148.

[17] From the report of Hellot and Geoffroy (registre des Procès Verbaux de séances de l'Académie Royale des Sciences (1741), p. 10): "Nous aurions fort souhaité que le Sr Charles Julien Quevanne, nommé par le Prevost General des Monnoyes et Mareschaussée de France, pour examnier cet étain, eut voulu dire par quels moyens il s'est trouvé en état dans un espace de temps de 7 ou 8 heures, de rapporter à cet officier qu'après plusieurs operations par lui faites sur une plaque d'étain tirée et coulée le jour d'auparavant vers les 8 heures du soir, d'un plus grand bain dans le Laboratoire [on the fringe: "Extrait du Procès Verbal"], du Sr de Kemerlin, que *cet étain est privé du cri qui fait connoître l'acide corrosif que l'étain contient ordinairement; qu'il est sans melange d'aucuns metaux ni d'aucune chose nuisible à la santé, ne contenant uniquement que sa propre substance epurées, que par la calcination, dudit étain on n'apperçoit aucun melange de teinture, et que les fleurs en sont d'une blancheur parfaite*. La route qu'il a suivie dans son examen, seroit elle aussi sûre qu'elle est courte. Quoi qu'il en soit nous allons donner le détail des experiences que nous avons precedemment annoncées."

Ménon[18]); and Macquer read of course his dissertation on this subject only the next April 16 in session, praising in passing Hellot's scientific qualities.[19] Macquer, by the way, inherited in his turn Hellot's activities after the latter's death, assuring for the State the continuity of his functions from at least 1730 (with Du Fay) to 1780.

2 Hellot and the Creation of a True Social Status for the Scientist

Except for the two mentioned books (which were actually ordered by the Controller-General of Finances) and a dozen or so published dissertations in the *Histoire et Mémoires de l'Académie Royale des Sciences*,[20] most of Hellot's scientific production is only accessible in various handwritten archives: the bulky annual *Registres de Procès-Verbaux de séances de l'Académie Royale des Sciences*, the 8552 notes and observations of the *Papiers* bought by Trudaine from Hellot's widow and today kept in the public library of Caen,[21] a thousand or so pages of notes and reports about the mining exploitations in France kept in Mazarine Library,[22] and the work documents about porcelain factory. All these documents are well known but they were only superficially studied by the historians of science who preferred to outline an image of Hellot from his public positions of organizer of mine, dyeing and porcelain industries. However, while Hellot entered the service of the State he achieved a status which exceeded his simple status of scholar. He became a man on whom the State relied to carry out some necessary transformations for economic development of the country and to allow the rise of French industry. Hellot embodied the new relations between sciences, technology and politics, taking part in a certain

[18] On this case, see Lehman C. (who nevertheless did not mention Hellot's role in the delay for Macquer's dissertation présentation in the *Académie*): Lehman (2012) L'art de la teinture à l'Académie royale des sciences au XVIIIe siècle, *Methodos* [on line], 12|2012, URL: http://methodos.revues.org/2874; DOI: 10.4000/methodos.2874.

[19] See the Registre of Procès-Verbaux de séances de l'Académie Royale des Sciences (1749), t. 68, pp. 86 and 204–208.

[20] 1735: "Analise chimique du zinc. Premier mémoire", pp. 12–31; "Analise chimique du zinc. Second mémoire", pp. 221–243; 1736: "Conjectures sur la couleur rouge des vapeurs de l'Esprit de Nitre & de l'Eau-forte"; 1737: "Le phosphore de Kunckel, analyse de l'urine", pp. 342–378"; 1738: "Sur du Sel de Glauber trouvé dans le Vitriol sans addition de matiére étrangere", pp. 288–298; 1739: "Sur la liqueur éthérée de M. Frobenius", pp. 62–83; 1740: "Théorie Chymique de la Teinture des Etoffes", pp. 126–148; with L. Lemery, C.-L. Geoffroy "Examen du Sel de Pecais", pp. 361–370; 1741": "Théorie Chymique de la Teinture des Etoffes. Second mémoire", pp. 38–71; 1746: with Camus "Sur l'Etalon de l'aune du Bureau des Marchands Merciers de la ville de Paris"; 1756: "Sur l'exploitation des Mines"; 1763: with Tillet and P.-J. Macquer "Mémoire sur les essais des matières d'or et d'argent", 1–14; with Duhamel and de Montigny "Sur les vapeurs inflammables qui se trouvent dans les Mines, de charbon de terre de Briançon".

[21] Jean Hellot, Ms. Varia 140, *Papiers*, 9.

[22] Jean Hellot, Ms. 2755: *Préface de la traduction de Sclutter et Recueil d'indication des mines de France*, 188p.; and Ms. 2756: *Notes et observations sur les mines de France*, 474.

decline of the scientific production of the chemists inside the *Académie des Sciences*[23] because of their increasing investment in many committees ordered by the *Compagnie* or in different businesses outside of the *Académie*. The scattered tasks of the academician chemists corresponded to the multiplication of chemical activities in the French Society where at that time more and more people not belonging to the *Académie* had chemical abilities thanks to the strong popularization of chemistry in particular by the means of the many public and private chemistry courses in France.[24] In the case of chemistry the *Académie des Sciences* was not anymore completely able to maintain its monopoly in the field of legitimate scientific authority, in control of scientific production and technical innovation. It seemed to have often followed, administratively, the evolutions.

Hellot was, above all, a servant of the State. Being a member of the *Académie des Sciences* since 1666—the creation date of this institution—meant sharing pure interests in Science; academicians were supposed to work only for the progress of their discipline and not for personal interests, unlike the common people who were always suspected of looking for profit.[25] However from Hellot's time, unselfish scientific work for an academician could mean work for the interests of the State, whose purpose was to use science to rationalize craft and factory practices. So Hellot worked to satisfy such interests. Hellot inaugurated changes in the social representation of the academician-scholar who now had a true function in French Society. No longer being simply isolated authorities in the field of sciences, nor a distinct and separate clan, nor an elite group existing far away from vulgar people and enclosed in their *Académie*, Hellot and the Academy's chemists now reached out into the social world and acquired a true social status in France. It was the consequence of the social recognition by the State of the relevance of their participation in the activities of French Society.

3 Hellot, Adviser to the Controller-General of Finances

Jean Hellot spent most of his time conscientiously studying all the dissertations and requests, on questions of crafts and trades, addressed to Ministers—in particular the Finance Minister—in order to determine their worth, to allot permissions to

[23] The decline in the number of chemical publications of the *Académie des Sciences* was quite strong in the middle of the century. Whereas the average over the century of the published chemistry dissertations in the *Histoire & Mémoires de l'Académie Royale des Sciences* represented around 13 % of the whole of the published dissertations, their number at the end of the 1740's and the beginning of the 1750's fell to a few units a year (before going up actually to 40 % of all the dissertations at the end of the 1770's with the polemical Lavoisier's work); see Halleux et al. (2001), *Les publications de l'Académie royale des sciences de Paris (1666–1793)*, t. 2 (étude statistique et index généraux), Brepols, Turnhout, pp. 24–25.

[24] See Perkins (2010) Chemistry Courses, the Parisian Chemical World and the Chemical Revolution, 1770–1790, *Ambix*, 57/1, 27–47.

[25] Heinich (1987) Arts et sciences à l'âge classique, *Actes de la recherche en sciences sociales*, 66–67, 73.

exploit a mine or to work ores, to evaluate the cost of such exploitations and even the scientific and technical competences of the applicants, as well as to carry out tests, to order inquiries, to get information about mining practices in foreign countries, to intervene in the training of mining engineers, to propose modifications in legislation,[26] to improve the security of workers, to supply them with a solidarity fund in case of accident or death, to limit pollution and consumption of wood, to fight against fraud by the manufactured goods or craftsmen productions, and to protect the public health. Hellot's work had one goal: working toward the economic development of France.

From the very first pages of Jean Hellot's manuscript entitled *Notes et Observations sur les Mines de France*, the expectations of the State for Hellot's work are clear. These pages are about the search for gold spangles accumulated in the sand of the rivers in the Cevennes and the Pyrenees, and correspond to Hellot's reports to the Chancellor from 08/10/1751, 31/05/1752, 03/12/1753, 19/06/1754 and 22/08/1754 about several requests of the abbot de Gua, Jean-Paul de Gua de Malves[27] who was at the beginning of the adventure of the famous *Encyclopédie ou Dictionnaire raisonnée des sciences, des arts et des métiers*. Hellot's function was here—as he said himself—"to help the Minister to make the wisest decisions".

First Jean Hellot received—not necessarily directly—a dissertation of de Gua on the exploitation of gold spangles from rivers. He presented the content of the text to the Chancellor. Then he explained in detail the historical and scientific hypothesis about the topic and went further into de Gua's arguments. Lastly Hellot expressed his favourable opinion:

> [...] I believe [...] that we can risk some expense to clear up an assumption which is not unreasonable [...]. I conclude in that way only by supposing that we could make sure first, at little expense, of the reality of the deposit of those gold materials in the pits [...].[28]

But he warned that the demand from de Gua, who wanted too many workers and too much equipment, should not be fulfilled to the letter: "what, in my opinion, presupposes a very considerable expense". So Hellot proposed to de Gua to work differently by carrying out some tests in a less expensive way, with a kind of drill:

> Consequently it will only cost the expenses for the drill and the trip of the author of the project who undoubtedly will be delighted if the Intendant appoints 2 or 3 people, who can certify the success of his operation, to accompany him.[29]

---

[26] On the evolution of mining legislation in France and the role played by Hellot, see Birembaut (1964) L'enseignement de la minéralogie et des techniques minières, *Enseignement et diffusion des sciences en France au XVIIIe siècle*; and Todericiu (1983).

[27] See also Condorcet (1786) Éloge de M. L'abbé de Gua, *Histoire et Mémoires de l'Académie Royale des Sciences*, 63–76.

[28] Hellot, *Notes*, 2. "[...] je crois [...] qu'on peut risquer quelque depense pour s'éclaircir sur une supposition, qui n'est pas déraisonnable [...]. Je ne conclus ainsi qu'en supposant qu'on pourroit s'assurer d'abord à peu de frais de la réalité du dépôt de ces matiéres aurifiques dans les fosses [...]".

[29] Hellot, *Notes*, 2 "Alors il n'en coutera que les frais de la sonde et les frais du voyage de l'auteur du projet qui sans doute sera charmé que Mr l'Intendant le fasse accompagner par 2 ou 3 personnes qui puissent certifier le succès de son operation".

Hellot recalled what Réaumur had written in his dissertations of the *Académie des Sciences* about orpailleurs' work and listed what was the most necessary to start the tests:

> Here is, as I believe, all that he needs, and 7 to 8 thousand pounds could be enough for the continuation of these experiments until the end of September[30].

Afterwards Hellot received, through the Chancellor and a tax farmers-general, another handwritten document with some samples from de Gua's tests. So he carried out some chemical experiments to assess how much gold they contained and he reported his conclusions to the Minister. The amount of gold in the samples was enough to start the exploitation of the spangles.

Hellot's following report was addressed to Trudaine, Intendant of Finances. He summed up the affair and expressed some doubts about de Gua's ability to find partners in order to form a company which had to be accepted by the Council of State.[31] After a new letter of de Gua sent to the Controller-General of Finances, Hellot proposed to allot him at last the permission to exploit only for 15 months a well-defined area provided that he and his partners compensated the owners of this area for the land and the rivers, secured their exploitation and sent every 3 months to the Council of State a certificate of a public officer about the progress of their work.[32] Only after this period, if the Intendant of Languedoc thought that their exploitation was really profitable, could they continue and go further. Then after 5 years, they would pay taxes to the king.

In other affairs reported in his manuscripts, Hellot considered on the contrary that the dissertations he received from the Controller-General of Finances were so ridiculous or suspect that he advised that no answer at all be sent to the applicants.

4 Hellot and Alchemy[33]

In the second third of the eighteenth century, the necessity to develop the economy of France was very strong. Nevertheless, alchemical ideas seemed still rather strongly widespread in French Society, which was why Venel made the report in 1753 of an ignorance of what chemistry was and of confusion among people, even educated, between a chemist and a researcher of the Philosophical Stone.[34] It became then

---

[30] Hellot, *Notes*, 3. "Voilà, à ce que je crois, tout ce qu'il lui faut, et 7 à 8 mille livres pourront suffire pour une suite de ces experiences jusqu'à la fin septembre".

[31] Hellot, *Notes*, 4.

[32] Hellot, *Notes*, 5.

[33] The topic of alchemy will be tackled here through the reports ordered by the Controller-General of Finances or other Ministers to Hellot, within the framework of his functions of adviser. The personal alchemical interest and practices of Hellot will be the subject of a following paper which will be published in *Almagest*, 2014.

[34] Venel (1753) Chymie, *in* Diderot D., Le Rond D'Alembert J., *Encyclopédie, Dictionnaire raisonné des Sciences, des Arts et des Métiers*, t. III, 408. See Rémi Franckowiak (2009) La chimie

urgent and important to put order in this confusion: that is to reduce charlatanism in order to protect the State and the public, to divert the people from hazardous alchemical investments towards reliable metallurgical and mining initiatives, while sometimes putting worthy alchemical discoveries to the test. In fact, there were not really at that time two different chemistries, but only one which affected the economic Power and, for this reason, no "useful discovery" was being refused. In a way the State could encourage alchemical research for which the hope was legitimate; alchemy then was especially judged by the measure of its economic profitability.

For example, we can read in Hellot's manuscripts of the following affair.[35] In March 1749, a chemistry laboratory was denounced at the Mint. It was suspected of producing false money. The laboratory was established in a large, rather beautiful, house, with a porte cochère, at number 27 of the faubourg St Lazare in Paris. Fourteen furnaces with fire were found. All was seized and soldiers were called to stand guard in front of the house. On Monday, March 31, Mr. Gouault, public prosecutor of the court of the Mint, came to lead Jean Hellot to that address, with the consent of Mr. Rouillé, the Chancellor of the Exchequer, to whom Jean Hellot was answerable. Jean Hellot found there on the right side of the ground floor a chemistry laboratory made up of three rooms. Mr. de Bazinghen, advisor and reporter of the court of the Mint, Mr Renard de Petiton director of the Mint of Paris, and Mr Mayol apothecary opposite to La Madeleine were already on the spot. Hellot then dictated to the clerk of the court of the Mint a description of the furnaces, hermetically closed vessels, utensils, retorts and matters that were there. He concluded his official report by affirming that nothing reprehensible was to be noticed: the people in this laboratory just worked the 'high chemistry', and more precisely they had the aim of extracting "the soul from March and Venus" for the improvement of imperfect metals. So this affair was not a big deal for Hellot, but the court of the Mint wanted to make it important. All the accused people were questioned: Jean-Baptiste Fleuron one of the craftsmen, Jean Thevenot a craftsman in charge of the register of the laboratory, an Italian Joseph Melavi director of the laboratory, Barthélémy Chéron owner of the laboratory and ex-banker imprisoned for five years in the prison of the Abbey of Saint Germain for debts, and his partners: the Knight Roch Eugène de Plessier d'Attencourt, Jean Charles Lucquet de Perceville, usher of the King and police chief of the gendarmerie, and finally Mr. Le Gendre d'Arményy, brother of Mme Crozat.

The question that actually had to be answered was the following one: Did this alchemical work contravene the privilege granted to Chéron on February 11th, 1738? Because the laboratory was perfectly legal in fact. An authorization in due form had been delivered for the conversion of iron into steel and especially for the production of genuine tin, similar to Cornish tin, but without using tin ores, in other words a production of artificial tin. Hellot did not hide his perplexity about the reason of the issue of such an authorisation because he refuted the idea that it was possible to produce tin starting from other things than tin ores. And saying that all the work

dans les dictionnaires et encyclopédies au XVIIIe siècle: une incuriosité peu philosophique, *Corpus, revue de philosophie*, 56, 37–57.

[35] Hellot, *Notes,* pp. 410–415.

carried out in this laboratory remained strictly within the framework of the 1738 authorisation, as Chéron defended himself, could not be a good enough argument because, as Hellot noticed: "this answer in order to divert the idea of an alchemical work is alchemical itself and full of pure gibberish, incomprehensible even for alchemists". And for Hellot it was necessary to denounce the "alleged alchemists".

Looking at the 20 letters, on philosophical gold, the universal medicine, the putrefaction of silver, etc., found in the laboratory which Hellot had also to appraise and explain, he showed that this laboratory was actually intended for the preparation of what "the authors of alchemy name Specifics,[36] that is operations supposed to make the alchemists rich, but much more slowly than with the Philosopher's stone." And Hellot continued by saying that "In the hands of suspect people, these processes are used to amuse the credulous ones and to allow these suspect people to live at the expense of the latters."

At the beginning, Hellot thought sincerely that Chéron had duped his partners into financing his alchemical research in the hope of all of them becoming rich. However, after having piled up some information, Hellot understood that the Knight d'Attencourt, Le Gendre d'Arményy and Perceville, certainly duped by Chéron, duped in their turn other people into giving them money. However, Hellot became more and more disappointed: actually, everybody was implicated in this affair:

- Further to the intervention of the public prosecutor Gouault, the judgment published by the Mint on May 19th, 1749 was not the judgment delivered by the first president of the court of the Mint. Chéron's furnaces, ingot moulds and crucibles were broken but vessels, matters, and utensils were returned to him.
- Around the end of the month, Chéron escaped from prison and left for Germany with the Knight d'Attencourt to live at a princess' expense.
- Perceville boasted that he would open soon another laboratory in a place where he would not be worried.
- Trudaine, Intendant of Finances, said to Hellot that: "It was necessary to let them do it without appearing to know; madmen as they are, these alchemists are able to find sometimes good things etc."
- Then, on Friday, May 31, Bazinghen, advisor of the Mint, "came to Hellot's home in order to be given back the retort full of cooked yellow vitriol which Hellot had taken with him, in order to examine it, the day of the inventory". Then Bazinghen confessed to him that the retort he returned him had cost him 1200 pounds.
- Finally, the following day, via the clerk, the public prosecutor of the Mint, Gouault, tried to corrupt Hellot. He wanted most probably Hellot not to make waves nor talk around him about this affair. No doubt, Gouault had invested also some money in Chéron's laboratory.

This story is derived from Hellot's handwritten copies of reports and letters during his activities as adviser of several Ministers of the mines of France. There are other such accounts, like Hellot's appraisal, ordered by the Chancellor of the Exchequer

[36] "Particuliers" in French.

himself, of an artificial silver production just based on the heating of mercury with very high temperature and strong pressure.[37] The appraisal concluded of course that it was a failure but the operation was worth carrying out because, according to Hellot, Wilhelm Homberg would have had a similar idea in one of his dissertations read at the *Académie des Sciences.*[38] Another account? This one about a certain Pillon who would have been in touch with a secret chemistry laboratory at Barrière de St Jacques in Paris and who referred to a charlatan and wine dealer, called Soulatre, in league with some Parisian goldsmiths, and who worked with Madame d'Urfé,[39] known for her propensity towards alchemy and especially for huge amounts of money she gave to some adventurers such as Casanova, the Count of Saint German and Cagliostro.

5 Conclusion

Hellot's involvement at the service of the State led to systematization of the use of scientific expertise and standardization of expertise procedures. The science advisor to business of the public had now become an almost official position. While the chemist had a recognized function in French Society—a true social status—the *Académie des Sciences* lost a part of its legitimacy on the matter of science. From now on, the chemist was not completely anymore in possession of the scientific authority because he was belonging to this institution but he could claim it independently.

Moreover, the standardization of the expertise procedures concerned indiscriminately a charcoal mine exploitation in Northern France (Anzin) for example and the production of artificial gold or silver in a discreet Parisian laboratory. In the French Society of the second third of the eighteenth century, alchemy was a still widespread practice in which the State could be interested if it was supposed profitable. At that time a more dangerous practice than chemical deceits was rather a badly-exploited mine. The State—with Hellot's help—had the will to regulate all craft and factory practices, alchemical as well as mining. Nevertheless it was much more worrying to the State to risk spoiling all its efforts in encouragement to exploit mines by the behaviour of incompetent people, than to let probably erroneous alchemical doctrines be studied and practiced. The necessity to develop the economy of France was very strong and thus it appeared urgent to put the crafts in order, to reduce the number of harmful projects and activities of interest to the State, to divert hazardous investments of swindlers to well-managed metallurgical and mining works, while on occasion it could be interesting to put to the test some metal transmutations judged possibly profitable.

---

[37] Hellot, *Notes*, pp. 407–410.

[38] We note here the specific interventions of Camus, academician of the class of Geometry, and Baron who republished 14 years later Lemery's *Course of Chymistry*.

[39] Hellot, pp. 415–416.

References

Birembaut A (1964) L'enseignement de la minéralogie et des techniques minières. In: Enseignement et diffusion des sciences en France au XVIIIe siècle. Hermann, Paris

Condorcet N (1786) Éloge de M. L'abbé de Gua. Histoire et Mémoires de l'Académie Royale des Sciences 63–76

d'Albis A (1983) Procédés de fabrication de la porcelaine tendre de Vincennes, d'après les livres de Hellot, 69/3–4. Museo Internazionale delle Ceramiche di Faenza, Faenza, pp 202–216

Demeulenaere-Douyère C, Sturdy DJ (2008) L'enquête du Régent 1716–1718. Sciences, techniques et politique dans la France pré-industrielle. Brepols, Turnhout

Duhamel H-L, Grosse J (1734) Recherche chimique sur la composition d'une liqueur tres-volatile, connuë sous le nom d'Éther. Histoire et Mémoires de l'Académie Royale des Sciences 41–54

Franckowiak R (2009) La chimie dans les dictionnaires et encyclopédies au XVIIIe siècle: une incuriosité peu philosophique. Corpus, revue de philosophie 56:37–57

Grandjean de Fouchy J-P (1766) Eloge de M. Hellot. Histoire et Mémoires de l'Académie Royale des Sciences 169–179

Hahn R (1994) L'Anatomie d'une institution scientifique: l'Académie des sciences de Paris, 1666–1803, (1st English edition in 1969). Éditions des archives contemporaines, Yverdon

Halleux R, Mc Clellan J, Bebariu D, Xhayet G (2001) Les publications de l''Académie royale des sciences de Paris (1666–1793), t. 2 (étude statistique et index généraux). Brepols, Turnhout

Heinich N (1987) Arts et sciences Ã l'âge classique. Actes de la recherche en sciences sociales 66–67:47–78

Hellot J (1750a) L'art de la teinture des laines et des étoffes de laine en grand et petit teint, avec une instruction sur les déboüillis. Veuve Pissot, Paris

Hellot J (1750b) De la Fonte des mines, des fonderies, & c. Pissot La Veuve, Hérissant JT, Pissot fils, Paris

Hellot J (1753) De la Fonte des mines, des fonderies, & c. Hérissant JT & Pissot JN, Paris

Hellot J Notes et observations des mines de France. Bibliothèque Mazarine, Paris, p 474. Ms. 2756

Hellot J Préface de la traduction de Sclutter et Recueil d'indication des mines de France. Bibliothèque Mazarine, Paris, p 188. Ms. 2755

Hellot J Papiers, Bibliothèque Municipale de Caen, Mss. Varia 140, vol 9

Lehman C (2012) L'art de la teinture Ã l'Académie royale des sciences au XVIIIe siècle, Methodos [on line from 10 April 2012], 12|2012, URL: http://methodos.revues.org/2874. doi:10.4000/methodos.2874. Accessed 21 Dec 2013

Monnet AG (1797–98), Démonstration de la fausseté des principes des nouveaux chimistes, pour servir de supplément au raité de la dissolution des métaux

Perkins J (2010) Chemistry courses, the Parisian chemical world and the chemical revolution, 1770–1790. Ambix 57(1):27–47

Procès-Verbaux de séances de l'Académie Royale des Sciences, Archives de l'Académie des Sciences, Mss, 1735, t. 54

Procès-Verbaux de séances de l'Académie Royale des Sciences, Archives de l'Académie des Sciences, Mss, 1739, t. 58

Procès-Verbaux de séances de l'Académie Royale des Sciences, Archives de l'Académie des Sciences, Mss, 1741, t. 60

Procès-Verbaux de séances de l'Académie Royale des Sciences, Archives de l'Académie des Sciences, Mss, 1749, t. 68

The record of the Royal Society of London for the promotion of nature knowledge, The Royal Society, London, 1940

Thuillier G, Birembaut A (1966) Une source inédite: les cahiers du chimiste Jean Hellot (1685–1766). Annales. Economies, Sociétés, Civilisations 21(2):357–364

Todericiu D (1975) Chimie théorique et appliquée au milieu du XVIIIe siècle. Œuvre et vie de Jean Hellot (1685–1766). PhD thesis, University of Paris-Sorbonne [handwritten 522p. and annexes]

Todericiu D (1976a) La bibliothèque d''un savant chimiste et technologue parisien du XVIIIe siècle. Livres et manuscrits de Jean Hellot (1685–1766). Physis 18(2):198–216

Todericiu D (1976b) Correspondance de Jean Hellot. Revue de Synthèse 97(81–82):129

Todericiu D (1977) Le traité de chimie inachevé de Jean Hellot (1685–1766). Physis 19(1–4):355–375

Todericiu D (1983) Les mines du pays de Liège dans les papiers du Savant Français Jean Hellot (1685–1766). Technologia: Histoire des techniques, archéologie industrielle 6(2):61–68

Venel G-F (1753) Chymie. In: Diderot D, Le Rond D'Alembert J (eds) Encyclopédie, Dictionnaire raisonné des Sciences, des Arts arts et des Métiers. Tome III, p. 408.

Wisniak J (2009) Jean Hellot. A pioneer of chemical technology. Revista CENIC Ciencias Químicas 40(2):111–121

Engineering Creativity: An Essay on Epistemological Analysis

Elena Alexandrovna Gavrilina

Abstract The article contains an evaluation of the place of technical reality and the role of engineering creativity that acts as its integral component. The place of creativity is represented in different phases of engineering activity; the factors that influence modern manifestations of creative activity in engineering are reviewed. The author proves that engineering activity in the present-day world acquired a number of peculiar features and specific characteristics. Also is substantiated the thesis statement on the involvement of human beings in designed technical systems not only as a subject of activity but as a structural element of designed system as well, i.e. as an object for manipulation. The necessity of understanding engineering practices as socially determined and socially significant is reasoned.

Keywords Engineering creativity · Epistemological analysis · Technical reality · Engineering activity · Technical systems · Engineering practice · Society

1 Introduction

Humankind (even at the level of collective consciousness) is already almost completely aware of and reconciled to the fact that, in the contemporary world, the functioning of a society as a whole, as well as of separate social institutions and a specific person, is performed as part of an unprecedented complex multi-component structure. This structure is a technical reality. The same structure also becomes one of the dominant factors of social and personal development. As a relatively independent sphere of society life it includes the following components:

1. subject of activity;
2. activity itself aimed at creation and usage of a technical environment;
3. substantive results of the activity (artifacts) as socio-cultural values;
4. system of relationships between subjects which arises in this sphere in connection with technical artifacts, methods of activity on their basis etc.;

E. A. Gavrilina (✉)
Bauman Moscow State Technical University, Moscow, Russian Federation
e-mail: gavrilina@bmstu.ru

R. Pisano (ed.), *A Bridge between Conceptual Frameworks,* History of Mechanism and Machine Science 27, DOI 10.1007/978-94-017-9645-3_11

5. system of relationships between the results of such an activity as artificial environment and a natural world;
6. system of relationships between subjects and objects of technical activity in the context of the feedback principle.

Technical reality is increasingly contrasted with both nature and human beings. But at the same time a certain mode of human existence and nowadays also the destiny of civilization are connected with technology. Therefore, an ever increasing role in this reality is played by engineering activity, which creates artificial technical (in the broadest sense of the word) systems. And creativity—creative work—is an integral part of engineering activity.

2 Creativity as an Integral Part of Engineering Activity

In order to prove the previous thesis statement let us review the role of creativity at different stages of engineering activity.

Recognition of a need and definition of design objectives as well as functions of created technical objects always are the first stage of any engineering activity. This stage is to a large extent definitive but at the same time it is the least formalizable. The step of function description for a technical object is performed intuitively in many respects, on the basis of the designer's experience. Development of design specification on the basis of the list of requirements serves as a result of the work. It is perfectly obvious that the role of creative component at this stage is extremely high. Quite often in this step works are performed without financing, frequently accompanied by absolute indifference or even opposition from milieu and further development of works is possible only provided that the author of an idea possesses enough emotional fervour, creative inspiration and enthusiasm to overcome these obstacles.

At the next stage additional engineering research is often needed for a solution of a formulated technical problem. This stage of an engineer's work is akin to scientific inquiry and basically solves the same problem of elaboration of new knowledge, yet in the sphere of technical sciences and engineering activity. This is undoubtedly a creative work, although for its implementation and support it is necessary to perform some routine tasks like information support, calculations in compliance with known methods, and standard experimental research.

3 Designing, Invention and Construction in Engineering Activity in the Light of Creativity

Upon availability of clearly-worded requirements for a technical object (outcome of the first stage) and results of engineering research (the second stage) it is possible to begin the design of a technical object as such. A design objective is to create an object that satisfies certain requirements and at the same time possesses necessary qualities in the presence of time and resource restrictions. Design to a wide extent organizes object-making activity.

However, design was not immediately singled out as an independent element of engineering activity. Originally an engineer's activity was mainly intended for handling of material objects. And the key stages of an engineer's work were invention and construction.

Invention is the most representative of all kinds of engineering activity. The word "engineer" for good reason originates from Latin *ingenium*, which means abilities, talent, inventiveness, as well as a smart idea (Latin-English Dictionary n.d.). Undoubtedly, this activity is creative but as contrasted with creativity in other scopes of activity, it is performed not only in the ideal space of abstract models and symbols but also to a large extent in the space of real technical objects, artifacts, life experience and physical laws.

However, one can't say that an inventive act is oriented only to pragmatic aspects and practical usefulness of each specific invention. Invention is always creation of something that wasn't before, that never existed earlier. Therefore, it is always a breakthrough to ideality.

Furthermore, any invention is based upon overcoming of some contradiction: between productivity and costs, between complexity and reliability, between manufacturability and quality factors of a technical object. In other words, there are always contradictions between operating efficiency of some technical equipment and available resources, between ends and means.

Overcoming of these contradictions requires implementation of creative action, insight. At all times it was possible only for a few, as analogous to creation of a piece of literature, art or music. However, in the twentieth century some general provisions of creative task solution theory were at last formulated (Altshuller 1973; Polovinkin 1988 etc.). Of course this does not mean mass training in creativity, but it is quite possible to help in mastering skills for creative labour. Since invention is not only a flight but also a huge amount of routine work, Edison described it as follows: "Genius is one percent inspiration and ninety-nine percent perspiration."

Construction is a sort of a bridge from modeling of a technical device or a system in terms of semiotics to technology of its manufacturing as an element of the material world. Although it still only happens on paper, designer already must give specific features to designed items. He defines its mass, dimensions, constructional materials from which it will be manufactured, technologies of their processing etc. At this point past experience (as a designer, technological, operating) as well as the creative component of activity play significant roles.

4 Producibility vs Creativity in Engineering Activity

Then starts the technological stage. An engineering technologist estimates producibility of a designed product, requirements for equipment and materials. Strictly speaking, only at this stage is further viability of a future device or system determined.

Besides, not all the information concerning new technology gets included in patents and descriptions, many nuances, small details, secrets, the so-called know-how turn out to be disguised. All of these of course prove the creative nature of an engineering technologist's labour.

Therefore, it can be concluded that engineering activity is inseparably linked to creativity.

> Creativity is the prerogative of a free person, capable of self-development. Creativity is a mode of personal existence, as opposed to impersonal action, which kills personality in its utterly "purified" kind[1].

In the broad sense creativity is the activity which gives rise to new knowledge or products. With such an approach creativity in engineering activity may be regarded as a process of intensive professional activity of an engineer during which his independence, initiative, need for fuller realization of his knowledge, skills and abilities, continuous professional advancement, use of unconventional solutions etc. are shown to the full.

5 Creativity: Activity Aspect

In the strict sense of the word creativity is the activity which gives rise to something totally new, original and inimitable, as well as notable for its social and historical uniqueness and usefulness.

In the context of the science and technology sphere, such activity results in new knowledge, creation of totally new samples of equipment and technology, discoveries and inventions.

From the philosophical point of view any human activity represents an act of creation, because both subject and object exit the act of activity being different from the state in which they entered it (Bergson 2005). However, for the convenience of analysis, differentiation of thinking activity to creative and reproductive is fully justified and it is carried out on the basis of both objective and subjective criteria:

1. An activity is called creative when it leads to a new result or new product.
2. Because a new result may be achieved accidentally or by means of continuous non-heuristic enumeration of possibilities, the criterion of novelty of process with the help of which this product was obtained (new method, approach, modus operandi) is usually added to the criterion of result novelty.

   The process or result of an act of thinking is called creative only if it can't be obtained as a result of a simple formal conclusion or action in accordance with an algorithm. In case of a genuinely creative act, a formal logic gap is overcome on the way from the statement of a problem to its solution. Overcoming of this gap is possible with the use of irrationality, intuition, and non-formal (dialectical) logic.

[1] Smirnov 2001, p. 145.

3. Creative thinking is usually associated with ability to singlehandedly reveal and formulate a problem rather than with solution of a problem that was already set by somebody else. Mathematicians state that mathematical talent manifests not in the skill of solving mathematical problems (practically anybody can be trained in that), but above all in the ability to formulate in mathematical language a problem taken from real life or other area of knowledge, i.e. to set it as a mathematical problem.
4. The presence of pronounced emotional experience preceding the moment of finding a solution of a problem is an important psychological criterion. The role of emotional experience, particularly sense of beauty that directs the solution search itself, was also noted by the famous French mathematician H. Poincaré (Poincaré 1913).
5. The act of creative thinking usually requires stable and lasting or more short-term but very strong motivation.

In general our view is close to the position of P. Tillich who stated that "a human being lives creatively in various spheres of meaning." By the word "creatively" he denotes not the innate ability natural for a genius but life in immediate action and reaction regarding culture content. The one who lives creatively inside meanings establishes oneself as a participant of these meanings. He establishes himself as the one who creatively perceives and alters reality (Tillich 1959).

6 New Technologies and Creation

It is possible to distinguish the following main directions of influence of new developing technologies (above all information technologies) on some aspects of human creativity:

- At present there appear new materials, means and modes of creative process implementation. Appearance and development of a technical base (writing system, magnetic recording, radio and computer technologies etc.) changes the forms of creative process. New technologies frequently significantly expand creative potential, increasing searching opportunities and expressiveness for writer, musician, artist, scientist, and engineer. But the same technologies often restrict creative pursuit by defining a corresponding "format" of creative activities and its products.
- Owing to the change of creative processes, forms and modes of existence for creative activity products also change. For example literary and art, information and even scientific works acquire forms of hypertext and electronic hypertext, accompanied by multimedia. Existing and emerging musical compositions are increasingly transferred and presented in digital form. Painting, photography and technical drawing are digitized. New modes of existence for textual, graphic, musical and other products make them widely and easily accessible in computer

networks. This accounts for growth of reproducing, copying and transportation of works of authorship, which in its turn leads to depersonalization of works, their increasingly independent objective and material existence.

- Capabilities of modern facilities for analog and digital transmission and storage of information facilitate to a great extent growth of audiences, popularization of some kinds of activity and increase of their social significance. The existing technology not only changes conditions and scope of works distribution but also makes them generally accessible, generally comprehensible, decreasing sometimes the level of expertise required and removing many "restrictions" from perception and comprehension. For example, one does not need to be literate to perceive TV and radio programmes. Due to changes in the structure of perception, greater information saturation, and greater accessibility of creative works, social consciousness is altered and changes occur in the concept of creativity such as notions of what kinds of activity and works can be creative. Besides, some kinds of activity in collective consciousness are strongly associated with the status of "creative" labour. Development of advanced information technologies made possible the appearance of new kinds of creativity (electronic music, video art, computer graphics, telecommunication art, web design etc.).
- Development of information technologies changes opportunities for mystification of creativity. Walter Benjamin (1968) stated that authenticity and originality lose their meaning when an artist starts to use methods to which multiplicity is inherent. For example, he noted that technically one can make an infinite number of photographs (prints) from photographic negatives and it is meaningless to recognize authenticity and originality of one of them. In the age of digital reproduction not only the idea of genuineness gets modified, but also the idea of authorship of work, the paradigm of creativity itself changes. We fill our life space with "simulacra" (J. Baudrillard 1994). Each product of creativity, each work ceases to be absolutely unique and becomes relatively unique because of the possibility to reproduce or interpret it infinitely. If previously an artist created work, text and meaning, now he or she creates context that allows recipients of informational messages to perform reconstruction of meanings to the best of his or her education and cultural range. Furthermore, under present-day conditions, results are much more quickly mastered socially, and also the "creative standard" forms more quickly. Human beings switch over from subject activity to subjective one (Batishhev 1997), but at the same time remains within the confines of the defined "standard" that becomes mass adopted very quickly.

Subject activity	A person's ability to put and to correct goals, to realize motives, to build actions independently and to assess their consequences.
Subjective one	Attitude to something defined by personal tastes, preferences, beliefs, fears, lack of objectivity.

- As follows from the above, creative processes become mass adopted; therefore, one can notice the appearance of a new paradigm of creativity. In a classic model of creative activity an artist possesses the privileged status of creator of a work—message (painting, text, new scientific theory etc.). Such is the modern reality that the artist acts only as an initiator of generation of some creative field, open for other individuals to join, in which creative interaction is performed. In the first case, work has a material, a sign and semantic integrity. And in the second case work itself doesn't exist, a system of meanings and senses is not represented, but demonstrates the progress of an existing constructive process. There is no concept of author in the new paradigm; it is replaced by the concept of authorship. As far back as in the 1960s, M. McLuhan (1995) pointed out the collective nature of the new information forms of creativity, comparing them with folklore and finding in them similar attributes (polyvariance, anonymity and absence of copyright claims).

The new paradigm of creativity is in opposition to the classic paradigm, based on the author-public dichotomy that emphasizes privileged position of creator-prophet, who is always unique, exceptional and special. The new paradigm partially demystifies the role of artist and reduces it to the function of initiator.

7 Influence of Technologies on Engineering Creativity

Engineering activity is not an exception. In the modern world it has acquired a number of new features and specific characteristics. Let us describe those that are conditioned by development of information technologies and their global nature. With the appearance and distribution of the latter, such an ontological property of engineering activity as modeling of the world around human being gained a new meaning. The conventional approach to the construction of technical systems, the boundaries of which were marked by the machine frame walls, underwent a number of qualitative modifications.

Separation of designing as an independent element within the confines of engineering activity became an essential factor for understanding of the nature of these modifications. With the development of the notions of designing in engineering activity there occurred a transformation from immediate direct material influence on the material world to modeling of this world in sign, semiotic systems (Rozin 1997).

In the late twenty—early twenty-first century due to accelerating development of information technologies the application area of system designing expanded. It already includes not only industrial production but also other spheres of social practice (service, consumption, teaching, management etc.). The new property of sociotechnical engineering activity product manifests in the fact that now it can't be "touched" as a single-piece item as it was in conventional engineering activity (Stjopin et al. 1995, p. 370).

8 The Human Being as an Element of Sociotechnical Systems and Engineer's Responsibility

A modern engineer frequently performs modeling of real-world objects through the use of new information systems by means of a computer that becomes in many cases an irreplaceable instrument for implementation of a number of intellectual and design operations (algorithmizable calculations and problem solutions, drawing activity etc.). The use of computer technologies allowed creation of new methods of information structuring and management, organization of new communication possibilities etc. Computer games and conventions are realities that not only immerse a user in a special eventful world almost without differences from artistic worlds, but also allows him to take an active part in the events.

Technologies of virtual reality offer new opportunities of handling sign systems by creation of human-computer-machine symbiosis in which human functions are paradoxical: on the one hand, the human being as a virtual user is an element of such symbiosis; on the other hand, he remains able to, as a human, use it for his own purposes. In other words, technologies of virtual realities allow creation of new semiotic systems by including in them aspects and fragments of live human activity and behaviour. There is a possibility of including the human being and his activity in machine systems and computers in human activity.

Transition to semiotic practices and "dematerialization" of engineering activity includes the human in designed systems not only as a subject of activity but also as an object for manipulation, i.e. as a structural element of a designed system.

However, it is the capability of distancing from one's own existence and objectification of oneself as an object of the external world that serves as a prerequisite for technical intervention in human life interconnections. One of the tendencies for such changes consists in attempting to improve quality of life by making the natural technically reproducible. But there appear a number of problems connected with conceptual incompleteness of models made by researchers. In its turn, it depends on both peculiarities of human intellectual apparatus and conceptual incompleteness of humanity's knowledge about the real world.

Indeed, aspiration for solution of new problems on the basis of more and more complicated models leads to necessity or receiving and processing of increasingly more complicated and less exact information. The more complicated a modeled system is, the quicker its capability to formulate precise meaningful statements about its behaviour decreases, up to some threshold beyond which preciseness and meaning become mutually exclusive. Imprecision, inconsistency, incompleteness of information is also explained by imperfection of measuring devices and the fact that in many cases an expert serves as the only source of information (Bagdasarian 1998). And considering that modern level of development technology is incomprehensible for a normal human: opening of a device does not reveal a system of rods and levers comprehensible for normal experience; the connection between pressing a key and a result reveals features of magic action, and we come to an understanding of the fact that the area of uncontrollable consequences will grow further.

On the other hand, in modern scientific society one can more and more often come across the ideas of methodological pluralism and as a result development of notions of multiobjectivity and multimodelity of studied and designed objects. In its turn it leads to conceptual incompleteness and limitedness of any object, among other things owing to the complexity of designed objects. Therefore, a design approach leads to limitation of recognition of an engineer's responsibility for objects, models, schemes that were designed by him and comes into conflict with the modern concept of social control over technology.

9 Virtualization and Dematerialization of Engineering Practices

"Dematerialization" of modern engineering practices now gains almost literal meaning. Previously engineers and designers chose material for use on the basis of its physical properties. New technologies of material processing, particularly nanotechnologies, make it possible to design and manufacture molecular material the properties of which can be defined theoretically and the product itself specifically designed for satisfaction of requirements that can occur afterwards[2]. Moreover, the principle of manufacturability is implied even in the names of the new spheres of engineering activity: nanotechnologies, biotechnologies, information and communications technologies etc.

One more characteristic feature of modern engineering activity is the creation of the so-called "distributed creative teams" for joint work on realization of engineering and engineering-scientific projects with real-time interaction of executors over the Internet. Actually it can be considered a quasi-personal work of engineers. Furthermore, the results of such researches are frequently documented as electronic publications on the same web or become the property of electronic magazines and libraries. Also the world supply of electronic scientific resources is headily increas-

[2] Example from practice: there are windows with self-cleaning glass Pilkington Activ. The secret of self-cleaning properties is that the exterior side of the glass has a thin transparent coating of titanium oxide. Under the influence of UV rays an oxygen chemical reaction takes place, as a result of which dust particles are detached from the surface. When raindrops fall on it, they form a uniform water film that easily slides off the glass and completely washes away dirt. After that the moisture quickly dries up without leaving streaks. The coating "works" permanently, doesn't peel off or fade over time. Dirt washes away each time when it rains. Coating strength is achieved by a special pyrolytic manufacturing method: a microscopically thin special layer is applied to one of the sides of a glass plate while it is still hot. After cooling this layer becomes an integral part of the surface and can be damaged only if the glass itself is scratched or broken.—http://www.ivd.ru/news.xgi?id=823.

ing. It puts into the forefront the new global problem of creation of effective tools for the search of necessary scientific information in electronic networks.

However, apart from the problem of information search there appears one more problem connected with the fact that in modern societies a significant role in formation of real-world notions and in construction of the meaning of social phenomena starts to be played by information networks and mass media. This process as well affects experts, due to their involvement in the society. And taking into account that people nowadays contact more with indirect notions of the physical and social world than with just the objective reality of their surroundings (the typical example would be proliferation of the so-called social networks: Facebook, LinkedIn, Myspace etc.), it must be acknowledged that a person often acts not on the basis of what happens in actual fact but on the basis of his notions of what happens. Undoubtedly it also leads to inaccuracies and distortions during the modeling of engineering objects.

10 Conclusion

The degree of penetration of virtual technologies in modern engineering activity is already quite large and continues to grow rapidly. Technologies of virtual reality have become a very effective instrument for scientific and engineering research. They allow one to clearly model the development of many real processes or phenomena and in such conditions or modes that are expensive or dangerous for reproduction in real conditions.

On the other hand, the possibility of quickening performance of the main stages of design process leads to quickening of introduction of technological and technical novelties. In its turn it requires an increase of consumed resources volume. Furthermore, it is commonly known that introduced novelties go out of date very quickly and are replaced by other technologies and technical systems. All of this results in unsustainable utilization of technical systems and resources and undoubtedly impacts the environmental situation in the world.

Therefore, it can be documented that engineering activity and engineering creativity are undergoing substantial transformations caused by global distribution of information technologies. These changes occur in the objective area of engineering as well as perhaps in the area of recognition of one's own activity an formation of a responsible attitude towards it. This fact makes it necessary for us to understand engineering practices as socially determined and socially significant.

References

Altshuller GS (1973) Algorithm of invention. ([Algoritm izobretenija], Moscow, Moscow worker. This book in English was published under the name: Altshuller G.S. (1999) Innovation algorithm). Technical Innovation Center, Worcester

Bagdasarian NG (1998) Professional engineering culture: mechanisms of exploration [Professional'naja kul'tura inzhenera: mehanizmy osvoenija]. Bauman Moscow State Technical University Press, Moscow

Batishhev GS (1997) Introduction to dialectics of creativity [Vvedenie v dialektiku tvorchestva]. Russian Christian Institute for the Humanities Press, Saint-Petersburg

Baudrillard J (1994) Simulacra and simulations. University of Michigan Press, Michigan

Benjamin W (1968) The work of art in the age of mechanical reproduction. In: Illuminations. Arendt H (ed), New York. http://brutishandshort.com/wp-content/uploads/2011/06/Hannah-Arendt-Introduction-to-Illuminations.pdf

Bergson H (2005) Creative evolution (New York, Cosimo. The first edition in English: Bergson Henri (1911) creative evolution). Henry Holt and company, New York

Latin-English Dictionary (n.d.). http://www.latin-dictionary.net/search/latin/ingenium

McLuhan M (1995) The Gutenberg galaxy: the making of typographic man. University of Toronto Press, Toronto

Poincaré H (1913) The foundations of science. Science, New York. http://archive.org/stream/foundationsscie01poingoog#page/n8/mode/2up

Polovinkin AI (1988) Basis of engineering creativity [Osnovy inzhenernogo tvorchestva]. Engineering Publishing, Moscow

Rozin VM (ed) (1997) Philosophy of technology: history and modernity [Filosofija tehniki: istorija i sovremennost']. IPhRAS Publishing, Moscow

Smirnov SD (2001) Pedagogy and psychology of higher education: from activity to personality [Pedagogika i psihologija vysshego obrazovanija: ot dejatel'nosti k lichnosti]. Academia, Moscow

Stjopin VS, Gorohov VG, Rozov MA (1995) Philosophy of science and technology [Filosofija nauki i tehniki]. IPhRAS Publishing, Moscow

Tillich P (1964) Theology of culture. Oxford University Press, Oxford

Galileo's "Technoscience"

Vitaly Gorokhov

Abstract Three main features of engineering thought have formed over the centuries: artistic, practical (or technical), and scientific. In the Renaissance time the relation between art and nature to each other were interpreted in three different ways. Galileo criticized the craftsmen's approach to technical activity that overlooked scientific knowledge and laws of physics in building machinery that would be impossible without them. He created more than a model of experimental activity; he demonstrated how to develop scientific knowledge so that it could be used for technical purposes. That is why "technoscience" is an appropriate name for Galileo's new science.

Keywords Galileo · Scientist-engineer · Scientific engineering education · Natural and engineering sciences · Technoscience

Although engineers resort more often to drawings and diagrams, and scientists to formulas and texts (e.g. papers, monographs, textbooks), modern engineering thinking is basically scientific engineering. Modern engineering thinking is at the same time scientific thinking in opposition of the technical thinking before scientific engineering. The drawing and the diagram (the elements of the engineer's language) are permeated with science and mathematics. The scientific picture of the world worked out during the seventeenth to eighteenth centuries began timidly and hesitantly to intrude on the practice of the ordinary engineer only in the nineteenth century. In the eighteenth century, Galileo's mathematics based on experimental science failed to exhaust engineering practice, which at that time remained an engineering art. We can distinguish already in this time three types of engineers: master-cum-engineer (artist-engineer), engineer-scientist, scientist-engineer (organizer of scientific engineering education and TA-expert).

V. Gorokhov (✉)
Research Center for Philosophy of Technology and Engineering Ethics of the Institute of Philosophy, Russian Academy of Sciences, Russian Federation, Moscow, Russia
e-mail: vitaly.gorokhov@mail.ru

R. Pisano (ed.), *A Bridge between Conceptual Frameworks,* History of Mechanism and Machine Science 27, DOI 10.1007/978-94-017-9645-3_12

1 The Emerging of the Scientific Education of Engineers

The first engineers appeared in the Renaissance. They came from the circle of scientists who turned to technology or from groups of self taught artisans interested in science. This period also saw appearance of the engineer, technician-and-expert, whose main and, later, only occupation was the construction of civil structures and military installations. There were also managers of large-scale technical projects and theoreticians of technology in the last decades of the Roman Empire, but both occupations were rigidly divided from one another. A technician remained either a foreman or a supervisor only, who needed no theory; a theoretician was mostly a philosophizing dilettante. Gradually, the engineer became a professional like teachers, doctors, lawyers and so on, although the social organization of engineering and the socio-economic mechanism for protecting priority and inventors' rights had not fully taken shape.

The first engineers of the Renaissance were at the same time artists and architects; consultant engineers specializing in fortification, artillery, and civil structures (Pisano 2009); alchemists and physicians; mathematicians; natural scientists; and inventors. The traditional guild-regulated crafts were gradually replaced by science-based engineering activity. Instead of anonymous craftsmen, more and more professional technicians enter the scene, the outstanding technical personalities whose fame extends far beyond the area of their immediate operation. However, the rapid and radically new development of technology required that its structure should be fundamentally changed. Technology comes to a point from which its further advance is impossible without its saturation with science. The need is felt everywhere for new technical theory, for codification of technical knowledge, for some general theoretical basis of that knowledge. Technology requires the application of science.

> The emergence of the figure of the engineer seen as a technician in some way educated in sciences, is a characteristic feature of the XV century and the first half of the XVI. Indeed this is perhaps the main feature of science, where the reduced creativity (real or apparent) of *pure* scientists, was counterbalanced by the great creativity of *applied* scientists. [...] Although there were no public funding to encourage scientists to devote their efforts to the study of technical applications and to the improvement of their knowledge, a common ground arose, particularly in Central and Northern Italy. The link between engineers and scientists emerged, at least in part, through the creation of some technical centres in the courts of the principalities which had been set up. This was the case of Medici's court in Florence, but also, and perhaps more importantly, the court of Milan under Francesco Sforza with its very rich library.[1]

Gradually, the engineer became a professional. Engineering had already broken away from the craft guild structure. The education of artist-engineer was in the artist workshop, in the Abaco schools and Academies. For example the Academy of the Art of Design (*Accademia dell'Arte del Disegno*) in Florence was the first official school of drawing in Europe and an Academy for Doing. The Academy became a model of the training of artists and engineers in Italy. Artists, engineers,

[1] Pisano 2013, p. 32.

and mathematicians were often equally expert in practical geometry, geodesy, perspective, technical drawing etc. (Pisano and Capecchi 2015; Valleriani 2010, pp. 7–12).

Galileo was directly associated with engineers and technicians of the Renaissance. His scientific career had a "technical" beginning; Galileo studied in Florence, where his teacher was Ostilio Ricci, an engineer and architect belonging to the Tartaglia school. Taking from him an interest in technical practice and engineering problems, Galileo maintained close ties with engineers all his life. The social need for technical innovations in Italy of that time stimulated many people to try their hand in some way at inventing things. Galileo was also caught by this fever. For years he built scientific instruments and carried out tests in a workshop in his house in Padua (Pisano and Bussotti 2015). This city was in the Republic of Venice, and Galileo maintained constant contacts with the Venetian arsenal.

Many medieval views and notions were assimilated during the Renaissance, but they took on a new meaning and conveyed a new emphasis; the comprehension of the divine plan began to be interpreted as the discovery of the laws of nature (acquisition of scientific knowledge), and technical activity in accordance with those laws was interpreted as a practical "engineering" action. As a result, the architect-cum-engineer and the technician-cum-inventor of that time considered nature, described in philosophy and science, to be the object of practical activity, and the latter was regarded as the art that followed the laws of nature. But in the Renaissance time the relation between art and nature to each other were interpreted in three different ways.

The Renaissance brought about a particular attitude towards the engineer differing from that towards the craftsman or technician of the Middle Ages. The engineer, like the Divine Creator, became a creator by creating reality. At the outset he imitates the Creator of the World and nature, and gradually he begins creating the world and another nature. The artist now imitates not so much God's creations, which, naturally, also takes place, as His process of creation: in God's creations, that is, in natural things, the artist strives to discern the law of their making. Man comes to the centre of the Universe; he is the best creation of God, and he rules over all other creatures. Meanwhile, the artist and engineer ceases to be a rank-and-file member of a trade guild and becomes a courtier, a "prince" of the arts, a bearer of the divine gift, equal in his art to God Himself. Giorgio Vasari wrote:

> The origin of the arts we are discussing was nature itself, and that the first image or model was the beautiful fabric of the world, and that the master who taught us was that divine light infused in us by special grace, which has made us not only superior to the animal creation but even, if one may say so, like God Himself (Vasari 1978, p. 19).

Galileo were busy with understanding physical phenomena. He criticized the craftsmen's approach to technical activity that overlooked scientific knowledge and the laws in building machinery:

> I have seen all engineers deceived, while they would apply their engines to works of their own nature impossible [...] (Crombie 1981, p. 277).

The main reason for those errors was that practical engineers who developed their inventions on false foundations deceived nature, failing to see its basic laws.

The rapid development, of the states and trade promoted improvements in military technology, mainly fortification and artillery (Pisano and Capecchi 2009); the construction of water works and civil engineering structures; the manufacture of machines, including ingenious mechanisms and automatic devices for entertainment. The development of artillery and fortification was essential to the existence of the cities and republics in Italy; their independence often relied on the accuracy and range of their cannons and the strength of their fortification. Therefore, engineering consultants were in demand everywhere and were valued by kings, dukes, and citizens.

But traditional artisan skills were no longer enough. That is why the first engineers and inventors turned to mathematics and mechanics, where they got knowledge and borrowed calculation methods. When that knowledge was insufficient, they tried to obtain new knowledge on their own, often becoming very productive scientists. Niccolò Tartaglia (Pisano and Capecchi 2015), for one, was a self-taught engineer and a "free-lance" consultant in mathematics to technicians. While tackling the problem of increasing the range of artillery fire (his tables for calculating missile trajectories were used for a long time by artillery officers) he published a book entitled *Nova Scientia*. The author explained that the book was necessary for every Christian to be better equipped to be offensive as well as defensive against Sultan Suleiman who was then threatening the Venetians.

Knowledge was then considered to be a real power, and the engineer its holder. Nevertheless the question remains, why did the Renaissance bring forth so many engineers and inventors who claimed rights and social status when there had been so few of such people in the Middle Ages (or, perhaps, we know nothing about them)? The spirit of invention and innovation pervaded all the strata of society at that time. Numerous impostors and pseudo-inventors appeared alongside genuine inventors (which means that to be an inventor was prestigious!). It has been considered normal since the Renaissance to claim an inventor's right, and if an invention bears the name of another man, the inventor suffers beyond all measure. On the contrary, a medieval man ascribed his creation (invention or treatise) to a divine or an indisputable human authority. Hence Polydore Vergil in his "*De Inventoribus Rerum*" (the first edition was published in Venice in 1499) complained that it was impossible to name the authors of many ancient inventions, such as cannon, mills, mechanical striking clocks. This is hardly surprising—they had kept themselves anonymous on principle. On the other hand, many things known in ancient times like gunpowder, for instance, which had been used by the ancient Chinese, were rediscovered, but attributed to individuals during the Renaissance. This too is understandable, because that was when one of the first kinds of engineering, viz. invention, and one of the first engineering professions, viz. the inventor, came into existence. Vergil's programme as far as inventors were concerned was to list those who first invented or began all things or arts.

To promote their innovations, medieval inventors often concealed their authorship or obscured it, ascribing them to some authority. Finally, many things in this respect can be accounted for by the psychology of the medieval craftsman who did not see himself as separate from his shop, guild, or corporation. While improving his products, the craftsman did not realize that he created something new, and he even did not try to realize that, because the whole socio-cultural situation hindered him from doing so.

In the fifteenth to seventeenth centuries the attitude towards innovations radically changed. The mark of the Master took on a personal significance, and he became a free creative individual. The social status of the Master and his treatment by society also changed.

Albrecht Dürer (1471–1528), a painter, engraver, architect and fortification builder, mathematician and optician is an example of the emancipation of the Master from the world of craftsmen. Born into the family of a Nuremberg jeweller, he belonged to the artisan class, but he managed to overcome their narrow-mindedness and became a learned man (jewellers had to master metal smelting, flattening, alloy mixing, coining, engraving, enameling—quite a lot of skills!), he was the first to sign not only his paintings and engravings, but also his drawings, as he believed that a sketch made by a great, master was more valuable than a carefully executed work of a craftsman. Dürer attached much importance to science, particularly to mathematics, primarily to geometry and perspective, passionately believing in the omnipotence of science. In fact, he worked as a master engineer even in painting, by actually designing portraits (Harnest 1996).

The engineers of the Renaissance did not canonize unattainable standards nor did they belong to a narrow circle of masters of a guild: rather, they tried to improve current technologies, to leave a personal imprint and make them public property, to associate the names of inventors with inventions so that they would bring fame to those people. That was not anything extraordinary in the Renaissance culture, something once created by an individual scientist to demonstrate the omnipotence of science, as it was with Archimedes. Ingenious machines like those developed by Archimedes were now built, by many people everywhere. They not merely amazed people, they became necessary, and their designers were paid by numerous customers and users.

In his letter offering his services to Lodovico Sforza, the Duke of Milan, young Leonardo da Vinci first enumerated his abilities as a military engineer and only then his achievements as a sculptor and artist (Hart 1961, pp. 22–23). Combining the activity of an artist (initially, he was a pupil at the studio of the painter, sculptor, and technician Verrocchio) with that of an experimenter, Leonardo spent a great deal of time on developing entertainment device and water works, on consulting and supervising fortification builders, on experiments with paints, not always successful (his painting of The Last Supper has been soon ruined for that reason). He used to work on his paintings very slowly spending most of his time designing scaffolds and mechanical accessories, which annoyed his pupils and exasperated his customers. And so Leonardo got most of his income as a military and civil engineer.

The list of engineering inventions and tasks offered by Leonardo in his letter was not an empty boast or impracticable "engineering" fantasy, although which of them he could really bring into effect is unknown. During his lifetime Leonardo managed to realize some of his promises, although many others could not have been realized in his times. His notes contain detailed descriptions and drawings, which, of course, are not addressed to anyone in particular, but which indicate a way to embody them in specific structures and devices. Some "draft projects" were based on careful studies of nature (Cianchi 1998).

Scholars believe that Leonardo da Vinci's notes contain descriptions of some apparatus and machines developed by other engineers as well as his own unrealized designs (Pisano and Bussotti 2014). His notes are indicative of a design approach he employed, although he enciphered them. Leonardo was active even in alchemy, but his notes show nothing of the mythical or mystical, they are basically scientific. Leonardo's "projects" were more than the "engineering" fantasies of Roger Bacon. Unrealized does not mean impracticable: Leonardo da Vinci (Pisano 2013) always dismissed empty dreams and fantastic chimeras. Invention creates things non-existent but possible in nature, while fantasy deals with things which are chimerical, impossible, and impracticable. An invention or even a painting was, for Leonardo da Vinci, not merely a product of imagination, a semiartistic inspiration, or a blind adherence to craft traditions; it resulted from a careful study of nature and its laws. He wrote:

> Those, who are not in love with principle or knowledge are like the sailor who goes into a ship without a rudder or compass and who never can be certain whether he is going. Practice must always be founded on sound theory (Parsons 1939, p. 36, p. 37).

Brunelleschi and Alberti did not simply dream about the possibility of creating miraculous enlarging devices; they (and some others) developed the theory of perspective (in effect, geometrical optics). This was used as a basis for building so-called perspective (optical) instruments and devices, primarily the camera obscura, which was the prototype of Galileo's telescope, initially called perspective.

The presence of both unrealized and realized designs is the first feature of a design culture, which came into being during the Renaissance. A design culture can be defined as placing emphasis on the ideal moments of existence, while the spiritual aspect of existence remains real, and material wealth is only a means and not an end; it prefers the reality of the possible to the possibility of the real. Unrealized designs are no less important than those realized.

> It now seems that both the traditional sharp contrast between the great inventor and his colleagues and the more recent attempts to continue Leonardo's engineering activity within the limits of practice, procedures and projects already fully developed by contemporary engineers and those of previous generations must be rejected as inadequate. [...] Leonardo was original also in his drawings which, even in their incompleteness, are correctly interpreted as the conceptual equivalent of the "model". [...] I this field Leonardo boasts a supremacy which is unrivalled and which places him at the very beginning of modern scientific illustration. Never before had anyone managed to demonstrate a complex technical design so effectively in a drawing (Leonardo 2005, p. 131, p. 132).

Galileo goes in the same way: compare two drafts of the same mechanism for the transform of the rotatory motion into progressive movement from Galileo Galilei and Leonardo da Vinci (Fig. 1).

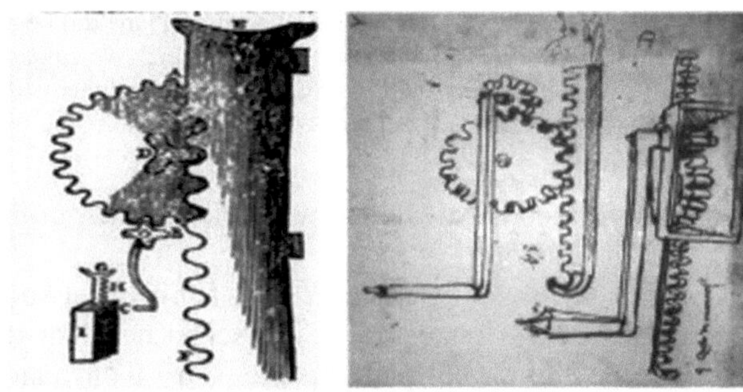

Fig. 1 On the *left* is Galileo's draft of a mechanism for the transform of rotatory motion in the progressive movement (Galilei 1634, p. 103), on the *right* is Leonardo's draft of the same mechanism (Cianchi 1998, p. 76)

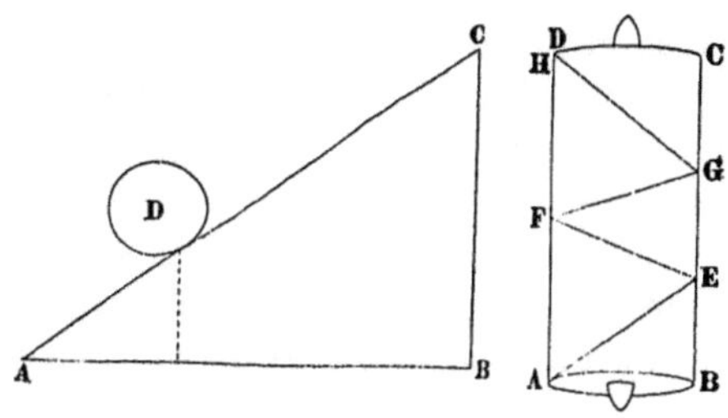

Fig. 2 Geometric representation of the inclined plane as the abstract object of the new science (on the *left*) and the explanation of the functioning of screw with help of this model (on the *right*) by Galileo (Galilei 1665, pp. 26–27)

But Galileo, contrary to Leonardo, reduced such drawings to the geometrical models. For example, he used the inclined plane (the abstract object of the new science) as the universal explanatory model for the functioning of all machines (see Fig. 2).

Galileo investigated in his *Mechanics* the nature of the screw with help of the ideal model of the inclined plane as *triangle*.

> Returning now to our first Intention, which was to investigate the Nature of the Screw, we will consider the Triangle ABC, of which the Line AB is Horizontal, BC perpendicular to the said Horizon, and AC a Plane elevated; upon which the Moveable D shall be drawn by a Force so much less than it, by how much the Line BC is shorter than CA: But to elevate or raise the said Weight along the said Plane AC, is as much as if the Triangle CAB standing still, the Weight D be moved towards C, which is the same, as if the same Weight never removing from the Perpendicular AE, the Triangle did press forwards towards H. For if it were in the Site FHG, the Moveable would be found to have mounted the height AI. Now, in fine, the primary Form and Essence of the Screw is nothing else but such a Triangle ACB, which being forced forwards, shall work itself under the Grave Body to be raised, and lifted it up, as we say, by the head and shoulders. And this was its first Original: For its first Inventor (whoever he was) considering how that the Triangle A B C going forwards raised the Weight D, he might have framed an Instrument like to the said Triangle, of a very solid Matter, which being thrust forwards did raise up the proposed Weight: But afterwards considering better, how that that same Machine might be reduced into a much lesser and more commodious Form, taking the same Triangle he twined and wound it about the Cylinder ABCD in such a fashion, that the height of the said Triangle, that is the Line CB, did make

> the Height of the Cylinder, and the Ascending Plane did beget upon the said Cylinder the Helical Line described by the Line AEFGH, which we vulgarly call the Wale of the Screw, which was produced by the Line AC. And in this manner is the Instrument made, which is [...] a Screw; which and insinuate with its Wales under the Weight, and with facility raised it (Galilei 1665, pp. 26–27).

With help of geometry Galileo was able to teach military engineers to use mathematical instruments.

Galileo as scientist-engineer was directly associated with engineers and technicians of the Renaissance. The social need for technical innovation in Italy of that time stimulated many people to try their hand in some way at inventing things. Galileo's military compass was a mathematics teaching instrument (Pisano and Bussotti 2015) for the art of war.

Therefore, Galileo demonstrated how to develop scientific knowledge so that it could be used for technical purposes. Galileo's works paved the way for the formation of engineering thinking and activity in practice as well as theory.

The medieval universities were unable to provide broad practical knowledge and they stood aloof from the developing engineering practice. However, the desire to obtain such knowledge together with mathematics was growing among the public. The scholastic university science exhausted its potential, but new knowledge was very slow to catch on. In time, of course, the fresh air of change swept away the traditionalism of university science, and it made a significant contribution to development of the theoretical basis of engineering. However, this breakthrough took place in the seventeenth to eighteenth centuries within other new social structures, such as scientific societies, academies and later high technological schools.

École Polytechnique (*Polytechnique*) was founded in 1794 in Paris by Gaspard Monge. The *Polytechnique* was oriented to the theoretical instruction of students from the initial period of its existence. The *Polytechnique* proved to be a standard for many engineering schools in Russia, Germany, Spain, Sweden, and the USA. For the first time students were introduced there to genuine mathematics and genuine theoretical science. The School's first graduates—polytechnic engineer (Poinsot, Poisson, Cauchy, Navier, and others)—made a great contribution to the development of experimental and engineering science. This was the first time that the curriculum of a higher technical school included a course in machine design.

Galileo not only related a geometrical scheme to physical reality, but also to the construction of different complex machines. But it was Euclidean geometry. The next phase of development of the theory of mechanisms (kinematics of machinery) as an engineering science was not elaborated by Euclidean geometry but by the descriptive geometry of Gaspard Monge. But in both cases, scientific engineering education was a decisive factor for development of the theoretical basis for codification and systematization of practical technical knowledge.

2 Galileo's New Science as "Technoscience"

2.1 The Structure of Natural-Scientific Theory

In the structure of natural-scientific theory we shall differentiate three basic components: theoretical schemes, the mathematical apparatus and the conceptual apparatus.

Theoretical (ontological) schemes are a set of ideal objects of theory, which are, on one hand, oriented to application of the appropriate mathematical apparatus and, on the other hand, oriented to the "mental or thought experiment" or conceptual modeling, i.e. to the design of possible experimental arrangements. At the same time they are developed so that mathematical operations (calculations) are possible with them. Theoretical schemes are special idealized models, often in the form of graphical (geometrical) representations. Theoretical schemes also express a particular "vision of the world" from a specific point of view given in theory (that is why they cannot be referred to as ontological). On one hand, they reflect some properties and aspects of real objects which are of interest to a given theory, and, on the other hand, they are its operational means for a particular idealized representation, which can then be practically realized in experiment (through eliminating side effects). Thus, it is the so-called ideal objects (in other words, abstract, idealized, objects of theory) that are of considerable importance, especially in mathematized scientific theories. They are specially constructed in theoretical knowledge as a result of a particular idealization and schematization of experimental objects. In a broader context the things concerned must be not only experimental objects, but also objects of e engineering activity—technical systems.

> Among ideal objects used in a scientific research at least two main types are traditionally singled out: empirical and theoretical objects. Empirical objects are abstractions which fix features of real objects of experience. They are a kind of schematization of fragments of the real world. Any feature—the "carrier" of which is an empirical object—can be found in corresponding real objects … Theoretical objects, unlike empirical ones, are idealizations, "logical reconstruction of reality". They may be provided not only with features corresponding to the features and connections of real objects, but also with features not proper for any such object (Stepin 2005, pp. 47–48).

The mathematical apparatus is necessary chiefly for analyzing experimental situations which are a means of substantiation and verification of obtained theoretical knowledge. Moreover, in a well-developed theory it is used for its evolutionary presentation or deductive transformation of ideal objects. The mathematization of ideal theoretical object transformation rules makes it possible to obtain new knowledge without resorting to experiment and observation, i.e. staying within the theoretical activity framework. One can say that one or another science is really mathematized only when it begins using mathematical methods for processing experimental research findings, but also in looking for new laws, constructing theories and developing a special formalized language of this science.

> In the language of science the syntactical aspect becomes most important when we make formal operations with symbols, for instance, with physical magnitudes (which enter mathematical expressions of physical laws), in accordance with rules of mathematics. Making such operations, a researcher disengages himself from the meaning of the linguistic terms and considers them only as signs which create formulae in their connections and then deduce other formulae according to the rules of the linguistic system given (Stepin 2005, p. 46).

To mathematize a scientific discipline, it is necessary to simultaneously or even preliminarily evolve an adequate conceptual apparatus. Theoretical schemes and the mathematical apparatus are always used in the context of a particular conceptual environment. In this sense the conceptual apparatus is necessary for the conceptual fixation of theoretical schemes and mathematical apparatus of a theory. At the same time each theoretical concept contains, in a non-evolved form, a corresponding theoretical scheme and mathematical procedure. For example, the physical sense of the concept of capacity is different from the common one. In physics, it is, on one hand, a theoretical scheme of a particular physical process (the flow of electric charges in the capacitor plates or of current through the capacitor) and, on the other hand, a mathematical operation of integration when it is considered in the operational calculus context.

> In the theoretical language a theoretical scheme can be characterized by means of at least two types of expressions. First, it may be pithy descriptions like those regarded above: "a material point is moving along the continuum of points of a spatial-temporal frame of reference", "the force changes the state of motion of a material point" etc. Such expressions describe connections and relations of abstract objects forming a theoretical scheme. At the same time these connections can be expressed as mathematical dependencies. This can be reached through mapping abstract objects of a theoretical scheme onto the objects of mathematics. For instance, a frame of reference may be connected with coordinates (the inertial frame of reference in mechanics can be identified—within certain limits—with a system of rectangular, spherical or cylindrical coordinates in Euclidean space) (Stepin 2005, p. 53).

In theory, several conceptual strata (levels) correlating to each other can be differentiated.

Three main levels in the theoretical (ontological) schemes of a natural scientific theory can be discerned. *The functional scheme* is oriented on the mathematical description and fixes the general idea about the object under investigation. The units of this scheme reflect only the functional properties of the elements of the object for the sake of which they are included in it to attain the general objective and reflect certain mathematical relations. In classical natural science it was a geometric scheme of physical processes. *Flow schemes* describe natural, for instance, physical processes taking place in the object and connecting its elements into a single whole. The units of such schemes reflect various operations performed in a natural process by the elements of the object. These are based on natural-scientific concepts. Finally, *structural schemes* reflect the structural arrangement of elements and linkages in the given experimental equipment. These schemes represent parameters of the projects of new experimental situations.

One of the best examples is Galileo's theoretical investigation in which he sequentially moved from a geometric scheme to physical processes and from the latter to a design diagram. Galileo staged an experiment mentally (he compared the rotation of two wheels—one small, one large—on the geometric drawing) and arrived at a radically new conclusion, which went against the first one, passing from the artificial, technical, model to an explanation of a natural phenomenon:

> […] it might be supposed that the whirling of the earth would no more suffice to throw off stones than would any other wheel, as small as you please, which rotated so slowly as to make but one revolution every twenty-four hours (Galilei 1952, p. 217).

This approach gives rise to an ideal and real, natural and mathematical, syncretic object, that is, an object that hides a mathematical (geometric) scheme. Galileo noted in this respect:

> […] because of the imperfection of matter, a body which ought to be perfectly spherical and a plane which ought to be perfectly flat do not achieve concretely what one imagines of them in the abstract… Then whenever you apply a material sphere to a material plane in the concrete, you apply a sphere which is not perfect to a plane which is not perfect, and you say that these do not touch each other in one point. But I tell you that even in the abstract an immaterial sphere which is not a perfect sphere can touch an immaterial plane which is not perfectly flat in not one point, but over a part of its surface, so that what happens in the concrete up to this point happens in the same way in the abstract. It would be novel indeed if computations and ratios made in abstract numbers should not thereafter correspond to concrete gold and silver coins and merchandise. Do you know what does happen, Simplicio? Just as the computer who wants his calculations to deal with sugar, silk and wool must discount the boxes, bales, and other packing, so the mathematical scientist (filosofo geometra), when he wants to recognise in the concrete the effects which he has proved in the abstract, must deduct the material hindrances, and if he is able to do so, I assure you that things are in no less agreement than arithmetical computations. The errors, then, lie not in the abstractness or concreteness, not in geometry or physics, but in a calculator who does not know how to make a true accounting. Hence if you had a perfect sphere and a perfect plane, even though they were material, you would have no doubt that they touched in only one point; on the other hand if it is impossible to have these, then it was quite beside the purpose to say that *sphaera alnea non tangit in puncto* (a bronze sphere does not touch a plane at a point) (Galilei 1952, pp. 207–208).

Moreover, Galileo believed there existed engineering methods that allowed imperfect material objects "bulging" from a perfect geometric form to be brought near to ideal, mathematically perfect objects. Examples are a grinding process or the use of an undeformable material for which insignificant deviations from an ideal shape can be neglected (this is the way he designed his telescope).

There is no doubt that Galileo, for all his abstract reasoning, was guided by the engineering practices of his time:

> Maybe these mathematical ratios which are true in the abstract do not exactly correspond when applied in the concrete to physical and elemental circles. Though it does seem to me that a cooper, in determining the radius of the bottom to be made for a barrel, makes use of the abstract rules of the mathematicians despite such bottoms being very material and concrete things (Galilei 1952, pp. 232–233).

In a well-developed science, theory formation is generally begun from using a theoretical model from some better-developed area of science, which serves as an initial one, and is corrected for a new class of phenomena. For example, Galileo has borrowed the geometro-kinematic scheme from astronomy where motions of celestial bodies along ideal curves were considered in the purest form, in accordance with the theorems and postulates of Euclidean geometry. The further restructuring of this model was by means of constructive introduction of new ideal objects.

At early stages of science development, theoretical models are evolved through the direct schematization of experience. Then these initial models are used, as means for construction of new theoretical models and this way becomes decisive in science development. Galileo realised in practice the purposeful application of scientific knowledge that formed the basis of engineering thought and engineering activity. This approach became possible because Galileo's new science had its roots in technical practice (which had progressed by his time and was urgently in need of generalisation) and was oriented to it.

2.2 The Structure Engineering Theory

In engineering theory, the same components (theoretical schemes, the mathematical apparatus and conceptual apparatus) can be differentiated. However, their content will be different from the natural-scientific theory.

Three main levels in the theoretical (ontological) schemes of an engineering or technological theory can be discerned. *The functional scheme* is oriented on a mathematical description and fixes the general idea about the *technical system*, irrespective of the method of its realization. The units of this scheme reflect only the functional properties of the elements of the technical system for the sake of which they are included in it to attain the general objective and reflect certain mathematical relations. *Flow schemes,* or schemes of performance, describe natural, for instance, physical processes taking place in the technical system. The units of such schemes reflect various operations performed in the natural process by the elements of the technical system while it is functioning. Finally, *structural schemes* reflect the structural arrangement of elements and linkages in the given technical system and presuppose its possible realization. The elements of the latter are regarded in them as having not only functional properties, but also properties of the second order, i.e. those undesirable properties which are added by a definitely realized element, for instance, non-linear distortions of the amplified signal in the amplifier. These schemes represent constructive-technical and technological parameters, i.e. they reflect specific problems cropping up in engineering practice.

The functioning of engineering theory is realized by the iteration method. At first a special engineering problem is formulated. Then it is represented in the form of the structural scheme of the technical system which is transformed into the idea about the natural process reflecting its performance. To calculate and mathematically model this process a functional scheme is constructed. Consequently, the engineering problem is reformulated into a scientific one and then into a mathematical problem solved by the deductive method. This path from the bottom to the top represents *the analysis of schemes.* The way in the opposite direction—*the synthesis of schemes*—makes it possible to synthesize the ideal model of a new technical system from idealized structural elements according to the appropriate rules of deductive transformation, to calculate engineering discipline can be considered as formed when a mathematized engineering theory is constructed in it. It should also clearly give the procedures of the transition from structural schemes to flow and functional schemes (schemes of analysis) and vice versa (schemes of synthesis). Only when an engineering science has worked out the means of the theoretical synthesis of engineering systems which make it possible to extrapolate the theoretical results obtained for the class of hypothetical technical systems (with the orientation on practical and methodological knowledge) can its generalized ontological scheme be considered universal in relation to the given class of objects.

Engineering disciplines are peculiar in that the engineering activity takes the place of experiment in them. It is in the engineering activity that theoretical conclusions are checked for adequacy and new empirical material is drawn. That is why theoretical knowledge must be brought here up to the level of practical engineer recommendations. In the natural science, the most important thing is solving theoretical problems in terms of the natural process reconstruction aimed at prediction and description of its future states. Here mathematical relationships and experimental findings are mere auxiliaries used for substantiation, analysis, validation, etc. The specificity of the engineering theory is based on that its findings are used largely for constructing technical systems rather than explaining natural processes. To solve this problem, the theory must feature clear-cut rules of correspondence and transition from some "model" levels to other ones. In the engineering science the problem of interpretation and empirical substantiation is formulated as a problem of realization. The specificity of the engineering theory is based on that its findings are used largely for constructing technical systems rather than explaining natural processes. The requisite condition of engineering theory productivity is the presence of practical methodological knowledge, i.e. engineering recommendations stemming from theoretical research, in its empirical basis.

Galileo's geometric-kinematic theoretical schematic model of the machines was a beginning and precondition of the application of the natural scientific theory to the first special engineering science—the theory of mechanisms and machines or kinematics. Franz Reuleaux so as Galileo

> Personified a new figure in the industrial age, *the engineer-scientist*; professor, kinematics theorist, head of a university, industrial consultant and confidant to capitalists, government expert and technical ambassador to the emerging global industrial world ... The machine, he said, consists of one or more mechanisms which can be separated into kinematic chains which in turn can be broken down into kinematic pairs or fundamental mathematical constraints. The tools of this reductionism is *analysis* ... Reuleaux believed there were scientific principles behind invention and the creation of new machines or what we call *synthesis* today (Moon 2001, p. 5).

Franz Reuleaux in his "Kinematics of Machinery" wrote that kinematics or phoronomy (pure kinematics or kinematics geometry) is "the study of geometric representation of motion".

> "For the practical mechanician who has made himself familiar with the modern Phoronomy, and still more for the theorist, the machine becomes instinct with a life of its own through

> the rolling geometrical forms everywhere connected with it". He said that "the geometrical abstraction of machine" is "the soul of machine" (Reuleaux 1876, p. 56, 84, 85).

On one hand, while using engineering means, Galileo reasoned and acted essentially as a natural scientist. He not only developed stricter scientific terms but also "designed" a peculiar plan of reasoning, an ideal mental experiment as a "project" of a real experiment, an idealised concept of natural objects which then could be actually realised in an experiment (by eliminating the influencing factors). In experimental science, however, the scientist must build up a logical theory explaining and predicting the run of a given natural phenomenon and also design a practical experiment reproducing that phenomenon artificially, in its "purest" form, ignoring its unessential properties and verifying the validity of the theory. Indeed, it is necessary, in order to conduct an experiment, to eliminate side effects and to reproduce a natural process in an engineering way under conditions that can hardly be found in nature in pure form. For example, when checking the law of free fall of bodies, Galileo selected a ball made from a hard material, which meant its deformation could be neglected. In addition, he did his best to eliminate friction in a slot cut on a board by gluing polished parchment to the surface.

On the other hand, the situations experimentally reproduced by engineering methods must be presented and described scientifically as certain idealised constructions. In such an experiment, the construction was represented by an inclined surface. The experimental situation thus obtained was then considered as some idealised natural process of motion of natural bodies on an inclined surface, that is, objectively. The theoretical scheme obtained could be extended to cover the whole class of real objects for which friction and elastic deformations can be neglected. In the way Galileo reasoned as an engineer (or better to say as scientist-engineer in the engineering science), and, quite naturally, he often appealed to craftsmen's technical practice rather than to pure observation and the contemplation of the obvious, which was typical of antique science. While proving that the Moon's surface was rough by the manner in which it reflects the sunlight, Galileo wrote:

> Burnished steel appears very bright from some viewpoints and very dark from others [...]. And note that the diversity of what is seen upon looking at a burnished surface causes such a different appearance that to imitate or depict burnished armor, for example, one must combine pure black and white, one beside the other, in parts of the arms where the light falls equally (Galilei 1952, p. 78, 79).

To prove his statements Galileo also resorted to observation of operating technical devices, for instance a pump. He wrote in his Dialogues Concerning the Two New Sciences:

> The stock of the pump carried its sucker and valve in the upper part so that the water was lifted by attraction and not by a push as is the case with pumps in which the sucker is placed lower down. This pump worked perfectly so long as the water in the cistern stood above a certain level; but below this level the pump failed to work. When I first noticed this phenomenon I thought the machine was out of order; but the workman whom I called in to repair it told me the defect was not in the pump but in the water which had fallen too low to be raised through such a height; and he added that it was not possible, either by a

pump or by any other machine working on the principle of attraction, to lift water a hair's breadth above eighteen cubits (an ancient measure of length, about 18–22 inches, originally the length of the arm from the end of the middle finger to the elbow); whether the pump be large or small this is the extreme limit of the lift (Galilei 1952, p. 137).

That some natural phenomenon could not be reproduced artificially was a weighty argument for Galileo. A mathematical object (e.g. a point) for him always corresponded both to a natural physical object (say, a stone) and to a man-made object (e.g. a cannon ball). Galileo not merely compared them, he idealised them, "designing" in theory particular "ideal objects" (in other words, the abstract, idealised objects of theory).

These objects are specially designed in theoretical knowledge as a result of a particular kind of idealisation and schematisation of experimental, and hence, technical systems. Such are Galileo's inclined surface or the mathematical pendulum—an idealised model of a gravity pendulum, which can be used to investigate the laws of free fall. Here, the action of one cause, the resistance of the air, is separated from the action of another cause, the pull of gravity. It is this experiment that, in his view, proved without any doubt that Aristotle's ideas prevalent at the time were invalid. Without such an idealisation, both experimental science and engineering science were impossible.

The modern engineer mostly has to do with drawings, diagrams, plans, rather than directly with real technical objects; he does not manufacture them with his own hands but directs the manufacturing process, plans for it, and organises their service and maintenance. The activity of the engineer is much closer to that of the experimenting research scientist than is often thought. Today the close connection between natural and engineering science is expressed in the development of technoscience.

The term 'technoscience' is increasingly being used to refer to such contemporary disciplines as information and communication technology, nanotechnology, artificial intelligence and also to biotechnology. The term technoscience' is thus not only a useful pointer to the highly commercialized setting in which modern biotechnology and many other contemporary undertakings are conducted, it also suggests that science and technology, which presumably were once distinct activities, have become so much intertwined as to be virtually indistinguishable nowadays. The present popularity of the term may thus reflect the historical process in which science and technology have become increasingly interwoven, but it may also partly reflect the dominant preoccupations and concerns of those who use the term (van den Belt 2009, p. 1311).

But such connection between science and technology is typical also for Galileo.

Galileo, son of the Renaissance, offered his contribution to architecture by means of an interdisciplinary style thinking, intertwining operational and theoretical skills, mathematics, physics and art. Besides facing theoretical thematic [...] he was busy with civil, military facts as he reported in his work on fortifications (Pisano and Capecchi 2009, p. 28; see also Pisano and Bussotti 2014).

2.3 The Structure of Technoscience

In technoscience, on the one hand, explanatory models of natural phenomena are drawn up and predictions of the course of certain natural events on the basis of mathematics and experimental data are formulated as in classical natural science, and as in the engineering sciences, on the other hand, not only experimental arrangements are constructed, but also structural plans of new technical systems previously unknown in nature and technology.

Galileo did more than just observe natural phenomena. He would first *construct an idealized experimental situation,* leaving aside the question of its technical feasibility (the situation itself, while not existing in nature, was, however, reproducible in principle). Then he would design an ingenious project of the technically feasible experimental situation, say a pendulum (a mass suspended from a string), where the gravity force was separated from the force applied to the solid. Based on this project, a real experiment could be devised and conducted.

Similarly interms of nanotechnoscience:

> Nanotechnology comprises *not only the manipulation of natural molecules, but also the creation of molecules not found in nature*. The multifariousness of the relationship between nanotechnology and nature is expressed in the fact that some nanotechnological objects are clearly distinct from comparable natural objects, while others are identical to natural objects. Nanotechnology, however, does not only create an artificial world that is distinct from nature. It also relates to natural processes and materials in a new way. In this respect it is difficult to separate it from nature (Schiemann 2005, pp. 77–96).

Galileo Galilei was one of those who created this new science oriented to technical needs. He established the relation between scientific knowledge and the objects of practice. His fundamental work *Dialogues Concerning the Two New Sciences* begins with a description of Venice's famous arsenal:

> The constant activity which you Venetians display in your famous arsenal suggests to the studious mind a large field for investigation, especially that part of the work which involves mechanics; for in this department all types of instruments and machines are constantly being constructed by many artisans, among whom there must be some who, partly by inherited experience and partly by their own observations, have become highly expert and clever in explanation… You are quite right. Indeed, I myself, being curious by nature, frequently visit this place for the mere pleasure of observing the work of those who, on account of their superiority over other artisans, we call 'first rank men'. Conference with them has often helped me in the investigation of certain effects including not only those which are striking, but also those which are recondite and almost incredible (Galilei 1952, p. 131).

Galileo created more than a model of experimental activity; he demonstrated how to develop scientific knowledge so that it could be used for technical purposes. This approach became possible because Galileo's new science had its roots in technical practice (which had progressed by his time and was urgently in need of generalization) and was oriented to it. Galileo constantly emphasized the practical orientation of his idea. In the foreword to his "Discourse on Bodies in Waters" he noted, for instance, that his work was useful "in occurrences of building bridges or fabrics on the water" (Galilei 1980, p. 3). In his new science, Galileo manipulated natural objects like the present-day engineer:

> [...] if the terrestrial globe were perforated through the centre, a cannon ball descending through the hole would have acquired at the centre such an impetus from its speed that it would pass beyond the centre and be driven upward through as much space as it had fallen, its velocity beyond the centre always diminishing with losses equal to the increments acquired in the descent; and I believe that the time consumed in this second ascending motion would be equal to its time of descent (Galilei 1952, p. 227).

However, Galileo's new style of scientific-engineering and engineering-scientific thought and action manifested itself mainly in the sphere of thought rather than in practical activity.

That is why "technoscience" is an appropriate name for Galileo's new science. This is a combination of the natural and engineering sciences. In relation to the observers of nature, Galileo is more of a practical man, who destroys and restructures a natural object in order to discover a universal principle underlying it, and in relation to practice, he is more of an observer, who sees in an engineering process the universal law it reveals, not a particular end to be achieved. It is the middle place between natural and engineering sciences that is characteristic for technoscience.

3 Conclusion

Galileo's works paved the way for the formation of engineering thinking and activity in practice as well as in theory. Galileo himself was not engaged in the building and designing of machines. But he was able to produce a new science, that is a new scientific approach to physical phenomena also taking into account mechanics, devices and techniques (i.e., applied to cantilevers and machines modelling).

The first engineers in the Renaissance came from the circle of scientists (*scientist-engineers*) who turned to technology or from self taught artisans (*artist-engineers*) interested in science. This period also saw appearance of the engineer, technician-and-expert, whose main and, later, only occupation was the construction of civil structures and military installations. The teacher of Leonardo da Vinci, Verrocchio, who had come from a handicrafts tradition, was profoundly occupied with mathematics and taught it to his pupils. According to Leonardo mathematical relationships are found everywhere in nature. Proportions are found not only in numbers and measures, but also in sounds, weights, times, and places, and in every force. Galileo was familiar with the theory of perspective put forward by Italian artists. He had a life-long friendship with Lodovico Cigoli, an outstanding painter of his time. Galileo even helped him (in a letter) to argue against those who stated that sculpture was superior to painting. This is a geometrical interpretation of nature, or, in other words, materialized geometry that enabled Galileo to develop a new science—a mathematized experimental natural science. The visual representation of natural objects by the Renaissance painters made it possible to describe them in terms of geometry in the science of modern times. Modern engineering also employs its methods: the use of drawings and schematic diagrams lays the groundwork for future engineering projects and graphic design documentation. In his notes

Leonardo da Vinci, being a genuine engineer, hotly insisted on the advantages of drawings over verbal descriptions:

> What poet can represent to you in words, oh lover, the true image of your ideal as faithfully as the painter will do? (Richter, I, p. 55).

Therefore, painting for the artists and engineers of the Renaissance was not merely a natural science but also a means for working out the rules of action based on the disclosed laws of nature.

> "If you ask me what these rules accomplish or what good they are," wrote Leonardo da Vinci, "I would answer that they keep a restraining hand on engineers and investigators, teaching them not to promise impossible things to themselves or others, in consequence of which they may be considered either crazy or impostors" (Parsons, p. 371).

Painting for the Renaissance artists was primarily designing a perfect image: if there was no such a thing in nature, the artist made it up from various things that actually existed in nature (as the image of a perfect man). But this approach again required science, the study of natural structures.

> The painter who draws by practice and judgement of the eye, without the use of reason, is like the mirror which reproduces within itself all the objects which are set opposite to it without knowledge of the same (Parsons, p. 24).

For the analysis of complex machines Galileo applied geometrical representation of their principle of operation (Fig. 3). He started his "Mechanics" with appeal as his program of the theoretical analysis of machines:

> mechanicians deceive themselves in going about to apply machines to many operations of their own nature impossible; by the success whereof they have been disappointed, and others likewise frustrate of the hope which they had conceived upon the promise of those presumptuous undertakers: of which mistakes I think I have found the principal cause to be the belief and constant opinion these artificers had, and still have, that they are able with a small force to move and raise great weights [...]. In the mean time, since I have hinted, that the benefit and help derived from machines is [...] to move those weights, which, without it, could not be moved by the same Force: it would not be besides the purpose to declare what the commodities be which are derived to us from such like faculties, for if no profit

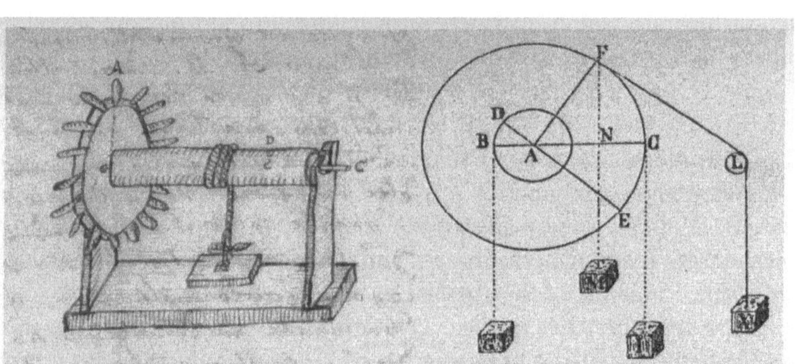

Fig. 3 Practical description of axle wheel on the *left* and geometrical illustration of the same instrument on the *right*. (Valleriani 2010, p. 101)

> were to be hoped for, all endeavours employed in the acquist thereof will be but lost labour (Galilei 1665, pp. 1–2).

Similarly, Robert Willis wrote in his "The Principles of Mechanisms" in 1841:

> My object has been to form a system that would embrace all the elementary combinations of mechanism, and at the same time admit of a mathematical investigation of the laws by which their modifications of motion are governed. I have confined myself to the Elements of Pure Mechanism, that is, … to reduce the various combinations of Pure Mechanism to system, and *to investigate them upon geometrical principles alone* (Willis 1870, p. xiii, p. 4; see below Fig. 4).

In engineering science (theory of mechanisms) it is important to reduce the constructive mechanism (or machine design) as a real technical system to the various combinations of pure mechanism (sometimes called kinematics of machinery) as an ideal model of this system.

Before Galileo, scientific studies followed the ancient standard of obtaining knowledge about an object that was regarded as unchangeable. It occurred to nobody to change practically the real object of investigation (as it would then be considered to be another object). On the contrary, scientists strove to improve their theoretical model so that it would fully describe the behaviour of the real object. In Galileo's view, the real object exactly corresponds to the ideal object but is interpreted as a distortion of the ideal object's behaviour under the action of various factors, for instance friction. This made it possible for Galileo to modify the real object by

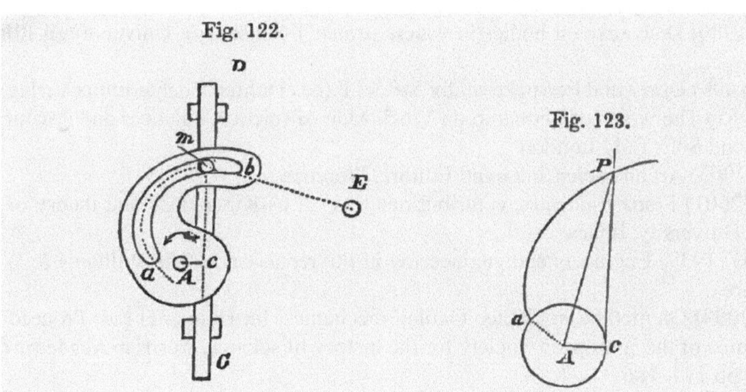

Fig. 4 "In fig. 122, *A* is the center of motion of a revolving plate in which a slit *a b* is pierced, having parallel sides so as to embrace and nearly fit a pin *m*, which is carried by a bar *CD* fitted between guides so as to be capable of sliding in the direction of its length. If the plate revolve in the direction of the *arrow* the inner side of the slit presses against the pin and moves it further from the center *A*, but when the plate revolves in the opposite direction the outer edge of the slit acts against the pin and moves it in the opposite direction. If the curved edges of the slit be involutes' of the circle whose radius is *Ac*, where *Ac* is a perpendicular upon the path *m c* of the bar, it appears from Art. 133 that the velocity ratio of plate and bar will be constant, and the linear velocity of the bar equal to that of the point *c* of the plate. But if any other velocity ratio be required, let *Pc* (fig. 123) be the path of the sliding bar, *P* the pin, *A* the center of the curve, *aP* the curve." (Willis 1870, pp. 152–153)

acting on it in a practical way. As a result, its “negative” properties, which prevented it from being identical to the ideal object, became neutralized. Galileo chose an approach unusual for scholastic science: technology began to lean on mathematical knowledge and models. The orientation towards both engineering practice and mathematical knowledge (obtained strictly analytically) largely determined the line of development of Galileo’s ideas.

Acknowledgments This study is part of a largely research supported by the project “From Galileo’s Technoscience to the Nanotechnoscience (Philosopical and Methodological Analysis)”, Nr. 13-03-00190, of the Russian Foundation for Humanities.

References

Cianchi M (1998) Leonardo da Vinci’s Machines. Becocci Editore, Florence

Crombie AC (1981) Philosophical presuppositions and shifting interpretations of Galileo. In: Theory change, ancient axiomatics and galileo’s methodology. Proceedings of the 1978 Pisa conference of the history and philosophy of science, vol 1. Heidel Publishing Company, Dordrecht

Galilei G (1634) Les méchaniques. http://archimedes.mpiwg-berlin.mpg.de/cgi-bin/toc/toc.cgi?step=thumb&dir=galil_mecha_047_fr_1634

Galilei G (1665) Mechanics. http://echo.mpiwg-belin.mpg.de/ECHOdocuViewfull?url=/mpiwg/online/permanent/archimedes/galil_mecha_070_en_1665

Galilei G (1952) Dialogues concerning the two new sciences. The great books of the Western world, vol 28. Encyclopaedia Britannica Inc., Chicago

Galilei G (1980) Discourse on bodies in waters (trans: T Salisbury). University of Illinois Press, Urbana

Harnest J (1996) Dürer und Perspektive. In: Stridel P (ed) Dührer. Bechtermünz Verlag, Augsburg

Hart JB (1961) The world of Leonardo da Vinci. Man of science, engineer and dreamer of flight. Purnell and Sons Ltd., London

Leonardo (2005) Art and science. Giunti Editore, Florence

Moon FC (2001) Franz Reuleaux: contributions to 19th C. Kinematics and theory of machines. Cornell University, Ithaca

Parsons WB (1939) Engineers and engineering in the renaissance. The Williams & Wilkins Co., Baltimore

Pisano R (2009) On method in Galileo Galilei’ mechanics. In: Hunger H (ed) Proceedings of 3rd Conference of the European Society for the history of science, Austrian Academy of Science, Vienna, pp 174–186

Pisano R (2013) Reflections on the scientific conceptual streams in Leonardo da Vinci and his relationship with Luca Pacioli. Advances in Historical Studies 2(2):32–45

Pisano R, Bussotti P (2014) Historical and Philosophical Reflections on the Culture of Machines around the Renaissance. How Science and Technique Work? Acta Baltica Historiae et Philosophiae Scientiarum 2/2:20–42

Pisano R, Bussotti P (2015) Galileo in Padua: architecture, fortifications, mathematics and “practical” science. Lettera Matematica Springer International 2(4):209–221

Pisano R , Capecchi D (2015) Tartaglia’s Science Weights and Mechanics in XVI century. Selection from Quesiti et inventioni diverse: Books VII–VIII. Springer, Dordrecht

Pisano R, Capecchi D (2009) Galileo Galilei. Notes on Trattato di Fortificazione. In: Altamore A, Antonini G (eds) Galileo and the renaissance scientific discourse. Edizioni Nuova Cultura, Roma

Reuleaux F (1876) Kinematics of machinery: outlines of a theory of machines (trans: Kennedy ABW). McMillan, London
Schiemann G (2005) Nanotechnology and nature. On two criteria for understanding their relationship. HYLE—International Journal for Philosophy of Chemistry 11:1. http://www.hyle.org
Stepin V (2005) Theoretical knowledge, vol 326. Springer, Berlin
Richter JP ([1939] 1970) ed Literary works of Leonardo da Vinci. 2 vols. New York, Oxford University Press
Valleriani M (2001) A view of Galileo's Ricordi autografi. Galileo practitioner in Padua. In: Montesinos J, Solís C (eds) Largo campo di filosofare. Fondación Canaria Orotava de Historia de la Ciencia, La Orotava, pp 285–288
Valleriani M (2010) Galileo engineer. Springer, Dordrecht
van den Belt H (2009) Philosophy of biotechnology. In: Philosophy of technology and engineering sciences: 9 (Handbook of the philosophy of science). Elsevier, Amsterdam, p 1311
Vasari G (1978) Artists of the renaissance. A selection from lives of the artists (trans: G Bull). George Bull, London
Willis R (1870) Principles of mechanism. Longmans, Green, and Co., London

Mathematical Language as a Bridge Between Conceptualization of Motion and Experimental Practice

Ladislav Kvasz

Abstract In this paper we try to distinguish two different styles of experimental practice—roughly speaking the Galilean and the Newtonian. They differ in the way they intertwine mathematics and experimentation. We offer a theoretical reconstruction of the transition from the Galilean to the Newtonian experimental practice. It seems that this transition was brought about by gradual changes of the conceptual framework for the representation of motion. The aim of the paper is to argue that in many of these changes Cartesian physics played an important role.

Keywords Galileo · Descartes · Newton · Force · Interaction · Experimental practice

1 The Galilean Style of Experimental Practice

In January 1610 Galileo Galilei (1564–1642) constructed the telescope and through it he made a series of important astronomical discoveries. He discovered mountains on the Moon, the satellites of Jupiter, the phases of Venus, the sunspots, as well as many new stars. Thus in one single month—January 1610—there occurred more changes in astronomy than during the whole preceding century. Galileo's discoveries played an important role in the defense of the Copernican theory (Swerdlow 1998; Shea 1998). Galileo published his astronomical discoveries in *Sidereus nuncius* (Galilei 1610). The book created a real storm. The reason for the intense reactions was not only the novelty and significance of the discoveries themselves, but also the fact that he made them using a telescope. His critics accused Galileo

L. Kvasz (✉)
Institute of Philosophy, Academy of Sciences of the Czech Republic, Prague, Czech Republic
e-mail: ladislavkvasz@gmail.com

R. Pisano (ed.), *A Bridge between Conceptual Frameworks,* History of Mechanism and Machine Science 27, DOI 10.1007/978-94-017-9645-3_13

of naivety. At that time, the telescope was considered to be an illusionist toy which shows the phenomena not as they really are, but altered. Therefore "observations" with a telescope are unreliable and cannot be a basis for a serious science. Science has to examine the phenomena as they really are. Galileo's grounding scientific theories on "observations" made with a telescope was considered to be similarly naive as to try to develop a theory from "observations" made by a curved mirror. Galileo wanted to persuade his opponents and therefore he sent them a telescope, so that they could see with their own eyes what he spoke about. "The majority of the natural philosophers simply did not think it worthwhile even to look through his telescopes." (Ronchi 1967, p. 201). And it was not mere reluctance. A lens creates images which are greater or smaller than the real object observed by the naked eye. The object seems to be nearer or more distant than it really is and sometimes it is even turned upside down. Thus the lenses do not show true images of the things, but create illusions.

With his telescope Galileo brought a fundamental change in the notion of observation. The classical astronomical instruments like the *sextant* or the *astrolabe* were only put alongside the axis that connects the eye with the object on the sky. They did not change the way in which a particular object is disclosed to our view in everyday experience. Thus these classical instruments only sharpen our natural experience. Their precision has limits given by the resolving power of our eye. On the other hand Galileo's *telescope* enters between the observed object and us. It makes it possible to see things which without its help we are unable to see, as for instance the satellites of Jupiter or the countless amounts of new stars whose magnitude is below the threshold of our eye's sensitivity. Thus the telescope expands in a fundamental way the scope of our experience. This *new kind of instrumentation of experience* is one of the characteristic features of Galilean experimental practice.

Instruments like the telescope broaden the realm of our experience. Nevertheless, they do not intervene into the constitution of the observed phenomena—they only change the resolving power of our senses. There is, however, a whole range of phenomena, for grasping of which the instrumentation of observation is insufficient. If we take for instance free fall, we are unable to see whether it is uniform or not. We are not able to perceive it as something ideal and perfect. And no instrumentation of perception can be of help here, because the problem lies not in the insufficient resolving power of our sight, but in the ambiguity of the perception of motion. Free fall is a motion, and therefore time enters in a fundamental way into its constitution. Nevertheless, we have no specific organ for the perception of time, the resolving power of which could be eventually enhanced by some instrument. Some phenomena are too complex and we cannot grasp directly the mathematical forms which determine them. Therefore, according to Galileo it is necessary to create simplified situations in which the phenomenon would disclose itself in its purity and would reveal its mathematical form. The creation of such simplified situations requires invention, and Galileo's analysis of free fall is a beautiful example of such an invention.

For Aristotle free fall and horizontal motion were qualitatively distinct. Free fall was a natural motion, because the body moved towards its natural position. On the other hand, horizontal motion (in the sublunary region, of course) was an unnatural motion, requiring an external mover. Galileo's idea was to consider these two motions from the point of view of the artificial motion on an inclined plane. Free fall is a motion on a totally inclined (i.e. vertical) plane, while horizontal motion is a motion on an inclined plane the inclination of which is zero. Therefore, by continuously changing the inclination of the plane we can pass from free fall to horizontal motion and back. In this way Galileo's imagination, using the artificial device of an inclined plane, connected two phenomena which apparently have nothing in common. This connection has considerable technical advantages, because the motion on an inclined plane is relatively slow and therefore it is more suitable for observation than free fall. What Galileo discovered was a regularity: the distance passed by the ball grew as the square of the time. After the first pulse the ball reached the first line, after the second pulse the fourth, after the third pulse the ninth line. If we increase the inclination of the plane, the motion will accelerate. Nevertheless, the basic regularity—distance proportional to the square of the time—will be preserved. From this we can derive the conclusion that in the limit case of the vertical plane the distance will still be proportional to the square of time. It is plausible, even if we cannot observe it directly. Thus an experiment, by creating an artificial situation in which the mathematical form of the phenomenon is accessible to observation, sheds light on situations, in which the mathematical form remains hidden. The motion down an inclined plane made it possible to discover the law of free fall. We arrived at a concept of experiment, which is central for Galilean physics. *An experiment is an inventive disclosing of the mathematical form of phenomena using artificial situations*. It is a characteristic feature of the Galilean style of experimental practice that *the experimentally studied phenomena are accessible to ordinary experience.*[1]

The aim of an experiment is to create by help of an artificial situation access to the mathematical form of phenomena. After achieving that, its task is usually finished. Nevertheless, when Torricelli created a vacuum in a glass tube, it was not the end of the story. The reason was that the phenomenon of atmospheric pressure, the mathematical form of which he disclosed in this way, is not accessible in any other way. In ordinary experience we are not aware of atmospheric pressure, and many cultures did not even suspect that there existed something like this. In this respect there is a radical difference between heat and pressure. Heat is disclosed to ordinary perception and therefore the thermometer can be interpreted as an instrument that only sharpens the perception of heat. With the atmospheric pressure the situation is different. Without a barometer we have no idea even of the existence of this phenomenon. That is why Torricelli's tube did not "end in a museum" (i.e. did not become of interest only to historians), but was transformed into the barometer, which is a device opening an access to the phenomenon of pressure. A measurement is a standardization of experiment. Thus in order to understand what measurement

[1] Even though air pressure (which we will discuss below) is not directly accessible to ordinary experience, there is nevertheless a directly observable phenomenon—the strange behavior of the water pumps that refuse to pump water from depths greater than a certain limit—that is accessible to ordinary experience.

is, one has to keep in mind what an experiment is. An experiment is the disclosing of the mathematical form of a phenomenon by means of artificial situations. *A measurement is based on the standardization of the artificial situation of the experiment i.e. of the objects, relations and procedures that constitute it.* For instance in the case of the barometer we fix the diameter and the length of the tube, the amount of mercury. We may also determine the number scale which we fix to the tube, choose the suitable physical units and determine the scope of temperatures at which the barometer gives reliable results. In this way we secure the reproducibility and so the intersubjectivity of the measurement.

As long as physics operates in the area of phenomena to which we have an immediate access through our senses it is possible to understand measurement as a process of refinement of the picture of reality, offered by the senses. For instance in the case of *free fall* we are not able to decide by the use of our senses, whether it is uniform or accelerated. Nevertheless, we know free fall from our experience and thus we are inclined to interpret the measurement as a device that only helps us to determine that free fall is accelerated. In the case of *temperature*, the interpretation of measurement as a process of refining the sensory image which we get by the immediate contact with the body whose temperature we are measuring becomes more problematic. The problem is that we are able to measure the temperature of bodies that are so hot that by touching them our hand would be carbonized, and so we cannot speak about any sensory image. In the case of *atmospheric pressure* it is even worse. The gradual decrease of pressure manifests itself first by a headache, and it ends with the explosion of our organism. To speak about the measurement of pressure as making our sensory impressions more precise is impossible. What impression corresponds to the pressure of 0.01 atmospheres? *Measurement extends reality beyond the boundaries of the phenomenal world.* So we cannot interpret measurement as a simple refinement of the phenomena.

But the situation is even worse. We cannot interpret measurement even as a prolongation of the phenomenal world. The reason is that the picture of the world offered by the measuring devices comes into a conflict with the picture based on everyday experience. For instance, all motions which we can see around us on the Earth have a natural tendency to stop. In contrast to this, the account of motion offered by physics says that all motions are inertial and their ceasing is only the result of friction. In this way the natural, everyday experience is deprived of legitimacy. It starts to be regarded only as an *inaccurate and distorted picture* of the "real" reality which presents itself in measurement. We can find this shift already in Galileo.

> Hence I think that tastes, odours, colours, and so on are no more than mere names so far as the object in which we place them is concerned, and that they reside only in the consciousness. Hence if the living creature were removed, all these qualities would be wiped away and annihilated. (Galilei 1623, p. 274).

Galileo tells us, that physically real is not the picture, disclosed to us by our senses, but only a part of it, which we are able to grasp by help of measurement.

2 Four Ways of Conceptualizing Motion

The most radical changes brought about by the Galilean experimental practice concern the concept of motion. So in the next four sections we will offer a reconstruction of the development of conceptualization of motion from Galileo till Newton. It will turn out that this development necessitated the transition to a different style of experimental practice. (For more details see Kvasz 2002, 2003, 2005.)

Generally speaking, it is possible to distinguish four different ways of representing mechanical motion. The Aristotelian description of local motion can be characterized as a *geometrical transition*.[2] According to Aristotle, everybody has its natural place determined by the geometrical structure of the universe, and motion is a transition from one place of this geometrical structure to another. Galilean theory of motion can be seen as a theory of a *geometrical flow*.[3] Thus Galileo has replaced the Aristotelian concept of motion as a transition from one place to another by the concept of motion as a flow along a particular trajectory. The nature of the motion is, however, still given by the geometrical properties of its trajectory, and the global structure of the universe remains a geometrical order.

The basic innovation of Descartes can be interpreted as the replacement of a *geometrical* theory of motion by a *dynamic* one. Motion cannot be understood in terms of geometry because geometry does not allow us to understand interaction. Descartes described an interaction as a collision, i.e. as a transition from the initial state (the state before the collision) to the terminal state (the state after the collision). Descartes' theory is based on the comparison of two states and so it can be described as a theory of *dynamic transition*. Being a theory of transition, it resembles the Aristotelian theory. But there is also a deep difference between the Aristotelian and the Cartesian concepts of motion. Motion according to Descartes is not a transition from an initial position to a terminal one in a geometrically ordered, static universe. It is a transition from an initial state to the terminal state in a mechanically united dynamic universe.

---

[2] *Geometrical transition* means that the basic representation of motion is *geometrical* (as opposed to a *dynamic* one): it describes motion as a change of *position* in space rather than as a change of *state*. Further, *geometrical transition* represents motion as *transition*, which means that the focus of attention is on the starting point and the terminal point of motion, while the process of motion itself does not enter the representation. (This does not mean that the change of *state* or the *process* of change is not thought about, but only that they are not included into the representation of motion. Thus even though Aristotle understands change as a becoming, in the definitions of (violent) motions these aspects do not enter.).

[3] *Geometrical flow* means that the representation of motion is still *geometrical*, it concerns a change of position in space and not a change of state. Nevertheless, it is a *flow*. It represents motion as a continuous passing along a trajectory (as opposed to a *transition* from the starting position to the terminal one).

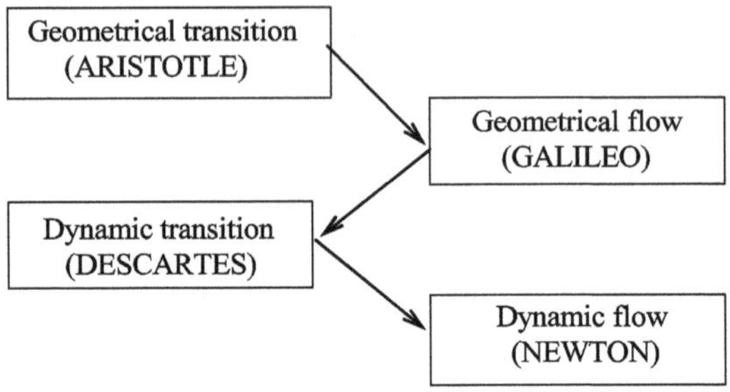

We can characterize Newton's theory as a similar transformation of Cartesian physics as Galilean theory was of Aristotelian physics. Galileo replaced the Aristotelian idea of motion as a transition from an initial position to a terminal one by a flow along a trajectory connecting the two positions. Similarly Newton's theory replaces the Cartesian idea of transition between states by a *dynamic flow* along a vector field. These resemblances and differences between the four mentioned theories can be summarized in the form of the diagram above. Its aim is to represent the inner tension of the Cartesian system. In one sense Descartes moves from Galileo backwards to Aristotle (as he employs the concept of transition instead of flow), but in another sense he is a step forwards in that he incorporates interaction into the representation of motion.

3 Aspects of the Galilean Conceptualization of Motion

Galileo's law of free fall is often described as the first scientific law, i.e. regular correlation between empirical quantities expressed in mathematical form. Nevertheless, there is a number of conflicting interpretations of the role of Galileo in the history of modern science. Historians differ in their interpretations of the core of the Galilean project. Some of them see the main contribution of Galileo in his experimental method (Drake 1973; Hill 1988), others in his mathematical Platonism (De Caro 1993), still others stress his use of the Aristotelian deductive method (Wallace 1984) or of a combination of experiment and deduction (Naylor 1990). It seems that they do not exclude each other but rather represent different aspects of Galileo's scientific work which existed side by side.

Despite his fundamental contributions, which are well known and so there is no need to deal with them here, Galileo's ideas had also some grave shortcomings which are the reason why modern science is not a direct continuation of the Galilean project. The problem is not with Galileo's mistakes (his conviction that inertial motions are circular or his reluctance to accept Kepler's discovery of the elliptical shape of the planetary orbits). Such mistakes can easily be corrected. When speaking about shortcomings of the Galilean system we have in mind several problems of Galilean conceptualization of motion. First of all, Galileo's concept of motion is to a great extent a geometrical concept. Galilean physics *lacks a framework for the representation of interaction.* His description of motion is always a description of the motion of a single, isolated body. The laws discovered by Galileo bear a witness to this. The law of free fall, the law of the isochrony of the pendulum, or the law of the trajectory of projectile motion, these are all laws describing bodies without interaction. Secondly, Galileo has a *too narrow concept of a natural law*. The laws mentioned above lack generality. Be it the law of free fall or the law of the pendulum, they are laws describing *particular phenomena.* In Galilean science for each phenomenon there is a special law that describes it. And finally, Galilean physics *lacks any description of states*. Galilean physics deals only with observable quantities and tries to discover regularities in them. Contemporary science, on the other hand, is based on the description of the state of a physical system (using a *Lagrangian* or a *Hamiltonian function*) and its temporal evolution (by means of *Lagrangian* or *Hamiltonian equations*).

It seems that these shortcomings have a common root. They are the consequence of the use of *geometry* as the language of science. Galileo believed that the book of nature is written in the language of mathematics:

> Philosophy is written in this grand book, the universe, which stands continually open to our gaze. But the book cannot be understood unless one first learns to comprehend the language and read the letters in which it is composed. It is written in the language of mathematics, and its characters are triangles, circles, and other geometric figures without which it is humanly impossible to understand a single word of it; without these, one wanders in a dark labyrinth. (Galilei 1623, p 237, p. 238).

This passage is often quoted, but its strange nature is rarely recognized. Modern science *is not* based on any triangles, circles, and other geometric figures but on functions and differential equations. The fact that the laws discovered by Galileo all describe isolated bodies without interaction can be brought into connection with his choice of geometry as the language of science. Splitting nature into isolated phenomena, reducing natural laws to mere phenomenal regularities, and sticking to observable quantities is closely connected to the role which Galileo has given in his scientific project to geometry. The language of synthetic geometry is too concrete. It does allow neither for universal laws nor for interaction.

4 Cartesian Conceptualization of Motion as a Foil to the Newtonian Conceptualization of Motion

While nobody would seriously question the great value of Galileo's contributions to the development of modern science, things are not nearly so simple with respect to Rene Descartes (1594–1650). Descartes is omitted in many expositions of the development of physics and Newton is seen as deriving directly from Galileo. It is sufficient to quote the words of Stephen Gaukroger:

> With the exception of the work in optics, his contribution to the development of classical physics is minimal. Insofar as kinematics is concerned, Cartesian physics accomplishes considerably less than had been achieved by Galileo in his Two New Sciences, and insofar as Descartes' physics can be considered a dynamical theory it is often hopelessly confused, particularly in comparison with Newtonian dynamics. (Gaukroger 1980, p. 123).

The views of Daniel Garber are rather similar:

> Descartes' intellectual program failed, of course; while pieces of the program may have proved important inspirations to later thinkers, as an approach toward understanding the natural world Descartes' program turned out to be a dead end. But while the design may have been faulty, and the edifice doomed from the start, it is fascinating to contemplate the entire structure as the architect planned it, [...] (Garber 1992, p. 2).

It seems that this interpretation of the rise of modern science is not adequate. Excluding Descartes from the main stream of the history of science prevents us from understanding the origins of two central features of physics: its ontological homogeneity, and descriptive universality.

Descartes, if he had wanted to, could have worked out the Galilean project much further than Galileo was able to, because he was a much better mathematician. Descartes was one of the creators of *analytic geometry* and he introduced the *algebraic notation*, which is still in use. He could therefore develop the ideas which Galileo arrived at by help of a cumbersome symbolism and a rudimentary idea of a coordinate system, in a much more elegant way using analytic geometry and symbolic algebra. Nevertheless, Descartes was not interested in the motion of isolated bodies as Galileo was, but in the interactions among bodies, a phenomenon Galileo never understood. From Descartes stems the idea that science should search for *universal laws*, that these laws should *describe interaction* and that this description of interaction should have an *ontological foundation*. Despite the fact that Cartesian physics is formulated in ordinary language, which seems to have misled many historians of science, *in a deeper sense Cartesian physics is algebraic*. Cartesian physics takes from algebra the universality of its laws. Even though these laws are formulated in ordinary language, they have the *same kind universality as algebraic formulas*. The language of algebra enabled Descartes to create the first description of interaction. He did it by introducing the notion of the quantity of motion, and formulating the law of conservation of the quantity of motion. It is important to realize that the quantity of motion is *an algebraic quantity* and the law of its conservation is *an algebraic equation*. And finally, we have to realize that the quantity of motion is not a phenomenal quantity. It cannot be measured directly. It is an ontological quantity; it is related to the state of a physical system and not to its appearance. Thus Cartesian physics receives its universality, its ability to represent interaction, and its ontological grounding from the universality, functionality, and abstractness of the language of algebra.[4]

---

[4] It seems that the *extended body* of Cartesian physics is closer to the *cosa* of the *Cossists* than to the Aristotelian *pragmata*. An extended body enters into mechanical interactions with other bodies just like a *cosa* enters into algebraic relations with other quantities. Furthermore, an extended body has only mechanical properties just like a *cosa* has purely algebraic ones.

I would like to show that it is a mistake to interpret Newtonian physics as a continuation of the Galilean "*non-metaphysical and problem-oriented conception of natural philosophy*". The Galilean system lacks universal laws as well as a description of interaction, and so it is too far away from the Newtonian system. Therefore the Galilean system cannot serve as a background for understanding of what Newton did when he created his physics. Only when we put Newtonian science against the backdrop of the Cartesian system will it be possible to understand some peculiar features of Newtonian physics. Therefore let us now turn to a reconstruction of the relations between the Cartesian and the Newtonian systems. It can be argued that Newton took the idea of universal laws, the idea of interactions, and the idea of ontology in physics from Descartes. Of course, the *particular* universal laws by which Newtonian physics described a physical system were very different from the laws which Descartes ascribed to it. Similarly, the *particular* way how Newtonian science described interactions among bodies was very different from the Cartesian description of interactions. And finally, the *particular* ontology on which the Newtonian system was based, differed substantially from the Cartesian ontology. Thus not the technical details, not the way in which Newton formulated his laws, described interactions, and introduced ontology were Cartesian. Nevertheless, the very idea that science should search for universal laws, that these laws should describe interactions and that this description of interactions should have an ontological foundation was Cartesian. That is the reason why, despite all the antagonism between the Newtonians and the Cartesians, I believe that there was a fundamental influence of Cartesian philosophy on Newtonian science.

If we take any of the laws discovered by Galileo—the law of free fall, the law of isochronous motion of the pendulum, or the law of the descent on an inclined surface—all these laws are laws describing particular phenomena. On the other hand the laws of the Newtonian system as the law of inertia, the law of force, or the law of action and reaction, are *universal laws*. We can take a falling body, a pendulum, or a body descending on an inclined surface, the tree Newtonian laws apply to all of them. The law of conservation of the quantity of motion, introduced by Descartes, seems to have been the first universal law in physics. This law was universal because, like Newton's laws, it applied to a falling body, to a pendulum, as well as to a body sliding on an inclined surface. Therefore, in using universal laws in the description of nature, Newtonian science was Cartesian rather than Galilean.

Another interesting common feature of the laws discovered by Galileo is that they are laws describing the behavior of single isolated bodies. The law of free fall describes the fall of one single body just like the law of the pendulum describes the periodical motion of one single body, and the law of the descent on an inclined surface describes the descent of one single body. Thus it is fair to say that Galilean physics lacked the notion of interaction. On the other hand, the fundamental aim of Newtonian physics was to *describe interaction among bodies*. The first theoretical description of interaction among bodies was in all probability given by Descartes, when he introduced the notion of a dynamic state (characterized by extension and motion) and described interaction as a change of this state (in collisions). Therefore in describing interactions among bodies Newtonian science was Cartesian rather than Galilean.

A third characteristic of Galilean science is that it lacked ontology. Galileo intended to develop science in a phenomenological way (I use the term phenomenology not in the Husserlian sense, but in the sense as this term is used in physics). The aim of Galilean science was to discover the mathematical regularities hidden in the phenomena. In contrast to this notion of science, Newtonian physics had a *clear (corpuscular) ontology*. The first who realized the necessity to base physics on ontological foundations was Descartes when he introduced extension as the ontological foundation of the physical description of phenomena. So by building its theories on explicit ontological foundations Newtonian science was Cartesian rather than Galilean.

Newton rejected the Cartesian laws of nature, the Cartesian description of interaction, as well as the Cartesian ontology. Nevertheless, he owes to Descartes the idea that the laws of nature must be universal, that they must describe interactions among bodies, and that these bodies must have some ontological status. These ideas are fundamentally Cartesian, and thus it is fair to say that Newton was closer to Descartes than to Galileo. Some influence of Descartes on Newton is visible already in the title of Newton's *Philosophiae Naturalis Principia Mathematica*. This title is an allusion to the title of Descartes' *Principia philosophiae*. Besides this allusion there are similarities also in the general structure of the two systems. Newton's system has three laws of motion just like the Cartesian system and Newton's formulation of his law of inertia is simply a juxtaposition of the first two laws of Descartes. Nevertheless, these are rather superficial similarities that do not touch the content of the two systems. If we go further, and leave these salient similarities aside, we will find that there is a much stronger sense in which the Cartesian system exerted influence on Newton. The point is that the *main problems solved by Newtonian physics were of Cartesian origin*. To see this we have to concentrate on the main shortcomings of Cartesian physics (just like we concentrated on the main shortcomings of the Galilean system when we wanted to see more clearly the contributions of Descartes). Among the shortcomings of the Cartesian system were: a too loose connection between the phenomenal and the ontological levels; causal openness of the description of motion and an unsatisfactory description of interactions. If we look from this perspective on the Newtonian system, we see that the main achievements of Newtonian science were in a sense answers to, or solutions of, the shortcomings of the Cartesian system (in a similar way as the main achievements of the Cartesian system can be interpreted as answers to, or solutions of, the main shortcomings of the Galilean system).

The first shortcoming of the Cartesian system was that it had only a loose connection between the phenomena and the explanatory models that were used to account for them. Thus for instance Descartes explained the phenomenon of gravity by his vortex model. He postulated a vortex of fine matter, but gave no clue how particular aspects of the vortex (its velocity, structure, orientation, etc.) relate to specific attributes of gravity (its homogeneity, direction, permanence, etc.). This was obviously a main weakness, which gave the whole theory a speculative flavor. In the Newtonian system the ontological level and the phenomenal level are tied together by a mathematical framework, which allows to derive from an attribute of the phenomenon a corresponding aspect of the ontology and vice versa. Thus Newton could for instance derive the inverse square law of universal attraction (i.e. a particular aspect of the force of gravity, which, as all forces, belonged to the ontological level) from Kepler's laws (which belonged to the phenomenal level). But even if these close ties between ontology and the phenomena are a non-Cartesian aspect of the Newtonian system, they can be seen as an answer to a deep tension of the Cartesian system—the unreliability of its explanatory models. And so in an indirect sense it is a Cartesian aspect of the Newtonian system.

Another weakness of the Cartesian system was that its description of motion was causally opened. The mind could, according to Cartesian physics, have a causal influence on the body. Thus *a physical process*, as for instance the lifting of my arm, *can be caused by a nonphysical event*, in this case by my decision to do so. Descartes' description of motion was not causally closed. This causal gap is closely related to the fact that according to Descartes velocity is a scalar quantity. Therefore a change of direction of motion does not influence the value of the quantity of motion. Consequently the law of conservation of the quantity of motion does not determine the changes of direction of motion and thus in the Cartesian system there is a gap where the mind can intervene. Newton closed this gap when he introduced the notion of velocity as a vector quantity. Therefore changes of direction of motion are in the Newtonian system changes of the quantity of motion, and so they must be caused by forces (i.e. by physical causes). Even though the Newtonian notion of velocity as a vector quantity is a non-Cartesian concept, Newton introduced it in order to solve a deep problem of the Cartesian system—its causal openness. And so in an indirect sense it is a Cartesian aspect of the Newtonian system.

A further shortcoming of the Cartesian system was that the notion of the quantity of motion was introduced for the universe as such and so it included the motions of all bodies in the universe. Therefore, strictly speaking, it was impossible to calculate its value. So even if Descartes introduced this notion in order to describe interactions, it could not be applied to any concrete situation. This made, of course, the law of conservation of the quantity of motion practically useless. It is true that Descartes used his law in the description of collisions of bodies. But all these descriptions were counterfactual because in reality, according to Descartes, all bodies were submerged in a vortex of fine matter, which took away portions of the quantity of motion. Therefore in the Cartesian system the law of conservation of the quantity of motion could hold only for the whole universe. Only when Newton turned to empty space as the background of the theory of motion, the conservation of the quantity of motion in smaller systems became possible. Thus by eliminating the Cartesian fine matter Newton opened the possibility to *describe restricted mechanical systems*. For the description of such systems he created a new mathematical tool—differential equations (or, more precisely, something which we today call differential equations). Newton's second law was perhaps the first differential equation in history. The notion of a differential equation, which is a mathematical tool that can describe the temporal evolution of the state of a mechanical system, is a non-Cartesian notion. But Newton introduced it in order to solve a tension in Cartesian physics—its inability to describe interactions in a restricted system. And so in an indirect sense it is a Cartesian aspect of Newton's physics.

We have seen that the most important achievement of Newtonian physics—the mathematical description of interactions in a causally closed mechanical system of finite extension—was an answer to problems inherent in the Cartesian system. Thus, although Newton's system of natural philosophy was very different from the Cartesian system, Newton developed some of the most important aspects of his system in reaction to deep conceptual problems and internal tensions of the Cartesian one.

5 Aspects of the Newtonian Conceptualization of Motion

In the previous chapter I presented several arguments in order to justify taking the Cartesian system as a contrasting background of the Newtonian system. I believe that only against this background will it become sufficiently clear that the fundamental innovations of Sir Isaac Newton (1643–1727) in physics are based on infinitesimal calculus. The infinitesimal aspect of Newtonian physics is hidden at the first sight. In this respect the Newtonian system is analogous to the Cartesian one—in both cases the linguistic surface and the epistemological structure are very different. In the case of the Cartesian system it is difficult to see the fundamental role of algebra in its representation of motion, because the whole system is formulated in ordinary language. Only when we contrast the Cartesian system with the Galilean, its *descriptive universality*, *dynamic unity* and *ontological foundations* become visible. All these hidden features of the Cartesian system are algebraic in their character. So despite being formulated in ordinary language, Cartesian physics is based on the algebraic framework.

After clarifying the connection between the Cartesian and the Newtonian systems it becomes obvious that the Cartesian system is a suitable background for Newton, and we can turn to the analysis of the technical details of the Newtonian description of interaction. On the basis of this analysis, I will argue that Newton's breakthrough in physics was made possible by a fundamental change of the mathematical framework—the creation of the infinitesimal calculus. For some readers this may seem obvious, but we have to keep in mind that Newton's *Principia* are written in the geometrical language and there are historians who dispute the relevance and even the legitimacy of the use of infinitesimal calculus in the historical reconstruction of Newtonian physics.

In the previous chapter we indicated that Newton's creation of his description of interaction had its roots in the conceptual problems of Cartesian physics. It is fascinating to see that many of the components which Newton used in building of his own theory were present already in the Cartesian system, even though they had there a different function than they have in the Newtonian system. We will concentrate on three Newtonian components: the description of interaction as a *transfer of momentum*; the notion of *force*, and the idea that the transfer of momentum is *governed by forces*. It is almost unbelievable, but all three of these fundamental components of the Newtonian theory were present already in Descartes. Descartes understood interaction as a transfer of a particular quantity of motion from one body to another; he used in his theory of interaction the notion of force; and he understood that the transfer of the quantity of motion is governed by forces. Thus Newton could find all the necessary components of his theory of interaction present in Descartes. But the way how these components were put together in the Cartesian and in the Newtonian theory are fundamentally different. Newton had to *exempt* these notions from the Cartesian context, *change* them in a substantial way, and then *reunite* them in a rather different order. The main point of the present section is that Newton was enabled to make these changes thanks to his invention of the infinitesimal calculus.

5.1 Interaction as Transfer of Momentum (Quantity of Motion)

The Cartesian notion of interaction is based on the idea of a *contest*, understood as the collision of two tendencies to preserve the previous state of rest and motion respectively (see Gabbey 1980). The result of the contest is the victory of one tendency at the expense of the other. The paradigmatic Cartesian model of interaction is a collision of two bodies. According to Descartes the greater body determines the outcome of the collision and thus also the further motion of both bodies. If the moving body is greater, then after the collision both bodies will move together in the direction of the original motion. Nevertheless, the description of interaction is *separated* from the description of motion. As long as a body can, it moves uniformly in a straight line. When such motion becomes impossible a collision occurs. So the motion of a body consists of periods of uniform motion in a straight line that are separated from each other by *singular events*—collisions, when the state of the body is changed.

Let us consider a body *B* moving with the velocity V_B colliding with a resting body *C*, and let us assume that the moving body is bigger. Before the collision the total amount of the quantity of motion was $B \times V_B$. After the collision both bodies will, according to Descartes, move together with a velocity that can be determined from the law of conservation of the quantity of motion as $V = \frac{B \times V_B}{B + C}$. We see, that the original velocity V_B is here multiplied by the factor $\frac{B}{B + C}$, which is smaller than 1. This means, that the body *B* was decelerated, i.e. it lost a particular amount of its original quantity of motion. On the other hand the resting body *C* started to move, and so it acquired a particular quantity of motion. Both of these quantities are equal to $\frac{C \times B \times V_B}{B + C}$. *Thus according to Descartes the interaction consists in an exchange of a particular quantity of motion between the interacting bodies.* Basically the same thing holds also about the Newtonian notion of interaction, even though Newton changed many aspects of the Cartesian notion of interaction (Fig. 1).

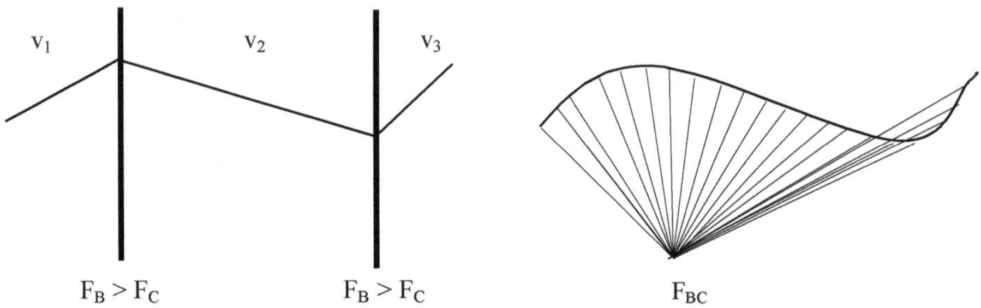

Fig. 1 **a** Descartes' notion of interaction. **b** Newton's notion of interaction

In contrast to Descartes, Newton understands interaction as *cooperation.* In the process of interaction the faster body accelerates the slower body and at the same time the slower body decelerates the faster one. The resulting motion is a compromise, it is the result of the action of both bodies. Thus the result of a collision is neither a simple re-bouncing from the obstacle, nor is it a coupling of the two bodies into one, but something in-between. A second, maybe even more important change is that motion and interaction happen *simultaneously*. They are not separated from each other as in Descartes. According to Newton the forces act all the time and their action accompanies the whole motion. A third change is that interaction is not a singular event, it happens not during isolated moments in time as in Descartes. According to Newton a body acts on the other body during an *infinitesimal time interval* d*t* (or *o*). It is true, that Newton still speaks about impulses of forces, but in all concrete calculations he makes a limit transition. In the limiting case the impulses are becoming infinitesimally dense, and the magnitude of each separate impulse becomes infinitesimally small, thus at the end we are getting a continuous picture. And it is this continuous picture that is important, because all the relations which Newton uses in his calculations hold only for this limiting case.

5.2 The Concept of Force

Another ingredient of the Newtonian theory of motion, which is clearly of Cartesian origin, is the notion of force. Galileo did not use this notion in his physics, because he considered the idea that one body could act on another at a distance as occult. Forces were introduced into physics by Descartes. The role of forces in the Cartesian system was to *preserve the state* of a body. They did not act between bodies; they rather bound each body to its present state of rest or uniform rectilinear motion. Therefore we can represent them using arrows oriented downwards. They play a role only during the moments of collision, when they decide about the direction and velocity of the next interval of uniform rectilinear motion of the body.

Let us consider again the moving body *B* colliding with the resting body *C*. Descartes defined the force for proceeding of the moving body *B* as $F_B = \dfrac{B^2 \times V_B}{B+C}$, that is as the product of the magnitude of the moving body *B* and the common velocity after the collision $V = \dfrac{B \times V_B}{B+C}$. On the other hand, the force of resisting of the

Fig. 2 **a** Descartes' notion of force. **b** Newton's notion of force

resting body C he defined as $F_C = \frac{C \times B \times V_B}{B+C}$, that is as the product of the magnitude of the resting body C and the common velocity after the collision. Depending on which of these two forces is greater—the force for proceeding F_B in the body B, or the force of resisting F_C in the body C—the outcome of the collision will be either that B imposes its motion on C, or that B will simply rebound, while C preserves its rest. Thus, *according to Descartes, the force of a body acts upon the body itself and it simply preserves its state of motion or rest* (Fig. 2).

According to Newton, a force is something, by the virtue of which *one body acts on another body*, and causes a change of its state. The role of the preservation of the state, which Descartes ascribed to forces, is in the Newtonian system played by masses. Newton by introducing the notion of mass liberated the forces from the role of binding bodies to their own states and thus opened the possibility to ascribe forces a new role—the role of changing the states of other bodies. Newtonian forces are forces of interaction; they act along the line connecting the two bodies.

5.3 *The Role of Force in the Exchange of a Quantity of Motion*

According to Descartes, the result of a collision depends on the question which force is greater, whether the force for proceeding in the body B, or the force of resisting in the body C. The collision is thus determined by a relation of the form

$$\frac{B^2 \times V_B}{B+C} > \frac{C \times B \times V_B}{B+C}.$$

The quantity on the left-hand side is the force for proceeding in the body B, the quantity on the right-hand side is the force of resisting in the body C. These quantities have the same denominators, and actually the only difference between them is the magnitude of the bodies. The other quantities cancel each other. Thus we arrived at the Cartesian result that the moving body B wins the contest if and only if $B > C$, i.e. if the magnitude of the body B is greater than the magnitude of the body C.

Now we see why Descartes maintained (in spite of contrary evidence) that a moving body can bring a resting body into motion only if it is greater. It is a necessary consequence of the formula. When this happens, the moving body B must pass a portion of its quantity of motion to the resting body C in order to start its motion.

We see that Descartes saw correctly that interaction consists in the transference of a particular quantity of motion from one body to the other. Nevertheless, *in the Cartesian system the passing of motion from one body to another was separated from the action of forces*. The forces decided about the outcome of the contest, they decided whether the result will be the re-bouncing of the moving body or a common motion of the two bodies. The forces did not play any role in the following transference of motion from one body to the other. This transference was governed only by the law of conservation of the quantity of motion. Descartes thus represented interaction on two levels. The first level consisted in the contest of the forces and it was governed by the above formula. The second level consisted in the transference of motion between the bodies, and was governed by the conservation law. But these levels were separated from each other.

The unfolding of the interaction from a singular moment into the time interval dt and the notion of forces as forces of interaction enabled Newton to *join* the action of forces with the transference of momentum. Descartes defined the force for proceeding as $F_B = \frac{B^2 \times V_B}{B+C}$, and the force of resisting as $F_C = \frac{C \times B \times V_B}{B+C}$. Even if these definitions seem similar (the force being in both cases defined as the product of the magnitude of the body and the common velocity after the collision), there is a remarkable *conceptual conflict* hidden in them. Descartes defines the force for proceeding as the *residual* momentum, which is left to the body B after the collision, while he defines the force of resisting as the *gain* of momentum, which the body C acquires in the collision. Thus it seems as if Descartes hesitated between two ways of connecting forces with momentum.

According to Newton, force is equal neither to the residual momentum nor to the gain of momentum, but to the *velocity of the change of momentum*. Descartes could not understand this connection, because he described interaction as a singular event in time. Therefore, in the Cartesian system, there is no way how to introduce the notion of velocity of change of momentum. We see the fundamental importance of the fact that Newton embedded the transference of momentum into the flux of time. Interaction is for him not a singular event, but it has temporal extension. This made it possible to connect force with the velocity of the change of momentum in his second law

$$\mathbf{F}\mathrm{d}t = \mathbf{d}p.$$

Thus one of the fundamental achievements of Newton was that he connected the change of momentum with the action of a force. The importance of this fact is often misunderstood, and Newton's second law is considered as a mere definition of the concept of force (for instance in Nagel 1961, p. 160). The fundamental conceptual work, which lies behind it, is thus veiled. Newton had to make profound changes in both, the concept of force and the concept of momentum, and above all he embedded the whole interaction into a continuous flux of time, to be able to connect the action of a force with the change of momentum. Only after all this conceptual work is done it became possible to use Newton's second law as an implicit definition of force.

5.4 Summary

We have seen that Newton's theory of interaction consists of the same components as the Cartesian theory: it understands interaction as an *exchange of momentum* (or quantity of motion), it *uses forces* in the description of interaction, and it *relates forces to the changes* of momentum. But these ingredients are in both theories understood quite differently. For Descartes an interaction is a *singular* event. This event has the form of a *contest*. The contest is governed by forces that are *forces of inertia*. These forces are equal to the *residual* or to the *transferred quantity of motion*. In contrast to this, for Newton interaction is a *continuous* process. This process has the form of *cooperation*. This cooperation is governed by forces that are *forces of interaction*. These forces of interaction are equal to the *velocity of the transfer of momentum*. Nevertheless, this very possibility to expose the whole Newtonian theory of interaction and in each fundamental point to be able to contrast it against the Cartesian background clearly shows the conceptual closeness of the Newtonian and the Cartesian theories. It is obvious that no such comparison would be possible say between the Newtonian and the Galilean theory of motion.

6 Conclusion. The Newtonian Style of Experimental Practice

The contribution of Newton to the experimental method consisted in finding a way of studying empirically the ontological basis of reality. Newton developed his new approach to experimental method during his study of colors in 1665–1667 (Hakfoort 1992, pp. 115–121). When Cartesian physics is criticized that it is purely speculative and does not care for empirical data, this criticism is not justified. For instance Hooke was one of the most prominent experimental scientists of his time and his style of experimental work was Cartesian. The Cartesian mechanical philosophy grew out, at least partially, from a criticism of the Galilean experimental method. According to Descartes, it is not enough to study different aspects of reality, but it is necessary to create a mechanical picture about its functioning. The core of Cartesian science is the endeavor to discover the mechanisms which are at the core of the experimental data. Thus the problem is not that the mechanical philosophy would not use experiments but rather that *the experiments are separated from the theoretical work* on the explanatory models which remained speculative. Descartes made experiments, but he used them only to activate his imagination. Theoretical work started after experimental work stopped.

Newton realized that for the further development of mechanical philosophy an experimental control of its theoretical models was necessary. The difficulty was that in experiments only the phenomena are accessible, while the theoretical models postulate ontological entities, which we cannot experience directly (for instance the vortex of fine matter, which is, according to Descartes, the cause of gravity). Newton's answer to this dilemma was his method of *inductive proof from phenomena.* I would like to call Newton's method of inductive proof from phenomena *analytical approach to the experimental method.* The analytical approach in the contemporary sense was born in algebra in 1591, when Viéte published his *In artem analyticam isagoge.* In 1637 Descartes transferred it to geometry. I would like to interpret Newton's contribution to the experimental method as a further step in this expansion of the analytic approach. According to Viéte, the core of the analytical method consisted in three steps. First, we *mark by letters* the known as well as the unknown quantities. The purpose of this step is to cancel the epistemic difference between the known and the unknown. In the second step we *write down* the relations which would hold between these quantities if the problem was already solved. In the third step we *solve* the equations and find the values of the unknown quantities. In contrast to the analytical method as we know it from algebra or geometry, where the basic difference is an *epistemic difference* between the known and the unknown quantities, in the analytical approach to the experimental method the fundamental difference is a *methodological difference* between the measurable quantity (position, velocity) and the non-measurable quantities (forces). Thus Newton first *marks by letters* the measurable quantities as well as the non-measurable ones. In this way he cancels their methodological difference. Then he writes down the equations that hold between these quantities. Finally he derives from these equations some relation in which only measurable quantities occur, which relation can be therefore checked experimentally.

In order to see the novelty of this method, let us compare it with the methods of Galilean and Cartesian physics. Galileo simply refused to speak about non-measurable quantities and thus he considered all theories which supposed for instance an influence of the Moon on earthly phenomena as unscientific. For Galileo the world of science was restricted to phenomenal reality. In this respect Cartesian physics was a step forward. It was able to conceive an influence of the Moon onto earthly phenomena. The vortex of the fine matter could in principle transfer such an influence. Nevertheless, about the physical characteristics of this vortex Descartes was not able to say anything specific and so, at the end, he was not able to say anything specific about this influence itself. This is so, because in Cartesian physics the world-picture is split into two parts. One part is formed by ordinary bodies accessible to experimental investigation, the other part is formed by hypothetical substances, by help of which the results of the experiments are explained. These substances are not accessible to experiments, but only to speculation. Thus Cartesian physics has its phenomenal and ontological levels of description unconnected. Newton realized that the relation between the phenomenal and the ontological levels of description in Cartesian physics is analogous to the relation of the known and the unknown quantities in algebra or analytic geometry. The non-measurable quantity (e.g. the force by which Earth attracts the Moon) has to be marked by a letter and this letter has to be inserted into the equations that hold for such forces. Then some consequence of these equations should be derived, in which only measurable quantities occur. Finally this empirical prediction should be experimentally tested. In this way a measurement becomes a test of the analytic relations, from which the prediction was derived. The greatest advantage of Newton's method is that we are not obliged to measure directly the quantity which we are interested in. The network of analytic relations, into which this quantity is embedded, makes it possible to apply the experimental techniques at that particular place of the network which is for the measurement most suitable.

The characterization of measurement as a standardization of an experiment makes it possible to derive from our interpretation of Newton's experimentation as an analytic approach to the experimental method a new understanding of Newton's innovations in the field of measurement. As an illustration of the possibilities disclosed for physics by Newton's approach to measurement I would like to take the "weighing" of the Earth. In 1798 Henry Cavendish measured the force of attraction between two heavy spheres using very fine torsion weights. When he compared this tiny force with the weight of the heavy spheres, that is with the force with which they are attracted to the Earth, he was able to calculate the mass of the Earth. Cavendish's measurement was therefore often called "weighing of the Earth". In order to understand the novelty of the Newtonian approach to measurement, employed by Cavendish, let us compare it with the Galilean approach. In his measurement Cavendish used fine torsion weights, that is an *instrument*. He created an *artificial situation* that was thoroughly designed to exclude all disturbing effects, which could distort the result of the measurement. So far everything is in accordance with the Galilean approach. Nevertheless, the Galilean scientist would stop here and he would add the new phenomenon of attraction between the spheres to the known phenomena in a similar fashion as he added to them more than a century earlier the atmospheric pressure. The novelty of the Newtonian approach lies in the network of analytical relations which makes it possible to relate the measured force of attraction between the spheres to their weights and from this relation to determine the mass of the Earth.

If we realize how tiny the force between the two spheres is, it becomes clear that a Galilean scientist had no chance to discover it by lucky coincidence. He had no reason to construct such an ingenious experimental equipment as Cavendish created in order to make the force measurable. In ordinary experience there is no clue which could lead him to the discovery of the attraction of bodies, analogous to the failure of the pumping of water from deep shafts that led Torricelli to the discovery of the atmospheric pressure. In everyday experience there is no phenomenon which would reveal that macroscopic bodies attract each other. Therefore, for a Galilean scientist (and for his positivist followers) the gravitational force would remain probably forever undiscovered. Newtonian science differs from the Galilean in that it embeds the phenomena into a framework of analytic relations. It can then use this framework to search for artificial phenomena suitable for testing its predictions. Cavendish used the law of universal gravitation when he planned the experimental situation, in which the forces predicted by this law would become measurable. The law gave him an estimate of the magnitude of the force, and thus also an estimate of the precision that his instruments must reach. Thus while Galilean science uses in an experiment the technical equipment only to alter the phenomena which already exist in our ordinary experience Newtonian science goes much further than that. It *constructs new phenomena, which have no parallel in ordinary experience*. Of course, by this I do not mean that, for instance, the forces of attraction between macroscopic bodies did not exist before Cavendish measured them. These forces existed, but they were not phenomena, they were not accessible to human experience.

Acknowledgements I would like to thank Gábor Boros for his support and encouragement. The paper was written in the framework of the *Jan Evangelista Purkyně Fellowship* at the Institute of Philosophy of the Academy of Sciences of Czech Republic in Prague.

References

De Caro M (1993) Galileo's mathematical platonism. In: Czermak (ed) Philosophie der Mathematik. Holder–Pichler–Tempsky, Wien, pp 13–22

Drake S (1973) Galileo's experimental confirmation of horizontal inertia. Isis 64:291–305

Gabbey A (1980) Force and inertia in the seventeenth century: Descartes and Newton. In: Gaukroger S (ed) Descartes: philosophy, mathematics and physics. The Harvester Press, Sussex, pp 230–320

Galilei G (1610) The starry messenger. In: Drake S (ed) Discoveries and opinions of Galileo (1957). Doubleday, New York, pp 21–58

Galilei G (1623) The Assayer. In: Drake S (ed) Discoveries and opinions of Galileo (1957). Doubleday, New York, pp 229–280

Garber D (1992) Descartes' metaphysical physics. The University of Chicago Press, Chicago

Gaukroger S (1980) Descartes' project for a mathematical physics. In: Gaukroger S (ed) Descartes, philosophy, mathematics and physics (1980) The Harvester Press, Sussex, pp 97–140

Hakfoort C (1992) Newtons Optik, Wandel im spektrum der Wissenschaft. In: Fauvel J (ed) Newtons werk. Die Begründung der modernen Naturwissenschaft (1993) Birkhäuser, Basel, pp 109–133

Hill DK (1988) Dissecting trajectories: Galileo's early experiments on projectile motion and the law of fall. Isis 79:646–668

Kvasz L (2002) Galilean physics in light of Husserlian phenomenology. Philosophia Naturalis 39:209–233

Kvasz L (2003) The mathematisation of nature and Cartesian physics. Philosophia Naturalis 40:157–182

Kvasz L (2005) The mathematisation of nature and Newtonian physics. Philosophia Naturalis 42:183–211

Nagel E (1961) The structure of science. Hackett Publishing Company, New York

Naylor RH (1990) Galileo's method of analysis and synthesis. Isis 81:695–707

Ronchi V (1967) The influence of the early development of optics on science and philosophy. In: McMullin E (ed) Galileo, man of science. (1967) Basic Books, New York, pp 195–206

Shea W (1998) Galileo's Copernicanism: the science and the rhetoric. In: Machamer P (ed) The Cambridge companion to Galileo. (1998) Cambridge University Press, Cambridge, pp 211–243

Swerdlow N (1998) Galileo's discoveries with the telescope and their evidence for the Copernican theory. In: Machamer P (ed) The Cambridge companion to Galileo. (1998) Cambridge University Press, Cambridge, pp 244–270

Wallace WA (1984) Galileo and his sources: the heritage of the Collegio Romano in Galileo's science. Princeton University Press, Princeton

'The Renaissance of Physics': Karl K. Darrow (1891–1982) and the Dissemination of Quantum Theory at the Bell Telephone Laboratories

Roberto Lalli

Abstract Karl K. Darrow was a central actor in the reception of quantum theory in the Bell Telephone Laboratories. He was the first industrial physicist to dedicate his entire working time to the dissemination of novel concepts and theoretical tools by means of long review papers. The present paper analyzes the evolution of Darrow's narratives of quantum theory and shows that Darrow's reviews aimed at substantiating the view that physics was an evolutionary process. The paper argues that this view was connected to Darrow's peculiar activity at the Bell Labs as well as to the contemporaneous attempts of leading American scientists to build an ideology of national science.

Keywords American ideology · Bell Telephone Laboratories · Quantum mechanics · Subcultures of physics · Erwin Schrödinger · Wave mechanics

1 Introduction

The evolution and establishment of quantum mechanics posed unprecedented challenges to those industrial laboratories that relied on physical research to develop innovative and competitive artifacts. Industrial laboratories had to implement novel intra-organizational communication strategies in order to acquire the new knowledge. They also had to reconfigure their organizational structure in order to apply the latest conceived theoretical tools to practical research problems. The introduction of quantum theory at the AT&T Bell Telephone Laboratories was one of the most interesting cases of this disruptive transformation in the application of fundamental physics. The acquisition of the concepts and theoretical tools of quantum theory, and their application to materials was a complex historical process, which involved at least two phases. The first was to retrain the research staff by communicating new theoretical developments to them—a process that lasted from the early 1920s to late 1930s. The second phase saw the hiring of young PhD physicists

R. Lalli (✉)
Max Planck Institute for the History of Science, Berlin, Germany
e-mail: rlalli@mpiwg-berlin.mpg.de

R. Pisano (ed.), *A Bridge between Conceptual Frameworks*, History of Mechanism and Machine Science 27, DOI 10.1007/978-94-017-9645-3_14

well-trained in quantum mechanics; this began in 1936, when the economic growth following a slight easing in the Great Depression allowed Bell Labs' director of research, Mervin J. Kelly (1894–1971), to appoint new members to the research staff (Hoddeson et al. 1992; Gertner 2012)[1].

Historian of science Lillian Hoddeson (1980) argued that the first step was unsuccessful. The lack of an adequate scientific training in modern physics made it impossible to apply the concepts and tools of quantum mechanics to the research needs of the Bell System. Nevertheless, it is just the first step that is of great historical value to understand how scientists tried to reconfigure their systems of beliefs in order to employ the new theoretical methods. This stage, indeed, shows the main cognitive struggles and interpretative difficulties. This process also reveals how communication schemes shaped the transmission of knowledge. Bell Labs' administrators adopted various strategies in order to disseminate the new physics, including maintaining an up-to-date library, organizing visits by leading theoretical physicists, encouraging researchers to attend courses and seminars in various universities, and publishing papers in the Bell System journals intended to explain the new physics in semi-popular fashion. Thanks to his stylistic ability in the exposition of scientific issues, the industrial physicist Karl K. Darrow became the central actor in the Bell Labs' efforts to retrain its researchers. Darrow published a number of long critical reviews in the *Bell System Technical Journal* (*BSTJ*) from 1923 to 1939, attended several symposia and courses, reported what he had learned to the scientific staff, and organized scientific conferences on quantum theory at the Bell Labs. Scientific dissemination shortly became Darrow's exclusive occupation at the Bell Labs, which made him the first industrial physicist to devote his whole working time to such an activity.

Various accounts suggest that beyond Bell Labs Darrow had a significant role in the reception of quantum theory in the United States, and several physicists recognized their debt to Darrow's efforts to disseminate a wide-ranging view of new theories through his critical review articles (Van Vleck 1967, p. 25; Weiner 1973, p. 27; Hoddeson 1980; Sopka 1988, pp. 87–88; Gertner 2012, pp. 41–42). Between the early 1920s and the late 1930s, Darrow occupied an exceptional position within the American scientific community as a synthesizer and interpreter of quantum theory at the juncture between university theoretical research and the American industrial environment. However, historians have so far given scant attention to Darrow's activity as a mediator between different subcultures of physics. The only scholarly study of Darrow's writings appears within philosopher of science Nancy Cartwright's analysis of the reception of quantum theory in the United States. Cartwright (1987) claimed that Darrow is a very fine example of the influence that the operationalist-pragmatist tradition had in the way in which American

[1] The present paper covers a period from the late 1910s to late 1930s. The AT&T Bell Telephone Laboratories were created only in 1925. Before that date, the personnel that would merge into the Bell Telephone Laboratories worked in the Engineering Department of the Western Electric Co., which was one of the member companies of the Bell System. For the sake of brevity, I will use the term *Bell Labs* to refer to all the laboratories of the Bell System during the period under consideration.

physicists dealt with the philosophical problems of quantum theory. The present paper aims at extending and historicizing Cartwright's analysis.

Darrow's beliefs did not remain fixed while sudden theoretical developments were altering the face of physics. He had to actively reconfigure his worldview in order to engage with new concepts and new theoretical tools. By following the chronological order of Darrow's reviews of topics related to quantum theory, one discovers that Darrow's views evolved differently from those of leading American theoretical physicists. I will show that Darrow's interpretations and translations of the developments of quantum theory were strongly shaped by his peculiar role within the American physics community as well as his ideological view of physics as a continuous evolutionary process, from measurement to knowing, and from knowing to invention. I suggest that the relationship between Darrow's ideological cast and his expositions of quantum theory might reveal deep interconnections between the changing social-epistemic context embodied by the Bell Labs milieu and Darrow's interpretative endeavors. In conclusion, I argue that the evolutionary technoscientific views held by Darrow were embedded in the American ideology of national science, which dominated the scientific landscape in the period 1919–1930 (Tobey 1971).

2 Creating a New Profession in the Bell Labs: Darrow's Unique Role in the Life of American Physics

Born in 1891 in Chicago, Karl Kelchner Darrow exhibited from an early age an impressive interest in several different disciplines. A local newspaper documented Darrow's talent in both humanities and mathematics, dedicating to the 7-year-old boy a long article, in which the journalist stated that the young Darrow was already "an indisputable authority in history, geography and mythology, an unparalleled mathematician, a poet and author, and an expert at operating the typewriter" ("Karl Darrow's Genius"). The article provided an extremely vivid image of the personal capacities that would make Darrow a unique interpreter of physics from the early 1920s onward. The boy showed a striking memory coupled with a broad interest in various branches of knowledge from literature to mathematics, from art to natural science. Such a vast range of interests would become the mark of Darrow's approach to physics writing. As John H. Van Vleck (1899–1980) described it, Darrow's "unique role in the life of American physics" depended on his particular style, which in turn stemmed from "his wide literary and cultural background" (Van Vleck 1967, p. 25) (Fig. 1).

Among his various intellectual interests, Darrow chose to pursue the professional study of physics, while preserving his passion toward arts and literature. The University of Chicago provided an ideal environment for Darrow to complete his undergraduate and doctoral studies. There, Darrow received his Bachelor's degree in physics in 1911 and his PhD in 1917 with an experimental dissertation concerning the measurement of the ratio of the specific heats of hydrogen under the supervision of Robert Andrews Millikan (1868–1953).

Fig. 1 Photograph of Karl K. Darrow. (AIP Emilio Segre Visual Archives, Darrow Collection)

Although he spent almost two years of study in Paris and Berlin between 1911 and 1913, Darrow's approach to physics was shaped almost exclusively by Millikan's teachings (Darrow 1964)[2]. Darrow's diaries of this formative period highlight Millikan's impact on Darrow's view of physics. Experiments were the centerpiece of the academic courses. The subordination of theoretical advancements to novel experiments shows that Millikan taught physics as an experimentally-driven discipline. Darrow's diaries also reveal that quantum theory was not part of the normal physics curriculum at that time. Darrow acquired some knowledge of the developments of quantum theory only *after* he completed his dissertation by attending a summer course taught by Millikan himself. The reading assignments of this course focused on experimental problems and instrumental methodologies confirming that Millikan's teaching agenda was built on the belief that experiments had primacy over theory (Darrow 1917a).

Not only did Millikan shape Darrow's views on physics; Millikan also had an active role in the continuation of Darrow's professional career. Soon after Darrow earned his PhD, Millikan helped him to obtain a research position at the Engineering Department of the Western Electric Company, which would become the AT&T Bell Telephone Laboratories in 1925 (Millikan 1917). When Darrow joined the research staff of the Bell System in 1917, the multidisciplinary research environment was in ferment. The 1917 entry of the United States into the war requested major efforts by U.S. industrial companies to serve the military needs of their country. The involvement of physicists in the military research agenda of the

[2] Unless otherwise reported, the biographical information contained in this section is taken from this interview.

American government during World War I was deeply transforming the relationship physicists had with their discipline. While before the war physicists conceived themselves as lone researchers, the coordinated and product-focused military efforts led several leading American scientists to appreciate the value of team-organized and interdisciplinary environments for pursuing pure research. Since the United States began organizing their industrial structures for the entry into the war, the research teams of the Bell System actively worked on military projects, and the industrial multidisciplinary environment became a model of research organization for several universities (Kevles 1971, pp. 102–138; Tobey 1971, pp. 20–61). Darrow was hired to increase the scientific manpower of the Bell System working on military projects.

The successes of Bell Labs' researchers in pure physics led several historians to study the essential elements of the Bell Labs environment. The various accounts provide a complex landscape of contradictory images. In particular, the role of pure research within the organizational strategies of the research directors is controversial. On the one hand, historian of science Steven Shapin convincingly argues that industrial laboratories were not morally inferior to academic environments in their commitment to pure research, providing several examples concerning the freedom enjoyed by Bell Labs' industrial physicists (Shapin 2008). On the other hand, certain Bell Labs' research directors pointed out that scientists' research freedom was rigorously subordinated to the commercial needs of the firm (Millman 1983, pp. xiii-xxi, 1–17). By epitomizing the various historical accounts, it is possible to deduce that the balance between pure and applied research within the Bell Labs milieu was historically contingent and depended on specific negotiations between research directors and the research staff.

An essential historical element that favored the Bell System involvement in pure research was Millikan's direct influence in shaping the views and actions of various research directors of the firm. Several of Millikan's students found jobs within Bell Labs and some reached eminent positions, including Frank B. Jewett (1879–1949)—who became Vice-President of the AT&T in 1921 and President of the Bell Labs in 1925—and Harold D. Arnold (1883–1933)—who became the Director of the Research Branch of the Engineering Department in 1921 and the first director of the AT&T Bell Telephone Laboratories in 1925 (Gertner 2012, pp. 9–40). Darrow, then, was following a route common to many experimental physicists who graduated under Millikan in the first decades of the twentieth century.

Darrow appreciated the lively environment and the spirit of research, and began helping with the Bell System's war efforts (Darrow 1917b). Soon after the war, however, a physical problem jeopardized his career at the Bell Labs: His trembling hands made him unsuitable to continue experimental research (White 1976). Nevertheless, thanks to Arnold and Jewett's managerial decisions, instead of losing his job, Darrow was asked to organize the unpublished literature of scientific and engineering memoranda. Later, Darrow became a sort of "intelligencer to the community at the laboratory" (Darrow 1964)[3], occupying a void in the organization

[3] This expression was used by the interviewer W. J. King.

of the industrial firm with a series of activities aimed at acquiring new knowledge within the Bell System research environment.

Such an activity was multifaceted and evolved with time. The deep changes due to the development of quantum physics highlighted the need for Bell Labs' researchers to maintain an updated knowledge of physics. To accomplish this task, 23 members of the Bell System research staff, mostly of the Physical Research Department, organized a society called Colloquium. This society organized meetings that enabled members "to keep in touch with recent thought and experiment in physics and allied sciences" (Darrow 1920a)[4]. In 1919, the "little republic of serious thinkers," as the society was later called, began gathering two evenings a month in which one of the members introduced a topic to be discussed by the participants (Darrow 1920b; "Modern Physical Theories"). Darrow's lack of involvement in active research allowed him a central role in the Colloquium, becoming its perpetual secretary ("First meeting").

The informal and voluntary character of the Colloquium demonstrates that this dissemination was more a need of individual researchers than a strategy developed by research directors. The beginning of Darrow's activities in disseminating quantum theory did not depend, then, on well-organized research strategies. They were an outcome of the interplay of individual decisions of research directors trained by Millikan and personal interests of Bell Labs' physicists. Only in 1923, when Darrow began publishing a series of papers in the *Bell System Technical Journal* (*BSTJ*), did Darrow's role within the Bell System research organization assume a more stable form.

While continuing his work on the organization of the unpublished literature, Darrow began using his literary talent to write typed reports about the meetings of the American Physical Society. Starting as a spontaneous activity, Darrow's role as a synthesizer and interpreter of new ideas became a real job. In 1923, the *BSTJ* editor Robert W. King asked him to write reports of the recent developments of physics that could be of interest for the Bell System community of researchers. The quarterly *BSTJ* had been created in 1922 with the intent to publish articles by the staff concerning scientific and engineering aspects of electrical communication. During the first year, the publishing policy maintained a strong link between the production needs of the Bell System and the issued topics. Darrow's papers were the first and unique exception to such a policy. Following King's request, Darrow began publishing a permanent section called "Contemporary Advances in Physics." The foreword of Darrow's first paper clearly cast the aim of the overall series: to make available to the Bell Labs' research staff and the broader readership of the journal reviews of recent researches in physics that Darrow considered of special interest. As the foreword pointed out, the publication was a way to institutionalize a work that Darrow was already doing in a more informal manner (Darrow 1923).

Some of the directors of research of the Bell System showed an uncommon intuition in allowing Darrow to use his literary talents in the behalf of the entire community of physicists. As Darrow and other physicists stressed, such an activity was

[4] The list of the topics of the first season clearly shows that the principal aim of the Colloquium was to engage with the development of quantum theory ("Meeting of the Colloquium").

essential for two reasons. First, Darrow was creating a bibliography in which novel research endeavours were epitomized in order to make them intelligible to researchers specialized in different areas of physics; second, in a more implicit way, Darrow's papers were an attempt to slow down the centrifugal forces of specializations within the growing American community of physicists (Van Vleck 1967; White 1976; Wooldridge 1976; Gertner 2012, pp. 41–42). In 1929, the American Physical Society (APS) recognized the vital importance of this kind of literature by creating the *Reviews of Modern Physics*, which shared the same targets of Darrow's "Contemporary Advances in Physics" (Lalli 2014).

Before the APS began publishing *Reviews of Modern Physics*, Darrow's reviews were almost unique in the American physics literature. Even after the APS institutionalized the need for long critical reviews, Darrow's writings maintained exceptional features. For these reasons, they represent fundamental documents for investigating the transmission of knowledge between different cultural and institutional settings. Darrow's interpretation of quantum theory, his choices of topics, his rhetorical style, his philosophical background are all elements that improve our knowledge of how the new physics was transmitted from European theoretical physics departments to an American industrial environment. To analyze Darrow's writings, it is necessary to consider both the two spatial features of the environment in which Darrow acted; namely, the specific fluctuating subculture of the industrial laboratory of the Bell System, and the broader philosophical underpinning of the American reception of quantum mechanics (Coben 1971; Schweber 1986; Holton 1988; Schweber 1990; Assmus 1992a, 1992b). This kind of approach will allow me to draw the strong connection between Darrow's views and the technoscientific ideology held by several exponents of the American physical community in the period in which American theoretical physics "came of age" (Sopka 1988, pp. xvii-xxiii).

3 Darrow's Interpretation of Old Quantum Theory: Waves, Corpuscles, and the Art of Model Building

Old quantum theory was the main topic to which Darrow dedicated his quarterly reviews in the period 1923–1926. In this period, Darrow's papers dealt almost exclusively with the atomic structure and spectral phenomena. Although Darrow preferred focusing on experimental issues, he did not elude conceptual problems and provided his personal judgments. In particular, the relationship between the corpuscular and undulatory pictures of light puzzled Darrow. The complex and perplexing features of this relationship led him to reveal his underlying philosophical beliefs when he first discussed the topic in 1925.

Darrow built his paper concerning the corpuscular and wave phenomena of light on the opinion that experiments had primacy over theory. Such a view is evident in Darrow's assessment of Planck's theory of the blackbody radiation. Darrow (1925) declared that Planck's theory "might practically be confined to the pages of the more profound treatises on the philosophical aspects of physics, if certain experiments had not been guided to seek and to discover phenomena so simple that none could fail to apprehend them, so extraordinary that none could fail to be amazed" (p. 286). Here, not only did Darrow maintain that experiments should guide theoretical research; he also extrapolated two features that experiments must have to be significant. The experiments should lead to the discovery of phenomena that are both simple and unforeseen. These kinds of phenomena were the best suited for stimulating the growth of reliable physical theories—a view that was deeply related to Darrow's overall perspective about what theoretical physics should be.

Darrow considered model building as the fundamental practice of the theoretical branch of physics. This belief became explicit in Darrow's discussion of the photon concept that Albert Einstein (1879–1955) had introduced in 1905. To demonstrate that the apparent lack of physical reality of theoretical models should not prevent one to formulate bold hypotheses, Darrow argued that the "absurdity" of the photon concept faded gradually out of view since Einstein's heuristic hypotheses led to explain more and more experimental phenomena (p. 287). Darrow used the expression "as if" to epitomize the relationship between observations and theoretical models. In the photoelectric effect, the emissions of electrons occurred *as if* the energy of light was concentrated in packets of amount $h\nu$. After having illustrated a variety of experiments bearing on the photon concept, Darrow admitted that they forced one to accept that radiation travels by means of corpuscles of specific energy and momentum. This view, Darrow stressed, was irreconcilable with the wave phenomena characterizing light behavior in different experimental settings. Moreover, the definition itself of the energy and momentum of the photon depended on wave concepts such as frequencies and wavelengths. To reconcile these two seemingly incompatible views, Darrow called for what he defined "a *revolutionary* extension of the art of thinking" (p. 326 my emphasis). Only such a revolutionary transformation could allow physicists to cope with the copresence of such different phenomena; but, unfortunately, Darrow did not try to define the factors that should characterize the modification in the way of thinking he was demanding.

In order to understand Darrow's view of revolutionary change one has to rely on the prolegomena to the first edition of his book *Introduction to Contemporary Physics*, in which Darrow (1926) provided a more clear assessment of both his philosophical views and methodological prescriptions. Cartwright (1987) has recognized the explanatory clarity and philosophical outlook of this text describing it as a "fine piece of philosophy" (p. 426). Darrow's prolegomena was one of the main sources of Cartwright's analysis of the American response to the philosophical queries posed by quantum theory. In Darrow's words, Cartwright argues, one can uncover the conceptual roots of the apparent lack of philosophical anxiety of the American physicists. According to Cartwright, the apparent disinterest of American physicists toward philosophical questions was related to the influence of Bridgman's operationalism as a prescriptive rule for ensuing reliable scientific research. In Darrow's prolegomena, however, Cartwright found the manifestation of a philosophical perspective wider than operationalism, which she epitomized in two methodological prescriptions:

> (1) The task of physics is to describe what it can as accurately as it can, at the same time striving for simplicity and economy of presentation. In particular, the task of physics is not to explain. (2) Physics should not postulate hypotheses, but should accept only what is experimentally verifiable[5].

These two philosophical theses, Cartwright maintained, shaped the American reception of quantum theory and Darrow was an intellectually clear example of a national philosophical tradition.

Cartwright is right when she affirms that Darrow exposed quite explicitly the first thesis. The second thesis, instead, seems to me not to adequately capture Darrow's thoughts. Moreover, the view Darrow articulated was not simply a mirror of the broader cultural milieu, neither can one say that it represented a coherent philosophical doctrine. Darrow's normative prescriptions came out of the historically contingent intellectual conflict between his views of physics as an evolutionary process and the specific status of theoretical physics in that period. When Darrow wrote the book, quantum theory was "a lamentable hodgepodge of hypotheses, principles, theorems, and computational recipes rather than a consistent theory", as Jammer (1966, p. 196) defined it. Darrow's response to such a hodgepodge was to stress the relevance of model-building activity in the advancement of knowledge. The criterion physicists should follow to build physical models was that the models be as simple as possible and describe the greatest number of phenomena—features well summarized by Cartwright's first prescription rule. For Darrow (1926), the belief that simple models governed also complex phenomena was an "act of faith" (p. xix). Such an act of faith had induced theorists to apply the models built on simple experimental phenomena to more complex ones. Darrow contended that the theory of quanta had revealed the limits of this procedure by leading to the coexistence of mutually irreconcilable models such as the corpuscular and wave theories of light, used to explain different phenomena related to the same physical entities.

Darrow's activity as a synthesizer and interpreter of new ideas helped him to find various examples to demonstrate the scientific value of accepting different, and often incompatible, theoretical models of the same supposed entity. To show that this methodology had been useful in the advancement of knowledge, Darrow contended that scientists had been drawing different atom models to describe different phenomena. The nineteenth-century chemists' atom model was different from the rigid elastic sphere early twentieth-century physicists conceived to describe gas features such as elasticity, pressure and specific heat. The latter model, in turn, was completely different from the Rutherford-Bohr atom model, which dealt with the radiation emitted by luminous gases. Each of the three models had its irreplaceable function explaining a particular range of phenomena, but they were barely compatible and did not lead to a coherent vision of the ultimate atom, interpreted as a real entity. Darrow advised the students to "adopt the practice of regarding atom-models as creations of the imagination, as the building stones of mental models designed to copy chosen phenomena of the environing world" (p. xxii), and maintained that this was the practice actually adopted by most physicists. In other

[5] Cartwright 1987, p. 425.

words, each model had its own validity only within the range of phenomena that it was designed to imitate. The issue that Darrow problematized seems to be, then, more the reality content of theoretical hypotheses than the indisputable role of hypothesizing in the practice of theoretical physics.

Philosopher of science Ian Hacking (1983, pp. 27–28) argued that it is fundamental to differentiate between two kinds of realism: realism about theories—according to which theories are either true or false and science aims at discovering how the world really is—and realism about entities, according to which several entities described by scientific theories exist objectively. Hacking made the point that experimentalists usually believe in physical entities because they can use them (p. 262). One is tempted to ask whether and how Darrow's philosophical perspective fits Hacking's criteria. Certainly, Darrow's view of theoretical physics as a model-building activity demonstrates that, for him, theories were not true and could not aspire to grasp the world as it is. But what about entities? In the prolegomena, Darrow implied that entities such as atoms and electrons existed independently of the theories built to describe their experimental features, but did not explain the reasons underlying this conviction. Darrow just took their existence for granted.

Darrow's main target was to give precise normative rules that theoretical physicists should follow to advance the knowledge of nature. Usually, Darrow argued, only a limited number of features of invisible entities were to be considered as experimentally verified. In order to explain broader phenomena, it was necessary to build models whose properties included and yet exceeded the ones empirically detected. Darrow prescribed a sort of freedom of stipulation about these exceeding properties. The only requirement was that these hypothesized properties should describe an ample range of phenomena. Darrow's advice was not to consider such surplus properties as physical realities and to use different models depicting the same entity in different situations.

In order to extend Cartwright's analysis, it is necessary to emphasize that the issue at stance in Darrow's reflections was the unity of physics as a discipline. Darrow was trying to find some criteria to counterbalance the conceptual and social forces that were undermining the coherence of the theoretical apparatus as well as the compactness of the physics community. In the prolegomena, not only did Darrow explicitly reject that theories might catch the real world; more fundamentally, Darrow was implying that physicists had to renounce to what Hacking (1996, p. 50) calls *local reductionism*; namely, the possibility that physics could be reduced to a set of fundamental principles, from which all the scientific laws might be deduced. In the history of physics, this practical precept has been playing a strong role in the self-perception of unity by practitioners of physics (Cat 1998). In order to articulate a different criterion on which to base a unitary view of physics, Darrow replied to the challenge posed by the conceptual incongruences of the wave-particle duality, by stressing the methodological unity of the model-building practice[6].

[6] My interpretation of Darrow's writings as a quest for the methodological unity of physics is analogous to the analysis philosopher of science Jordi Cat (1998) made of the debate between emergence and reductionism in physics as implicitly referring to two different models of unification.

Darrow's book (1926) was a recasting of those articles he had been publishing in the *BSTJ* that assessed the structure of the atoms and related phenomena in the framework of the old quantum theory. One might consider Darrow's prescription about the net separation between theoretical models and the underlying reality as a response to the "revolutionary extension of the art of thinking" he was asking for to cope with quantum inconsistencies. In Darrow's opinion, quantum dominion had definitely ruled out the local reductionist program of physics. To reconfigure their theoretical work, physicists had to consider different ways in which to define their activity. Darrow, then, linked the revolutionary change to the re-definition of theoretical work as a research practice limited to the building of models. Of course, Darrow's thoughts show several elements of the pragmatist perspective that had a great relevance in the philosophical attitude of several American physicists. However, they show more than that. Darrow's peculiar role makes it evident the tension between the theoretical evolution of quantum physics and his underlying ideological conviction. As I will show in the next sections, Darrow's subsequent reviews of the advancement of quantum mechanics demonstrate that Darrow continued to reconfigure his philosophical beliefs in order to intellectually engage with novel theoretical developments.

4 Wave Mechanics and the Heuristic Value of Visualizability in Physics

Darrow could not imagine that while his book was being printed, Max Born (1882–1970), Werner Heisenberg (1901–1976), Pascual Jordan (1902–1980), and Erwin Schrödinger (1887–1961) were discovering the basis of quantum mechanics. To assess the novelties produced in the other side of the Atlantic Ocean, Darrow followed a path quite common in the early reception of both matrix and wave mechanics. He read with anxiety the publications of Heisenberg, Born and Jordan, while embracing with relief Schrödinger's elaboration of De Broglie's wave interpretation of particles' behaviour. In his *BSTJ* reviews, Darrow completely bypassed the publications concerning matrix formalism. Darrow (1927) dedicated, instead, the entire first review of the new physics to wave mechanics. He spoke as a representative of the enthusiasm rose in the world of physics about Schrödinger's theory, which seemed "to promise a fulfilment of th[e] long-baffled and insuppressible desire" of those physicists who "yearn[ed] for *continuity* in their images of science" (p. 653 my emphasis). In July 1926, he had got the impression that Schrödinger's theory was widely regarded as superior to the concurrent approach pursued by the theoretical physicists in Copenhagen and Gottingen. Canadian-American spectroscopist Arthur Jeffrey Dempster (1886–1950), had offered him a vivid image of the reception of Schrödinger's papers in Munich:

> Schrödinger's recent developments seem to have put Born and Heisenberg's papers into eclipse (to everyone's apparent satisfaction). At Munich there seems to be the hope that we would possible formally get a satisfactory formulation of the quantum theory in the way[7].

Although Darrow tried to maintain his role of impartial observer of what was going on in physics, he did not hide his preference for Schrödinger's approach. Darrow's personal predisposition is unsurprising because a great number of physicists favorably received wave mechanics[8]. Not only did wave mechanics provide theoretical tools more practicable to resolve specific problems than those of matrix mechanics, it also seemed promising for the possibilities to expand its range of applicability. Moreover, Schrödinger's return to a quasi-classical conception based on the continuity of nature entailed a reinstatement of the visualizability of nature, which the abstractness of matrix mechanics seemed having ruled out permanently. The wave formalism allowed for a descriptive link between microphysical phenomena and more familiar ones, which physicists could use as examples and inspiration (Jammer 1966, pp. 274–84; MacKinnon 1980; Wessels 1983; Beller 1983, 1997; Mehra and Rechenberg 1987, pp. 577–868).

It was precisely the possibility to build a visualizable account of Schrödinger's theory that shaped Darrow's method of exposition. Darrow pointed out that wave mechanics allowed a multiplicity of approaches and that the one he chose was substantially different by those followed by both Schrödinger and Louis De Broglie (1892–1987). His method of exposition, Darrow (1927) believed, was "the one which Schrödinger meant when he wrote 'I had originally the intention of establishing the new formulation of the quantum conditions in this more visualizable (*anschaulich*) way, but preferred a neutral mathematical form, because it makes the essence clearer'" (p. 655).

With his focus on the visualizability of the new description, Darrow was touching one of the most controversial topics in the developments of quantum physics. The notion of *anschaulichkeit* and its relation to the quantum world has led to deep philosophical disagreements between the main actors of the theoretical development of quantum mechanics (Miller 1984, pp. 125–183; Camilleri 2009, pp. 48–53; Jähnert and Lehner 2015). In these controversies, Schrödinger maintained that a visualizable account of the quantum phenomena was essential. However, it is controversial the exact role that this concept played in Schrödinger's development of wave mechanics as well as the connection of such a concept with his broader philosophical position. While some physicists and philosophers claim that Schrödinger's quest for visualization stemmed from his ontological realist position (see, e.g., Harré 1992), philosopher of science Henk de Regt (1997) has convincingly argued that Schrödinger was a methodological realist. For Schrödinger, so De Regt's argument goes, *anschaulichkeit* meant both visualizability and intelligibility because understanding was equivalent to forming space-time pictures. However, De Regt concludes, Schrödinger did not believe that these space-time pictures reproduced

---

[7] Dempster 1926.

[8] Historian of science Mara Beller has convincingly argued that the success of Schrödinger's approach was widespread and not limited to "the conservative quarterlies of the physics community" (Beller 1983, p. 470).

reality; Schrödinger thought that scientists should work as if they did (for the complexity of Schrödinger's philosophy, see also Wessels 1983; Bitbol 1996; Beller 1997).

Darrow's own approach to such a controversial issue led him to expose his personal interpretation of what visualization meant. While he seemed to follow what De Regt interprets as the Schrödinger's program for visualization, Darrow's account is curiously different from Schrödinger's. The differences between them gave a clue to Darrow's understanding of Schrodinger's theory as well as his personal program based on the conviction that physics was an evolutionary and continuous process. Darrow did not find it helpful to address wave mechanics from the point of view of the Hamilton analogy between mechanics and optics, which constituted a fundamental part in the evolution of Schrödinger's approach to wave mechanics (Schrödinger 1928, pp. ix, 13–40; for recent in-depth historical analyses see Joas and Lehner 2009; Renn 2013). He insisted, instead, that physical common vibratory processes were essential in order to understand the theory. For him, the word visualization was connected to the possibility of relating the most recent theoretical advancements to mechanical processes well known by a broad section of the scientific community, including the applied physicists and engineers of the Bell Labs.

In order to provide a visualizable account of wave mechanics, Darrow relied on the theory of acoustics whose characteristic problems, Darrow argued, corresponded to the characteristic problems of wave mechanics. From the theory of acoustics, Darrow selected three vibratory phenomena that could be used as models to visualize the atomic stationary states as stationary wave patterns whose natural frequencies corresponded to the quotient between the stationary states' energies and the Planck constant h. The three "familiar examples of stationary wave-patterns," as Darrow (1927, p. 664) called them, were the stretched string, the tensed membrane, and the ball of fluid confined in a spherical shell. The first phenomenon represented the simplest instance of a vibratory phenomenon in one spatial dimension and two variables x and t. The tensed membrane corresponded to the case of a physical system with two spatial dimensions and three variables; while the third, and more complex, example represented a physical vibratory system in three spatial coordinates and four variables x, y, z, and t. The latter, Darrow contended, "present[ed] the closest analogy to the atom-model for the hydrogen atom in wave-mechanics" (p. 673).

Darrow showed that the imposition of boundary conditions and the correlated choice of the coordinate system rigorously determined the frequencies of the permitted vibrations in all the three cases. The mathematical treatment of the three processes allowed Darrow to calculate the formulas of the eigenvalues and eigenfunctions in term of the boundary conditions. The reader might build, then, a clear image of wave mechanics solutions by relating the wave-mechanical formalism to familiar phenomena. Of course, the phenomenon most difficult to visualize was the three-dimensional case of the fluid in a spherical shell. Contrary to the other two acoustic models, the term ψ appearing in the associated wave equation of the three-dimensional fluid could not be interpreted as a perpendicular displacement of the medium with respect to its static position. Darrow, however, did not renounce to the possibility to visualize the wave motion and suggested the reader to consider ψ "as a condensation or a rarefaction, after the fashion of sound-waves" (p. 675).

Once Darrow had completed the discussion of the three examples of wave patterns, his aim turned into relating the mathematical formalism and its associated physical pictures to the cases to which Schrödinger had applied wave mechanics, including the one-dimensional harmonic oscillator and the hydrogen atom. To accomplish this correlation Darrow asked the reader for an imaginative effort: "It would suffice to imagine strings and fluids not uniform like those of the simple theory of vibrating systems and sound but varying from point to point in a curious and artificial way" (p. 655). Darrow was thus seeking to describe through images of physical processes those mathematical terms that made wave mechanics different from ordinary acoustic models. The general differential equation of wave motion was:

$$u^2\nabla^2\Psi = d^2\Psi / dt^2 \tag{1}$$

where ∇^2 was the Laplacian differential operator, and ψ stood for the quantity transmitted by the wave, which in acoustic models was the displacement of the medium from its equilibrium position. In acoustic models, u was the velocity of propagation of the wave front, and u^2 was, of course, always a positive constant. In wave mechanics, the same did not hold. There were cases in which the wave-mechanical correspondent to u^2 varied in function of the coordinates, and cases in which it could take negative values. This meant that one should imagine a string or a fluid whose deformation traveled with an imaginary wave speed. Although Darrow recognized the danger in attaching to mathematical terms words "devoid of any physical meaning," he did not renounce to use visualizable analogies. For him, the acceptable solutions of Schrödinger's wave equations of the harmonic oscillator and the hydrogen atom represented respectively "imaginary strings" and "imaginary fluids." More specifically, Darrow pointed out that any stationary states of the hydrogen atom might be represented by an imaginary fluid, each pervading the whole space.

Darrow's efforts to build an analogy between wave mechanics and more visualizable phenomena led him to believe that "the vibrating imaginary fluid [would] furnish the customary symbolism for expressing the data of experiment," for many years to come (p. 685). Darrow's confidence derived from his belief in the heuristic power of visualizable models in the advancement of knowledge. Although still incomplete, Schrödinger's theory represented a model very close to Darrow's own ideal of physical theory. Not only did the theory capture in a single picture a number of different phenomena, including the Davisson-Germer experiment on the diffraction of electrons recently performed at the Bell Labs; it also provided a visualizable scheme to relate invisible phenomena to more familiar ones. For Darrow, the physical models should be as simple as possible, and his discussion of wave mechanics demonstrates that visualizability was a fundamental element in Darrow's conception of simplicity in physics.

That his discussion of wave mechanics was immersed in Darrow's view of theoretical physics as a model-building activity is demonstrated by his discussion of the physical meaning of the wavefunction ψ. At the end of the paper, Darrow distanced himself from Schrödinger's tentative electromagnetic interpretation of ψ as a measure of the density of the electric charge. In the four part of his paper "Quantization as a Problem of Proper Values," Schrödinger had suggested that the term

$$\rho = e\,\psi\psi^* \tag{2}$$

was the electron charge density; where e was the charge of the electron, ψ was the wavefunction, and ψ^* was its complex conjugate. Even recognizing the heuristic value of considering Eq. (2) as a measure of the electric density, Darrow exposed the difficulties to regard it as a realistic picture of what was going on inside the atom. How could one accept the dissolution of the electron in space and, at the same time, use in the wave equation the Coulomb potential that described the force between two localized particles? Darrow used this striking contradiction to confirm his own view that physical theories did not aim at the truth, while he seemingly considered Schrödinger's interpretation as a step in the direction of an ontological explanation. Darrow wanted to make it clear that the success of Schrödinger's model depended only on the fact that it was "unrivalled [as] a *device* for *picturing* the radiation-process" (p. 700 my emphasis). The relationship between Schrödinger's model and the reality of microscopic phenomena was, in Darrow's views, out of reach of theoretical analysis. This perspective led Darrow to distance himself from Schrödinger's views in the following years by embracing the statistical interpretation of the wave function, while retaining his preference for Schrödinger's formalism.

5 Ensemble-Statistical Interpretation of the Wavefunction

After 1927, Darrow faced the task to deal with the developments and the wide acceptance of a more abstract quantum mechanics. Darrow went along a very personal route by continuing to believe that the wave picture had an epistemic and physical priority over the particle picture, while at the same time embracing the statistical interpretation of the wave function. The first of Darrow's attempt to cope with the new formulation of quantum mechanics concerned Heisenberg's indeterminacy principle, which Darrow (1930) called the "principle of indefiniteness." Darrow disagreed with the usual translation of Heisenberg's word *Unbestimmtheit*. As Darrow saw it, both uncertainty and indeterminacy did not represent the very idea underlying the mathematical formulation of the principle. Although Darrow did not explain the reason of his semantic choice, one might relate it to his understanding of the principle. Darrow stressed that Heisenberg's principle was a successful attempt of "harmonizing contradictory ideas," by fusing together the corpuscular and the wave pictures (p. 188). Seemingly, Darrow followed those physicists who, like Niels Bohr (1885–1962), saw in the Heisenberg's principle a partial solution to the problem of the wave-particle duality (Jammer 1966, pp. 66–79). However, Darrow held a very personal interpretation by stressing that the fusion of the two apparently irreconcilable pictures was made at the expense of the corpuscular picture.

The way in which Darrow illustrated the principle adhered to this general perspective, by challenging Heisenberg's exposition of the principle. Darrow criticized the use of the γ-rays *Gedankenexperiment* to prove the principle because *thought* and *experiment* were, for him, two contradictory words. Darrow preferred to review actual experiments in which the relationship between the "indefiniteness" principle and the observations was factually evident. The banning of thought experiments from the methodology of physics was an important indication of Darrow's beliefs concerning the prescriptive rules that should have governed reliable scientific research. Methodology and theories were indissolubly linked in a general view of physics in which experiments and visualizability continued to have a privileged epistemic value. Darrow called *illustrations* the actual applications of the principle he discussed in his paper. Illustrating and visualizing were the actions that allowed Darrow to maintain a contact between a tangible and visible world, and the new physical theories, which presented a high degree of abstractness. This attempt underlay Darrow's interpretation of Heisenberg's principle. For him, the principle was "a startling way to save the waves, affirming in substance that what they cannot describe cannot exist" (Darrow 1930, p. 189).

In a subsequent paper on quantum mechanics, the merging of epistemological perspectives and theoretical descriptions became even more explicit. Four years later, Darrow (1934) published a long review of various aspects of quantum mechanics, including wave mechanics, matrix algebra and quantum operators. In this account, Darrow clarified what was the epistemological perspective guiding his assessment of the new advancements in physics. The contradictory copresence of corpuscular and wave conceptions of microphysical phenomena had been Darrow's main concern about quantum physics. Darrow could not figure out how to link such contradictory images of the physical world and he interpreted all the theoretical advancements as a response to this concern. In 1934, after quantum mechanics had reached a level of completeness unpredictable ten years earlier, Darrow evaluated what had been accomplished. The result of this evaluation was to distance his exposition from the way in which quantum mechanics was usually discussed and taught.

Darrow (1934) expressed his distaste for "[t]he trend toward perfect abstraction, which for several years ha[d] been dominant in quantum mechanics" (p. 44). As Darrow explicitly stated, he belonged to the group of scientists who "crave[d] to retain, for as long as possible, as many as possible of the links with the past" (p. 44). This attempt to maintain a clear connection with past achievements in physics led Darrow to restate his preference for Schrödinger's approach. For Darrow, corpuscular and wave pictures had not the same epistemic place in quantum mechanics, but "corpuscles must be subordinated to the waves" (p. 44).

In this paper, Darrow seemed even more inclined to follow Schrödinger's exposition than he had been in his previous paper on wave mechanics. Here, indeed, Darrow accepted Schrödinger's optical-mechanical analogy, which he had dismissed in "Introduction of Wave Mechanics" (Darrow 1927). However, this propensity toward Schrödinger's approach remained embedded in the strong conviction that theoretical physics was a model-building practice as Darrow's interpretation of the wavefunction demonstrates. Darrow accepted what is usually called the ensemble-statistical interpretation of quantum mechanics, according to which wavefunctions do not apply to an individual system (Jammer 1974, pp. 38–44, 440–443; Beller 1990). Darrow (1926) believed it to be a strength of the theory that it did not deal with individual atoms "for no statement about the individual behavior of an atom [could] ever be subject to an experimental test" (p. 319).

In order to understand the singularity of Darrow's approach, it is useful to evaluate it against the views held by American theoretical physicists at the time. As Cartwright showed, the statistical interpretation, which Darrow accepted, was widely accepted by the American community of theoretical physicists between the late 1920s and the early 1930s (Cartwright 1987, p. 419). Cartwright used the widespread acceptance of the ensemble-statistical interpretation to confirm that the operationalist-pragmatist philosophical perspective had shaped the American reception of quantum theory. However, the nuanced differences between the various approaches were even more revealing than their similarities. To make these distinctions emerge, one can compare Darrow's views with those of Van Vleck, who was to become a champion of the statistical interpretation of quantum mechanics in the United States.

In 1929, Van Vleck declared that the choice among the various formulations of quantum mechanics depended on the mathematical training:

> [T]he wave formulation of the new mechanics is apt to appeal most strongly to mathematical physicists who have been trained primarily along the line of classical nineteenth century mathematical physics, whereas the matrix formulation appeals more strongly to those who have been trained in the mathematics of the old quantum theory, especially the correspondence principle[9].

Van Vleck was describing a generation gap depending on the different mathematical training of younger American theoretical physicists with respect to the older generation. Like many theorists of his generation, Van Vleck stood for the more abstract formalism of the transformation theory of Paul Dirac (1902–1984) and Jordan, interpreted as a statistical description of the behavior of several particles. The transformation theory pleased him because it was the most comprehensive formulation of quantum mechanics and included all the other formulations as special cases. In Van Vleck's views, the statistical interpretation became the physical viewpoint that allowed for the unification of all the various mathematical approaches. As for Schrödinger's theory, Van Vleck (1929b) contended that it was strongly related to what he called its "extreme hydrodynamical interpretation" (p. 480); namely, the view that the electron was a fluid-like substance whose actual motion was described by the wave equation. According to Van Vleck, this interpretation appealed only to those physicists who were "rather loath to accept the revolutionary philosophy of Heisenberg's indeterminism principle" (p. 479).

Historical studies corroborated Van Vleck's opinion that different interpretations of quantum mechanics depended on different mathematical trainings (Sopka 1988; Servos 1986). However, I have shown that Darrow accepted the statistical interpretation of the wave equation and discarded what Van Vleck called its "extreme

[9] Van Vleck 1929a, p. 485.

hydrodynamical interpretation." Darrow preferred Schrödinger's theory because it was more visualizable than the alternative approaches. More deeply, Schrödinger's formalism allowed Darrow to maintain "the links with the past." The particular role of wave mechanics in Darrow's view of the progress of physics suggests that the choice between different formalisms touched a more profound level than the generation gap Van Vleck emphasized. The divergence between Darrow and Van Vleck's attitudes stemmed from different opinions about how physics evolved, which in turn depended on their daily activities. Van Vleck was a theorist who actively engaged with the physical problems of quantum mechanics, while Darrow had a completely different role. He tried to make sense of these fast theoretical changes for a broader community of physicists. These changes were undermining the unity of physics because conceptual problems were distancing the assumptions held by quantum theorists from those used, for example, by the experimenters who provided the experimental basis of quantum mechanics (see, e.g., Chang 1995). In a period of theoretical instability, Darrow chose to explore and highlight the coherence of the scientific tradition. The different activities of Van Vleck and Darrow led to conflicting descriptions of quantum mechanics even though the proponents shared the same broad philosophical position and accepted the same physical interpretation. As I will show shortly, Darrow's choice depended, in fact, on precise ideological views on the progress of physics.

6 The Evolutionary Character of Physics

Schrödinger's equation was the mathematical tool that allowed Darrow to bridge novel theoretical developments with classical physics, and to demonstrate the evolutionary character of physics. When Darrow (1925) first tried to review the wave-corpuscle duality, he concluded his paper asking for a revolutionary change of mind in order to deal with the contradictory results of different experiments. Ten years later, he challenged the revolutionary perspective by pointing out, instead, that "what [was] happening in modern physics [was] a tremendously rapid evolution" (Darrow 1936, p. 16). The program of unifying the wave and corpuscular pictures of both radiation and matter accomplished by quantum mechanics was only apparently revolutionary. This appearance was fueled by "many among the workers in quantum mechanics [who] have helped to confirm that impression, by writing or speaking of the downfall, the overthrow, or the repudiation of classical theories" (Darrow 1934, p. 24). But, Darrow stressed, there had never been a revolution "more gradual, more cautious, more tenacious of all the virtues of the old regime" (p. 24). The radicalism of single new ideas did not touch the "immense conservatism of the scientific mind," which continued to drive the progress of physics (p. 24). Although physicists were making "strange things" with the Hamilton equation, the very equation and, above all, the relationship between mathematical computations and observations showed a clear continuity with classical physics (p. 24). In Darrow's views,

Maxwell and Lorentz were not superseded. Quite the contrary, they provided the unmodified bases of the new advancements.

The clearest exposition of Darrow's belief in the evolutionary character of physics is in the collection of the lectures Darrow (1936) gave before the Lowell Institute in Boston, published with the revealing name of *The Renaissance of Physics*. The introduction of the book was an ode to the conservatism of physics. For Darrow, contemporary innovators moved in a very definite path whose borders had been solidly built by the works of their predecessors. Darrow's thesis of the evolutionary process of physics, of course, required that something remained constant in what might seem a twisting of the ordinary notions of physics, such as particles and waves, or space and time. When Darrow tried to define what he called the "royal line" of physics (p. 16), he contended that the "historic continuity of physics" depended on the permanence of the tools, theoretical as well as material (p. 78). Centuries of research have provided the physicists with all the "instruments for the hand" and the "instruments for the mind," which they used to do their daily work (p. 14). Without such instruments, Darrow implied, physicists could do nothing. And the tools did not change in an abrupt revolutionary way. On the contrary, they persisted and made possible novel combinations of the representation of nature.

While the gradual transformation of experimental tools seemed self-evident to Darrow, the development of quantum mechanics required an in-depth analysis of the evolutionary character of theories. To demonstrate that intellectual tools resisted revolutionary changes as much as measuring instruments did, Darrow again referred to the normative usefulness of the model-building activity. Darrow took as the most relevant example the copresence of corpuscle and wave theories in contemporary explanations of quantum phenomena. As Darrow saw it, physicists "ha[d] spent centuries trying to reject either the wave theory or the corpuscle theory; and [they] ha[d] ended by keeping them both, tacitly consenting to whatever violence must be done to our habits of thought" (p. 15). But, how should one interpret such a coexistence of apparently irreconcilable theories? In the chapter on the duality of waves and corpuscles, Darrow eventually gave up any hope of resolving the mystery in any definite manner. He expressed his scepticism about all the metaphysical explanations that had so far been proposed, and advised the reader not to blindly follow any particular school of thought.

With *The Renaissance of Physics*, Darrow's personal struggle with quantum mechanics, and, in particular, with the particle-wave duality, came to an end with a strong reaffirmation of his earlier philosophical position on the necessary copresence of different theoretical models. By the late 1930s, Darrow claimed, the intellectual war between the two conceptions had "died away without victory" (p. 168). All physicists had come to embrace, or at least, to tolerate the duality. All they needed to know were De Broglie's rules of correlation linking energy and momentum of the particles with the frequencies and wavelength of the associated waves. The experimental evidence for the different phenomena was overwhelming. While Darrow considered the various interpretations to be inconclusive, he claimed total confidence in the experimental verifications of both wave and corpuscular phenomena. In the conclusion of his ten-year personal conceptual fight with the wave-particle duality, Darrow reconfirmed his belief that experiments had priority over theories in the evolution of physics. He contended that physicists had to believe in waves because of the phenomena of diffraction, and had to believe in corpuscles because they had seen their splashes and tracks in the Wilson chamber. Darrow's apparent endorsement of the positivist motto that "seeing is believing" represented actually a view quite different from those held by the leading exponents of positivist and logical-positivist doctrines (Hacking 1983, p. 63). Darrow was perfectly aware that there were different ways of seeing. When he stated that his generation was the first to have seen the atom, he stressed that physicists had not seen it in the same way in which they saw macroscopic objects, such as balls and pebbles. Physicists had seen atomic tracks. For Darrow, there was no epistemic difference between seeing an object and seeing its tracks; both constituted conclusive evidence for the ultimate reality of the physical entities. Such a view was consistent with a strong tradition in experimental physics, which historian of science Peter Galison (1997, p. 22) called "image tradition"—an epistemic tradition that experimentalists upheld throughout the twentieth century. Although Darrow was not an experimental physicist, he was one of the most receptive exponents of this tradition and his focus on visualizable theoretical models seems to be in line with this tradition. In all of his writings on quantum mechanics, he consciously tried to build a hermeneutics of quantum mechanical problems that would allow the continuity of this tradition to emerge out of the changing world of theoretical physicists.

7 Conclusion. American Ideology of National Science and the "Renaissance of Physics"

Darrow was neither a theoretical nor an experimental physicist. He was a mediator between different subcultures of physics. Darrow believed in the need for synthetic reviews of complex developments in physics and dedicated his stylistic talent to this purpose. Nevertheless, he did not receive passively what was going on in theoretical physics. On the one hand, he had to reconfigure his set of beliefs in order to cope with the duality of the wave-corpuscle pictures of light and matter as well as with the theoretical responses to this problem. On the other hand, he interpreted and translated the new concepts and mathematical tools in a very personal way. Cartwright's interpretation of Darrow's words as embedded in the American operationalist-pragmatist approach to quantum theory is valuable, but it does not consider the several nuances of this case as well as its historical evolution. Darrow shared various common epistemological beliefs with other American physicists, but he questioned the inclination for abstractness of matrix algebra and transformation theory, which many American theoretical physicists showed (Van Vleck 1929a, 1929b; Sopka 1988, pp. 221–302). Darrow's personal accounts of quantum theory ultimately depended on the strong belief that physics was essentially an evolutionary discipline—a view deeply different from that maintained by those theoretical physicists who stressed the revolutionary character of the new theories.

Although Darrow gave equal relevance to the theoretical and material tools in the continuity of physics, his picture of how physicists acquire knowledge focused on the experimental side of physics. He affirmed his admiration for Kamerlingh Onnes's famous motto "through measuring to knowing" (Darrow 1936, p. 51). For Darrow, measurement was the only way through which conceptions of invisible entities acquire meaning. Since nature did not present herself in such a way that simple laws could be deciphered, so Darrow's argument went, physicists needed to arrange for situations that allowed them to observe the underlying simplicity. Thus Darrow demonstrated his full adherence to a Baconian view of experimental physics: Only by forcing nature into a specific artificial setting could she reveal a part of her secrets (Hacking1983, p. 246).

One is tempted to trace the connections between Darrow's full commitment to the Baconian view of experimentation and his working milieu. The Bell Labs were a multidisciplinary setting in which physicists and engineers worked closely together in a goal-oriented practice. Since the late nineteenth century, in such an environment the borders between pure and applied science were contingently shifting and, often, disappeared (Shapin 2008). Darrow argued that the engineer was nothing but an applied physicist. He also stated that the pure and the applied physicist made essentially the same thing. They both modified nature: the former to understand her; the latter to make her serve people. What Darrow was implicitly providing was a description of his own perception of how scientists acted and thought in a techno-scientific environment[10].

Experimentation, in Darrow's words, became a means for dividing the atom and using its constituent parts to create new elements. Even though Darrow was probably influenced by the recent discoveries of new particles, such as the neutron and the positron, the major factor influencing Darrow's explicit connection between pure and applied physics strongly depended on the ideological trend that American physicists were following in the 1920s. Historian of science Ronald Tobey (1971) argued that leading American scientists tried to popularize an ideology of national science in the post-World War I period. The conservative character of such an ideology implied that scientific progress was evolutionary, not revolutionary. Tobey maintained that this ideology considered experimental physics and goal-oriented industrial research groups as an alternative model of inquiry to the solitary and esoteric practices of theoretical physics. In their struggle for recognition and funds, certain American scientists were trying to build an image of pure sciences as the basis of applied sciences. Millikan, one of the principal actors in developing such an ideology, claimed that American industrial development would have suffered if pure research were not supported sufficiently. This ideological hierarchy served as an argument for the funding campaign for the National Research Council among American industrial firms, which started in 1926.

[10] Darrow's description shares common elements with the idealized framework the historian of science John Pickstone (2000) has created about the ways in which scientists know nature and make artifacts.

AT&T was one of the few industrial firms that substantially contributed to the fund. The reciprocal influence between Millikan's views on physics and the research policies of the Bell Labs was enormous and complex (Kevles 1971; Gertner 2012). Millikan also had a strong influence in shaping Darrow's view of physics as well as his career. Darrow had been a student of Millikan and the latter strongly supported the appointing of the former within the research staff of the Bell System. Moreover, Darrow's position at the juncture between experimental and industrial physics was in line with the ideology professed by many American scientists. It is unsurprising that the politically conservative Darrow sincerely shared the set of values popularized by Millikan in the 1920s. *The Renaissance of Physics* was, indeed, the summa of this ideological view of physics. In the introduction, Darrow (1936) explicitly asked the reader to accept that engineering was "a province of physics" (p. 6). By subordinating applied physics and engineering to pure physics, Darrow was giving his contribution to the widespread endeavor to provide funding for the National Research Council.

Seen in this ideological perspective, the "historic continuity of physics" stood as the central argument on which Darrow tried to demonstrate the genetic role of physics with respect to related applied sciences. The diachronic continuity between classical physics and quantum theory corresponded to the synchronic continuity of methods between different disciplines. Darrow disliked specializations. For him, the unification of physics was to be social as well as epistemic. A community of pure and applied physicists should have continued to be able to discuss all the topics of the various subdisciplines of physics. His work to spread the new knowledge and the way in which he coped with various topics had just this aim: to preserve as long as possible the continuity of physics. While he thought of himself as a champion of the temporal continuity of physics, his implicit purpose was to challenge the impending fragmentation of the physics community (Weart 1992; Kaiser 2012). From this perspective, Darrow's reading of quantum theoretical developments and his commitment to Schrödinger's formalism were just some of the strategies through which he struggled for epistemic and social continuity within physics.

Acknowledgments The archival research was supported by a grant-in-aid from the friends of the Center for the History of Physics, American Institute of Physics. I gratefully thank Marie Burks and Max Levis for the English revision of the paper and Cristopher Lehner for the useful discussions on Darrow's interpretation of wave mechanics. I am also grateful to Peter Galison, Michael Gordin, Gerald Holton and the other participants of the Phunday 2013 workshop where I presented a previous version of the paper. I especially thank David Kaiser, Skúli Sigurdsson and an anonymous referee for their helpful comments and suggestions.

References

Assmus A (1992a) The molecular tradition in early quantum theory. Historical Studies in Physical and Biological Studies 22:209–231

Assmus A (1992b) The Americanization of molecular physics. Historical Studies in Physical and Biological Studies 23:1–34

Beller M (1983) Matrix theory before Schrödinger: philosophy, problems, consequences. Isis 74:469–491
Beller M (1990) Born's probabilistic interpretation: a case study of 'Concepts in Flux'. Studies in History and Philosophy of Science Part A 21:563–588
Beller M (1997) 'Against the stream'—Schrödinger's interpretation of quantum mechanics. Studies in History and Philosophy of Science Part B 28:421–432
Bitbol M (1996) Schrödinger's philosophy of quantum mechanics. Kluwer, Dordrecht
Camilleri K (2009) Heisenberg and the interpretation of quantum mechanics: the physicist as philosopher. Cambridge University Press, Cambridge
Cartwright N (1987) Philosophical problems of quantum theory: the response of American physicists. In: Krüger L, Gigerenzer G, Morgan M (eds) The probabilistic revolution, vol. 2: ideas in the sciences. MIT Press, Cambridge, pp 417–435
Cat J (1998) The physicists' debates on unification in physics at the end of the 20th century. Historical Studies in the Physical and Biological Sciences 28:253–300
Chang H (1995) The quantum counter-revolution: internal conflicts in scientific change. Studies in History and Philosophy of Modern Physics 26:121–136
Coben S (1971) The scientific establishment and the transmission of quantum mechanics to the United States, 1919–32. American Historical Review 76:442–460
Darrow KK (1917a) (Diary 1917). in: KDP (Box 6, Folder 14)
Darrow KK (1917b) (Letter to father and mother, 15 September). In: KDP (Box 2, Folder 25)
Darrow KK (1920a) (Letter to G. S. Gorton, 15 March). In: LHP (Box 2, Folder 30)
Darrow KK (1920b) (Letter to F. B. Jewett, 15 March). In: LHP (Box 2, Folder 30)
Darrow KK (1923) Contemporary advances in physics. Bell System Technical Journal 2:101–115
Darrow KK (1925) Contemporary advances in physics—VII waves and quanta. Bell System Technical Journal 4:280–326
Darrow KK (1926) Introduction to contemporary physics, 1st edn. Van Nostrand, New York
Darrow KK (1927) Contemporary advances in physics—XIV introduction of wave mechanics. BSTJ 6:653–701
Darrow KK (1930) Contemporary advances in physics—XIX fusion of wave and corpuscle theories. Bell System Technical Journal 9:163–188
Darrow KK (1934) Elementary notions of quantum mechanics. Reviews of Modern Physics 6:23–68
Darrow KK (1936) The renaissance of physics. MacMIllan, New York
Darrow KK (1964). Oral history interview by Henry Barton and W. J. King. AIP. http://www.aip.org/history/ohilist/4567_1.html. Accessed 12 Nov 2012
De Regt H (1997) Erwin Schrödinger, Anschaulichkeit, and quantum theory. Studies in History and Philosophy of Modern Physics 28:461–481
Dempster AJ (1926) (Letter to K. K. Darrow, 27 July). In: KDP (Box 12, Folder 8)
First meeting of the Colloquium, season 1919–1920 on 20 October 1919. In: LHP (Box 2, Folder 30)
Galison P (1997) Image and logic: a material culture of microphysics. University of Chicago Press, Chicago
Gertner J (2012) The idea factory: Bell Labs and the great age of American innovation. Penguin, New York
Hacking I (1983) Representing and intervening: introductory topics in the philosophy of natural science. Cambridge University Press, Cambridge
Hacking I (1996) The disunities of the sciences. In: Galison P, Stump DJ (eds) The disunity of science: boundaries, contexts, and power. Stanford University Press, Stanford, pp 37–74
Harré R (1992) Bohr and Schrödinger. In Bitbol M, Darrigol O (eds) Erwin Schrödinger: philosophy and birth of quantum mechanics. Editions Frontières, Gif-sur-Yvette, pp 143–50
Hoddeson L (1980) The entry of the quantum theory of solids into the Bell Telephone Laboratories, 1925–1940: a case-study of the industrial application of fundamental science. Minerva 38:422–447
Hoddeson L, Braun E, Teichmann J, Weart S (eds) (1992) Out of the crystal maze: chapters from the history of solid-state physics. Oxford University Press, New York

Holton G (1988) On the hesitant rise of quantum physics research in the United States. In: Holton G (ed) Thematic origins of scientific thought: Kepler to Einstein, 2nd edn. Harvard University Press, Cambridge, pp 147–187
Jähnert M, Lehner C (2015) The early debates about the interpretation of quantum mechanics. (In press)
Jammer M (1966) The conceptual development of quantum mechanics. McGraw-Hill, New York
Jammer M (1974) The philosophy of quantum mechanics. Wiley, New York
Joas C, Lehner C (2009) The classical roots of wave mechanics: Schrödinger's transformations of the optical-mechanical analogy. Studies in History and Philosophy of Modern Physics 40:338–351
Kaiser D (2012) Booms, busts, and the world of ideas: enrollment pressures and the challenge of specialization. Osiris 27:276–302
Karl Darrow's a genius: seven-year-old boy is a marvel of learning—he is writing a monster novel on the world's fair. (1898 ca.). (Newspaper clip). In: KDP (Box 1, Folder 3)
KDP [Archival Collection Karl K. Darrow Papers], American Institute of Physics (AIP). Center for History of Physics. One Physics Ellipse College Park, Maryland 20740, USA
Kevles DJ (1971 [1995]) The physicists: the history of a scientific community in modern America. Harvard University Press, Cambridge
Lalli R (2014) A new scientific journal takes the scene: the birth of Reviews of Modern Physics. Annalen der Physik 526:A83–A87
LHP [Archival Collection Lillian Hoddeson Papers], American Institute of Physics (AIP). Center for History of Physics. One Physics Ellipse College Park, Maryland 20740, USA
MacKinnon E (1980) The rise and fall of the Schrödinger interpretation. In Suppes P (ed) Studies in the foundations of quantum mechanics. Philosophy of Science Association. East Lansing, pp 1–57
Meeting of the Colloquium during its first season 1919–1920. Attached to (Darrow 1920b)
Mehra J, Rechenberg H (1987) The historical development of quantum theory, vol 5, part 2, the creation of wave mechanics: early response and applications 1925–1926. Springer, Berlin
Miller AI (1984) Imagery and scientific thought: creating 20th-century physics. Birkhäuser, Boston
Millikan RA (1917) (Letter to K. K. Darrow, 6 September). In: KDP (Box 12, Folder 2)
Millman S (ed) (1983) History of engineering and science in the bell system. vol 4: physical sciences (1925–1980) Bell Telephone Laboratories, Murray Hill
Modern physical theories—time table. In: LHP (Box 2, Folder 30)
NBA [Niels Bohr Library & Archives], American Institute of Physics (AIP). Center for History of Physics. One Physics Ellipse College Park, Maryland 20740, USA
Pickstone JV (2000) Ways of knowing: a new history of science, technology and medicine. University of Chicago Press, Chicago
Renn J (2013) Schrödinger and the genesis of wave mechanics. In Reiter WL, Yngvason J (eds) Erwin Schrödinger—50 years after. ESI Lectures in Mathematics and Physics, European Mathematical Society Publishing House, Zürich. http://www.mpiwg-berlin.mpg.de/Preprints/P437.PDF. Accessed 20 March 2013
Schrödinger E (1928) Collected papers on wave mechanics. Blackie & Son, London
Schweber SS (1986) The empiricist temper regnant: theoretical physics in the United States 1920–1950. Historical Studies in Physical and Biological Science 17:55–98
Schweber SS (1990) The young John Clarke Slater and the development of quantum chemistry. Historical Studies in Physical and Biological Science 20:339–406
Servos JW (1986) Mathematics and physical sciences in America, 1880–1930. Isis 77:611–629
Shapin S (2008) The scientific life: a moral history of a late modern vocation. University of Chicago Press, Chicago
Sopka K (1988) Quantum physics in America: the years through 1935. American Institute of Physics, New York
Tobey R (1971) The American ideology of natural science, 1919–1930. University of Pittsburgh Press, Pittsburgh
Van Vleck JH (1929a) The new quantum mechanics. Chemical Review 5:467–507

Van Vleck JH (1929b) The statistical interpretation of various formulation of quantum mechanics. Journal of the Franklin Institute 207:475–498

Van Vleck JH (1967) Karl Kelchner Darrow—writer, councillor and secretary. Physics Today 20:23–26

Weart SR (1992) The solid community. In: Hoddeson L, Braun E, Teichmann J, Weart S (eds) Out of the crystal maze: chapters from the history of solid-state physics. Oxford University Press, New York, pp 617–669

Weiner C (1973) How the transistor emerged. IEEE Spectrum 10:24–33

Wessels L (1983) Erwin Schrödinger and the descriptive tradition. In: Aris R, Davis HT, Stuewer RH (eds) Springs of scientific creativity: essay on founders of the modern science. University of Minnesota Press, Minneapolis, pp 254–278

White A (1976) Interview by L. Hoddeson. AIP. http://www.aip.org/history/ohilist/4960.html. Accessed 28 Jan 2013

Wooldridge D (1976). Interview by L. Hoddeson. http://www.aip.org/history/ohilist/4981.html. Accessed 15 Jan 2013

The Historical Development of X-ray Absorption Fine Spectroscopy and of Its Applications to Materials Science

Annibale Mottana and Augusto Marcelli

Abstract This essay sketches the development of X-ray Absorption Fine Spectroscopy (XAFS) ever since the second half of the twentieth century. At that time, synchrotrons started competing with X-ray discharge tubes as the sources of the excitation able to show the pre- and near-edge structures (XANES) and extended oscillations (EXAFS) that characterize the X-ray absorption edge of solid matter. Actually, modern XAFS began to be used after 1975, when the hard-X-ray synchrotron radiation derived from storage rings took over. Ever since, XAFS have greatly contributed to both technical refinement and to theoretical development of Materials Science. Although a unified theory of X-ray fine absorption has not been reached yet, many XAFS advancements benefited from theoretical models and complex calculations made possible by continuous growth of computing power, while contributing to developing new or previously never used materials.

Keywords X-rays · Absorption edge · Synchrotron radiation · EXAFS · XANES · XAFS · Multiple scattering · Single scattering · History of science

1 Introduction

Wilhelm Conrad Röntgen (March 28, 1845–February 10, 1923) detected absorption of X-rays by matter on the very evening he discovered X-rays (November 8, 1895). Some days later, he also realized that different materials display different degrees of transparency, as he took

A. Mottana (✉)
Department of Geosciences, University of Roma Tre, Roma, Italy
e-mail: annibale.mottana@uniroma3.it

A. Marcelli
Istituto Nazionale di Fisica Nucleare, Laboratori Nazionali di Frascati, Frascati, Italy
e-mail: augusto.marcelli@infn.lnf.it

R. Pisano (ed.), *A Bridge between Conceptual Frameworks,* History of Mechanism and Machine Science 27, DOI 10.1007/978-94-017-9645-3_15

> [...] an image of his wife's hand which showed the shadows thrown by the bones of her hand and that of a ring she was wearing, surrounded by the penumbra of the flesh, which was more permeable to the rays and therefore threw a fainter shadow.[1]

His first, simple observations were crucial not only for the discovery of X-rays but also for their quick expansion throughout the world. It is well known, unfortunately, that Science is considered by most people to be of little use ("pure") if it does not produce practical results at once. The transparency noticed by Röntgen found a practical application almost immediately in clinical Medicine: numberless clinical transparency images were taken, making Radiology the most productive diagnostic method to date.[2]

Soon X-rays were also tested in detail for scientific reasons. However, reviewing all branches of Science that benefited from X-rays is not the aim of this contribution. Rather, we will describe the development of a field of X-ray-based research on solid matter that started over half a century after Röntgen's discovery: X-ray Absorption Fine Spectroscopy (XAFS).

XAFS is the modern development of X-ray Absorption Spectroscopy (XAS), a branch of X-ray-based Physics that began over 10 years later than Röntgen's X-ray discovery, and a couple of years later than X-ray Diffraction (XRD), the first widespread form of application of X-rays for scientific purposes. For a long time XAS was poorly considered or even under-esteemed, overshadowed as it was by XRD extraordinary success. Indeed, XRD had made a solid-state research take off because it could determine the lattice structure of ordered matter, but it could do nothing when such an order was not present, as in amorphous materials, highly disordered systems and liquid solutions. This was indeed XAS' deed, but this took a long time to be understood and even more to attain successful results. Only during the second half of the last century did XAS take off too, and in the new form of XAFS. In order to reach the front-line of Science, it had to wait for the availability of powerful and brilliant X-ray sources such as the electron synchrotrons first, and then the much more stable and brilliant storage rings. In addition, the availability of powerful computers and advanced computational methods pushed to better insight into theory (an essential part of Science), by allowing simulations that would both validate and crosscheck the results of experiments.

Now over 50 % of the requests for time reaching synchrotron laboratories from industrial customers involve one or another X-ray absorption method, making XAFS the choice field for any study on innovative materials. XAFS is also the best tool to evaluate their technological applications, particularly for Chemistry, which is the science that creates new materials; for Biology to characterize biomolecules, both natural and synthetic; finally, yet just as importantly, for the Earth's scientists,

---

[1] This sentence is in the autobiography Röntgen wrote at the time of the Nobel award, later translated into English and published in the *Nobel Lectures* (1967).

[2] As early as on February 3, 1896, at Dartmouth Medical School and Hospital, Hanover, NH, USA, the first photographic image of the fractured wrist of a boy was taken (Spiegel 1995, p. 242) and the injured bone could be appropriately compounded: first clinical application of X-rays in the whole world.

to characterize amorphous or even glassy materials, where XRD is of little help, if any.

The historical development of XAFS has been reviewed several times, but only by bits and pieces. Actually, there is one authoritative general review (Stumm von Bordwehr 1989), which however stops in the year 1975, when synchrotrons started replacing discharge tubes as X-ray sources. In addition, there are short review papers published in the proceedings of the many international conferences taking place ever since 1982 (e.g., the series "EXAFS and Near Edge Structure", or "Advances in X-ray Spectroscopy", etc.), and there are also accounts by or about people who contributed to work in progress (e.g., Citrin 1986; Doniach et al. 1997; Lytle 1999; Stern 2001; Mottana 2003; etc.). Nevertheless, there is no appraisal yet of the pros and cons that Materials Science has derived from XAFS during 50 odd years of life and development. This is indeed the purpose of our still limited (in terms of space) review.

1.1 How XAS Came into Being

1913 was a most extraordinary year for Physics. It saw the birth of some aspects of this science that are basic for its modern state-of-art, such as the first formulation of the atomic theory (Niels Bohr); the first crystal structure solution by XRD (W. Henry Bragg and W. Lawrence Bragg); the essentials of X-ray fluorescence (XRF) analysis (Henry G.J. Moseley). It was the year too when (August) Julius Hedwig (October 17, 1879 – January 26, 1936) in Germany and (Louis-César-Victor-) Maurice de Broglie (April 27, 1875 – July 14, 1960) in France, working independently but both using a modified type of spectrometer in a reverse way to impinge a metal sheet with monochromatic X-rays, detected on photographic plates some sharp or diffuse lines and bands that appeared next to the strong, enhanced edge of the investigated metal atom (platinum or tungsten) in the region of shorter wavelengths: a spectrum, then. On this rather weak basis, M. de Broglie (1913) dared proposing a structure for X-rays that was similar to that—long since known—of light spectra. Neither Herweg nor de Broglie were completely confident in their observations[3]. The former scientist (whom priority should be assigned, as he submitted his paper on June 30) gave up with this kind of research. The latter one (who presented his first note on November 17) was heavily criticized and even had to admit he had misinterpreted his first photos. Nevertheless, he kept on experimenting and communicating new results for some years, thus amply justifying his reputation as the man who "*invented X-ray spectroscopy*" (Lytle 1999, p. 123). The existence of modulations at the X-ray absorption edge of metals was first established without doubt between 1918 and 1920 by [Karl] Wilhelm Stenström (January 28, 1891–November 7, 1973) and Hugo Fricke (August 15, 1892–November

[3] *Eine Aneinanderreihung von Flecken, die verschiedenen Wellenlängen entsprechen* (Herweg 1913, p 556*); un véritable spectre de raies, ayant tout à fait l'aspect des photographies de spectre lumineux, avec des raies fines or diffuses, des bandes, etc.* (de Broglie 1913, p 925).

10, 1972), both working at Lund University under the supervision of [Karl] Manne (Georg) Siegbahn (December 3, 1886–September 26, 1978), who earned the 1924 Nobel Prize because of it.

Actually, XAS experienced a great upgrade at the beginning of the second decade of the twentieth century, when it was instrumental in probing the electronic structure of atoms and in the successful development of quantum theory. Many people contributed to this early phase of basic research, and by the final years 1920s, the fundamentals of XAS were firmly set. Then, limitations inherent in the X-ray sources and detectors restricted its further development for many years especially on what the method was most in the need for: the interpretation of the fine structures detectable around the absorption threshold (known, at that time, as the "*Kossel structure*"). They had been understood in their generality, but not really worked out in detail. Moreover, other features at higher energy (known as the "*Kronig structure*") had been detected, studied and explained, however without reaching real consensus on their meaning among scientists, mainly because experimental results and calculated data rarely matched.

For almost 30 years, it was clear that in order to really reach the front-line if Science XAS needed more intense X-ray sources, much better detection methods, and—last but not least—a better theoretical background than the theories formulated by Walther Kossel (1920) and Ralph deLaer Kronig (1931, 1932a, b).

2 XAFS Moves to Synchrotron Radiation Sources

2.1 Improvements in the Apparatus

A first upgrade in the experimental part of XAS studies that would eventually lead to XAFS occurred only after the Second World War (WWII), when Norelco designed a novel type of diffractometer for powder XRD use. Indeed, a Norelco vertical powder diffractometer was the instrument modified by Robert Van Nostrand (November 28, 1918–January 20, 2012) to measure the absorption coefficient of the catalysts produced by the company he was employed in. As monochromator he used a single crystal of Si or quartz set at the centre of the Rowland circle, i.e., where the sample to be analysed by XRD is, and he recorded the reflected beam intensity using a counter that could be rotated stepwise. He first was doing it for the beam passing free (I_0), then for the beam filtered by a finely powdered sample (I_1). The absorbance, $\ln(I_0/I_1)$, was plotted as a function of θ, the rotation angle of the diffractometer, the intensity at each θ angle being a fraction of the maximum occurring at the absorption jump, which he normalized to one irrespectively of the operation conditions, thus introducing a normalized absorption coefficient (Van Nostrand 1960).

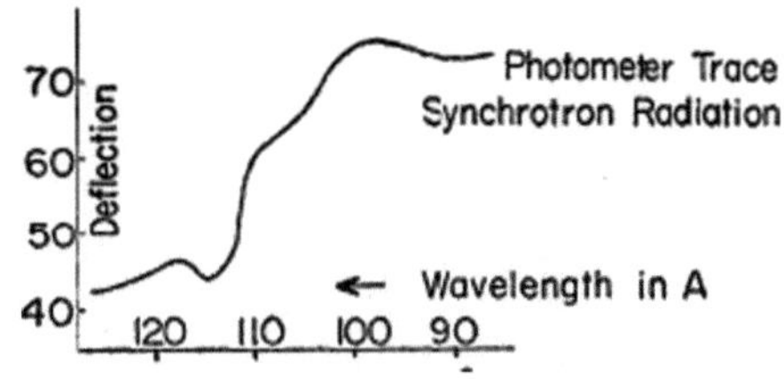

Fig. 1 The first absorption spectrum recorded using synchrotron radiation: the B K-edge using the 300 MeV electron synchrotron. (Johnston and Tomboulian (1954, p. 1589), Fig. 5, strongly reduced)

In this way, which mimics most of the powder-XRD practice, he introduced into XAS the manner of presenting data that has now become standard[4] not only in the industry but also among scientists. Indeed, he could use the absorption data not only for internal comparison, but also to exchange with outside laboratories, thus creating the first XAS data bank. Despite this substantial improvement, which prompted not only other experimental studies but also reconsideration of the entire theory (see later), nothing highly significant would have really happened with XAS had a new intense and brilliant source of radiation not entered physical sciences: the synchrotron.

Particle accelerators had been studied theoretically during WWII and were built for research in the early years 1950s. Using these early accelerators the first *K*-edge absorption spectrum ever-recorded was that of Be metal (Johnston and Tomboulian 1954, Fig. 1) using the soft X-rays emitted by the Cornell synchrotron. In Europe, the first ones were the Al and Cu *K*-edges recorded in May 1963 at the Frascati electron synchrotron (Cauchois et al. 1963; Mottana and Marcelli 2013).

The electron synchrotrons, despite their limited energy (~ 1 GeV) and large instabilities, made detailed studies in the soft X-ray energy range possible. Actually, such studies were few (e.g., Sagawa et al. 1966; Jaeglé and Missoni 1966; Balzarotti et al. 1970), and were limited to the edges of gases and of certain metal atoms. Yet, they promoted not only the use of synchrotron radiation (SR), but also the reactivation of studying XAS basic theory. Unfortunately, it is a historical fact that, in spite of all advantages offered by orbit-derived radiation (Parrat 1959), SR as the X-ray source did not become really popular among the XAS community until storage rings were introduced i.e., 10–15 years after installing the first electron synchrotrons. Such a long time delay had several good reasons: the conventional source was much cheaper, and had the advantage of being stable and reliable, besides being at home.

The approach to Science changed only in the middle of the 1970s, and compelled people to move from their laboratories to large facilities operating stable storage rings i.e., to radiation sources much more suitable for high-quality research.

[4] *For the study of the chemistry of catalysts and other non-crystalline systems this technique may have a role comparable to that of X-ray and electron diffraction in crystalline systems* (Van Nostrand 1960, p. 184). Indeed, he was wrong only in that he did not foresee that the technique could be applied just as successfully to natural samples that are significant for the oil and mining industries, and more recently to such environmental problems such as pollution by dust and aerosols.

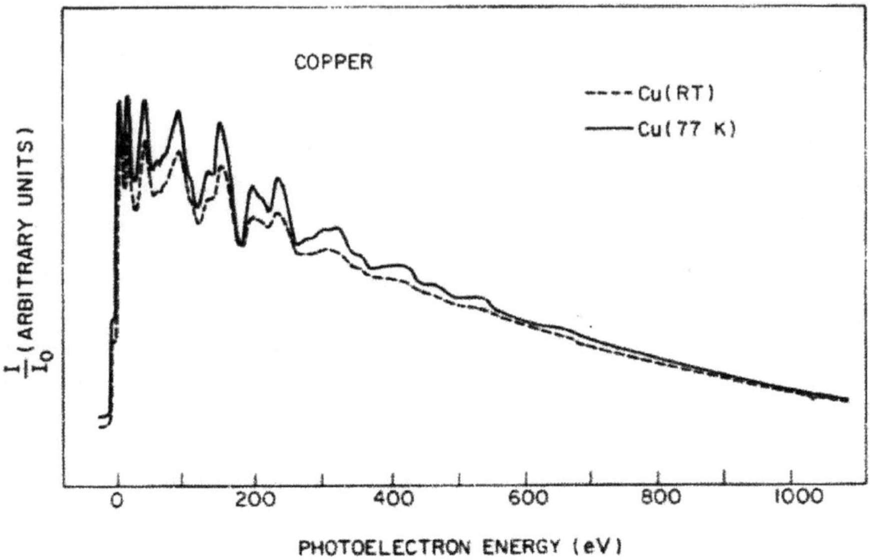

Fig. 2 The first published XAS spectrum recorded using the synchrotron radiation extracted from the SPEAR I storage ring. (Eisenberger et al. (1974, p. 806), Fig. 1)

Their highly increased brilliance, combined with the production of hard X-rays, made it possible to measure the *K*-edges of heavy metal atoms, and these in turn widened the use of XAFS to branches of Science well outside Physics, such as Chemistry, Geochemistry, Biophysics, Metallurgy, etc.

The first published experimental spectrum that used SR extracted from a storage ring by a bending magnet was recorded on metallic copper at the Stanford Synchrotron Radiation Facility (SSRF, now SSRL; cf. Doniach et al. 1997), the orbiting storage ring being SPEAR I (Fig. 2).[5] Irrespectively on whether those experiments were done at room or liquid nitrogen temperature, the improvement in intensity over a conventional source was beyond any expectation. It was calculated to be 5×10^4, together with a 10^4 enhancement in the signal-to-noise ratio, so that the recording time of a complete EXAFS spectrum could be reduced to 30 min (Kincaid and Eisenberger 1975, p. 1361). The following statement by one of the founding fathers of modern XAFS is worth quoting, as it points out effectively the change of paradigm: *In one trip to the synchrotron we collected more and better data in 3 days than in the previous 10 years. I shut down all three X-ray spectrometers in the Boeing laboratory. A new era had arrived* (Lytle 1999, p. 132).

Over a short period of time substantial improvements were conceived and put into operation aiming to record even better spectra, not only by new techniques such as fluorescence detection based on solid-state detectors and optimized for diluted systems (Jaklevic et al. 1977), but also by an increase of the SR flux that bending

[5] The same paper (Eisenberger et al. 1974) shows in addition the Cu XANES spectrum recorded on a Cu-porphyrin: this is the first spectrum of an organic molecule ever recorded.

magnets extracted from the source. Wiggler magnets were developed jointly in the early years 1980s, and first used at SSRL (Winick et al. 1981), then at Frascati National Laboratories (LNF), where the ADONE storage ring had started operation in 1978. They proved to be immensely useful because they produced a SR spectrum with a much higher critical energy, thus providing a high-enough intensity for all absorption experiments requiring high-energy photons (>20 keV). Consequently, many beam lines optimized for spectroscopy moved from bending magnets to wigglers devices. Further energy gains were obtained by developing undulators, which increase power and brilliance of several orders of magnitude.

Because of all this upgrade, modern XAFS blossomed, and a great number of data became available to renew and test the theory, which was largely lying still on Kronig's one (1931). Nevertheless, there were other technical limitations in the apparatus and particularly in the computing power. They made so that two specialized techniques developed independently, in both experiment and calculation: (a) EXAFS, which provides a wealth of quantitative bond distance information based on single scattering processes, and (b) XANES, which implements the quantitative structural information deduced from EXAFS with information on the electronic properties, and also recognizes the local coordination and gives bond angles information. The time-maps show that the developments of these two techniques were not the same despite both used the same experimental apparatus. EXAFS progressed rapidly; XANES lagged behind hampered by its own complexity. The use of new acronyms underlines a new theoretical and practical approach. In fact, XAFS main purpose shifted towards developing a structure-determination technique that would stand side-by-side or even compete with the XRD-dependent science of Crystallography, by then mature and able to determine the average crystal structure with accuracies in bond lengths ±0.001 Å and angles ±0.1°. During the years 1980s, it was clearly understood that determining the atomic structure with all its local defects is the reason behind any study on solid-state matter, as number and distribution of defects determine and/or modify physical properties. Such a result can be best attained by XAFS, which is a local probe, and is reached even better when its results are combined with information on the average crystal structure. The change of paradigm of Materials Science was complete, and XAFS was finally in the front line of Science.

For a rather long period of time scientists working with XAFS could make use of SR from storage rings only parasitically, but the evident rapid progress they made and the usefulness of the technique for fundamental research and industrial applications suggested first creating dedicated beam lines, then entirely dedicated laboratories. Intensity and tunability were the properties most required and best made use of. First generation storage rings had been designed for researches on particle physics, but were implemented with partly dedicated or parasitic beam time and with beam lines especially designed for XAFS already in the early 1970s. Second generation storage rings designed to run as fully dedicated sources started operation around 1975 (the first one being SOR in Tokyo, Japan, a 380 MeV ring). There are now plenty of second- and third-generation storage rings in operation, where all types of XAFS experiments can be performed. The largest, most efficient and best-equipped facilities are ESRF at Grenoble in Europe (6 GeV), APS II at Argonne in USA (7 GeV) and Spring-8 in Japan (8 GeV).[6]

The continuous upgrade of the synchrotron radiation facilities, such as the increased current and stability and, more recently, the availability of topping up modes, may explain the continuous increase in number and quality of XAFS studies, which now involve all possible kinds of materials and phenomena, and operate even under extreme P, T conditions and high magnetic fields. To achieve these results, and in particular to perform experiments on extremely diluted systems, the availability of new solid-state multi-element detectors was absolutely needed. Computer Science also contributed with new methods that make the old cumbersome ones simply legendary. The extensive use of computers is also responsible for the software packages now available for EXAFS and XANES analyses.

2.2 Development of Theory

R. Stumm von Bordwehr (1989) closed his review by reaffirming (p. 444) why he decided to finish his work with the state-of-art of XAS in 1975: *at this date the theory of XAFS was established ... and* [the year 1975] *marks the beginning of an explosive growth in the field* (p. 379). In fact, the change from XAS to XAFS and all related growth were propelled not only by the introduction of synchrotron as the radiation source of choice, but because the obsolete Kronig's theory was profoundly revised and implemented. The poor calculations resulting from this theory forced researchers to develop new methods and strategies, and a virtuous loop started of experimental data requiring theoretical refining.

For clarity, we anticipate that modern XAFS theory describes the entire absorption spectrum as resulting from the scattering of a photoelectron, which progressively damps out in energy moving from a multiple-scattering (MS) regime, which entails many scattering contributions generated by the photoelectron with neighbouring atoms around the photoabsorber, containing plenty of information on the local geometrical structure, i.e. bond distances and bond angles, to a single-scattering (SS) regime, which involves only couples of neighbouring atoms and contains information on bond distances (Fig. 3).

2.2.1 EXAFS Spectroscopy

In his recollection of the EXAFS first steps (1999), Farrel W. Lytle (November 10, 1934) makes a strong point at stating that, before going to SSRL, *all the data had been obtained with the conventional X-ray sources in our laboratory* (p. 132).

[6] To date, there are 67 synchrotron radiation facilities (www.lightsources.org/) in operation across the world, with around 152 beam lines optimized for XAFS measurements: 61 in the Americas, 54 in Europe and 37 in Asia (H. Oyanagi, personal communication).

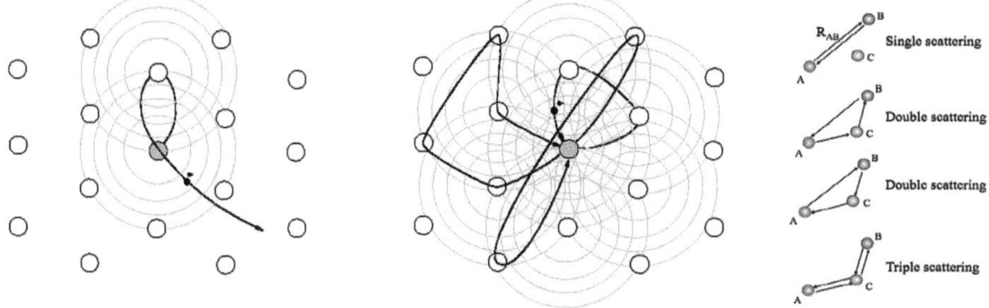

Fig. 3 Pictorial view of the photoelectron scattering processes: (*left*) single scattering by a single neighbour atom = EXAFS; (*right*) simple multiple scattering pathways associated to processes generated by interactions with neighbouring atoms set at different distances and angles = XANES (*centre*). All events generate waves and contribute to the absorption cross section, before the photoabsorber atom returns to its fundamental state in ca. 10^{-15} s

Moreover, he stresses that the new theory of EXAFS[7] was conceived by him and first presented as nothing more than a preliminary hypothesis (Lytle 1965) because it was totally different (*a break*: p. 129) with respect to Kronig's theory, which at that time was still the reference one, although it was 35 years old. Lytle gradually improved his theory by working in close cooperation with Dale E. Sayers (1944–November, 2004) and Edward A. Stern (s.d) (Sayers et al. 1970, 1971, 1972). Actually, it is clear that he had felt the need of developing a new theory because he had a wealth of experimental data gathered using a Siemens automatic diffractometer with conventional X-ray source that he had to justify to his company heads. He had modified his instrument, which operated in a horizontal arrangement, essentially mimicking the vertical spectrometer set up by Van Nostrand. To Lytle's luck, 10 years later his horizontal orientation turned out to fit perfectly the requirements of storage rings, so that the entire EXAFS technique his group had developed for home studies was quickly transferred to the newly available, powerful and time-sparing SR sources.[8] Additional work (especially for the correct use of the Fourier transform) was required to perfect the quantitative interpretation technique and this work too the three scientists performed jointly. The final EXAFS package covered all three major points: theory (Stern 1974), experiment (Lytle et al. 1975), and determination of structural parameters (Stern et al. 1975). The complete EXAFS formula was:

$$\chi(k) = -\sum_j \frac{N_j}{kR_j^2}\left|f^j(k,\pi)\right| \exp\left(\frac{-2R_j}{\lambda}\right) \sin\left[2kR_j + \delta_j(k)\right] \exp\left(-2k^2\sigma_j^2\right)$$

[7] This acronym is credited explicitly by him to J.A. Prins (Lytle 1999, p. 130).

[8] *Synchrotron radiation revolutioned the experimental side of EXAFS, making it accessible to non-experts and attracting the largest number of users at synchrotron sources* (Stern 2001, p. 51). Unfortunately, over-interpretation, in particular over-estimation of the spatial resolution, is the negative outcome of EXAFS enormous growth among such inexperienced people (ivi).

It may appear strange that EXAFS (formerly known as a weak "secondary absorption" and noticed several years later than the "fine structures" near the absorption edge; cf. Mottana, in preparation) was clarified earlier and quantitatively solved more deeply than the "fine-structures" near the absorption edge, known from 1918 to 1920. This historical fact can be easily explained. EXAFS structures had been disregarded by early researchers essentially because of the limitations inherent in their photographic detectors, but they could be recorded well by the counters of modified diffractometer, which also span over much wider angles. Moreover, both Van Nostrand and Lytle started working on this region of the XAS spectrum because they looked for few diagnostic features that could be usable for practical purposes, and simple enough as to require only a simple mathematical analysis.

The theoretical innovation that Lytle's group proposed for EXAFS is based, in fact, on phenomenological rather than abstract theoretical grounds. It considers the structures in the EXAFS region as due to interference of electrons: when an X-ray photon impinges an atom, it excites a core electron, which moves away through the atomic lattice as a photoelectron having energy equal to that of the incoming photon minus the core binding energy. Alternatively, the photoelectron may be seen to propagate through the lattice as a spherical wave, its wavelength decreasing with increasing the energy of the impinging photon. This spherical wave interacts with waves arising from the atoms neighbouring the absorber within a certain distance, and each atoms reacts behaving as a point-scatterer, so that the propagating total wave is the sum of all scattered waves. The total absorption (and absorption coefficient) is determined by the dipole transition matrix element between the initial core state and the photoelectron final state, which in turn is determined by the superimposition of the outgoing and incoming spherical waves. Inter-atomic distance and phase relationships between these two waves determine final state amplitude and modulation of the effects; the frequency of the damped oscillations in the spectrum depends on the distance between absorber and back-scatterer atoms; the amplitude is proportional to the number of the back-scatterer atoms, i.e. to the coordination around the absorber. EXAFS is indeed a local probe i.e., it is insensitive to the Long Range Order (LRO), as the photoelectron has a mean-free-path typically less than 10 Å (Müller et al. 1982). Nevertheless, EXAFS is a powerful method to study the Short Range Order (SRO) and is chemically selective. In fact, the large energy separation between different inner shells makes it possible to tune the incident X-ray photon energy of a particular core level corresponding to the chosen atom edge; the spectrum is then scanned up to *c.* 1000÷2000 eV above edge, provided there are no interfering contributions by any edges of other atoms present in the sample.

Modern EXAFS, now conceived as, and actually being, an almost unique local structural probe (e.g., Stern 1978; Lee et al. 1981), contributes significantly to understanding the physical behaviours of non-ordered systems such as defective and semi-amorphous materials that have important applications in modern life e.g., catalysts, semiconductors, biomolecules, etc. Perhaps the EXAFS equation may appear to many researchers to be *formidable* (Lytle 1999, p. 131), and yet an ordered sequence of mathematical steps performed by *ad hoc* computer programs allows extracting easily (albeit still by trial and error, occasionally) from the experimentally recorded signal all the intrinsic structural information: (a) interatomic distance R (down to a $0.01 \div 0.02$ Å); (b) coordination number N (with a ~10% approximation); (c) Debye-Waller factor; (d) asymmetry due to thermal expansion. The major drawbacks in the EXAFS state-of-art concern the main approximations non-accounted for by the theory, and consequently neglected (or even ignored) by most users: it is not sensitive to bond angles, and is complex to interpret when the system is too ordered (cf. Gunnella et al. 1990). Moreover, because of the large error inherent in the determination of coordination atoms, the EXAFS analysis fails when the absorber coordination is high: typically for $N > 8$.

Nevertheless, the continuous improvement of EXAFS methods allowed also identifying shake-up and shake-off channels that are associated with multi-electron transition processes (Bernieri and Burattini 1987), and which for a long time had been considered invalid speculations. Much experimental and theoretical effort has been dedicated to determine amplitudes and positions of those excitations and to calculate their relative cross sections (cf. Chaboy et al. 1994, 1995). Although the intensities of such multi-electron excitations are very low, and the unambiguous identification of their energy position and shape is difficult in the presence of the large EXAFS oscillations, their presence affects the EXAFS data, leading to errors in the determination of interatomic distances, coordination numbers and even first-shell anharmonicity (D'Angelo et al. 2004).

2.2.2 XANES Spectroscopy

The "fine structure" across the absorption edge, although discovered first and interpreted properly on its general lines, has proved to be a rather difficult subject to study in detail. It still has many obscure points, which often make it unfit to be made use of even when powerful computers and modern methods are available. Despite the overall picture being unclear for several reasons, XANES spectra are frequently used in the "fingerprinting" mode e.g., to recognize in the most straightforward way a tetrahedral coordination from an octahedral one (e.g., Mottana et al. 1997), etc. It is physically senseless, although customary, to distinguish between XANES[9] and NEXAFS (cf. Bianconi 1988; Stöhr 1992), as they are the same spectroscopy method. They differ only in that hard and, respectively, soft X-rays are used to scan the sample. In other words, some researchers prefer to use the acronym NEXAFS (Near Edge X-Ray Absorption Fine Structure) to present and discuss X-ray absorption data recorded while doing surface experiments (see later) or the K-edges of low Z atoms such as C, N or O.

When the absorption effects of atoms impinged by SR started being recorded accurately in their whole extent and complexity, two different approaches to near-edge data presentation, i.e. to XANES, developed: (a) a phenomenological—"fingerprinting-like"—description with the simple interpretation of the main

[9] This acronym was used first by Antonio Bianconi (April 1, 1944) in 1980 while working at SSRL, Menlo Park, USA. It was published by him two years later (Bianconi et al. 1982).

features occurring across the main absorption edge, given as E_0 (thus from 10 eV to +30 eV with respect to E_0); (b) an analytical, theoretical study of the MS contributions generated by the photoelectron in relationships to the local geometry of the material probed and to the disorder and defects shown by its atom arrangement around the absorber. The last approach may open up as far as to include the atoms located in high-order coordination shells, in this trying to reach a unified picture of the entire absorption (cf. Stern 1978; Natoli and Benfatto 1986). In fact, the two approaches are interconnected; while the first one is typically followed by EXAFS users because it complements their quantitative information on the relative distance among atoms with qualitative data on the coordination and electronic state of the absorber, the second one is followed only by small, theoretically-gifted groups that try deciphering by appropriate calculations the overall physical reactions undergone by the photoelectron displaced from the absorber in the studied material impinged by an electromagnetic radiation of suitable energy. The MS approach, therefore, is essentially a method to calculate the structure of polyatomic matter from first principles, working in the real space (Lee and Pendry 1975; Lee and Beni 1977; Kutzler et al. 1980; Natoli et al. 1980; Durham et al. 1982; Natoli 1983, 1984; Bunker and Stern 1984; Rehr and Albers 1990; Ankudinov et al. 1998; etc.) and searching for an accurate description of the wave function.

Thus, the MS approach proposes and tests algorithms having both theoretical and computational insight. Underneath its application to XANES, there is the assumption that, in the proximity of an edge, the X-ray absorption coefficient depends upon the photoelectron transitions to low-lying unoccupied states in the conduction band of the absorber atom. In the MS framework, both the initial state, which is characterized by a well-defined core level with a localized wave function having precise angular momentum symmetry, and the continuum part, which is represented by modulations of the density of non-occupied electronic states, are calculated. These structures, in fact, are the features near the edge generated by the dipole selection rule, which samples the density of the final states projected onto the angular momentum and convoluted with the core-hole energy width.

In the MS framework, a XANES calculation starts with an approximate molecular potential obtained by partitioning the cluster of atoms considered to be representative of the studied material into distinct atomic and interatomic regions, with each atom enclosed in a sphere of specific radius ("muffin-tin") and an outer sphere enveloping the entire atomic cluster whose size depends by the mean-free-path (typically <10 Å) and the lifetime ($\sim 10^{-15}$ s) of the X-ray excitation. The Coulomb and exchange part of the potential are approximated via the total charge density of the cluster and may or may not consider the existence of dynamical effects. The photoelectron path is made of the several segments that connect the atoms involved in the collision. Calculations are iteratively performed by increasing the size of the cluster and different potentials are considered until reaching calculated spectra that closely resemble the experimental spectrum.

The EXAFS theory had taken about 10 years to reach completeness (see above) and from that moment on the EXAFS technique gained wide appreciation among both the scientists and the industrial world. By contrast, although there is general convergence on the MS formalism to explain XANES spectra, a comprehensive, easy to use and universally accepted practical approach to XANES is still lacking, mainly because of the relevant, but still incomplete and somewhat unclear electronic information present in the edge region. At present, there are several formulations, each one based on different algorithms. They underline a gigantic effort at understanding (cf. Natoli et al. 1990, 2003; Rehr and Albers 2000; Ravel and Newville 2005; etc.) and, in addition, reflect the increased computer power that makes these calculations possible in large atomic clusters consisting of up to ~200 atoms or more.

A slow, but steadily increasing theoretical development has taken place ever since the first application of the MS method, which was proposed in the analogy of the Fano's (1935) concept of "shape resonances", which he had conceived for nuclear physics. Indeed, the MS interpretation is its extension to condensed systems. Such "shape resonances" are localized states in the continuum observed in the spectra of diatomic molecules like N_2, which could be easily interpreted by the MS theory (Dill and Dehmer 1974; Dehmer and Dill 1975). Only later, it was recognized that there is no true physical reason to make use of different algorithms for XANES and EXAFS, as the MS calculation method may reproduce the entire spectrum, if properly applied (Benfatto et al. 1986).

The first major step ahead in this direction was taken by Calogero Rino Natoli (May 3, 1941-), who first showed that through MS calculations one can determine the absorber to scatterer distance (Natoli 1983, 1984). Then, Benfatto et al. (1986) showed how the MS theory applied to liquid solutions connects the lower-energy XANES region to the EXAFS region at high energy via an intermediate energy region where only a few MS contributions contribute to the XAFS spectrum (see also Natoli and Benfatto 1986). Nowadays, it is well recognized that, neglecting the small structures before the edge (i.e., the "pre-edge"), a XAFS spectrum consists of three regions undergoing different scattering modes: a) a full multiple scattering (FMS) region close to the main edge E_0 i.e., with $10/20 \leq E \leq +20/25$ eV, which is nothing else but the former XAS "fine structure" or "Kossel structure", now called XANES; b) an intermediate multiple scattering (IMS) region ($+20/25 \leq E \leq +100/150$ eV); and a single scattering (SS) region, which extends to +500 eV or +1000 eV or even more above the edge value E_0 and is nothing else but the former "secondary absorption" or "Kronig structure", now called EXAFS (Fig. 4).

Within this theoretical framework, the transitional IMS region plays a critical role for the quantitative evaluation of XAFS spectra (Pendry 1983), although it has rarely been investigated in detail so far (e.g., Bugaev et al. 2001; Brigatti et al. 2000, 2008). It is now acknowledged that a full XAFS spectrum is the sum of all these regions (Stern 2001), and yet separate mathematical treatments of the XANES and EXAFS regions are still normally done, since the latter one, particularly, is quite simple to apply and is straightforward for most applications, whereas the former requires the cumbersome application of complex computer codes.

As a matter of fact, no unified theory nor computer program have been proposed so far that are able to account for the entire spectrum by a single algorithm, although the scenario is wide and in continuous evolution (EXCURVE: Binsted et al. 1991;

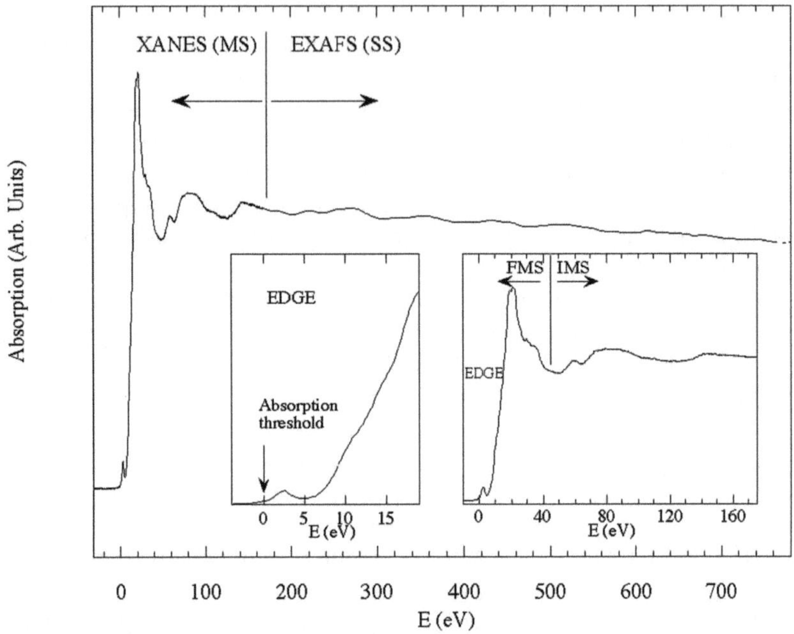

Fig. 4 A modern schematic picture of the XAFS spectrum of the *K*-edge of a transition metal atom showing the MS or XANES region (with in the *left* inset the FMS sub-region with the pre-edge and the edge rise, and in the *right* inset the transitional IMS sub-region) and the region of SS or EXAFS. (After Mottana (2004, p. 478), Fig. 4, largely modified)

CONTINUUM: Natoli et al. 1990, 2003; GNXAS: Di Cicco 1995; Filipponi and Di Cicco 2000; FEFF: Newville 2001; Ravel and Newville 2005); MXAN (Benfatto et al. 2001, 2003; WIEN2k: Blaha et al. 2001; SPRKKR: Ebert 2000; and several others optimized for special cases).

The contribution of the upgrade of computing methods (both hardware and software) to XANES development through calculations performed by the MS method cannot be overestimated. Calculations are now possible within both the "muffin-tin" and "non-muffin-tin" approximations (Joly 2001; Hatada et al. 2010), although a major, unfortunately still incompletely worked out problems, first evidenced by Natoli et al. (1986) and studied by Gunnella et al. (1990), lies in the use of potentials (Joly et al. 1999).

2.3 Advances in Experimental Work

2.3.1 EXAFS

The algorithms developed in 1974/1975 by Lytle's group (see above) proved their worth in a variety of applications that were previously believed to be possible only by XRD, and are now to be considered classical. They allow data retrieval from experimental recordings performed practically under any condition: at low- as well as high-temperature and pressure (e.g., Itié 1992; Itié et al. 1997; Comez et al. 2002), down to near 0 K and up to 1 Mbar, especially when using the diamond-anvil cell (DAC) in the energy dispersive setting (Bassett et al. 2000). Quick- and Turbo-EXAFS (Dartyge et al. 1986; Pascarelli et al. 1999) provide data in milliseconds that allow live recording reactions while they are in progress, although extracting their structural information then requires some additional time. The polarization of the SR beam has been carefully analysed (Benfatto et al. 1989; Brouder 1990; Pettifer et al. 1990; Benfatto and Della Longa 2001) and made advantage of to measure the atom distribution in materials displaying layered structures (e.g., Manceau et al. 1988; Cibin et al. 2010), and also the distribution and orientation of dopants epitactically growing over certain chemical compounds (e.g., Asakura and Ijima 2001).

Another major step ahead in XAFS due to the power of third-generation synchrotrons is the development of the XAFS-microprobe (SmX: Sutton et al. 1995, 2002; μ-XAS: Mosbah et al. 1999). This apparatus allows recording spectra at micron- and even pixel-size, thus also permitting imaging and quantitative mapping of the atom distribution in the studied materials, and consequently also exposing their zoning and all their intrinsic and extrinsic defects. SmX may take advantage either of the fingerprinting method or, more carefully, of quantitative algorithms. It is particularly interesting for metamorphic petrologists, as it sorts out compositionally different generations of the same mineral, thus allowing a well-planned use of geo-thermobarometers (Dyar et al. 2002; cf. Sutton et al. 2002). A combination of XANES and XRF makes it possible to draw pixel-size single images as well as large maps showing composition and speciation for selected elements (Muñoz et al. 2006).

A field where EXAFS has shown all its worth is in the study of metamict materials (e.g., Greegor et al. 1984; Nakai et al. 1987), which are particularly important in the modern world because they are a common by-product of nuclear power plants. Moreover, EXAFS showed its worth also on quasicrystals (Sadoc 1986) i.e., a new category of synthetic materials having non-crystallographic properties that only recently sufficient evidence to attain the Nobel Prize in Chemistry (2010). Nevertheless, the most exciting results of EXAFS research concern geological materials, particularly those related to the oil and coal industries, and environmental problems produced both by the mining industry and the metallurgical and chemical industrial use of various geological materials. Such studies as determining the size contamination of sediments and mine tailings by As, Pb, U, etc. are to be considered extremely valuable contribution to the integrated study of Man Living Space (cf. Fenter et al. 2002). Indeed, EXAFS is at its best when it studies semi-amorphous ores and ore derivatives, where XRD cannot give information. Moreover, the exact location and ordering of selected trace elements in crystalline hosts (e.g., diamond) and even the measurement of nanometric metal inclusions (e.g., Clausena et al. 1994) are possible by EXAFS, while XRD often cannot even detect their presence. Finally, it is interesting to note that the Continuous Cauchy Wavelet Transform (CCWT) analysis (Munoz et al. 2003), which some time ago seemed to supersede the classical EX-

AFS treatment because it gives a much wider description of the radial distribution functions and in three dimension, in contrast with the classical method which is limited to a one-dimensional representation, has declined in the general appreciation: the classical method is too well rooted among users to be easily replaced.

From the very first stages of SR extraction from storage rings, EXAFS was optimized also for surface experiments. This special technique, called SEXAFS (Surface Extended X-ray Absorption Fine Structure), concerns the study of electronic transitions from core levels of atoms located at the surface of solids. The peculiar local character of core level excitations in the X-ray absorption processes makes it attractive to study the atoms at the surface of solids and to investigate both their local structure and localized electronic states. The knowledge of the atomic arrangement of neighbour atoms around a selected atom on a surface is important in many industrial studies such as chemisorption, oxidation processes and catalysis. SEXAFS has gained wide appreciation because the local structure of chemisorption sites and the local structure of surface amorphous oxides are basic information that are not directly given by any other technique based on diffraction methods e.g., Low Energy Electron Diffraction (LEED).

SEXAFS was born in the middle of seventies as soon as the tunable and intense SR X-rays of storage rings become available. Indeed, in order to efficiently start, it had high flux requirements, owing to the very low concentration of surface atoms: the main experimental problem of all surface X-ray spectroscopies was and still is how to enhance the sensitivity at surface. First Lukirskii and Brytov (1964), using the continuum bremsstrahlung of a standard X-ray tube, and later Gudat and Kunz (1972), using SR, demonstrated that the total electron yield (TY) by the sample surface is proportional to the bulk absorption coefficient, while having a sampling depth of the order of magnitude of few tens of Å. Therefore, TY is fit for surface investigations, but it could not be proficiently used till a strong enough SR would come in. The first SEXAFS experiment having high surface sensitivity was carried out at the Stanford Synchrotron Radiation Facility (SSRF, now SSRL) by detecting the Auger electron yield (Bianconi et al. 1977). The absorption spectrum of the surface Al atoms in the top monolayer of an aluminium crystal could be distinguished from the aluminium bulk spectrum: contrast was increased by selecting an Auger line originating from the Al atoms at the surface, which interacted with chemisorbed oxygen producing an inter-atomic, Al–O Auger transition. Both the total electron yield (Citrin et al. 1978) and the Auger electron yield (Stöhr et al. 1978) techniques were then made use of to measure the signal of different atomic species chemisorbed on solid materials. These experiments brought with them, as an immediate consequence, that the relaxation undergone by the atoms in the exposed substrate started being studied too (Bianconi and Bachrach 1979; Bianconi et al. 1979).

As shown above, SEXAFS concerns the study of the absorption coefficient modulations over a range of photoelectron wave-vectors above ~ 3 Å^{-1}. In this range, the experimental data can be analyzed in the framework of the well-established EXAFS theory (Stöhr 1986; Norman 1986), and contribute to policy-making particularly on matters related to environmental remediation (e.g., Brown et al. 1999; Sherman and Randall 2003). By contrast, the XAFS structures present in the low-energy range (few tens of eV), which contain information on the local geometry i.e., on bond distances and bond angles, should be evaluated and discussed in the framework of the XANES theory (Bianconi and Marcelli 1992).

2.3.2 XANES

In the near-edge fine structures of XANES, the higher-order terms of the correlation function of the atomic distribution become much more important than the pair correlation function of the atomic distribution probed in the EXAFS range. They too have been intensively studied using SR ever since the first application of this as the X-ray source. However, due to the lack of a reliable theoretical method of analysis, early XANES studies on local geometry determination have been limited to a fingerprint approach using model compounds. This compelled scientists to perform a preliminary step that was considered useful, i.e., to re-record, using SR as the source and updated instrumental detection setups, most if not all the near-edge XAS features of the many materials that had been collected during the previous 50 years. As pointed out before, many of them were actually very accurate, and yet they all underwent confirmation. Among the best results on this line of work, those on Mn (Belli et al. 1980), Fe (Waychunas et al. 1983), and Ti (Waychunas 1987; Paris et al. 1993; Farges et al. 1996) are worth mentioning: they all concern transition-metal atoms of high significance for industry. These studies also updated certain already well-known XANES properties such as, e.g., the chemical shift, or angular dependence, or coordination dependence (cf. Mottana, in preparation). Moreover, a completely new emphasis was given to the role of the pre-edge features, since instrumental upgrade made it possible to record them with resolution and intensity much higher than ever done before. Important advances towards a quantitative theoretical analysis of the surface XANES data have also been carried out (Bianconi and Marcelli 1992). In particular, priority has been given to biomolecules (e.g., Shulman et al. 1976; Bianconi 1983), to soft semi-amorphous compositionally inhomogeneous compounds (e.g., bitumen and coal: Wong et al. 1983; catalysts), and to glassy materials (e.g., Greaves et al. 1984; Greaves 1985), because their very low crystallinity hinders studies using XRD, while allowing XAFS, which does not requires LRO for detection and study.

Practically, two different lines of approach to XANES have developed, which both persist up to now. The first and easiest one makes use of the fingerprinting method and extracts from the XANES region the electronic information required to best interpreting EXAFS quantitative results, e.g., the oxidation state of the absorber. Typical examples of this trend are the many studies concerning iron and other transition atoms, starting with the first realization that energy and shape of the pre-edges of transition-atom-bearing minerals could bear quantitative information of the relative amounts of e.g., coexisting Fe^{2+} and Fe^{3+} (Bajt et al. 1994). Furthermore, it was shown that the intensity of the pre-edge brings information on the relative coordination e.g., of $Ti^{[4]}$ and $Ti^{[6]}$ (Paris et al. 1993). This trend has been pursued particularly by geochemists and is in general universally acknowledged for transition metals and metalloids, its best example of application being on the iron-bearing natural materials and on the oxidation state of arsenic in waters and soils. Indeed, studies on pre-edge energy and shape have shown that these two features account for both coordination and oxidation state of Fe in known minerals (e.g., Wilke et al. 2001; Berry et al. 2008) and may foresee conditions that are not yet realized on Earth, but which might occur on the Moon and in meteorites. Similarly, Sherman and Randall (2003), although using a somewhat crude fingerprinting approach, could demonstrate how reacting As^{5+}-bearing solutions with a Fe^{3+}-bearing oxide- and hydroxide-rich substrate can contribute to reclaiming heavily polluted environments, essentially by reducing poisonous $As^{5+}$to almost harmless As^{3+}. Environmental studies benefit the most of XANES because this kind of spectroscopy is utterly independent on the state of the absorber, while being able to reveal very low contents of it (ppb). Thus, recently, several manuscripts appeared that deal with aerosols and nanoparticles suspended as fine dust in the low layers of the atmosphere, or even permanently trapped in continental ice, such as in Antarctica (Lama et al. 2012; Marcelli et al. 2012; Schlaf et al. 2012; etc.).

Chemical reactions also occur at the interface between two solid layers of different composition and/or structure, which often are packed together to form multilayers. These have lately become a kind of artificial material highly interesting for advanced industry and in many scientific applications. As a matter of fact, for multilayers, the use of XAFS in the total reflection mode (ReflEXAFS: Davoli et al. 2003) gives results that are far more useful than SEXAFS, especially for samples that are grown by molecular beam epitaxy, the added layers of which have nanometric thickness (López-Flores et al. 2009).

Another major step ahead in XAFS due to third-generation synchrotrons is the further upgrade of XAFS-microprobes, which allow recording spectra at micron- and smaller scales limited only by the pixel size of the available detectors. They permit imaging and quantitative mapping of a sample showing intrinsic and extrinsic defects and compositional variations. Indeed, for all applications from biology to environmental science it is always more important to fully understand the concentration profile and the chemical state of the major number of elements present in a given material.

In the last decade, several two-dimensional (2D) X-ray imaging methods have been developed for *in situ* analysis of large sample areas. These data show the distribution of particles in a sample, while XAFS reveals the kind of elements present and their chemical state. Moreover, X-ray microscopy and tomography can be also combined with XAFS resulting in a much more effective analytical technique. A three-dimensional (3D) computed tomographic XANES approach has been recently demonstrated (Meirer et al. 2011). It uses a CCD detector having a space resolution of a few μm. The ability to simultaneously probe morphology and phase distribution in complex systems at multiple length scales may unravel the interplay of nano- and micrometre-scale factors that are at the origin of a material macroscopic behaviour. The newest X-ray layouts combine full-field transmission X-ray microscopy (TXM) with XANES spectroscopy, to follow 2D and 3D morphological and chemical changes with a resolution of tens of nanometres in thick samples, allowing the analysis of large areas over a time from few min to few hours.

This mass of new information prompted the in-depth theoretical studies on XANES outlined previously and the formulation of the new software based on the MS theory extensively described there. It is now customary, even among industrial users, to accompany their experimental results with structural (or pseudo-structural) interpretations based on theoretical simulations performed using anyone of the software packages listed above, most of which are freely available (e.g., Wu et al. 2004; Xu et al. 2011). Therefore, XAFS in all its various techniques of application is now one of the most specialized techniques of largest industrial interest. Modern industry—particularly chip technology, which is essential for both computers and communication tools—needs to know as precisely as possible the state of the surface of the material and of interfaces rather than its bulk. In fact, most reactions involved in technological operations (e.g., lithography of circuits; epitaxial growth of either required dopants and of unwanted impurities, these ones to be removed) occur at the surface or in the first few layers of the material, and are further complicated by the relaxation effect any well-ordered bulk structure undergoes when it is suddenly interrupted. Consequently, both fingerprinting and MS techniques typical of SEXAFS and ReflEXAFS have been used to cope with these assignments on a variety of materials, starting from ores up to proteins, in a practically continuous, numberless sequence.

3 Conclusion

XAFS i.e., X-ray absorption fine spectroscopy (now intended to be XANES+EXAFS, performed together at high resolution), is certainly one of the most advanced techniques of material investigation particularly among industrial and applied scientists. Indeed, SR facilities are flooded by requests of users willing to apply XAFS methods to different materials and phenomena. XAFS-related methods such as SEXAFS (Surface EXAFS) and ReflEXAFS (Reflectivity EXAFS), or others too recent or too poorly pushed to be mentioned here such as e.g., XES (X-Ray Emission Spectroscopy), RIXS (Resonant Inelastic X-Ray Scattering), TXRF (Total Reflection X-ray Fluorescence), XMCD (X-ray Magnetic Circular Dichroism, e.g. Fukui et al. 2001), and DAFS (Diffraction Anomalous Fine Structure), are successfully applied to complex systems, the deep knowledge of which is requested by modern research aiming to industrial applications. New instruments are available, which allow performing polarized experiments on single crystals and, recently, even μ-XAFS, i.e., XAFS performed on micrometre-size spots or microparticles, is increasingly developed, thus preparing the further evolution towards nanoparticle studies, which still mostly rely on high-resolution transmission X-ray microscopy (TXRM) with a spatial resolution in the range ~30–50 nm. Nowadays, time-resolved experiments using either dispersive or QUICK-XAFS layouts permit following the advance of physical phenomena or of chemical reactions with a time resolution from seconds down to few tens of microseconds.

The continuous growth of XAFS documented here pays a particular attention to new ideas and their theoretical extension. Early XAS contributed to the understanding of the atomic theory of matter and to the establishing of Materials Science (cf. Mottana, in preparation). Since 1975, XAFS contributed to its advanced understanding; starting from the characterization of complex or extremely diluted materials up to the interpretation of order-disorder phenomena under the peculiar aspect of electronic interactions among neighbouring atoms in a complex chemical environment. The explosion of modern XAFS opened new fields and generated unexpected results. All this historical sequence of projects, tests, and events brings support to the opinion, widespread among the scientific community, that much greater results are to be expected, such as those offered in the near future by the incoming availability of fourth-generation SR coherent X-ray sources.

However, now, the front line of XAFS research is at trying to make the best out of FEL (Free-Electron Laser). This novel, extremely sophisticated energy source, allows studies on the near-edge fine structure of solids by means of femtosecond soft X-ray pulses. So far, only pioneering attempts on model materials have been successfully carried out (Bernstein et al. 2009). Such preliminary experiments could be carried out only under many strong constraints on experimental conditions such as sample geometry, spectrographic energy dispersion, single shot position-sensitive detection. In addition, they used a data normalization procedure that eliminates the severe fluctuations of the incident intensity in space and in photon energy present in FEL, which resemble those occurring in the very first synchrotrons. Nevertheless, all these pioneering works gave convincing results, thus opening new potential applications to XAFS that will meet with the increasing demands of Materials Science as the relevant branch of Physics and, without any doubt, of the most advanced, high-tech industry, which is definitively oriented to make the best out of its nanometre-size products.

Acknowledgments Our attempt at sketching the development of XAS and at clarifying more extensively its transition to XAFS up to the present time is based on more than two decades of readings and discussions with many people, whom we thank greatly for contributing to our effort. Among them, we particularly thank Antonio Bianconi, Maurizio Benfatto, Jesús Chaboy Nalda, Giannantonio Cibin, Ivan Davoli, Rino Natoli, Hiroyuki Oyanagi, Eleonora Paris, Piero Pianetta, Trevor Anthony Tyson, and Ziyu Wu. Nevertheless, the responsibility for this historical account, which necessarily involves selection and evaluation of persons and works, lies entirely upon us. We certainly have missed some related important research papers, but both the capacity of our memory and the length allowed to this contribution have limitations. We sincerely apologize to all colleagues whose worthy papers we neglected to quote.

References

Ankudinov AL, Ravel B, Rehr JJ, Conradson SD (1998) Real space multiple scattering calculation of XANES. Physical Review B 58:7565–7576

Asakura K, Ijima K (2001) Polarization-dependent EXAFS studies on the structure of Mo oxides dispersed on single crystals. Journal of Electronic Spectroscopy and Related Phenomena 119:185–192

Bajt S, Sutton SR, Delaney JS (1994) Microanalysis of iron oxidation states in silicates and oxides using X-ray absorption near edge structure (XANES). Geochim Cosmochochim Acta 58:5209–5214

Balzarotti A, Piacentini M, Grandolfo M (1970) Measurement of the spectral distribution of the Frascati electron synchrotron radiation in the (80 ÷ 1200) Å region. Lettere al Nuovo Cimento s. I 3:15–18

Bassett WA, Anderson AJ, Mayanovic RA, Chou I-M (2000) Hydrothermal diamond anvil cell XAFS study of first-row transition elements in aqueous solutions up to supercritical conditions. Chemical Geology 167:3–10

Belli M, Scafati A, Bianconi A, Mobilio S, Palladino L, Reale A, Burattini E (1980) X-ray absorption near edge structures (XANES) in simple and complex Mn compounds. Solid State Communications 35:355–361

Benfatto M, Della Longa S (2001) Geometrical fitting of experimental XANES spectra by a full multiple-scattering procedure. Journal of Synchrotron Radiation 8:1087–1094

Benfatto M, Natoli CR, Bianconi A, Garcia J, Marcelli A, Fanfoni M, Davoli I (1986) Multiple scattering regime and higher order correlations in x-ray absorption spectra of liquid solutions. Physical Review B 34:5774–5781

Benfatto M, Natoli CR, Brouder C, Pettifer RF, Ruiz Lopez MF (1989) Polarized curved wave EXAFS: theory and application. Physical Review B 39:1936–1939

Benfatto M, Congiu Castellano A, Daniele A, Della Longa S (2001) MXAN: a new software procedure to perform fitting of experimental XANES spectra. Journal of Synchrotron Radiation 8:267–269

Benfatto M, Della Longa S, Natoli CR (2003) The MXAN procedure: a new method for analysing the XANES spectra of metallo-proteins to obtain structural quantitative information. Journal of Synchrotron Radiation 10:51–57

Bernieri E, Burattini E (1987) Multielectron transitions above the krypton *K* edge. Physical Review A 35:3322–3326

Bernstein DP, Acremann Y, Scherz A, Burkhardt M, Stöhr J, Beye M, Schlotter WF, Beeck T, Sorgenfrei T, Pietzsch A, Wurth W, Föhlisch A (2009) Near edge x-ray absorption fine structure spectroscopy with x-ray free-electron lasers. Applied Physics Letters 95:134102:1–3

Berry AJ, Danyushevsky LV, O'Neill HStC, Newville M, Sutton SR (2008) Oxidation state of iron in komatiitic melt inclusions indicates hot Archaean mantle. Nature 455(07377):960–964

Bianconi A (1983) XANES Spectroscopy of local structures in complex systems. In: Bianconi A, Incoccia L, Stipcich S (eds) EXAFS and near-edge structure. Proceedings of the international conference, Frascati, Italy, September 13–17, 1982. Springer, Berlin, pp 118–129

Bianconi A (1988) XANES spectroscopy. In: Koningsberger DC, Prins R (eds) X-ray absorption: principles, applications, techniques of EXAFS, SEXAFS and XANES. (Chemical analysis 92). Wiley, New York, pp 573–662

Bianconi A, Bachrach RZ (1979) Al surface relaxation using surface extended X-ray absorption fine structure. Applied Physics Letters 42:104–108

Bianconi A, Marcelli A (1992) Surface X-ray absorption near-edge structure: XANES. In: Bachrach RZ (ed) Synchrotron radiation research. Advances in surface and interface science. Techniques, vol 1. Plenum, New York, pp 63–115

Bianconi A, Bachrach RZ, Flodström SA (1977) Study of the initial oxidation of single crystal aluminum by inter-atomic Auger yield spectroscopy, Solid State Communications 24:539–542

Bianconi A, Bachrach RZ, Flodström SA (1979) Oxygen chemisorption on Al: unoccupied extrinsic surface resonances and site-structure determination by surface soft X-ray absorption. Physical Review B 19:3879–3888

Bianconi A, Dell'Ariccia M, Durham PJ, Pendry JB (1982) Multiple-scattering resonances and structural effects in the x-ray-absorption near-edge spectra of FeII and FeIII hexacyanide complexes. Physical Review B 26:6502–6508

Binsted N, Campbell JW, Gurman SJ, Stephenson PC (1991) EXCURVE. SERC Daresbury Lab, Warrington

Blaha P, Schwarz K, Madsen GKH, Kvasnicka D, Luitz J (2001) WIEN2k, an augmented plane wave + local orbitals program for calculating crystal properties. Schwarz, Wien

Brigatti MF, Lugli C, Cibin G, Marcelli A, Giuli G, Paris E, Mottana A, Wu Z (2000) Reduction and sorption of chromium by Fe(II)-bearing phyllosilicates: chemical treatments and X-Ray absorption spectroscopy (XAS) studies. Clays and Clay Minerals 48:272–282

Brigatti MF, Malferrari D, Poppi M, Mottana A, Cibin G, Marcelli A, Cinque G (2008) Interlayer potassium and its surrounding in micas: crystal chemical modeling and XANES spectroscopy. American Mineralogist 93:821–835

Broglie M de (1913) Sur un nouveau procédé permettant d'obtenir la photographie des spectres des rayons Röntgen. Comptes Rendus de l'Académie des Sciences 157:924–926

Brouder C (1990) Angular dependence of x-ray absorption spectra. Journal de Physique: Condensed Matter 2:701–738

Brown GE Jr, Henrich V, Casey W, Clark D, Eggleston C, Felmy A, Goodman DW, Gratzel M, Maciel G, McCarthy MI, Nealson KH, Sverjensky D, Toney M, Zachara JM (1999) Metal oxide surfaces and their interactions with aqueous solutions and microbial organisms. Chemical Review 99:77–174

Bugaev LA, Sokolenko AP, Dmitrienko HV, Flank A-M (2001) Fourier filtration of XANES as a source of quantitative information of interatomic distances and coordination numbers in crystalline minerals and amorphous compounds. Physical Review B 65:024105-1–024105-8

Bunker G, Stern EA (1984) Experimental study of multiple scattering in X-ray-absorption near-edge structure. Physical Review Lettersers 52:1990–1993

Cauchois Y, Bonnelle C, Missoni G (1963) Premiers spectres X du rayonnement d'orbite du synchrotron de Frascati. Comptes Rendus de l'Académie des Sciences de Paris 257:409–412

Chaboy J, Marcelli A, Tyson TA (1994) Influence of double-electron transitions on the EXAFS *L* edges of rare-earth systems. Physical Review B 49:11652–11661

Chaboy J, Tyson TA, Marcelli A (1995) Relative cross sections for bound-state double-electron $LN_{4,5}$-edge transitions of rare earths and nonradioactive elements of the sixth row. Prensas Universitarias, Zaragoza

Cibin G, Mottana A, Marcelli A, Cinque G, Xu W, Wu Z, Brigatti MF (2010) The interlayer structure of trioctahedral lithian micas: an AXANES spectroscopy study at the potassium *K*-edge. Am Mineral 95:1084–1094

Citrin PH (1986) An overview of SEXAFS during the past decade. In: Lagarde P, Raoux D, Petiau J (eds) EXAFS and near edge structure IV, Abbaye Royale de Fontevraud, France, July 7–11, 1986. J Phys 47(12):C8, supplement, pp C8-437–472

Citrin PH, Eisenberger P, Hewitt RC (1978) Extended x-ray absorption fine structure of surface atoms on single-crystal fine structure of surface atoms substrates: Iodine adsorbed on Ag(111). Physical Review Letters 41:309–312

Clausena BS, Topsøe H, Hansen LB, Stoltze P, Nørskov JK (1994) Determination of metal particle size from EXAFS. Catalysis Today 24:49–55

Comez L, Di Cicco A, Itié JP, Polian A (2002) High-pressure and high-temperature X-ray absorption study of liquid and solid gallium. Physical Review B65:014114-1–014114-10

D'Angelo P, Petit P-E, Pavel NV (2004) Double-electron excitation channels at the Ca^{2+} *K*-edge of hydrated calcium ion. Journal of Physical Chemistry B108:11857–11865

Dartyge E, Fontaine A, Tourillon G, Cortes R, Jucha A (1986) X-ray absorption spectroscopy in dispersive mode and by total reflection. Physics Letters 113A:384–388

Davoli I, Thanh HN, d'Acapito F (2003) ReflEXAFS technique: a powerful tool for structural study in new materials. In: Bianconi A, Marcelli A, Saini NL (eds) CP652, X-ray and inner-shell processes. 19th international conference on X-ray and inner-shell processes, American Institute of Physics, Melville, NY, pp 388–394

Dehmer JL, Dill D (1975) Shape resonances in K-shell photoionization of diatomic molecules. Physical Review Letters 35:213–215

Di Cicco A (1995) EXAFS multiple-scattering data analysis: GNXAS methodology and applications. Physica B 208–209:125–128

Dill D, Dehmer JL (1974) Electron-molecule scattering and molecular photo-ionization using multiple scattering method. Journal of Chemical Physics 61:694–699

Doniach S, Hodgson K, Lindau I, Pianetta P, Winick H (1997) Early work with synchrotron radiation at Stanford. Journal of Synchrotron Radiation 4:380–395

Durham PJ, Pendry JB, Hodge VH (1982) Calculation of X-ray absorption spectra based on the multiple scattering theory. Computer Physics Communications 25:193–205

Dyar MD, Lowe EW, Guidotti CV, Delaney JS (2002) Fe^{3+} and Fe^{2+} partitioning among silicates in metapelites: a synchrotron micro-XANES study. American Mineralogist 87:514–522

Ebert H (2000) Electronic structure and physical properties of solids. In: Dreysse H (ed) Lecture notes in physics 535. Springer, Berlin, pp 191–246

Eisenberger P, Kincaid B, Hunter S, Sayers DE, Stern EA, Lytle F (1974) EXAFS measurements at SPEAR. In: Koch EE, Haensel R, Kunz C (eds) Vacuum ultraviolet radiation physics. Proceedings of the 4th international conference on vacuum ultraviolet radiation physics, Hamburg, July 22–26, 1974. Vierweg, Braunschweig, Pergamon, pp 806–807

Fano U (1935) Sullo spettro di assorbimento dei gas nobili presso il limite dello spettro d'arco. Nuovo Cimento 12:154–161

Farges F, Brown GE Jr, Reher JJ (1996) Coordination chemistry of Ti(IV) in silicate glasses and melts: I. XAFS study of titanium coordination in oxide model compounds. Geochimica et Cosmochimica Acta 60:3023–3038

Fenter PA, Rivers ML, Sturchio NC, Sutton SR (2002) (eds) Applications of synchrotron radiation in low-temperature geochemistry and environmental science. Reviews of Mineralogy and Geochemistry 49. The Mineralogical Society of America, Washington, D.C.

Filipponi A, Di Cicco A (2000) GNXAS: a software package for advanced EXAFS multiple-scattering calculations and data-analysis. Task Quarterly 4:575–669

Fukui K, Ogasawara H, Kotani A, Harada I, Maruyama H, Kawamura N, Kobayashi K, Chaboy J, Marcelli A (2001) X-ray magnetic circular dichroism at rare-Earth L2,3 edges in RE2Fe14B compounds (RE = a, Pr, Nd, Sm, Gd, Tb, Dy, Ho, Er, Tm, Yb, and Lu). Physical Review B64:10440

Greaves GN (1985) EXAFS and the structure of glass. Journal of Non-Crystalline Solids 71:203–217

Greaves GN, Binstead N, Henderson CMB (1984) The environments of modifiers in oxide glasses. In: Hodgson KO, Hedman B, Penner-Hahn JE (eds) Near edge structure III. Proceeding of an international conference, Stanford, CA, July 16–20, 1984. Springer, Berlin, pp 297–301

Greegor RB, Lytle FW, Ewing RC, Hasker RF (1984) EXAFS/XANES studies of metamict materials. In: Hodgson KO, Hedman B, Penner-Hahn JE (eds) Near edge structure III. Proceeding of an international conference, Stanford, CA, July 16–20, 1984. Springer, Berlin, pp 343–348

Gudat W, Kunz C (1972) Close similarity between photoelectric yield and photoabsorption spectra in the soft-x-ray range. Physical Review Letters 29:169–172

Gunnella R, Benfatto M, Marcelli A, Natoli CR (1990) Application of complex potential to the interpretation of XANES spectra. The case of Na K-edge in NaCl. Solid State Communications 76:109–111

Hatada K, Hayakawa K, Benfatto M, Natoli CR (2010) Full-potential multiple scattering theory with space-filling cells for bound and continuum states. Journal of Physics: Condensed Matter 22(18):185501

Herweg J (1913) Über das Spektrum der Röntgenstrahlen. Verhandlungen der Deutschen Physikalischen Gesellschaft 15:555–556

Itié JP (1992) X-ray absorption spectroscopy under high pressure. Phase Transition 39:81–98

Itié JP, Polian A, Martinez-Garcia D., Briois V, Di Cicco A, Filipponi P, San Miguel A (1997) X-ray absorption spectroscopy under extreme conditions. Journal de Physique IV 7:C2–31

Jaeglé P, Missoni G (1966) Coefficient d'absorption massique de l'or dans la région de longueur d'onde de 26 à 120 Å. Comptes Rendus de l'Académie des Sciences de Paris 262:71–74

Jaklevic J, Kirby TA, Klein MP, Robertson AS, Brown GS, Eisenberger P (1977) Fluorescence detection of exafs: sensitivity enhancement for dilute species and thin films. Solid State Communications 23:679–682

Johnston RW, Tomboulian DH (1954) Absorption spectrum of beryllium in the neighborhood of the K edge. Physical Review 94:1585–1589

Joly Y (2001) X-ray absorption near-edge structure calculations beyond the muffin-tin approximation. Physical Review Letters B63:125120-1–125120-10

Joly Y, Cabaret D, Renevier H, Natoli CR (1999) Electron population analysis by full-potential X-ray absorption simulations. Physical Review Letters 82:2398–2401
Kincaid BM, Eisenberger P (1975) Synchrotron radiation studies of the *K*-edge photoabsorption spectra of Kr, Br2, and GeCl4: A comparison of theory and experiment. Physical Review Letters 34:1361–1364
Kossel W (1920) Zum Bau der Röntgenspektren. Zeitschrift für Physik 1:119–134
Kronig RL (1931) Zur Theorie der Feinstruktur in den Röntgenabsorptionsspektren. Zeitschrift für Physik 70:317–323
Kronig RL (1932a) Zur Theorie der Feinstruktur in den Röntgenabsorptionsspektren. II. Zeitschrift für Physik 75:191–210
Kronig RL (1932b) Zur Theorie der Feinstruktur in den Röntgenabsorptionsspektren. III. Zeitschrift für Physik 75:468–475
Kutzler FW, Natoli CR, Misemer DK, Doniach S, Hodgson KO (1980) Use of one electron theory for the interpretation of near edge structure in *K*-shell X-ray absorption spectra of transition metal complexes. Journal of Chemical Physics 73:3274–3288
Lama PJ, Ohnemus DC, Marcus MA (2012) The speciation of marine particulate iron adjacent to active and passive continental margins. Geochimica et Cosmochimica Acta 80:108–124
Lee PA, Beni G (1977) New method for the calculation of atomic phase shifts: application to extended x-ray absorption fine structure (EXAFS) in molecules and crystals. Physical Review B15:2862–2867
Lee PA, Pendry JB (1975) Theory of extended x-ray absorption fine structure. Physical Review B11:2795–2811
Lee PA, Citrin PH, Eisenberger P, Kincaid BM (1981) Extended X-ray absorption fine structure—its strengths and limitations as a structural tool. Reviews of Modern Physics 53:769–806
López-Flores V, Ansell S, Ramos S, Bowron DT, Díaz-Moreno S, Muñoz-Páez A (2009) Development of ReflEXAFS data analysis for deeper surface structure studies. In: 14th international conference on X-ray absorption fine structure (XAFS14). Journal of Physics: Conference Series 190:012110-2-012110-6
Lukirskii AP, Brytov IA (1964) Investigation of the energy structure of Be and BeO by ultra-soft X-ray spectroscopy. Soviet Phys—Solid State 6:33–41
Lytle FW (1965) X-ray absorption fine structure in crystalline and non-crystalline material; In: Prins J (ed) Physics of non-crystalline solids. North-Holland, Amsterdam, pp 12–30
Lytle FW (1999) The EXAFS family tree: a personal history of the development of extended X-ray absorption fine structure. Journal of Synchrotron Radiation 6:123–134
Lytle FW, Sayers D, Stern E (1975) Extended X-ray-absorption fine-structure technique. II. Experimental practice and selected results. Physical Review B11:4825–4835
Manceau A, Bonnin D, Kaiser P, Frétigny C (1988) Polarized EXAFS of biotite and chlorite. Physics and Chemistry of Minerals 16:180–185
Marcelli A, Cibin G, Hampai D, Giannone F, Sala M, Pignotti S, Maggi V, Marino F (2012) XRF-XANES characterization of deep ice core insoluble dust in the ppb range. Journal of Analytical Atomic Spectrometry 27:33–37
Meirer F, Cabana J, Liu Y, Mehta A, Andrews JC, Pianetta P (2011) Three-dimensional imaging of chemical phase transformations at the nanoscale with full-field transmission X-ray microscopy. Journal of Synchrotron Radiation 18:773–781
Mosbah M, Duraud JP, Métrich N, Wu Z, Delaney JS, San Miguel A (1999) Micro-XANES with synchrotron radiation: a complementary tool of micro-PIXE and micro-SXRF for the determination of oxidation state of elements. Application to geological materials. Nuclear Instruments and Methods in Physical Research B158:214–220
Mottana A (2003) X-ray absorption spectroscopy in mineralogy: a review. In: Bianconi A, Marcelli A, Saini NL (eds) CP652, X-ray and inner-shell processes. 19th international conference on X-ray and inner-shell processes, American Institute of Physics, Melville, NY, pp 415–437
Mottana A (2004) X-ray absorption spectroscopy in mineralogy: theory and experiments in the XANES region. In: Beran A, Libowitzky E (eds) Spectroscopic methods in mineralogy. EMU Notes in Mineralogy, 6. Eötvös University Press, Budapest, pp 465–552

Mottana A (2014) 1913–2013—The centennial of X-ray absorption spectroscopy (XAS): evidences about a question still open. Journal Electron Spectroscopy Related Phenomena. doi:org/10.1016/j.elspec.2013.12.004
Mottana A, Marcelli A (2013) Fifty years since the first European synchrotron radiation-derived XAFS spectrum (Frascati, 1963). Journal of Synchrotron Radiation 20:811–815
Mottana A, Robert J-L, Marcelli A, Giuli G, Della Ventura A, Paris E, Wu ZY (1997) Octahedral versus tetrahedral coordination of Al in synthetic micas determined by XANES. American Mineralogist 82:497–502
Müller JE, Jepsen O, Wilkins JW (1982) X-ray absorption spectra: K-edge of 3d transition metals, L-edges of 3d and 4d metals and M-edges of palladium. Solid State Communications. 42:365–368
Munoz M, Argoul P, Farges F (2003) Continuous Cauchy Wavelet Transform (CCWT) analyses of EXAFS spectra: a qualitative approach. American Mineralogist 88:694–700
Muñoz M, De Andrade V, Vidal O, Lewin E, Pascarelli S, Susini J (2006) Redox and speciation micromapping using dispersive X-ray absorption spectroscopy: application to iron in chlorite mineral of a metamorphic rock thin section. Geochemistry, Geophysics, Geosystems 7(11)
Nakai I, Imafuku M, Akimoto J, Miyawaki R, Sugitani Y, Koto K (1987) Characterization of the amorphous state in metamict silicates and niobates by EXAFS and XANES analyses. Physics and Chemistry of Minerals 15:113–124
Natoli CR (1983) Near edge absorption structure in the framework of the multiple scattering model. Potential resonance on barrier effects. In: Bianconi A, Incoccia L, Stipcich S (eds) EXAFS and near edge structure. Springer series in chemical physics 27. Springer, Berlin, pp 43–47
Natoli CR (1984) Distance dependence of continuum and bound state of excitonic resonance in x-ray absorption near-edge structure (XANES). In: Hodgson KO, Hedman B, Penner-Hahn JE (eds) EXAFS and near-edge structure III. Springer, Berlin, pp 38–42
Natoli CR, Benfatto M (1986) A unifying scheme of interpretation of x-ray absorption spectra based on the multiple scattering theory. In: Lagarde P, Raoux D, Petiau J (eds) EXAFS and near edge structure IV, Abbaye Royale de Fontevraud, France, July 7–11, 1986. Journal de Physique 47(12):C8, supplement., pp 11–23
Natoli CR, Misemer DK, Doniach S, Kutzler FW (1980) First-principles calculation of X-ray absorption edge structure in molecular clusters. Physical Review B22:1104–1108
Natoli CR, Benfatto M, Doniach S. (1986) Use of general potentials in multiple scattering theory. Physical Review A 34:4682–4694
Natoli CR, Benfatto M, Brouder C, Ruiz Lopez MZ, Foulis DL (1990) Multichannel multiple-scattering theory with general potentials. Physical Review B42:1944–1968
Natoli CR, Benfatto M, Della Longa S, Hatada K (2003) X-ray absorption spectroscopy: state-of-the-art analysis. Journal of Synchrotron Radiation 10:26–42
Newville M (2001) EXAFS analysis using FEFF and FEFFIT. Journal of Synchrotron Radiation 8:96–100
Nobel Lectures, Physics 1901–1921 (1967). Elsevier, Amsterdam
Norman D (1986) X-Ray absorption spectroscopy (EXAFS and XANES) at surfaces. Journal de Physique C 19:3273–3311
Paris E, Mottana A, Della Ventura G, Robert J-L (1993) Titanium valence and coordination in synthetic richterite—Ti-richterite amphiboles. A synchrotron-radiation XAS study. European Journal of Mineralogy 5:455–464
Parrat LG (1959) Using of synchrotron-orbit radiation in x-ray physics. Review of Scientific Instruments 30:297–299
Pascarelli S, Neisius T, De Panfiliis S (1999) Turbo-XAS: dispersive XAS using sequential acquisition. Journal of Synchrotron Radiation 6:1044–1050
Pendry JB (1983) The transition region between XANES and EXAFS. In: Bianconi A, Incoccia L, Stipcich S (eds) EXAFS and near-edge structure. Proceedings of the international conference, Frascati, Italy, September 13–17, 1982. Springer, Berlin, pp 4–10
Pettifer RF, Brouder C, Benfatto M, Natoli CR, Hermes C, Ruiz López MF (1990) Magic-angle theorem in powder x-ray-absorption spectroscopy. Physical Review B42:37–42

Ravel B, Newville M (2005) ATHENA, ARTEMIS, HEPHAESTUS: data analysis for X-ray absorption spectroscopy using IFEFFIT. Journal of Synchrotron Radiation 12:537–541
Rehr JJ, Albers RC (1990) Scattering-matrix formulation of curved-wave multiple-scattering theory: application to X-ray-absorption fine structure. Physical Review B41:8139–8149
Rehr JJ, Albers RC (2000) Theoretical approaches to x-ray absorption fine structure. Reviews of Modern Physics 73(3):621–654
Sadoc A (1986) The EXAFS of quasicrystals. In: Lagarde P, Raoux D, Petiau J (eds) EXAFS and near edge structure IV, Abbaye Royale de Fontevraud, France, July 7–11, 1986. J Phys 47(12):C8, supplement, pp 1003–1008
Sagawa T, Iguchi Y, Sasanuma M, Nasu T, Yamaguchi S, Fujiwara S, Nakamura M, Ejiri A, Masuoka T, Sasaki T, Oshio T (1966) Soft X-ray absorption spectra of alkali halides. I. KCl and NaCl. Journal of the Physical Society of Japan 21:2587–2598
Sayers DE, Lytle FW, Stern EA (1970) Point scattering theory of X-ray K-absorption fine structure. In: Henke BL, Newkirk JB, Mallet GR (eds) Advances in X-ray analysis 13. Plenum, London, pp 248–271
Sayers DE, Stern EA, Lytle FW (1971) New technique for investigating noncrystalline structures: fourier analysis of the extended X-ray absorption fine structure. Physical Review Letters 27:1204–1208
Sayers DE, Lytle FW, Stern EA (1972) Structure determination of amorphous Ge, GeO_2 and GeSe by Fourier analysis of extended X-ray absorption fine structure (EXAFS). Journal of Non-Crystalline Solids8–10:401–407
Schlaf R, Höpfner B, Figueroa J, Tridas E, Welter E, Köhler T, Lauermann I, Fischer Ch-H (2012) X-ray absorption measurements on an ultrasonic spray aerosol. Journal of Synchrotron Radiation 19:126–128
Sherman DM, Randall SR (2003) Surface complexation of arsenic(V) to iron(III) (hydr)oxides: structural mechanism from ab initio geometries and EXAFS spectroscopy. Geochimica et Cosmochimica Acta 67(22):4223–4230
Shulman RG, Yafet Y, Eiserberger P, Blumberg WE (1976) Observation and interpretation of x-ray absorption edges in iron compounds and proteins. Proceedings of the National Academy of Science 73(5):1384–1388
Spiegel PK (1995) The first clinical X-ray made in America—100 Years. American Journal of Radiology 164:241–243
Stern EA (1974) Theory of the extended X-ray-absorption fine structure. Physical Review B10:3027–3037
Stern EA (1978) Development of XAFS into a structure determination technique. In: Haase A, Landwehr G, Umbach E (eds) Röntgen centennial: X-rays in natural and life sciences. World Scientific, Singapore, pp 323–340
Stern EA (2001) Musings about the development of XAFS. Journal of Synchrotron Radiation 8:49–54
Stern EA, Sayers DE, Lytle FW (1975) Extended X-ray-absorption fine-structure technique. Determination of physical parameters. Physical Review B11:4836–4846
Stöhr J (1986) SEXAFS: everything you always wanted to know about SEXAFS but were afraid to ask. In: Prinz R, Konigsberger D (eds) X-ray Absorption: principles and techniques of EXAFS, SEXAFS and XANES. Wiley, New York, pp 443–571
Stöhr J (1992) NEXAFS Spectroscopy. Springer series in surface sciences 25. Springer, Berlin
Stöhr J, Denley D, Perfetti P (1978) Surface extended x-ray absorption fine structure in the soft-x-ray region: Study of an oxidized Al surface. Physical Review B18:4132–4135
Stumm von Bordwehr (1989) A history of the X-ray absorption fine structure. Annales de Physique 14:377–465
Sutton SR, Bajt S, Delaney J, Schulze D, Tokunaga T (1995) Synchrotron x-ray fluorescence microprobe: quantification and mapping of mixed valence state samples using micro-XANES. Review of Scientific Instruments 66:1464–1467
Sutton SR, Bertsch PM, Newville M, Rivers M, Lanzirotti A, Eng P (2002) Microfluorescence and microtomography analyses of heterogeneous earth and environmental materials. In: Fenter PA,

Rivers ML, Sturchio NC, Sutton SR (eds) Applications of synchrotron radiation in low-temperature geochemistry and environmental science. Reviews in Mineralogy and Geochemistry 49. Mineralogical Society of America, Washington DC, pp 429–483

Van Nostrand R (1960) The use of X-ray K-absorption edges in the study of catalytically active solids. In: Eley DD (ed) Advances in catalysis 12. Elsevier, Amsterdam, pp 149–187

Waychunas GA (1987) Synchrotron radiation XANES spectroscopy of Ti in minerals: effects of Ti bonding distances, Ti valence, and site geometry on absorption edge structure. American Mineralogist 72:80–101

Waychunas GA, Apted MJ, Brown GE Jr (1983) X-ray K-edge absorption spectra of Fe mineral and model compounds: I. Near-edge structure. Physics and Chemistry of Minerals 10:1–9

Wilke M, Farges F, Petit P-E, Brown GE Jr, Martin F (2001) Oxidation state and coordination of Fe in minerals: an Fe K-XANES spectroscopic study. American Mineralogist 86:714–730

Winick H, Brown G, Halbach K, Harris J (1981) Wiggler and undulator magnets. Physics Today 34(5):50–63

Wong J, Maylotte DH, Lytle FW, Greegor RB, Peters RFSt (1983) EXAFS and XANES investigations of trace V and Ti in coal. In: Bianconi A, Incoccia L, Stipcich S (eds): EXAFS and near-edge structure. Proceedings of the international conference, Frascati, Italy, September 13–17, 1982. Springer, Berlin, pp 206–209

Wu ZY, Xian DC, Hu TD, Xie YN, Tao Y, Natoli CR, Paris E, Marcelli A (2004) Quadrupolar transitions and medium-range-order effects in metal K-edge x-ray absorption spectra of 3d transition-metal compounds. Physical Review B70:033104

Xu W, Chen D, Chu W, Wu Z, Marcelli A, Mottana A, Soldatov A, Brigatti MF (2011) Quantitative local structure determination in mica crystals: ab initio simulations of polarization XANES at the potassium K-edge. Journal of Synchrotron Radiation 18:418–426

Mathematics and Technology at the University of Tartu

Peeter Müürsepp

Abstract In the nineteenth century the University of Tartu was among the leading universities in Europe. The story of its success started with the reopening of the university in 1802 as a result of the efforts of Georges Frédéric Parrot. Due to Parrot's personal connections, the university managed to employ several outstanding personalities, mostly mathematicians, astronomers and natural scientists, whose works enabled development of a number of mathematical tools in close connection with their practical applications. Those applications benefited astronomy most directly. For this reason our focus is on the achievements of Wilhelm Struve. At the same time, we relate an amazing episode that had already happened in the history of the university during the late seventeenth century.

Keywords University of Tartu (Dorpat) · Dimberg · Newton · The Geodetic Arc · Struve · Parrot · Mathematics · Gauss · Astronomy

1 Introduction—Newton Involved

The forthcoming analysis will provide a novel insight into some important developments at the University of Tartu, then Dorpat, in the late seventeenth and nineteenth centuries that show an interesting interaction between mathematics, science and technology in the form of equipment that is important for research, thus contributing to our epistemological understanding of the structure and composition of the universe. Obviously, this is mostly about astronomy. However, the amazing role of the importance of the involvement of technology, even in the history of such an abstract discipline as mathematics, becomes one central focus of our forthcoming story.

The story starts in the last decade of the seventeenth century. In 1690 when the University of Tartu reopened after a long break, an interesting individual arrived in Tartu. He came across the Baltic Sea from the mainland of Sweden to the eastern

P. Müürsepp (✉)
Tallinn University of Technology, Tallinn, Estonia
e-mail: peeter.muursepp@ttu.ee

R. Pisano (ed.), *A Bridge between Conceptual Frameworks,* History of Mechanism and Machine Science 27, DOI 10.1007/978-94-017-9645-3_16

part of the Empire to work as a Professor of Mathematics at a newly reopened easternmost university in Europe. His name was Sven Dimberg (1661–1731).

As we know, in 1687 Newton's *Philosophiae naturalis principia mathematica* appeared in London. Dimberg is famous exactly for his early interest in Newton. There are good reasons to believe that he was one of the first to teach the revolutionary method to university students, perhaps the very first one. True, Dimberg had a strong competitor, David Gregory (Gregorie) of the University of Edinburgh. It is not the task of this paper to settle the dispute over this priority. The fact is that there is enough evidence supporting Sven Dimberg's position, namely that he really did teach Newton's method at the University of Tartu in the early 1690s.

We find the evidence in preserved parts of the curricula that describe the content of the courses of mathematics Dimberg taught in Tartu. True, there is always the theoretical possibility that not everything stated in the curricula resulted in reality. Still, concerning some historical data we are justified in taking the position that the evidence gives us enough ground to believe the statements concerning Newton's method. There are several mentions of it from different angles. However, materials other than the curricula provide no further information. Studying the curricula opened up new paths to getting the story right. Before, there was only the following paragraph written by G. von Rauch:

> It was namely Dimberg who brought Newton to Tartu! In 1695–1696 he lectures on mathematics according to *Contemplationes Neutoniae* (curr. 1696, announcing this as a continuation to the lectures begun in the previous year). From 1697 to 1699 he continues these lectures penetrating "even deeper into the analysis of the mathematical principles of Newton's natural sciences (*Naturphilosophie*) and into his higher mathematics". Simultaneously he also approaches the queen of science of those times—astronomy—and here he has followed Newton as well (curr., 1697 and 1699. Quatations translated from Latin). First of all, it should be emphasised what it means, to get acquainted with Newton in Tartu, as early as in 1695, when Newton was little known in Uppsala as well. Annerstedt (1909, p. 323, 318), thinks that in Uppsala Newton's law of gravitation could be discerned in a disputation only in 1703; as for the lectures, an impact of Newton's and Leibniz's teachings can be noticed in those of Johan Vallerius as late as in 1711. Dimberg's indisputable merit is that he duly understood the enormous significance of this pioneering researcher and started to spread Newton's revolutionary ideas during his lifetime already, making it possible to participate directly in the spiritual life of the Occident at such a distant place as Tartu.[1]

The correction of the dates became possible thanks to the curriculum record for the academic year 1694/1695 that reappeared just some 15 years ago. This curriculum

[1] While compiling the translation of the paragraph into English, the original footnotes where changed into parentheses in the text by Ülo Lumiste and Helmut Piirimäe; the abbreviation "curr." means "curriculum". The data on applied curricula in use appeared in Rauch (Rauch 1943, p 456). In this connection G. von Rauch's note that he had translated the quotation concerning the year 1699 from Latin, i.e. from the original, is misleading. At the same time he shows the corresponding curriculum only via (Backmeister 1764), which is provided with a German translation, most probably not a very exact one, because Dimberg could have hardly used in the original the term "higher mathematics", which came into use only later. Actually, the quotation under discussion is the first sentence of the curriculum of 1697/1698 and a free paraphrase of the little that we know about 1698/1699. Later this transferred also to the years 1695–1696, creating some confusion over the issue.

notes that Newton's "Astronomical hypotheses" was already part of the curriculum during the previous academic year. This actually brings us even further back in history meaning that Sven Dimberg perhaps started to teach Newton's method even in his first year in Tartu. The very first curriculum Dimberg compiled for his teaching in Tartu contains a reference to the 'recent method' (Lumiste and Piirimäe 2001, p. 7). The name of Newton does not appear here but it is difficult to come up with a reasonable guess what that recent method might have been if not Newton's. Still, if we require the names to appear then we have to admit that Dimberg taught Newton's world system in his lectures for the first time in the academic year 1694/1695. Still, as we shall see below, there are good reasons to believe that Newton's method was included into the curriculum in Tartu in the academic year 1693/1694 already. The corresponding reference is simple and clear enough.

It is not just the plain fact of teaching Newton's method that the curricula enabled us to determine. There is more interesting information that is almost as important. First, the range of time of the teaching of the new method seems clear. However, this is not the main issue. More than that, some historians of science claim that Dimberg was actually not physically present in Tartu on all these occasions mentioned in the curricula. As far as it is known, however, being absent from Tartu does not concern the most 'crucial' academic year of 1693/1694.

It is much more interesting, however, to take a good look at the content, at the topics from *Principia* that Dimberg considered appropriate for his students at the end of the seventeenth century. It is obvious that Dimberg did not follow Newton blindly. He did not teach the whole *Principia*.

In order to be precise, let us begin with the curriculum of the academic year 1692/1693. The teaching plan says: "Sven Dimberg, Professor of Mathematics, after having completed the foundations of mathematics according to Euclid's Elements as he promised, will evidently start at 2 o'clock explaining publicly in general order the use of globes and the theory of calendar computation (*computi Ecclesiastici*); in addition he will give private classes in different areas of mathematics to those who request it" (cited from Lumiste and Piirimäe 2001, p. 8). There is no mention of Newton yet and not even to any 'recent method' here but the quote from the curriculum gives us a relatively good idea of how Dimberg built up his courses. One can obtain an even better idea by studying the interests of Dimberg's students. For instance, Nicolaus Limatius, a student of Sven Dimberg, defended his thesis entitled "On Mathematical Proof" (*De Apodixi mathematica*). One can find a synthesis of mathematics, physics and theology of the seventeenth century in the thesis. The defence took place on December 10, 1692 (Lumiste and Piirimäe 2001, p. 8). We don't now if Limatius was aware of Newton's works while writing his thesis, but the composition of his work bears the stamp of a time that was quite typical for Sir Isaac as well. It is common knowledge that the great contributor to the classical scientific method was a believer who looked at the world system as God's creation and searched for the laws put in place by the creator himself.

In the academic year 1694/1695 Sven Dimberg was the Rector of the University of Tartu. In the same academic year Newton's name was first mentioned in Dimberg's curricula:

> Sven Dimberg, Professor of Mathematics, having completed the treatment of analysis of Significant Principles as they are called and which he intermittently deals with, sets out to explain these Astronomical hypotheses of I. Newton, which he emphasised last year. For the sake of such contemplations *astrophilum*, the hours usefully spent on analysis and conic sections can hardly be regretted. And in addition to that, he would immediately take up the theory of closed conic sections he promised to discuss last year, if not the wishes of those who are more interested in the observation of Sphaerica prevent him from doing so. At home he continues private classes in Geometry, architecture and other subjects begun earlier. His public lectures will be delivered at the usual time and place.[2]

Here we have a direct reference to something happening the year before. Although the name of Isaac Newton does not appear in that particular context, the reference is clear enough. Thus, the start of teaching the method of Newton at a university could have been 1 year earlier. It is obvious enough, however, that Sven Dimberg had been teaching Newton's method in the academic year 1693/1694 already.

Very interesting new information appears in the curriculum for the academic year 1697/1698:

> Sven Dimberg, Professor of Mathematics, continues publicly and in general order the analysis of mathematical principles of Newton's Natural science; here the definitions and axioms (*Definitiones & Axiomata*) of the previous year will be presented, thereafter the basic theorems of deducible statics (*Statica*) and theory of levers (*Mochlica*); the author announces that the lectures will also include general lemmas on primary and final relations (*de Methodo Rationum Primarum & Ultimarum*) and certainly these sentences which concern the centripetal forces (*de Virium Centripetarium*); a theory on the movement of bodies along the circle (*in Perimetris cyclicis*) and along the eccentric conic section (*in Sectionibus Concis Excentricis*); from among the remaining part of these principles, he explains the world system (*Systema Mundanum*) from the third book. In the small lecture-room at 2 pm. At home, however, he explains simpler mathematics [...].[3]

One cannot be sure, however, to what extent this plan was actually realised. On September 25, 1697 Sven Dimberg was reported to have been among the three professors who applied for permission to travel to Stockholm to attend a meeting of the Senate of the University. The professors left quite early in the academic year. Thus, substitutes took their place. Unfortunately, there was no one who was able to substitute the Professor of Mathematics. The lectures simply did not take place. The minutes of the Senate meeting do not specify why the three professors were keen on leaving for the mainland of Sweden. As for Dimberg, however, the explanation shows up in the minutes of the Consistory of Uppsala University (Lumiste and Piirimäe 2001, p. 11). He was applying for the vacant post of Professor of Uppsala University.

It seems that in 1998 Dimberg had the intention to return to Tartu. He submitted his curriculum well in time. The original of the curriculum, however, exists. In the translation into English, one can read the following: "Dimberg will continue the public explanation of axioms of Newton's higher mathematics" (Backmeister 1764).

---

[2] Programmata Regiae Academiae Dorpatensis & Pernavienses 1653–1709. Catalogi Praelecionum Acad. Dorpat & Pernav. 1690–1707. Library of Uppsala University. Catalogued under the call number: Univ. progr. Ryssland. (Translation into English by Ülo Lumiste and Helmut Piirimäe).

[3] Riksarkivet, Livonica II 469. (Translation into English by Ülo Lumiste and Helmut Piirimäe).

Still, Dimberg did not return to Tartu that year. He was finally dismissed from the university after a long absence. Still, Dimberg did return to Livonia. This happened only in the year 1706. This time he came to Riga as an assessor of the highest court. The University of Tartu was evacuated to Pärnu at that time to get away from the hostilities of the Northern War between Sweden and Russia. Conrad Quensel occupied the post of Professor of Mathematics in Pärnu. Dimberg left Riga for Sweden in 1709 because the war was getting close to the capital of Livonia (Westerlund 1923).

There is still the interesting question, what were Dimberg's priorities concerning the original presentation of the new method by Isaac Newton? It is quite evident from the preserved curricula that Dimberg relied solely on Newton's main work *Mathematical Principles of Natural Science (Philosophiae naturalis principia matematica)* that appeared in London in 1687. It is known that Dimberg visited England shortly after the publication of Newton's book. The professors at Uppsala got hold of the book only after Dimberg had used it in Tartu already (Nordenmark 1934, p. 9).

If we compare the structure of Newton's book with Dimberg's curricula then it appears that Dimberg planned to deal with the beginning of Newton's book until the third section of Book I. Next, he intended to move on to the third book *De Mundi Systemate Liber Tertius*. The second book, however, where Newton presents the theory of the movement of the resistance of bodies, was to be out altogether (Lumiste and Piirimäe 2001, p. 13).

There is a good rationale for setting up just such plan. In the preface to Book III Newton explains the general composition of his work. In order to enable understanding of his work, the explanation of the world system and the Law of Gravitation presented in Book III, it is necessary to go through the preceding parts. Nevertheless, this did not necessarily concern all of them. Newton writes:

> I do not really insist that everyone should necessarily study each sentence in these books, because there are so many of them that it would require a great deal of time even from a reader well versed in mathematics. It is sufficient if one acquaints himself carefully with the definitions, laws of motion and the three first parts of Book I, then one may proceed with the book dealing with the world system.[4]

Obviously, Dimberg used Newton's own suggestions while compiling his curriculum. It is clear that he had studied Newton's book with special care. This helps to understand the curriculum of the academic year 1694/1695 as well. It has become possible to decipher the phrases "Important principles" and "Astronomical hypotheses" as well as that "emphasised last year".[5] These are all findings of Book III of *Principia.*

[4] Newton, I. Philosophiae naturalis principia mathematica. London, 1726—the last edition that appeared during Newton's lifetime. There is a new edition with addenda and comments: Newton, I. Philosophiae naturalis principia mathematica. Reprint of the third edition (1726) with variant readings. Vol. I, II. Harvard Univ. Press, Cambridge, Mass., 1972.

[5] There is a tendency to accuse Dimberg of the inaccurate use of the word "hypotheses" in its present-day sense. Newton used his famous sentence *Hypotheses non fingo* in the last general

The latter evidence enables us to sum up: "S. Dimberg dealt with Newton's *Principia* in Tartu in the academic years 1693/1694–1694/1695 and 1696/1697–1697/1698 (during the latter approximately till the end of September), presenting the three first parts of Book I (together with addenda) and the world system from book III" (Lumiste and Piirimäe 2001, p. 14). Nevertheless, the mystery of the 'recent method' is still with us.

Obviously, there are good reasons to believe that Dimberg did not follow Newton blindly. It is most probable that he had listened to lectures on conic sections, mechanics and statics at Uppsala University. It is obvious that Dimberg had compiled his curricula in such a way that his students were to become prepared for the introduction into Newton's method before teaching of the method itself was actually started. Dimberg used knowledge acquired in Uppsala during his first academic year in Tartu. All this looks very much like Dimberg was aware that he is going to teach the method of Isaac Newton in a year or two. This could really be the case. By all evidence, Dimberg was already in possession of a copy of Newton's *Principia* by then.

2 The Case of the 'Mathematical' Instruments

Normally, the discussion observed above is the main issue that is considered interesting concerning the personality of Sven Dimberg and his activities at the University of Tartu. In our current context, however, it is perhaps even more important to emphasize Sven Dimberg's conviction that mathematics and astronomy should develop hand in hand. Thus, one of his first moves in Tartu was to buy a telescope and an astrolabe for the university. By the way, back then these devices were often called mathematical instruments. Obviously, mathematics was still in those days considered an important part of empirical research in the context of natural philosophy, not some kind of specific language closed up into itself as it is often believed today. Even the title of Newton's fundamental treatise exemplifies this attitude more than convincingly.

Already at the meeting of the senate of the University of Tartu on Dec. 1, 1691 Dimberg made the proposal to purchase a telescope (*tubus opticus*) for the university since there existed an account of extraordinary funds (Lumiste and Piirimäe 2001, p. 7). According to the existing minutes, all other professors enthusiastically supported Dimberg's initiative using several emotional expressions. The Rector, Professor of Medicine Lars Micrander has been reported saying that the study of

explanation to the second edition of the *Principia* (1713). It is important to note that from here one could conclude that by the time Dimberg had not fully understood the contents of the third book of the *Principia*.

philosophy (which included mathematics, physics and astronomy back then) and medicine without instruments was like body without life.[6]

The Rector Micrander travelled to Stockholm to apply for additional funds for the University of Tartu for various purposes. His list included Dimberg's order for "mathematical instruments". In the list the instruments were called astrolabe (*astrolabium*) and telescope (*tubus opticus*).[7] The application of Micrander was convincingly substantiated, stating the need to give the students a better grounding in mathematical and physical sciences and bringing fame to the university.

Still, the petition was not convincing enough for the King. He took the position that each Professor should himself purchase the requisites and instruments necessary in his speciality.[8]

It appears that Sven Dimberg had not counted on generosity from the King. He was perhaps aware of the main interests of the latter that did not concern intellectual matters. Dimberg had already ordered a telescope from England at his own risk. The telescope reached Tartu in spring 1692. On May 4 Dimberg demonstrated the instrument to his colleagues.

Dimberg had in 1691 already purchased an astrolabe for the university. He had spent 50 rix-dollars in total on the purchases. At the meeting of the university senate on Dec. 16, 1692 Dimberg asked the university to buy the telescope from him. At the senate meeting of Aug. 14, 1693, however, it appeared that the Professor of Mathematics had remained not compensated for his purchases for the university. The Chancellor of the university was present at the meeting. The Senate decided to pay the costs to Dimberg from extraordinary funds that included everything that could be saved. It is clear, however, that in 1694 the Chancellor issued a new order that the Professor of Mathematics should get the money. Still, Lumiste and Piirimäe are sure that Dimberg never got the money (2001, p. 8).

Thus, one can conclude that the 'marriage' of mathematics and technology at the University of Tartu of the Swedish period was arranged as the result of the efforts of one enthusiastic individual, Sven Dimberg, who was a real innovator in his field. His role in introducing Newton into the curricula of universities cannot be overestimated, be it an absolute priority or not.

The University of Tartu closed down at the end of the Northern War in 1710 and was reopened only in 1802 when this part of the world, Livonia, had already been a province of the Russian Empire for nearly a century. Fortunately, the reopening became possible as the result of activities of educated men who were under the influence of the spirit of the Enlightenment and implemented by people who believed in the capacity of human reason to develop natural science with the aim of making human life better.

---

[6] Minutes of the Consistory of the University of Tartu for 1690–1709. Department of Manuscripts and Rarities, Library of the University of Tartu, stock 7, entries 24–28.

[7] Riksarkivet. Livonica II 469.

[8] Riksarkivet. Livonica II 460. Letter of Charles XII to the University of Tartu, June 6, 1692.

3 The Restart of the University

The University of Tartu was founded as a Swedish Lutheran university in the seventeenth century. Then Livonia was expected to remain a province of the Swedish Empire forever. The Russian conquest in early eighteenth century ended this dream. For a long time it seemed that the University of Tartu would remain a short undertaking of the Swedish period.

Fortunately, the tables turned when the Enlightenment reached the western provinces of the Russian Empire. The university reopened in 1802 mostly thanks to the efforts of an outstanding Frenchman, Georges Frédéric Parrot (1767–1852) who had arrived in Livonia in 1795. Thus, there were real people who brought the new cultural spirit to Russia. The university could have been located in what is currently Latvia. Tartu prevailed as the choice of location mostly due to the historical tradition created by the Swedes.

The story of the reopening of the University of Tartu is an integral part of our story as the champion of the event, Georges Frédéric Parrot, made his mark on the mathematical natural science of the time. Parrot's coming forward with the idea of reopening the University of Tartu and launching the building process dates back to the year 1799 when the Tsar was Paul I. But the Tsar changed his mind concerning the location of the university. Under the influence of the Curonian nobility the Tsar decided in favour of *Academia Petrina* in Mitau (Jelgava in Latvia) (Tohvri 2011, p. 355). Then Alexander I became the Emperor and decided to open the university still in Tartu.

The Baltic German nobility had the aim of achieving the status of *Baltische Landesuniversität* for the newly reopened university. They managed to have the majority in the Board of Trustees (Curatorium) of the university. Parrot, the first Rector of the university, was in opposition with the Board of Trustees. The latter wanted to control the activities of the faculty and denied peasants' sons admission to the university. Admitting females was not even an option back then.

Parrot's idea was that the University of Tartu had to be subordinated directly to the central authorities of the Russian Empire. The university had to be open to all social ranks. Parrot was a proponent of abolishing serfdom in the Baltic provinces of the Russian Empire and emancipating the peasants through education to launch a process in which the talented ones among the peasants could become medical doctors, lawyers or state officials (Parrot 1803, p. 13). One can feel Parrot's own educational background behind this idea. He had been studying cameralistics, a kind of German type of public administration of the time, in Stuttgart. It is interesting that in addition to such 'social scientific' background, Parrot had managed to become very well versed in both natural science and mathematics as well. In a way, it is a normal case if we consider the strong affinity that Parrot was feeling towards the entire Enlightenment movement.

Still, in the context of the reopening of the University of Tartu, Parrot did not give up even concerning education for women. He was active in the foundation of co-educational parish schools and girls' schools (Tohvri 2011, p. 358). He held the view that the Enlightenment reached the province of Livonia through school reform creating an educational system on a ladder principle, which means that every step in education is a preparation for the next one (Parrot 1804, pp. 1–16), something that is normally taken for granted today but not necessarily in Livonia of the late eighteenth century.

Parrot's administrative ability was remarkable. His political views were liberal, even very much so for early nineteenth century. His views materialised in the Statute of the University of Tartu. Through the Statute Parrot claimed complete academic autonomy and juridical immunity for the university with a legal subordination to the university's own court (Tohvri 2011, p. 358). The Statute was a radical document in its time. In principle, it called for establishing the so-called "scholarly republic", an expression that has begun to come into usage again lately.

According to the Statute compiled by Parrot, the University of Tartu should be an institution of balanced mental and material domains. The aesthetical and architectural sides should be almost as important as the academic content working to the benefit of the latter of course. In order to accomplish the architectural part of the campus of the reopened university Parrot invited his acquaintance Johann Wilhelm Krause. Parrot was looking for building a "Temple of Wisdom, with its exterior sides altered into a Temple of Nature". Although Parrot's vision never accomplished in full and the University of Tartu never became a real campus type university, Toome Hill evolved into one of the first public urban parks in Estonia. The opening such public urban space even works as evidence of the first signs of democratization in the Baltic provinces of the Russian Empire (Tohvri 2009, pp. 171, 172).

Parrot's main scientific interests were in chemistry and physics, rapidly developing fields in the age of the Enlightenment that still were based on mathematics, the language of science that makes natural science both exact and rigorous. Jānis Stradinš has given a good overview of Parrot's experimental research during his Riga period in his paper written in the Russian language (Stradinš 1968). Being a true son of the Enlightenment Parrot had the idea of making human life better. As he was living in Riga, the capital of the province of Livonia, the main goals of his research came from the need to contribute to the wellbeing of the inhabitants of this city. Actually, even the research problems and hypotheses had their origin in the local environment. Among the main issues were the quality of water in the river Daugava and the quality of air in the hospitals of Riga.

During reopening of the University in Tartu, Parrot took special care of the technological disciplines. His university should not be a place for dry theoretical contemplation but rather an environment for technological innovation. The Chair of Agricultural Technology and Architecture started within the Faculty of Philosophy. Among the principal courses of the faculty, led by Professor Johann Krause, agricultural technology should be specially mentioned (Mägi 2013, p. 78).

Parrot was involved in teaching general and solid state physics at the Faculty of Theoretical and Experimental Physics. He addressed topics like electricity, magnetism and galvanism, complemented by meteorology and physical geography as additional courses. Vahur Mägi has claimed that Parrot was the founder of the first Cabinet of Physics in Tartu (Mägi 2007). It is quite probable, however, that the Swedes had something like this in the seventeenth century already. Parrot's own research papers normally concerned very practical issues like illuminating rooms, constructing ship masts or making gunpowder.

Although an empirical researcher by his nature, Parrot understood the importance of mathematics as the foundation of exact science. He even attempted to make his mark in mathematics as well by presenting an original theory of the tides.

There is a special story about this mathematical aspiration of Parrot. In 1826, Parrot wrote two letters to the great mathematician Carl Friedrich Gauss (1777–1855) himself where he explained his theory of the tides. These letters are the only existing source about this theory. Gauss was quite well aware of the mathematical studies of the tides. In his response to Parrot Gauss managed to convince Parrot about the superiority of his own gravitational theory in this respect (Reich and Roussanova 2011, p. 142).

Coming back to the case of the reopening of the University of Tartu it is crucially important, however, that the cultural tradition of the original university was preserved. The University of Tartu remained a Lutheran university. Now, it became a German university on the territory of the Russian Empire. It is not surprising because the official language of the whole province of Livonia as well as that of Estonia was German. The language of instruction was an important reason why German academics were quite eager to come to live and work in this faraway Northern Province. By all evidence, the language was not the only reason for this strife. Interestingly, compared to Germany, the living conditions in Tartu did not differ much. "... the atmosphere was homely, teaching was carried out in German, scientific literature, textbooks and study aids published in Germany were used also in Tartu" (Tankler 2001, p. 28). Some of the faculty of the university were German nationals and some were Baltic Germans. There was still a third group of Germans among the faculty members. They were Germans who had been born on the main territory of Russia. The University of Tartu had relations with every university and institution of higher education in Germany. The closest contacts the university had were with Berlin University. A number of the faculty members of the University of Tartu had graduated from there. Some of them, like F. Minding and H. Bruns, had also been teaching at Berlin University.

The relations with Göttingen University were also close throughout the nineteenth century. Even Carl Friedrich Gauss himself, perhaps the greatest mathematician of all times, had close links to the University of Tartu. The same holds for the universities of Heidelberg, Halle, Jena, Königsberg and Leipzig, all established centres of higher learning.

4 Wilhelm Struve and the Geodetic Arc

Real fruitful interaction of Mathematics and Technology at the University of Tartu connects inevitably to the name of Wilhelm Struve (1793–1864). Wilhelm Struve was born in 1793 in a place where a famous observatory is located, in Altona near Hamburg. He studied in Tartu since 1808 and finished his thesis in 1813. The title of his thesis was *De geographica positione speculae astronomicae Dorpatensis*. It was published in Jelgava (Mitau) (Struve 1813). It is an interesting fact that a copy of the thesis exists in the Gauss Library in Göttingen[9]. This fact is remarkable as it testifies that there were perhaps contacts between Struve and Gauss at the time of the defence.

In 1813, at the age of just 20 years Struve became Extraordinary Professor of mathematics and astronomy at the University of Dorpat (Tartu). In 1820 he became Ordinary Professor of astronomy. This is the time when mathematics started to separate from astronomy becoming slowly a self-containing discipline without any natural link to the material world. Quite naturally, and fortunately still, astronomy became Struve's main aspiration. He became the person who transformed the Observatory of Tartu into one of the very best in Europe. First, Struve established an astronomical journal *Observationes astronomicae institutae in specula Universitatis caesareae Dorpatensis*. The journal was published in Latin. Struve himself published eight volumes. The first one appeared in 1817, the last one in 1839 (Reich and Roussanova 2011, p. 143).

Due to his influence, Struve could afford to buy expensive instruments. Among these instruments was the refractor constructed by Joseph von Fraunhofer (1787–1826). This instrument was the biggest telescope in the world at the time. The excellent instrument enabled Struve to make his most successful observations. The results mostly appeared in catalogues of double stars. Besides discoveries concerning the double stars, Struve measured the parallax of the Lyrae. Some of these observations were even earlier than Bessel's corresponding ones (Lawrynowicz 1995, pp. 191–214). Struve's publications, however, came later.

Another important activity of Struve was surveying the Baltics. In 1820 Gauss had begun surveying the Kingdom of Hanover. Gauss and Struve met during the measurement of the basis in Braak, near Hamburg.[10] Heinrich Christian Schumacher (1780–1850) managed the measurement process. Schumacher was a Danish citizen who lived in Altona. This is perhaps the main reason why the Danish King Fredrik VI agreed to pay the expenses of this very expensive procedure.

The Struve Arc belongs to the list of the UNESCO World Heritage sites. There is just one more site on the territory of Estonia that belongs to the prestigious list as well. This is the historical Old Town of the capital city Tallinn. It is interesting to note the main criteria of the inclusion of the Arc into the list. First, it is the first accurate measuring of a long segment of a meridian. This is important from the perspective of establishing the exact size and shape of the Earth. In addition to this, UNESCO has singled out an extraordinary interchange of human values in the form of scientific collaboration among scientists from different countries for a scientific cause.

---

[9] Gauss Library 1370, with a personal dedication from Struve: "*Dem Herrn Professor Gauß hochachtungsvoll, der Verfasser*".

[10] Struve's letter to Gauss from November 11 (October 30), 1821.

The Struve Geodetic Arc is an outstanding technological ensemble. It presents the triangulation points of the measuring of the meridian as the not movable and not tangible part of measurement technology.

Measuring of the arc and the results received are directly associated to the important problem of the shape and size of the Earth. The results help to build empirical evidence for the hypothesis of Isaac Newton that the Earth is not an exact sphere.

These were the main points that the UNESCO officials were impressed by. The theory behind the practical endeavour of arc measurement deserves still deeper investigation.

The understanding that the Earth is not flat but has some kind of spherical shape dates back to about the year 500 BC. At first, it was just a theoretical hypothesis of course. In the third century BC, however, Eratosthenes developed a new surveying technique and came forward with a theory for determining the size of the Earth. The theory of Eratosthenes was about to enable determining the size of the Earth using length measurement and angles determined by observations of stars. The theory could work potentially. Still, the measurements themselves were inaccurate due to inadequate methods and equipment.

Better measurement equipment became available only in the seventeenth century. Invention of the new method of triangulation occurred about the same time. The new method enabled one to avoid the need to measure long distances directly, which increases the probability of inaccuracy. The new method was based on measuring a shorter distance. As a result, measurement of longer distances became possible by means of covering the landscape with a chain of triangles. The triangles could span several hundreds of kilometres. Each side (base line) of the triangles was 100 km long. Each triangle in the chain had one common base line with at least one other triangle and two common corners (station points) with another triangle.

The new method helped to establish the true shape of the Earth by means of long arcs in Peru and Lapland in the 1730s and 1740s.

However, the problem of the size of the Earth remained unresolved and became even more complex in a way. Now it was clear that the Earth is not a perfect sphere. The arcs applied in the eighteenth and early nineteenth century had different shortcomings that did not allow an accurate solution of the problem of the size. The issue became even more acute by political rather than scientific reasons. Napoleon had been defeated by 1815. There was a decision made to establish internationally agreed boundaries of Europe. Tsar Alexander I of Russia was deeply interested in the project. He provided Wilhelm Struve with all the necessary resources for a new long geodetic arc.

The role of Struve's measurement in the history of science can hardly be overestimated. It appears as the first step in the development of modern geodetic framework and topographic mapping.

Struve's idea was that the arc should follow a line of longitude (meridian) passing through the observatory of his home university, the University of Tartu (Dorpat). Carl Tenner had measured a shorter arc in Lithuania in 1840. The long Struve arc was formed by connecting earlier, shorter arcs to the southern one measured by Tenner and by extending then to the north and south. As the result, the arc covered a line connecting Fuglenæs near Hammerfest in the north with Staro-Nekrasowska, near Ismail on the Black Sea. The distance covers 2800 km.

The World Heritage site consists of 34 of the original station points established by Struve and his colleagues between 1816 and 1851—four points in Norway, four in Sweden, six in Finland, one in Russia, three in Estonia, two in Latvia, three in Lithuania, five in Belarus, one in Moldova and four in Ukraine. This corresponds to the political borders of today of course. The marks appear in different forms. There are small holes drilled into rock surface and sometimes filled with lead. There are cross-shaped engraved marks on rock surface, solid stone or brick with a marker inset, rock structures (cairns) with a central stone or brick marked by a drilled hole and single bricks. There are also specially constructed monuments to commemorate the point and the arc.

The Struve Geodetic Arc is a great achievement not alone on its own as a scientific and technological achievement. It is a wonderful example of interchange of human values in the form of collaboration among scientists of different countries not just for the sake of knowledge but for practical reasons as well.

5 People Around Struve

One of the most interesting personalities at the University of Tartu in the early nineteenth century was Magnus Georg Paucker (1787–1855). Paucker was born in a small Estonian village called Simuna (St Simonis). He studied astronomy and physics at the University of Tartu with Parrot and Johann Wilhelm Andreas Pfaff. Paucker had remarkable achievements in the field of technological innovations. In 1808, he took part in surveying of the Emajõgi (Embach) river that flows through Tartu. In 1809, Paucker contributed to the construction of the first optical telegraph in Russia in Tsarskoye Selo near St Petersburg. In 1811, he took over the tasks of an observer at the University of Tartu, after the death of astronomer Knorre. An important year is 1813. That year Paucker completed his thesis in Tartu. It had the title *De nova explicatione phaenomeni elasticitatis corporum rigidorum*. The thesis is in physics but Paucker's supervisor was the mathematician Huth who was also the supervisor of Struve's thesis. Struve completed his thesis in the very same year.

It was Struve who became assistant and later Professor in Tartu. Therefore, Paucker had to leave Tartu in 1813 and became teacher of mathematics at the Gymnasium Illustre in Mitau (Jelgava) what is Latvia today (Reich and Roussanova 2011, p. 146).

The Gymnasium Illustre has quite an interesting history. Peter von Biron, Duke of Courland, founded it in 1775. Initially, it got the name Academia Petrina. In 2015, there will be the 240 anniversary of the academy. For that reason, the next Baltic Conference on the History of Science in Latvia will take place in that year in Riga and Jelgava. By all evidence, the next Baltic conference in Estonia in 2017 will be dedicated to the 250th anniversary of the birth of Georges Frédéric Parrot.

The name Gymnasium Illustre became official in 1806. The gymnasium had a very big library and a small observatory. In 1815, the Kurländische Gesellschaft für Literatur und Kunst was founded in Mitau. This is another and even a more significant reason to hold the Baltic conference in Mitau (Jelgava) in 2015. Paucker became the perpetual secretary of the newly founded Society. In 1819, Gauss became an honorary member of the Society. The diploma, dating from June 1, 1819 and signed by Paucker is still preserved in Brunswick.[11]

The connection between Gauss and Paucker is not just formal. In the same year (1819) Paucker published his work *Ueber die Anwendung der Methode der kleinsten Quadratsumme auf physikalische Beobachtungen*. In his treatise Paucker quoted in detail Gauss' and Legendre's contributions to the subject (Paucker 1819). It is not the only work where Paucker has referred to Gauss.

But Paucker was not just a follower of the greatest mathematician. He had remarkable and somewhat unrecognized achievements of his own. They concern both geometry and number theory. In 1822, Paucker published his *Geometrische Verzeichung des regelmäßigen Siebzehn-Ecks und Zweihundertsiebenundfünfzig-Ecks in den Kreis* (Paucker 1822). As the original title suggests, Paucker analysed the theory and the construction of the regular polygons with 17 and 257 sides. Normally, the theory and construction of polygons with 257 sides has been attributed to Friedrich Julius Richelot (1808–1875) (Richelot 1832) who was a student of Carl Gustav Jacob Jacobi (1804–1851). It is easy to notice that Paucker's publication came out 10 years earlier. Paucker was in correspondence with Gauss during his research. Gauss shared his own findings with him. In a letter dating from Jan. 2, 1820 Gauss expresses his thanks to Paucker for his work on the polygons and mentions that there was nobody in Germany who was able to achieve further developments in number theory (Reich and Roussanova 2011, p. 147). Paucker became really an expert on Gauss' work and the two men were close penfriends.

Besides his studies of theoretical mathematics, Paucker was interested in practical applications. For instance, he published several papers on the Russian system of measurement. As the result of these studies he wrote a big handbook: *Handbuch der Metrologie Rußlands und seiner deutschen Provinzen*. This work brought him the Demidov Prize, the highest scientific award in Russia of the time. The originator of the award was Pavel Nikolayevich Demidov (1798–1840), a philanthropist and a member of the rich Demidov family. Paucker received the Demidov Prize in 1832. It included the full prize. It was unusual at the time because the contributions of the winners of the prize had to be in Russian. The entire voluminous work of Paucker, however, was entirely in German (Mezenin 1987).

Struve's direct successor in Tartu was Johann Heinrich Mädler (1794–1874) (Eelsalu and Herrmann 1985). In collaboration with Wilhelm Beer (1797–1850) in Berlin Mädler worked out accurate maps of the Moon. There is the hypothesis that again Gauss himself was impressed. When the chair of Struve became vacant, Gauss offered Mädler his special support. By all evidence, this was decisive. Struve himself had supported the candidacy of Karl Eduard Senff (1810–1849). Gauss'

[11] Stadtarchiv Braunschweig, shelfmark G IX 21. 44 Nr. 7.

support, however, decided the competition in Mädler's favour. As expected, Mädler continued Struve's work. His contribution was significant but it does not match that of Struve's. Mädler retired in 1866 and returned to Germany. Thomas Clausen (1801–1885), another astronomer and mathematician whose credentials impressed Gauss himself succeeded Mädler. Just like his predecessors, Clausen was active in surveying, astronomy and mathematics. The most remarkable results Clausen achieved in the latter. He showed that the Fermat number $2^{64}+1$ did not have a prime number as the result as it was expected.

Several other individuals contributed to the preservation and development of the glory of the University of Tartu in the nineteenth century. As far as men in or around of Mathematics are concerned, one could mention Martin Bartels (1769–1836) and Friedrich Parrot (1791–1841), the son of Georges Frédéric (Georg Friedrich) Parrot. True, the latter is rather famous for his achievements in mountaineering and direct exploration of nature during his many expeditions.

Unfortunately, mostly due to the world wars and the political turmoil they brought about, the next century was not that bright for the university any more. We have to hope that the twenty-first century will be more peaceful and stable again favouring the flourishing of academic activities.

6 Conclusion

The University of Tartu has been an important centre of knowledge production since its foundation in 1632. In the late seventeenth century, it became an outstanding centre of learning as Newton's method was introduced here earlier than anywhere else, the University of Edinburgh perhaps excluded.

Throughout the nineteenth century, the University of Tartu was among the leading centres of academic research and the applications of the research results in the world. As far as mathematized natural science is concerned, the main achievements of the university remain connected with the activities of Wilhelm Struve. His observations of the double stars with the biggest telescope of the time, the Fraunhofer refractor, were at the very edge of astronomy of the time.

The other major achievement of Struve is as significant. The measurement of the Geodetic Arc reaching from the northern coast of Norway to the Black Sea is a major achievement concerning application of natural scientific knowledge and cooperation of scientists and politicians of different countries.

Above the achievements of the researchers, however, the organisational genius of Georges Frédéric Parrot deserves special merit. It is legitimate for Parrot to claim the leading role as he was not just an academic administrator and organiser but a researcher in his own right who came out with some original discoveries. It is important to notice that Parrot was fully aware of the fundamental role of mathematics in contemporary natural science. More than that, he even did research in mathematics concerning the theory of the tides. The great Carl Friedrich Gauss himself overruled Parrot's theory but Parrot managed to get the attention of the greatest mathematician of all times. Parrot was even close to bringing Gauss to Tartu as Professor of Mathematics. Despite the failure in the effort of bringing the greatest mathematician on board, the University of Tartu of the ninteenth century became one of the most significant academic centres of the era.

In the twentieth century, the situation changed and due to the world wars, the University of Tartu lost its leading position. Today it is a normal university with remarkable achievements in different fields but it is somewhat difficult to believe that the glorious days of the nineteenth century will ever come back.

Acknowledgements The current paper was written thanks to the support from the Estonian Science Foundation Grant No. 9362.

The final version was completed during the author's Kone Fellowship at the Helsinki Collegium for Advanced Studies.

References

Annerstedt C (1909) Upsala Universitets historia. Andra delen, 1655–1718. W. Schultz, Uppsala

Backmeister HLC (1764) Nachrichten von den ehemaligen Universitäten zu Dorpat und Pernau. Sammlung russischer Geschichte, IX. Kaiserliche Akademie der Wissenschaften, St. Petersburg

Eelsalu H, Herrmann DB (1985) Johann Heinrich Mädler (1784–1874): Eine dokumentarische Biographie. Akademie, Berlin

Lawrynowicz K (1995) Friedrich Wilhelm Bessel 1784–1846, Vita mathematica series, vol 9. Birkhäuser, Basel

Lumiste Ü, Piirimäe H (2001) Newton's *principia* in the curricula of the University of Tartu (Dorpat) in the early 1690s. In: Vihalemm R (ed) Estonian studies in the history and philosophy of science (Boston studies in the philosophy of science), vol 219, pp 3–18. Kluwer Academic Publishers, Dordrecht/Boston/London

Mägi V (2007) From the 17th century to the present-day in Estonia: evolution of exact and natural sciences, The Global and the Local: the History of Science and the Cultural Integration of Europe. In: Kokowski M (ed). Proceedings of the 2nd International Conference of European Society for the History of Science, Polish Academy of Science, Cracow, pp 920–924

Mägi V (2013) Estonian language of technology as a factor supporting the evolution of engineering thinking. Acta Baltica Historiae et Philosophiae Scientiarum 1(1):75–94

Mezenin NA (1987) Laureaty Demidivskikh premii Peterburgskoi Akademii nauk. Nauka, Leningrad

Nordenmark MVE (1934) Svensk astronomi och svenska astronomer 1700–1730. Arkiv för matematik, astronomi och fysik 24A:2

Parrot GF (1803) Rede bei Gelegenheit der publication der Statuten der Universität und der Abgabe des rectorats am 21. Sept. 1803 gehalten. M. G. Grenzius, Dorpat

Parrot GF (1804) Rede bei Eröffnung der Dörptschen Töchter-Schule: über weibliche Schul-Bildung. Am Geburtstage Ihrer kaiserl. Majestät Maria [...] Fedorowna. M.G. Grenzius, Dorpat

Paucker MG (1819) Ueber die Anwendung der Methode der kleinsten Quadratsummeauf physikalische Beobachtungen. Steffenhagen, Mitau (Gauss Library 1279)

Paucker MG (1822) Geometrische Verzeichnung des regelmäβigen Siebzehn- und Zweihundertsiebenundfünfzig-Ecks in den Kreis in Jahresverhandlungen der Kurländischen Gesellschaft für Literatur und Kunst (Mitau), vol 2. Steffenhagen u. Sohn, Mitau, pp 160–219

Rauch Gv (1943) Die Universität Dorpat und das Eindringen der früher Aufklärungin Livland 1690–1710. G. Olms, Essen

Reich K, Roussanova E (2011) Carl Friedrich Gauss' correspondents in the Baltics. In: Müürsepp P (ed) Baltic Journal of European Studies 1(1):139–156
Richelot FJ (1832) De resolution algebraica aequationis x257 = 1, sive de divisione per bisectionem anguli septies repetitam in partes 257 inter se aequales commentatio coronate. Journal für die reine und angewandte Mathematik 9:1–26, 146–161, 209–230, 337–358
Stradinš J (1968) 'Akademik G.F. Parrot i ego deyatelnost v Rige' ('Academician G.F. Parrot and his Activities in Riga'). On the history of natural science and technology in the Baltic States, vol. I (VII). Zinatne, Riga
Struve W (1813) De geographica positione speculae astronomicae Dorpatensis. J. F. Steffenhagen et filii, Mitau (Gauss Library 1370)
Tankler H (2001) A university between two cultures. On the development of Tartu/Dorpat University in the 19th and early 20th centuries. In: Vihalemm R (ed) Estonian Studies in the history and philosophy of science. Boston Studies in the Philosophy of Science, vol 219. pp 19–34
Tohvri E (2009) Valgustusideede mõju Tartu Arhitektuurikultuurile 19. sajandi alguses. Dissertationes Historiae Universitatis Tartuensis. Tartu Ülikooli Kirjastus, Tartu
Tohvri E (2011) Some new aspects of Georges Frédéric Parrot's visions about the institutional and architectural establishment of the University of Tartu in the early nineteenth century. In: Müürsepp P (ed) Baltic Journal of European Studies 1(1):354–362
Westerlund AW (1923) Åbo Hovrätts presidenter, ledamöter och tjärstemän 1623–1923. H. Edgrens bokhandel, distr, Åbo

Highlights of King Sejong's Astronomical Project: Observatory *Ganui-dae* and Calendar *Chiljeong-san*

Moon-hyon Nam and Il-seong Nha

Abstract The fourth monarch King Sejong (r. 1418-50) of the Joseon dynasty (1392-1910) launched a program to equip the Royal Observatory *(Ganui-dae)* during 1432-38 to establish Neo-Confucian political ideology, and to promote agriculture and social culture. The project was focused on adjusting the Season-granting system fitting to the latitude of Hanyang (now Seoul). And it was also aimed on making observational equipments for calculating astronomical constants fitting to Hanyang as well as timekeeping devices for civil services. Sejong constructed five kinds of astronomical equipment in the main palace grounds: the Simplified Armilla and its platform, the forty-*cheok* -high (828 cm) Template and Bronze Tall Gnomon with Shadow Aligner, the Direction-determining Square Table and the Water-operated Armillary Sphere and Celestial Globe. To do this, the Korean version of the Chinese *Zhou* Foot-Rule (*Ju-cheok*) which is equivalent to 20.7 cm employed as standard measure for scaling. In 1442, the first Joseon native calendar Calculations on the Celestial Motions of Seven Regulators (*Chiljeong-san*) came into use. Moreover, the Rain-gauge (*Cheugu-gi*), the oldest known philosophical instrument for measuring precipitation, was invented at this time and put into use. The new system of observing the rule of heaven enabled the king to implement the ruling by Neo-Confucian rites and virtues following the practices of the Yao and Shun. The Observatory provided the momentum to lay the groundwork for scientific and social norm in fifteenth century Korea. Sejong's achievements not only helped to advance East Asian astronomy, but also set a milestone in the world history of astronomy.

Keywords Neo-confucianism · King Sejong · Royal observatory · Astronomical equipment · Calendar · Standard measure · Rain-gauge · Guo Shoujing

---

M.-h. Nam (✉)
Konkuk University and Jagyeongnu Research Institute, Seoul, South Korea
e-mail: monroe@konkuk.ac.kr

I.-s. Nha
Yonsei University and The Nha Museum of Astronomy, Yecheon, South Korea
e-mail: slisnha@chol.com

R. Pisano (ed.), *A Bridge between Conceptual Frameworks,* History of Mechanism and Machine Science 27, DOI 10.1007/978-94-017-9645-3_17

1 Introduction

The development of science and technology was particularly prominent in fifteenth century Korea, thanks in large measure to the contributions by Sejong the Great (r. 1418–1450).[1] Koreans already had made important advancements from the Three Kingdoms Period (BC 57–AD 935). The Goguryeo kingdom, which ruled the northern part of the Korean peninsula and much of what is now Manchuria employed Chinese calendars, and painted murals of the Four Directional Deities—blue dragon (east), white tiger (west), red phoenix (south) and black warrior (north)—in the interior of the burial-chamber of the great tombs Gangseo Daemyo built during the sixth–seventh century in Nampo, North Korea. In the Tomb of Wrestlers, which was built in the early fifth century at present-day Jian, Jilin Province, China, the burial chamber ceiling painted to represent sky, including a sun disc with the three-legged crow inside, and a lunar bearing the image of a toad, and various constellations.[2] In Pyeongyang, which was one of the Goguryeo capitals in ancient times, are relics of an ancient astronomical observatory (*cheomseong-dae*) and a star map carved in stone. In Silla kingdom, Queen Seondeok (r. 632–647) built an astronomical observatory (*cheomseong-dae*) in 671 at the capital city of Gyeongju, and installed a water-clock at the Hwangryong-sa (temple) in 718. Moreover, a "sacred bell" was cast in 771 to honour the memory of King Seongdeok (r. 702–737) at Bongdeok-sa (temple), which bell tolled to announce the time for morning and evening prayers. Silla imported the "Great Expansion" (*Dayen-li*), "Chimer Virtue" calendar (*Linte-li*), and "Extending Enlightenment" calendar (*Xuanming-li*) from Tang.[3] From AD 554, Baekje kingdom sent calendar and astronomy experts to Japan to supervise calendar-and clock-making. Among the works of such missions, was the water-clock made in 671 during the Emperor Tenji (r. 661–671). In the *History of Three Kingdoms* (*Samguksagi*), various heavenly phenomena of astronomical and astrological significance are recorded, to include solar and lunar eclipses, comets, meteors, novae, Venus in the daytime and sunspots. Astronomical activities such as observations of the heavenly phenomena, calendar making and divination during the Goryeo period (918–1391) can be found in the "Treatises of Astrology, Mathematical Astronomy and Five-star Conjunction" in the *History of Goryeo* (*Goryeosa*). They imported the "Season-granting" system (*Shoushi-li*) from Yuan dynasty in 1281 and the "Great Concordance" (*Datong-li*) from the Ming in 1370, respectively. They also built an astronomical observatory near Man-wol-dae, a palace in the capital of Gaegyeong (now Gaeseong) and the Altar for Star (*Chamseong-dan*) at Mt. Mani in Ganghwa island, where the Goryeo government resisted Mongolian invaders between 1216 and 1259[4]. They ran the Astro-calendric Office (*Seoun-gwan*) which took charge of the astronomical, calendar and clepsydral timekeeping matters.

[1] Needham et al. (1986), *HHR*, p. 17.

[2] Kim L (ed) (2004), *Koguryo Tomb Murals*, p. 15 and 74.

[3] Nam MH (2008), Time in Korea.

[4] Rufus WC (1936), p. 18 and Figs. 13 and 22.

The founders of the Joseon dynasty (1392–1910) adopted Neo-Confucianism as the state ideology and initiated the notion that the rule of heaven was the ideal politics of royal statesmanship. The Neo-Confucian tenet of "observing the heaven and granting the seasons to the people" led to the production of calendars for farming and the reckoning of the time of day, which was regarded as the ruler's most important tasks.[5] The Joseon rulers took over the Astro-claendric Office from the former dynasty and assigned astronomical, calendrical, meteorological, divinations, and clepsydral timekeeping duties. The Prime State Councilor was appointed as the head of the office. Astronomy became the major discipline of the royal family. Taejo (r. 1392–1398) ordered the inscription of a star map (*cheonsang-yeolcha-bunya-jido*)[6] based on a rubbing from the Goguryeo star map at Pyeongyang to ensure dynastic legitimacy, and he affirmed the duties of the ruler as "venerating the heaven and working diligently on behalf of the people". He also ordered the casting of a big bell and hung it in a bell tower in the center of the capital. It rang at dawn and dusk to control civilian timekeeping. He also ordered the casting of a night-watch clepsydra (*gyeong-nu*) and had the clepsydra office erected next to the bell tower in 1398. The third monarch, Taejong (r. 1400–1418), revived metal type and printing technology of the former dynasty to print books for the government and his people. In 1402, he ordered production of a world map named (*honil-gangni-yeok-dae-gukdo-jido*)[7] which depicted thirteenth century view of the Korean peninsula, China, Japan and the Arab and Islamic land as well as Africa and Europe. It might be the first Afro-Eurasian map ever drawn in East Asia by combining contemporary Korean and Japanese maps to compensate improperly drawn Chinese maps on Manchuria and Japan. The cartographic work provided Joseon with a vision of what the world looked like.

The fourth monarch, Sejong was strongly influenced by the Neo-Confucian education he received during his formative years. He sought to establish an ideal Confucian state and focused his policy on creating a new system of rites, music, the calendar, weight and measures. As mentioned above, people of Joseon inherited the Astro-calendric Office and employed the Extending Enlightenment calendars, Season-granting system and the Great Concordance calendars as a basis of calendar-making and accordingly time-measurement. The calendar-makers did not master the calculation methods using the Season-granting system with Extending Enlightenment calendars and they frequently erred in predicting solar and lunar eclipses. As the sun was traditionally the symbol of the king in East-Asian culture, the darkening of the sun's light during a solar eclipse implied the ruler's temporary absence, necessitating a ritual ceremony to recover the sun from the eclipse.[8] After ascending the throne, Sejong commissioned his ministers to correct the calendar by comparing the Extending Enlightenment and Season-granting system, and by sending calendar-makers to China to learn how to predict the solar and lunar eclipses

[5] Yi TJ, Jeon SW (1998) Science, Technology, and Agriculture in Fifteenth Century Korea.

[6] Rufus WC (1913), p. 27.

[7] Yi TJ (2012), op cit, pp. 277–278.

[8] Jeon SW (1998), *A History of Science in Korea*, p. 124; Park SR (1998), *Portents and Politics in Korean History*, p. 43.

more accurately. As suzerain, the Ming Emperor promulgated the almanac used in Joseon, and almanac-makers in Hanyang (now Seoul) had difficulties in time-measurement with Nanjing-based almanac[9]. Joseon calendar-makers did not recognize the time differences resulting from the longitudinal discrepancy, which caused errors in predicting the time of the eclipses. Moreover, the Great Concordance did not consider the lengthening and shortening of the Year Numerator[10] and declared 1368 as the beginning of the new epoch, the establishment of Ming, rather than the year 1281, the epoch of the Season-granting system. Sejong decided to adjust the Season-granting system according to the latitude of the Joseon capital and perceived that it required installation of the instruments to observe heavens.

The king appointed Supervisor on Calendrical Matters Jeong In-ji (1396–1478) to oversee the project to equip the Royal Observatory, and commissioned him "to research the classics, to create instruments and to provide experiments and verifications" in the fall of 1432. Thereafter the project proceeded under the King's leadership and was given high priority. Ultimately, five astronomical installations and instruments were constructed and ten kinds of timekeeping devices newly created. Several instruments were installed on the west garden of the main palace compound, and others at the Astro-calendric Office near the palace for regular observations. The Striking Palace Clepsydra at Announcing Clepsydra Pavilion began to herald standard time adjusted to the times of sunrise and sunset on the first day of the seventh lunar moon of 1434, a milestone in Korean time-measurement history. The constructions and the equipment of the Respectful Venerating Pavilion, which housed Striking Heavenly Clepsydra, ended on the main palace—Gyeongbokgung—grounds in the first lunar moon of 1438, thus finishing the work on a Royal Observatory (*Ganui-dae*). Calendrical undertakings ended in 1444 after more than 20 and more years of effort: the correction of the Season-granting system fitted to the latitude of Hanyang ended with publishing of the *Calculations on the Celestial Motions of Seven Regulators* calendar *Inner and Outer Volume* (*Chiljeong-san nae-oe-pyeon*).[11] A royal secretary Lee Soon-ji (Before 1465) edited *Collections of the Chinese History and Writings on the Astronomy, Calendar, Instrument and Timekeeper* (*Jega-yeoksang-jip*) and *Star Catalogues and Astrological Illustrations* (*Cheonmun-yucho*). The *Computational Case Study on the Solar and Lunar Eclipse on eighth lunar moon in 1447* (*Jeongmyonyeon gyosik-garyeong*) employing *Chiljeong-san nae-oe-pyeon* enabled them to predict the solar and lunar eclipses more accurately.[12] In addition to the astronomical achievements, the crown

[9] Ming China used Nanjing-based calendar and time after moving the capital to Beijing until the adoption of Temporal Model Calendar (*Shixian-li*) in 1656.

[10] 365.2425 days.

[11] Outer Volume *Chiljeong-san oe-pyeon* was compiled based on the Chinese-Islamic calendar (*Huihui-lifa*) and *A Gateway to the Islamic Method for the Computational Case-study of the Sun* (*Weidu Taiyang Tongjing*). (See Yu GR 2001, p. 43) After the compilation of new calendars, it seems that the Astro-calendric Office made the astronomical almanacs according to the *Chiljeong-san* under the name of the Great Concordance (*Datong-li*) which was distributed by the Ming dynasty. (See *Sejong sillok* 101: 4a).

[12] Outer Volume (*Oe-pyeon*) proved to be superior to predict eclipses and employed as a supplementary calendar. (Lee EH and Hahn YH (2012), *Korean Journal History of Science*, Vol. 34, No. 1, pp. 36–70).

prince (later Munjong, r. 1451–1452) invented a unique meteorological instrument Rain-gauge (*Cheugu-gi*) and put it into use in the capital and provincial areas. A Water-level Marker (*yangsu-pio*) was installed on the main stream of the capital to measure the effects of rainfall and on the Han-gang River.

Many of Sejong's accomplishments, from the invention of the Korean alphabet (*Hangeul*) to the equipment of the Royal Observatory, reflected his Neo-Confucian values and their implementation in accordance with the doctrines of the Chinese sage-rulers Yao and Shun. Sejong's 32-year reign marked the heyday for Joseon social[13] and scientific[14] advancement as the state demonstrated creativity. The success was not accidental but rather accomplished through close collaboration between a strong king and his most talented scholar-officials. This tradition was maintained for generations, and later kings oversaw the production and use of armillary spheres and calendars.

In this chapter, major achievements of King Sejong's astronomical undertakings palace observatory (*Ganui-dae*) and the first native calendar (*Chiljeong-san*) were introduced by citing the Records of Observatory (*Ganuidae-gi,* hereinafter referred to as "Observatory Records") in Chap. 77 of *Annals of Sejong* (*Sejong sillok*)[15]. In addition, recent studies on the observatory and its reconstructions as well as relics were introduced.

2 The Equipment of the Astronomical Observatory in the Palace

As mentioned above, King Sejong executed an astronomical project to measure local Joseon time independent of that used in Ming China. Moreover, Sejong aimed to establish an ideal Confucian state by implementing Neo-Confucian state ideology and reforming the ways of the preceding Goryeo dynasty to conform to Neo-Confucian values in accordance with the doctrines of Yao and Shun as well as contemporary systems. Sejong consolidated various libraries into the Hall of Worthies[16] (*Jiphyeon-jeon*), a royal institute functioning as an advisory body for the king to ponder pending national policies and issues. The Jiphyeon-jeon scholars researched ancient and contemporary Chinese rites, music, linguistics, military affairs, astronomy and calendar, law, geography. Their major achievements were Five Rites (*Orye*), Music Score (*Akbo*), Treatise on Geography (*Jiri-ji*) and Calcula-

[13] Kim YS (1998), OSIRIS, Vol. 13, p. 56.

[14] Lee PH and Theodore de Bary W (eds) (1997), Sources of Korean Tradition, Vol. 1, p. 262.

[15] The Veritable Records of Sejong issued in 1454 where every significant occurrence from the time when King Sejong ascended the throne was entered chronologically by year, month and day in that order. The Observatory Records is an article on the fifteenth day of the fourth lunar moon of the nineteenth year (1437) after ascending to the throne.

[16] Lee PH and Theodore de Bary W (eds) (1997), op cit, p. 293.

tions on the Celestial Motions of the Seven Regulators (*Chiljeong-san*)[17]. The Hall was responsible for most of the research on classics and designs of the astronomical instruments and clocks created during the Royal Observatory Project and other calendrical matters. The astronomical undertakings were meant to provide scientific knowledge for calendar-makers, examining the major astronomical constants needed to adjust an almanac to the latitude of the Joseon capital.

Table 1 summarizes the astronomical instruments and timekeeping devices made during a 7-year project according to the Observatory Records. In addition to the project achievements, the star chart inscribed in stone, publications from the calendar studies, the Rain-gauge and Water-level Marker are described below the end of the main text. The equipment and installation were arrayed on the west garden of the Gyeongbok-gung-palace, centering on the platform of the Simplified Armilla, which was north of Banquet Area (*Gyeonghoe-ru*). The Sun-and-Star Time-determining Instrument was installed east of the platform as shown in Fig. 1b. The site was chosen because of the water-ways, which provided drainage for the water-clocks and water-operated armillary sphere. The installations also needed to be near the Royal Council Hall (*Cheonchu-jeon*) and easily accessible to the people who worked at the observatory. The Astro-calendric Office was between the main palace (*Gyeongbok-gung*) and the auxiliary palace (*Changdeok-gung*) as shown in Fig. 1a. King Sejong moved the Simplified Armilla, Template and Bronze Tall Gnomon to the northwest corner of the palace ground in 1443 as shown in Fig. 1.

2.1 Restoration of the Yuan Observatory and its Instruments

Astronomical instruments described in the Observatory Records are the Simplified Armilla, Direction-determining Square Table, Template and Bronze Tall Gnomon with Shadow Aligner, Water-operated Armillary Sphere with house shared by Celestial Globe and these came from the Chinese history records. Among these, Korean versions of Guo Shoujing's (1231–1316) (hereinafter, "Guo") Simplified Instrument, Template and 40 *chi* Tall Gnomon with Shadow Aligner, Direction-determining Table, Celestial Globe and the Imperial Observatory (*Sitian-tai* of the Astrology Commission's headquarters *Taishi-yuan*) were directly modeled after the Chinese.[18]

[17] Sejong sillok 156–158: Inner Volume, 159–163: Outer Volume; Yu GR (2001), pp. 33–57.

[18] *Sitian-tai* was called ancient observatory *ling-tai* where the Guo Shoujing's Simplified Instrument, Tall Gnomon, Shadow Aligner, Observing Table, Ingenious Instrument, Upward-facing Instrument, Direction-determining Table, Celestial Globe, Star-dial Time-determining Instrument, Lantern Clepsydra at Daming Hall and other instruments were installed. Guo might have employed a sky measuring ruler *tianwen-chi* equivalent to 24.5 cm per one *chi*, in contrast to King Sejong who employed Korean *Ju-cheok* (*Zhou-chi* in Chinese) 20.7 cm. (Ref. Wang D 2011, pp. 300–301) So the scales of the equipment of Sejong's observatory were smaller than Guo's observatory *sitian-tai.*

Table 1 Equipment made for King Sejong's Observatory during 1432–1438

Name	Translation	Number of instruments	*Sejong sillok* and others	Remark
1. *Ilseong jeongsi-ui*	The Sun-and-Star time-determining instrument	4-set	77: 7a–8b	King Sejong's memoirs, Kim Don's inscriptions with prefaces
2. *So gan-ui*	Small simplified Armilla	2-set	77: 9a, 10a	Jeong Cho's inscriptions
3. *So Ilseong jeongsi-ui*	Small Sun-and-Star time-determining instrument	several	77: 9ab, 10a	Portable
4. *Gan-ui* and *Ganui-dae*	Simplified Armilla (or simplified instrument) and its platform	1-set	77: 9b	Kim Don's records
		1-set	*Jega yeoksang-jip*[b] 3: 25a–28b	Guo Shoujing's[a] *jian-yi*
5. *Jeongbang-an*	Direction-determining square table	1-set	77: 9b	Guo Shoujing's *zhengfang-an*
			Jega yeoksang-jip 3: 30a-31a	
6. *Dongpio* (*Gyu-pio*) with *Yeongbu*	Template and bronze tall Gnomon with shadow aligner	1-set(40 c)	77: 9b	Guo Shoujing's 40 *chi gao-biao*
		1-set(8 c)	*Jega yeoksang-jip* 3: 31a-32b	
7. *Hon-ui* in the Honui honsang-gak	Water-operated Armillary Sphere with house shared by Celestial Globe	1-set	77: 9b	Wu Cheng's *hun-yi*
		1-housing	60: 38b	
			61: 24a	
8. *Hon-sang*	Water-operated Celestial Globe	1-set	77: 9b	Guo Shoujing's *hun-xiang*
9. *Borugang-nu* in the Boru-gak	Striking Palace Clepsydra in the Annunciating Clepsydra Pavilion	1-set	65: 1a–3b	Kim Don's memoirs, Kim Bin's inscriptions with prefaces
		1-housing	77: 9b–10a	
10. *Heumgyeong gang-nu* in the Heumgyeong-gak	Striking Heavenly Clepsydra in the Respectful Veneration Pavilion	1 set	77: 10a	Kim Don's memoirs
		1-housing	80: 5a–6a	
11. *Angbu-ilgui*	Scaphe Sundial	2-public use	66: 1a	Kim Don's inscriptions
		Assume several in the palaces and the observatory	77: 10a	Guo Shoujing's *yang yi*
			Jega yeoksang-jip 3: 28b-30a	
12. *Hyeonju-ilgui*	Plummet Sundial	several	77: 10a	portable

Table 1 (continued)

Name	Translation	Number of instruments	*Sejong sillok* and others	Remark
13. *Haeng-nu*	Traveling Clepsydra	several	77: 10ab	Guo Shoujing's *xing-lou*
14. *Cheonpyeong-ilgui*	Horizontal Sundial	several	77:10b	portable
15. *Jeongnam-ilgui*	South-determin-ing Sundial	several	77: 10b	portable
16. *Ju-cheok*	*Zhou*-Foot-Rule	several	77: 11ab	Korean version of the Chinese *Zhou*-Foot-Rule
17. *Cheonmun-do* made in 1433[b]	Star map *Cheon-sang yeolcha bunya-jido* made in 1395?	1-set	107: 21b *Jeungbo- mun-heon -bigo* 1, *Sangwigo* 2: 31b	

[a] Treatise on Astrology 1 in the *History of Yuan* (Song Lian, 1976) 48 (*Yuan shi tian wen zhi*). (Same as other notations in the table)
[b] Lee (1445) *Collections of the Chinese History and Writings on the Astronomy, Calendar, Instrument and Timekeeper*

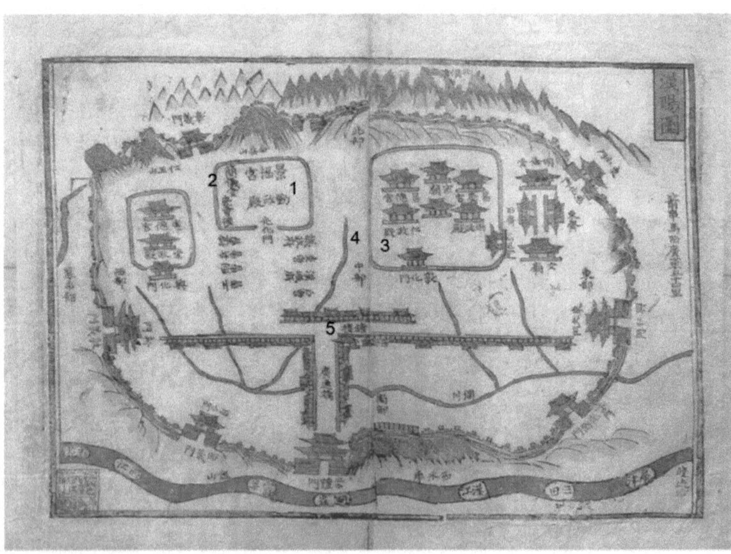

Fig. 1 **a** Map of the capital city Hanyang (now Seoul). *1*: Main Palace; *2*: Royal Observatory; *3*: Auxiliary Palace; *4*: Astro-calendric Office; *5*: Bell Tower. (Adapted from Nam (1995), Figs. 2–9). **b** Map of Gyeongbok-gung palace where the Royal Observatory placed. *1*: Observatory; *2*: Bronze Tall Gnomon; *3*: Banquet Area; *4*: Royal Council Hall; *5*: Respectful Veneration Pavilion; *6*: Announcing Clepsydra Pavilion; *7*: Palace Branch of the Astro-calendric Office; *8*: Sun-and-star Time-determining Instrument Platform; *9*: Small Simplified Armilla Platform; *10*: Water-operated Armillary Sphere and Celestial Globe House; *11*: waterway. (Adapted from Nam (2002), Fig. 7)

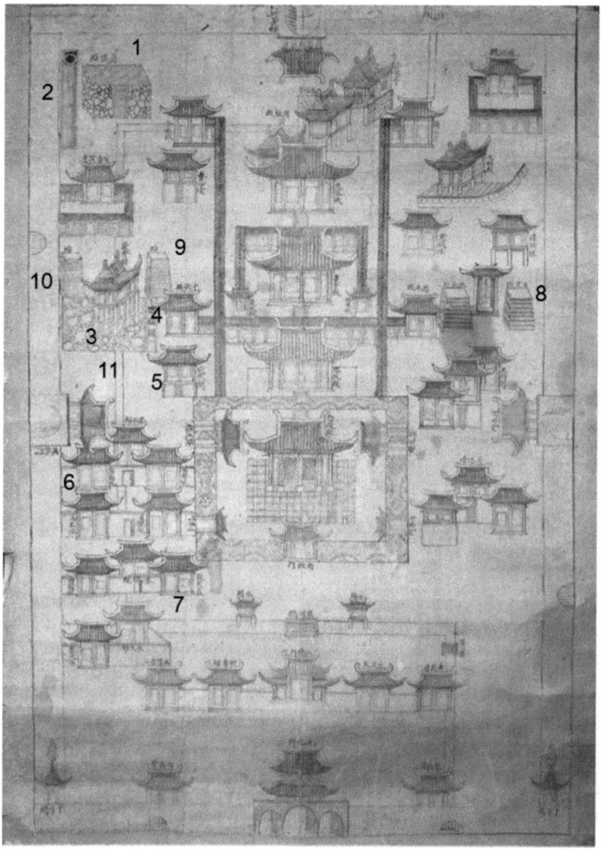

Fig. 1 (continued)

2.1.1 Simplified Armilla and its platform

Simplified Armilla Descriptions on the Simplified Armilla in the Observatory Records are as follows and the details can be found elsewhere[19].

> [...] first made a wooden model, and measured polar elevation to be 38.25^d, coinciding with the value recorded in the *History of Yuan* (*Yuan shi*)[20], afterwards made the instrument in bronze by castings, ..., The Simplified Armilla was placed on the platform [...]

We can infer how the Simplified Armilla was made: first, they made a wooden prototype model following an article on Guo's Simplified Instrument on the Treatise on Astrology in the *History of Yuan* (herein after "*Yuan Treatise*") and checked the polar elevation, i.e. geographic latitude of Hanyang and found it 38.25^d, the same

[19] Pan N and Xiang Y (1980), pp. 30–39; Pan N (ed) (2005), pp. 157–59; Needham et al. (1986) *HHR*, pp. 64–68; Sivin N (2009), Granting the Seasons, pp. 194–198, etc.

[20] *Yuan Treatise* 1: 1.

as recorded on the Yuan Treatise[21]. They then cast a bronze Simplified Armilla. The king ordered Ahn Soon (1370–1440) to build an observation platform for the new instruments. Finally, the Simplified Armilla was placed on the platform.

The Simplified Instrument was made first by Guo in 1279 reforming the traditional Chinese armillary sphere and was a core instrument among the equipment invented by him, contributing much to the development of astronomy and calendrical science. Sejong and his scholar-officials experimented with a wooden-model of Guo's Simplified Instrument on the platform, verified the latitude of the Joseon capital as 38.25^d, the same as the recorded value in the Yuan Treatise[22]. Then, Sejong commissioned casting in bronze and had it installed on the platform. Five observatory watchers made nightly observations starting from the fourth of the third lunar moon of the king's twentieth reign year (1438).[23] This was a new instrument for the king. The observatory was christened as the "Simplified Armilla Platform" (*Ganui-dae*) and the principle of operation was applied to make smaller versions and to develop the Sun-and-Star Time-determining Instrument and various other timekeepers. As Needham et al. (1986) argued, the casting of a new Simplified Instrument marked the first step in Sejong's ambitious program of re-equipping the Royal Observatory.[24]

As its name implied, Guo's Simplified Instrument came from the unpacking of traditional three-layer armillary sphere rings (or circles) into one-layer, discarded the ecliptic components to broaden the observer's visual field, and mounted the equatorial and the azimuthal assemblies separately. In contemporary Korea, reconstructions were made twice based on the Yuan Treatise: the first by the Cultural Properties Office of Korea in 1994 and the second by the Korea Astronomy & Space Science Institute (KASI) in 2000.[25] Figure 2a shows the Ming reconstruction of Guo's in 1437 and Fig. 2b shows the components of the reconstruction by KASI.

Platform The article on the Simplified Armilla in the Observatory Records cited above continues to describe the platform as follows:

> [...] all was ready the king ordered to the Minister of Taxation Ahn Soon to build a stone observatory-platform north of Banquet Area (*Gyeonghoe-ru*) in the rear garden. The height of the platform was 31 c, 47 c long and 32 c wide and provided with a stone banister [...]

---

[21] Actually 38.25^d was the polar elevation of *Gaeseong* the capital of Goryeo dynasty, not Seoul. Thus they made bronze instruments employing $38\ 1/6^d$ following Lee Soon-ji. (See *Sejo sillok* 36: 14a; Lee and Kim (1458), *Calculation of Luni-solar Eclipses*).

[22] In the Season-granting system, the circumference of the circle was 365.25^d (Chinese *tu*, Korean *do*) compared with the Islamic calendar's 360 deg.

[23] *Sejong sillok* 80: 26b.

[24] Needham et al. (1986), *HHR*, p. 68; Nha IS and FR Stephenson (eds) (1997), Oriental Astronomy from Guo Shoujing to King Sejong.

[25] On February 13–14, 1993, three Korean Professors Nha Il-seong (an author of this chapter), Park Seong-rae and Nam Moon-hyon (an author of this chapter) had a field trip to Purple Mountain Observatory, Academia Sinica, in Nanjing, China guided by late Professor Xu Zhentao and examined Ming reconstructed Simplified Instrument, Armillary Sphere and Template and Bronze Gnomon, and next year Professor Lee Yong-sam and Nam Moon-hyon had second tour.

Fig. 2 Simplified Instrument. **a** Ming reconstruction of Guo's in 1437. Purple Mountain Observatory, Academia Sinica, Nanjing, China. (Photo courtesy of Nam MH). **b** Simplified Armilla. Contemporary Korean reconstruction by Korea Astronomy & Space Science Institute, KASI), Daejeon. E-1: equatorial assemblies. *1*: equatorial ring reading for right ascension; *2*: alidade of the equatorial ring; *3*: hundred-marks ring divided into 12 double-hours with hundred-marks; *4*: alidade of the hundred-marks ring; *5*: four displacement (mobile declination) double-ring for reading declination; *6*: cross-struts; *7*: centrally pivoted alidade; *8*: sighting vanes with holes, *9*: pole-observing assemblies parallel to hundred-marks ring; *10*: pole-determining ring; A-1: azimuthal assemblies. *11*: standing movable ring for observing altitude; *12*: zenith, *13*: cross-struts; *14*: centrally pivoted alidade; *15*: eum-azimuth ring for reading azimuth; *16*: indicator-rod for reading azimuth; *17*: water-level groove; *18*: dragon column; *19*: south cloud purlin for supporting hundred-marks (diurnal) ring. *20*: north cloud purlin for supporting pole-observing assemblies. (Adapted from Kim SH et al. 2013, p. 33)

We can infer the scale of the platform as follows; a stone observatory-platform was 6.4 m high, 9.7 m long and 6.6 m wide. King Sejong moved the platform to the northwest corner of the palace grounds 10 years after its completion.[26] The map of the palace in Fig. 1b shows new site of the platform.

2.1.2 Direction-Determining Square Table

On the Direction-determining Square Table, the Observatory Records described briefly as "On the south of the Simplified Armilla, Direction-determining Square Table is installed". The table was portable to orient the north–south alignment of the fixed equipment. According to Pan and Xiang (1980)[27], the Direction-determining Square Table was shaped like a tabletop as shown in Fig. 3a, which was 4.0 *chis* (96 cm) square and 0.1 *chis* (2.4 cm) thick. A series of 19 circles are radiated from the center, the first circle with a 0.1 *chis* radius and the radii of the others increasingly progressively by 0.1 *chis*, reaching a maximum radius of 1.9 *chis*. The outmost circle was graduated in the 365.25 *tu* of the sidereal circle. In the center were erected three gnomons: 1.5 *chis* for near equinoxes, two *chis* for summer solstice and 1 *chi* for winter solstice. From sunrise to sunset, it marked the two points where it crossed each circle. In the morning and afternoon were marked the points where it crossed the same circle, and halving the distance between the points produced a point on the line running north and south from the center of the circle. The Direction-determining Square Table was employed for setting up the instruments.

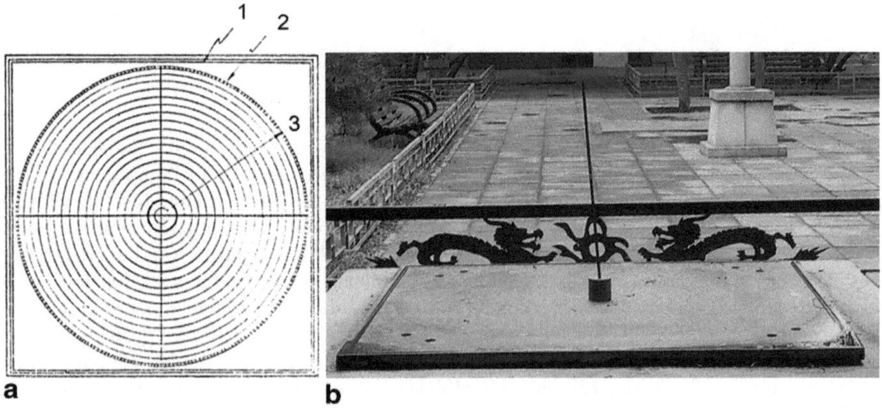

Fig. 3 Direction-determining *Square* Table. **a** Reconstruction drawing. *1*: water-leveling groove; *2*: sidereal circle graduations; *3*: maximum radius, 1.9 *chis*. (Adapted from Pan and Xiang 1980, Fig. 18). **b** Direction-determining Square Table. Reconstruction of the Beijing Ancient Observatory. (Photo courtesy of Nam MH)

[26] On moving the platform, Ref. *Sejong sillok* on 3rd, 14th, 22nd of the first lunar moon and 4th of the second lunar moon of the year.

[27] Pan and Xiang (1980), *Guo Shoujing*, p. 63.

Figure 3b shows the Direction-determining Square Table reconstruction drawing from Guo's design.

2.1.3 Template and Bronze Tall Gnomon with Shadow Aligner

The Template and the Gnomon calculated the length of a solar year by measuring the shadow of the sun at noon throught the year, and the data were used for making the calendar. Thus it was an essential astronomical instrument for calendar-makers from ancient times in China. In this instance, descriptions in the Observatory Records say they made Tall Gnomon according to Guo's design, 40 c (828 cm) tall and installed it with Stone Template which was a graduated decimal unit, i.e., *jang* (10 c), *cheok* c (20.7 cm), *chon* (0.1 c), *bun* (0.01c) on it.[28] They also made use of the Shadow Aligner for focusing the sun's shadow on the template for precise reading. In contemporary Korea, Cultural Heritage Administration (CHA) and KASI reconstructed 8 c (165.6 cm) Gnomon, in 1995 and 2011, respectively. Figure 4 shows the reconstruction made CHA and KASI, respectively.

Fig. 4 Template and 8 *cheok* (828 cm) Gnomon (reconstructions). **a** King Sejong Relics Office of CHA, 1995. (Photo: Nam 2003). **b** Korea Astronomy & Space Science Institute (KASI 2011). (Photo courtesy of Nam MH)

[28] According to the records on the length of shadows measured on the winter solstice in 1568, it was 130 years later first made, length of shadow of the Tall Gnomon was 67.52 c and Smaller Gnomon was 14.52 c on the 27th day of the eleventh lunar moon, eighteenth year throne of the King Myeonjong (r. 1546–1567) (*Myeonjong sillok* 29). This suggested King Sejong made traditional 8 c Gnomon in the Royal Observatory additionally.

2.1.4 Water-Operated Armillary Sphere and Celestial Globe

King Sejong made an observational armillary sphere first, compared it with the Simplified Armilla for measuring heavenly bodies at the platform, and made a celestial globe in conjunction with the construction of a water-driven armillary clock around 1435.[29] In this instance, we have descriptions in the Observatory Records as follows:

> West of the Tall Gnomon they built a small pavilion to house the armillary sphere and celestial globe, set up the sphere in the east and the globe in the west. The system of armillary sphere has not been same all through the generations, so this time the rings of the sphere made of lacquered wood relying on the *Annotation of Book of History* (*Shuzuan*) by Yuan scholar Wu Cheng. The celestial globe was made of lacquered cloth, round as a firelock-bullet, having circumference of 10.86 c. Drawing celestial degrees on it lengthwise and cross-wise; the Red Way (equator) was in the middle with Yellow Way (ecliptic) crossing it at an angle of 24^d weak. All over the lacquered cloth surface were drawn the constellations south and north of the equator, it made one revolution in one day with an addition of one *do* (1 degree). A model sun tied with a thread connecting it on the Yellow Way and moved by one *do* each day, accurately corresponding to the motion of the sun in the sky. All of the ingenious mechanisms of water-driven wheel were hidden invisible [...].

We can infer the general view of the armillary sphere that consisted of three-layer circles: Six Cardinal Points Assemblies i.e., north, south, east, west in the horizon plane, zenith and nadir, and the Arrangers of Time Assemblies i.e., meridian, equator, ecliptic and the Mobile Solar-declination Assemblies i.e., pole-mounted declination-ring, carrying sighting-tube. This was a typical type of armillary sphere made in the later Northern Song dynasty[30] and Fig. 5a shows reconstruction based on the *Annotation of Book of History* by Yuan scholar Wu Cheng. The celestial globe made by King Sejong revived diameter 6 *chi* made by Guo by enlarging the diameter, and Fig. 5b shows a reconstruction based on the Observatory Records.[31]

2.2 *Standard Measure*

In the making of all instruments described above, the Korean version of the Chinese *Zhou* Foot-Rule employed as standard measure for scaling.[32] Kim Don added a long paragraph discussing the measure as an appendix following the conclusive remarks. He discussed the history of the correction of the *Zhou* Foot-Rule in early Joseon by quoting the different printed edition in the Zhuzi's *Family Rites* for the making of ancestral tablets and the model tablets made by government officials.

---

[29] Hahn et al. (2001), Korean Historical Studies, Vol. 113, pp. 57–83.

[30] Su (1937), Essentials of the Method for the New Instruments; Needham et al. (1960) Heavenly Clockwork (HC).

[31] Nha IS (1994), *Celestial Globe of the King Sejong.*

[32] According to the recent studies, Guo Shoujing used Astronomical Rule *tianwen-chi*, one *chi* equivalent to 24.5 cm to make his instruments at the Yuan observatory *ling-tai* in Dadu (now Beijing). (Ref. Wang D (2011), *Studies History of Natural Sciences,* Vol. 30, No. 3, pp. 297–305).

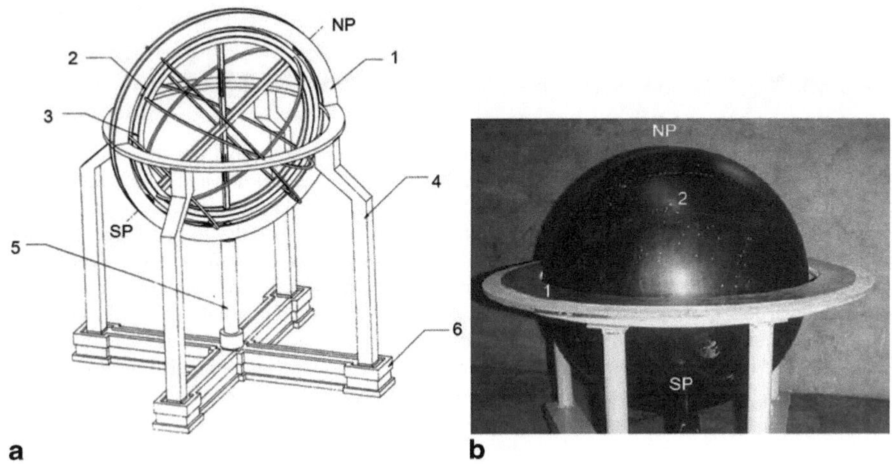

Fig. 5 Armillary Sphere and Celestial Globe. **a** Armillary Sphere of the King Sejong (reconstruction drawing). *1*: six cardinal points assemblies i.e., north, south, east, west in the horizon plane, zenith and nadir; *2*: arrangers of time assemblies i.e., meridian, equator, ecliptic; *3*: four displacement (mobile declination) assemblies i.e., pole-mounted declination-ring, carrying sighting-tube; *4*: four supporting dragon columns; *5*: cloud-and-tortoise column, concealing the transmission shaft; *6*: crosspiece of the base, incorporating water-levels; NP: north pole, SP: south pole. (Drawing courtesy of Nam MH). **b** Celestial Globe of the King Sejong (a reconstruction). *1*: single-ring horizontal *circle* of outer nest; *2*: globe; NP: North Pole; SP: South Pole. (Photo courtesy of Nha IS)

Finally, they determined the length of the Korean version of the Chinese *Zhou* Foot-Rule (*Ju-cheok*) by reducing 25 % of the contemporary Joseon official rule as standard measure, which corresponded to the 75 % of the *sheng-chi* in the *Notes of the Family Rites*. This rule applied to making ancestral tablets, mileposts, target distance on the archery range and astronomical instruments. The law specified the ratios between the five standard measures, distributed to local government officials and kept in the history repository. During the Imjin War with Japan (1592–1598), the standard measures were lost, and King Yeongjo (r. 1725–1776) revived them in 1740 as shown in Fig. 6.[33] According to the 1740s ruler, the length of one *Ju-cheok* is equivalent to 20.7 cm and Yeongjo promulgated the new ruler as the standard measures and used *Ju-cheok* to make instruments such as the rain-gauge and water-level marker.

2.3 Others

2.3.1 Star Map

In 1433, a star map was carved in the stone during the astronomical project.[34] It might have been made when they were preparing to draw the constellations on the Celestial Globe mentioned in Sect. 17.2.1.4.

[33] Nam MH (1995) Korean Water-clocks, pp. 271–304.

[34] *Sejong sillok* 107: 21b; *Jeungbo munheon bigo* 1, *Sangwigo* 2: 31b.

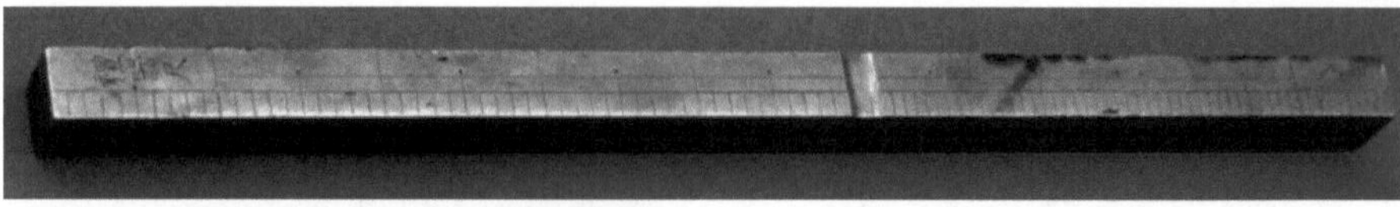

Fig. 6 The Standard Measure. Reconstructed in 1740 under Yeongjo from the relics made in Sejong's reign in 1437. It was made of bronze, 24.6 cm long, 1.2 cm wide and 1.5 cm high rectangular and used until the end of nineteenth century. It is composed of the graduations in half-scale and uses of five standard measures on four-sides, two measures on the front surface; *Ju-cheok* 10.35 cm for making the astronomical instruments and *Yegi-cheok* 13.7 cm for making ceremonial vessels; *Pobaik-cheok* 24.6 cm for clothing and trading on the second side; *Yeongjo-cheok* 15.4 cm for the constructions on the third side; *Hwangjong-cheok* 17.3 cm for making musical instruments on the fourth side. Doubling the length of each ruler make up one *cheok*. The graduation of a foot *cheok* divided into ten *chon*, and a *chon* into ten inch *bun* following the decimal system. Collection of the National Palace Museum of Korea, Seoul. (Photo courtesy of Nam MH)

At that time the Astro-calendric Office referred to Guo's star catalogue and provided new information to draw the constellations on the Water-operated Celestial Globe. Nha (1996) postulated that the upside-down side of the Star Map *Cheonsang yeolcha bunya-jido* made in 1395 by Taejo and now in the National Palace Museum of Korea might have been inscribed in 1433.[35]

2.3.2 Rain-Gauge

The Rain-gauge was invented in 1441 after the astronomical undertakings by the crown prince, later Moonjong (r. 1451–1452) and put into use from 1442.[36] It was a result of Sejong's meteorological studies, which aimed at improving agricultural administration and thus benefiting his people. The king ordered his officials to report the rainfall infiltration into soil by measuring the depth of wet soil all over the country.[37] The crown prince developed the rain-gauge in collaboration with court scientists after conducting an "experiment to measure precipitation", whereby a bronze vessel was set up to collect rainfall in the royal courtyard. He might have been inspired from the development of the water-clock of the Striking Palace Clepsydra, which determined the time by pouring water steadily into a cylinder marked with time scales. The difference between the clepsydra and the rain-gauge was that the latter used the standard *Ju-cheok* to measure the level of rainfall collected in a cylinder after rain, while the former used a floating indicator-rod with time scales. A year later, the Astro-calendric Office made official rain-gauge as a 1.5 c (31.0 cm) high bronze cylinder, diameter of 0.7 c (14.4 cm). The government distributed it to local authorities throughout the state, together with a standard ruler *Ju-cheok*, one *Ju-cheok* is equivalent to 20.7 cm, and a manual for its usage and manufacturing.

[35] Nha I S (1996) *Dongbanghak-ji* 93: pp. 69–101.

[36] *Sejong sillok* 92: 24b; Lee PH and Theodore de Bary W (eds) (1997), op cit, p. 308.

[37] It was called "reporting the amount of precipitation—the gratitude from the heaven."

The local authorities had to record the depth of precipitation and report it to the Department of Taxation. Officials were required to write down the date, duration and amount of precipitation as decimal system *cheok*, 0.1 *cheok* and 0.01 *cheok*. Rain-gauge became a major instrument of the Astro-calendric Office *Seoun-gwan* (later changed its name to *Gwansang-gam*) for recording rainfall as an astronomical data and the data were reported to the king for state affairs regularly.[38] Rainfall data were put to use at the local government for assessment of agricultural land taxes. There remain several articles about the amount of precipitation in the *Annals*.[39] The execution of rainfall recording and reporting, however, had been halted after the Japanese invasions of Korea (1592–1598, so called Imjin War)[40] and resumed during Yeongjo's reign (r. 1725–1776) in 1770 and his grandson Jeongjo (r. 1776–1800) in 1782 and Heonjong (r. 1835–1849) in 1837, respectively. As a result, records on rainfall in the Seoul area cover some 140 years since the 1770s, which is a unique achievement in Joseon. The only extant *cheugu-gi* is the one used at the Office of the Chungcheong provincial governor as shown in Fig. 7d, and it consists of three parts for accuracy in measurement and convenience in handling: a base cylinder and two separate tubes above that. The rain-gauge in Sejong's day might have been in the same shape. The diameter of the rain-gauge 14.4 cm is appropriate for collecting rain drops and it is similar to the average diameter of modern rain gauges used in various countries. If the size is bigger than that, inaccuracy is high for measuring light rainfall, and if smaller, collecting raindrops is difficult when wind is blowing. In short, Sejong's rain-gauge was the optimal size and the state-wide rainfall reporting began on the 19th day of the fifth month in 1442. In contemporary Korea, the Ministry of International Trade and Industry designated May 19 as the "National Day of Invention" in commemoration of the beginning of the system.

Japanese meteorologist Wada Yuji (1859–1918) first introduced the *Cheugu-gi* to the Western world in 1910.[41] Figure 7a shows a platform for the rain-gauge made in Sejong's reign, (b) a platform for Yeongjo made in 1770, (c) the platform made under Jeongjo in 1782 and (d) the vessel made in the reign of Heonjong in 1837, respectively.

2.3.3 Water-Level Marker

The Water-level Marker (*Supyo*) played a similar role as the rain-gauge. This marker, a long stick marked with length scales such as a *cheok*, 0.1*cheok* and 0.01 *cheok*, was erected in front of a bridge (then called Majeon-gyo, later renamed Supyo-gyo) and attached to a rock wall of the Han-gang river.

---

[38] Seong JD (1818) Treatise on the Astro-calendric Office, Vol. 1, p. 30a.

[39] For example, "From 28th day to today, rainy or shinny, height of the rain gauge was 0.05 c (1.0 cm)." (*Joonjong sillok* 98: 41b), etc.

[40] Jeon SW (1998), A History of Science in Korea, p. 143.

[41] Wada (1910) had written a report on the Korean Rain-gauge in French, and sent it to his academic acquaintances in France. In the report, he argued that the *Cheugu-gi* was developed about 200 years before Italian Benedetto Castelli (1578–1643) invented a rain gauge in 1638.

a b c d

Fig. 7 The Rain-gauge, the platforms and a model-ruler. **a** The platform for the rain-gauge made in 1442 and used at the Astro-calendric Office: 92 (W) × 58 (L) × 68 (H) cm, granite, a hole 16.5 cm diameter, 4.7 cm deep for inserting the rain-gauge vessel on the surface. Collection of the Korea Meteorological Administration (KMA). (Photo courtesy of Nam MH). **b** A platform for the Rain-gauge made in 1770 reign of Yeongjo, inscribed "Platform for the Rain-gauge (*cheugu-dae*) on both sides and "Made in May of 1770 Qianlong": 37 (W) × 37 (W) × 46 (H) cm, granite, a hole 16 cm diameter, 4.3 cm deep for inserting the rain-gauge vessel on the surface. Collected of the KMA. (Photo: Nam 1994). **c** Platform for the rain-gauge of the Royal Institute (*Gyujang-gak*) made in 1782 reign of Jeongjo, engraved the inscriptions of the rain-gauge on four sides. Collection of the National Palace Museum of Korea, Seoul. (Photo courtesy of Nam MH) **d** Rain-gauge made in the reign of Heonjong in 1837, inscribed "*Cheugu-gi* (Rain Gauge)" at the Governor's Office of Chungcheong Province, the man in service measures and report the official, 1.5 c (31.0 cm) high, 0.7 c (14.4 cm) diameter, made in 1837, 11 *geun* (6.2 kg) weigh, three pieces. Wada (1917) reported the ruler was *Ju-cheok* equivalent to 20.7 cm and it seemed to be lost in Japan and has not yet returned to Korea. Collection of the KMA. (Photo: Nam 1994; Source: The inscriptions made in May of 1770 Qianlong caused Chinese scholars to make mistakes about the invention of the rain-gauge as China and the Qing government distributed it to Joseon at that time. (Ref. n. 62, PL. I) (See Wang (1985), *Studies History of Natural Science*, Vol. 4, No. 1, pp. 237–46; also Kim SS (1990) *Korea Journal*, Vol. 30, No. 7, pp. 23–32; Park SR (1999), In: Kim and Bray (eds), pp. 9–11.)

On the rocks by the Han River, the *cheok* and *chon* scales were engraved and the official in charge of managing ferries monitored and reported the water-levels regularly.[42] A Water-level Marker made of granite was erected in front of the Supyo-gyo (i.e., Water-level Marker Bridge) in order to alert for floods in Cheonggye Stream across the capital in 1773 during the reign of Yeongjo, who resumed the rain-gauge in 1770. Figure 8 shows relics of Water-level Marker made in 1773.

2.4 Achievements and Leading Figures

Kim Don concluded his Observatory Records by summarizing the equipping of the Royal Observatory undertakings.

[42] *Sejong sillok* 93: 22a.

a b

Fig. 8 Water-level Marker made of granite stone in 1773. **a** Front surface of the marker, ten *Ju-cheok* were inscribed each spaced a *cheok* about 21 cm and marked numbers from 2 to 10 *cheok*. **b** The holes bored into the back marker indicate the water-level as "flood," "moderate," "draught" from top down. Also inscribed "Dredge again at 1773", 3 m high and 20 cm wide. Collection of the King Sejong Memorial Society, Seoul. (Photo courtesy of Nam MH)

In the spring of 1438, the project ended under the initiative of the king who considered the creation of instruments for observation and edification as a top priority for correcting the calendar. He commissioned his scholar-officials to study Guo's Simplified Instrument, Upward-facing Instrument and Tall Gnomon thoroughly and made the several models by comparing with Guo's instruments. However, most came from the king and resulted in new creations. The Season-granting system was corrected, and five kinds of astronomical instruments were made, following the time of heavens above and servicing the works of the people below. Calendar studies came to an end in 1442 after 20 and more years. The correction of the Season-granting system fitted to the latitude of Hanyang ended with the compiling of the *Calculations on the Celestial Motions of Seven Regulators*, so-called first Korean native calendar. Lee Soon-ji edited *Collections of the Chinese History and Writings on the Astronomy, Calendar, Instrument and Timekeeper* and *Star Catalogues and Astrological Illustration*. These were the major outcomes from King Sejong's astronomical project.

Fig. 9 Leading figures. From *left*: Jang Yeong-sil (Photo courtesy of the Scientist Jang Yeong-sil Memorial Society, Seoul, Korea), Yi Cheon (Photo courtesy of the Yean Yi Clan Society, Seoul), Statue of the King Sejong (Photo courtesy of the King Sejong Memorial Society, Seoul) and Lee Soon-ji (Photo courtesy of the Yangseong Lee Clan Society, Seoul)

Sejong and his court scholar-officials, engineers and artisans were involved with the observatory undertakings. The king initiated the project and contributed to the research, guiding the scholars Jeong In-ji, Kim Don (1385–1440), Lee Soon-ji and Kim Bin who belonged to the Hall of Worthies and royal court engineer Jang Yeong-sil for construction and engineering works. Moreover, he was a great thinker, a philologist who invented the Korean alphabet and a cultural hero who is revered for fostering cultural creativity, revitalizing intellectual life, and encouraging technological and scientific advances.[43] Sigimondi (2012) asserted that King Sejong the Great and Sylvester II of the Roman Pontiff were strikingly similar in terms of their achievements in astronomy, music, philosophy, medicine and various scientific disciplines.[44] Jeong Cho (Before 1434) took charge of calendar matters and classics, and Jeong In-ji oversaw calendar matters, equipping the Royal Observatory. He taught mathematics to the king. Yi Cheon (1376–1451) was in charge of engineering and construction affairs with the chief royal court engineer Jang Yeong-sil (Unknown), and they collaborated on the making of the equipment with Kim Don and Kim Jo (former name Kim Bin, Before 1445). Kim Don wrote the records and memoirs about most of the instruments and buildings produced during the project. Lee Soon-ji and Kim Dam (1416–1464) were the experts on mathematical astronomy and they took charge of adjusting the Chinese and Islamic calendar fit to the latitude of Hanyang compiling these into the first Korean calendars *Chiljeong-san*. Figure 9 shows the figures of major contributors to the Royal Observatory project.

[43] Baker D (1998) King Sejong the Great, p. 174.

[44] Sigimondi C 2012, Proc. 1st Int. Conf. Korea Science and Culture Agency, pp. 37–41.

3 Conclusion

The astronomical undertakings launched from King Sejong's court in 1432 aimed to establish bureaucratic organizations for making accurate calendars and observing heavenly bodies to provide astronomical constants for future use by improving the Astro-calendric Office. The project was focused on adjusting the Season-granting system fitting to the latitude of Hanyang (now Seoul). And it was also focused on making observational equipments for calculating astronomical constants fitting to Hanyang, which resulted in equipping the Royal Observatory successfully in 1438. In the course of the undertakings, several young scholars of the Hall of Worthies who were involved with doing research on astronomy and calendars played direct roles in equipping the Observatory and making the first native calendar, and they furthered the development of scientific endeavors in early fifteenth century Joseon. Consequently, the new system of observing the rule of heaven enabled the king to implement the ruling by Neo-Confucian rites and virtues following the practices of the Yao and Shun. The Observatory provided the momentum to lay the groundwork for scientific and social norm in fifteenth century Korea. Sejong's achievements not only helped to advance East Asian astronomy, but also set a milestone in the world history of astronomy. The research and development policies promoted by King Sejong in equipping the Royal Observatory can be appraised as follows: state-sponsored R&D projects; collaborative research environment; benchmarking of advanced scientific models; modeling-experimentation-prototype making. The Royal Observatory founded by Sejong in 1437 was one of the best-equipped observatories in the world in its days, incorporating the long tradition of contemporary Chinese astronomy with medieval East-West science and technology.

Acknowledgements The author wishes to thank the King Sejong Memorial Society, National Palace Museum of Korea, Korea Astronomy & Space Science Institute (KASI), Korea Meteorological Administration (KMA), Purple Mountain Observatory and the Institute for the History of Natural Science of the Academia Sinica, Beijing Ancient Observatory, Scientist Jang Yeong-sil Memorial Society, Research Center for the Korean History of Technology of Konkuk University. The author is grateful to Professors Timothy Atkinson at Seoul University of Foreign Studies and Dr. Yoonkey Nam at Korea Advanced Institute of Science and Technology (KAIST) for reviewing the manuscript.

References

Annals of Sejong (Sejong sillok) (1454), Annals of Sejo (Sejo sillok), Annals of Jungjong (Jungjong sillok), Annals of Myeongjong (Myeongjong sillok), Diaries of Gwanghaegun (Gwanghaegun ilgi), Annals of Yeongjo (Yeongjo sillok), Annals of Jeongjo (Jeongjo sillok), Annals of Heonjong (Heonjong sillok), National Institute of Korean History), reprinted 1986, Seoul

Baker D (1998) King Sejong The Great: bringing Heaven and Earth into harmony. In: Kim-Renaud YK (ed) King Sejong the Great: the light of fifteenth century Korea. The International Circle of Korean Linguistics. George Washington University, Washington

Park YD (eds) (1910) Comprehensive Study of Korean Civilization (1980) Jeungbo-munheon-bigo (Korean Translation) revised and expanded, vol 1. The King Sejong Memorial Society, Seoul pp 142–150
Hahn YH, Nam MH, Lee SW (2001) A study on the astronomical clock of Joseon Dynasty—water-driven Armillary clock. Korean Historical Studies 113:57–83
Jeon SW (1998) A history of science in Korea. Jimoondang, SeoulKim SS (1990) Comments on the Chinese claim for the invention of Rain Gauges. Korea Journal 30(7):23–32
Kim SS (1990) Comments on the Chinese claim for the invention of Rain Gauges. Korea Journal 30(7):23–32
Kim YS (1998) Problems and possibilities in the study of the history of Korean science. OSIRIS 13:56
Kim L (ed) (2004) World heritage Koguryo Tomb murals. YEMACK, Seoul
Kim SH, Min BH, Ahn YS, Yang HJ, Lee YS (2013) Apparatus for Construing the Heavens. Korea Astronomy and Space Science Institute (KASI), Daejeon
Lee SJ (1445) Chinese history and writings on the astronomy, calendar, instrument and timekeeper. Collection of Kyujanggak Institute for Korean Studies, Seoul National University, Seoul
Lee EH, Hahn YH (2012) The study of Kyosik-Garyong (交食假令) published in the early Joseon dynasty. Korean Journal of the History of Science 34(1):36–70
Lee SJ, Kim SJ (1458) Calculation of Luni-solar eclipses. Collection of Kyujanggak Institute for Korean Studies. Seoul National University, Seoul
Lee PH, Theodore de Bary W (eds) (1997) Sources of Korean tradition, vol 1. Columbia University, New York
Nam MH (1995) Korean Water-Clocks: Jagyongnu, the Striking Clepsydra & the history of control & instrumentation engineering. Konkuk University Press, Seoul
Nam MH (2002) Jang Yeong-Sil and the Striking Clepsydra Jagyeongnoo—revival of the time measuring history of the Chosun Period. Seoul National University Press, Seoul
Nam MH (2008) Time in Korea. In: Selin H (ed) (2008) Encyclopedia of history of science, technology, and medicine in non-western cultures. Springer, Heidelberg. http://www.springerlink.com/content/tt722t7002h86885/fullt.... Accessed 22 Sept 2010
Nam M (2015) Innovations on the timekeeping devices at King Sejong's observatory Ganui-dae. In: Pisano R (ed) A bridge between conceptual frameworks, sciences, society and technology studies. History of mechanism and machine science 27. Springer, Dordrecht, pp 344–366
Needham J, Wang L, Solla Price DJ (1960) Heavenly Clockwork (HC): The great astronomical clocks of mediaeval China, 2nd edn. Cambridge University Press, Cambridge. (supplement by Combridge JH, 1986)
Needham, J, Lu GD, Combridge, JH, Major, JS (1986) The Hall of Heavenly records (HHR): Korean astronomical instruments and clocks. Cambridge University Press, Cambridge, pp 1380–1780
Nha IS (1994) Celestial Globe of the King Sejong. Cultural Heritage Administration of Korea, Daejeon
Nha IS (1996) Cheonsang-yeolcha-bunya-jido (Star Map made by King Taejo in 1395 Korea), Dongbanghak-ji (Korean Studies, Yonsei University) 93:69–101
Nha IS, Stephenson FR (eds) (1997) Oriental astronomy from Guo Shoujing to King Sejong—곽수경에서 세종대왕까지의 동양 천문학. Proceedings of an international conference on oriental astronomy (ICOA-1), Seoul, Korea, 6–11 October, 1993. Yonsei University Press, Seoul
Pan N (ed) (2005) The history of ancient astronomical instrument of China. Shanxi Education, Taiyuan, pp 157–159
Pan N, Xiang Y (1980) Guo Shoujing. Shangahi People's, Shanghai, pp 30–39
Park SR (1998) Portents and politics in Korean history. Jimoondang, Seoul
Park SR (1999) Pride and prejudice in the historiography of science in East Asia. In: Kim YS, Bray F (eds) Current perspectives in the history of science in East Asia. Seoul National University Press, Seoul, pp 9–11

Rufus WC (1913) The Celestial Planisphere of King Yi Tai-Jo [Yi T'aejo]. Transaction Korea Branch Royal Asiatic Society IV(III):23–72
Rufus WC (1936) Korean astronomy. Transaction Korea Branch Royal Asiatic Society XXVI:1–48
Seong JD (1818) Treatise on the Astro-calendric office Seoungwan-ji, vol 1, p 30a
Sigimondi C (2012) Sylvester II—King Sejong the Great Parallel. Proceeding of the 1st International Conference of Korea Science and Culture Agency, 15–17 August, 2012, Yecheon, Korea
Sivin N (2009) Granting the seasons—The Chinese astronomical reform of 1280, with a study of its many dimensions and an annotated translation of its records. Springer, Heidenberg
Song Lian (1976) Standard History of the Yuan. China Publishing Company, Beijing
Su S (1937) Essentials of the method for the new instruments. Commercial Press, Shanghai
Wada Y (1910) Les Pluviometres coreens du XVe siècle, scientific memoirs of the Korean Meteorological Observatory. Jemulpo, Korea
Wada Y (1917) Scientific Reports of the Ancient Korean Observation Records. Joseon-General Government
Wang D (2011) The measurement proof of Su Song's astronomical clock tower. Studies History of Natural Science 30(3):297–305
Yi TJ (2012) A new history of Korea. Kachibooks, Seoul
Yi TJ, Jeon SW (1998) Science, technology, and agriculture in fifteenth century Korea. In: Kim-Renaud Y-K (ed) King Sejong the Great—The light of 15th century. The International circle of Korean Linguistics, Washington, pp 233–238
Yu GR (2001) Publication of Chil Jung San Nae Oe Pyun by the Great King Sejong. In: Astronomy in the time of King Sejong. Proceeding of international conference on commemorate of 600th anniversary of King Sejong's Birth (1997, 1st September). Daejeon

Innovations on the Timekeeping Devices at King Sejong's Observatory *Ganui-dae*

Moon-hyon Nam

Abstract In the course of seven year's astronomical project launched by King Sejong (r. 1418-50) of the Joseon dynasty (1392-1910); beginning in 1432, the Royal Observatory installed fifteen kinds of astronomical equipment and timekeeping devices in the main palace grounds. Among these, five kinds were astronomical instruments. Ten kinds were timekeeping devices. Jang Yeong-sil (Unknown) invented two Striking Clepsydras: Ball-powered Striking Palace Clepsydra employed as a standard timekeeper, and Water- and ball-operated Striking Heavenly Clepsydra as an instrument to edify Neo-Confucian ideology. The Sun-and-Star Time-determining Instrument and its smaller version consisted of a sun- and star-dial mounting equatorial-polar alignment and thread gnomons. Two Small Simplified Armilla were used to observe celestial bodies and/or timekeeping at the palace observatory and Astro-calendric Office, respectively. The sundials, mounting an equatorial-polar alignment and/or all thread-gnomons Scaphe-, Plummet-, Horizontal-, and South-determining, were new and used at the observatory, Astro-claendric Office, military camps, public places and palaces. A portable water-clock, Traveling Clepsydra, was used for the royal family, the Astro-claendric Office and military camps. The Striking Clepsydras, the Sun-and-Star Time-determining Instrument and the Small Simplified Armilla were evaluated as Korean originals in the world history of astronomical instruments and clocks. Inventions and innovations in constructing the timekeeping device at the Royal Observatory attribute in large measure to King Sejong's state-supported science policies. Time-measuring devices are reviewed from the contemporary official records with extant relics and reconstructions.

Keywords King Sejong · Royal Observatory · Sundial · Star-dial · Striking Clepsydra · Sun-and-Star Time-determining Instrument · Small Simplified Armilla · Traveling clepsydra

M.-h. Nam (✉)
Konkuk University and Jagyeongnu Research Institute, Seoul, South Korea
e-mail: monroe@konkuk.ac.kr

R. Pisano (ed.), *A Bridge between Conceptual Frameworks,* History of Mechanism and Machine Science 27, DOI 10.1007/978-94-017-9645-3_18

1 Introduction

Koreans have had a long history of clock making since the Three Kingdoms Period (BC 57–AD 935).[1] In the Silla kingdom, they ran the Office of Water-clock and installed a water-clock in the Hwangryong-sa (temple) in 718. In 554 Baekje sent calendar and astronomy experts to Japan to supervise calendar- and clock-making. Among the works of such missions were the water-clock made in 671 during the reign of Emperor Tenji (r. 661–671). According to Xu Jing (1091–1153) of Northern Song who visited Goryeo dynasty (918–1391) on a mission from China in 1123 and compiled his works describing Goryeo and their customs, he briefly described time-telling.[2] The Goyro government ran the Astro-calendric Office (*Seoun-gwan*) which took charge of the astronomical, calendar and clepsydral timekeeping matters. The Joseon dynasty (1392–1910) founders took over the Astro-claendric Office from the former dynasty and assigned it to astronomical, calendrical, meteorological, divinations, and clepsydral timekeeping duties. In 1398, Night Clepsydra (*Gyeong-nu*) was installed and the Clepsydra Office was erected next to the bell tower in the center of the capital. In 1424, King Sejong (r. 1418–1450) ordered the casting of the night Clepsydra (*Gyeong-jeom-jigi*) and replaced 1398 night-watch clepsydra.

King Sejong commissioned Jeong In-ji (1396–1478) who was supervising calendar studies "to research the classics, to create instruments and to provide experiments and verifications" in the fall of 1432 to equip a new observatory in the Gyeongbok-gung palace compound. In the course of the astronomical project, the Striking Palace Clepsydra at Announcing Clepsydra Pavilion began to herald standard time adjusted to the times of sunrise and sunset on the first day of the seventh lunar moon of 1434. And then, the Striking Heavenly Clepsydra and Traveling Clepsydra; the Scaphe-, Plummet-, Horizontal-, and South-determining-sundials; the Small Simplified Armilla, the Sun-and-Star Time-determining Instrument and the Small Time-determining Instrument were created and installed at the Royal Observatory (*Ganui-dae*), the Astro-calendric Office, a number of palaces, public places, with several copies of them being sent to military camps.

Sejong and his court scholar-officials, engineers and artisans were involved with the observatory project. The king initiated the project and contributed to the research, guiding the scholars Kim Don (1385–1440) and Kim Jo (former name Kim Bin, Before 1445) and royal court engineer Jang Yeong-sil (Unknown) for creating the Striking Clepsydras, and inventing the Sun-and-Star Time-determining Instrument. Moreover, he was a great thinker, a philologist who invented the Korean alphabet *Hangeul* and a cultural hero who was revered for fostering cultural creativity, revitalizing intellectual life, and encouraging technological and scientific advances.[3] Sigimondi (2012) asserted that King Sejong the Great and Sylvester II of the Roman Pontiff were strikingly similar in terms of their achievements in astronomy, music, philosophy, medicine and

[1] Nam (2008).

[2] "Reaching the hour, a time-teller beats the drum and suspends a time-tablet on the column. In the middle of a double-hour an official dressed in a red uniform stands in the left bearing a time-tablet, an official dressed in a green uniform bows and says "It is xx hour" and suspends the time-tablet on the column and leaves". (Xu 1123).

[3] Baker (1998), p. 174

various scientific disciplines.[4] Jeong Cho (Before 1434) took charge of classics, and Jeong In-ji oversaw equipping the Observatory. Yi Cheon was in charge of engineering and construction affairs with Jang Yeong-sil the chief royal court engineer, and they collaborated on the making of the equipment with Kim Don, Lee Soon-ji (Before 1465) and Kim Jo.[5]

In this chapter, the timekeeping devices created in the course of the astronomical project during 1432–1438 to equip the King Sejong's Observatory were introduced by citing the Records of Observatory (*Ganuidae-gi,* hereinafter referred to as "Observatory Records") in Chap. 77 of *Annals of Sejong* (*Sejong sillok*)[6]. In addition, recent studies on the timekeeping devices are reviewed from contemporary official records with extant relics and reconstructions.

2 Creation of the Time-Determining Instruments and Clocks

The astronomical project was launched to measure local Joseon time independent of that of Ming China. Sejong consolidated various libraries into the Hall of Worthies[7] (*Jiphyeon-jeon*), a royal institute functioning as an advisory body for the king to ponder pending national policies and issues. The Hall was responsible for most of the research on the classics and designs of the astronomical instruments and clocks created during the observatory construction project and calendar studies.

Table 1 summarizes the timekeeping equipment made during the 7-year project according to the Observatory Records. The water-clocks were arrayed on the west side of the Gyeongbok-gung palace, centering on the south of Banquet Area (Gyeonghoe-ru). The site was near the waterways, which provided drainage for the water-clocks. The Sun-and-Star Time-determining Instrument was located at the eastern part of the palace.

In the making of all instruments described above, the Korean version of the Chinese *Zhou* Foot-Rule was employed as the standard measure for scaling.[8] During the Imjin War with Japan (1592–1598), the standard measures were lost, and Yeongjo (r. 1725–1776) revived them in 1740 as shown in Fig. 1.[9] According to the 1740s ruler, the length of one *Ju-cheok* is equivalent to 20.7 cm and the Yeongjo promulgated the new ruler as the standard measures and used *Ju-cheok* to make the instruments such as the rain-gauges and water-level marker.

4 Sigimondi (2012), pp. 37–41.

5 Nam and Nha (2015).

6 The records is an article on the fifteenth day of the fourth lunar moon of nineteenth year (1437) after throne. An academician of the Hall of Worthies Kim Don wrote the records.

7 Lee and Theodore de Bary (1997), op cit, p. 293.

8 According to recent studies, Guo Shoujing used Astronomical Rule tianwen-chi, one chi equivalent to 24.5 cm to make his instruments at the Yuan observatory ling-tai in Dadu (now Beijing). (Ref. Wang (2011)).

9 Nam (1995), pp. 271–304.

Table 1 Timekeeping equipment made for King Sejong's Observatory during 1432–1438

Name	Translation	Number of instruments made	*Sejong sillok* and others	Remark
1. *Ilseong jeongsi-ui*	The Sun-and-Star Time-determining instrument	4 (1-Royal Obs., 1-Astro-calendric Office, 2-Field Army HQ)	77: 7a–8b	King Sejong's memoirs, Kim Don's inscriptions with prefaces
2. *So gan-ui*	Small simplified armilla	2-set	77: 9a, 10a	Jeong Cho's inscriptions
3. *So Ilseong jeongsi-ui*	Small Sun-and-Star Time-determining instrument	Several	77: 9ab, 10a	Portable
4. *Borugang-nu* in the Boru-gak	Striking Palace Clepsydra in the Annunciating Clepsydra Pavilion	1-set 1-housing	65: 1a–3b 77: 9b–10a	Kim Don's memoirs Kim Bin's inscriptions with prefaces
5. *Heumgyeong gang-nu* in the Heumgyeong-gak	Striking Heavenly Clepsydra in the Respectful Veneration Pavilion	1 set 1-housing	77: 10a 80: 5a–6a	Kim Don's memoirs
6. *Angbu-ilgui*[a]	Scaphe sundial	2-public use, assume several in the palaces and the observatory	66: 1a 77: 10a *Jega yeoksang-jip* 3: 28b-30a[b]	Kim Don's inscriptions Guo Shoujing's *yang yi*
7. *Hyeonju-ilgui*	Plummet sundial	Several	77: 10a	Portable
8. *Haeng-nu*	Traveling Clepsydra	Several	77: 10ab	Guo Shoujing's *xing-lou*
9. *Cheonpyeong-ilgui*	Horizontal sundial	Several	77:10b	Portable
10. *Jeongnam-ilgui*	South-determining sundial	Several	77: 10b	Portable

[a] Treatise on Astrology 1 in the *History of Yuan* (Song Lian, 1976) 48 (*Yuan shi tian wen zhi*). Same as other notations in the table

[b] Lee (1445) Collections of the Chinese History and Writings on the Astronomy, Calendar, Instrument and Timekeeper

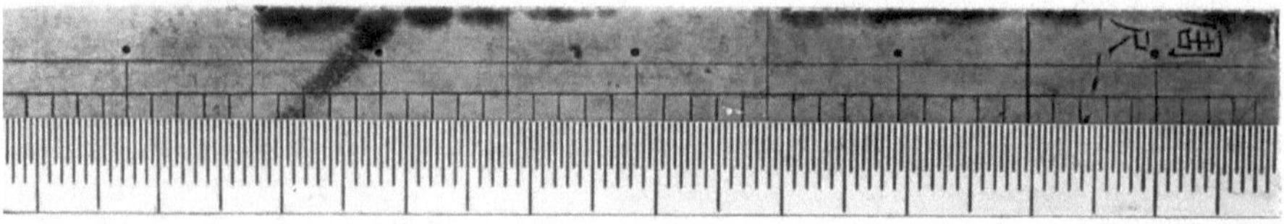

Fig. 1 Half-length of revived *Ju-cheok* (*upper part*). *Lower part* is modern scale. Doubling the length of ruler makes up one *cheok* equivalent to 20.7 cm. The graduation of a foot *cheok* was divided into ten *chon*, and a *chon* divided into 10-in. *bun* following the decimal system. (Collection of the National Palace Museum of Korea, Seoul). (Photo courtesy of Nam MH)

2.1 *The Sun-and-Star Time-Determining Instrument*

Until the Sun-and-Star Time-Determining Instrument was invented under Sejong, the nighttime measurement in Joseon mostly relied upon observations of the 28 lunar mansions to measure five night-watches. A star-dial, which was capable of measuring nighttime without knowing the seasonal meridian star, has not yet developed in East Asia.[10] Whereas sundials have been used for measuring daytime from remote antiquity, nighttime-measuring instruments were only developed after the Royal Observatory project was launched. In this instance, the Observatory Records describe the creation of the Sun-and-Star Time-Determining Instrument for measuring both daytime and nighttime in detail. According to the Preface of Sun-and-Star Time-Determining Instrument in the Observatory Records, Sejong got the ideas, "after dark, stars are used to divide the night" from the book *Zhou Rites* and "to determine the time by observing the stars, but they do not depict exactly the technique to measure with it" from the *History of Yuan* (*Yuan Shi*). Then a project was launched to develop a compound instrument of a star-dial and a sundial, which was christened as the Sun-and-Star Time-Determining Instrument.

Although there were devices called the "Time instrument for both day and night (*Urstrolabe*)" that are cited among the Islamic instruments (*Xiyu yixiang*) in the *Yuan Treatise* and a "Star-dial Time-determining Instrument (*Xing-gui ding shi-ui*)" in the *Biography* of Guo Shoujing (1231–1316) in the *History of Yuan*, no mention was made of how to make one.[11] Because it was very difficult to develop a new instrument from brief articles on *Zhou Rites* and *Guo's Biography*, Sejong studied the *Yuan Treatise* thoroughly and applied the components of the Simplified Instrument, to include equatorial ring, diurnal rings, alidade, pole-determining ring to create the Sun-and-Star Time-Determining Instrument for time-determining both the length of both daylight and nighttime hours. He made four instruments: one which was decorated with a cloud-dragon column for the palace (Fig. 2a), one for the Astro-calendric Office for regular observation, two for both military headquarters in Pyeong-an and Ham-gil -Province to support the work of guarding the frontier.

Kim Don continued the Preface, which was written by Sejong on the design and usage of the instrument at length. In this instance, the design and usage were illustrated through the Inscriptions followed by the Preface[12], and main content is as follows:

[10] Night time could be measured with the Simplified Instrument by observing a meridian star, but it requires knowing the meridian stars by seasonal time zone. Thus we don't call the Simplified Instrument a star-dial.

[11] Guo Shoujing, Biography, History of Yuan 164:51, p. 989.

[12] Needham et al. (1986), pp. 46–54.

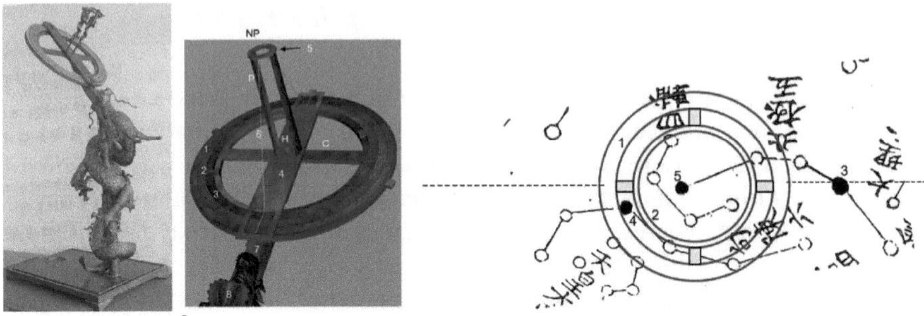

a **b** **c**

Fig. 2 The Sun-and-Star time-determining instrument. **a** Reconstruction by Moon-hyon Nam (author) in 1996. (Photo courtesy of Nam MH). **b** Enlarged components of the observation part. *1*: mobile celestial-equator ring graduated 365 $1/4^{d}$, each of four fractions set on the equatorial-wheel (external diameter 41.4, 0.8 cm thick, 6.2 cm wide) beneath; *2*: fixed sundial hundred-marks ring (external diameter 37.2, 0.8 cm thick, 1.6 cm wide) is graduated for solar-time determinations, with 12 double-hours and 100-marks of day-and-night; *3*: mobile star-dial (external diameter 34.7, 0.8 cm thick, 1.6 cm wide) is graduated same as sundial; *4*: 43.4 cm long pivoted alidade; *5*: pole-determining ring comprised of outer and inner rings; *6*: sighting-thread; *7*: equatorial wheel and its handle; *8*: dragon-column; *C*: cross-struts of the equatorial wheel; *H*: hole pierced in the pivot; *P*: 20.7 cm high dragon-pillar erected on each side of pivot for supporting the pole-determining rings; *NP*: North Pole. **c** Orienting the polar axis and equatorial plane using the star-dial pole-determining rings for observing Chinese fifteenth century circumpolar constellations. *1*: outer ring (diameter 4.7, 0.6 cm wide, 0.4 cm thick); *2*: inner ring (diameter 3.0, 0.8 cm wide, 0.4 cm thick); *3*: Imperial Throne (βU. mi.); *4*: Angular Arrangers (αU. mi.); *5*: Heavenly Pivot, the last star of North Pole constellations which begins with γU. mi

> [...] first, made a round wheel with cross-struts and a handle. It is set in the equatorial plane by making the north higher than the south. The coiled-dragon standing on the platform holds the protruding handle of the wheel in its mouth. The platform has a water-groove for leveling when needed. The three rings set on top of the wheel are concentric. The outer-most one is celestial circumferential ring, and degrees and fractions are graduated on it[13]. There are two rings separately within celestial ring: sundial ring, graduated (12 double-hours and) 100-marks on it, and star-dial ring, likewise graduated. The time on the star-dial corresponds to the celestial degrees. As the outer and inner rings are mobile, the middle one is fixed. The surface of the rings has an alidade horizontally, and there is a pivot at the center. The pivot is penetrated by a hole no bigger than a mustard seed. The degrees, fractions and the times (of the rings) can be read through the slotted ends of the alidade. Two dragon pillars on each side of the pivot uphold the pole-determining rings. The stars are seen between the outer and inner rings. What stars are these? They are the Angular Arranger and Heavenly Pivot, used to locate south north and thereby determine the east west properly. How do they observe with it? They use threads. First, threads are extended from the pole-determining (outer) rings above and pass through the ends of alidade below. For measuring the sun, they use two threads, and only one for observing the stars. The constellation Imperial Throne is red and bright, and is near the celestial pole. So one can know the pole and the time also. The water-clock being used first to ascertain the midnight, and this is marked on the wheel and ring. Moreover, this is where the celestial circumference ring's tour of the heavens starts. Every night it passes through the celestial circumference, and the degrees and fractions ends and starts. The instrument is simple but sophisticated, and its use is comprehensive and full of detail [...].

[13] The circumference is 365 1/4d, i.e., same as the length of the days in a year.

We can infer the design of the instrument from the above-cited Inscriptions, Fig. 2a shows a reconstruction comprised of components for observation and a dragon-column standing on a rectangular base. Figure 2b shows an enlarged observation part of the instrument. The outer adjustable celestial-equatorial ring which was graduated 365 1/4^d (d is Korean notation of the Chinese degree *tu*) is intended to be shifted 1^d anti-clockwise every 66 years to compensate for slow processional movements of the equinoxes, with the North Pole. For daytime use, the fixed sundial ring was employed with the alidade and oblique threads to constitute a sundial. The alidade was set to the shadow of a sighting-thread falling along the center-line of the alidade and the time was then shown by the position of the thread on the sundial's 100-marks ring. For night time-determinations, the alidade was set through the axial hole in the pivot, a sighting-thread was aligned with the star βU. mi. as shown in Fig. 2c and solar time was shown by the position of the thread on the inner ring. This is for allowing the cumulative difference between solar time and sidereal time, so the star-dial ring was shifted 1^d clockwise per day relative to the circumference ring. A relic of the Observational Instrument made in the17th year (1486) of Seongjong (r. 1470–1494) patterned to King Sejong's are collected at the King Sejong Memorial Hall, Seoul.[14] Rufus (1936) took a photo of its original instrument named as "Oblique Sundial and Moondial, Yi".[15]

Up to now some Chinese scholars hypothesized that the "Star-dial Time-determining Instrument" is comprised of a Star-dial and a Time- determining Instrument[16]. Bo (1997) stated that all these past conjectures were wrong and the Xing-gui ding shi-ui is the same as the Korean *Ilseong-jeongsi-ui* and *Xing-gui* is not a star-dial; *Xing* means star, *Gui* means the solar shadow.[17] *Xing-gui ding shi-ui* depicted in the Biography of Guo is as follows:

> The sky has its Red Way [i.e., the celestial equator]; a circle serves in its place. The two poles rise and sink below it, with graduations marking [positions]; thus he made a Star-dial time-determining Instrument (*hsing-kuei ting shih i*).[18]

The Sun-and-Star Time-Determining Instrument is derived directly from the mobile equatorial ring and associated components of Guo Shoujing's Simplified Instrument. Sejong's astronomers did not merely duplicate the old instruments but also extended and improved upon them.[19] Sejong also was inspired by the Islamic instruments "Time-determining Instrument for both day and night (*Urstrolabe*)" and "Star-dial Time-determining Instrument (*Xing-gui ding shi-ui*)" of Guo.

[14] Nam (1995), pp. 105–108.

[15] Rufus (1936), vol 26, plate 17, Fig. 33.

[16] Chen (1955), p. 133.

[17] Bo (1997), p. 21.

[18] Translation taken from Sivin (2009), p. 204.

[19] Needham et al. (1986) pp. 52–54.

2.2 Small Sun-and-Star Time Determining Instrument

Kim Don wrote in the Observatory Records, "The Sun-and-Star Time-Determining Instrument is heavy and inconvenient for military use, prompting the construction of the Small Sun-and-Star Time-Determining Instrument. The design was same as the larger model, with slight differences. For example, the pole-determining rings were omitted to light the instrument and facilitate use in the field". This instrument was very similar to the full-sized version, but the Observatory Records did not detail the size and materials used. No other records on this instrument and no relics are extant. Needham et al. (1986) has studied this in detail by comparing it with a Chinese equatorial sundial with double-ended alidade and triangular-gnomon thread.[20]

2.3 Small Simplified Armilla

Jeong Cho's Inscriptions on the Small Simplified Armilla says, "The best instruments are simplest. The old Simplified Instrument was bulky and ladder-like in its purlin and columns. New one has been made to be portable. So it is even simpler than the old Simplified Instrument". This smaller instrument is made with a bronze column, and uses equatorial assemblies i.e., components of the equatorial ring, hundred-mark ring and mobile solar-declination ring, of the Simplified Armilla as both the equatorial and azimuthal assemblies. The mobile solar-declination ring and hundred-mark ring employed as the standing movable ring for observing altitude and Eum azimuth ring for reading azimuth which constitute the azimuthal assemblies. When the equatorial assemblies are held by a column in a slanting position parallel to the equatorial plane, the mobile declination ring is set along the polar axis. Thus, when the mobile declination ring is set straight upward on the column, the equatorial assemblies can be used for all positional measures. Several columns and supporters of the Simplified Instrument are omitted, and only one column supports equatorial assemblies. Two Small Simplified Armilla were made, one on the platform near the Royal Council Hall and the other on the Small Simplified Armilla platform (*So-ganui-dae*) in the Astro-calendric Office, which remained at the original location (See Fig. 4). Figure 3 shows the Small Simplified Armilla reconstructed by the KASI and Fig. 4 shows the relics of the Small Simplified Armilla platform on the later Astro-calendric Office (The name *Seoun-gwan* was changed to *Gwansang-gam* in 1466).

[20] Op. cit. pp. 60–63.

Fig. 3 The small simplified Armilla (reconstruction). **a** Components. *1*: equatorial ring graduated position of the stars and constellations north and south of the equator regard to 28 lunar mansions and celestial circumference degrees 365 1/4^{d}; *2*: hundred-mark ring graduated 12 double-hours and 100-mark; *3*: four displacement (mobile declination) ring can turn east and west and carries a sighting alidade; *4*: cross-struts; *5*: alidade can move north and south; *6*: center of the earth; *NP*: North Pole; *SP*: South Pole. **b** When the small simplified Armilla is used as the equatorial assemblies. **c** When the small simplified Armilla is used as the azimuthal assemblies. (Adapted from Kim et al. 2013, pp. 36, 38)

Fig. 4 Relics of the small simplified Armilla platform. **a** Photo of the relics around 1913. (From "A Short History of Astronomical Observation", *Daily Handbook*, Incheon Observatory of Joseon Government-General, 1913). **b** Relics of the small simplified Armilla platform on the premises of the later Astro-calendric Office. (Photo: Nam 1995)

2.4 The Striking Palace Clepsydra in the Announcing Clepsydra Pavilion

The Striking Palace Clepsydra (*Jagyeok-gung-nu*) was developed by Chief Royal Court Engineer Jang Yeong-sil under the guidance of Sejong in 1433, and it began to announce standard time in Hanyang from the first day of the seventh month of the 16th year (1434) of Sejong's reign. The account on the making of the Striking Palace Clepsydra was written by Kim Don and Kim Bin as the Records of the Announcing Clepsydra Pavilion. The text includes Memoirs by Kim Don, Inscriptions with Preface by Kim Bin. The new clepsydra was named after its housing as the Clock of the Announcing Clepsydra Pavilion (*Borugang-nu*) officially or "Palace Clepsydra" (*Geum-nu*) literally.

The uniqueness of the Striking Palace Clepsydra was its capability to announce dual-times automatically with visual and audible signals: 12 double-hours (1 double-hour, or *si*, corresponds to two of today's hours), and five night-watches and their points (each watch, *gyeong*, was divided into five *joem*, resulting in twenty-five time subdivisions per night from dusk to dawn. Each watch corresponds to about 2 h today.) Thus the clepsydra was used as a standard timekeeper to announce the time of the curfew alert *Injeong* ("people at rest") in the evening and the removal of curfew *Paru* ("quit the clepsydra") in the morning, and mid-day drum *Ogo* ("quit morning services") following the sunset and sunrise in the capital Hanyang. To change the clepsydra's indicating-rods corresponding to the length of the night-watches to the seasonal variations, *Manual on the Calculation and Usage of Clepsydra Indicator-rods* (*Nujutongui*) were compiled according to sunrise and sunset hours of Seoul in 1437. Eleven pairs of indicator-rods with Fortnightly Periods different graduations were used during a year.[21]

The story of the Striking Palace Clepsydra in the Observatory Records is very concise, so more expanded descriptions are cited from the Memoirs of Kim Don on the clepsydra as follows:

> The king worried about potential errors and mistakes made by the time-announcing officials, so he commissioned Chief Royal Court Engineer Jang Yeong-sil to make wooden immortal figures to announce the time automatically, without human involvement. The construction of the time-announcing machinery was preceded by building of the three-pillar-wide pavilion, and a two-story box was built in the space between the eastern pillars. The three wooden puppets were erected on the pedestal of the upper story as sounders, announcing double-hours by a bell, night-watches by a drum and the number of watch-points by a gong. Below the lower story was set up a horizontal wheel arraying a line of 12 double-hour figures around the circumference. Each figure was carried on thick iron rods that allowed it to pop up and down bearing a time-tablet to announce one of twelve double-hours in sequence.

According to Kim Don's Memoirs, the Striking Palace Clepsydra consisted of six modules (the labels M-x indicate the module number as shown in Fig. 5) as follows. M-1: An inflow-type water-clock system with a dual-time measuring and dual-time

[21] Details See Nam (2014).

Fig. 5 The Striking Palace Clepsydra of King Sejong created in 1434, **a** South elevation, **b** Inside of the dual-time announcing system. (Rear view) The labels are same as in the text. Exhbition of the National Palace Museum of Korea, Seoul. (Photo courtesy of Nam MH)

small-ball releasing mechanisms to mark the beginning of each double-hours and each night-watch and their points; M-2: Transmission routes connecting the two systems M-1 and M-3 via dual-time small-ball receiving funnels and guides; M-3: A dual-time announcing system operated by dual-time large-balls at the beginning of each double-hour, and at each night-watch and their points.; M-4: A dual-time small-ball-to-large-ball converter; M-5: The double-hour machine; M-6: The night watch-point machine.[22]

To reinstate the Striking Palace Clepsydra, the Cultural Heritage Administration (CHA) of Korea launched a national project in 2004–2005 commemorating the scheduled opening of the new National Palace Museum of Korea supposed to be inaugurated in 2006. The construction of the replica was carried out from 2004–2005 under contract with Konkuk University, with which the author of this chapter (Moon-hyon Nam) has been affiliated. The reconstructed instruments worked well and announced dual-times exactly. It has launched daily operations starting in November 2007 for permanent exhibition as shown in Fig. 5.[23]

Ball-driven mechanisms enabled the time-announcing mechanisms to work discretely while securing effective energy transmission and accurate time-keeping through the reconstruction studies.[24] Importantly, the night watch-point announcements by pre-programmed float-rods enabled the Striking Palace Clepsydra to operate as a kind of programmable computer. This clepsydra operated on a closed-loop system whereby it continued to work as long as smaller balls were loaded into the small-ball-holders above the inflow-vessels and the rest into the large-ball-channels of the ball-rest-release mechanisms. As such it is similar to the Candle and the Elephant Clocks of Al-Jazari (1206).[25] As Needham et al. (1986) argued, the elaborate time-annunciating mechanism using ball-operated jack-works was inspired by Al-Jazari's 1206 clocks, which were not copied but instead were adapted to fit the Korean time-keeping tradition by the eminent mechanical engineer Jang Yeong-sil and his supporters.[26] The Striking Palace Clepsydra was a key machine in the history of clock-making, mechanism and machine science.

2.5 *The Striking Heavenly Clepsydra in the Respectful Veneration Pavilion*

The descriptions of the Striking Heavenly Clepsydra in the Observatory Records are very concise and provide an ample view of what the clepsydra looked like. In this instance, Kim Don introduced the makers of the clepsydra, which was not mentioned in the Observatory Records, in his memoirs as follows:

> Respectful Veneration Pavilion (*Heumgyeong-gak*) has been completed [on the seventh day of the first lunar moon of the twentieth year (1438) of Sejong's reign]. Daehogun Jang Yeong-sil installed it; however, its scales and exquisiteness of the systems resulted from the

---

[22] Nam (2012), pp. 91–94.

[23] National Palace Museum of Korea (2007).

[24] Nam et al. (2010).

[25] Nam (1995), pp 185–186; Al-Jazari (1206), pp 58–70; Romdhane and Zeghloul (2010), pp. 1–20.

[26] Needham et al. (1986), p. 42.

Fig. 6 Image of the Striking Heavenly Clepsydra. (South elevation). *1* model sun; *2* four jade female immortals; *3* four guardian gods; *4* hour-jacks and three warriors; *5* 12 Spirit Generals of the Zodiac; *6* Dragon double-hour immortal; *7* figure dressed as an official; *8* inclining vessel; *9* rural scenery sculpture. Numbers are same as that of the text. (Adapted from Nam 2012)

> king's wisdom. The pavilion is located at the vicinity of the royal council hall at the main palace [*Gyeongbok-gung*]. The king ordered Kim Don to write a memoir [...].

We can infer that Chief Royal Court Engineer Jang installed it and the main ideas came from the king who wanted to have a splendid heavenly clock capable of displaying a revolving model of the sun, time-announcing jackwork figures, rural scenes of four-seasons and an inclining-vessel. In this instance, we have descriptions of the making of the heavenly clepsydra in the Observatory Records. The entry is summarized as numbers in Fig. 6 as follows: A model sun (1) attached to a mechanism revolves the mountain, four jade female immortals (figures) (2) are erected on the multi-colored cloud, each facing in one of the four cardinal directions, and strikes a golden bell with a wooden mallet to announce the double-hours. At the same time, the Four Guardian Spirits (Eastern Blue Dragon, Southern Red Bird, Western White Tiger and Northern Black Snake) (3) face in the four cardinal directions. At the southern foot of the mountain are an hour-jack and three warriors (4) that announce the double-hours, night-watches and points with respective musical instruments similar to those devices of the Striking Palace Clepsydra. On the ground are the 12 Spirit Generals (figures) of the Zodiac (5), occupying their respective positions, and behind them are holes. At 5th double-hour, the Hour of the Dragon, the hole behind the Dragon opens, the immortal (6) with a time-tablet comes out, and the Dragon figure stands still. The process continues with other hours in succession. At the south of the mountain, an officially-dressed (Fig. 7) with a silver bottle pours water into an inclining vessel (8)[27]. Rural scenery sculpture (9)

[27] The vessel lies on its side when empty, stands upright when half full of water, and falls over again when full. This vessel was understood to serve as an instrument to remind of moral standard, i.e., "humility receives benefit, full brings upon loss".

cited from the *Ode of Bin* depicts peasants working around the mountain in the four seasons.

The uniqueness of the clepsydra was its capability to demonstrate celestial movements and to announce the 12 double-hours and five night-watches and their points automatically with various visual and audible signals. As its name implies, it was used in conjunction with edification rather than observation. It was reconstructed in 1554 and again in 1614 after the destruction by fire. The last version destroyed in 1656. It was maintained for 218 years, and the technical details might have been handed down and innovated upon to make subsequent armillary clocks from the seventeenth century onwards in Korea. Information related to making astronomical clocks made it possible to conjecture plausible technical features.[28]

2.6 *Traveling Clepsydra*

According to Kim Don's Observatory Records, it was small and simple in design and consisted of a water-delivering vessel, a water-receiving vessel and a siphon for water flow and changeover. The receiving vessel was changed between the Rat-double-hours (*Jasi*) and Horse-double-hours (*Osi*), or between Rabbit- (*Myosi*) and Rooster-double-hours (*Yusi*), i.e., they would run for only 6 double-hours at a time. It seems to be an inflow-type water-clock graduated 6 double-hours on the wall of the vessels or with a float and indicator-rod. A floating-siphon might be used for dripping the water to provide a constant flow, instead of an ordinary siphon as shown in Fig. 7.[29] Recently, Hua (2004) argued that the Traveling Clepsydra might come from Li Lan's steelyard clepsydra employed floating-siphons.[30]

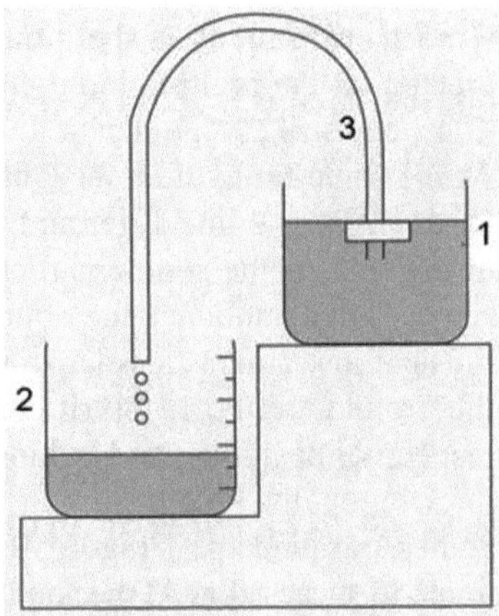

Fig. 7 Traveling Clepsydra (reconstruction drawing). *1* water-delivering vessel; *2* water-receiving vessel; *3* floating-siphon. (From Nam 2002, p. 72, Fig. 1–34)

[28] Hahn et al. (2000).

[29] E.g., Heron-type floating-siphon described in Mayr (1970).

[30] Hua (2004).

Guo Shou-jing made traveling clepsydra (*xing lou*) during the reign of Temur (r. 1294–1307) for the occasions when the emperor journeyed for the suburban and ancestral sacrifices.[31] The scholars working with Sejong might have been inspired from Guo's clepsydra and made similar ones for the royal journey for suburban royal tomb's and shrines. Several of those were sent to both military headquarters to support guard duties, and those kept at the Astro-calendric Office.

2.7 Scaphe Sundial

Kim Don described the sundial very briefly in his Observatory Records: "Ordinary people were ignorant about time-reckoning. Therefore two Scaphe Sundials were made, with the (12) double-hour Guardians of Zodiac—Rabbit, Dragon, Snake, Horse, Sheep, Monkey, Rooster, Dog, and others—drawn inside them, to enable the common people to tell the time. One was set up in front of the Benevolent Government Bridge and the other on the south street of the Royal Ancestral Shrine". Actually they were set up during the observatory project in the tenth lunar moon of the 16th year (1434) of Sejong's reign, and Kim Don wrote inscriptions on the sundial at that time. According to the inscriptions, it was made of bronze in the shape of a kettle. A circular gnomon was set up at the South Pole axis toward the North Pole, and graduated with a reticular scale—12 longitudinal hour-circle lines and 13 latitudinal solar-declination lines—inside the semi-circular scaphe.

Figure 8a shows a reconstruction of Sejong's sundial and the example of the relics from the seventeenth century onwards. It was assumed that this sundial used the equatorial coordinates, the celestial circumference 365.25^d, the 12 double-hours with 100-mark system in a full day and the latitude of Hanyang $38\ 1/6^d$ depending on the Great Concordance calendar (Datong-li). On the other hand, seventeenth century's revivals used the same coordinates, 360°, the 12 double-hours with 96-mark system and the latitude of Hanyang as 37° 39′ 15″, based on the Temporal Model calendar (*Shixian-li*) and latitude of Joseon capital. On the winter solstice, the shadow of the tip of the gnomon falls on the uppermost solar-declination line and on the lowermost line on the summer solstice. Following the movement of the sun, the shadow moves from the left to right, i.e., clockwise (CW), and it falls on the center-line at high-noon. One can tell the time of day by reading the hour-circle lines and the season in Fortnightly Periods by solar-declination lines. As shown in Fig. 8b Sejong made the sundial by transforming the armillary sphere and celestial globe by removing the upper part of the horizontal ring (lip), and held up the hemisphere by four dragon columns standing on the cross-water-grooves for orienting and leveling. On the lip was inscribed the locations corresponding to the 12 double-hours.

The Scaphe sundials were made of ceramic, marble, and ivory in various types and sizes until the end of the twentieth century and the royal court sent several

[31] Famous Servitor of the Yuan dynasty, Grand Astrologer Guo, p. 17b.

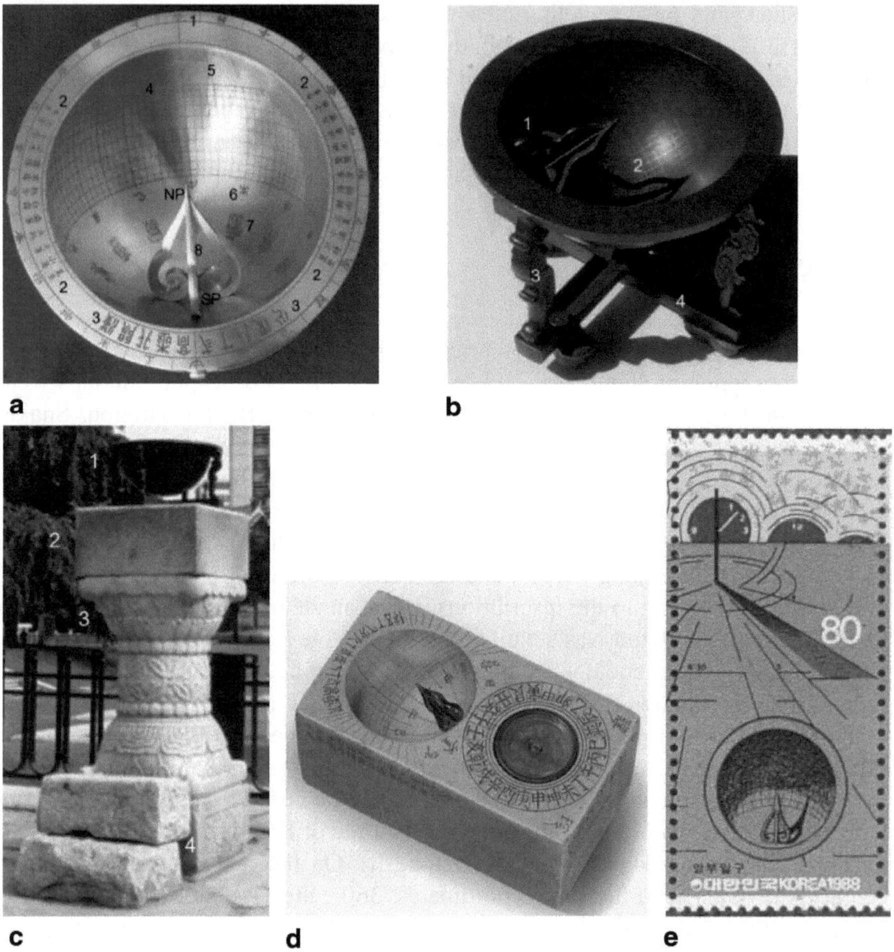

Fig. 8 **a** The Scaphe Sundial of the King Sejong following Kim Don's inscriptions on the *Angbu-ilgui* (a reconstruction): *1* the locations corresponding to the double-hours are inscribed on the lip; *2* names of the half of the Fortnightly Periods are inscribed from the winter solstice (*top*) to summer solstice (*bottom*) on the right side and from the summer solstice (*bottom*) to winter solstice (*top*) are on the other side; *3* latitude of Hanyang 38 1/4^{d}; *4* thirteen solar-declination lines; *5* twelve double-hour-circle lines; *6* names of the double-hours from Rabbit *myo* to Rooster *yu*; *7* double-hours guardian spirits inscribed; *8* burning flame-like gnomon set up at South Pole axis toward North Pole. (Reconstructed by Nam 1996 and donated to the Museum of the Konkuk University, Seoul). **b** Relics of the eighteenth century revivals of the Sejong's Scaphe Sundial *1*: gnomon; *2*: shadow of the gnomon; *3*: dragon columns; *4*: cross-water-groove for orienting and leveling on the stand. The diameter of the lip is about 35 cm and the radius of the hemisphere is about 30 cm. The characters and graduations are silver inlaid. (Collection of the National Palace Museum of Korea, Seoul). (Photo courtesy of Nam MH). **c** The Scaphe Sundial on the stand: *1*: Scaphe sundial; *2*: hexahedral stand; *3*: lotus and lotus leaves; *4*: sky-peach pattern. (Height: 136 cm) (Photo courtesy of Nam MH, *Deoksu-gung* palace, Seoul). **d** Portable Scaphe Sundial made of marble in 1908 by Kang Moon-su (After 1862). Collection of the Seoul Museum of History. The length 7.2 cm; height 3.1 cm; width 3.8 cm. (Photo courtesy of Nam MH). **e** A stamp with a Scaphe Sundial (*Angbu-ilgui*) was issued to commemorate '1988 Seoul Olympic Games by the Department of Post and Communication, Korea

marble-made portable ones to Qing's court as gifts similar to those shown in Fig. 8d[32]. Figure 8e shows a stamp with a Scaphe Sundial (Angbu-ilgui) that was issued to commemorate the 1988 Seoul Olympic Games by the Department of Post and Communication, Korea. Needham et al. (1986) said that it was innovative and a conspicuous tribute to the public-spirited beneficence of the ruler.[33]

2.8 Plummet Sundial

Kim Don wrote about the sundial in his Observatory Records and his words can be summarized as follows: A rectangular stand is 0.63 c (13.0 cm) long, a column set up in its northern part. There is a pool on the southern part, north of the column was engraved with a cross and suspend a weight by a plumb line from the top of the column, above it. In this instance, it was unnecessary to fill the water in the groove for leveling, because the sundial adjusted itself automatically. The (12 double-hours) with 100-marks were inscribed around a small disc with a diameter 0.32 c (7.0 cm); this had a protruding handle to fit into a dovetail slot positioned obliquely from the column. The hole in the disc had a thin thread through it that connected to the top of the column above, and to the southern part of the stand below. The thread casts a shadow on the disc, so the time was given by the locations of the shadow in the 12 double-hours with one hundred-marks graduations.

We can depict the general view of the sundial following the above-descriptions as follows: The column is placed near the north end of the base's centerline. The sundial was a portable; 13.0 cm long base and set up a 7.0 cm diameter disc, inscribed with 12 double-hours and 100 marks on both faces. The protruding handle of the disc was inserted into the dovetail slot on the base obliquely to fit to the equatorial plane of Hanyang 37.5° to the vertical. A thin thread was used to connect to the top of the column behind the slot and the small hole mimicking the South Pole in the southern edge of the base through a hole at the center of the disc at right angles, thus forming a pole-pointing thread-gnomon. In the base, there is a pool for a floating compass-needle for orientation. In this instance, a bronze sundial as shown in Fig. 9a was discovered at Buddhist temple (*Haein-sa*) where the *Tripitaka Koreana Goryeo Daejanggyeong-pan*[34] housed from the early fifteenth century. It was revealed that the sundial was the Sejong's.[35] Figure. 9b shows a reconstruction. It was a good case of the equatorial type instrument innovating from Guo Shoujing's Simplified Instrument and is acknowledged as the oldest thread-gnomon pole-pointing portable sundial known to exist.

[32] I ST (1986); Jeon (1998), pp. 72–73; Sivin (2009), p. 200.

[33] Needham et al. (1986).

[34] Goryeo Daejanggyeong-pan (80,000 Tripitaka Koreana woodblocks), acknowledged as the best and oldest intact version of Buddhist canon in Chinese script ever assembled. As such it has provided a very important contribution to the advancement of the Buddhist faith worldwide and was designated as a World Cultural Heritage by UNESCO in 1995.

[35] Nam (1995), pp. 108–113, figure-appendix 3–29.

a b

Fig. 9 Plummet Sundial. Collection of the Haein-sa Buddhist temple, Hapcheon, Gyeongnam Province, Korea. **a** Relics intact; a time-dial inscribed with 12 double-hours and 100-marks inserted into the dovetail slot, and a square hole *4* for the column (lost) on the north of the base. (Base 129.3 cm long, 8.3 cm wide, 2.5 cm high, water-groove for leveling and a pool for floating compass-needle for orientation). It seemed that Yejong's (r. 1469–69) Queen Ansun donated it to the temple for merit from the Buddha. (Photo courtesy of Nam MH). **b** The Plummet Sundial (a reconstruction). *1* time-dial (front face); *2* time-dial (back face); *3* thread; *4* column; *5* handle of the time-dial; *6* weight and the incised cross; *7* pool; *8* water-groove; *9* base; *NP*: Celestial North Pole; *SP*: Celestial South Pole. A threaded hole was pierced in the center of the time-dial and a thread stretched out to the north and south perpendicular to the dial. (Adapted from Nam 1995, p. 114)

2.9 Horizontal Sundial

This sundial was similar to the Plummet Sundial on its design in general, but it had a pool more to the north and a column set up to the center of the base. A thread ran from the south to the top of the column, holding up and pointing to the south. It was used to tell the time on horseback and for journeys. Figure 10 shows a conjectural reconstruction of the sundial, the two pools on the base; one in the south for floating compass-needle for orientation, and the other for folding the dial and the column when out of use.

Fig. 10 Horizontal Sundial (reconstruction drawing). (Perspective view) The dial and the column are folded into the pool below when out of use. (From Nam 1995, p. 115)

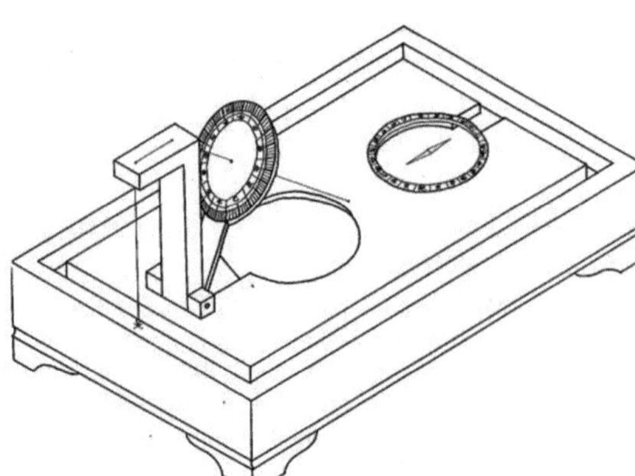

2.10 South-Determining Sundial

Kim Don wrote in his Observatory Records about the construction and the use of the South-determining Sundial. He said that traditionally one who wanted to examine the heavens to know the time used a compass certainly, but the South-determining Sundial could align the south-north axis automatically for determining south without a compass. The components of the Four Displacements ring [mobile ring] of an armillary sphere were transformed to make this instrument, passing the sighting-alidade that has a round hole on one side and a square aperture on the other side between the split rings of the mobile solar-declination ring. On the inside of the solid ring was carved a meridian line. The sighting-alidade reset according to the sun's north-polar distance of every day. And the rays of the sun formed a circle on the equatorial half-ring which was inscribed with day time 'marks' opposite the square aperture. Therefore, the sundial determines the south and tells the time naturally. This portable instrument might be used for fixing the south of various instruments equipped at the Royal Observatory. Figure 11 shows a reconstruction of the South-determining Sundial according to Kim Don's Observatory Records.

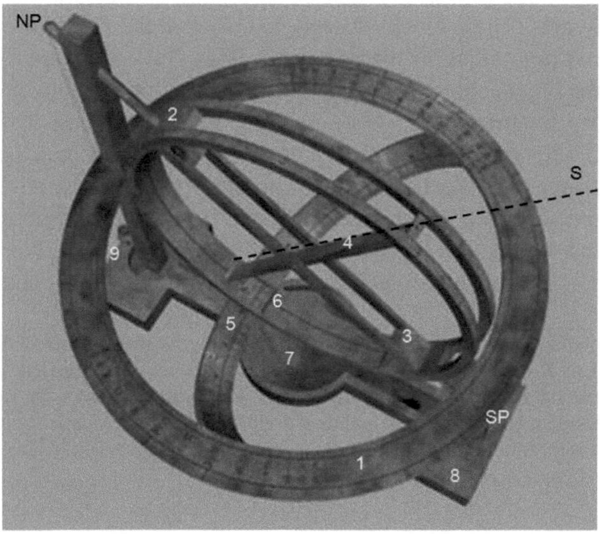

Fig. 11 South-determining Sundial (reconstruction drawing). *1*: Fixed Horizontal ring is graduated to indicate the 24 azimuth directions as well as the azimuths of sunrise and sunset during the 24 Fortnightly Periods. It was held up by the top of the 0.59 c (12.2 cm) high south column (hidden) and used for equalizing sunrise and sunset at the summer solstice; *2*: Four Displacements ring which received its pivots in the 0.11 c (2.2 cm) below the top of 1.1 c (22.7 cm) long north column and 0.038 c (0.7 cm) below the top of the south column. And it was graduated with the degrees of the celestial circumference, from North 16^{d} to 167^{d} there was a slot in the ring and engraved a meridian line on the inside of the rest; *3*: Cross-struts carrying a sighting-alidade; *4*: Alidade having a round hole one side and a square aperture the other side; *5*: Equatorial half-ring inscribed day time 'marks'; *6*: Square aperture for reading the 'marks' of the equatorial half-ring; *7*: Round pool 0.26 c (5.3 cm) in diameter having water-grooves reaching to the both ends and encircling the columns; *8*: Stand 1.25 c (25.8 cm) long, 0.4 c (8.2 cm) wide at both ends for a distance of 0.2 c (4.1 cm), and 0.1 c (2.0 cm) wide and 0.85 c (17.5 cm) long waist; *9*: Weight for leveling opposite to the cross below. *NP*: celestial North Pole; *SP*: South Pole; *S*: Solar rays. (Adapted from Nam 2002, p. 71)

3 Conclusion

The Neo-Confucian tenet of "Observing the Heaven and Granting the Seasons to the people" prompted the rulers to produce calendars for farming and the reckoning of the time of day. Objects of the astronomical project were initiated by King Sejong during 1432–1438 in Joseon dynasty, aiming to facilitate celestial observation and timekeeping equipment at the newly constructed palace observatory. Astronomical instruments and clocks of Chinese History Books and medieval Islamic science and technology inspired King Sejong's court scholars and engineers to construct the observatory and devise equipment. This equipment included two monumental timekeeping installations Striking Clepsydras, a compound of a sun- and star-dial mounting an equatorial-polar alignment and thread gnomons Sun-and-Star Time-determination Instrument and its smaller version, the celestial observation and/or time-keeping instrument Small Simplified Armilla, the sundials mounting an equatorial-polar alignment and/or all thread-gnomons Scaphe-, Plummet-, Horizontal-, and South-determining-sundials, a portable water-clock Traveling Clepsydra were created. In the course of the program, chief court engineer Jang Yeong-sil invented liquid-driven ball-falling and the ball-driven discrete motion-controlling mechanisms inspired by the medieval Islamic technology and employed these to make Striking Clepsydras. These clocks deserve to be called the first hydro-mechanically engineered dual-time clock in history of horology. Inventions and innovations in constructing the timekeeping installations and making of portable clocks at the palace observatory is attributed in large measure to King Sejong. His research and development (R&D) policies on equipping the observatory can be appraised as follows: state-supported R&D programs; portability and automation; practicality, adaptability and standardization. The Striking Clepsydras, the Sun-and-Star Time-determining Instrument and the Small Simplified Armilla have been evaluated as Korean originals in the world history of astronomical instrument and clocks.

Acknowledgements The author wishes to thank the King Sejong Memorial Society, National Palace Museum of Korea, Korea Astronomy & Space Science Institute (KASI), and Haein-sa (temple), Scientist Jang Yeong-sil Memorial Society, Research Center for the Korean History of Technology of Konkuk University. The author is grateful to Professors Timothy Atkinson at Seoul University of Foreign Studies and Dr. Yoonkey Nam at Korea Advanced Institute of Science and Technology (KAIST) for reviewing the manuscript.

References

Al-Jazari I (1206) *Kitáb fi ma'rifat al-hiyal al-handasiyya* (The Book of Knowledge of Ingenious Mechanical Devices), translated and annotated by Hill DR 1974. Reidel Publishing Company, Dordrecht

Annals of Sejong, (Sejong sillok), Annals of Jungjong (Jungjong sillok), Diaries of Gwanghaegun (Gwanghaegun ilgi), Annals of Yeongjo (Yeongjo sillok), National Institute of Korean History, reprinted 1986, Seoul

Baker D (1998) King Sejong The great: bringing heaven and earth into harmony. In: Kim-Renaud YK (ed) King Sejong the Great: the light of fifteenth century Korea. The International Circle of Korean Linguistics. George Washington University, Washington

Bo S (1997) An outline of Guo Shoujing's astronomical work. In: Nha IS, Stephenson FR (eds) Oriental astronomy from Guo Shoujing to King Sejong – 곽수경에서 세종대왕까지의 동양 천문학–: Proceedings of an international conference oriental astronomy (ICOA-1), Seoul, Korea, 6–11 October, 1993. Yonsei University Press, Seoul, pp 15–24

Chen TK (1955) Brief history of Chinese astronomy. Shanghai Ren-min Publishing Company, Shanghai

Famous Servitor of the Yuan dynasty 9, Biography of Grand Astrologer Guo. In: Chinese collections of science and technology classics series—Technology Volume. Reprinted Henan Education Publishing House, Zhengzhou, p 17b

Hahn, YH, Nam MH, Lee SW, Yi MK (2000) Astronomical clocks of Chosun dynasty—King Sejong's Heumgyonggak-nu. Technology and History 1:99–140

Hua T (2004) The Steelyard Clepsydra: its structure and principle of steady flow. Studies History of Natural Science 23(1):23

I ST (1986) Upward-facing bowl-shaped Scaphe sundial and upward-facing instrument. Studies History of Natural Sciences 5(1):41–48

Jeon SW (1998) A history of science in Korea. Jimoondang Pub. Co., Seoul

Kim SH, Min BH, Ahn YS, Yang HJ, Lee YS (2013) Apparatus for Construing the Heavens. Korea Astronomy and Space Science Institute (KASI), Daejeon

Lee SJ (1445) Collections of the Chinese history and writings on the astronomy, calendar, instrument and timekeeper. Collections of Seoul National University Kyujanggak Institute for Korean Studies, Seoul

Lee PH, Theodore de Bary W (eds) (1997) Sources of Korean tradition, vol 1. Columbia University, New York

Mayr O (1970) The origins of feedback control. M.I.T. Press, Cambridge

Nam MH (1995) Korean water-clocks: Jagyongnu, the Striking Clepsydra & the history of control & instrumentation engineering. Konkuk University Press, Seoul

Nam MH (2002) Jang Yeong-Sil and the Striking Clepsydra JAGYEONGNOO—revival of the time measuring history of the Chosun period. Seoul National University Press, Seoul

Nam MH (2008) Time in Korea. In: Selin H (ed) Encyclopedia of history of science, technology, and medicine in non-western cultures. Springer, Heidelberg. http://www.springerlink.com/content/tt722t7002h86885/fullt... Accessed 22 Sept. 2010

Nam MH (2012) Jang Yeong-sil: inventor of the Striking Clepsydra during the reign of King Sejong in Joseon. In: Koetsier T, Ceccarelli M (eds) Explorations in the history of machines and mechanisms. International symposium on the history of mechanism and machine science, VU University Amsterdam, Amsterdam, 7–11 May. Proc. HMM2012, pp 83–106

Nam M (2014) Yeong-sil Jang (Unknown). In: Ceccarelli M (ed) Distinguished figures in mechanism and machine science: their contributions and legacies, vol 3. Springer, Dordrecht

Nam MH, Nha, IS (2015) Highlights of King Sejong's Astronomical Project: Observatory *Ganuidae* and calendar *Chiljeong-san*. In: Pisano R (ed) A bridge between conceptual frameworks, sciences, society and technology studies. History of mechanism and machine science 27. Springer, Dordrecht, pp 321–343

Nam MH, Hahn YH, Seo MH, Lee JH, Seo J (2010) The recreation of King Sejong's Striking Clepsydra Jagyeongnoo at the National Palace Museum of Korea. Chinese Journal of History of Science and Technology 31:461–470

National Palace Museum of Korea (ed) (2007) Jagyeongnoo (self-striking water clock). In: Catalog of National Palace Museum of Korea. National Palace Museum of Korea, Seoul

Needham J, Lu GD, Combridge JH, Major JS (1986) The hall of Heavenly records (HHR): Korean astronomical instruments and clocks. 1380–1780. Cambridge University Press, Cambridge

Romdhane L, Zeghloul S. (2010) AL-JAZARI (1136–1206). In: Ceccarelli M (ed) Distinguished figures in mechanism and machine science: their contributions and legacies, Part 2. Springer Science + Business Media B.V., pp 1–20. doi:10.1007/978-90-481-2346-9_1

Rufus WC (1936) Korean astronomy. Transaction Korea Branch Royal Asiatic Society XXVI:1–48
Sigimondi C (2012) Sylvester II—King Sejong the Great Parallel. Proceeding of 1st international conference in Korea Science and Culture Agency, 15–17 August, 2012, Yecheon, Korea
Sivin N (2009) Granting the seasons–The Chinese astronomical reform of 1280, with a study of its many dimensions and an annotated translation of its records. Springer, New York
Song Lian (1976) Standard History of the Yuan. China Publishing Company, Beijing
Wang D (2011) The measurement proof of Su song's astronomical clock tower. Studies in the History of Natural Sciences 30(3):297–305
Xu J (1123) Illustrated Account of Goryeo (Goryeo-dogyeong). Chap. 22: Water-clock, reprinted by Ewha Historical Institute, Ewha Woman's University, 1970, Seoul

Lazare Carnot and the Birth of Machines Science

Agamenon Rodrigues Eufrásio Oliveira

Abstract In this paper, we are mainly concerned with the development of the first theory applied to machines made by Lazare Carnot (1753–1823). The two Carnot's essays on machines previously published before *Fundamental Principles* have in common the use of the concept of work as a fundamental step to build that theory of mechanics within the framework of Rational Mechanics. Carnot accomplished this task by starting to develop his new theory applying d'Alembert's (1717–1783) principle and crowning his project with *Principes Fondamentaux de l'Equilibre et du Mouvement* published in 1803.

Keywords History of work · History of energy · History of applied mechanics

1 Introduction

The first author to reference Carnot's work appears in the first decades of the nineteenth century. He was André Guenyveau (1782–1861), who in 1810 published the *Essay on Science of Machines*, different from Carnot's works. He studied the equilibrium of machines and presented a series of practical applications, using Coulomb's memoir in which machine motion can appear by means of shock or pressure. This distinction also indicates that continuous transmissions provide the best efficiency. But Guenyveau does not attributes to Lazare Carnot the origin of the principle of living forces used in spite of the fact that, in the preface of his *Essay,* Carnot is referred to as an author of a general treatise on machines.

Another author who cites Carnot is Jean-Nicolas-Pierre Hachette (1769–1834). He refers to Carnot's (1803) *Fundamental Principles of Equilibrium and Motion* appearing in the preface of his *Elementary Treatise of Machines*, published in 1811. Hachette remarks that Carnot in the last chapter of *Fundamental Principles* studies the whole theory of machines and the moving applied forces. In addition he mentions Carnot's work as *the most profound savant and experienced engineer.* Paradoxically Hachette uses few of Carnot's achievements. In fact the course taught

A. R. E. Oliveira (✉)
Polytechnic School, Federal University of Rio de Janeiro, Rio de Janeiro, Brazil
e-mail: agamenon.oliveira@globo.com

R. Pisano (ed.), *A Bridge between Conceptual Frameworks,* History of Mechanism and Machine Science 27, DOI 10.1007/978-94-017-9645-3_19

by him is much more a collection of drawings of particular machines, studying also machine elements such as gears, pulleys, etc.

Alexis Petit (1791–1820) developed an investigation in the same sense of others that applies the work concept. In 1818 he published *About the Employment of the Living Forces Principle in the Calculation of the Effect of Machines* which is a very interesting work and coincides with our line of investigation. Petit was a physician who together with Pierre Dulong (1785–1838) postulated a law with his name and died prematurely. He presented general properties for motion as the conservation of living forces sponsored by him as the most efficient approach to machine calculation. According to him it is the living force that permits in each particular case the best natural evaluation of a motor and the effect produced. The equation expressing the relation between these two quantities can provide the direct solution to a machine problem. In fact what Petit was proposing is a method of energy balance to solve this problem.

With respect to further developments of applied mechanics, Coriolis' (1792–1843) textbook is the result of writings accumulated since 1819 (Coriolis 1829). His main objective was to develop the Lazare Carnot project of creating a general theory of machines but in general as a product of the tradition of Newtonian mechanics (Coriolis 1832). What is new in Coriolis' approach is the extensive use of the concept of work with its mathematical formalism and applying in different situations to machines the Newtonian concepts. The new approach adopted by him can be better understood if we look at the two general approaches used to solve a given mechanical problem. One is to consider the variations of motion as the result of forces, in a Newtonian way. Alternatively, knowing the amount of work generated by the system, try to find the "living forces" involved (Poncelet 1827). In other words, we can use Newton's second law or work-kinetic energy principle. In 1844, 1 year after Coriolis' death, his master work was published with the title: *Treatise on mechanics of solid bodies.*

2 Lazare Carnot's Biographical Note

Lazare-Nicolas-Marguerite Carnot was born on May 13th 1753 in a bourgeois home in Bourgogne in France, who family occupied a remarkable position in local society (Charnay 1990). Claude Carnot, his father, was a lawyer and public notary in Nolay, a small city close to Côte-d'Or. The place where Lazare Carnot was born has not changed significantly as of now and still belongs to Carnot's family (Fig. 1).

Carnot's father took upon himself the education of his three older children and they, on the other hand, helped the younger brothers as tutors or preceptors. Among the six sons, two demonstrated early ability in mathematics and technical questions: Lazare, the second son and Claude-Marie (1755–1876), because their characteristics were compatible with a military engineering career.

During the eighteenth century it was usual in order to assure the son's education, mainly when the families were richer, to ask for the help of a priest or a preceptor.

Fig. 1 Lazare Carnot (1753–1823)

There were only two ways, the classic or humanistic studies and the military schools. At the Universities one could prepare for a career in law, medicine or theology.

Carnot completed his humanistic course in the Autumn School, where after the Jesuits had left the school, the Oratories replaced them. Ten years later Napoleon Bonaparte (1769–1821) studied there. Lazare Carnot finished these studies at the age of 16.

With the purpose to save his father's money, Carnot decided to study alone in order to submit to the examinations of Mézières School of Engineering. Because of his unsuccessful attempt to be admitted to that School at the end of 1769, his father went with him to Paris. In addition, Carnot obtained the patronage of the Duke of Aumont and thus he gained entry into one of the most famous Paris institutions which was directed by Louis-Siméon de Longpré, located in Marais. Longpré supported himself thus early by giving instruction in mathematics, and soon after becoming an author. Being well versed in mathematics, he conceived the plan of erecting a school for the preparation of young men destined for service in the army.

From the beginning of his classes' period, Carnot had the opportunity to meet Charles Bossut (1730–1814) and d'Alembert, two figures of great influence in Carnot's approach to mechanical problems. Bossut, who directed Mézières's examinations, published in 1763 the first edition of a very popular mechanics handbook entitled: *Elementary Mechanical Treatise and Applied Dynamics mainly to Machines Motions*. He also published other algebraic, geometrical and arithmetic works. In spite of his main interest in mathematics, he was particularly concerned with mathematical education. He wrote *A General History of Mathematics from the Earliest Times to the Middle of the Eighteenth Century* which was published in 1802.

At that time d'Alembert was an old man and maintained a strong friendship with Longpré. The great mathematician frequently met the students around him to suggest mathematical problems. Later on when Carnot became also an old man he enjoyed telling how d'Alembert had influenced his approach to solving difficult questions and how d'Alembert predicted a successful career for him.

After passing his examinations, Carnot entered the Mézières's School of Engineering, on January of 1771, to follow 2 years of basic formation which is the fundamental step to the future regular formation of engineers. At that time, the Engineering School required from candidates proof that they belong to a noble class or, to a family living in a noble way. In other words what was required of families was a non-degrading life. It is important to say that the majority of the students were of bourgeois origin and Carnot had no difficulty in obtaining the certificate.

Unlike Carnot, Gaspar Monge (1746–1818), one of the greatest talents among those students, had suffered cumbersome situations because of these requirements. Monge's family had modest resources and he had to overcome this situation by means of his intellectual qualities. He managed to overcome all these difficulties and became a lecturer of mathematics and physics (Belhoste 1989).

With respect to Monge's influence on Carnot, there are divergences among the biographists. One did not refer to the other and there is no indication that Monge had been interested in Carnot's work. Indeed there were profound differences of personality between the two men. Monge was a mathematician and a pedagog without any tendency to politics. Always when he was confronted with political issues, he demonstrated a theoretical spirit, became very emotional and incapable of making decisions, exhibited no confidence in his own judgments and paid no attention to details. On the other hand, Carnot behaved inversely, being an engineer with clear and objective ideas, with ability to grasp a conception and execute his assigned tasks with uniformity of character and preferring to judge with practical spirit and adapt to circumstances.

Carnot left Mézières on first January 1773, having developed military aspirations, and served in several military posts, initially in Calais, then in Havre, Béthume, Arras and Aire. In Calais he worked in its port fortification. Later on, he moved to Havre and worked there for 3 years in order to participate in the building of Cherbourg's port, an advanced technical project being considered one of the most modern undertakings in military French engineering (Chatzis 1999). With this and other participations, Carnot was considered among his peers a well-known and reputed engineer.

In order to understand Carnot's scientific work perhaps we should remember Coulomb's quotation made in 1776, about the situation of everyone that left engineering school. He refers to them as: *young men who studied and after left the School had no other thing to do, to face tedious and monotony into the corporative situation that is the dedication to any science branch or to literature absolutely strange to his study.*

Obviously it happened to Carnot after he finished the course in Mézières's School. He spent part of his time pursuing mechanical and mathematical studies, achieving a higher level of knowledge when compared with other students from

Mézières. These courses had provided the following: the study of four books of Charles-Etienne-Louis Camus (1699–1768) on arithmetic, geometry and statics; in addition Bossut's treatises on dynamics and hydrodynamics; as complements, annual courses on industrial design, perspective or descriptive geometry, as well as experimental physics. Camus was an important French mathematician and mechanician who studied also civil and military architecture and astronomy. He became secretary of the Academy of Architecture and fellow of the Royal Society of London.

Probably when Carnot left Mézières's School, facing the solitude of the military post life, he began to read again d'Alembert, Bossut, Bélidor, Euler's mechanics and Daniel Bernoulli's hydrodynamics. This is an easy conclusion when we read his early works, where he demonstrated a good knowledge in the above subjects. Another argument to corroborate this supposition is his distance from Paris's scientific circles in a different way followed by Charles Augustin Coulomb (1736–1806) and Charles Meusnier de la Place (1754–1793). Sometimes Carnot spent his vacations in Nolay or Dijon. Documents from the Paris Sciences Academy and the German Sciences Academy also confirm this supposition. Other manuscripts belonging to Carnot's family delineate in some way the circumstances and the context in which he wrote his early works on mechanics and mathematics.

Carnot's first studies were addressed to his participation in conferences sponsored by scientific societies. In 1777, the Paris Academy proposed a conference in the year of 1779, but it was only accomplished in 1791, because the referees were not satisfied with the works submitted for the first time. Consequently Carnot wrote two of his first studies on machines theory. Both entitled: *Essay about the machines*; he concluded the first in Cherbourg on March 1778 and the second on July 1780 in Béthume. At that time a memoir by Coulomb was the winner and Carnot received an honorable mention.

With respect to his interest in mathematical foundations the motivation is quite similar. His well-known text *Considerations about the metaphysics of infinitesimal calculus was* a memoir submitted to the Prussian Royal Academy in 1785. Yet, in keeping with the interests of the planned conference, the Paris Academy asked for anyone interested to write about the first human flight. Carnot had addressed the topic in a letter of January 17th, 1784 entitled *Letter about the aerostats*. We should remember that on June 5th, 1783 the Mongolfier brothers had sent into space a hot-air balloon that achieved 1800 m of height. This event led to the solution of several technical problems about motion and stability of flight.

Carnot had first studied the air-ship problem and proposed building a very unusual and fantastic propeller which displaced itself like a medusa working according to systole and diastole motions created inside the balloon by heat dissipation.

In the above description appear the most remarkable facts related with the beginning of Carnot's scientific career. In the next lines we will see his participation in politics mainly at the beginning of the French Revolution.

If we wish to truly characterize Carnot and portray his real character, we must begin by identifying him as an authentic republican. He was involved with important actions of the French revolutionary period, for instance voting for the death of

Louis XVI, unmistakably confirming his republican convictions. On the other hand, his military position contributed to the view of the movements by the left as creating the main menace to the recently established Republic. Hence, he fought against the revolutionary movement called *Conspiracy of Equals,* commanded by Gracchus Babeuf (1760–1797).

Babeuf's name is a tribute to the Roman tribunes of the people and reformers, the Gracchi brothers. He was a French revolutionary political and journalist of the Revolutionary period. In spite of the efforts of his Jacobin friends to save his life, Babeuf was arrested, tried and convicted for his participation in the *Conspiracy of Equals.* In that time neither anarchist nor communist were existing words. *Conspiracy of Equals* was an armed rising fixed for Floréal 22, year IV (11 May 1796). One day before Babeuf was arrested, many of his associates were gathered by the police on order from Lazare Carnot. Babeuf was guillotined at Vêndome Prairal 8 (27 May 1797) without appeal. This episode is important from the viewpoint of revolutionary movements in general, because some historians see the French Revolution as the beginning of a proletarian movement within a bourgeois revolution.

Carnot also participated in regrouping of the French army after the disorganization caused by the Thermidorian movement that maintained control of their military administration. However the French Revolution delivered plenty of surprises, with victories and defeats of the republican forces and movements for reestablishing the monarchy.

In spring of 1797, legislative elections showed the possibility for resurrection of the monarchy. The left could not accept the defeat of the Revolution and thus tried to reject the elections' verdict and attempted to eliminate the reactionary members from assemblies. In the *coup d'état* of 18 Frutidor (September, 4th) Carnot was convinced that it was not possible to maintain the Constitution, and as a result did not participate in this movement. Upon receiving a communication about the directory's intention, he left Luxembourg and obtained refuge in Paris during some weeks before moving to Switzerland.

During this type of political unrest, Carnot frequently spent part of his time in researching mechanics and mathematics, according to his son Hippolyte. In a similar situation in 1797, during his participation as a member of the directory, his book *Considerations about the metaphysics of infinitesimal calculus* appeared.

With the taking of power by Napoleon Bonaparte in 1799 (18th Brumaire), Carnot came back to France with the benefit of a general amnesty in favor of Termidor's victims. He realizes that, although this regime has created difficulties, it is not sufficient that he be hostile to the Republic but must also pay attention to his own affairs. During that time a number of his works were published. In 1800, at the end of the year, he wrote his *Letter to citizen Bossut containing new visions about trigonometry*; in 1801, he published *From the correlation of the geometry's figures;* in 1803 his *Essay on machines is published* with the new title *Fundamental principles of equilibrium and motion* which occupies, for the purpose of this paper, a central position. In addition, in this year he also published *Geometry of position,* considered by him his masterpiece; in 1806 there appears his *Memoir about the relation that exists among the respective distances of any five points took in space*, followed by *An essay about the theory of transversals.*

The two sons of Lazare Carnot, Nicolas Léonard Sadi and Lazare Hippolyte were born in 1796 and 1801, respectively. Sadi was the founder of thermodynamics and we come back to him in order to analyze the influences received by him from his father (Gillispie and Pisano 2014).

Another scientific activity developed by Carnot must be emphasized but it is necessary to look back to the time of the French Revolution (Rashed 1988). The beginning of terror, August 8th 1793, the Sciences Academy as well as other French institutions representing the *ancient régime* has been abolished or closed, because they were considered as founded on privileges and not adequate to the spirit of the Republic. Only on August 7th did Carnot became a member of the Committee of Public Safety, thus we cannot attribute to him the decision to close the Sciences Academy. As we know, with the end of the Academy was created the French Institute with Carnot appointed as a member and with the purpose to give continuity to the old Academy's works. The new institution would undertake to unify the arts, letters and science, as branches belonging to the same culture, which was considered the best symbol of the Republic.

In 1805, Carnot became president of the French Institute. His main activity was to work for the progress of science, to publish its memoirs and to recommend editions of other studies submitted to him and considered of sufficient importance to be divulgated. In addition he sometimes worked as a public consultant, analyzing and judging technological problems, studying the conception and design of new machines, and the viability of industrial processes. At that time it was common for inventors and machine constructors to have their designs submitted to the government in order to obtain financial resources.

Yet, frequently Carnot participated in technical commissions within the attributions of the Institute in order to evaluate projects of practical characteristics. One example of this case was the Niepce motor. In an enthusiastic report he emphasized that this machine was the first way, never before imagined, to obtain primary energy by means of heated air expansion, or in other words, air combined with caloric. Conversely to the steam machine which required a big consumption of caloric to heat and vaporize water before using steam expansibility, an air motor presented a great advantage in not using combustibles except to produce expansion where energy arises.

In 1809, Carnot wrote, and addressed to the Institute, a new report about a new model of a thermal motor which was conceived by an inventor named Cagniard de Latour. This motor had attracted the attention of Sadi Carnot according to Thomas Kuhn. Like the Niepce machine, the fluid used is the air, but its utilization was more subtle. Some characteristics of this motor are described by Sadi Carnot in his *Considerations*. In addition Kuhn refers to another important aspect of Cagniard's motor that is its reversibility. This concept appears in Sadi Carnot's works and we come back to it later on.

The last years of Carnot can be briefly outlined. When Napoleon came back from Elba Island, Carnot allied to him, being his interior minister in the well-known government of 100 days. However because Carnot had voted for Louis XVI death, confirming again his republican convictions, he never was forgiven by the restored monarchy. Carnot went into exile again with his son Hippolyte. After passing by

Brussels, Munich, Wien and Cracow they arrived at Warsaw where he tried to stay and live. It was not possible to stay there and, prohibited from living in the Rhineland that is close to France, finally they went to Magdeburg where he died on August 2nd, 1823.

3 Application of Living Forces Conservation Principle to Machines

According to Claude Henri Navier (1785–1836), the first study where we find the principle of living forces conservation applied to machines is *Hydrodynamics* of Daniel Bernoulli (1700–1782), published in 1738. It is also the first time where the relation between hydrodynamics and that principle is established. Daniel Bernoulli shows that, if we abstract frictions, losses and we consider an incompressible fluid, an increase in the speed of the fluid occurs simultaneously with a decrease in pressure or a decrease in the fluid's potential energy, relating speed with height and pressure in a similar way that was done in the elevation of a weight to determine the capacity of the machine. Bernoulli performed his experiments on liquids, so his equation is valid for incompressible flow. Thus, in an arbitrary point along a streamline the sum of living forces, weight multiplying height and pressure divided by fluid density is constant.

Unfortunately, this important achievement made by Bernoulli, which expresses actually an energy balance, was completely forgotten in the most important works on mechanics that looked for an application, mainly that addressed to engineers. Among others we can cite *Physics* of Jean Théophile Désagulliers (1683–1744) and *Hydraulic Architecture* of Bernard Forest de Bélidor (1698–1761). Even the physicians and mathematicians that worked on more theoretical mechanics also did not realize the importance of Bernoulli's principle. Euler, for instance, didn't use these Bernoulli's ideas, when he studied the reaction wheel, the centrifugal wheel and Archimedes' screw. Only with the memoir of Jean-Charles Claude Borda (1733–1799) entitled *Memoir on the hydraulic wheels*, published by the Sciences Academy in 1767, did the principle of living forces conservation start to be known and to be applied to machines. Borda was the first to apply this principle to hydraulic wheels.

Some years later, in 1781, Coulomb published a memoir on windmills and used the same principle to study this machine. In addition he included the losses caused by shocks. Coulomb's text was entitled *Theoretical and experimental considerations on the effect of windmills*, also published by Sciences Academy (Coulomb 2002).

To Navier, the contributions of Borda and Coulomb are fundamental steps and remarkable progress when we compare these developments to Bernoulli's *Hydrodynamics* in spite of their restricted particular applications of the living forces conservation principle. Navier also emphasizes that after these studies mentioned above there is a need for the creation of a general theory involving that principle with the capacity to calculate machines efficiency. It is exactly in this context that he affirms that this theory was created by Lazare Carnot in his *Fundamental Principles of*

Equilibrium and Motion. Navier attributes to Carnot the general demonstration of the theorem which calculates the loss of living force due to shocks between non-elastic bodies (hard bodies). This fact was not observed, neither by Borda nor by Coulomb, unless in very particular cases.

After these achievements, which made an important contribution to the development and application of living forces conservation principle, came the work of Navier. In a note published by the *Proceedings of Chemical and Physics*, he communicates his intention to submit to the Academy some notes and additions prepared for a new edition of Bélidor's *Hydraulic Architecture*. It is a kind of update of this important book. In his remarks Navier demonstrates the principle of living forces conservation to a single mass but he also generalizes this result to a system of n particles through the d'Alembert principle. This study is made by means of virtual velocities and Navier derives Carnot's later theorem from the conservation of living forces. According to him this theorem fulfills the principle of living forces conservation for the case of a sudden change as follows: *The addition of living forces which arise in a system after a sudden change, is lesser than that which occurs previously, and the system has lost an amount of living force equals to that which would happen if the bodies were animated of velocities that are lost since this change*. As usual, Carnot uses the decomposition of velocities proposed by d'Alembert's principle before applying the living forces balance.

However, a great evolution in the living forces principle occurs with the publication of Coriolis' book—*Du Calcul de l'Effet des Machines*, published in 1829. This work is considered one of the most important in mechanical engineering to appear in the nineteenth century. In addition this book is a fundamental step in the history of work concept, because the term *work* was coined in this book, becoming adopted by the polytechnicians engineers as well as by the future technical literature. Thus, the word *work* was progressively used and replaced the previous denominations as mechanical power, quantity of action or mechanical effect, etc.

4 Carnot's Theory of Machines

4.1 Carnot's Essays on Machines

Before the appearance of his *Fundamental Principles* in 1803, Carnot wrote two introductory studies known as *Essay on Machines* (Gillispie and Youschkevitch 1979). The fundamental assumption to the development of an applied mechanics theory is that these two E*ssays* were preliminary studies to his masterpiece. As we are focusing our attention on a general theory of machines instead of a general analysis on Carnot's mechanics, only the aspects directly associated to this theory are considered. Obviously, the passages from the E*ssays* to *Fundamental Principles* will be studied from an evolutive viewpoint.

The problem to be solved by Carnot, which is the motion of any machine, is quite similar to the dynamics of a system of particles discussed by d'Alembert in his famous principle. In fact it is a system of **n** bodies or particles, with the possibility of most of them being constrained or connected among them, and the system as a whole constrained in any manner, and to this system we apply any initial condition to one or more particles. The solution is to find the subsequent position for any particle of the system. If we call **m** the mass of any corpuscle or particle of the system, **V** the velocity that it would have if free, i.e. without reactions from the other parts of the system, **u** the velocity that it actually has and **y** the angle between **V** and **u**, we will have: $\mathbf{S\ mu\ (V\cos y - u) = 0}$, that is, the summation of the products of quantities of motion for each corpuscle by the lost velocity, calculated in the direction of **u**, is zero.

This can be proved by decomposing **V** into two components, **u** and the other which is the lost velocity in shocks, then the velocity estimated in **u** direction is $\mathbf{V\cos y - u}$. This is the proposition which initiates the first memoir part related to machines and enunciated in paragraph 27. In paragraph 29, Carnot postulates what he calls a fundamental theorem, in the following terms:

> If any system of hard bodies acting among them of any manner being immediately or by means of a machine, calling **m** any molecule of the system, **V** the velocity that it would have free from interactions in a given instant, **u** the velocity which it actually acquired due to reciprocal action of different parts of the system, **y** is the angle between the directions of velocities **V** and **u**; I affirm that we have: $\mathbf{S\ mu\ (V}\cos\mathbf{y)=0}$.

Regarding that the decomposition **V** cos **y** is **u** itself we have in fact $\mathbf{S\ mu^2=0}$, which means the conservation of total kinetic energy of the system of particles. What we have to take into account is that Carnot doesn't make any difference between integral and summation symbols, as well as discrete and continuous systems. Other remark is that the above equation actually represents an energy balance, where the system communicates motion to other parts of it such that the process is conservative.

Following Carnot's first memoir, he made many particular applications, including cases where the system is in equilibrium situation, but always starting from motions situations. In item 48, he discusses the quantities which vary when the machine is in motion such that we can take advantage if we know this form of variation. Carnot affirms: *The advantage of machines is that we can vary the factors **F, u, t**, but the product must be ever **M g H**, since that it is a weight or a similar quantity that is involved, even being other kind of resistance.*

What Carnot explains many times by different ways is the conservation of energy within the field of mechanics. Thus, it is possible to vary in an appropriate form the three quantities force, velocity and time, but the work or energy used in any process never overcomes the constant quantity **M g H**.

The second part of this memoir deals with the equilibrium of simple machines and we will go to the second memoir also called *Essay*, where a general theory of machines is going on.

Theorem 2: General Principle of Machines in Motion If we impress suddenly motion to a machine with another motion being any geometric one and if we release the machine to itself, the conservation of living forces will happen immediately and at each elapsed instant of motion for any alteration of the driven forces.

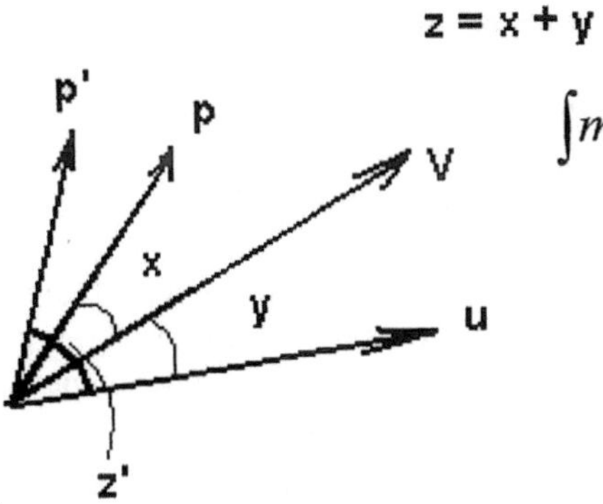

$$\int mu\, du = \int mu\, dt\, p' \cos z'$$

Fig. 2 Equivalence between living forces and work

The following nomenclature will be used (Fig. 2):

M = mass of each corpuscle; **V** = actual velocity; **P** = driven force
U = velocity after the variation of actual motion in another geometric one
X = angle between **V** and **p; Y** = angle between **V** and **u; Z** = angle between **u** and **P**
p′ = driven force after the arbitrary variation; **z′** = angle between **p′** and **u**

The virtual velocity of **m** estimated in **u** direction is **V cos y+d (V cos y)** and the lost velocity by **m** during **dt** estimated in **u** direction is **p dt cos z−dt (V cos y).**

With respect to Theorem 2, it is a kind of application of the living forces conservation principle having in the equation shown in the figure on one side living force and on the other work. Geometric motion also cited in the theorem is related to the invariability of system constraints and is not referred to application of the virtual works principle to a dynamical system.

4.2 Fundamental Principles of Equilibrium and Motion

Starting from page 227, paragraph 252, Carnot treats directly the questions related to machines. It is the final part of his *Fundamental Principles*, from where we obtain the same subtitle, in which he studies deeply the problems arisen from the machines operation. The sequence followed by him starts with machines in equilibrium, its motion analysis, and the forms to increase machines efficiency. Obviously these are problems emerged from the needs of industry and engineering at that moment and therefore the polytechnician engineers of the first decades of the nineteenth century since the publication of *Fundamental Principles* will develop new concepts and a new theory of machines. It is important to point out the role of the concept of work to these developments. As we will see this concept occupies a central position in the majority of the investigations of this period.

Carnot starts by defining a machine as a body or a system of bodies which are in-between two or more powers (forces), with the condition to fulfill a given objective. He emphasizes that, in general, these bodies are considered without mass because of its small effect on the applied system of forces, independently if these forces are driven or inertia forces. He states that this abstraction simplifies the problem.

At this beginning Carnot also discusses questions of modeling a machine by the use of a convenient representation with a great quantity of corpuscles separated by strings and bars, through which motion is transmitted from one element to its neighbors and so on.

Carnot is aware of the general character of his investigation and said that his intention is not to search for particular properties of each machine, what he has remarked before, but to offer some considerations about the machines, and the common properties to every machine.

For the case of motion, not only the weight must be considered, but the height to be elevated, making these two operations with the machine, quite different. For the equilibrium, the machine could centuplicate the force effect while for the machine in motion we have an invariable quantity which is always the product of a force by the path described, estimated in the force sense. In other words, Carnot discusses exactly the question of energy in the field of mechanics, which is expressed by mechanical work. To eliminate any doubt he uses an example of a horse which is two times stronger than another, meaning that it can elevate a new quantity of water, for instance, a double height of the second at the same time, or a double quantity of water for the same height, also to the same time. It is important to say that the definition of mechanical power appears here in this context, in a very simple form.

Another important question related to economic concerns is the capacity of a machine to elevate a given weight to a certain height. This is the method used to calculate the machine capacity to provide work as well as the labour of workers and its estimation in order to pay salaries. What Carnot supposes is that the concept of work from physics is also useful to calculate salary's value as referred in the *Introduction*. Although Carnot's concerns with economic questions which also are related to mechanics is a remarkable fact.

In the context of the above discussion Carnot enunciates the famous principle: *In any machine in motion, one loses always in time or velocity what one wins in force.* Carnot then analyzes the true meaning of this principle, discussing the effect produced by a machine in motion. If we call **P** the weight to be elevated to a given height **H**, this effect will be represented by **PH**. Assuming that the force used to produce that effect is **F**, **V** the velocity estimated in the force sense, **T** the time during which the operation occurs and supposing that for the sake of simplicity the motion is uniform, one has **FVT** = **PH**. As mentioned before, Carnot calls the work "moment of activity", which is equal to **FVT**.

Carnot comes to the discussion of machine's utilization and states: the advantage which machines present is not to produce a great effect with small means, but to provide the choosing among different means which we can consider equivalent, to the most convenient circumstances. In other words, a machine is not a mere tool for the forces multiplication, but mainly, an apparatus which has the availability of some amount of work that can be used in a great variety of forms. Carnot completes this discussion saying that it is always necessary that the moment of activity spent by the driven forces be equal to the effect of the motion absorbed at the same time by the resistant forces. These considerations seems to be sufficient to finish the illusion that machines having assemblies of levers mysteriously linked can transform a weak agent capable to produce huge effects, by transposing the reasoning adopted to the equilibrium situation to the motion condition. Indeed, for a machine in motion it is always limited and thus, the machine cannot overcome the spent moment of activity produced by the correspondent agent. The difference is for the equilibrium condition, that motion is obstructed; for the motion situation the objective is to facilitate its appearing and, therefore to maintain it. This condition means one more consideration that is to know the actual velocity of each point of the system.

Carnot describes in details, for the equilibrium and motion situations the internal processes of machines related to dissipation of motion by obstacles. For the dynamical case, the fixed points and obstacles, of any type, are forces of passive nature, which absorb motion, of any intensity but never could provide its birth, even for a small one, when the body is in equilibrium. Although a small power cannot annihilate a great power, but it means that only the resistance imposed by fixed points. In other words, a small power is capable only to annulate a small part of a great one and the obstacles do the rest.

Carnot studies then the problem of the transformation of work in motion by considering all the parameters involved. From this viewpoint it means to establish convenient variations of the terms of the quantity **FVT**, i.e., the moment of activity, later on denominated work by Coriolis. Thus, if time is the most important parameter and we should minimize it, the effect must be produced in a very short time. It is possible to generalize these reasoning for the case of a system of forces, for instance; if we have the forces **F**, **F**′, **F**″ with the velocities **V**, **V**′, **V**″, acting during the times **T**, **T**′, **T**″, respectively, then one reads:

$$\mathbf{FVT = F'V'T' = F''V''T'' = PH}$$

If the motion of each one of the forces is variable, we will take the quantity: $\int \mathbf{(FVdt + F'V'dt' + F''V''dt'')}$, or if we have the forces directions with respect to velocities, one has:

$$\int \mathbf{[FVdtcos(F \wedge V) + F'V'dt'cos(F' \wedge V') +}$$
$$\mathbf{F''V''dt''cos(F'' \wedge V'')]}$$

This is the definition of work done by all forces.

The quantity **PH**, the effect to be produced by machine, is, by Carnot called latent living force. If we call **M** the mass of the weight **P**, and **V** the velocity correspondent to a height **H**, one reads:

$$\mathbf{PH} = \tfrac{1}{2}\mathbf{M}^2.$$

Following Carnot's concerns the problem of a machines' efficiency comes back. This is a question of central importance not only to mechanics of that time, but to development of applied mechanics, and, later on to industrial mechanics. Carnot remembers that previous considerations about machines in motion, all of them were made without taking into account shocks and sudden variations of velocities. The variations considered are always by means of insensible changes otherwise we have a great loss of living forces. He states: *In order to obtain from machines the best possible effect, it is important that it has been built such that the motion does not vary unless by insensible degrees. The exceptions are that with possibility to support different percussions, as the majority of mills. But even in this case, obviously we should avoid sudden variation unless that essential to the machine structure and operation.*

Carnot concluded the discussion above of how to obtain the best possible effect in a hydraulic machine, moved by water flow, denying that the solution is to adapt a water wheel to a machine, because the shocks against water appear necessarily. Then, two reasons can impede that the maximum effect be obtained: the fluid percussion itself and when this shock occurs, there is always a residual velocity associated to the pure loss, and in some cases could be employed to produce yet a new effect added to the first. To design an improved hydraulic machine, with capacity to produce the maximum possible effect, we should look to fulfill the following conditions:

1. For design considerations to oblige that fluid could lose all motion by its action over the machine, or at least that the remainder motion be precisely that necessary to escape.
2. That fluid loses totally the motion by insensible degrees without the occurrence of percussion from the fluid as well as from some machine component. This is independent of the type of machine.

The machine that better fulfills the conditions above will produce always the greater possible effect. Because it is so difficult to achieve the above objectives, mainly the water wheels, where shocks and percussions appear, it is possible at least to fulfill the second condition.

When Carnot discusses, in the following lines, what should be taken into account in order to produce the greatest possible effect, he affirms that this problem depends upon particular circumstances and therefore the problem does not admit a general solution to be applied for any situation.

The effect produced by a machine is a real or latent living force, always referred to the product **PH** of one weight **P** and a height **H**; we should call **q** this effect. On the other hand, to produce it we need that all driven forces spend a moment of activity **Q**, which cannot be lesser than **q**; this means that there is no loss moment of activity to be consumed by a driven force, or that **Q = q**. But the moment **Q** of activity consumed by the force **F** in a time **T**, moving with velocity **V**, if we suppose a simplified form, **F** and **V** constants, and that the angle between **F** and **V** can be designated by **(F ^V)**, we have **FVcos (F^V)** as the quantity that must be maximized.

The equation above is, obviously the moment of activity or, that is the same as the work done by the driven force. It depends upon four quantities: **F**, **V**, **T** and **(F ^V)**; one form to maximize this product is to insure that the direction of force coincides with that of velocity, i.e., that the force is in phase with velocity, using other words. With respect to the quantities force, velocity, and time during which the force acts, it is much more difficult to determine in an absolute form its intensities. If we could calculate approximately, the problem of optimization should be tried.

Carnot comments the case of the work done by a man where fatigue is also involved. The knowledge of his physical constitution is fundamental. In general this data could only be obtained by experience.

Within the work's study undertaken by man considered as a machine which originates ergonomics and physiology, Coulomb pointed out the importance of previous studies as, for instance, Daniel Bernoulli's: *Resultat de plusiers experiences destinées à determiner la quantité d'action que les homes peuvent fournir par leur travail journalier, suivant les differentes manières don't ils emploient leurs forces*; Bossut's *Mécanique* is referred to as containing important considerations about machines and also an Euler's memoir entitled: *De Machine in Genere.*

If we are looking at the maximization effect obtained by the machine, it is important to minimize the effect of passive forces, such as the friction force, string stiffness, air resistance, etc. Because of the impossibility of eliminating all these resistances and passive forces, which impose the progressive decreasing of machine velocity, this implies the impossibility of a perpetual motion. If percussion does exist the motion will decrease rapidly and the addition of living forces always decreases when percussion appears. In this context it is important to quote Carnot about the impossibility of the perpetual motion: *Obviously we cannot produce a perpetual motion, if it is true that all the driven forces that exist in nature are attractions and that this force have the general property, of being always the same to equal distances, between given bodies, i.e., a function that does not vary unless for the case where the distance of these bodies varies itself.*

Again Carnot discusses the importance of the concept of work within the theory of machines, which is recognized by him as follows: *A general consideration about what was said, is that kind of quantity called moment of activity, plays a great role in the theory of machines in motion: because is in general this quantity that is necessary to save the maximum possible in order to obtain from one agent all the effect which can be done.*

If we are studying a machine at rest, where it is necessary to overcome a body's inertia, but if we wish to give rise to a motion, the moment of activity which must be spent will be equal to the half-add of living forces which should be born.

5 Conclusion

In Carnot's book *Fundamental Principles,* as analyzed above, the concept of work plays a fundamental role in the building of his new general theory of machines. Once more, this theory is completely integrated into the conceptual framework of Rational Mechanics, because the used tools are based on its fundamental concepts including the concept of geometric motion which is a modified version and an attempt to generalize the principle of virtual velocities. Therefore, the structure of *Fundamental Principles*, contains a mechanical review emphasizing the problems involving shocks and sudden variations of velocities which is precisely the transmission of motion inside the machine. The originality of Carnot's contribution is clearly different from Lagrange's mechanics, in the sense of continuous variations, or in insensible forms, as characterized by him but presenting a simple approach which uses geometry and trigonometry, where a cosine law is transformed into an equation of energy conservation. At last, the new theory of machines proposed by Carnot is also up dated with the current practice of engineering of his time.

If we compare the famous Coriolis' work with Carnot's *Fundamental Principles*, it is easy to see that it represents a great progress from the technical point of view, as well as by its completeness, style and language. The 25 years that separate these two fundamental books in the mechanical engineering history does not present important works, except the notes and remarks of Navier about Bélidor, as previously mentioned. This advancement in the mechanical field, which means the development of applied mechanics in the industrial progress context, is also a remarkable aspect of engineering development. This process had a great influence in engineering education due to the need to prepare the technicians in order to give continuity to the technical progress itself and thus also provide a fundamental requirement to the Industrial Revolution at this moment spreading along the European continent. As we know, Coriolis was a lecturer in a Polytechnic School, Navier did the same in a School of Bridges and Highways.

References

Belhoste B (1989) Les origines de l'École polytechnique. Histoire de l'Education 42:13–53
Carnot L (1803) Principes fondamentaux de equilibre et du mouvement. Èditions Deterville, Paris
Charnay JP (1990) Lazare carnot ou le savant-citoyen. Presses de l'Université de Paris-Sorbonne, Paris
Chatzis K (1999) Theory and practice in the education of French engineers from the middle of the 18th century to the present. American Engineering Education, Reforms and Changes, 1920–1980 5:40–63
Coriolis GG (1829) Du calcul de l'effet des machines. Carilian-Goeury, Librairie, Paris
Coriolis GG (1832) Sur le principe des forces vives dans le mouvements relatifs des machines. J l'Ècole Polytechnique, vingt-nième cahier, tome XIII, Paris
Coulomb CA (2002) Théorie des machines simples. Librairie Scientifique et Technique Albert Blanchard, Paris

Gillispie CC, Pisano R (2014) Lazare and Sadi Carnot. A scientific and filial relationship (2nd edn). Springer, Dordrecht

Gillispie CC, Youschkevitch AP (1979) Lazare carnot et sa contribuition a la théorie de l'Infini mathematique. Librairie Philosophique J. Vrin, Paris

Poncelet JV (1827) Mémoire sur les roues hydrauliques a aubes courbes, second mémoire. Librairie de Me. Ve. Thiel, Place St. Jacques, No. 4, Metz

Rashed R (1988) Sciences a l'Èpoque de la revolution Française (Recherches Historiques). Librairie Scientifique et Technique Albert Blanchard, Paris

On the History and Engineering of the *Human Factor*

Dominique Pécaud

Abstract Human labor has been a subject of research for almost a 130 years. The study of human labor typically aims for a technical rationalization of labor by simultaneously applying its insights to production tools, to their use, and to human behavior. Over time, this field of research has given birth to the concept of the *human factor*. Studies of the human factor aim to improve the reliability of the manufacturing process and to reduce industrial risks. The history of the concept of human factor, from the manufacturing process to the engineering of man–machine interfaces sheds light on the evolution of the political definition of men and their management.

Keywords Human factor · Human labor · Practical reason · Rationalization of social life

1 Introduction

Here, we bring the human factor to the fore to explain the manner in which a maintenance team responds to a system breakdown, the consequences of which threaten to be highly consequential. We seek, namely, to call attention to the fact that machine operators often forgo electrical safety measures in order to save time. In the field of industrial safety, the human factor can be nebulous in meaning, even if its improvement appears to be of critical importance. But what exactly does the human factor mean, especially in the context of political policy? If the human factor historically originates in the man–machine interface, its use now extends to reliability issues within vast systems, industrial or otherwise. Indeed, it can now be applied to a discussion of areas of human activity in the broader sense.

We would like to address how this concept was born and is now used, and to observe, across various field studies in which the concept of the *human factor* arises,

D. Pécaud (✉)
Centre François-Viète, University of Nantes, Nantes, France
e-mail: dominique.pecaud@univ-nantes.fr

R. Pisano (ed.), *A Bridge between Conceptual Frameworks,* History of Mechanism and Machine Science 27, DOI 10.1007/978-94-017-9645-3_20

the importance of a movement of *reification*[1]. Stéphane Haber points, in the perspective of Georg Lukacs (1967), to reification as "the act of treating something like a simple manipulable entity when that is absolutely not the nature of the thing at all [...], in a way that engenders the possible forgetting of what we have already recognized about it" (Haber 2009, p. 189).

According to Axel Honneth (2008, pp. 15–18), reification extends to four different domains today: (1) the application of economics values to the ensemble of human activities; (2) a form of behavior marked by the privileging of opportunistic attitudes; (3) an instrumental understanding of human beings, corresponding to the meaning of the term as Luckacs originally understood it; (4) the domination of emotions through the naturalistic analysis of neural connections.

Ultimately, use of the term would appear to correspond to two main objectives: to organize a rational policy of risk prevention on the one hand, and to increase the effectiveness of these policies by preserving a dimension of humanity on the other. Here, we define the concept of humanity as inter-subjectivity in action, discovering expression though cultural forms that are sensible, cognitive, political, and oriented toward an ethos as an "expression of a culturally standardized system of organization of the instincts and emotions of the individuals" (Bateson 1958, p. 119).

This ethos is a considered to be a compassionate one when it demonstrates an extreme attention to suffering (and) a singular willingness to listen (Fassin 2007).

Our hypothesis is as follows: in both cases, use of the term *human factor* can lend itself to the overriding of regulatory codes grounded in an anthropological understanding of individual and collective *precautions*, which are then reconstrued as ways of organizing collective activity geared toward safety.

Per Sandin enumerates three big differences between *prevention* and *precaution*.

> First, prevention is neutral with regard to value. Good things can be prevented as well as bad ones, while precaution is about avoiding something undesirable. Secondly, prevention implies certainty. Just as it is strange to say 'x caused y, but y did not happen', it would be strange to say 'x prevented y, but y happened nevertheless'. This not the case with precaution. It is quite reasonable to say that in spite of all precautions, the unwanted event occurred nevertheless. Thirdly, talking of precaution implies talk of actions. Taking precautions is something agents do intentionally. This is not necessarily the case with prevention. Even inanimate objects may prevent things from happening, but they cannot take precautions.[2]

As Pascal Boyer and Pierre Lienard observe, "ritualization may be seen as an occasional by-product of specific precaution systems and action-parsing capacities in humans" (Boyer and Lienard 2006, p. 18).

After clarifying the conceptual and historical context in which the concept of the *human factor* arises, we describe the *explanatory practices* normally implemented to explain the functioning of technical systems. We will identify the consequences of the roadblock to these explanatory practices that follows from the application of the concept of the *human factor*. Then we will analyze the ways in which the concept of the human factor can be used to justify choices made in the design and

[1] Mainly: Lukacs (1967); Polanyi (2001); Honneth (2008).

[2] Sandin (2004, p. 15).

implementation of corrective actions applied to the conception and regulation of technologies and their uses. Finally, we will take a look at the distinction Hannah Arendt (1906–1975) Arendt ([1959] 1998) made between labor, work, and action and the concept of *practical reason* as Paul Ricœur (1913–2005) (Ricœur 2007) defines it to elucidate the different reductions which the *human factor* has as its goal.

2 A Concept with Multiple Meanings

The human factor arises with the existence of objects or *technical systems*, the elements of which exist in a situation of interdependence (Gille 1979; Stiegler 1998). The functioning of a technical system is contingent upon its fidelity to the laws of nature that define, direct, and ensure its forward march, but also on the activity of those humans who are directly or indirectly responsible for its functioning. A technical system is therefore of a socio-technical nature, bringing non-human objects, human beings, and environments into interaction (Akrich 1987, 1989). It could be described as a structural-phenomenological organization (Drăgănescu 1996).

As a *structure*, it mobilizes elements that are physically and symbolically ordered. As a phenomenon, it is constituted from multiple interactions, and defines a point of view to the degree that there are human beings involved in its functioning.

The activity of manipulating technical systems implies the existence of a *techne,* one that is traditionally defined as an intelligence engaged in practice (Détienne and Vernant 1978).

Currently, the *techne* refers rather to a set of methods and tools derived from an instrumental rationality that recalls both the traditional sciences and their applications within the sciences of the engineer and the sciences of management. Since Greek antiquity, it had been believed that *metis* (cunning) and *phronesis* (prudence) accompanied all technical activity. Use of the *human factor* concept would seem to revive this tradition of thought, but only by appearance. It calls upon the *human factor* primarily to complete a vision of the technical system reduced to the means of material production. Thus, the *human factor* designates either a class of phenomena surrounding system malfunctions or the activities undertaken by men to stop them. It is, depending on the case, the source of (as opposed to merely a factor in) the correct functioning of a technical system, or an area of research concerned with the performance of such a system, or a principle of efficiency assuring the safety or the correct use of the system, or a source of mechanical equilibrium or disequilibrium between a technical system and men at work.

One can trace the human factor

> back to the nineteenth century, in military circles, where it designated the baseline operator who activates the technical devices. The concept of human factors established itself as a current of thought during the Second World War, when the American army conducted studies aiming to improve weapons systems that were becoming more and more complex.[3]

[3] Fanchini (1999, p. 67).

This use of the term is older, and anchored in a determinist way of thinking capable of giving rise to a science of war, not merely an art. Thus, in 1911:

> By careful collation of past events it becomes clear that certain lines of conduct tend normally to produce certain effects; [...] We can even go further. By pursuing an historical and comparative method we can detect than even the human factor is not quite indeterminable. We can assert that certain situations will normally produce, whether in ourselves or in our adversaries, certain moral states on which we may calculate.[4]

In the 1950s, use of the term became more widespread with the creation of the Human Factors and Ergonomics Society:

> Between the two wars, the Human Factors and Ergonomics Society (HFES) directs the current of thought principally toward performance research. Then, in integrating ergonomics and the idea of adapting work to man, the HFES sets about the goal of creating and disseminating a body of knowledge dedicated to the creation of systems where men and machines interact in various environments, with the aim of improving not only their efficiency, but also their safety and ease of use.[5]

The following excerpt from this first issue of the publication sets forth the society's objectives:

> In this study, in which the cross-fertilization between life sciences and engineering is examined, the human factor is considered in relation to the machines and environments in which man works and plays. The ultimate aim of each human factors effort is toward the optimal utilization of human and machine capabilities to achieve the highest degree of effectiveness of the total system. The human is the stable element in each system, since he does not undergo drastic changes. Although each advance in machines and environments will have a bearing on the human component, the essential principles governing human behavior and performance capacity will not be altered. Thus, the elucidation of human factors principles is of ever lasting benefit, and it is the purpose of the Human Factors Society of America and this Journal to contribute to the advancement of this knowledge.[6]

Today, we see this same preoccupation with the reliability of the man–machine interface, or that of man himself. This preoccupation draws inspiration from a determinist approach to work situations.

> To speak of human factors is to admit, then attempt to comprehend, control, and if possible reduce the variability of the human being, so that this composite can be treated like a non-aleatoric domain, compatible with technical and organizational theories of the industrial world.[7]

The determinist project seeks to comprehend the functioning of the technical system and the human factor according to the same heuristic models, often in order to place machine and man on a plane of equality.

Implicitly, its aim is to defend of the notion of the *human factor*.

---

[4] Corbett (2004, p. 6–7).

[5] Fanchini (1999, p. 67).

[6] Morehouse (1958, p. 1).

[7] Amalberti (1998, p. 2).

3 Use and Industrial Consequences of the Human Factor

The *philosophy of human factors* has a great influence on the functioning of technical systems (Pécaud 2006). Many empirical studies we have conducted in different sectors of activity (the metallurgical industry, industrial maintenance, transport companies) reveal the presence of three main modes of technical or social regulation within technical systems. The contemporaneity by which a single technique applies to both man and machine speaks to the presence of *social scripts*, understood as the practical intentions contained within every technical system, whatever the intention explicitly declared by its creators or users may be (Dodier 1995, 1997; Foot 2003). However, the coexistence of man and machine is not always harmonious. The diversity of the social practices to which these modes of regulation can refer adds a layer of complexity to the safety and smooth functioning of systems.

The first type of system can be described in terms of three, interrelated stages: design, *normal* functioning, and maintenance and control. Maintenance and control do not function to call into question the design of a system. If the system functions well and produces what it is supposed to, it's because it was designed as such, and regular inspections should be enough to ensure its correct functioning. Maintenance, which is defined from the moment of design, erases or anticipates the effects of the system's inevitable degradation. If each *unit of time* in the life of a system undergoes a phenomenon of entropy, the consequences are cancelled out. Preventive maintenance encompasses the control, regulation, and replacement of a system's components.

The *human factor* helps justify the design options at hand, notably through ergonomic considerations pertaining to the prediction of uses (Stiegler 1998).

Furthermore, preventive maintenance aims to reduce, or has the effect of reducing, the need for inventive, human intervention.

This system is based on a naturalist approach to the technical, its design purporting to offer *a priori* mastery of the system. It is expected that the *real* functioning of the system will match its predicted functioning. The question of humanity comes to the fore in two ways. One the one hand, technology can be defined as a traditional kind of knowhow, thereby reflecting the humanity of the system. "Techniques are to be defined as traditional actions combined in order to produce a mechanical, physical, or chemical effect, these actions being recognised to have that effect" (Mauss 2007, p. 24).

But on the other hand, the procedural elaboration of the system can also lead to the exclusion of humanity. For Martin Heidegger (1889–1976) (Heidegger 1977), the *techne* designates not only the *doing* of the craftsman. The *techne* is always production, *poiesis*. Modern technology appears to be a threat, a *seizing* of nature so that it delivers up to man the energy it possesses.

The second kind of system inspires a philosophy in which the technical becomes more or less estranged from humanity. It is based on a distinction between the smooth, mechanical functioning of the system and the disturbances it can undergo as a result of human intervention. An unexpected humanity reveals itself, then, when the design is undermined by a repeated malfunction or an accident capable of shedding doubt upon the design.

Regulation of the system takes place through a neo-functionalist process of *integrative spill-over*. It takes potential hazards into account when they threaten the system's reliability. The analysis of potential hazards can bring the *human factor* to the fore as the cause of their appearance (Pécaud 2011). Implicitly, such analysis provides the basis for a principle. Every technical system can be disturbed by the *human factor*, even when the *human factor* is designed and described by its creators or users in a naturalist fashion, through structural and functional means.

The third kind of system regards the *human factor* as an element that is completely independent of and external to its functioning. Regarded from a neo-functionalist perspective, the human factor can be assimilated into an *integrative continuum* (Soldatos 1989, 116 sq.) that is defined by two facets of the system's forward march: a functional sequence of events and a voluntary transformation of this sequence, or of the system, itself. However, any questioning of a system's design is rarely undertaken in the name of underlying economic or social concerns, such as those regarding the *division of social labor* (Durkheim 1997) and the methods of symbolic domination that accompany it.

Within this more determinist perspective, the structural role of the *human factor* can be conceived in line with dehumanized, technical logics. The *human factor* could refer to the notion of behavior to circumscribe and technically define a regulatory element of the technical system. This approach can affirm a mechanistic vision of human activity within the system, even as the aims that guide this activity and comprise its humanity are reduced to the mental or political objective of executing a predetermined procedure.

The distinction established here between technical and behavioral procedures upholds, in illusory fashion, the humanity at the heart of the system. However, if it publicly reveals a humanity reduced to its behavioral forms, it impoverishes the symbolic and formative experience of human labor as it is understood as *work* and *political action* in public space, grounded in an assumption of the equality of all the subjectivity of each (Arendt 1998). Because the *human factor* suggests a false complementarity between a dehumanized technical universe and a social universe reduced to its behavioral normativity, it precipitates the exclusion of cultures of labor. It promotes a technological politics that seeks to associate the experience of human labor with a succession of dehumanized technical activities under procedural control and geared toward improving the safe functioning of the system.

Generally, therefore, use of the concept of the *human factor* privileges the instrumentalization of labor. It reinforces man's enslavement to deliberately dehumanized, technical systems. A simultaneous vision of technical and behavioral excellence is born. The instrumentalization of human labor (Pécaud 2005) thus expresses itself through a finer and finer elaboration of the activities necessary to the correct functioning of technical systems. It is also evident in the proliferation of a vocabulary dedicated to the human movements of labor, then extending progressively to the ensemble of human movements.

We note an evolution of forms of control. Human labor, reduced to the objective expression of activities that are useful to the functioning of systems, would appear to be designed and controlled from the outside. A system of control integrates itself wholly into a technical system. Moreover, technical control threatens to infiltrate even psychic life. Everyone is invited to practice self-control according to previously established forms.

The method of instrumental reasoning applied to human labor provides a pretext for causal analyses guided by logics of attribution or incrimination (Dodier 1995; Kouabenan 1985, 1999; Pécaud 2001). Finally, the description of external behavior excludes the role of individual subjectivity and collective references, which provide labor with meaning, both in itself and for others. We bear witness to the reign of material labor for the sake of consumption, characteristic of the *Homo laborans* and his survival, which the behavioral labor that facilitates this material labor is meant to achieve, to the detriment of the subjectivity at work in human labor. Hannah Arendt (1998) describes the labor of the *Homo laborans*, oriented toward the production of ephemeral consumer goods capable of ensuring survival. This labor is anonymous and collective in the sense that it does not allow for the expression of individual subjectivity. In this respect, Arendt does not believe it to be specifically human. By contrast, she describes the work of the *Homo faber*. His production is meant for use and not only consumption. It inscribes itself in institutional time and participates in a collective culture; it is the work of the man who acts, and thereby participates in the political elaboration of, collective action. This action is open to deliberation; it takes place in a public space that brings together humans according to the principle of equality while distinguishing between them in the name of their subjectivity.

4 Domestic Use of the Human Factor and its Consequences

If use of the *human factor* historically originates out of a need to characterize components in the man–machine interface, it is now called upon in attempts to identify and understand the ensemble of human attitudes expressed in response to the implementation of new technologies.

To illustrate this use of the concept, we turn to a study we conducted concerning the rehabilitation of low income housing units and questions of energy efficiency.

By *technologists*, we mean to designate the technicians vested with the task of planning and executing the housing rehabilitation project, opening a discourse on the indispensability of the role of technology in carrying it out and offering a better quality of life to its tenants. To understand the political use that is made of such expressions, we will examine how *technologists* envision their activity and its consequences. This activity presents itself in the form of planned project management. Here, the housing rehabilitation project is conceived in view of a timeline that starts with a situation that is identified as undesirable (the waste of energy, mediocre results) and ends with a desirable one, obtained through the use of technologies (or technical systems) that are intended to improve, in a predetermined manner, the energetic performance of the units targeted for rehabilitation.

The *human factor* refers to the different attitudes adopted by the technologists or the residents of the rehabilitated housing. The technologists' attitudes are criticized, at least in appearance, by the sponsors of the study, who often fret over the lack of respect they feel coming from *their* technicians, notably when these technicians point to the tenants' ignorance in all technical matters. According to the technologists, but also according to the sponsors, comprehension of these matters is commensurate to a political acceptance of the technical solutions implemented to rehabilitate the housing. Both parties point to the fact that meetings were organized to explain the process, that the tenants received clear answers to all of the questions they asked. They are surprised, then, when they see that certain tenants have an uncooperative attitude vis-à-vis the procedure to be followed and its intended results—and also when they see that tenants are manipulating the technologies they are presented with (for example, central heating and air, mechanically controlled ventilation systems) "in a manner that runs contrary to common sense," common sense here amounting to the sort of technical rationality that the technologists boast of perfectly mastering, including when they try to explain it to others.

But this lack of consideration has also been observed when the *technologists*, much like the sponsors themselves, note tenants' resistances to change in relation to proposed and implemented renovations. Here, in the recognition of these different attitudes, the *human factor* designates a cognitive, symbolic, cultural and political heritage crystallized through the expression of an ensemble of justifications put forth by tenants to contend with a technological advance that is either desired or feared.

Human attitudes and their movement toward greater technical rationality are at the heart of analyses concerned with situations in which technological artifacts are installed in domestic spaces that are governed by multiple forms of rationality. Once more, the political function of the *human factor* would seem ambiguous. One the one hand, it points to the existence of something that needs to be taken into account when one seeks to dictate a person's manner of doing something but also his or her manner of existing within technological rationality. On the other, it designates the quasi-technical means by which one ensures the smooth functioning of planned, behavioral and social transformations.

We discover three major dimensions governing tenants' attitudes toward the two-speed technical system that is the housing rehabilitation campaign (Oskamp 1977). The first concerns the cognitive dimension of their collective evaluation of the rehabilitation process and of the new technical instruments they are now invited to manipulate in their housing units. What do these objects, installed in their homes, represent to them? Opinion is divided: for residents, technology is as constraining as it is helpful, as much a source of difficulty as it is a source of ease, whereas for the technologists, it is the sign of progress, of improvement.

The second involves the affective dimension of any attitude. Is technology good for mankind? On the contrary, is it harmful? The presence of new technical devices in domestic space can be either a relief or a threat. Neither impression constitutes the dominant stance, and when such impressions arise, it is clear that they stem from multiple interactions with the technical system. Among these, we will point to the importance of tenants' perspectives on social housing landlords and the state, and the importance attributed to neighborly relations in the egalitarian implementation of housing renovation initiatives, the assessment of social status, cultural codes, and the merits attributed by tenants to other tenants.

The final dimension has to do with the behavioral component of the attitudes adopted and the origins of their direction. The belief that fuels any attitude will determine the direction of the activity undertaken by the person who adopts it. If this attitude projects itself along the behavioral path of individual engagement in the activity, the belief that fuels it exists only if it is shared with a group united by a common vision of lived experience.

5 Returning to Practical Reason to Rehumanize Activity

The widespread use of the concept of the *human factor* today goes hand in hand with the rationalization of individual and collective action under the guise of improving the functioning of socio-technical systems. But the study of labor and its attendant *precautions* reveals the existence of other avenues for facilitating this functioning and ensuring the integrity of the individuals whose job it is to ensure it. The concept of *practical reason* affords us a look at these avenues.

The uses of the *human factor* we've enumerated thus lead to two major impasses. The first is the product of a calculating conscience seeking to guide human action through a kind of *voluntary deliberation.* Pierre Bourdieu (1930–2002) denounces the "half-learned philosophy of action" (Bourdieu 2000, p. 137).

He describes the pernicious importance that it traditionally accords to a

> philosophy of conscience [...] that cannot conceive of spontaneity and creativity without the intervention of a creative intention, or finality without a conscious aiming at ends, regularity without observance of rules, signification in the absence of signifying intention.[8]

The rationalizing illusion it engenders is largely nourished by the standardized self-evaluations completed by technical operators when they are asked to describe and justify the work they do (Pécaud 2005). The second is the product of a hypothesis grounded in a determinism of the body, structured by past experiences and giving rise to "systems of perception, of appreciation, and action" that exist outside of a voluntary semantics of action (Bourdieu 2000, p. 137).

According to Max Weber (1864–1920) (Weber 1978), Paul Ricœur proposes a theory of practical action based on the collective intelligibility of every human action. For Ricœur, human action can be "neither dumb nor incommunicable" (Ricœur 2007, p. 189).

[8] Bourdieu (2000, p. 137).

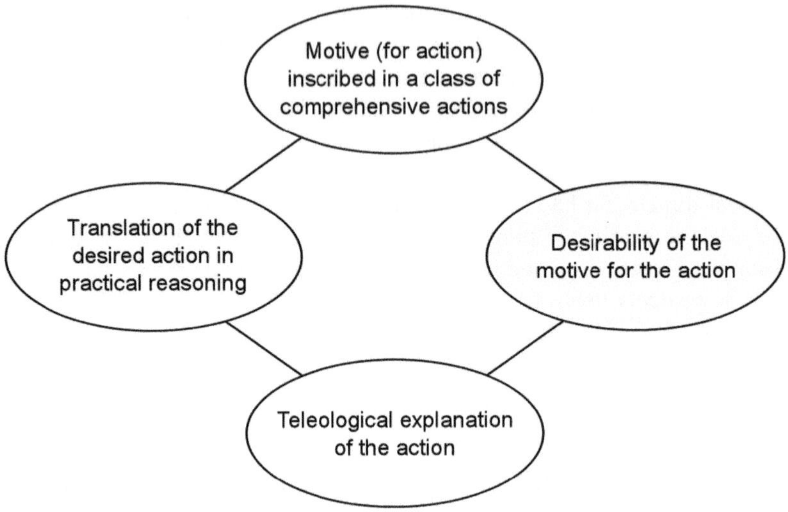

Fig. 1 The modalities of practical reason

This intelligibility expresses itself whenever the action in question satisfies the requirements of a lexicon encompassing a shared language and system of values.

This makes it possible for one to explain and justify the actions enumerated in responses to questions such as, what are you doing? Why are you doing it? Why did you do it?

The concept of practical reason is divided into four semantic and lexical modalities that seemingly double as descriptions of social activity and its construction (Fig. 1).

The first concerns the motive for the action and the acceptability of this motive for the person who is acting. If this motive can be emotional or rational in nature, it is freer from constraint. Paul Ricœur borrows the concept of *desirability* from Gertrude E. M. Anscombe (1919–2002) (Anscombe 1957), to describe the motive behind an action as the result of a hierarchy of "*heterogeneous motives*," originating out of desire. "For how indeed could one course of action be deemed preferred to another if we could not say how one seemed more desirable than the other?" (Ricœur 2007, p. 190).

Vincent Descombes explains that:

> To identify the intention behind something an agent does is to re-describe his action (in the present) with the help of an expanded description (in time, or in space). It follows that, as long as we stick to descriptions that the agent might himself give for his action, stating what he does and why are two ways of describing the same thing, the same event in the history of the world. "What are you doing?" The subject can respond, "I am looking for my camera," or "I am going upstairs to get my camera." If these descriptions from the agent are one and the same from the point of view of reality, are they one and the same from the point of view of the action? No, because they form an intentional framework. [...] The descriptions are

> sequentially linked following an A-B-C-D scheme: the subject has the intention of doing A under the present circumstances, he has this intention because he wants to do D, and to do A under such circumstances, is to do B, and to do B under these circumstances, is to do C, and so on and so on until reaching the final destination of D, which gives the broadest description of the action by indicating its outcome.[9]

The second modality relates to the relationship between desirability and socially available modes of explanation. If it is possible to generalize the motives behind people's actions, this is not simply a consequence of weighing individual perspectives against dominant, collective perspectives, but also of the forms of expression through which such motives are communicated. The terms of this generalization can be found in the pre-generalizable character of an expressed motive. In other words, the motives for action encompassing the inherent desirability of *singular* action partly reflect the pre-constructed meaning of that action, thereby rendering its expression generalizable.

The third modality relates to the intended goal of an action. This teleological aim is not to be confused with the reasons for or origins of an action, which are explanations of a causal nature. Rather, it corresponds to "an explanation using available terms" (Ricœur 2007, p. 190).

If it is acceptable by others, it's because it's socially acceptable.

The fourth modality aims for a lexical translation of an action within the framework of *practical reasoning*, itself socially available and accepted (refer to *practical syllogisms*). Invoking Charles Taylor (b. 1931) (Taylor 1964), Ricœur observes that the explanation of an action's final result "is an explanation in which the global configuration of events is itself a factor in its own production" (Ricœur 2007, p. 191). It attributes to available, semantic and lexical forms the capacity to enable the social actor to produce a teleological explanation of his own action, grounded in reasoning.

The use of practical reason can counterbalance the processes of reification threatening to infiltrate the concept of the human factor. From this point of view, the implementation of practical reason offers a new approach to the risk prevention program. It can aid with system design as well as regulation. Surely, the justifications abound. Firstly, practical reason inscribes all security-oriented measures within a signification that is collectively shared. Second, it goes beyond the divide between a rationalization of the world that responds to the laws of nature and the multiple motives invoked by those who act. It mobilizes the ensemble of available significations in the justification of the action in question, thereby bypassing those neglected or devalued modes of explanation that do not correspond to the canons of scientific or technological rationality. Thirdly, it reconciles action with the discourse that surrounds it, guaranteeing effective feedback capable of giving meaning and direction to practices that are dictated by precautions.

[9] Descombes (preface to Anscombe 2002, pp. 16–17).

6 Conclusion

Ricœur notes, referring to the modality corresponding to the concept of practical reason, that it would seem impossible to clearly distinguish desire from reason. Referring to Book 6 of Aristotle's (384–322 BC) *Nichomachean Ethics* (Aristotle 2002), he suggests that desire can permeate the sphere of language and of social relations as a whole without the need of an intermediary. It's this possibility of expressive, factual and linguistic practice that we refer to by the idea of the precaution (Pécaud 2010).

Technical reason is never a substitute for desire. Yet, proponents of the human factor's exploitative misuse would have us believe as much. They seek to develop rational control over the socio-technical system in order to put an end to practices grounded in deliberative reason or deliberative desire. But these deliberative practices remain crucial in understanding the factory of precautions inherent in situations of labor or domestic living conditions.

The significance of Paul Ricœur's proposition resides in its phenomenological foundation, which shelters us from the scholarly illusion of a theory of human attitudes reduced entirely to their behavioral expression, a theory indispensable to the political project of controlling human beings. It also shelters us from the realist illusion of an absolute determinism of action, defined with reference to the functioning of technical systems.

The notion of intention contributes to the idea of an actor who is capable of making choices, but who must simultaneously answer to obligations both internal (as the product of practical reasoning) and external (as the product of the instruction and publication of practical reasoning skills corresponding to available explanations). Without denying the question of determinism, it opens a discussion of the field of relations between human interaction, action, and language. It raises the question of human labor, but also that of risk prevention, within the political perspective of man acting for himself, but also with and for others.

If the hypothesis of practical reason doesn't shelter him or her from technical rationality's jurisdiction over preventative practices established in the name of the human factor, it nevertheless provides the basis for a program of political resistance to this control, which fuels mass society and its industrial avatar as they progressively infiltrate all our ways of being together.

References

Akrich M (1987) Comment décrire les objets techniques. Technique et culture 9:49–64

Akrich M (1989) La construction d'un système socio-technique. Esquisse pour une anthropologie des techniques. Anthropologie et sociétés 13(2):31–54

Amalberti R (1998) Les facteurs humains à l'aube de l'an 2000. Phébus, Paris

Anscombe GEM (1957) Intention. Basil Blackwell, Oxford

Anscombe GEM (2002) L'intention. Gallimard, Paris

Arendt H ([1959] 1998) The human condition. University of Chicago Press, Chicago

Aristotle (2002) Nicomachean ethics. Oxford University Press, Oxford

Bateson G (1958) Naven: the culture of the Iatmul people of new guinea as revealed through a study of the Naven ceremony. Stanford University Press, Stanford

Bourdieu P (2000) Pascalian Meditations (1997). Stanford University Press, Stanford
Boyer P, Lienard P (2006) Why ritualization behavior? Precaution systems and action parsing in developmental, pathological and cultural rituals. Behavioral and Brain Sciences 29:1–56
Corbett JS (2004) Principles of maritime strategy (1911). Dover Edition, Dover
Détienne M, Vernant JP (1978) Cunning intelligence in Greek culture and society (1974). The Harvester Press (Humanities Press), Hassocks (Atlantic Highlands)
Dodier N (1995) Les hommes et les machines, la conscience collective dans les sociétés technicisées. Métailié, Paris
Dodier N (1997) Remarques sur la conscience du collectif dans les réseaux sociotechniques. Sociologie du travail 2:131–148
Drăgănescu M (1996) Ontological Universality of Information. Editura Academiei, Bucarest
Durkheim E (1997) The division of labor in society (1930). Free Press, New York
Fanchini H (1999) Paradoxes et nouvelles orientations du facteur humain en sûreté de fonctionnement. Revue de l'Armement 67:67–72
Fassin D (2007) Un ethos compassionnel. La souffrance comme langage, l'écoute comme politique. In: Soulet MH (ed) La souffrance sociale. Nouveau malaise dans la civilisation. Academic, Fribourg, pp 51–72
Foot R (2003) Une approche pragmatique "des positions inouïes du sujet." In: Actes du colloque "L'individu hyper moderne", ESCP-EAP/LCS, Université Paris 7, II, pp 22–32
Gille B (1979) La notion de système technique, essai d'épistémologie génétique. Technique et culture 1:8–18
Haber S (2009) L'homme dépossédé. Une tradition critique de Marx à Honneth. CNRS éditions, Paris
Heidegger M (1977) The question concerning technology and other essays (1954). Harper & Row, New York
Honneth A (2008) Reification: a new look at an old idea. Oxford Universirty Press, Oxford
Kouabenan DR (1985) L'analyse des attributions causales des accidents. Le travail humain 48(1):1–17
Kouabenan DR (1999) Explication naïve de l'accident et prévention. Presses Universitaires de France, Paris
Lukacs G (1967) History and class consciousness (1923). Merlin Press, London
Mauss M (2007) Manual of ethnography (1947). Berghahn Books (The Durkheim Press), New York (Oxford)
Morehouse LE (1958) A human factors philosophy. Human Factors: The Journal of the Human Factors and Ergonomics Society 1(1):1
Oskamp S (1977) Attitudes and opinions. Prentice Hall, Englewood Cliffs
Pécaud D (2001) L'usage des notions de facteur comportemental et de facteur humain dans l'analyse des accidents de travail. Prévenir 40:217–234
Pécaud D (2005) Risques et précautions. L'interminable rationalisation du social. La Dispute, Paris
Pécaud D (2006) Améliorer la sûreté de fonctionnement: un projet de rationalisation de l'usage de l'expérience. Atelier de l'Association Internationale de Management Stratégique (AIMS). Le projet dans l'action collective, Regards pluriels, Nantes, June 1–2 (not published)
Pécaud D (2010) Ingénieries et sciences humaines. La prévention des risques en dispute. Lavoisier, Paris
Polanyi K (2001) The great transformation. The political and economic origins of our time (1944). Beacon Press, Boston
Ricœur P (2007) From text to action. Essays in hermeneutics, vol 2 (1986). Northwestern University Press, Evanston
Sandin P (2004) Better safe than sorry: applying philosophical methods to the debate on risk and the precautionary principles. Dissertation, Royal Institute of Technology, Stockholm
Soldatos P (1989) Le système institutionnel et politique des communautés européennes dans un monde en mutation. Théorie et pratique. Bruyland, Bruxelles
Stiegler B (1998) The fault of Epimetheus (1994). Stanford University Press, Stanford
Taylor C (1964) The explanation of behavior. Routledge & Kegan Paul, London
Weber M (1978) Economy and society. An outline of interpretative sociology, vol 1. University of California Press, Berkeley

The Emergencies of Mechanics and Thermodynamics in the Western Technoscience-Society during Eighteenth–Nineteenth Century

Raffaele Pisano and Paolo Bussotti

Abstract In this paper, we present a brief history of the development of mechanics and mechanical machines theory (particularly in Lazare Carnot) and consequently the birth and advanced studies in thermodynamics and heat machines (particularly in Sadi Carnot) that influenced science & technology (as technoscience) in Western society between the eighteenth and nineteenth centuries. We focus on theoretical aspects of thermodynamics efficiency—for improvements of practical studies from cannons to civil steam engines. Considerations on mechanical and thermodynamic machines and correlated topics like the role played by the impossibility of perpetual motion in mechanical and heat machines are discussed.

Keywords Mechanics · Machines · Carnot · Thermodynamics · Cannons · Steam engines · Physics–mathematics–geometry relationships · Western civilization & science in context · Technoscience

1 An Introduction

1.1 An Outline

In between the seventeenth and eighteenth Centuries, military weapons achieved a consistent efficiency of cannons and guns, particularly in France and England. The latter produced interesting practical and early theoretical studies of heat engines due to technological and social industrial events. The main interests concentrated on the research of a source of unlimited power, evidently correlated with theoretical

R. Pisano (✉)
Department of Physics, Lille 1 University Science and Technology, Villeneuve d'Ascq, France/Centre François Viète, University of Nantes, Nantes, France
e-mail: pisanoraffaele@iol.it

P. Bussotti
The Alexander von Humboldt Foundation, Berlin, Germany
e-mail: paolobussotti66@gmail.com

R. Pisano (ed.), *A Bridge between Conceptual Frameworks,* History of Mechanism and Machine Science 27, DOI 10.1007/978-94-017-9645-3_21

studies on the conversion of heat to work: *How?* In effect, a total conversion[1] was, and is, an ideal idea within the formulation of the first law of thermodynamics and its correlation with conceptualization of the more general energy conservation law[2]. That was mathematically confirmed later, mainly thanks to results of James Prescott Joule (1818–1889) (Fox 1969; Joule 1965, pp. 277–281; 1847, pp. 173–176), Rudolf Clausius (1822–1888) and William Thomson (1824–1907) (Clausius 1850, 1865, 1898; Thomson 1848, 1851a, 1851b). In spite of that Lazare Nicolas Marguérite Carnot (1753–1823)—already in 1778 (Carnot 1778, 1780, §§ 149–160)—stated this *chimerical dream* within his studies on general mechanical machines[3] (Carnot 1786).

In the nineteenth century, the great variety of heat engines was initially studied by theoretical physicists who took—as the best instance—the most powerful engine i.e. a cannon, which was considered in its impressive phase only, the explosion; in particular steam engines were preliminarily used where water was lacking.

Nicolas Léonard Sadi Carnot (1796–1832), son of Lazare and inventor of thermodynamics (Gillispie and Pisano 2014, pp. 176–183), already in his unpublished works (*Ivi*, Chap. VI) proposed a theory which radically revolutionized the foundations and paradigms of machine science during the nineteenth century. His analysis (Gillispie and Pisano 2014; Carnot 1978) explicitly referred to a civil heat machine. The *puissance motrice du feu*, is not caused by a chemical explosion; as his father had done with mechanical machines (Carnot 1778, 1780, 1786, 1803; see also Girard 1824), the calculation of the efficiency is independent from whatever *working substance*. Here his famous theorem (Carnot 1978, p. 38) is proved. From the beginnings of his book, Sadi Carnot faced the technological and theoretical problems related to power production from a given quantity of heat between two reservoirs.

> We draw a distinction here between the vapor [steam] engine and heat ["à feu"] engine in general ["*en général*"]. The latter may be employed with any working substance, steam or anything else, in order to develop the motive power of heat.[4]

He introduced important scientific concepts: i.e. *state of a system, reversible processes*, *cycle*, phases (from three to four phases), and his above cited theorem demonstrated by an *ad absurdum* proof (untypically for a physical science at that time). His physical and technological arguments were based on the impossibility

[1] Briefly, just to specify that the total conversion *work–heat* is only produced within particular ideal conditions (i.e, thermodynamics cyclic process) in which the internal energy can change (ΔU) during the course of a cyclic process; when the cyclic process finishes the system's energy is the same as the energy it had when the process began ($\Delta U = 0$). This also means that the ideal equivalence (idealistic total conversion) work–heat ($W = Q$) is obtained. Conventionally, within the loop of the cyclic process, W is positive, then it represents the *heat engine* case-study; if W is negative, then it represents a *heat pump* case-study.

[2] We make precise that the conservation law is a crucial concept in between the history of mechanics and thermodynamics because, i.e., it was a concept faraway from Leibniz (*mv*) Descartes (mv^2) and Newton (*motion lost*) science.

[3] Carnot (1786, pp. ix–x, pp. 89–94, pp. vij–ix, pp. 86–87; see also 1780, §§ 151–152); Gillispie (1971, Appendix C, §§ 151–152, pp. 328–329).

[4] Carnot (1978, p. 8, ft. 1, line 1).

of perpetual motion. Finally, the emergence of two Carnots' sciences, mechanics and thermodynamics, produced an *anomaly* because their foundations were incompatible both with a cannon —the practical/technology use of science— and with theoretical science (i.e., by Laplace, Fourier, Lamé, Ampère, etc.) at that time. *Thus, how describe the theoretical causes of these emergencies? When and how tension between technosciences explicitly practised?* From a modern standpoint: *what was the conceptual bridge between theoretical scientific research and the organization of technological researches in the development of the industry applications?* Therefore, how to recognize the *bridge between conceptual frameworks, sciences, society and technology studies?* (Pisano 2014).

1.2 Science and Technology in Context: Fourteenth–Eighteenth Century

In the late Middle Ages, the *heavy artilleries* began to be organized. These weapons had fanciful names such as *serpentine, bastard, culverin, falconets, falcon.* A standard vocabulary did not exist, and, sometimes, the same name was attributed to different weapons. In an initial phase (fifteenth–sixteenth century), the weapons were assembled and furnished with gunpowder and projectiles during the battle. They were generally utilized during the sieges. It was the famous *Mons Meg*, from the Belgian city Mons, that was constructed and presented (1457) to King James II by Phillip the Good, Duke of Burgundy. This cannon fired projectiles having a mass of 330 pounds (149.6 kg) with a range of two miles (more than 3 km). As to small-calibre artillery, guns and mortars are known. Both of them were breech-loading until the end of the sixteenth century. They were composed of an iron-chamber kept by a stirrup that reached the breech. During the late Middle Ages and the early modern time, these weapons had many technical problems: they could explode (this happened frequently for small-calibre artillery), they could lose gases during the shooting and this caused a limited shoot-energy and, consequently, a shorter range.

The mortar-bombs were constructed as a cave sphere with a touch-hole holding the match. A serious danger was caused by the bomb, which, if inserted with the lighted match, could explode inside the short mortar-barrel. Scientific studies on these weapons did not exist. However, numerous pictures and illustrations show continuous attempts to improve the technique. Around the end of the fifteenth century there were improvements in the transportation of guns and in aiming technique. The so-called *orillions* were constructed on the muzzle. They allowed a quick rotation and an aiming-change in elevation. The improvement in foundry technique held an important role: in the Middle Ages, guns and mortars were likely constructed with copper-alloys, but, in the fifteenth–sixteenth centuries, weapons of cast iron were constructed thanks to improvement of the smelting furnaces. As to this, Vannoccio Biringuccio (1540; see also [1943] 1990) (1480–1539?) presented a detailed description of the methods used in metallurgy, in particular with regard to the bronze-cast. He supplied further particulars as the most suitable ratio between copper and tin in the bronze (ca. 1/10). Likely, for the iron-guns, a technique similar to that used for bronze-guns was utilized. The construction was based on artisan and experiential-techniques. The metal was not worked and afterwards cast; it was only cast. Hence, the piece obtained was structurally weak. Furthermore, every new muzzle needed a new construction-process. Thus, no weapon was similar to another one. The muzzles were not correctly bored, but irregularly perforated. During the battle, the gunner had to choose the right projectile for his weapon. Nevertheless, neither were the projectiles homogeneous. Most historians agree as to the increasing importance of firearms, both for defence and attack. It is however necessary to be cautious when terms such as *siege* and *artillery* recur in documents of the thirteenth–fourteenth centuries.

The damages caused by the projectiles were far-inferior to the ones the men expected the guns could produce, given the supposed energy they believed these weapons could generate. The gunpowder, composed by saltpetre (ca. 40 %) and by sulphur and small charcoal (ca. 30 %), was not refined and its energy was slight, enough. An important problem was transportation of heavy artilleries. Around the Renaissance period, the feudatories left armory to the kings and the princes, the only ones who could bear the heavy costs of new arms. Improvements in the composition of gunpowder (France, ca. 1425) depended on the different proportions of its components and on the use of wet solutes. This permitted production of a more powerful gunpowder with homogeneous grains. It is known that the king of France, Charles VIII (1470–1498), when he invaded Italy in 1494, possessed a powerful artillery that could sway the battles in his favour in a relatively rapid manner. In 1503 Consalvo Pietro Navarra (1460–1528), Count of Oliveto, used the technique of mines in the battle for Castel Dell'Ovo in Napoli. In 1509, in Padua, the besieged side placed mines under the batteries of the enemy who was bombarding them with cannons. In the seventeenth century, the technique for artillery was further improved.

> All our engineers were men of war. Such statements of the obvious have the uncomfortable habit of often being true. Yet the sixteenth century had passed beyond warlike preoccupations and had constructed a complete technical system, just as it had built a new scientific system. More than their quest for deadly power, more than the amusements and the love of images, what has attracted us in these men is the difficult apprentice ship they served in a new world. Much remains to do before we understand the processes of their thought, before we appreciate their hesitations and grasp the nature of their ignorance and their failures. We must underline their gradual distortions of accepted truths, their difficult departures from the traditional paths, in order to give them credit for having [...] unique advance in the history of thought. [...]. But the enquiry remains open: it might bring to light other works still languishing in the dust of libraries, it might also provide a more precise analysis of the notebooks which have never been published and which are full of information.[5]

By the sixteenth century, cannon were made in a great variety of lengths and bore diameters, but the general rule was that the longer the barrel, the longer the range. Some cannons made during this time had barrels exceeding 10 ft (3.0 m) in length, and could weigh up to 20,000 pounds (9071.8 kg). Another notable effect of cannon on warfare during this period was the change in conventional fortifications. Niccolò Machiavelli (1469–1527) wrote: "There is no wall, whatever its thickness

[5] Gille 1966, p. 240 (see also Gille 1964, French original text).

that artillery will not destroy in only a few days" (Machiavelli 2005, p. 74). Instead of majestic towers and *merlons*, the walls of new fortresses were thicker, angulated, and sloped, while towers became lower and stouter; increasing use was also made of earthen, brick, and stone breastworks and redoubts. These new defences became known as *star forts* (Pisano 2008) after their characteristic shape. By the end of the fifteenth century, several technological advancements were made, making cannon more mobile.

Another interesting argument concerns the use of guns on ships in the seventeenth-century, particularly in England and correlated topics (Knox 1939; Heath 2005).

2 Mechanical Science and Machines in Context

Between the end of the seventeenth century and the beginning of the following one, physicists dealt with a series of complex problems: for example, the search for the oscillation centre of a rigid body, the study of vibrations of a chain or a thread. The search for the centre of oscillation was quite a relevant and difficult problem. This problem was historically discussed as solved by Christian Huygens (1629–1695) in his *Horologium oscillatorium* (Huygens 1673) by means a formulation of the theorem of living forces. Jakob Bernoulli revisited the subject in 1703 through a completely different and promising approach in (Bernoulli 1703). In this paper, one can find the roots of both D'Alembert's principle and the angular moment equation. Johann Bernoulli also addressed these problems at the end of the seventeenth century; his considerations, published in *Opera Omnia* (Bernoulli 1742a, 1742b) where he first introduced the concept of angular acceleration, are relevant. The problem of the vibrating chain was studied by scientists such as Euler, D'Alembert, Johann and Daniel Bernoulli. The motion of bodies on mobile surfaces, such as the motion of a heavy body upon an inclined plane without friction were among the main problems scientists dealt with in the eighteenth century. Johann Bernoulli (1742b) studied the motion of a material point through a Newtonian perspective by introducing the constraint reactions among the external forces, referring to them as *immaterial forces*, inasmuch as they were outside of the touching bodies.

In the first part of the eighteenth century, thanks to Euler and Johann Bernoulli, some general principles were affirmed, such as those regarding living forces and the *principles of least action*. On the *principle of living forces* we should mention the works by Johann Bernoulli (1727) and Daniel Bernoulli ([1748]1750) that followed chronologically the contributions made by Huygens and Leibniz. Afterwards, mainly D'Alembert (1743) and Lagrange (1764) provided the solutions to crucial questions. The *principle of living forces* alone was applicable to problems limited to one degree of freedom since it only gives a scalar equation. The *principle of least action* generally is attributed to Pierre de Fermat (1601–1665); he concerned himself in his work with the refraction of light (1662). One may consider the letter addressed to de La Chambre ("Sunday, January 1, 1662") where Fermat referred to the principle, according to which "[…] nature *always acts along the shortest paths* […]"[6] (Fermat 1891–1922, vol II, CXII [D., III, 5r.] p. 458). Pierre-Louis Moreau de Maupertuis (1698–1759) named the principle and extended it to mechanics (Maupertuis 1746). In fact, in the *De vera notione virium vivarium* (1735) Johann Bernoulli (1667–1748) modelled thermal phenomena by means of elastic and moving atoms, whose impacts obey the conservation law of kinetic energy (Bernoulli 1742c, pp. 239–260; see also 1742a, 1742b). Yet, his clever contribution stood ignored even by those scholars like John James Waterston (1811–1883) and Rudolf Clausius (1822–1888) (Clausius 1850, 1865a, b, 1868–1869; Mendoza 1959, 1960, 1982) who, a century later, rediscovered such a theoretical attitude.

Leonhard Euler (1707–1783) completed his *Mechanica sive motus scientia analytice exposita* (Euler 1736) and then published it in two volumes in 1736 (Pisano and Capecchi 2013). He offered a sophisticated model for theorizing still unresolved physical systems; his results on mechanics of fluids led theorists to think of heat as a new fluid. Nevertheless, this suggestion was misleading for the final theoretical result. By resurrecting a Newton's law (Bussotti and Pisano 2013a), in 1807 Fourier started some celebrated investigations on the mathematics of heat conduction (Fourier 1807). By means of a relevant number of works Fourier (1808, 1822, 1829) mathematically studied heat transport using differential equations (and trigonometric series), i.e his famous law (Fourier 1822, pp. 134–135). He accurately concluded that thermal phenomena are irreducible to mechanical theory (Fourier 1822, pp. xiv–xvij; see also 1807, 1808, 1829; Grattan-Guinness 1969, 1972; Grattan-Guinness and Ravetz 1980–1990). As to the physical principles, Leonhard Euler attempted to provide a more precise formulation of the least action principle. However, he did not reach a general formulation, while leaving to metaphysics the decision whether it could be assumed as a general principle or decided by the laws of mechanics. Lagrange (Lagrange [1762] 1973, I, pp. 363–468) gave an interesting proof of the least action principle based solely upon the laws of mechanics. In spite of some important successes, a feeling of disappointment emerged amongst scientists, who strove intellectually towards more simple and general principles. In the second half of the eighteenth century, this effort began to yield some results, with Euler's and Lagrange's achievements towards a nearly complete form of vector and analytical mechanics respectively. The great development of theoretical mechanics in the eighteenth century (Pisano and Capecchi 2013) was of little aid in suggesting the new thermodynamic notions. In fact, Newton's theoretical approach (Newton 1687; Bussotti and Pisano 2013a, 2013b, 2014) to the study of heat dominated physicists' minds along almost two centuries. According to this approach, *a body is composed of hard, fixed point-mass atoms, which mutually interact by means of gravitational force* (Newton [1704] 1730, III, Query 31, pp. 350–382; see also 1666). The conservation of mechanical energy had no theoretical citizenship, since the ideal hard body maintains its shape in an impact, hence no deformation energy was considered. In the whole history of technology, there is an interplay between military technology and science and civil technology[7]. The cultural influence of

[6] Author's italics.

[7] E.g., Leonardo da Vinci (Schneer 1973; Pisano 2009, 2014).

military technology on science was hardly noticed. Newton suggested the notion of a hard body, i.e. an ideal body that is capable of preserving its shape in an impact, irrespective of how large its impulse was Newton ([1704] 1730). Hard bodies were ideal for military technology studies, i.e. a cannon ball like body that produces an immense devastation without changing its physical nature. However, a theoretical viewpoint to put as an ideal case of a hard body implied in theoretical physics that two equal, centrally impacting, hard bodies stop instantaneously. Hence, no principle of conservation of energy is possible (Scott 1971). Thus, this suggestion of hard bodies, since supported by Newton's authority, led scientists to consider the conservation laws as a mere heuristic rule (Mendoza 1963). Not all scientists accepted this attitude. Among Newton's contemporaries, Leibniz was the most famous scientist to believe that the conservation principles played an important role in physics. Finally, the correct path for reaching a final thermodynamic theory was mislead by the Newtonian paradigm and attitude in theoretical mechanics.

2.1 On the Impossibility of Perpetual Motion in Mechanics and Mechanical Machines

2.1.1 Mechanics

Generally speaking, the history of science studies reports successful applications of science to technology (Singer 1954–1958). However, the relation science-technology is not uniform: it can happen that theoretical developments, which could be useful for technology, are exploited many years after their discovery and that, in contrast to this, some practical functioning instruments are constructed without a sufficient theoretical support (Grattan-Guinness 1990a, 1990b). Let us see some examples.

In 1638 Galileo Galilei (1564–1642) presented his celebrated formula for falling bodies (naturally accelerated motion; 3rd: Galilei 1954, p. 169; 1890–1909. III, p. 205). It was not successfully applied to military ballistic science (Capecchi and Pisano 2010) until two centuries later, when eventually new sophisticated theoretical calculations took into account a series of cumbersome corrections caused by air resistance of materials like friction (Pisano 2009a, 2009b). In fact, technology greatly influenced the theoretical process of construction of a new science. In particular, at the beginnings of the theoretical effort, military technology offered the most relevant case to be considered by theorists and hence their cultural address. *What about technologies in the history and science? How can we re-arrange, scientific knowledge (history of science) and its technology (History of science and technology) within a social dimension (technosciences)?* This particular history is interesting since some events exist in which military technology has been converted till reaching improvement of both civil society and even theoretical science; this is the case of the birth of thermodynamics by Sadi Carnot. It certainly has been pushed ahead by the widespread dream of gaining unlimited power by means of a least effort. Such a leading idea addressed the development of mechanical technology towards improbable dreams of obtaining a perpetual motion. In 1734 the French *Académie Royale des Sciences* decided to reject without examination any new claims to have obtained a perpetual motion. Paradoxically, the birth of modern theoretical physics was not a motivation for abandoning such impossible projects. Truly, in order to develop an *ad absurdum* proof on the equilibrium of a massive chain moving around a suspended solid triangle, Simon Stevin (1548–1620) first made use of the *impossibility of perpetual motion.* Stevin wrote:

> It is not true [*falsum*] that the globe moves by itself and the motion has not end [*aeternum*].[8]

Lazare Carnot also made large use of this principle, i.e.,

> […] LVII. what is finally the veritable purpose of moving machines? […] the machines in motion, always lose time and velocity, what is they gained in force.[9]
> […] It is therefore evident that we must absolutely give up the hope of producing that which we call perpetual motion, if it is true that all of the motive forces that exist in nature […].[10]

Joseph-Louis Lagrange (1736–1813) discussed cases in which scientists used the principle of a perpetual motion in theoretical mechanics (Lagrange 1813, p. 7). On the contrary, the birth of mechanics theory gave the impression to support such a hope, since it had been founded by Newton on a first sentence—i.e. Descartes' inertia principle (Descartes 1897–1913, *Inertia law*: VIII–1, II part, § XXXVII, pp. 62–63; see also Descartes 1897–1913; the conservation of movement: *Ivi*, § XXXIX and § XL, pp. 63–65; Bussotti and Pisano 2013a) which as a matter of fact considers a perpetual, uniform motion as a natural motion, which does not requires explanations. Newton's version of inertia principle stresses (as well known according to the three axioms or laws of motion) that *per accidens* only a force may change a moving body with constant velocity. As Alexandre Koyré (1892–1964) stated, René Descartes (1596–1650) and Isaac Newton (1642–1727) explained the real through the impossible (Koyré 1966, p. 276). Thus, after Newton the *idealistic* design to construct some machine providing a perpetual motion did not disagree with the highest theoretical arguments of mechanics. This is why in the eighteenth century (Pisano and Capecchi 2013; Pisano 2011), and in spite of the great advancement of eighteenth century technology, the dream of self-moving machines—as well as machines of unbounded power—was still so much uprooted in the minds of most inventors. In retrospect, such a leading idea obstructed the theoretical work by the scholars, who gave rise to the basic tenets of this new physical theory. Finally, the birth of thermodynamics originated from a radically different attitude from Newtonian (*Ibidem*) attitude in theoretical physics as it apparently induced theoretical laws from a reflection on technology.

---

[8] "[…] ipsique globi ex sese continuum et aeternum motum efficient, quod est falsum". Stevin (1605, p. 35; see also Stevin 1605–1608a, b, c, 1634).

[9] Carnot (1786, pp. 88–89) (Authors' *Italic* style).

[10] Carnot (1786, p. 95; see also p. 94). On principle of perpetual motion and virtual laws, see Pisano (2014).

2.1.2 Mechanical Machines

The essential notions, e.g., force, power and motion are differently used in practical science and in science. In some cases within practical science advanced uses of notions are applied in spite of theoretical science (e.g., the concept of trajectory, weight, friction, etc…). In this regard, Reuleaux claims:

> A machine as a combination of resistant bodies so arranged that by their means the mechanical forces of nature can be compelled to do work accompanied by certain determinant circumstances.[11]

Thus, a machine[12] can be, more or less, imperfect. Scientists do not study machines in themselves, rather the notions, which can be useful, in particular machineries applied to machines such as an imperfect object. Therefore, taking into account Reuleaux's quotation, one can add that a *machine* is an apparatus more or less complex (instruments and mechanism) consisting of one or more parts able to produce *Work (positive, negative, null, mechanical equilibrium)*. It can be mechanically, chemically, thermally, electrically etc… related with various applied sciences. From a physical point of view, we can ask: *what is the relationship between equilibrium, distance and work (positive and/or negative)?* Principle of virtual work laws (displacements, velocities)? A machine may be thought of as a device that helps to make work easier to perform by accomplishing one or more of the following functions:

- Transferring a force from one place to another,
- Changing the direction of a force,
- Increasing the magnitude of a force, or
- Increasing the distance or speed of a force.

But, it is also useful to think of a machine in terms of the *input force* (the force you apply) and the *output force* (force which is applied to the task). Therefore, when a machine takes a small input force and increases the magnitude of the output force, a *mechanical advantage* has been produced. *What kind of theoretical relationship can be drawn between applied force, machinery and motion*? A *mechanical advantage* is the ratio of output force divided by input force. If the output force is bigger than the input force, a machine has a mechanical advantage (greater than another machine). In machines that increase distance instead of force, the mechanical advantage is the ratio of the output distance and input distance. Thus *mechanical advantage* is an output/input from a machines geometry without focusing on the physical quantities (Pisano and Bussotti 2014); that is no measurements were necessary. *It was not possible to build a machine that increases both the physical quantity and the physical distance of a given "force" at the same time.* A boundary divides machines and particular functions (machinery) of particular built machines (crafts) (*Ibidem*) (Figs. 1 and 2).

[11] Reuleaux (1876, p. 35).

[12] On that, For sake of brevity we remind a deepen discussion to our papers (Pisano and Bussotti 2014, 2015).

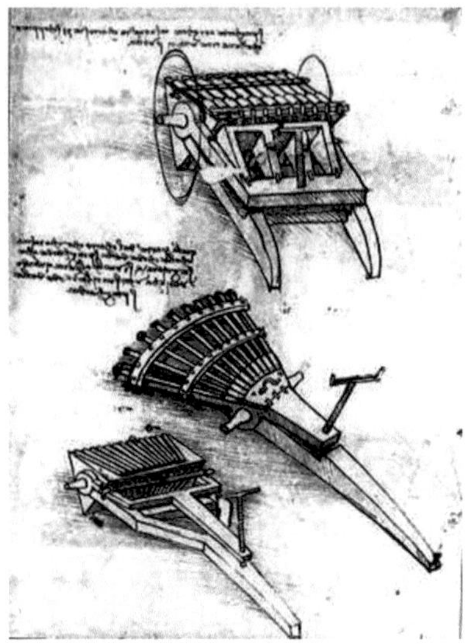

Fig. 1 da Vinci, studies on guns machine with array of horizontal barrels (c. 1481–1482) (da Vinci, *Codex Atlanticus*, f. 157r. It consisted of three sets of machine guns, set on a rotating drum. When the first set is fired, the force of the explosion would pivot the guns down, bringing the next set of guns to the top, ready to be fired. Source: *Biblioteca Ambrosiana*, Italy)

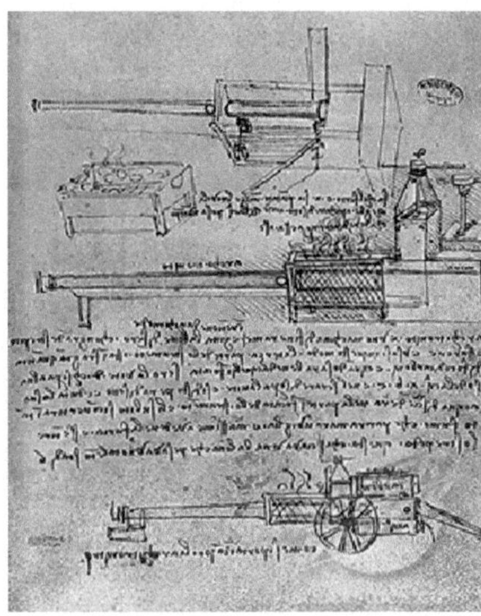

Fig. 2 da Vinci's Steam Cannon (1488–1497) (da Vinci. *Ms. B*, f. 33v. A tentative attempt to produce a steam cannon by copper and a breech. A small quantity of water should be injected into the chamber for instantly creating steam able to quickly power the cannon ball through the barrel. Source: *University of technology Sydney*, Australia.)

The animals and their natural movements were also studied as *new simple machines*: The human body is similar to a complex hydraulic machine[13]. Moreover, animal spirits behave as running water in the nerves and bones as levers with fulcra

[13] Hydraulic machines have been built and theorized from different standpoints. Generally speaking and related to heat machines and *analogies* at that time. We mainly refer to waterwheels for their large and civil use at that time. Particularly, undershot and overshot waterwheels.

in various articulations. Thus, they focused on the machineries of a machine (Figs. 3 and 4):

The relation between the practical construction of machines and the scientific aspect of technology became closer in the eighteenth century. In particular, this first happened in England in relation to hydraulic engines.[14]

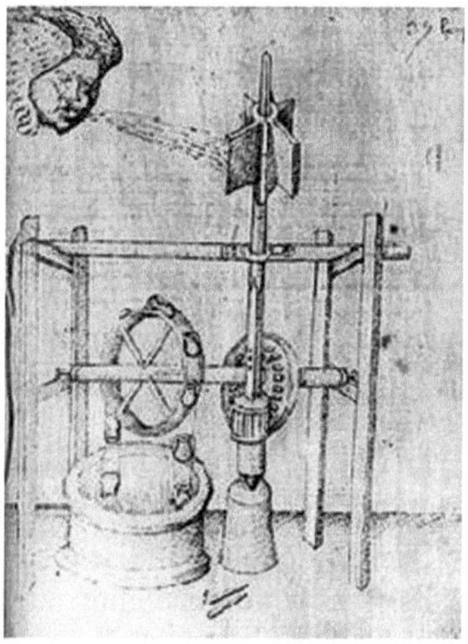

Fig. 3 Taccola, Wind-operated chain-pump (ca. 1405). (*Ivi*, f. 37r. Source: *Ibidem*)

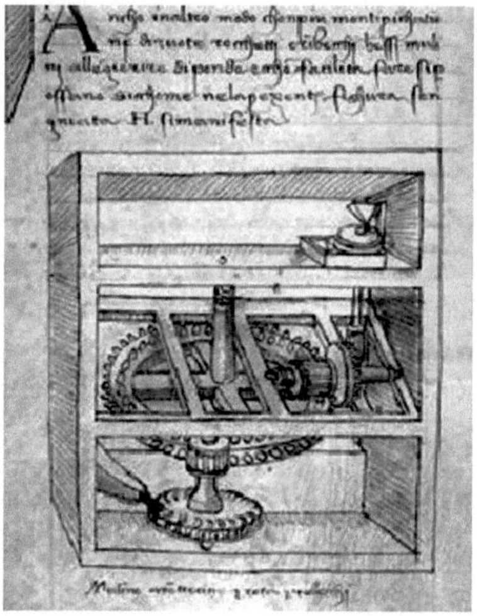

Fig. 4 di Giorgio Martini, Horizontal-waterwheel mill (1480–1482) (Di Giorgio Martini, *Ms. Saluzziano* 148, c. 34v. Source: *Ibidem*)

[14] Di Giorgio Martini, *Ms. Saluzziano* 148, c. 34v. Source: *Ibidem.*

2.2 Mechanics and Early Mechanical Machines Theory

The employment of hydraulic and thermal machines contributed to England's efforts to overcome the energy crisis of the eighteenth century. At the same time, the way of thinking of hydraulic and thermal energy radically changed. At first glance, calculation of the efficiency of waterwheels was necessary. Among many other items, the early contributions were within the development of hydrodynamics theory: i.e., on the force and water stream (Mariotte 1686, pp. 204–206; Parent 1745, pp. 116–123, 323–338).

It is known that in the industrial revolution, of which England was the epicentre in the eighteenth century, the improvements of machines played a prominent role in improving both production and productivity. Early studies, dating back to the beginning of the modern age, focused on paddle wheel and waterwheel as the main machinery system of the water machine. Among the first machines whose functioning was studied in a scientific manner, there are the hydraulic machines. A reasoning might be simply based on a volume rate (V) of water flow transported through a given cross-sectional area (A) with velocity (v):

$$V = Av.$$

A question could be raised: *what about the advantage from horizontal or vertical waterfall?* Until the eighteenth century, a precise physical answer was hard because of the lack of an adequate knowledge of relationship between physical phenomena and mathematical interpretation (Pisano 2011, 2013a); particularly a relationship between sections of the conductors and motive power applied was proposed. The sections of the conductors become smaller at the extremities. A theoretical and scientific enquire on the functioning of machines became hence necessary in order to get a general law of their efficiency. Significant contributions given by John Smeaton (1724–1792; Smeaton 1759) were important because he was able to improve of the efficiency of existing water wheels, a development that had been requested by English industrialization at that time. By means of several experiments and modelling (concerning *positive work* and *negative work*) he compared different hydraulic wheels through the method of *evaluative improvement*. Smeaton varied a single parameter and, *coeteris paribus*, checked the possible improvements of the machine. In the textile industry, which was the basis of the first industrial revolution, it was necessary that looms act quickly, in order to increase productivity. Time acquired a completely new role in modern industry, which was different from the role it had in the artisan-workshop. Hence, Smeaton chose the product of the force developed by the wheel times the wheel-speed to characterize the optimal functioning of the wheel itself. He called such a product *power* (in what follows indicated by P). He rolled a rope around the axis of a hydraulic wheel. This rope could elevate a weight $F_p = mg$ by means of a pulley. Smeaton defined the *power* (also called *original power*) of the water as the product between a certain quantity of water (during a time t) and the elevation from which water comes down. In his words:

> The word *Power*, as used in practical mechanics, I apprehend to signify the exertion of strength, gravitation, impulse, or pressure, compounded with motion, to be capable of producing an effect; and that no effect is properly mechanical, but what requires such a kind of power to produce it. The raising of a weight, relative to the height to which it can be raised in a given time, is the most: proper measure of power; or, in other words, *if the weight raised is multiplied by the height to which it can be raised in a given time, the product is the measure of the power raising it*; […] But note, all this is to be understood in case of flow or equable motion of the body raised; for in quick, accelerated, retarded motions, the *vis inertae* of the matter moved will make a variation.[15]

In other terms, we have:

$$P = mgv.$$

That is, the following power is obtained:

$$P = \frac{mgh}{t} = \frac{W}{t},$$

where *mgh* is equal to work *W*. In addition to these empirical studies, Smeaton concentrated his scientific researches on the causes concerning a greater loss of *mechanical power* (nowadays *energy*). Particularly he focused his attention on the undershot wheel and concluded that it was consumed in some sort of turbulence of the water ("nonelastic bodies"), when:

> […] the effect therefore of overshot wheels, under the same circumstance of quantity and fall, it is at medium double to that of the undershot: and, as consequence thereof, that non-elastic bodies, when acting by their impulse or collision, communicate only a part of their original power; the other part being spent in changing their figure in consequence of the stroke.[16]

Hence, the power is the work performed in the time-unity. This result referred to the idea of *mechanical work*: in the undershot wheel only the force due to the speed of the water that hits the buckets acts, while in the overshot wheel the work of the weight force of the water had to be added[17]. Smeaton obtained brilliant results in this field, but he was not able to extend them to the thermal machines, which continued to have a low efficiency (1 % for the Newcomen's machine).

[15] Smeaton (1759, pp. 105–106). Author's capital and italic. Particularly interesting are his reasonings concerning an early conceptualization of energy (*Ivi* 1782, pp. 341–344). We have no room to describe correlated conservation principles with regard to the emergencies of new sciences, i.e., like heat and electricity. However, we mention that "*probably a vibration of the corpuscles of bodies tending to separate them*" (1799) is how Humphry Davy (1778–1829) defined mechanical contrivances and heat.

[16] Smeaton (1782, p. 130). Author's italic. See also Smeaton (1776, pp. 455–458). Author's italic.

[17] Generally, the motive power for an *undershot wheel* was 33–50 %; and *Vis viva* was consumed by friction. The motive power for an *overshot wheel*: 52–76 %. It may be higher in studying an array of wheels and speeds. In the same experimental conditions and waterfall, an overshot effect paddle wheel is (on average) double an undershot effect paddle wheel. A larger wheel-total waterfall produced larger effects.

2.3 On Lazare Carnot's Mechanics

Lazare Carnot studied systematically the relation between theoretical physics and its applications to machines. He was not only a scientist but also a strategic theorist and a political man; one of the main leaders of the French Republic (Dhombres and Dhombres 1997; Gillispie 1971). His novelty in theoretical physics paralleled his well-known novelty in strategy. He was a follower of Gottfried Wilhelm von Leibniz (1646–1716), who in the analysis of the impact of bodies followed the ideal of an elastic body. Before the advent of the French revolution, in opposition to the theorists of *Total war,* he theorized *Total defence* (Gillispie 1960). Then in 1796, being the supreme chief of the French People's Army he, just by applying his defensive strategy, obtained *victoire* on European monarchist Armies (Reinhard 1950–1952, I, Chap. IX). The main tool of his defensive strategy was the stronghold, because—he wrote—*that it is preferable to constrain the enemy to fight walls and terre pleins than to consume men* (*Ibidem*). At first, a new theory of impact (Carnot 1786, p. 61, ft. 1) also included elastic bodies; notwithstanding his words, Lazare Carnot's basic notion for analysing the body impact was in fact the elastic body. Secondly, the first law of thermodynamics obliged theoretical physicists to include conservation of energy in theoretical mechanics, too. Therefore, the innovation of the law of conservation of energy came from a—commonly unrecognised—minor tradition in theoretical mechanics.

Lazare Carnot's science is strongly relevant in this tradition and he is recognised as the father of technical physics (Gillispie and Pisano 2014). His works on calculus and geometry are general in nature (Gillispie 1971); no technical link may be recognised (White 1962; Bernal 1973) and so different from Newtonian mechanical foundations (Bussotti and Pisano 2013b; Pisano 2010). He obtains a universal theory (Carnot 1786, 1803) concerning a general definition of a machine, mathematical calculus without infinitesimals, and new geometrical accounts. He did not consider forces on bodies, but interactions only, so no forces imposing compulsory behaviours to bodies, but mutual interactions only of the bodies[18].

> Given the virtual motion of any system of bodies (i.e., that which each of the bodies would describe if it were free), find the real motion it will assume in the next instant in consequence of the mutual interaction of the bodies considered as they exist in nature, i.e., endowed with the inertia that is common to all the parts of matter.[19]

Particularly, as mentioned, his topic was the problem of impact in all cases, both plastic bodies and elastic bodies; at first he solves it by introducing an index of plasticity. That provides a theory that in principle applied to all mechanical phenomena (provided that only contact forces are considered; a continuous force is considered as the limit of a series of jerks given by impact). His famous, *first* (Carnot 1786, p. 32) and *second* equations (*Ivi*, p. 33), generalized for multi-body systems are:

[18] On Lazare and Sadi Carnot we refer (sometime directly) to: Gillispie and Pisano (2014). Recently one of us wrote an advanced work on the development of principle of virtual laws taking into account Lazare Carnot's mechanics as case study (Pisano 2014). See also Dhombres and Dhombres (1997).

[19] Carnot (1786, § X, p. 21), see also Reinhard (1950–1952).

$$\sum mVU\cos\left(<\vec{U},\vec{V}>\right)=0\,(\mathrm{E})$$

$$\sum muU\cos(<\vec{U},\vec{u}>)=0\,(\mathrm{F})$$

$m=$ mass of the body
$W=$ velocity before interaction
$V=$ velocity after interaction
$U=$ W–V
$u=$ arbitrary geometric motion

In short[20]:

- The mass of the parts of a machine.
- Global magnitudes, abstracting from the mass of the mechanism.
- Kinematics first, then dynamics; and statics as a special case of dynamics.
- A theory of machines concerns a theory of the *communication of motions*.
- A machine is a connected system of (hard) bodies.
- The connections between the bodies constrain the *communication* of motion of the bodies.
- The theory of interaction–collisions by means of insensible degrees (e.g., see Carnot 1803 § 293, pp. 261–262) as the result of a sequence of infinitesimally small percussions.

Particularly, from its 2nd main equation—which is a generalization of the principle of virtual velocities (Pisano and Drago 2013; Pisano 2014)—conservation laws are derived through a notion—i.e., *geometric motions*[21]—which introduces groups of transformations; that is, for the first time—before Nöther's theorem—his physical theory is governed by a symmetry technique. Owing to the fact that conservation laws are solved for momentum, its mathematics has to solve merely algebraic equations; in other terms, his theory dispenses with differential equations. From a technological standpoint, Lazare Carnot mathematically stated the impossibility of a perpetual motion both for the seven simple machines and for whatsoever machine: no chimerical source of unlimited power is possible. Briefly, equilibrium was provided by an *applied power* to a machine taking into account a *resistant power*, i.e, simply friction. However, to formulate an adequate equation of motion, rather than that of equilibrium, including all resistant physical quantities at that time known (calculated/measured), was hard, but not impossible.

In his first *Mémoire sur la théorie des machines* of 1778 (Carnot 1778; Gillispie 1971, Appendix B, § 60) Lazare Carnot treated force of friction and other resistance forces (*Ivi*, § 50) as "active" forces (*Ibidem*) exhibiting "displacements by insensible

---

[20] For an extensive discussion see: Gillispie and Pisano (2014), Pisano (2014).

[21] Carnot (1786, pp. 28–34, 41–45, 1780, § 113); Gillispie (1971, Appendix C, § 113, pp. 308–309); Gillispie and Pisano (2014, Chap. 2 and Chap. 11).

degrees" (Carnot 1786, § X, p. 21, 92; 1803a, § 293, pp. 261–262). Very important was his conceptualization—frequently for emerging theories at that time—his model of the *communication/production of motion* (e.g. falling water in a hydraulic wheel) and reversibility processes[22] (Fox 1970; see also 1971a, 1971b, 1976, 1988).

3 Thermodynamics Science and Heat Machines in Context

Starting from the late Middle Ages, the two major accredited theories of heat were: the mechanical theory, which conceived of heat as an *accident of the matter*, taken from Roger Bacon (1214–1294) and Johannes Kepler (1571–1630), and the *fluidists* theory which conceived of heat as a *substance* (a fluid), while the scholars before Lavoisier (Pisano and Franckowiak 2014) associating heat with the element *feu*. The mechanical theory of heat (Callendar 1910) cohabited for a long time with the fluidist theory of heat based on the previous theory of the *phlogiston* studied by Georg Ernst Stahl (1660–1734). The fluidist theory was welcomed by scientists and philosophers because they found it coherent. Leonhard Euler, Henry Cavendish (1731–1810), Joseph Priestley (1733–1804), Adair Crawford (1749–1795), Johann Mayer (1752–1830), Immanuel Kant (1724–1804), and Alessandro Volta (1745–1827), considered *phlogiston* neither heat nor temperature; but as something which produced heat when it freed itself into the bodies. We remark that research on heat phenomena mainly implicated magnitudes such as heat Q and temperature t. The interest was focused on the ratio dQ/dt, because it seemed the basic theoretical physical magnitude. The experiments by Joseph Black (1728–1799) were crucial for development of the theory (Black 1803; Guerlac 1970–1990, pp. 173–183). He started with Georg Wilhem Richmann's[23] (1711–1753) formula[24], extended it to fluids different from water and made primal studies on what today we call thermometry and calorimetry, latent heat and specific heat. In 1789, chemistry produced a real revolution and Lavoisier, as well as other chemists of his time, searched for basic principles of this new theory in a revolutionary fashion.

Lavoisier assumed (1789; 1862–1893) new elements: *chaleur, calorique*[25] (et *lumière*). The research carried out by Lavoisier and Pierre Simon Laplace (1749–1827) (Lavoisier and Laplace 1784) is also of remarkable importance. In 1802

[22] In this context, one may remember the decisive role played—according to authoritative historians—by the notion of efficiency in order to assure the validity of the law of energy conservation (Kuhn 1959; see also 1960, 1962).

[23] Richmann (1750, pp. 171–172). He wrote "caloris" that correspond to our modern magnitude temperature.

[24] "De quantitate caloris, quae post miscelam fluidorum, certo gradu calidorum, oriri debet, cogitations", *Ivi.*

[25] Lavoisier attempted an early distinction between heat and temperature. (Lavoisier 1789, pp. 12–17, line 37).

Joseph-Louis Gay-Lussac (1778–1850) formulated a law of gases; then Pierre Louis Dulong (1785–1838) and Alexis Thérèse Petit (1791–1820) showed in 1816–1819 that specific heat depended on temperature (in some cases). However, we have to wait for the second half of the 19th century to acquire verification, not only experimental, of the laws of gases known at that time and studied by theoretical physicists (and chemists). In particular, the adiabatic law (Poisson 1823, pp. 5–16; Laplace 1822) caused some confusion and various versions of it were formulated; even when Siméon Denis Poisson (1781–1840) formulated the right equation, most scientists did not consider the issue resolved. Scientific knowledge on the matter took two paths, one based on the properties of gases (kinetic model of gases) and another one based on the efficiency of heat machines (Zeuner 1869), which naturally included the gas theory which would later become thermodynamics. In this context Sadi Nicolas Léonard Carnot (Payen 1968, pp. 18–30; Fox 1971a, pp. 67–72; Taton 1976; Mendoza 1959, pp. 377–396; Carnot 1986) relying on his father Lazare's (Carnot 1786; Gillispie 1971, 1970–1979) theory of mechanical machines, opted for the latter. His *Réflexions sur la Puissance Motrice du Feu*[26] (hereafter *Réflexions*) and his last work on the theory on gases, *Recherche d'une formule propre à représenter la puissance motrice de la Vapeur d'Eau*[27] (Carnot 1978, pp. 223–225) appeared in quite a short period of time, approximately three years.[28] It was a total overturning of the physics–mathematics relationship (Pisano 2013a; Barbin and Pisano 2013; Drago and Pisano 2007, pp. 497–525), a new theory starting from engineering practices. However, a strong analytical approach had already been adopted by Jean Baptiste Joseph Fourier (1768–1830). He formulated his differential equation for heat propagation in solids (1807), which was presented before the contribution of Sadi Carnot, and when the difference between heat and temperature was still subject of debate. He introduced many innovations. In the wake of Fourier (Pisano and Capecchi 2009) some 50 years later, Gabriel Lamé (1795–1870; *Ivi*) wrote *Lecons sur la theorie de la chaleur* (1861). At this point, the theory of heat was a well-defined science and several important conclusions had been reached on thermodynamics, whose field of applicability in physics was even studied by mathematicians.[29] Let us note that Lamé does not seem interested in thermodynamics and devotes himself to heat transmission, a topic that now has its own framing in a new branch of physics.

---

[26] In 1978, Robert Fox published an excellent critical edition of the *Réflexions sur la puissance mostrice du feu*, including crucial interpretative analyses. Here, he suggests that a technological study of Sadi's book cannot explain the originality of either theory or reasoning (Carnot 1978; English: Carnot 1986; Fox 1988).

[27] See also: Fox (1971b), Reech (1853, pp. 357–378), Lamé (1836).

[28] Sadi Carnot wrote a mathematical (only) footnote (Carnot 1978, ft. n. 1, pp. 73–79; Gillispie and Pisano 2014).

[29] Some of them: Liouville (1836), Dirichlet (1837), Reech (1853), Riemann (1861 [1868] cited in Weber 1892, pp. 391–404). Riemann's lecture (1861) "on the conduction of heat" was not published until 1868 (Spivak 1979, pp. 179–182). For Eugenio Beltrami, the titles were: 1871–1874 *Ricerche sulla cinematica dei fluidi*; 1882 *Sulle equazioni generali dell'elasticità*; 1884 *Sulla rappresentazione delle forze newtoniane per mezzo di forze elastiche*; 1889 *Note fisico–matematiche*.

3.1 The Impossibility of Perpetual Motion in Thermodynamics

In the progress of thermodynamics and its applications, the researches connected to the impossibility of a perpetual motion played a fundamental role. A step was to recognise that gas' behaviour can determine the behaviour of a great and heavy machine. It was necessary to study the specific laws for the gases.

Lazare Carnot stated that fuel can be generated by engines in three ways only: by passing to a gas, as gunpowder does; by heating water till it changes into steam; by raising the temperature of a gas. Consequently, there exist three fire engines only. No surprise if theorists again followed the ideal of cannon as the best machine. Before Sadi Carnot some theorists only tried to compute the efficiency of heat engines. By following the *chimera* of an unlimited power, they theorized engines as particular cases of a cannon. Accordingly, they considered two phases only; of course, the cannon suggests the explosion phase and, a phase of the expulsion of the ball. In fact, de Prony, Hachette, Petit, Coriolis, Navier and others calculated the *expansive force* of a gas moving a piston in a cylinder—an apparent scheme of a cannon rather than of a cumbersome heat engine (Scott 1971, p. 224; Hachette 1819, pp. 315–316; Grattan-Guinness 1990a, § 16.5.5.6; 1990b). When Sadi Carnot's theory put a bound in the efficiency of steam engines, it was the first time that theoretical physics met a bound to its dream of omniscience and omnipotence. This point suggested to Meyerson to qualify the resulting theory as *a defeat of the mechanistic reason* (Meyerson 1904). Moreover, one has to stress the crucial roles played by both notions of reversibility—the old "displacements by insensible degrees"—and the *impossibility of perpetual motion* in both Lazare Carnot's (i.e., Carnot (1786), § XLI, pp. 75–76; § XLII, pp. 77–78; LIX, pp. 91–92) mechanics and Lazare Carnot's strategy. Without these notions Sadi Carnot (i.e., 1978, p. 41) would have been unable to start his thermodynamics theory. The cycle, more than other notions severs civil engines from cannons. In fact the cycle represents a restoration of initial conditions; it underlines the idea of preservation, while the explosive power of a cannon underlines the idea of an irretrievably move (of destruction). In his theory Sadi Carnot achieves this notion of a cycle after some approximations to it. In the first paper the real steam engine, as subject of study, suggested to him a three stages cycle (vaporization of water, expansion and condensation of water). In the first pages of the book he describes this incomplete cycle. Nevertheless, a fourth stage is necessary for closing the cycle (Pisano 2010), as later by Sadi Carnot showed using the steam. He suggests the fourth and last phase closing the cycle, i.e. an isochor transformation—which results to be reversible since the interval of temperatures is made an infinitesimal.

Finally, Sadi Carnot, by passing to consider an air engine, describes the cycle of four operations with an adiabatic instead of the isochors, as in the present time version. A retired officer of French Army, Sadi Carnot took into account—contrarily to his colleagues of *École Polytechnique*—nothing of the military weapons.

3.2 Thermodynamics and Mechanical/Heat Machines: analogies

In the following we report the main common conceptions between mechanics and thermodynamics on machine in the case-study of Lazare and Sadi Carnot (Gillispie and Pisano 2014) (Table 1).

Sadi Carnot chooses variables considering the physical system globally:

$$W_{\max} / Q = f(Q, V, t).$$

By means of them he obtains a simple mathematical formula:

$$W = p\Delta V,$$

which, also has theoretical merit: in the exemplary thermal case, which is a gas in a cylinder, it formally belongs both to mechanical theory and heat theory. For, generally speaking we have

$$W = F\Delta s = (F / S)\, S\Delta s = P\Delta V.$$

It should be noted that when Lazare Carnot extends his mechanical theory to hydraulic machines he uses the case of a fluid in a cylinder as an example. Therefore, the concept of work used by Sadi Carnot in thermodynamics has the same mathematical formula as mechanical work in a particular case of the old theory. Neverthless, it is the conceptual core of the new theory; in this way, a formal element of continuity is given to the two theories. Just after the first announcement of his theorem (Carnot 1978, pp. 21–22) the analogy (Carnot 1978, pp. 8–9, 28–29) between the falling of water on a hydraulic wheel and the falling of caloric in a heat machine is proposed. According to Gillispie (1971, pp. 96–98), this analogy between two apparently different physical systems also reveals the scientific sensibility and attitude of the young French scholar.

Lazare Carnot was also interested in finding out the criteria (Carnot 1986, ft. 23, p. 124) for obtaining the greatest possible effect from machines (Carnot L 1780, §§ 149–152, §§ 155–157; Carnot 1786, pp. 89–94; Gillispie 1971, pp. 327–330, 332–333). His reference to the hydraulic engine (Carnot 1786, pp. ix–x, pp. 88–81; Carnot 1803, pp. xxi, p. 149, pp. 247–250) is interesting for our aim. As previously stated in the first paragraph, the product of *communication of motion* (Carnot 1780, ft *, § 148; Carnot 1786, p. iii–iv, 44; 1803, pp. xiii–xvi) and power should always be transmitted without percussion and all stocks should be avoided (Carnot 1786,

Table 1 On the common conception of *work* for a machine and its working substance

Lazare Carnot (1786)	Sadi Carnot (1824)
Work as a primary magnitude	Work as a primary magnitude
No intermolecular forces for the calculation of the work	No intermolecular forces for the calculation of the work
W_{max} and $W = (F/S)S\Delta s$	W_{max} and $W = p\Delta V$

pp. 45–48, 91–95, pp. 89–91, pp. 93–99; Carnot 1780, §§ 146–147, § 152, § 157; Gillispie 1971, pp. 323–325, pp. 328–329, p. 333). In particular,

> [Sadi] Carnot here remarked that the performance of work by heat is quite analogous to that by waterfall. By the fall of heat (*chute du calorique*) the performance of work is determined in quite similar manner to that performed by the fall of water (*chute d'eau*).[30]

Sadi Carnot substantially defined that concept when he utilized heat at a constant[31] temperature, under certain conditions, Q/t (Carnot 1978, p. 32). Nevertheless, that assumption is still weak. However, presuming that this assumption (or hypothesis) is valid, within analogy, the *falling of water* should be considered analogous to the *falling of the quantity of heat* at constant temperatures. That is to say:

$$Q_1/t_1 - Q_2/t_2 = \Delta S \text{ between the two temperatures.}$$

> [*chute du calorique* is quite similar *chute d'eau*]. But whilst for the water the performance of work is simply proportional to the height of fall, we may not put this performance in the case of heat proportional to the difference of temperature without a closer investigation.[32]

While in the hydraulic wheel W_{max} is proportional only at Δh, in the heat machine W_{max} can depend on unspecified variables. However, as Sadi Carnot remarks (Carnot 1978, p. 29) this work is not proportional to Δt since W_{max} seems to have greater experimental values at low temperatures (Carnot 1978, p. 72). Therefore, this is a new type of function. Sadi Carnot's theoretical effort reaches a standstill at this last difficulty, although he even attempted a calculation in his previously discussed famous footnote (Carnot 1978, ft. 1, pp. 73–79) to determine the efficiency function. However, let us note that the analogy is more persuasive than it appears in modern times. Moreover, around the XVIII century, it was common to consider machines by performing an abstraction from the masses of bodies (Carnot 1786, § XXX, p. 60) and separate the wire of a pendulum from the mass. Therefore, the water that falls on the hydraulic wheel could have been thought of without mass, that is, as a weightless fluid, as caloric fluid was conceived. This way of envisioning mechanical machines allowed (still within the caloric hypothesis) for the consideration of the analogy as a true connection of heat machines to the theory of mechanical machines which includes the case study of falling water and the hydraulic wheel. Therefore, for Lazare Carnot, this is a fundamental analogy and for Sadi Carnot, who doubts (see also Carnot L 1990) it is merely striking; for us it plays a relevant but not essential role.

3.3 On the Cannons

In eighteenth century, many amazing and innovative civil machines were constructed: steam engines to extract water from mines, new ships, traditional transports, and alternative studies like Mongolfier's balloon (Gillispie 1983). At the same time, problems related to unexpected explosion were partially solved. It is well-known

[30] Mach (1986, p. 201, line 37). (Author's italics and "( )").

[31] Carnot (1978, p. 32).

[32] Mach (1986, pp. 201–202, line 40). (Author's *italic*).

that in 1769, 1500 Spanish cannons have been tested; a large percent of them result as inefficient. Later new strategist' school advanced and the enormous power delivered by big cannons became the symbol of a new social power capable to mobilise the population oppressed by the *anciens regimes* (Reinhard 1950–1952). Nevertheless an organized production of civil and gun machines was weak. Therefore many studies addressed to improve, i.e., outward stroke of a heat engine, trying to conceive a heat machine as case study of a cannon. Both White and Bernal (White 1962; Bernal 1973, pp. 166–167) agree that

> […] cannon is not important in itself, [only] as a war machine consuming energy; it represents also one cylinder, inner combustion engine and all modern engines of this kind derived from the cannon.[33]

Among the new weapons, the carronade was adopted by *The Royal Navy* in 1779 (Manigault and Warren 1996).

In the 1810s and 1820s, great emphasis was placed on the accuracy of long-range gunfire, and less on the weight of a broadside. The carronade disappeared from the *Royal Navy* in the 1850s, after the development of jacketed steel cannon, by William George Armstrong (1810–1900) and Joseph Whitworth (1883–1887). Nevertheless, carronades were used in the *American Civil War* (Manigault and Warren 1996). The Great Turkish Bombards of the siege of Constantinople, after being on display for four centuries, were used to battle a British fleet in 1807, in the Dardanelles Operation (Wallechinsky and Irving 1975). The Western cannon, during the nineteenth century, became larger, more destructive, more accurate, and could fire at longer range (Hazlett et al. 2004). The cannons were crucial in Napoleon Bonaparte's (1769–1821) rise to power, and continued to play an important role in his army in later years[34]. The practice of rifling—casting spiralling lines inside the cannon's barrel—was applied to artillery more frequently by 1855, as it gave cannon gyroscopic stability, which improved their accuracy.

Generally speaking, the efficiency of cannons was improved by several inventions which attracted the attention of scientists. On that, Louis-Christophe-François Hachette (1800–1864)—a friend of Lazare Carnot—celebrated his second account(Hachette 1811, 1819) on machines devoting a special section to steam guns as suggested by scientific progress (Carnot 1803, pp. 49–51)[35]. The progression of the study (early by Petit 1818) of the efficiency proceeded by steps, preliminarily underlying the elastic nature of fluids, i.e. the gases since it was out of the perfectly hard bodies within Newtonian paradigm. An operative approach for calculating all referring to the gases in a running heat (Hachette 1819) and mechanical machines (Prony 1790, 1796, Prony et al. 1829) was developed. Particularly they also concentrated on the *expansive force*[36] of a gas in a machine, without necessarily showing

[33] White (1962, pp. 161–162).

[34] See Gillispie (1960), Chandler (1995), Adkin (2002), Wilkinson-Latham (1975).

[35] Let us note that the experimental nature of this set of principles was also emphasised by Dugas (1956).

[36] The term *expansive force of vapor* or *expansive force of heat* or *expansive force of caloric* was largely used at the time. Sadi Carnot used this term, too (Carnot 1978, p. 5). For details see Gillispie (Gillispie 1976) and Gillispie and Pisano (2014, pp. 400–403, and references in the note 67, p. 401).

interest for theorising of a working machine. It is also important to mention that, firstly Hachette (1811, 1819) and then Jean-Victor Poncelet (1788–1867) stressed (Poncelet 1845) the difference (heat, gas) produced by gunpowder. Of course, all inventions and scientific studies should take into account eventual financial sources and cost of the economical production; consequently, heat engines were welcomed because of social and economical reasons. However, most studies concentrated on the work of the expanding gas, like i.e. manifested impressively by big cannons. Petit calculated it (Gillispie and Pisano 2014, pp. 94–96) with the following formula

$$mv^2 = 2g\delta hab\log\frac{x}{a}.$$

It is related to geometry of piston and geometric characteristic of both cylinder and the elasticity of the fluid. Some scientists reiterated such formulas, sometimes without taking into account the expansion of the gas (*Ibidem*).

3.4 On Sadi Carnot's Thermodynamics

Sadi Carnot was a pupil of the *École Polytechnique* and invented thermodynamics theory (Pisano 2010; Gillispie and Pisano 2014). The *École Polytechnique* was militarised by Napoléon and in 1814 Sadi Carnot bravely fought to defend France from the attack of European monarchist Armies. Then he was a career officer; yet after a few years he retired. He received the suggestion of studying heat engines by his father. He devoted two years to collecting information, attending courses in *Conservatoire des Arts et Métiers* and eventually writing an early book[37] about him. In his study of heat machines Sadi Carnot took in account civil engineering only. It is remarkable that in the first pages of his unique published book, *Réflexion sur la puissance motrice du feu* (Carnot 1978, pp. 2–5), he compared the energy supplies of England with its Navy; he concludes that nothing is worse for a Nation than an energy crisis—i.e. a civil catastrophe, instead of a military catastrophe. Then he foresaw the starting of a new era as determined by the employment of engines in society (*Ibidem*). Recent studies confirmed the previous hypothesis by Gillispie (1971) concerning the scientific relationship between Sadi Carnot's thermodynamics and his father's, Lazare Carnot, mechanic theory as a *scientific and strong filiation*. Almost all the basic notions upon which Sadi Carnot founded thermodynamics may be recognised in Lazare Carnot's theoretical mechanics; i.e. maximum of efficiency, impossibility of a perpetual mobile, reversibility of operations, cycle of operations, state of the system, work (Gillispie and Pisano 2014). According to

[37] Sadi Carnot's *Recherche d'une formule propre à représenter la puissance motrice de la vapeur d'eau–unpublished manuscript* (Carnot S–EP s.d; see also Carnot 1986, pp. 167–180). Generally speaking, the correct date of publication still lacks historical decisive evidence. More or less the most recent Carnot historians agree on a date before 1824 (Gillispie and Pisano 2014, Chap. 3).

Challey[38] some historians thought that Sadi Carnot's theory was too abstract to be considered an essay on engineering: especially because the work did not easily and clearly defined how the function for heat machines' efficiency depended on temperature (Pisano 2001; Drago and Pisano 2005). Furthermore, Carnot's theoretical efficiency was disproportionately higher (some percentage) than that which was achievable at the time.

Other reviews suggested the opposite, considering Sadi Carnot's thermodynamics too practical and therefore less valuable for theoretical physics in general; that was also the opinion of Sadi Carnot's contemporaries, who were accustomed to infinitesimal analysis. They concluded that Carnot's book was naïve because it lacked the basic concepts for a nineteenth century scientific theory (Fox 1971a, 1971b, 1974), as absolute space and time, cause-force, differential equations, material-point, etc. were missing from it. In 1873 it was revealed that in some posthumous manuscript (Gabbey and Herivel 1966; Carnot 1978, pp. 223–225) Carnot had been in favour of the mechanical theory of heat[39] (Callen 1974, 423–443; Callendar 1910) and that he had even obtained the value of the heat equivalent—twenty years before Mayer (1814–1878) who did so in 1842. In this sense, he left the theory of conservation of caloric and presented the conservation of (energy) "puissance mécanique" (Carnot 1978, p. 248; Fox 1988, p. 299). In spite of all this, Carnot's theory seemed of little interest for over a century, essentially, because it was based on caloric theory, which was considered erroneous. Other studies however, have considered the young French scientist's thermodynamics fascinating, both for the simplicity of the language used and for the air of *mystery* produced by the lack of mathematical calculations in this work[40].

> But the fact remains that the central problem of the *Réflexions* is an engineering problem, and that the most of the solutions that are offered on the book have their roots firmly in the engineering tradition. This helps to explain why the argument has none of the elaborate mathematics that Carnot's contemporaries were accustomed to finding in the work of such as Laplace, Poisson and Fourier, all of whom wrote on the theory of heat, as mathematical physicists, between 1800 and the 1830s. […] For although the *Réflexions* lacks the mathematical sophistication of much of the physic of the period, its unassuming presentation conceals an argument of extraordinary depth and, in parts, of baffling complexity. [41]

---

[38] Challey (1971). "Sadi Carnot" in Gillispie 1970–1980, vol III, pp. 79–84. See also Kuhn (1962), Takayama (1979). Just a few lines to remark that Sadi Carnot knew James Watt's works (1736–1819) as one of us (RP) saw in his dossier at the *Ecole polythétique* de Paris. The Scottish engineer proposed a cycle of working operations of the entire engine but not of the working substance, for which he only proposed a prolongation of the isothermal expansion with a phase of adiabatic expansion. Finally, just two of the four phases were proposed by Carnot's cycle (Gillispie and Pisano 2014, Chap. 8).

[39] Moreover in "Notes sur les Mathématiques, la physiques et autre sujets" he clearly introduced his "thèse générale" on the energy: "[…] la puissance motrice est en quantité invariable dans la nature, qu'elle n'est jamais à proprement parler ni produite, ni détruite." ("Notes sur les Mathématiques, la physiques et autre sujets" III, folios 7, 44, line 9 in: Picard 1927, p 81, line 14).

[40] Sadi Carnot only wrote a mathematical footnote (Carnot 1978, pp. 73–79, ft. n. 1). Both calculus and reasoning mathematical footnote are published (into Italian) by: Pisano (2001, 205–230), Drago and Pisano (2005, pp. 37–58).

[41] Robert Fox in Carnot (1986, pp. 2–3, line 7).

Among those historians sympathetic to Carnot was the scientist Friedrich Wilhelm Ostwald[42] (1835–1932). He observed that Carnot used two different words to indicate heat: "chaleur" and "calorique". Particularly, Ostwald (1892) affirmed that in Sadi Carnot's theory, "calorique" is used to mean "heat changed without variation in temperature". Thus, Carnot's "calorique" is equivalent (due to the constant value of temperature) to entropy. Carnot's simple language had left his historical hypothesis appearing unsound: Ostwald proposed replacing *calorique* with *entropy*, thereby placing Carnot's theory within modern thermodynamics.

In the 1940s, through the careful re-construction of Sadi Carnot's biography (Aries 1921; Picard 1927; Kerker 1957, pp. 143–149), Léon Rosenfeld (1904–1974) presented an important interpretation of Carnot's and his own principles (Rosenfeld 1941, pp. 197–212). Rosenfeld also tentatively proposed a correlation between Carnot's theory and the technological developments of the industrial revolution (Singer [1954–1958] 1993; Gillispie 1960, Chaps. 9–10). One can certainly notice and acknowledge engineering in Carnot's theory, focusing as it did on the fundamental problem of finding the upper limit to machines' efficiencies. We know that Sadi Carnot based his work on the analogy between the gravitational fall of water in the hydraulic wheel and the caloric fall during the cycle of heat driven machines.[43] Additionally, it is significant that Sadi Carnot also sought to provide a general theory, which could be applied to any machine and could work with any fluid. The theoretical physical content of such a theory is actually very high, going far beyond the engineering study of machines. Carnot constructed a theoretical model, based on equations of state and functions of state, which helped him to reach some important theoretical results on gas laws, known at the time only through empirical observations. During the 1970s and 80s, Truesdell (1970) and Hoyer (1976, pp. 221–228) tried to interpret the foundations of thermodynamics using this new historical-foundational approach. The different historical interpretations of Sadi Carnot's theory appear similar to the two main existing interpretations of the birth of modern science: optics and mechanics in the 17th century.

Another historical interpretation, one which appears to be more far-reaching, is based on the connection (Gillispie and Pisano 2014) between Sadi Carnot and his father Lazare (Gillispie 1971, 1976; Drago and Pisano 2007, pp. 97–525). The most significant works to have arisen from this path of research include those by Koenig, Scott and Gillispie, their theories all being based on analysis of the same themes within Carnot's writings: discussion of gas theory, impossibility of perpetual motion and efficiency of a machine. For instance:

---

[42] He translated Carnot's book into German, aiding its dissemination. We also remark that in his famous *Traité élémentaire de chimie* Lavoisier had already dedicated some pages to the definition *calorique* and *chaleur* heat which he defined in a different way, according to different context (Lavoisier [1789] 1973, pp. 12–17, line 37; see Pisano and Franckowiak 2014).

[43] This is only valid if the heat, as well as the caloric, remains constant. For Carnot's cycle one may refer to: Kuhn (1955, pp. 91–94; *Id.*, 1961, pp. 567–574); Mendoza (1963, pp. 262–263), Klein (1976, pp. 213–219); Montbrial (1976, pp. 333–335).

> Sadi's work can be seen as an extension (of the father's mechanical theories) to the heat machine inspired by Lazare's thought.[44]

Ten years later, in 1971, Scott (1971) noted that the concept of *cycle* was already implied by Lazare Carnot's theory of mechanical machines, when these include springs.[45] In fact, according to Lazare one can calculate efficiency only after the springs have returned to their original; state; that is, when the entire machine has completed a full cycle of operation. In *Réflexions* Carnot attributes the concept of *cycle* with no significant application in theoretical physics until the nineteenth century, while it played an important role in his own work. It is noteworthy that, in later work, Carnot's father Lazare referred to a state in a mechanical system in which the resting position was restored, when the springs had returned to their initial positions (Scott 1971, pp. 82–103, 212–241). Then, *where did the cycle idea come from*? Previously, researchers have hypothesized that the concept of *cycle* in S. Carnot's theory could have been inspired by the typical working of the electric current in Volta's pile (Pisano 2003). Moreover, since Lazare Carnot had a clear idea of kinetic momentum and energy, he probably understood that the work can be carried out independently from the specific job performed: especially when the resting position at the beginning and at the end are the same in regards to gravity[46].

3.5 Sadi Carnot's Efficiency Formula

As already mentioned, the fact that Sadi Carnot calculated (from Gillispie and Pisano 2014) a cycle between only two isotherms, t and $t + dt$ can leave us perplexed because the (final) result is still valid from a modern point of view.

> Carnot now goes on to calculate the motive power obtained in the cycle [4 phases] for air, using the convenient assumption that the "fall" of temperature [between two isotherms] is so small that the effect of the adiabatic can be ignored.[47]

Nevertheless, nothing proves that this is also valid for a finite adiabatic. Let us see the possible calculus. Sadi Carnot utilized the following adiabatic formula[48]:

$$t(V) = \frac{K + \tau \log[\ln] V}{K' + \tau' \log[\ln] V}. \tag{1}$$

[44] Koenig (1959, pp. 57–111).

[45] Carnot (1786, 1803, 1813). See also "Note" in Carnot (1813, pp. 213–257). Poisson (1781–1840) also wrote about a cycle in *Traité de Mécanique*: Poisson (1833, pp. 5–16; 1823 (II), pp. 552–554); one can also see critical discussions in Truesdell (1970, pp. 36–38; 1980, pp. 208–235). But all these studies are occasional facts that do not seem to be inspirited by S. Carnot.

[46] Naturally, as long as there are no collisions and the shifting is done "par degrees insensible", that is, in Sadi Carnot's (Carnot 1978, p 33, line 9) reasoning, "rétablissement […] d'équilibre dans le calorique", reversibility, in modern times.

[47] Robert Fox in Carnot (1986, p 147, En. 78, line 1). (Authors' quotations).

[48] Carnot (1978, p. 66, line 10). Since there is no room here, we refer to Gillispie and Pisano's book (2014) for other contributors to completion of the history of mechanics and thermodynamics after Sadi Carnot's formulation of the efficiency of a heat machine.

Experimentally, the coefficient τ' of *log V* is a very small quantity, therefore, in order to simplify the calculations, as Sadi Carnot also suggested (see above) at the end of the mathematical footnote, $\tau' = 0$ can be introduced. In this case, Eq. (1) can be expressed in the following way:

$$t(V) = \frac{K}{K'} + \frac{\tau \log V}{K'},$$

from which, if we consider

$$\frac{K}{K'} = M \quad \text{and} \quad \frac{\tau}{K'} = \Omega, \tag{2}$$

it follows that

$$t(V) = M + \Omega \log V. \tag{3}$$

Let us consider Sadi Carnot's *ad hoc* cycle. For example, let us calculate a work $W_{\gamma\beta}$ on an adiabatic expansion along a path $\gamma\beta$ of an ideal gas:

$$W_{\gamma\beta} = \int_{\beta}^{\gamma} p dV. \tag{4}$$

From Sadi Carnot's state equation of ideal gases $pV = Nt$ (Carnot S–EP folios 5–6; Carnot 1986, pp. 172–173; Gillispie and Pisano 2014, Chap. XI) we obtain $p = N\frac{t}{V}$, which substituted in Eq. (4), gives:

$$W_{\gamma\beta} = N\int_{\beta}^{\gamma} \frac{t(V)}{V} dV. \tag{5}$$

From Eq. (3) we can write

$$\log V = \frac{t(V) - M}{\Omega},$$

obtaining the following expression for volumes V and dV, which depend on t–variable temperature only:

$$\begin{aligned} V &= e^{\frac{t-M}{\Omega}} = e^{\frac{t}{\Omega}-\frac{M}{\Omega}} = e^{-\frac{M}{\Omega}} e^{\frac{t}{\Omega}} = Be^{\frac{t}{\Omega}} \\ dV &= e^{\frac{t}{\Omega}} \frac{B}{\Omega} dt. \end{aligned} \tag{6}$$

Thus, by substituting Eq. (6) in Eq. (5) we have:

$$W_{\gamma\beta} = N\int_{\beta}^{\gamma} \frac{t}{e^{\frac{t}{\Omega}}} \frac{1}{\Omega} e^{\frac{t}{\Omega}} dt = N\int_{\beta}^{\gamma} \frac{t}{\Omega} t dt = \frac{N}{2\Omega}(t_{\gamma}^{2} - t_{\beta}^{2}). \tag{7}$$

Executing the same calculation on the other adiabatic-work $W_{\delta\alpha} = \frac{N}{2\Omega}(t_{\delta}^{2} - t_{\alpha}^{2})$, and since $t_{\gamma} = t_{\delta}$ and $t_{\beta} = t_{\alpha}$, in the end we conclude that the two works differ only by a sign; that is, they are equal and opposite:

$$W_{\gamma\beta} = -W_{\delta\alpha} \tag{8}$$

From a modern standpoint, the two adiabatic works do not give any contribution to the calculation of the efficiency within a Carnot's cycle. Their numerical values are equal and opposite; such as the result expressed by (8). Therefore, in modern terms, the reversible efficiency of a heat machine in a Carnot cycle is expressed as follows:

$$\eta_{rev} = \frac{Q_2 - Q_1}{Q_2} = 1 - \frac{Q_1}{Q_2} = 1 - \frac{T_1}{T_2}.$$

Sadi Carnot's footnote in *Réflexions sur la puissance motrice du feu* discussed by means of his original level of mathematics and the application of his father's *synthetic method* can contribute to clarify an enigmatic and crucial aspect of the birth and development of both Sadi Carnot's theory and the historical knowledge of thermodynamics.

> I believe that analysis of Carnot's description raises questions whose answers have both physical and historical interest.[49]

This is an aspect whose importance has been underestimated, since the possibility that Sadi Carnot could have applied his father's *method* was ignored (Gillispie and Pisano 2014). This cultural parentage strengthens the hypothesis of a strict scientific and filial relationship between Sadi Carnot and his father's science, especially concerning mathematics (*Ivi*, Chap. XI).

---

[49] Klein in Taton (1976, p. 214, line 19).

4 Conclusion

4.1 *Scientists in Context in Nineteenth Century*

From the second half of the eighteenth century, the developments of scientific studies (advanced waterwheels related to the new industries processes) and the increasing necessity of new source of energy, opened the door to new applied studies and professions where the relationship between science and technology became more decisive. A new link characterized theoretical and practical science. This link was far stronger the one existing in the previous centuries.

In the nineteenth century, in Europe the figure of the scientific engineer is emerging. In Paris the *Grandes Écoles* were founded, where the most distinguished mathematicians worked, writing treatises and teaching students. In 1794 the *École polytechnique de Paris*, a military school for the training of engineers, was founded. Joseph-Louis Lagrange (1736–1813) and Gaspard Monge (1746–1818) were among the first professors of mathematics. On 30 October 1794 (9 *brumaire,* a. III) the *École normale supérieure de Paris*[50] was also born (Dhombres 1992). The school aimed at training scholars and teachers of both science and humanities (Dhombres 2013). The attention of the French mathematicians toward applications was therefore, at least in part, due to the need that educational institutions train technicians for the new state; the situation was different in other European countries.

Particularly the core material in the scientific engineers' curricula was mainly mathematics and physics. This means analysis (*calculus*) in the first half of the eighteenth century up to include geometry (*projective geometry*) in the second half. Great importance was also attributed to purely theoretical disciplines, such as *number theory* and *abstract algebra*. Mechanics was developed within other disciplines becoming, ie., *Mécanique céleste* (stars, comets, planets satellites, stability of planetarium system), *planet* (geodesy, cartography), *mathematical physics applied to Engineering* (mechanics, instruments, frictions, structures, analytical theory of heat), *corporeal mechanics* (statics, dynamics, hydrodynamics, crystallography), *Molecular mechanics* (Elasticity). On the other hand, the interaction between scientists, engineers and society crossed the new abilities of machines and prototypes of civil engines. Therefore, a question continue to exist, i.e., *how to describe the scientific conversion from cannons to civil heat machines?*

The two Carnots scientific approaches[51] and sciences represent a clear contra attitudinal example—with respect to military attitudes—and social emergency of the use of science in society according with civil technology. Particularly the change

[50] The school was closed on 1795. Thus it was re–founded on 1808, 17th March. On this model, the *Scuola Normale Superiore in Pisa* (Italy) was founded (1810) as a branch of the *École normale supérieure* and later gained independence.

[51] Sadi Carnot stressed (Carnot 1978, p. 18, ft. 1, pp. 75–76; see also Gillispie and Pisano 2014, Chaps. 7–8.) that his own is the same method of infinitesimal analysis; really the method and almost the same words are those of the father Lazare in his celebrated work on calculus (Carnot 1813; Pisano, Capecchi and Lukešová 2013).

of attitude was also emphasized by the long oblivion[52] (25 years) of Sadi Carnot's book. The revolutionary birth of (Lavoisier's chemistry first and then thermodynamics; see Pisano and Franckowiak 2014) performed a new approach to science with respect to Newtonian physics–mathematics foundations (Pisano 2014c; Pisano and Bussotti 2012). In fact the interpretation cumulated by previous scientist was unable to give scientific justifications for the coming out of some crucial physical items like, i.e., heat, reversibility, state and cycle. From industry standpoint, we should underline important factor, the manufacturing. Major or minor defects made sure that a running machine should be corrected (a) after a theoretical re-examine of the scientific problem (e.g., friction, efficiency, conservation of living force), and (b) correlated to machineries and technology studies. Paraphrasing Pierre Duhem (1861–1916) on a side, the amount of technical improvements proposed by artisans (Duhem 1905–1906) as part of a scientific knowledge process for generating a modern view on the world—to use an expression by Koyré —was insufficient[53] (Koyré 1966, 1961, 1957) to develop the crucial theoretical conceptualizations and to define the technological role played by science in the Western society of the nineteenth century, on the other. Let us think about bounded efficiency, reversibility, cycle, impossibility of perpetual motion. The science in context is hard to develop by means of a unique path.

In the nineteenth century a significant case study regards the conversion from cannons (considered a heat machines) to civil steam engines (considered a complex heat machines). It was a crucial problem. In our opinion the translation from heat machines to steam civil engine followed a path concerned with the:

a. Ability to (physically and mathematically) calculate a general efficiency independent from the working substance used.
b. Ability to establish efficiency as depending on the difference of temperature only.
c. Clausius and Kelvin's accounts.

The first two items were (among other studies and other many social and scientific factors) physically collected by Lazare Carnot within his theories on *machines en général* (Carnot 1780, 1786). Later it was supplemented by Sadi Carnot(1978). He extended to the functioning of heat engines the analysis that his father had developed in his study on ordinary machines (Gillispie and Pisano 2014). The father and son in separate—but scientifically filial—works explained the various ways in which the sciences allowed us to conceive advanced modelling as well as, and the possible development of new technological ideas. Particularly,

[52] Paolo Ballada Count de Saint–Robert (1815–1888) was an Italian and military mechanicist and thermodynamicist who recuperated the useful information for writing a first sketch of the life and the works of Sadi Carnot (Pisano 2007a, 2007b; Gillispie and Pisano 2014, pp. 319–335).

[53] For historical accounts on that: Pisano (2009b); Pisano and Gaudiello (2009a, 2009b).

d. an emphasis on the role played by *geometric motion*[54], impossibility of perpetual motion, substance work and collision mechanisms related to the conservation of living force and impact for insensible degrees in Lazare's theory of mechanical machines,
e. and the reversible process, the novelty of concept of cycle and the dependence from temperature only within efficiency formula in Sadi's heat machine

provided a benefit to the scientific discourse on *machines en general*. The socio-economical implications were positive, as well. Sadi Carnot produced works, which derived from his training as engineer. His *Réflexions* represents the foundation of the science of thermodynamics and also the final item in a series of Carnot's memoirs on the science of machines, in which the analysis derived from engineering rather from the rational mechanics of the eighteenth century, culminated in Lagrange's *Mécanique analytique* (Lagrange 1788; Pisano and Capecchi 2013).

4.2 Concluding Remarks

The purpose of this paper is to identify and consequently examine the development of the interaction between theoretical sciences and technology within *mechanical* and *heat machines* studies in the eighteenth–nineteenth century) as technosciences. Basically we concentrated on the emerging sciences proposed by Lazare and Sadi Carnot. After Lazare Carnot's works on *mechanical machines* (Carnot 1778, 1780, 1786), Sadi's thermodynamics (Carnot 1978) originated by introducing the novelty of a cycle, a physical reasoning-process that restored the initial conditions on *heat machines* versus *steam civil engines*. This concept was out of artisan and military previous conceptualizations. A strong *scientific relationship* (Gillispie and Pisano 2014) between Lazare Carnot and Sadi Carnot's sciences was presented and—among other scholars—they offered a great contribution to theoretical and subsequently technological developments in society.

Finally, it is important to (historically) highlight the existence of two main branches of applied mechanics as technoscience. The first one deals with the machines (nowadays we can refer to theory of machines and machineries, industrial engineering); the second one examines the structures of constructions (today "civil engineering"). The main problem faced in this paper has been the relationship between physics and mathematics concerning the conversion of work to heat between history of science and history of science and technique. On these subjects, an emblematic example of how the effects of theoretical science on technology are detectable only many years after the achievement of a theoretical result, is offered by the studies published by the *Académie des sciences* and by the *École polytechnique de Paris*.

[54] Carnot (1786, pp. 28–34, pp. 41–45; see also 1780, § 113); Gillispie (1971, Appendix C, § 113, pp. 308–309).

References

Adkin M (2002) The Waterloo companion. The complete guide to history's most famous land battle. Stackpole Books, Mechanicsburg

Aries E (1921) L'Œuvre scientifique de Sadi Carnot. Introduction a l'étude de la thermodynamique. Payot, Paris

Barbin E, Pisano R (eds) (2013) The dialectic relation between physics and mathematics in the XIXth Century. Springer, Dordrecht

Bernal JD (1973) The social function of science. MIT Press, Cambridge

Bernoulli J (1703) Démonstration générale du centre de balancement et d'oscillation, tirée de la nature du levier. Mémoire de l'Académie Royale des Sciences, Paris, pp 78–84

Bernoulli J (1727) Theoremata selecta pro conservatione virium vivarum demonstranda et esperimenta confirmanda. Novi commentarii Academiae Scientiarum Imperialis Petropolitanae 2, pp 200–207

Bernoulli J (1742a) Propositiones variae mechanico-dynamicae. In: Bernoulli 1742a, vol IV, pp 253–386

Bernoulli J (1742b) Opera omnia, tam antea sparsim edita, quam hactenus inedita […] quibus continentur ea, quae ab A. 1690 usque ad A. 1727 prodierunt. Accedunt Lectiones Mathematicae de Calculo Integralium atque Anedokta. I-IV vols. Bousquet, Lausannae & Genevae (Facsimile 1968. Reprint Curavit Hofman JE. Georg Olms Verlag Hildeshe; very useful are The Collected Scientific Papers of the Mathematicians and Physicists of the Bernoulli Family by Patricia Radelet de Grave editor (Birkhäuser. Basel)

Bernoulli J (1742c) CXLV. De vera notione virium vivarum. In: Opera Ominia. III. Sumptibus Marcr–Michaelis Bosquet & Sociorum, Lausannae & Genevae, pp 239–260

Bernoulli D ([1748]1750) Remarques sur le principe de la conservation des forces vives pris dans un sens général. Mémoire de l'Académie Royale des Sciences et des Lettres de Berlin, vol IV, pp 356–364

Biringuccio V (1540) De la Pirothecnia. Libro X. Roffinello, Venezia

Biringuccio V ([1943] 1990) De la Pirotechnia. The Pirotechnia of Vannoccio Biringuccio, the classic sixteenth-Century treatis on metals and metallurgy, tradotto da Cyril Stanley Smith e Martha Teach Gnudi. Dover, New York

Black J (1803) Lectures on the elements of chemistry delivered in the University of Edinburgh by the late Joseph Black, M.D, 2 vols. John Robison, Edinburgh

Bussotti P, Pisano R (2013a) On the conceptual frames in René Descartes' *physical works*. Advances in Historical Studies 2(3):106–125

Bussotti P, Pisano R (2013b) On the *Jesuit Edition* of Newton's *Principia*. Science and advanced researches in the western civilization. In: Pisano R (ed) Isaac Newton and his scientific heritage. New studies in the history and historical epistemology of science. Advances in Historical Studies Special Issue 3(1)

Bussotti P, Pisano R (2014) Newton's Philosophiae Naturalis Principia Mathematica "Jesuit" Edition: The Tenor of a Huge Work. Accademia Nazionale Lincei-Rendiconti Matematica e Applicazioni 25:413–444

Callen H (1974) Thermodynamics as a science of symmetry. Foundations of Physics 4:423–440

Callendar HL (1910) Caloric theory of heat and Carnot's principle. Proceedings of the Physical Society–London 23(1):153–189

Capecchi D, Pisano R (2010) Scienza e tecnica nell'architettura del Rinascimento. CISU, Roma

Carnot L (1778) Mémoire sur la théorie des machines pour concourir au prix de 1779 propose par l'Académie Royale des Sciences de Paris. The manuscript is conserved in the Archives de l'Académie des sciences, Institut de France, and consists of 85 sections in 63 folios. Sections 27–60 are reproduced. In: Gillispie (1971), Appendix B, pp 271–296

Carnot L (1780) Mémoire sur la théorie des machines pour concourir au prix que l'Académie Royale des Sciences de Paris doit adjuger en 1781. The manuscript is dated from Béthune 15 July 1780. It is conserved in the Archives de l'Académie des sciences, Institut de France, and consists of 191 sections in 106 folios. Sections 101–160 are reproduced. In: Gillispie (1971), Appendix C, pp 299–343

Carnot L (1786) Essai sur les machines en général. Defay, Dijon

Carnot L (1803) Principes fondamentaux de l'équilibre et du mouvement. Deterville, Paris

Carnot L (1813) Réflexions sur la métaphysique du calcul infinitésimal. Courcier, Paris

Carnot S (1824) Réflexions sur la Puissance Motrice du Feu sur les machinés propre à développer cette puissance. Bachelier, Paris

Carnot S (1978) Réflexions sur la puissance motrice du feu sur les machinés propre à développer cette puissance, édition critique par Fox Robert. Vrin J, Paris

Carnot S (1986) English edition: Reflexions on the motive power of fire: a critical edition with the surviving scientific manuscripts. (trans and ed: Robert Fox). The Manchester University Press, Manchester

Carnot L (1990) Rapport sur une nouvelle machine inventée par MM. Niepce, et nommée par eux Pyrélophore; par MM. Berthollet et Carnot. Lu le 15 décembre 1806, imprimé en 1807, 1er semestre. In: Charnay JP (ed) Lazare Carnot ou le savant-citoyen, La Herne, Paris, Presses de L'Université de Paris–Sorbonne, 1990–Centre d'études et de recherches sur les stratégies et les conflits, Colloque en Sorbonne, 1988

Carnot S–EP (s.d.) Recherche d'une formule propre à représenter la puissance motrice de la vapeur d'eau. © Collections Archives de la Bibliothèque de l'École polytechnique, Paris

Challey F (1971) Sadi Carnot. In: Gillispie 1970–1980 III:79–84

Chandler DG (1995) The campaigns of Napoleon. Simon & Schuster, New York

Clausius RJE (1850) Über die bewegende Kraft der Wärme und die Gesetze, welche sich daraus für die Wärmelehre selbst ableiten lassen. Annalen der Physik und Chemie 79:368–397, 500. English edition: On the motive power of heat and on the laws which can be deduced from it for the theory of heat itself (trans: *Id.*), In: Mendoza 1960, pp 73–74, 109–152)

Clausius R (1865) Sur la détermination de la disgregation d'un corp et de la vraie capacité calorifique. Bibliothèque universelle de Genève. Archives des sciences physiques et naturelles XXV:117–124

Clausius, R (1898) Ueber die bewegende Kraft der Wärme, und die Gesetze, welche sich daraus für die Wärmelehre selbst ableiten Lassen. Leipzig. Engelmann, Leipzig (First published in Poggendorff's Annalen der physik und chemie 79, 1850)

Clausius RJE (1868–1869) Théorie Mécanique de la Chaleur, 2 vols. French édition (trans: Folie). Lacroix éditeur, Paris

D'Alembert JLR ([1743]1758) Traité de dynamique. Nouvelle édition, revue et fort augmentée par l'Auteur Chez, David Libraire, Paris

Descartes R (1897–1913) OEuvres de Descartes. Par Charles Adams et Paul Tannery, Paris, 12 vols; also see Discours de la méthode et Essais, Specimina philosophiae, vol VI; Physico–mathematica, vol X, Le Monde ou Traité de la lumière, vol XI (Id, 1964–1974 by Rochot B, Costabel P, Beaude J et Gabbery A, Paris)

Descartes R (1964–1974) Œuvres. Adam J et Tannery A. Nouvelle présentation par Rochet E, et Costabel P, 11 vols. Vrin, Paris

Dhombres J (1992) (ed) L'Ecole normale de l'an III. Leçons de mathématiques. Dunod, Paris

Dhombres J (2013) The role played by mathematics during the revolutionary and imperial Paris up to the restoration in the education of engineers with Lazare Carnot as a witness. In: Pisano R, Capecchi D, Lukešová A (2013) (eds), pp 11–24

Dhombres J, Dhombres N (1997) Lazare Carnot. Fayard, Paris

Dirichlet JPGL (1837) Über die Darstellung ganz willkürlicher Funktionen durch Sinus–und Cosinus–reihen. Repertorium der Physik 1:152–174

Drago A, Pisano R (2005) La nota matematica nelle Réflexions sur la puissance motrice du feu di Sadi Carnot: interpretazione del calcolo col metodo sintetico. Quaderni di Storia della Fisica–Giornale di Fisica 13:37–57

Drago A, Pisano R (2007) La novità del rapporto fisica–matematica nelle Réflexions di Sadi Carnot. Atti della Fondazione Giorgio Ronchi 62(4):497–525

Dugas R (1956) Histoire de la Mécanique. Griffon, Neuchatel

Duhem PM (1905–1906) Les origines de la statique, 2 Tomes. Hermann, Paris

Euler L (1736) Mechanica sive motus scientia analytice exposita. (Mechanis or science of motion exposed analytically). Auctore Leonhardo Eulero academiae imper. scientiarum membro et matheseos sublimioris professore. Tomus I. Instar supplementi ad commentar. acad. scient. imper. Petropoli. Ex typographia academiae, scientarum

Fermat P (1891–1922) Œuvres, four volumes plus Supplément aux tomes I–IV. Gautheir-Villars, Paris

Fourier JBJ (1807) Théorie de la propagation de la chaleur dans les solides. Nouveau Bulletin des Sciences par la Société philomathique de Paris, Tome I(6):112–116 (n. 6, mars 1808), Bernard, Paris

Fourier JBJ (1808) Théorie de la propagation de la chaleur dans les solides. Nouveau Bulletin des Sciences par la Société philomathique de Paris I(6):112–116

Fourier JBJ (1822) Théorie analytique de la chaleur. Firmin Didot, Paris

Fourier JBJ (1829) Mémoire su la théorie analytique de la chaleur dans les solides. Mémoires de l'Académie royale des sciences de l'Institut de France pour l'année 1825. Firmin Didot, Paris VIII:581–622

Fox R (1969) James Prescott Joule, 1818–1889. In: North J (ed) Mid-nineteenth-century scientists. Pergamon, New York, pp 72–103

Fox R (1970) Watt's expansive principle in the work of Sadi Carnot and Nicolas Clément. Notes and records of the Royal Society of London 24:233–253

Fox R (1971a) The intellectual environment of Sadi Carnot: a new look. In: Proceedings of XII Congres international d'histoire des sciences. Blanchard, Paris, vol IV, pp 67–72

Fox R (1971b) The Caloric theory of gases: from Lavoisier to Regnault. The Clarendon, Oxford

Fox R (1974) The rise and fall of Laplacian physics. Historical Studies in the Physical Sciences 4:89–136

Fox R (1976) The challenge of new technology: theorists and the high-pressure steam engine before 1824. In: Taton (ed) (1976) pp 149–168

Fox R (1988) Les Réflexions sur la puissance motrice du feu de Sadi Carnot et la leçon se leur éditions critique. La vie des sciences - Comptes rendus, série générale 5(4):283–301

Gabbey AW, Herivel JW (1966) Un manuscrit inédit de Sadi Carnot. Revue d'histoire des sciences et de leurs applications 19(2):151–166

Galilei G (1890–1909) Le opere di Galileo Galilei: Edizione nazionale sotto gli auspici di sua maestà il re d'Italia. Favaro A (ed). Barbèra, Firenze

Galilei G (1954) English edition: Dialogues concerning two new sciences (trans: Crew H, de Salvio A and an introduction by Favaro A). Dover, Mineola (see Galilei (1890–1909), VIII, pp 39–448)

Gille B (1964) Les ingénieurs de la Renaissance. Hermann, Paris

Gille G (1966) The renaissance engineers. MIT Press, Cambridge

Gillispie CC (1960) The age of objectivity. The Princeton University Press, Princeton

Gillispie CC (1970–1979) Carnot Lazare–Nicolas–Marguerite. In: Gillispie (ed) 1970–1980 III:70–79

Gillispie CC (ed) (1970–1980) Dictionary of scientific biography. Charles Scribner's Sons, New York

Gillispie CC (1971) Lazare Carnot Savant. A monograph treating Carnot's scientific work, with facsimile reproduction of his unpublished writings on mechanics and on the calculus, and an essay concerning the latter by Youschkevitch AP. Princeton University Press, Princeton

Gillispie CC (1976) The scientific work of Lazare Carnot, and its influence on that of his son. In: Taton (ed) (1976), pp 23–33

Gillispie CC (1983) The Montgolfier Brothers and the invention of aviation. Princeton University Press, Princeton

Gillispie CC, Pisano R (2014) Lazare and Sadi Carnot. A scientific and filial relationship (2nd edn). Sprinter, Dordrecht

Girard PS (1824) Réflexions sur la puissance motrice du feu sur les machinés propre à développer cette puissance, par S. Carnot, ancien élève de l'école polytechnique [...] prix 3 fr. Revue Encyclopédique XXIII–Juillet:414
Grattan-Guinness I (1969) Joseph Fourier and the revolution in mathematical physics. Journal of the Institute of Mathematics and its Applications 2(5):230–253
Grattan-Guinness I (1972) Joseph Fourier (1768–1830): A survey of his life and work. MIT Press, Cambridge
Grattan-Guinness I (1990a) Convolutions in French mathematics. Birklausen, Basel
Grattan-Guinness I (1990b) Work for the workers: engineering mechanics and its instruction. In: Grattan–Guinness (ed) 1990a, pp 1046–1121
Grattan-Guinness I, Ravetz J (1980–1990) Fourier Jean Baptiste Joseph. In: Gillispie 1970–1990, V:93–99
Guerlac H (1970–1990) Black Joseph. In: Gillispie (ed) (1970–1980), II:173–183
Hachette JNP (1811) Traité élémentaire des machines. Klostermann fils, Paris
Hachette JNP (1819) Traité élémentaire des machines 2 édition. Courcier, Paris
Heath B (2005) Discovering the Great South Land. Rosenberg, Kenthurst
Hazlett JC, Edwin O, Hume P (2004) Field artillery weapons of the American Civil War. The University of Illinois Press, Illinois, pp 88–108
Hoyer U (1976) La théorie de Carnot première et seconde approximations de la thermodynamique. In Taton (ed) (1976), pp 221–228
Huygens C (1673) Horologium Oscillatorium sive de motu pendolorum ad horologia aptato demonstrations geometricae. Muguet, Paris
Joule JP (1847) On the mechanical equivalent of heat, as determined by the heat evolved by the friction of fluids. Philosophical Magazine, 3rd series XXXI(207):173–176
Joule JP (1965) The scientific papers of James Prescott Joule, vol I. Dawson's, London
Kerker M (1957) Sadi Carnot. The Scientific Monthly 85:143–149
Klein MJ (1976) Closing the carnot cycle. In: Taton (ed) (1976), pp 213–219
Knox DW (1939) Naval documents related to the United States Wars with the Barbary Powers. I. The United States Government Printing Office, Washington
Koenig F (1959) On the history of science and the second law of thermodynamics. In: Evans HM (ed) Men and moments in the history of science. The University of Washington Press, Seattle, pp 57–111
Koyré A (1957) From the closed world to the infinite Universe. John Hopkins, Baltimore
Koyré A (1961) La révolution astronomique: Copernic, Kepler, Borelli. Hermann, Paris
Koyré A (1966) Études galiléennes. Hermann, Paris
Kuhn ST (1955) Carnot's version of cycle. The American Journal of Physics 23:91–94
Kuhn ST (1959) Engineering precedent for the work of Sadi Carnot. In: Proceedings of IX Congrès international d'Histoire des sciences, Barcelona, Madrid 1959, (Barcelona, Paris 1960), pp 530–535
Kuhn ST (1960) Engineering precedent for the work of Sadi Carnot. Archives Internationales d'Histoire des Sciences 13:251–255
Kuhn ST (1961) Sadi Carnot and the cagnard engine. Isis 52:567–574
Kuhn ST (1962) The structure of scientific revolutions. The Chicago University Press, Chicago
Lagrange JL ([1762] 1973) Application de la méthode exposée dans le mémoire précédente à la solution des différents problèmes de dynamique. In: Lagrange 1870–1873, I, pp 363–468
Lagrange JL (1764) Recherches sur la libration de la Lune dans lesquelles on tache de résoudre la question proposes par l'Académie Royale des Sciences, pour le Prix de l'année 1764. Prix de l'académie Royale des Sciences de Paris, IX (1777), pp 1–50. See also In: Lagrange 1870–1873, Tome VI, pp 5–61
Lagrange JL (1788) Mécanique Analytique. Desaint, Paris
Lagrange JL (1813) Théorie des fonctions analytiques. Courcier, Paris
Lagrange JL (1870–1873) Œuvres de Lagrange (2nd édn). In: Courcier (ed) I–XIV vols. (in X) Paris Gauthier–Villars, Paris

Lamé G (1836) Cours de physique de l'école polytechnique, vol I. Bachelier, Paris (*Id*, Cours de physique de l'école polytechnique, 2nd edn. Bruxelles)
Laplace PS (1822) Notes sur la vitesse du son. Annales de chimie et de physique XX:266–268
Lavoisier AL ([1789] 1973) Traité élémentaire de Chimie. Gauthier–Villars, Paris
Lavoisier AL (1862–1893) Œuvres de Lavoisier publiées par le soin de Son Excellence le Ministre de l'instruction publique et des cultes, 6 vols. Imprimerie Impériale, Paris
Lavoisier AL, Laplace PS (1784) Mémoires sur la Chaleur. Histoire de l'Académie Royale des Sciences. Mémoires, Paris, (1780 and first reading on 18th June 1783)
Liouville J (1836) Thèses de mécanique et d'astronomie. Bachelier, Paris
Mach E (1986) Principles of the theory of heat, historically and critically elucidated. In: McGuinness B (ed) vol 17. Reidel, Boston
Machiavelli N (2005). The art of war. III. The University of Chicago Press, Chicago, p 74
Manigault E, Warren R (1996) Siege train: the journal of a confederate artilleryman in the defense of charleston. The University of South Carolina Press, Charleston
Mariotte E (1686) Traité des eaux et autres corps fluides. Michallet, Paris
Maupertuis PL (1746) Les loix du mouvement et du repos déduits d'un principe métaphysique. Histoire de l'Académie Royale des Sciences et des Belles Lettres, pp 267–294
Mendoza E (1959) Contributions to the study of Sadi Carnot and his work. Archive International d'Historie des Sciences XII:377–396
Mendoza E (ed) (1960) Reflections on the motive power of fire and other papers on the second law of thermodynamics by E Clapeyron and R Clausius. Dover, New York
Mendoza E (1963) Sadi Carnot and the cagnard engine. Isis 54:262–263
Mendoza E (1982) The kinetic theory of matter: 1845–1855. Archive Internationales d'Histoire des Sciences 32:184–220
Meyerson (1904) Identité et Réalité. Alcan, Paris
Montbrial AL (1976) Interprétation économiques suggérées par le théorème de Carnot. In: Taton (ed) (1976), pp 333–335
Newton I (1666) De gravitatione et aequipondio fluidorum. Ms Add. 4003. The Cambridge University Library, Cambridge. http://www.newtonproject.sussex.ac.uk/view/texts/normalized/THEM00093
Newton I ([1704] 1730) Opticks: or, A Treatise treatise of the Reflectionsreflections, Refractionsrefractions, Inflections inflections and Colours colours of Lightlight. (4th ednition corrected). William Innys, London
Newton I ([1687] 1803) English edition: The mathematical principles of natural philosophy by Sir Isaac Newton (trans: Motte A). Symonds, London
Parent A (1745) Sur la plus grande perfection possible des machines. Histoire de l'Académie Royale des Sciences de Paris, année 1704. Martin–Coignard–Guerin, Paris, pp 323–338
Payen J (1968) Une source de la pensée de Sadi Carnot. Archive international d'Historie des Sciences XXI:18–30. Physik und Chemie 155:368–397; 500–524 (English trans: In: Mendoza, (1960), pp 73–74; 109–152)
Petit AT (1818) Sur l'emploi du principe des forces–vives dans le calcul de l'effet des machines. Annales de Chimie et de Physique par Gay–Lussac et Arago (eds), Tome 8:287–305
Picard E (1927) Sadi Carnot, biographie et manuscrit. Gauthier–Villars, Paris
Pisano R (2001) Interpretazione della nota matematica nelle Réflexions sur la puissance motrice du feu di Sadi Carnot. In: Proceedings of XX SISFA congress. Osservatiorio Astronomico di Capodimonte, Napoli, pp 205–230. http://www.brera.unimi.it/sisfa/atti/index.html
Pisano R (2003) Il ciclo di S Carnot e la pila di A Volta. In: Garuccio A (ed) Proceedings of XXIII SISFA congress. Progedit, Bari, pp 327–348
Pisano R (2007a) Note sui Principes de Thermodynamique di P de Saint Robert. In: Proceedings of XXIV SISFA congress. Biblipolis, Napoli–Avellino, pp 129–134
Pisano R (2007b) A history of chemistry à la Koyré? Introduction and setting of an epistemological problem. Khimiya 17(2):143–161

Pisano R (2008) Il ruolo della scienza meccanica nella progettazione degli architetti e degli ingegneri del Rinascimento. Ph.D. Dissertation, University of Roma La Sapienza, Roma. 2 vols. available in.pdf format via: International Galilean Bibliography, Istituto e Museo di Storia delle Scienze, Firenze; in hard copy at the Direction des collections, Département Sciences & Techniques, Bibliothèque Nationale de France, Paris, référence: DDD-TOL-2011-94

Pisano R (2009a) On method in Galileo Galilei' mechanics. In: Hunger H (ed) Proceedings of ESHS 3rd conférence. Austrian Academy of Science, Vienna, pp 147–186

Pisano R (2009b) Continuity and discontinuity. On method in Leonardo da Vinci' mechanics. Organon 41:165–182

Pisano R (2010) On principles in Sadi Carnot's thermodynamics (1824). Epistemological reflections. Almagest 2:128–179

Pisano R (2011) Physics–Mathematics relationship. Historical and epistemological notes. In: Barbin E, Kronfellner M, Tzanakis C (eds) Proceedings of the ESU 6 European Summer University History and Epistemology in Mathematics. Verlag Holzhausen GmbH–Holzhausen, Vienna, pp 457–472

Pisano (2013a) Historical reflections on physics mathematics relationship in electromagnetic theory. In: Barbin and Pisano (2013) (eds), pp 31–58

Pisano R (2013b) Reflections on the scientific conceptual streams in Leonardo da Vinci and his Relationship with Luca Pacioli. Advances in Historical Studies 2(2):32–45

Pisano R (2014) A Development of the Principle of Virtual Laws and its Conceptual Framework in Mechanics as Fundamental Relationship between Physics and Mathematics. Submitted to *Foundations of Physics*

Pisano R, Bussotti P (2012) Galileo and Kepler. On Theoremata Circa Centrum Gravitatis Solidorum and Mysterium Cosmographicum. History Research 2(2):110–145

Pisano R, Bussotti P (2014a) Historical and philosophical reflections on the culture of machines around the renaissance. How science and technique work? Acta Baltica Historiae et Philosophiae Scientiarum 2/2:20–42

Pisano R, Bussotti P (2015). Historical and philosophical reflections on the culture of machines around the renaissance: machines, machineries and perpetual motion. Acta Baltica Historiae et Philosophiae Scientiarum 3(1):65–85

Pisano R, Capecchi D (2009) La Théorie Analytique de la Chaleur. Notes on Fourier and Lamé. Bulletin de la Sabix 44:83–90

Pisano R, Capecchi D (2013) Conceptual and mathematical structures of mechanical science in the western civilization around the eighteenth Century. Almagest 4(2):86–121

Pisano R, Drago A (2013) The Modern Thermodynamics as based on the principle of virtual laws. In: Pisano R, Capecchi D, Lukešová A (eds), pp 345–352

Pisano R, Franckowiak R (2014) Lavoisier and Sadi Carnot. Chemical–and–Physical Sciences as Dating Back to Two Survived Scientific Revolutions: 1789 And 1824. In: Bilek M (ed) Proceedings of the 9th IOSTE symposium for Central and Eastern Europe. Hradec University, IX, Gaudeamus Králové, pp 42–54

Pisano R, Gaudiello I (2009a) Continuity and discontinuity. An epistemological inquiry based on the use of categories in history of science. Organon 41:245–265

Pisano R, Gaudiello I (2009b) On categories and scientific approach in historical discourse. In: Hunger H (ed) Proceedings of ESHS 3rd conference. Austrian Academy of Science, Vienna, pp 187–197

Pisano R, Capecchi D, Lukešová A (eds) (2013) Physics, astronomy and engineering. Critical problems in the history of science and society. International 32nd congress for The SISFA–Italian Society of Historians of Physics and Astronomy. The Scientia Socialis UAB & Scientific Methodical Centre Scientia Educologica Press, Šiauliai University, Lithuania

Poisson SD (1823) Sur la vitesse du son. Annales de Chimie et de Physique XXIII:5–16

Poisson SD (1833) Traité de Mécanique. Bachelier, Tome II, Paris

Poncelet JV (1845) Traité de mécanique appliquée aux machines. Deuxième partie, Leroux, Liège

Prony G (1790) Nouvelle architecture hydraulique. Contenant un traité de mécanique à l'usage de ceux qui se destinent aux constructions de tous les genres, et des Artistes en général. Première Partie. Firmin Didot, Paris

Prony G (1796) Nouvelle architecture hydraulique. Contenant la description détaillée des machines à feu. Seconde Partie. Firmin Didot, Paris

Prony G, Girard S, Navier CLMH (1829) Rapport suivant sur leMémoire de M. Coriolis, intitulé *du calcul de l'effet des machines*. In: Procès–verbaux des séances de l'Académie, 22 Jun, Institut de France. Académie des sciences. Imprimerie de l'observatoire d'Abbadia, Hendaye Basses Pyrénées [publication 1921] Année 1828–1831, Tome IX, pp 266–272

Reech F (1853) Théorie générale des effets dynamiques de la chaleur. Journal de Mathématiques pures et appliquées XVIII:357–378

Reinhard M (1950–1952) Le grand Carnot. De l'ingénieur au conventionnel, 1753–1792, 2 vols. Hachette, Paris

Reuleaux F (1876) Kinematics of machinery. Outlines of theory of machines. Mcmillan, London

Richmann GW (1750) Formulae pro gradu excessus caloris supra gradum caloris, mixti ex nivee et sale amoniaco. Novi commentarii Academiae Scientiarum Imperialis Petropolitanae I:171–172

Riemann GFB (1861 [1868]) Memoir on the conduction of heat. In: Weber 1892

Rosenfeld L (1941) La genèse des principes de la thermodynamique. Bulletin de la Société Royale de Liège 10:197–212

Schneer CJ (1973) Leonardo da Vinci. In: Gillispie 1970–1980, VIII:213–214

Scott WL (1971) The conflict between atomism and conservation laws 1640–1860. Elsevier, New York

Singer C ([1954–1958] 1993) A history of technology, 5 vols. The Clarendon, Oxford

Smeaton J (1759) An experimental enquiry concerning the natural powers of water and wind to turn mills, and other machines, depending on a circular motion. The Philosophical Transactions of the Royal Society of London 51:100–174

Smeaton J (1776) An experimental examination of the quantity and proportion of mechanic power necessary to be employed in living different degrees of velocity to heavy bodies from a state of rest. The Philosophical Transactions of the Royal Society of London 66:450–475

Smeaton J (1782) New fundamental experiments upon the collision of bodies. The Philosophical Transactions of the Royal Society of London 72:337–354

Spivak M (1979) A comprehensive introduction to differential geometry, vol 2. Perish, Berkeley, pp 179–182

Stevin S (1605) Tomus quartum mathematicorum hypomnematum de statica. Lugodini Batavorum [Translation into Latin by Willebrord Snel van Royen]. Ex Officina Ioannis Patii, Academiae Typographi

Stevin S (1605–1608a) Memoires mathematiques: contenant ce en quoy s'est exercé le très-illustre, très-excellent Prince & Seigneur Maurice, Prince d' Orange, Conte de Nassau, Catzenellenboghen, Vianden, Moers. [Translated into French by Jean Tuning]. Chez Ian Paedts Iacobsz. Marchand Libraire, & Maistre Imprimeur de l'Université de ladite Ville

Stevin S (1605–1608b) Hypomnemata Mathematica, hoc est eruditus ille pulvis, in quo se exercuit […] Mauritius, princeps Auraïcus […] a Simone Stevino [Translation into Latin by Willebrord Snel van Royen]. Lugodini Batavorum. Ex Officina Ioannis Patii, Academiae Typographi

Stevin S (1605–1608c) Wisconstige Gedachtenissen. Tot Leyden, Inde Druckerye van Ian Bouvvensz

Stevin S (1634) Les Œuvres Mathematiques de Simon Stevin de Bruges. Le tout reveu, corrigé, & augmenté Par Albert Girard. Samielois, Mathematicien. Bonaventure & Abraham Elsevier Imprimeurs ordinaires de L'Université, Leyde

Takayama S (1979) Development of the Theory of steam engines. Japanese Studies in History of Science 18:101–116 p 106

Taton A (ed) (1976) Sadi Carnot et l'essor de la thermodynamique, Table Ronde du Centre National de la Recherche Scientifique. École Polytechnique, 11–13 Juin 1974. Éditions du Centre National de la Recherche Scientifique, Paris

Thomson W (1848) On an absolute thermometric scale founded on Carnot's theory of the motive power of heat, and calculated from Regnault's observations. Proceedings of The Cambridge Philosophical Society 1(5):66–71

Thomson W (1851a) On an absolute thermometric scale founded on Carnot's theory of the motive power of heat, and calculated from Regnault's observations. Mathematical and physical papers, vol I. The Cambridge University Press, Cambridge, pp 100–106

Thomson W (1851b) On the dynamical theory of heat, with numerical results deduced from Mr Joule's equivalent of a thermal unit, and M. Regnault's observations on steam. Mathematical and physical papers, vol I. The Cambridge University Press, Cambridge, pp 175–183

Truesdell CA (1970) The tragicomedy of classical thermodynamics. Springer, Berlin

Truesdell CA (1980) The tragicomical history of thermodynamics. 1822–1854. Springer, Berlin

Wallechinsky D, Irving W (1975) The people's almanac. Doubleday, New York

Weber H (ed) (1892) Bernhard Riemann's Gesammelte Matheamtische Werk. Collected mathematical works, 2nd edn. Leipzig, Teubner

White L Jr (1962) Medieval technology and social change. The Oxford University Press, London

Wilkinson-Latham R (1975) Napoleon's artillery. Osprey, Oxford

Zeuner G (1869) German edition: Théorie mécanique de la chaleur, avec des applications aux machines (2nd édn) (trans: Arnthal Maurice ad Cazin Achille). Gauthier–Villars, Paris

A Critical Approach to Open Access Sources with a Focus on the History of Technology

Birutė Railienė

Abstract Open access sources (OAS) used by researchers are becoming increasingly demanded as well as widely available in a large number of subject areas because they allow the user to save time and they provide a highly convenient access to scholarly information. The article deals with the various aspects of OAS including the role of traditional academic libraries and librarians. Categories including primary, secondary and tertiary OAS are discussed. Selected items of history of technology in each category are presented for illustration purposes; also the types of open access providers are overviewed including open access journals, institutional repositories and specialised webpages/databases. Technical, organisational/economical, political and content quality challenges of open access sources are considered from the provider and end-user perspective. The library's role in educating end-users of OAS is emphasized.

Keywords Open access · History of technology · Primary sources of information · Secondary sources of information · Open access journals · Institutional repositories · Libraries

1 Introduction

Communication technologies have erased country barriers, and the combination of the local and the global further defines world history. The history of technology itself forms an essential part of society, providing tools with which to see the past and envision the future.

In a research culture, the shift from manuscript to printed sources was followed by the shift to digital information, which in the twenty-first century also became open, as an alternative to access by subscription (closed). One remarkable example is Project Gutenberg. It started from digitising the Declaration of Independence

B. Railienė (✉)
Head of Scholarly Information Department, The Wroblewski Library of the Lithuanian Academy of Sciences, Vilnius, Lithuania
e-mail: b.railiene@gmail.com

R. Pisano (ed.), *A Bridge between Conceptual Frameworks,* History of Mechanism and Machine Science 27, DOI 10.1007/978-94-017-9645-3_22

in 1971 and by 2013 having posted on the world-wide web 42,000 free e-books (Project Gutenberg 2013). This, obviously, gives new possibilities for both scholars and information specialists to find in libraries.

Reference service in a research library benefits from a policy that saves users time. The Indian mathematician and librarian Shiyali Rammarita Ranganathan (1892–1972), whose contributions have had worldwide influence, introduced five library laws, the fourth of which says: "Save the time of the reader" (Gorman 1998). This law is common among professionally devoted librarians. Library processes, even if they are labour consuming, intend to save time of the user. Heritage institutions such as academic libraries are encouraged to innovate, including retrospective digitization of archival documents. Libraries are extending open access to their catalogues, holdings, bibliographical and reference material. Depending on the financial resources, libraries subscribe to commercial databases, participate in institutional repositories and digitisation projects. However, this intention became very challenging in the information age.

Open access to the various sources of information via the internet (so-called electronic or soft documents) is nowadays becoming more and more popular in comparison with historically established use of printed sources in libraries or private collections or even access to the commercial databases available by subscription. In a survey of more than 200 academic librarians performed by InTech, an open access publisher, 95% of respondents were certain about the tangible benefits of open access (Assessing 2012, p. 3).

For authors, open access brings increased visibility, usage and impact of their work; it provides excellent means to improve their online presence in the research community. Since author scholars are the producers of the content, they represent the main part of the open access facility. At the same time, scholars also are the primary users of open access sources (OAS).

The paper intends to provide a tutorial over the OAS scene in the area of history of technology. The rest of the paper is organized as follows: Sect. 1 introduces services for open access; Sect. 2 considers categorization of primary, secondary and tertiary information sources; Sect. 3 illustrates the providers of open access sources; challenges related to open access sources are discussed in Sect. 4; Sect. 5 presents our conclusions. Appendix A is a glossary, Appendix B—selected items of history technology.

2 Information Service for Open Access

Researchers in the history of technology have increased their expectations of librarians mostly during the last few decades and the collaboration between historians and librarians offers new challenges and opportunities (Daniel 2012, p. 270). A famous information expert, a prophet of paperless society F. W. Lancaster (b. 1933) described a future role of a librarian as a freelance expert, a consulting information officer whose expertise could be highly valued in an information society, if only librarians themselves seized the coming opportunity (see Sapp 2003,

p. 559). However, the overflow of irrelevant information still is one of the greatest challenges for researchers. The assistance of information service professionals could be of a great value when a challenge to shortlist the relevant information from a vast choice of alternatives is met. Even teachers of history are rather pessimistic about abilities of their students "to use electronic resources responsibly" (Daniel 2012, p. 261).

Although instructional support is still on demand, libraries are expected to provide more research training in electronic resources (Daniel 2012, p. 274). Librarians face the challenge to provide training in information evaluation and scientific thinking, not only in information literacy. Thus, there is a need to promote and advocate for a critical approach to the OAS in order to distinguish between high quality and "other" sources. While introducing latest technologies in their work, librarians suspected the gap between their knowledge of open access and the knowledge their research patrons have. In a survey, performed by InTech, a concern of research librarians was revealed that almost half of their author and reader communities (54 %) were not familiar enough with open access (Assessing 2012, p. 2).

At the same time, information literacy is also crucial in the searching process and currently it is really a "hot topic" among research librarians in the academic community. Librarians tend not to use the term "bibliographical instruction" any more—instead they prefer the term "information literacy". The latter includes the whole world of information and is not limited to just library services. Librarians have their own expertise in finding, evaluating and using sources and are well prepared to keep up with fast changing information technologies (Daniel 2012, p. 267) and assist educators in their goal of helping their students to develop critical thinking skills (Kent-Drury 2005, p. 1).

3 Primary, Secondary and Tertiary Information Sources

Providing information for research or education, libraries have adopted a common system to classify information sources as *primary* and *secondary* (Archer et al. 2009, p. 410; Lombard 2010, p. 251). In the case of digital information or open access, the term *tertiary* becomes common, to describe a source that compiles or digests primary and secondary sources (Primary 2013, p. 1).

For example, in the paper (Primary 2013, p. 1) is stated: "academic research is based primarily on the analysis of primary sources, guided by perspectives on the topic which already exists via secondary sources". Some reference materials and textbooks are considered as "tertiary" sources when they are intended to list, summarize, or "simply repackage ideas or other information" (Primary 2013, p. 3).

However, categorisation depends on the topic being covered and the field of research. The line between primary and secondary sources is not always clear, since different fields define primary sources differently (Kent-Drury 2005, p. 1). As a result, some document may be categorised as "primary" in one study, but can be considered as "secondary" or even "tertiary" in another study. For example, an article in Wikipaedia on some technical subject can be considered as tertiary in the list of references presented in the student's coursework (because it served as a starting point for looking at the topic in question) but it also can be primary in the PhD research thesis in the area of Philology which is focusing on analysis of some peculiarities of the language usage in the World Wide Web documents (because it serves as a "raw material" for the research).

Typical categorisation can be presented as follows (see definitions of certain terms in Appendix A):

1. Primary sources:

   - Egodocuments (diaries, letters, memoirs, interviews, personal documents, autobiographies, etc.);
   - Current information of the period under investigation (newspaper articles, statistics, audio-visual information, oral history, art works, documentary or historical films, technical drawings, maps etc.);
   - Artifacts (instruments, tools, weapons, pottery, carving, building, etc.).

2. Secondary sources:

   - Scientific writings (monographs, books, dissertations, patents);
   - Cumulative works (bibliographies, catalogues, directories, indexes, library catalogues, etc.).

3. Tertiary sources:

   - Web encyclopaedias, web pages of learned societies, etc.

More detailed illustration of various OAS is shown in Appendix B where 42 primary, 24 secondary and 6 tertiary sources with their URLs are presented. The selected items are in the English language.

4 Types of Providers of Open Access Sources

Typical providers of the OASs include: open access journals, institutional repositories and specialised webpages/databases.

4.1 Open Access Journals

Scholarly writings, mainly articles and reviews as well as abstracts are available from the open access journals, and from some partly-free access commercial databases. Monographs are available from digital book projects. Directory of Open Access Journals (DOAJ) aims to increase the visibility and ease of use of open access scientific and scholarly journals. Participating editorial boards take responsibility to include the correct bibliographical and digital information for their journals.

The DOAJ, which in 2013 held 9060 journals, aims to be comprehensive and cover all open access scientific and scholarly journals that use a quality control system to guarantee the content (DOAJ 2013).

During the last decade, there has been an ongoing debate about the negative influence on scientific publishing by the open access journals. A survey to inform this debate was carried out. 610 open access journals from the Directory of Open Access Journals (DOAJ) were compared to the data of subscription journals 7609 subscription journals. A summary stated the rapid grow of open access publishing, though open access journals that fund publishing with article processing charges (APC) are on average cited more than other open access journals. The open access journals indexed in Web of Science and/or Scopus are approaching the same scientific impact and quality as subscription journals (Bo-Chister and Solomon 2012, p. 9).

4.2 Institutional Repositories

Institutional repositories may include peer-reviewed journal articles and conference proceedings, research data, spreadsheets, photographs, audio files, video files, representations of artwork, diagrams, monographs and books, book chapters, theses, dissertations and other research-related outputs, such as presentations (Swan 2013b). Many of them were open access resources, which are listed in the Directory of Open Access Repositories (OpenDOAR 2013). Access to the OpenDOAR system is based on the use of a Google Custom Search engine. Quality assured approach applied in OpenDOAR minimises (but does not eliminate!) spurious or junk results, and leads more directly to useful and relevant information. Full texts are available for most items.

4.3 Specialised Webpages/Databases

The idea of crowdsourcing to collect data is becoming more acceptable for scholarly communities, transforming former methods of data collecting (Doan et al. 2011, p. 90). Specialised webpages/databases, maintained by academic societies or research institutions offer the widest variety of information on open access. The level of membership-based access varies depending on the efforts required from the provider to create and maintain the data and implement the policy of the governing body. Though we understand a big variety of subjects and fields of interests among learned societies, we limited our survey to only a few international bodies: ICOTHEC, EASST and SHOT (see Appendix B).

5 Challenges Related to the Open Access Sources

Organising of the process of operation and usage of the OAS is facing a number of challenges, which relate to historical practise.

5.1 *Technical Challenges*

Before opening access to any source of information, one has to answer a number of technical questions. First, "is it possible from the technical point of view?" Open access starts from preparing soft documents (e.g. new text typing or converting from the previous text formats or image scanning, etc). A survey will help to formulate requirement analysis (including potential data storage volumes and potential data flows which will affect choice and configuration of the actual data storage facilities), performing system design process (in-the-house or by contractor) and, finally, implementing a target system which includes information editing and updating facilities and populating it by the target information sources.

The directories of OAS demand a continuous editing and updating (in order to keep access to the items, which may be migrating to different URL addresses). This Confederation of Open Access Repositories (COAR) within an Interoperability Project (Current State of Open Access Repository Interoperability 2012, p. 13) took this challenge on the international level.

The technical side of opening or closing access is easy, but it has a tremendous effect on users and subscription policies. Free access is usually still limited by the requirements to register and install certain browsing or downloading/viewing software on the user side in order to get proper access. The challenge of access also includes new generation of sources (e.g. those conforming to Web 2.0 line) as well as new end-user devices (e.g. smart phones, tablet computers, etc.).

5.2 *Organisational and Economical Challenges*

Tradition has a strong influence in a scholarly practise. Scholars were accustomed to scientific journals in paper format, but recently libraries are shifting to electronic formats mainly because of economical reasons. However, a subscription fee is a challenge for libraries (Noorden 2013, p. 1); it may lead to serious changes in an information service system. Sometimes decision to create their own electronic resources or to join the existing systems of open access are considered as the better alternative.

For the commercial source provider it means the investigating of a need for the new resource, potential market demand as well as the ways to market it. The consequence is a need for proper budgeting of the project, implementing it on time as well as providing long-term viability and return on investments from the resource after initial spending.

For the free source provider it usually means that demand was identified by the funding body (e.g. government or non-government organisation such as charity, learned society, etc.). If a project involves digitising of a pre-specified number of X items from the Technical Museum Y using defined technology Z, then there is no need for updates and any follow-ups after finishing of the project for which was spent allocated amount of money.

For the library it includes an annual increase of the subscription costs of the title, a demand to keep paper-based subscription, also a risk of title migration from one "package" to another. Assessment of the realities was published in "The Survey of Library Database Licensing Practices" (2011) by Primary Research Group, which includes responses from almost 70 academic, special and public libraries from the United States, the UK, continental Europe, Canada and Australia. The reports cover the impact of digital repositories and open access publishing on database licensing, article databases and directories, perceptions of price increases on various types of subjects areas, etc. Among other findings, the general conclusion on price increase was stated (The Survey of Library Database Licensing Practices 2011, p. 1). The new issue of the report stated: more than a quarter of sampled libraries make the "extensive use" of free access to back issues now offered by many journal publishers (The Survey of Library Database Licensing Practices 2013, p. 1).

5.3 *Political Challenges*

Open access became an international issue in a form of Budapest Open Access Declaration (2002), followed by Berlin Declaration on Open Access to Knowledge in the Sciences and Humanities (2003) and Bethesda Statement on Open Access Publishing (2003). The benefits of open access is also advocated by libraries, which join not only local, but also international consortiums and projects: OASIS (OASIS 2013), Electronic Information for Libraries Open Access (EIFL-OA 2013), etc.

Increasing legal support is coming from the national governments as well as from the international bodies. For example, the press release of the European Commission stated: "In science terms, because data is often not shared at all, there are risks of parallel research that wastes brainpower, time and money. Greater data transparency will also help reduce academic fraud" (Open Access 2012, p. 1). The issue of availability of data from publicly funded research projects and digitised books in the libraries are also mentioned in the Digital Agenda for Europe of the European Commission (Open Data 2012, p. 1).

5.4 *Content Quality Challenges*

The aspect of content quality of the accessed OAS is important to both, providers and users, i.e. supply and demand sides of the process.

High standards and high quality is easily compatible with open access. All parties such as editors, repository managers, web-page owners, etc., are charged with responsibility for the quality of contents they provide. In the case of scholarly journals, quality control is a responsibility of an editorial board. Then a decision to open access is a technical step usually. In the case of open access repositories the material will be either unrefereed preprints or refereed postprints (Is open 2013, p. 1). In all cases, institutions issue their own regulations to control the quality. The question of archival copies becomes more essential, since the publisher is expected to keep the archival copies.

6 Conclusion

Open access to information should be considered an appropriate policy just as free and open access to other important things such as a breath of air, a smile, a word of comfort…

In the report of the Working Group on Expending Access (2012) it was noted that research communications are already in a period of transition towards open access (Finch 2013, p. 1). If put more technically, open access is both an alternative and a choice, it is vital in science, it also encourages/forces commercial providers to introduce at least partly open access to scholarly content.

Nevertheless, when speaking about open access it is necessary to keep in mind all the parts, which include data sources, hardware, software, technical solutions and creativity. The latter is optimising a search interface and a process as a whole. The education of end-user should not be left without attention. Moreover, we should not forget to educate end-user how to use instruments and sources within the all-embracing information universe.

Appendix A. Glossary

Artifact is a simple object (such as a tool or weapon) that was made by people in the past (Merriam-Webster 2013).

Egodocument is an autobiographical writing, such as memoirs, diaries, letters and travel accounts (Center 2013).

Institutional repositories are digital collections of the outputs created within a university or research institution. While the purposes of repositories may vary (for example, some universities have teaching/learning repositories for educational materials), in most cases they are established to provide open access to the institution's research output (Swan 2013b).

Open access research literature is composed of free, online copies of peer-reviewed journal articles and conference papers as well as technical reports, theses and working papers. In most cases there are no licensing restrictions on their use by readers (Swan 2013a).

Primary source of information is a document or physical object, which was written or created during the time under study. These sources were present during an experience or time period and offer an inside view of a particular event. These can be original documents (egodocuments, official records), creative works, relics or artifacts (What 2013).

Secondary sources of information interpret and analyze primary sources. These sources are one or more steps removed from the event. Secondary sources may have pictures, quotes or graphics of primary sources in them (Ibid.).

Tertiary sources of information provide only general, simplified background on a topic, they would seldom be used in the capacity of a "primary" source, usually are not acceptable to base an academic research. These sources usually are not credited to a particular author. They provide a superficial overview of what the topic includes, its basic terminology, and often references for further reading (which would usually be Secondary sources, produced by established experts on the topic). Tertiary sources, such as dictionaries, can be used to get a fuller sense of definitions and meanings of the field's terminology (Primary 2013).

Appendix B. Examples of the Open Access Information Sources According to the Categories, Groups and Types

The Table below is organised as follows:

- Examples of the sources are given by categories starting from primary and followed by the secondary and tertiary;
- In each category groups of relevant documents (e.g. egodocuments, letters, etc.) are given. Not all groups are applicable to all categories;
- Examples of the three individual document types such as diaries in the egodocuments in the primary sources, etc. are given at each group.

Source category/ group/type	Source description and URL
1. Primary	
1.1. Egodocuments	
Diaries	Diaries of Court Ladies of Old Japan by Murasaki Shikibu, Izumi Shikibu and Sugawara no Takasue no Musume (1920) In: A celebration of women writers. The University of Pennsylvania. http://digital.library.upenn.edu/women/. Accessed 28 April 2013
	Taft HN The diary of Horatio Nelson Taft. In: American memory project. The Library of Congress. http://memory.loc.gov/ammem/index.html. Accessed 28 April 2013
	Narrative of De Sotoæs expedition based on the diary of Rodrigo Ranjel, his private secretary. In: American Journeys. Wisconsin Historical Society. http://www.americanjourneys.org/aj-062/index.asp. Accessed 28 April 2013

Source category/ group/type	Source description and URL
Letters	Butler (Hendricks) S, Hendricks AW (1856–1859) Letters. In: Hanover historical texts collection. http://history.hanover.edu/project.php#modern. Accessed 30 April 2013
	Goethals GW (George Washington) (1913) Letter from George W. Goethals, Chairman and Chief Engineer, Isthmian Canal Commission, at Culebra, to W. G. Comber, Resident Engineer, March 15, 1913: correspondence. In: Woodson Research Center, Rice University, Americas Collection, 1811–1920, MS 518. http://scholarship.rice.edu/handle/1911/22014. Accessed 30 April 2013
	Wright LE (1908) Letter from the secretary of War, Estero Bay, Florida, December 12, 1908. In: Florida environments online, the sciences and technologies general collection. http://ufdc.ufl.edu/UF00004582/00001. Accessed 30 April 2013
Memoirs	Hewitt KH (2004) Memoirs of the Admiral Kent H. Hewitt. Naval War College Press, Newport, Rhode Island. Scanned and electronically published by American Naval Records Society, Bolton Landing, New York, 2010. In: Ibiblo. http://www.ibiblio.org/anrs/docs/1004hewitt_memoirs.pdf. Accessed 30 April 2013
	Koller C (2011) Representing otherness: African, Indian, and European soldiers' letters and memoirs. In: Das S (ed) Race, empire and first world war writing. Cambridge, pp. 127–142. In: Zurich Open Repository and Archive. http://www.zora.uzh.ch/54507/1/Koller-Representing_Otherness.pdf. Accessed 30 April 2013
	Green FH and Eldridge E (1838). Memoirs of Elleanor Eldridge. Providence, B.T. Albro. In: North American Slave Narratives, Documenting the American South. http://docsouth.unc.edu/neh/eldridge/summary.html. Accessed 30 April 2013
Interviews	Achilles T (1972) Interview, November 13, 1972. In: The foreign affairs oral history collection of the Association for Diplomatic Studies and Training. American Memory Project, Library of Congress. http://memory.loc.gov/ammem/collections/diplomacy/author.html. Accessed 30 April 2013
	Benzer S (1991) Interview by Heidi Aspaturian. Pasadena, California, September 11-February 1991. In: Oral history project. California Institute of Technology Archives. http://oralhistories.library.caltech.edu/27/1/OH_Benzer_S.pdf. Accessed 30 April 2013
	Postman N (S. a.) Home alone with technology: an interview With Neil Postman: interview by Clark Norman. In: Iowa Research Online, University of Iowa's Institutional Repository. http://ir.uiowa.edu/cgi/viewcontent.cgi?article=402&context=ijcs. Accessed 30 April 2013
Personal documents	Vinci Ld The Codex Arundel: notebook. In: Digitised Manuscripts, the British Library. http://www.bl.uk/manuscripts/FullDisplay.aspx?ref=Arundel_MS_263. Accessed 28 April 2013
	Crophill J (1430–1485) Commonplace book including astrological prognostications, cookery recipes, medical and alchemical treatises and recipes. In: Digitised manuscripts. the British Library. http://www.bl.uk/manuscripts/FullDisplay.aspx?index=3&ref=Harley_MS_1735. Accessed 28 April 2013
	Galileo's biography. In: The Galileo Project: a source of information on the life and work of Galileo Galilei. http://galileo.rice.edu/. Accessed 28 April 2013

Source category/ group/type	Source description and URL
Autobiographies	Taussky-Todd, Olga (1906–1995) (1980) Autobiography, 1979–1980 Pasadena, California, 1980. In: Oral history project. California Institute of Technology Archives. http://oralhistories.library.caltech.edu/43/1/OH_Todd.pdf. Accessed 28 April 2013
	Smith H (1787–1860) (1903) The Autobiography of Lieutenant-General Sir Harry Smith, John Murray, London, 1903. In: A celebration of women writers. University of Pensylvania. http://digital.library.upenn.edu/women/hsmith/autobiography/harry.html. Accessed 28 April 2013
	Howard OO (1830–1909) (1908) Autobiography of Oliver Otis Howard, Major General, United States Army, vol 1. In: Perseus Digital Library. http://www.perseus.tufts.edu/hopper/text?doc=Perseus%3Atext%3A2001.05.0173. Accessed 28 April 2013
1.2. Current information from the period under investigation	
Newspaper articles written at a time of the event	Kentucky Gazette: [image] August 31, 1793. In: Kentucky Digital Library. http://kdl.kyvl.org/catalog/xt7xgx44rq00_1?. Accessed 30 April 2013
	Stamp Selling Machine. Examiner, Monday, 3 June, 1912. In: Digitised newspapers and more, Trove. http://trove.nla.gov.au/ndp/del/article/50654633. Accessed 30 April 2013
	Over Fifty Craft in the Raid. Te Puke Times, 10 July 1917, p. 3. In: PAPERPAST. http://paperspast.natlib.govt.nz/cgi-bin/paperspast. Accessed 30 April 2013
Statistics	Bridges, bridge sections, towers and lattice masts, of iron or steel (2003–2006) In: Industrial commodity statistics database. United Nations Statistics Division. http://data.un.org/Data.aspx?q=bridges&d=ICS&f=cmID%3a42110–0. Accessed 30 April 2013
	European Union indicators in 2005–2012. In: EUROSTAT. European commission statistical data. http://epp.eurostat.ec.europa.eu. Accessed 30 April 2013
	World Development Indicators. In: The world development indicators. http://databank.worldbank.org/data/home.aspx. Accessed 7 Aug 2013
1.3. Audio-visual information	
Items of oral history	The great fire of London (1666) In: Voices from the twentieth century: Sounds from the past. http://www.eyewitnesstohistory.com/londonfire.htm. Accessed 30 April 2013
	Smith, Tom (Part 1 of 1). An oral history of British photography. In: Oral history of British science. http://sounds.bl.uk/Oral-history/Observing-the-1980s/021M-C0459X0204XX-0001V0#sthash.Ysfs75zP.dpuf. Accessed 7 August 2013
	Banes L (1988) Lensing for Alfred Hitchcock and film equipment at Gaumont Studios interview by Peter Sargent on 28 July 1988. (A transcript of the interview). In: Oral history: BFI ScreenOnline. http://www.screenonline.org.uk/audio/id/952231/index.html. Accessed 7 Aug 2013
Art works	National galery of art. NGA images. https://images.nga.gov. Accessed 30 April 2013
	Metropolitan museum of art. http://www.perseus.tufts.edu. Accessed 30 April 2013
	WebMuseum, Paris. Famous artworks exhibition. http://www.ibiblio.org/wm/paint/. Accessed 30 April 2013

Source category/ group/type	Source description and URL
Documentaries or historical films	The Pirate Bay: away from keyboard (2013) In: Top documentary films. http://topdocumentaryfilms.com/pirate-bay-away-from-keyboard/. Accessed 30 April 2013
	The great train robbery (1903) In: History in motion, eye witness to history. http://www.eyewitnesstohistory.com/himtrainrobbery2.htm. Accessed 30 April 2013
	Titanic‘s Achilles heel (2007) In: Top documentary films. http://topdocumentaryfilms.com/titanics-achilles-heel/. Accessed 30 April 2013
Technical drawings	Floor plan of the Liner Lab, including room functions. Austin, Field & Fry, Architects Engineers, 22311 West Third Street, Los Angeles 57, California (1962) In: The prints and photographs online catalog, Library of Congress. http://www.loc.gov/pictures/item/ca3006.photos.192902p/. Accessed 30 April 2013
	General arrangement drawing of Pekin Syndicate Railways ‘2–6-0’ locomotive Order No 9141/ Draughtsman: F le Manquais. Image Online. In: Museum of Science and Industry. Manchester. http://www.mosi.org.uk/collections/explore-the-collections/images-online/display.aspx?irn=33310&row=2. Accessed 7 Aug 2013
	Da Vinci L (1452–1519) Drawings of water lifting devices I(image). In: Mystudios Art Galery. http://www.mystudios.com/artgallery/. Accessed 7 August 2013
Maps	Earliest known map, a town plan from Catal Hyuk, 6200 BCE. In: Cartographic Images Net. http://www.cartographic-images.net/100_Town_Plan_from_Catal_Hyuk.html. Accessed 30 April 2013
	Map of Texas, showing the line of the Texas and New Orleans Rail Road, and its connections in the US and adjacent territories [n.p.], (1860) In: Transportation and communication, American memory. http://memory.loc.gov/cgi-bin/query/D?gmd:1:./temp/~ammem_7gd4. Accessed 30 April 2013
	Ptolemaic world map in the Liber Chronicarum (1493) In: Henry Davis map collection. http://www.henry-davis.com/MAPS/AncientWebPages/119A.html. Accessed 30 April 2013
1.4. Artifacts	
Instrument, tool, weapon, pottery, carving, building, etc	Weight-driven clock (1774) Collection online. In: British Museum. London. http://www.britishmuseum.org/research/collection_online/collection_object_details.aspx?objectId=3344392&partId=1&searchText=oak&images=true&page=2. Accessed 7 Aug 2013
	Perseus: art and archeology artifact browser. http://www.perseus.tufts.edu. Accessed 30 April 2013
	Photographic negative showing 34,000 kW low pressure turbo-alternator set, 3000 rpm, 33 kV at Brimsdown A Power Station Customer. Image Online. In: Museum of Science and Industry. Manchester. http://www.mosi.org.uk/collections/explore-the-collections/images-online/display.aspx?irn=33695&row=22. Accessed 7 August 2013

Source category/ group/type	Source description and URL
2. Secondary	
2.1. Scientific writings	
Articles	OpenDoar (Directory of Open Access Repositories). http://www.opendoar.org/. Accessed 30 April 2013
	DOAJ (Directory of Open Access Journals). http://www.doaj.org/. Accessed 30 April 2013
	The Diversity in Engineering (DinE) online bibliography (a collection of journal articles and conference papers published 2005–2010). http://inesweb.org/dine/about. Accessed 30 April 2013
Books, monographs	Free Ebook Search Engine. http://ebookbrowse.com/. Accessed 30 April 2013
	E-Books Directory. http://www.e-booksdirectory.com/. Accessed 30 April 2013
	Project Gutenberg. http://www.gutenberg.org/. Accessed 30 April 2013
Dissertations	Networked Digital Library of Theses and Dissertations (NDLTD). http://www.ndltd.org/. Accessed 30 April 2013
	DART-Europe: European Research Theses. http://www.dart-europe.eu. Accessed 30 April 2013
	EThOS Electronic Theses Online Service. http://ethos.bl.uk/About.do. Accessed 30 April 2013
Patents	Espacenet. Patent Search. http://worldwide.espacenet.com/singleLineSearch?locale=en_ep. Accessed 30 April 2013
	US Patent and Trademark Office Full-text and Image Database. http://www.uspto.gov/patents/process/search/. Accessed 30 April 2013
	European Patent Office. Free Online Service. http://www.epo.org/searching/free.html. Accessed 7 Aug 2013
2.2. Cumulative works	
Bibliographies	Technology in world history: a basic bibliography of works in the field (2005) http://www.historyoftechnology.org/bibliography.html. Accessed 30 April 2013
	Scientific instrument commission cumulative bibliography (an electronic database listing the titles of books, pamphlets, catalogues and articles on or connected with historical scientific instruments). http://www.sic.iuhps.org/bibliography/index.shtml. Accessed 8 Aug 2013
	Isis current bibliography (history of science (1913–)). In: History of Science Society. http://www.hssonline.org/publications/current_bibliography.html. Accessed 30 April 2013
Catalogues	English short title catalogue (from the collections of the British Library and over 2000 other libraries hold records for the material published between 1473 and 1800). http://estc.bl.uk. Accessed 30 April 2013
	The nineteenth century (online catalogue of over 34,000 nineteenth-century works first published between 1801–1900, available on microfiche). http://c19.chadwyck.co.uk/. Accessed 30 April 2013
	Royal Society of London: Catalogue of scientific papers, London, 1867–1925, vol 15–16. In: Galica, a free digital library. http://gallica.bnf.fr/. Accessed 30 April 2013

Source category/ group/type	Source description and URL
Directories, indeces	Directory of museums. In: Artefacts, an association of historians of science and technology, mostly in museums and academic institutions. http://www.artefactsconsortium.org/Museums/MuseumsDirectoryF.html. Accessed 30 April 2013
	History of science and technology index. In: The open door website. http://www.saburchill.com/HOS/index.html. Accessed 7 August 2013
	Repertorium Commentationum a Societatibus Litterariis Editarum, Secundum Disciplinarum Ordinem. Compiled by Reuss, JD (1750–1837) (Index of articles published by scholarly societies, arranged by discipline) Gottingae, 1801–1821, vol 15–16. http://archive.org/details/repertoriumcomm11reusgoog. Accessed 30 April 2013
Library catalogues	The European library (links to 49 national Libraries in Europe) http://www.theeuropeanlibrary.org/tel4/discover/contributors. Accessed 7 Aug 2013
	The library of congress online catalog. http://catalog.loc.gov/. Accessed 30 April 2013
	WorldCat (world's largest network of library content and services). http://www.worldcat.org/. Accessed 7 Aug 2013
3. Tertiary	
3.1. Web encyclopaedias	Encyclopedia Astronautica. http://www.astronautix.com/. Accessed 30 April 2013
	History of Technology. In: Wikipedia. http://en.wikipedia.org/wiki/History_of_technology. Accessed 30 April 2013
	ScienceWorld, also Known as Eric Weisstein's World of Science (2002) http://scienceworld.wolfram.com/. Accessed 30 April 2013
3.2. Webpages of learned societies	ICOTHEC (International Committee for the History of Technology, since 1968, a part of IUHPS/DHST). http://www.icohtec.org/. Accessed 30 April 2013
	EASST (European Association for the Study of Science and Technology, since 1981). http://easst.net/. Accessed 30 April 2013
	SHOT (Society for the History of Technology, since 1958). http://www.historyoftechnology.org/. Accessed 30 April 2013

Acknowledgments I would like to express my sincere gratitude to Dr. Marius Naris for his kind suggestions during review process.

References

Archer J, Hanloan AM, Levine JA (2009) Investigating primary source literacy. The Journal of American Librarianship 35(5):410–420

Assessing the role of librarians in an Open Access world (2012) / prepared by TBI Communications on behalf of InTech.. http://www.intechopen.com/js/ckeditor/kcfinder/upload/files/Role%20of%20the20Librarian_Survey_Findings_Jun12.pdf. Accessed 24 April 2013

Berlin Declaration on Open Access to Knowledge in the Sciences and Humanities (2003) http://oa.mpg.de/lang/en-uk/berlin-prozess/berliner-erklarung/. Accessed 24 April 2013

Bethesda Statement on Open Access Publishing (2003) http://legacy.earlham.edu/~peters/fos/bethesda.htm. Accessed 24 April 2013

Bo-Chister B, Solomon D (2012) Open access versus subscription journals: a comparison of scientific impact. BMC Medicine 10(1):73–82 (Academic Search Complete/EBSCO Publishing. http://ebscohost.com)

Budapest Open Access Declaration (2002) http://www.budapestopenaccessinitiative.org/read. Accesed 24 April 2013

Center for the Study of Egodocuments and History (2013) http://www.egodocument.net/egodocument/index.html. Accessed 24 April 2013

Current State of Open Access Repository Interoperability (2012) COAR. http://www.coar-repositories.org/files/COAR-Current-State-of-Open-Access-Repository-Interoperability-26-10-2012.pdf. Accessed 30 April 2013

Daniel D (2012) Teaching students how to research the past: historians and librarians in the digital age. The History Teacher 45(2):261–282

DOAJ: Directory of Open Access Journals (2013) http://www.doaj.org/. Accessed 30 April 2013

Doan A, Ramakrishnan R, Alon HY (2011) Crowdsourcing systems on the world-wide web. Commun ACM 54:86–96. doi:10.1145/1924421.1924442 (EBSCOHost Business Source Complete. Accessed 14 Aug 2013)

EIFL-OA. Open Access (2013) http://www.eifl.net/openaccess. Accessed 24 April 2013

Finch D (2013) Accessibility, sustainability, excellence: the UK approach to open access. Information Services & Use 33(1):11–18. doi:10.3233/ISU-130687

Gorman M (1998) The five laws of library science: then & now. Li-brary Journal 44(7):20–23

Is Open Access Compatible With High Standards and High Quality? (2013) EIFL Open Access. http://www.eifl.net/faq/open-access-compatible-high-standards-and-high. Accessed 23 Aug 2013

Kent-Drury R (2005) Using internet primary sources to teach critical thinking skills in world literature. Libraries Unlimited, Westport, p 1

Lombard E (2010) Perspectives on primary and secondary sources. The Journal of American Librarianship 36(3):250–253

Merriam-Webster Learners Dictionary (2013) http://www.learnersdictionary.com/search/artifact. Accessed 24 April 2013

Noorden RV (2013) Open access: the true cost of science publishing. Nature 425(7442) http://www.nature.com/news/open-access-the-true-cost-of-science-publishing-1.12676. Accessed 24 April 2013

OASIS: Open Access Scholarly Information Sourcebook (2013) http://www.openoasis.org/. Accessed 24 April 2013

Open Access to Scientific Data—Communication and Recommendation—Background (2012) Press Release of 2012-07-17, Memo/12/565. http://europa.eu/rapid/press-release_MEMO-12–565_en.htm. Accessed 30 April 2013

Open Data (2012) Digital agenda for Europe. http://ec.europa.eu/digital-agenda/en/public-sector-information-raw-data-new-services-and-products. Accessed 9 Aug 2013

OpenDOAR (2013) The directory of open access repositories. http://www.opendoar.org/. Accessed 24 April 2013

Primary, Secondary, and Tertiary Sources (2013) ENGA14 Finnish Institutions research paper. http://www.uta.fi/FAST/FIN/RESEARCH/sources.html. Accessed 30 April 2013

Project Gutenberg (2013) http://www.gutenberg.org/wiki/Main_Page. Accessed 30 April 2013

Sapp G, Gilmour R (2003) A brief history of the future of Academic libraries: predictions and speculations from the literature of the profession, 1975 to 2000 /part two, 1990 to 2000 In: Portal: Libraries and the Academy 3, Nr 1, pp 13–34

The Survey of Library Database Licensing Practices (2011) Primary research group: selected findings from new report. ResourceShelf by FreePrint. http://web.resourceshelf.com/go/resourceblog/63300. The complete report (2011). http://www.primaryresearch.com/view_product.php?report_id=283. Accessed 9 Aug 2013

The Survey of Library Database Licensing Practices (2013) Primary research group. http://www.primaryresearch.com/news.php?id=107. Accessed 23 Aug 2013

Swan A (2013a) Open access: a briefing paper. http://www.openoasis.org. Accessed 30 April 2013

Swan A (2013b) Institutional repositories: a briefing paper. http://www.openoasis.org. Accessed 30 April 2013

What is a Primary Source? (2013) (Princeton University Library). http://www.princeton.edu/~refdesk/primary2.html. Accessed 30 April 2013

"Method and Much Scientific Probity": Hugo de Lacerda (1860–1944) and the Chair of Hydrography of the Lisbon Naval School (1897–1907)

Pedro Miguel Pinto Raposo

Abstract In 1897 Hugo de Lacerda (1860–1944), a Portuguese naval officer and hydrographic engineer, took charge of the chair of hydrography at the Lisbon Naval School. This chapter analyses how Hugo de Lacerda sought to overcome longstanding predicaments and controversies surrounding hydrographic activity in Portugal, by bringing together pedagogical strategies, observatory techniques and military values, under a patriotic rhetoric of scientific prowess.

Keywords Hydrography · Hydrographical engineers · Observatories · Military · Navy · Portugal · Empire · Pedagogy · Textbooks · Surveying · Astronomical Observatory of Lisbon · Naval School of Lisbon · Polytechnic School of Lisbon · Infante D. Luiz Meteorological Observatory · Hydrographical Mission of the Portuguese Coast

1 Introduction: a Plea for Hydrography

On 15 August 1885 the young naval officer Hugo de Lacerda (1860–1944) addressed a request to King Luiz of Portugal pleading to be admitted in the course of Hydrographic engineering (Historical Archive of the Portuguese Navy, Box 730). Lacerda had become a full-fledged officer in March that year and was now deployed in Eastern Africa, travelling the coasts of S. Tomé and Principe, Cape Verde and Angola. It had come to his knowledge that there was still one vacant place for that academic year. Only up to two officers were admitted each year, which led him to hurry his application. Lacerda was eager to enter the hydrographic profession. He was already involved in colonial surveys, and had found a vocation for this activity. As he proudly remarked in the letter, some charts bearing his name had been published by the Lisbon Geographical Society (Sociedade de Geografia de Lisboa, henceforth SGL). Lacerda's request was unsuccessful though. There were

P. M. Pinto Raposo (✉)
Inter-University Centre for the History of Science and Technology, University of Lisbon, Lisbon, Portugal
e-mail: pmraposol@fc.ul.pt

R. Pisano (ed.), *A Bridge between Conceptual Frameworks,* History of Mechanism and Machine Science 27, DOI 10.1007/978-94-017-9645-3_23

three more candidates and one of them won the place. Lacerda did not give up: claiming that the chosen officer was not experienced in hydrography, over the next months he successively presented three letters of complaint to the king, but always without success. He had to wait for the next academic year; in June 1886 he was finally admitted as a student of hydrography. After five years spent between the Polytechnic School of Lisbon (Escola Politécnica de Lisboa, henceforth EPL), the Naval School (Escola Naval), and several periods of practical training, Lacerda finally obtained the title of hydrographic engineer (Engenheiro Hidrógrafo, henceforth EH), which gave him access to the selective Hydrographic Corps of Portugal. There's a scent of irony to this story, as in 1897 Lacerda became the holder of the chair of hydrography at the Naval School, and in the decades that followed, one of the most influential figures in Portuguese hydrography. The irony gets starker if we take into account that, by that time, the shorelines of Portugal's mainland and overseas territories were yet to be properly charted, and that, during the existence of the Hydrographic Corps—from 1869 to 1895—practically none of its members was deployed to colonial service. With her longstanding maritime tradition and vast overseas empire, Portugal bore the shame of having to resort to foreign charts, especially British ones, to navigate her own coasts, metropolitan and colonial alike. Hugo de Lacerda believed that the teaching of hydrography at the Naval School could help reverse this state of affairs. This chapter focuses on the strategies he employed to fulfil this goal. First, the relevant historical background will be provided. Lacerda's pedagogical action must be placed in the context of local debates on the training, status, command and role of naval officers and EHs. The importance of military values, practices and codes in the development of surveying sciences (astronomy, geodesy, topography) has been highlighted in recent studies concerning, for instance, France and Sweden (Schiavon 2010; Widmalm 2010; see also Pyenson 1996). This was equally true of hydrography in Portugal, but we shall see that in his case the organization and control of hydrographic activity and its practitioners was subject to frequent tensions and controversies between the Navy, the Land Army, and the civil institutions, and above all amongst naval officers themselves. Such tensions and controversies were often related to the issue of how naval officers and hydrographers should be educated and trained. Historians of science have been unfolding the constructive role of pedagogy and training in the making of scientific knowledge (Kaiser 2005). We shall see that Lacerda's action as lecturer of hydrography represented precisely an attempt to use this constructive power in order to transform every naval cadet into a proficient surveyor. A related topic of paramount importance is the study of textbooks. In a recent survey of the concerned literature, M. Vicedo highlighted several themes whose historical approach has benefitted from inquiries on textbooks: pedagogical and training practices, the formation of new disciplines, the development of ideas, priority disputes, epistemological concerns, and more generally the social context of science (Vicedo 2012). Lacerda's use of textbooks at the Naval School of Lisbon invites us extend the scope of these investigations to topics such as the national appropriation of disciplines, the crafting of techno-scientific/military identities, and the formation of military expertise.

Astronomical and geophysical observatories also play an important role in Lacerda's story. D. Aubin, C. Bigg and O. Sibum propose a definition of 'observatory techniques' that entails, on the one hand, techno-scientific methods and procedures, and, on the other hand, social strategies used to organize labour and to articulate observatories with their wider contexts (Aubin et al. 2010). This two-fold concept will be particularly useful to approach Lacerda's appropriation and recontextualization, at the Naval School, of his previous experience as a trainee EH in the astronomical and meteorological observatories of Lisbon.

Finally, it must be underlined that Lacerda's major ambitions were to produce a national hydrographic chart, and to empower naval officers to survey the empire. This chapter is not specifically concerned with the making, content or use of Portuguese hydrographic charts. However, insights in the history of cartography presented by J. B. Harley and developed by historians such as M. Edney , namely that maps were used to legitimize the reality of conquest and empire (Harley 1988; Edney 1997), are very suitable to frame Lacerda's agency, and generally the Portuguese efforts to foster hydrography in the period under study. More than a response to practical needs of navigation, these were struggles for colonial legitimacy, and for the recognition of Portugal as an equal player in the arena of imperial powers. Let us start then with a general picture of how these ambitions emerged and developed in nineteenth-century Portugal.

2 Empire and the Convoluted Shaping of a Profession

2.1 Nineteenth-Century Portugal: War, Regeneration and The 'Scramble For Africa'

Portugal spent the first half of the nineteenth century in turmoil. It was successively assaulted by Napoleonic troops, plagued by a civil war between absolutists and liberals and, after the triumph of liberalism in 1834, beleaguered by frequent conflicts between the conservative and progressive sects of the new liberal monarchy. By the middle of the century an urge for political stability, social peace and modernization pervaded the country (Bonifácio 2002, 2009; Mattoso et al. 1993). On May 1, 1850, the Duke of Saldanha (1790–1876) led an upheaval that was meant to eradicate all upheavals (Cabral 1976, p 25). Thus began the Regeneration, a period that lasted roughly until 1890. Railways and telegraphic networks became badges of its extensive programmes of infra-structural enhancement. Resources were focused on the consolidation of an internal market so that industrial capitalism could acquire a solid footing in the country (Mónica 2009; Telo 2004).

However, Portugal also had to deal with an overseas empire whose effective domain was much above its capabilities. Upon the independence of Brazil in 1822, attentions started to divert towards Africa, especially Angola and Mozambique. In the 1870s, when the so-called 'Scramble for Africa' spurred the colonial ambitions of other European nations, the hinterlands of these regions remained largely unexplored (Lucas 1993). The Lisbon Geographical Society was founded in 1875 to change this state of affairs. It promoted extensive explorations of the African territories and their resources, seeking to imprint a mark of Portuguese domain (Medeiros 2004; Guimarães 1984). Traditionally, Portugal branded the argument of historical occupation to justify its sovereignty. After the Berlin conference of 1884–1885, it was effective occupation that counted. Over the next decades Portugal had to craft her colonial policies within a tapestry of imperial powers largely dominated by England. Issues with the empire had a tremendous impact in the life of the metropole. On 11 January 1890, British authorities compelled the Portuguese to retreat from disputed territories in Mozambique. The episode, known as the British Ultimatum, shook the very foundations of the Portuguese liberal monarchy. It triggered perceptions of imperial fiasco and international abashment that gave way to the implantation of a Republican regime in October 1910 (Proença 2010; Teixeira 1990). It was in this particularly tense period of Portuguese history that Hugo de Lacerda implemented his pedagogical programme at the Naval School. But the problems he was to tackle had much older roots.

2.2 Shaping the Profession

In 1834, a section for hydrography was established within the Commission for the Geographical Map of the Kingdom (Comissão da Carta Geográfica do Reino). Pedro Folque (1744–1848) and his son Filipe Folque (1800–1874), both Army officers, were the leaders of this cartographical endeavour. Filipe became the harbinger of geodesy and practical astronomy in Portugal. In 1837, he was appointed professor of Astronomy and Geodesy at the Polytechnic School of Lisbon (Carolino 2012a). The EPL was part of a new set of higher education institutions with which the liberal authorities expected to produce an elite of military officers and civil servants capable of modernizing the state. Techno-scientific training, with an emphasis on practical applications, was deemed the key to achieve this goal (Simões et al. 2013).

Hydrography was never taught at the EPL, but one year before its foundation, in 1836, the Minister of the Navy Sá da Bandeira (1795–1876) entrusted Filipe Folque with the elaboration of a geodesy course for four naval officers deployed in the above-mentioned commission (Decree of 15 April 1836, *Diário do Governo* no. 91, 18 April 1836). This has been referred as the first course in hydrographic engineering in Portugal, and as one of the first engineering courses in the country (Aguilar et al. 2001, p 15). However, placed against the extant documentation (File 'Hidrografia', Box 241–4, Historical Archive of the Portuguese Navy), this claim seems exaggerated. The officers were, most likely, just taught the elements of geodesy and spherical astronomy that constituted the core of Filipe Folque's course in Astronomy and Geodesy at the EPL.

The first official attempt to define a proper track of study for EHs dates from 1851 only. A decree issued on 21 March that year (*Diário do Governo* no. 69, 22 March 1851, p 1) created a Hydrographic Section,[1] this time within the Navy.

[1] "Secção Hidrográfica" in Portuguese. The English expression Hydrographic Section is used with capitals in order to avoid confusion with the preceding section of hydrography mentioned earlier in this chapter.

The Section comprised six positions for naval officers who were required to command the "theoretical and practical knowledge of hydrography" (*Ibidem*). In 1845 the general training of Navy officers had been reformed through the foundation of the Naval School. Prior to enrolling at this school, cadets had to obtain approval in the chairs of mathematics and physics of the EPL.[2] Then they would move on to the officers' course[3] of the Naval School. Approval in its five chairs,[4] together with three years of overseas on-board service, gave access to the post of Naval tenant.

Not all senior officers approved this long track of study and training, which took, at least, four years to complete. Nevertheless, a tendency for a heavy techno-scientific syllabus would prevail, and even more so for hydrographers. The officers' course provided only elementary notions of hydrography, as part of a wider course that also covered topics in artillery, fortification and geography (Eça 1892). To apply for the Hydrographic Section, aspiring EHs had to go back to the EPL and obtain approval in all of its courses, except those already attended as preparation for the Naval School. Besides mathematical and physical sciences, the EPL courses covered a wide array of topics: chemistry, zoology, botany, mineralogy, and political economy, among others.[5] Connection with hydrographic practice was evidently scant. Furthermore, the classes were generally bookish and centred upon the lecturer. Astronomy and Geodesy, in principle the most useful course for a future hydrographer, overlapped with some topics of spherical astronomy already taught in the officers' course.

In addition to this painstaking plan of study, aspiring EHs were also expected to be familiar with Hydrographic instrumentation, proficient in the drawing of coastal

---

[2] The so-called Full Course (Curso Completo) of the EPL originally comprised ten chairs, which can be summarized as follows: 1st and 2nd—algebra and geometry, calculus, probabilities; 3rd—mechanics and its applications; 4th—astronomy and geodesy; 5th—experimental and mathematical physics; 6th—general chemistry and its applications; 7th—mineralogy, geology, and principles of metallurgy; 8th—comparative anatomy and zoology; 9th—botany and principles of agriculture; 10th—political economy and principles of administrative and commercial law (*Diário do Governo* no. 13, 16 January 1837, p 70). As in the case of Naval cadets, there were several combinations of specific courses as preparation for further military and techno-scientific studies in other institutions.

[3] The Naval School also offered courses for pilots and other naval specialties. The expression "officers' course" will be used hereafter to refer to the general course that gave access to the career of Naval officer.

[4] The five chairs were the following: (1). Elements of mechanics; spherical and nautical astronomy; (2). Principles of optics, practice of astronomical observations and navigational computing, performance of a complete journey; (3). Theoretical and practical artillery, principles of provisional fortification, geography and hydrography; (4). Naval architecture and technical drawing; (5). Naval manoeuvres and tactics. Complementary activities included fencing, swimming and military exercises. It must be noted that this arrangement of courses and complementary activities (as well as the preparatory studies at the EPL) underwent several reformations over the following decades. It is impossible to give detailed notice of all of them here. For an overview see Morais (1945) and Eça (1892). I am indebted to Ana Patrícia Martins for giving me access to unpublished material of her doctoral research on the mathematician and lecturer of the Naval School, Daniel Augusto da Silva (1814–1878).

[5] (See note 2).

plans and charts, and experienced in hydrographic surveys. A year of practice in fieldwork was required, but experience in geodesy or topography was also accepted. In any case, practice had no primacy in the admission to the Hydrographic Section. Preference was given to the candidates who had obtained the best marks at the EPL.

This scheme of selection and training was clearly misadjusted to the pressing needs of Portuguese hydrography. However, other reasons dictated the limited productivity of the Hydrographic Section. It was too small, some of its members diverted to functions with no direct relation to hydrography, and those who performed actual fieldwork worked essentially in the triangulations of the mainland. At best, they would survey the metropolitan harbours. Moreover, in practice it was Folque who commanded the Hydrographic Section. Some naval officers and EHs were infuriated to see a land army officer taking control of a naval department.

2.3 An Imperial Hydrographic Corps Wedged in the Metropole

By the late 1860s the reorganization the Hydrographic Section started to be discussed in the context of a wider reformation of the Portuguese Navy. New professional categories were established in an attempt to modernize the corporation. By the mid-1860s the Naval School offered courses for naval construction engineers, pilots of the merchant navy, naval machinists, and administrative officers. A course for naval doctors was eventually added. As regards EHs, there was never a specific degree. In different guises, their training was always approached as complementary to the officers' course.

In 1869 the Hydrographic Section was replaced with the Hydrographic Corps (Decree of 24 April 1869, Diário do Governo no. 93, 27 April 1836, p 528). The number of staff members was increased to eight officers, and significant changes were introduced in their training. The long track of complementary study at the EPL was recognised as excessive and unnecessary. Henceforth aspiring EHs would enrol only in the courses of Mechanics, Descriptive Geometry, and Astronomy and Geodesy. Then they would attend the courses of Practical Geodesy, Topography and Drawing, and Canals and Rivers, taught at the Army School.[6] A second major change was the establishment of a 1-year advanced course in hydrography exclusively for EHs, to be taught at the Naval School by a senior hydrographer. The programme is not specified in the decree that established the Corps, but it was certainly meant to function as a transmission of expertise based on the proximity between the students (no more than two per year) and an experienced EH.

Notwithstanding a greater focus on hydrography proper, EHs were still regarded and employed as multivalent experts. A two-year apprenticeship was now required. Future EHs could engage in hydrographic surveys, carry out astronomical, magnetic and meteorological work in the observatories of the country, or serve in the recently

[6] Similarly to the EPL, the Army School (Escola do Exército) resulted from the liberal reformations of higher education implemented in the mid-1830s. It played a very important role in the training of state engineers who conducted important infra-structural projects (Macedo 2012).

created War Depot (Depósito de Guerra). After obtaining the title of EH they could stay in these functions, and also take teaching positions at the Naval School. As we shall see in Sect. 3, hydrographers developed an especial liaison with the astronomical and meteorological observatories of Lisbon. But these observatories had no official ties with the Navy. And the War Depot was just a temporary avatar of the Directorate of Geodesic Works,[7] controlled by Folque until his death in 1874. After publishing a comprehensive geographical map of Portugal in 1865, Folque set out to produce a complementary chorographic map, for which he needed all available personnel, hydrographers included.[8] Moreover, the enhancement of the national ports, especially the Port of Lisbon, required high-scale surveys that were assigned to EHs. Consequently, a Hydrographic Corps that should have gone charting the colonial coasts ended up wedged in the metropole surveying the inland and the harbours—when its officers were not acting as astronomers, meteorologists, or lecturers.

Folque and a later director of the geodesic works, Army General Carlos Arbués Moreira (1845–1923), were often used by naval officers as scapegoats for the incipient state of Portuguese hydrography. Folque's immediate successor in the leadership of the Directorate for Geodesic Works was his former student Francisco Pereira da Silva (1813–1891). Silva was a naval officer and EH, but there were no significant improvements in hydrography during his tenure. Furthermore, views and opinions about the hydrographers' training, status and sphere of action varied amongst naval officers and EHs themselves.

2.4 *"Books and One's Own Intelligence": The 1886 Debate on the Hydrographic Profession in Portugal*

After the Berlin conference of 1883–1844 charting colonial coasts and rivers became a pressing need. It was obvious that the Hydrographic Corps and Portuguese hydrography in general still needed significant reformations. In 1886 a debate emerged around this issue. Its protagonists were Francisco Pereira da Silva, the Geographical Society of Lisbon, and Admiral Luiz de Moraes e Sousa (1845–1924). The debate broached in the sequence of a proposal presented by Pereira da Silva (1886), which was analysed by an especial commission appointed by the SGL (Eça 1886). EH Frederico Augusto Oom (1830–1890), director of the Astronomical Observatory of Lisbon (Observatório Astronómico de Lisboa, henceforth OAL), presided over this commission. Amongst the other members was EH Artur Baldaque da Silva (1852–1915), a keen defender of an authoritative intervention of hydrographers in the on-going enhancement of the Port of Lisbon (AB Silva 1893). Moraes e Sousa also partook the commission, but published his own views separately (Sousa

[7] This directorate underwent several denominations. It is generally referred as Direcção Geral dos Trabalhos Geodésicos. For a complete chronology of its official denominations and institutional configurations, see Alegria and Garcia (2000; p 17).

[8] This project was only completed by the beginning of the twentieth century (Alegria and Garcia 2000).

1886). All these men shared the opinion that Portugal, pioneer of hydrography, had failed to keep abreast of its development, and that using foreign charts to navigate Portuguese coasts was shameful. Besides, they complained that Portuguese names were often obliterated from those charts, in which, they claimed, cartographic mistakes abounded. It was also consensual that the Ministry of the Navy should exert a greater control over Hydrographic matters. There were, nevertheless, several points of divergence.

The few members of the Hydrographic Corps who did actual surveys worked for the Ministry of Public Works (Ministério das Obras Públicas, Comércio e Indústria, henceforth MOP), the overlord of the Directorate of Geodesic Works. Pereira da Silva stood for a clear separation between the hydrographic services carried out under the MOP, and those performed under the Ministry of the Navy. The SGL commission criticised this stance, arguing that EHs were officially under the aegis of the Ministry of the Navy, and stressing their value for the Directorate of Geodesic Works. So far they had been deployed to geodesic surveys, but were also expected to work in undertakings such as the improvement of commercial harbours and the domestication of watercourses. In brief, the SGL regarded EHs as important actors in the infrastructural enhancement of the metropole promoted by the cabinets of the Regeneration. Moraes e Sousa contested this position. He also wanted naval officers to play an active role in the development of the country, but in his opinion specialized EHs should act, above all, for the benefit of navigation.

As far as the overseas territories were concerned, Pereira da Silva wanted EHs to take full control not only of hydrography, but actually of all colonial cartography. He deplored the absence of EHs in the Commission for Nautical and Maritime Cartography of the Ministry of the Navy,[9] which had been publishing colonial maps, plans and charts since 1883. Pereira da Silva further suggested that a new Hydrographic Depot replaced this Commission. The idea was starkly criticised by the SGL, whose general opinion was that EHs lacked the expertise to carry out extensive surveys in the African hinterlands. Moraes e Sousa went even further: EHs, he claimed, sought access to the Corps precisely to escape colonial service whilst enjoying the bonuses conceded to its members. This was a mischievous remark but in fact not a single EH had, so far, participated in an overseas survey.

Pereira da Silva had suggested the introduction of two distinct categories of hydrographers: 1st class hydrographers with advanced training, and 2nd class hydrographers with a shorter complementary course and a practical apprenticeship. The Admiral held a more radical view. In his opinion the plan of study for EHs was "absurd". Most of the courses that aspiring EHs were compelled to attend, he argued, were too theoretical and overlapped with the officers course of the Naval School. Moraes e Sousa believed that a solid training in mathematics at the Naval School would prepare every officer to master a wide range of practical applications. All that was needed, as the Admiral put it, was "books and one's own intelligence"[10] (Sousa 1886, p 283). Therefore, the Naval School should provide all officers with the basic

[9] Usually mentioned as Comissão de Cartografia.

[10] "[precisam de] livros e da própria inteligência".

knowledge necessary to carry out hydrographic surveys. For those interested in this career path, the title of EH would be conceded simply upon a 2-year apprenticeship, entailing no especial privileges.

Hugo de Lacerda's action as lecturer of hydrography a decade later resonated considerably with Moraes e Sousa's ideas. There was an important nuance though: Lacerda believed that practice was as important as theory in the officers' basic training. And, as far as practice was concerned, observatory techniques constituted a precious asset.

3 Surrogate Naval Observatories: The Astronomical Observatory of Lisbon and the Infante D. Luiz Observatory

In France, naval observatories were established together with hydrography schools that were founded in the country from 1825 onwards (Boistel 2010). In Spain, hydrography holds close historical ties with the San Fernando Observatory (Gonzales 1992). In other countries, observatories originally shaped to support seafaring became prominent representatives of the astronomical sciences: the cases of the Royal Observatory, Greenwich, and the U.S. Naval Observatory immediately come to mind (Dick 2003; Smith 1991).

In 1798, the short-lived Royal Maritime Society (Sociedade Real Marítima) was founded in Lisbon with the aim of centralizing all cartographic activity, hydrography included.[11] Concomitantly, Portuguese authorities sought to bring together assistance to navigation, the training of seafarers, and the advancement of astronomy in a single observatory. In the same year, the Royal Observatory of the Navy (Real Observatório da Marinha, henceforth ROM) was officially established for this purpose. Its ensuing history, though, was particularly unfortunate (Reis 2009). In 1807, upon the invasion of Portugal by Napoleonic troops, the court moved to Brazil, taking most of the observatory's instruments to Rio de Janeiro. Over the next decades the ROM's location changed several times. The observatory was eventually installed in improvised facilities at the Navy Arsenal, where the few available instruments were exposed to the humidity of river Tejo and to the arsenal's fumes. In 1855 Folque was appointed director. He sought to refurbish and reequip the observatory, but with little success. The ROM was eventually extinct in 1874. In the following year, a new teaching observatory started to take shape in the yards of the EPL (Carolino 2011).

By that time, another observatory was under completion in the Portuguese capital. The construction of the Astronomical Observatory of Lisbon (OAL) had begun in 1861, under the patronage of King Pedro V. The OAL was essentially designed

[11] The society, whose complete name was Sociedade Real Marítima, Militar e Geográfica para o Desenho, Gravura e Impressão das Cartas Hidrográficas, Geográficas e Militares, was dissolved c. 1807, upon the Napoleonic invasions of Portugal.

to collaborate with Central Observatory of Russia in Pulkovo in the measurement of stellar parallax, but its development was hampered by a longstanding dearth of qualified personnel (Raposo 2010). Aspiring astronomers were expected to be skilful in mathematics and adroit in the use of instruments. Besides, exclusive dedication was mandatory. There were some attempts to shape the statutes of the observatory so that professors of the University of Coimbra and the higher-education schools of Lisbon could take posts at the OAL in a more flexible regime. EH Frederico Augusto Oom, who had practised in Pulkovo to become the OAL's first director, opposed them vehemently (Oom 1875). Oom feared what he perceived as the risk of the OAL becoming a reservoir of sinecures for bookish dons. The statutory decree of the OAL, issued in 1878 (*Lei Orgânica do Real Observatório Astronómico de Lisboa* 1878) vindicated Oom's position. It also shortened the number of prospective astronomers. Scientific staff comprised five positions, but vacancies were frequent. The advantage was that those who stayed were usually efficient and dedicated (Fig. 1).

During its first decades of activity, the OAL was essentially worked by Oom and two other EHs: César de Campos Rodrigues (1836–1919) and Augusto Alves do Rio (1845–1905). The parallax project was set aside, in favour of timekeeping. The renewal of the Port of Lisbon and the growth of railway and telegraph networks intensified the need for a national timekeeper. This function was included in the OAL's statutes as an ancillary assignment but it soon became the observatory's core activity.

Fig. 1 The Astronomical Observatory of Lisbon in the nineteenth century. (Courtesy of the Astronomical Observatory of Lisbon)

From 1885 onwards the OAL controlled the drop of a time-ball installed in the Port of Lisbon, via the telegraphic network. The apparatus of the time-ball was also used to relay time signals to other institutions and services.

The exactness of the OAL's signals became a badge of the observatory's commitment to precision. Work at the observatory was grounded on solid discipline and a rigorous division of tasks. For each day, preparatory work (choice of stars, determination of the instruments' constant errors and the observers' personal equation), the observation of stars with a transit instrument, the ensuing reduction of observations, and the telegraphic transmission of time signals, were carefully distributed by the three astronomers. They also tackled most administrative paperwork. Affinity and compliance to military discipline was crucial to keep the OAL running efficiently, but there was another pivotal element: Campos Rodrigues's thorough study of instruments, observing techniques, and computing methods. Rodrigues managed to introduce several improvements in the apparatus of the OAL, and also developed a number of slide-rules and graphs that eased the computing work (Raposo 2008). In 1904 he was awarded the Valz Prize of the Académie des Sciences de Paris, a distinction that confirmed him as the leading astronomer in Portugal and scientific hero of Portuguese hydrographers.

Several EHs, including Hugo de Lacerda (he did his apprenticeship at the OAL between 1890 and 1891), practised under Rodrigues's guidance. Their training included: observations with sextants and other portable instruments; transit observations by eye and ear, and by the American method;[12] the study of observing errors and, in this context, personal equation measurements. Personal equation has been discussed in relation to topics such as the moral probity of the scientists, physiological investigations on reaction times, and the disciplining of observers (Canales 2009, pp. 21–58; Hoffmann 2007; Daston and Galison 1992; Schaffer 1988). This last aspect was crucial in the training of EHs at the OAL. Personal equation measurements were used to assess their progression in the observation of transits. But here the discipline of observing did not mean solely an external imposition of rigour and self-restraint. It also instilled self-confidence, an important resource to deal with the loneliness and often-awkward conditions of fieldwork. As EH and former OAL trainee Joaquim Patrício Ferreira (1847–1925) put it: "aware of his own value and having to rely solely on his own devices, [the hydrographer can] tackle all kinds of observations, even the most delicate ones".[13]

Another institution of reference for Portuguese EHs was the Infante D. Luiz Meteorological Observatory (Observatório Meteorológico do Infante D. Luiz, henceforth OMIDL), where, as trainees, they were assigned meteorological and magnetic

[12] In the eye and ear method, the observer listened attentively to the beats of a clock and estimated the fraction of second corresponding to the passage of the star by a reticule thread. In the American method, an electric chronograph was used to record both the signals from the clock and the signals from the observer, who pressed (or released) a key in the moment he saw the star crossing the thread.

[13] "já cônscio de quanto vale e a não ter que apelar senão para os seus próprios recursos, [o hidrógrafo pode] proceder a todos os géneros de observações por mais delicadas que sejam (…)" (Ferreira 1886, p 62).

observations. The OMIDL was founded in 1853, upon the initiative of Guilherme Dias Pegado (1804–1885). Pegado was a lecturer of physics at the EPL, to which the observatory was appended. The OMIDL was meant to function as a centre of calculation (Latour 1987) for meteorology in the country and the overseas empire. It was also assigned the coordination of meteorological observations carried out aboard Portuguese vessels, following the guidelines issued from the Brussels conference of 1853 (Simões et al. 2013, pp 106–111; Raposo 2012; Tavares 2009; pp 55–61). Similarly to the OAL, the OMIDL was a civilian institution but developed a strong link with the War Navy. In 1855 naval officers Fernando da Gama Lobo (1829–1879) and João Carlos de Brito Capelo (1831–1901) were engaged as assistant observers. In 1875 Capelo was appointed director, a post he held until his death in 1901. Capelo became an internationally renowned pursuer of astronomical, meteorological and magnetic investigations. His work covered topics such as the patterns of winds and currents in the Atlantic Ocean, the deviations of the compass needle at sea, and the relation between sunspots and terrestrial magnetism (Bonifácio et. al. 2007; Canas and Silva 1999). To the eyes of EHs and naval officers in general, Capelo was in geophysics what Rodrigues was in astronomy: the model of an accomplished naval scientist.

Thus the OAL and the OMIDL functioned as surrogate naval observatories, providing EHs with training and technical advice, and imposing the mark of the Portuguese Navy in national and international science.

4 "The True Sense of Being a Seaman": Lacerda and the Chair of Hydrography of the Lisbon Naval School

4.1 Textbooks and the Nationalization of Hydrography

It 1897, Hugo de Lacerda (Fig. 2) became the holder of the chair of hydrography at the Naval School.

Fig. 2 Hugo de Lacerda. (Undated, courtesy of the Instituto Hidrográfico)

Fig. 3 Frontcover of the hydrography textbook for the academic year 1904–1905. (Courtesy of Biblioteca Nacional de Portugal)

An independent chair of hydrography had been introduced in 1887 and implemented from 1895 onwards.[14] Partaking feelings of wounded national pride, common in Portugal after the British Ultimatum, Lacerda set out to affirm Portuguese hydrography. The Hydrographic Corps had been dissolved in 1895. Lacerda would now use the chair of hydrography to create a wide basis of hydrographic expertise and pinpoint the better candidates for the title of EH. To implement this plan, Lacerda counted on the background of observatory techniques acquired at the OAL and the OMIDL, on his own experience in hydrography, and on three additional resources: a printing press, textbooks, and legislation. A lithographic workshop was installed at the Naval School to print textbooks, as well as hydrographic plans and preliminary charts based on the students' surveys. Textbooks were elaborated regularly since the first year of Lacerda's tenure. This practice was continued by EH Victor Hugo de Azevedo Coutinho (1871–1955), who in 1903 succeeded Lacerda in the chair. The National Library of Lisbon keeps a copy of the first part of the textbook for the academic year 1904–1905 (Fig. 3). It is entitled *Apontamentos para um Curso Elementar de Hydrographia* (Notes for an elementary course in hydrography) (Escola Naval 1904). The content is organized into eight parts: (1) preliminary notions, including the theory of errors and the description of instruments; (2) angle measurements; (3) distance measurements; (4) simultaneous measurements of angles and distances; (5) spirit levels and their sights; (6) further topics on topographic instruments and devices; (7) instruments employed in hydrography proper; (8) instrument for drawing and copying hydrographic works.[15]

[14] It was the 9th chair in the 1887 reform, and the 8th in the 1895 reform, with the full title of Hydrography and principles of geodesy (Hidrografia e princípios de geodesia). In 1903 its full name changed to Hydrography, lighthouses, oceanography and maritime routes (Hidrografia, faróis, oceanografia e derrotas) (Morais 1945, pp 42–43; Silva 1945, pp 186).

[15] The second part should be dedicated to the actual processes of surveying, as it can be inferred from a later version cited below.

The *Apontamentos* were significantly based on English and French reports and textbooks on hydrography, topography, oceanography and related matters.[16] But more than appropriating hydrography and reinforcing its disciplinary contours, there was an assumed intention to shape a national version of this subject. The sections addressing errors and the measurement of angles and distances clearly benefited from teachings received at the OAL. The bibliography also included the Portuguese textbook *Curso de Topographia* (Course in Topography), written by the lecturers of the Army School Mendes de Almeida (1854–1943) and Rodolfo Guimarães (1866–1918). This book provided Hugo de Lacerda with several descriptions of techniques and devices developed by Portuguese officers, deceased and alive, who had been involved in geodesic, topographical and hydrographic surveys. The science conveyed in the *Apontamentos* was intended to be, as much as possible, recognised as a Portuguese science. This point was clearly stated by Lacerda in the preface to a more polished version of the *Apontamentos*, revised by Azevedo Coutinho and issued in 1906 (Coutinho 1906). Lacerda introduced it as a "hydrography book written in Portuguese, following methods established by Portuguese masters such as Folque, Batalha, and Brito Limpo, already deceased, and also by Campos Rodrigues, fortunately still alive for the glory and profit of Portugal". Such a book was necessary, he added, "not only for the cadets of the Naval School, but in fact for the majority of the War Navy" (preface to Coutinho 1886, p VII).[17] The textbooks would thus stimulate a patriotic assimilation of hydrography; then practice, Lacerda believed, would give the cadets "the true sense of being a seaman" (*Ibidem*).[18]

4.2 "Method and Much Scientific Probity": Hydrographic Practice as a Rite of Passage

Lacerda was certainly aware of how the statutes of the OAL had been important in shaping the astronomers' profile. Similarly, he crafted a bid that became a decree in January 1901, and which established that the promotion of midshipmen to the

[16] The following works are included in the bibliography: Frochot, *Marées—Campaigne du "Ougnay-Frouin", 1903–1904*; A. Germain (1882), *Traité d'hydrographie*, Paris; Thompton S. Lecky (1884), *Wrinkles in practical navigation*; Le Bail, *Hydrographie: Campaigne du "Ougnay-Frouin", 1903–1904*; Eugène Prévot (1898), *Topographie: la topographie expediée*, Paris; M. J. Thoulet (1896), *Oceanographie dynamique*, Paris; M. J. Thoulet (1890), *Oceanographie statique*, Paris; William J. L. Wharton (1898), *Hydrographic surveying: A Description of the Means and Methods Employed in Constructing Marine Charts*, London.

[17] "Um livro de hydrographia escripto em portuguez, segundo processos estabelecidos por mestres portugueses, como Folque, Batalha, e Brito Limpo, já fallecidos, e ainda por Campos Rodrigues, felizmente ainda vivo para glória e proveito de Portugal, não era só preciso para os alumnos da Escola Naval; era-o para a maioria da Corporação da Armada." Caetano Maria Batalha (1810–1881) was among the first group of hydrographers who studied geodesy with Folque in the 1830s. Francisco António de Brito Limpo (1832–1891) was a Land Army engineer and surveyor who gained local prestige as a designer of geodesic and topographic instruments.

[18] "o verdadeiro sentido do marinheiro."

post of Naval tenant implied the successful participation in a hydrographic survey (Decree of the Ministry of the Navy, 25 January 1901). The decree constituted a legal basis to enforce hydrographic practice, and also conveyed a moral portrait of the ideal naval officer, whose virtues were essential to pass the test of fieldwork. These virtues included "method and much scientific probity" (*Ibidem*),[19] solicitude, prudence and resoluteness, together with physical stamina. Since hydrographic surveys were often carried out in remote and dangerous areas, the possible perils added to the challenge. Fieldwork would thus function as a ritual of passage for aspiring officers, putting their braveness and dexterity to test.

Following the decree, Hugo de Lacerda elaborated a stern set of rules for the conduction of the students' surveys. Cadets were demanded to comply with a disciplinary regime in which the ship functioned as centre of calculation (Latour 1987) and study room, and where the observing discipline and data-management methods of observatories had a noticeable influence. In the beginning of the survey, cadets were divided into two teams, each of them with a cadet acting as leader. Time management was strict. There were no Sunday or holyday breaks. Surveying works started by dawn and extended through dusk. There was only a daily two-hour break for lunch and rest, with some occasional allowance for sea swimming. Travel between different areas took place by night in order to make full use of daylight for surveying activities. When appropriate, the period of travel was used to perform in-depth soundings and other operations carried out from the ship's deck. By the end of each surveying day, cadets returned to the ship. From 8 to 10 pm, they revised their notes and discussed difficult and unclear points with the survey's director and the team leader. An individual report was elaborated and, after signed and dated, delivered to the head of the survey (i.e. the hydrography lecturer). Data collection and management were carefully regulated. In each day two cadets were assigned the synthesis of all accomplished work. Data was recorded in standardized forms and, like the individual notes, presented to the director by the end of the day. Intensive note taking during fieldwork was recommended to all cadets. Besides an officer's knife, their personal weaponry should include paper, pencil and rubber, to be carried all the time. There were also 24-hour shifts for the tide gauge readings. In each day (in this case, counting from midday to midday) two cadets would camp near the place where the gauge was installed and alternate in the readings. In the end of the shift all observations were copied to the ship's log, and then used to trace a tidal curve.

After the whole survey was completed, each ensign had eight days to present his final report. Its assessment decided the promotion to Naval tenant. The head of the survey also produced a general report. Published extracts from Lacerda's report for the 1901–1902 academic year (Lacerda 1903b) provide us with an insight on how the student surveys were actually implemented. The survey described in this report was undertaken between 15 and 31 August 1902. It covered several locations, from the mouth of the Lima River (in the northwest coast of Portugal) to the river Sado.[20]

[19] "método, e muita probidade científica."

[20] The River Sado runs through the southern province of Alentejo and meets the ocean near the city of Setúbal, roughly 30 km to the south of Lisbon.

The class was divided into two groups of ten elements. Each group spent roughly one week in the field. Practical works included a hydrographic and topographical study of the Lima's mouth; a geodesic survey in S. Martinho harbour; an elementary survey of the Farilhões islets, to evaluate mooring and shelter conditions; and studies of navigability in the rivers Douro and Sado. These activities were performed in very short periods of time. Moreover, as Lacerda acknowledged, the ship used in the survey[21] was not fully adapted to this kind of function. Accommodation for the cadets was improvised and rough, officers could not avoid mingling with them whilst moving on-board, and discipline often lessened. The intense summer heat and the long exposure to the sun also affected some students. In other occasions, episodes of bad weather limited the observance of the survey's rules.

Nevertheless, the survey yielded useful cartographic material. Data were amassed for the elaboration of preliminary charts in the scale 1/5000.[22] The cadets were also required to draw coastal features. A complete photographic set was taken aboard, so that they could take pictures too. These elements were collected for the eventual elaboration of a coastal guidebook. Several warnings to navigation were also issued on the basis of the students' observations. Their reports were essentially descriptive, but outstanding work occasionally appeared. For instance, in 1902 a cadet named Anahory Athias developed an investigation on the siltation of S. Martinho bay that Lacerda deemed publishable. Yearly student surveys were carried out until 1907, when another reorganization of the Portuguese Navy placed hydrographic matters under the aegis of the Chief Naval Command (Majoria General da Armada). Taken together, the student surveys resulted in the publication of at least 14 elementary charts (Aguilar et al. 2001, p 18) (Fig. 4).

4.3 *Republicanizing the Shores: the Hydrographic Mission of the Portuguese Coast*

The next step was to elaborate a comprehensive hydrographic chart of Portugal. This chart would complement the chorographic maps originally coordinated by Folque, and thus fill in what Lacerda deemed "an unacceptable gap" (Lacerda 1903b, p 213).[23] These ideas were implemented in earnest after the Republican revolution of 1910, in which naval officers played a major role. In 1912, Lacerda was placed in charge of the Hydrographic Mission of the Portuguese Coast (Missão Hidrográfica da Costa de Portugal, henceforth MHCP). Several officers, who, as cadets, had partaken his student-surveys, formed a crew that, in Hugo de Lacerda's

[21] It was *Bérrio*, a supply vessel constructed in 1886 in which some changes were introduced to accomodate the cadets.

[22] According to the decree of 1901 cited above, fluvial and coastal surveys should cover at least half a mile of terrain and adopt the scale 1:5000; for preliminary surveys made from the ship at least a full mile and the scale 1:10,000 were required.

[23] "uma lacuna inqualificável".

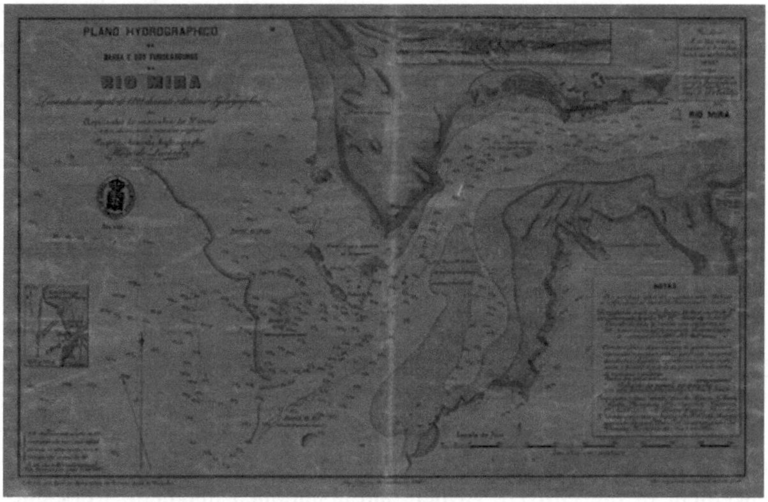

Fig. 4 Hydrographic plan of the River Mira's mouth, based on the student survey of 1898–1899. (courtesy of the Biblioteca Nacional de Portugal)

own words, constituted no especial team (Lacerda 1914, p. 473), that is, none of the officers apart from Lacerda was a specialized hydrographer. This crew went aboard the *Aviso 5 de Outubro*, formerly a royal yacht named Amélia, in honour of Queen Amélia, wife of King Carlos.[24] King Carlos had used it between 1898 and 1903 in oceanographic campaigns. The king was a keen participant in the culture of yachtism that helped foster oceanographic studies in the nineteenth century (Ramos 2006, p. 180; Rozwadovski 1996, p. 421). Under Lacerda's command and the name *5 de Outubro* (October 5th, the date of the Republican revolution) the ship would now sail the coasts of the mainland and tame them on paper. The first campaign of the MHCP took place between 1913 and 1914 and covered the northwest coast of Portugal, between Caminha and Espinho. It rendered a navigational chart in the scale 1/150,000, hydrographic plans of several ports, and a lithological chart. This first survey was also used to initiate magnetic and meteorological investigations, to collect elements for a costal guidebook, and to evaluate the coverage and range of existing lighthouses (Ministério da Marinha 1915). Lacerda eventually passed on the command of the MHCP to Vítor Hugo de Azevedo Coutinho. The MHCP remained active until 1935.

In 1919, Coutinho and another member of the MHCP, Henrique Baeta Neves (1890–1956), keenly represented Portugal in the hydrographic conference held in London between 24 June and 16 July that year (Coutinho and Neves 1920). The

[24] King Carlos (1863–1908) was entroned in 1889. He was murdered by radical Republicans during a public appearance on February 1st, 1908. The throne was then occupied by his son Manuel II (1889–1932) until the revolution of October 1910.

conference set forward the basis for the International Hydrographic Bureau, founded in 1921. The United Kingdom, the USA and France took the lead, but Portuguese involvement constituted an important moral achievement. Besides providing the groundwork for a national hydrographic chart, the MHCP had also instilled Portuguese hydrographers with the confidence they needed to play an active role in the international arena.

5 Conclusion

In 1925 the training of EHs was revised again. Their specialization was now called Hydrography and Navigation. It maintained the basic structure of the previous study plans, with the addition of short-term apprenticeships aboard hydrographic and oceanographic vessels, practice at the Geo-Magnetic Observatory of Coimbra in replacement of the OMIDL (which had abandoned magnetic work due to the introduction of electric traction in Lisbon), and some courses distributed by the Faculty of Sciences of Lisbon (Faculdade de Ciências de Lisboa, former EPL) and the new Instituto Superior Técnico (a school for advanced engineering studies). Essentially, the revision represented an adaptation to the Republican reforms of higher education, and an effort to prepare EHs to engage in oceanographic studies. During the MHCP Lacerda had already promoted the convergence between hydrography and oceanographic investigations, for the sake of national fisheries. EHs effectively came to play an important role in the development of such investigations. The foundation of the Hydrographic Institute (Instituto Hidrográfico, IH) in 1960 sealed the institutional commitment of the Portuguese Navy to the techno-scientific exploration of coasts and sea. With a noteworthy delay, but benefiting from the imperialist agenda and centralizing policies of the New State (Estado Novo, the dictatorial regime implanted in 1933), the IH managed to unify metropolitan and imperial hydrography under a single naval department.

These reforms were significantly influenced by Lacerda's pedagogical action at the Naval School of Lisbon. By imposing hydrographic training and practice to all aspiring officers, Lacerda created a critical mass of hydrographic expertise. Mandatory practice boosted the development of surveying skills, and highlighted promising candidates to the title of EH. Since all officers were expected to command at least a basic level of hydrographic expertise, those who pursued advanced training could now legitimately be submitted to a more demanding and diversified track of complementary study and practice. And regardless of further specialization, the inscription of hydrographic practice in the ethos of the naval officer helped strengthen the association between the Navy and the cultivation of maritime sciences.

Hugo de Lacerda, of course, did not solve all the predicaments of Portuguese hydrography. The surveying of overseas coasts progressed slowly, with hydrographic brigades being sent at spaces to Portuguese Guinea (1912), Timor (1937), Cape Verde (1945), Angola and S. Tomé (1953), and Macao (1960s) (Aguilar et al.

2001).[25] Furthermore, Lacerda's positions and strategies were often contested and rebutted by his peers. EH Jaime Wills de Araújo (1872–1941), for instance, was particularly critical of Lacerda's valorisation of practice. More important than enforcing hydrographic practice to all officers, he argued, was to assure the availability of EHs well versed in the concerned sciences and able to command surveying work (Minute nº 28, 28 August 1930, Historical Archive of the Portuguese Navy, Box 1380-A-10). The MHCP's modus operandi was not consensual either. Another influential EH, Pires de Matos (1901–?) played it down for the lack of standardized practices and uniformity of criteria. Because the choice of instruments and methods rested upon the decisions of the officer in command of each survey, the MHCP had resulted, in Matos's words, in "chaos, Babel Tower, the organized disorder" (Matos 1932; p 166).[26] It was only with the foundation of the IH that these problems were definitely settled. Whilst in the United States, for instance, the institutional control of hydrography required a careful negotiation between civil and military interests (Manning 2004; Slotten 1993), divisiveness within the Navy and among hydrographic engineers were probably the major hindrance of Portuguese hydrography. Military stamina and devices of patriotic rhetoric did not suffice to overcome it, but by blending them with pedagogy, scientific values, and observatory techniques into fieldwork, Lacerda managed to boost hydrographic cartography in Portugal and to reinforce the place of science in the ethos of Portuguese naval officers.

Acknowledgements Research presented in this paper was carried out under the post-doctoral grant SFRH/BPD/73373/2010, awarded by the FCT—Portuguese Foundation for Science and Technology. I would like to thank Luis M. Carolino and Teresa Salomé for inviting me to present an earlier version of this paper at the workshop A Experiência Politécnica (University of Lisbon, March 2013), Ana Simões for her comments and suggestions, and Ana Patrícia Martins for giving me access to unpublished material of her doctoral research on the Lisbon Naval School.

References

'Acta nº 28' (28 August 1930) Historical Archive of the Portuguese Navy, Box 1380-A-10

Aguilar T et al (2001) A Marinha na Investigação do Mar. Instituto Hidrográfico, Lisboa

Alegria MF, Garcia JC (2000) A cartografia hidrográfica de Portugal continental na 2ª metade do século XIX e início do século XX. In: Amorim I, Polónia A, Osswald H (eds) O litoral em perspectiva histórica (Séc. XVI a XVIII): um ponto da situação historiográfica. Instituto de História Moderna, Porto, pp 9–20

Aubin D, Bigg C, Sibbum HO (eds) The Heavens on Earth: observatories and astronomy in nineteenth-century science and culture. Duke University, Durham

[25] Mozambique stands out in this picture. Due to the strategic importance of its seaports for traffic in south-eastern Africa, several surveys were undertaken there between 1870 and 1914. After leaving the Naval School in 1903, Hugo de Lacerda went on to coordinate the enhancement of the harbour in Lourenço Marques nowadays Maputo (Lacerda 1906, 1907).

[26] "É o caos, a torre de Babel, a desordem organizada (…) o sistema em que o interesse pessoal pelo trabalho, que pode existir ou não, supre completamente a ordem e o método."

Boistel G (2010) Training seafarers in astronomy: methods, Naval Schools, and naval observatories in eighteenth- and nineteenth-century France. In: Aubin D, Bigg C, Sibbum HO (eds) The Heavens on Earth: observatories and astronomy in nineteenth-century science and culture. Duke University, Durham, pp 148–173

Bonifácio MF (2002) O século XIX português. Imprensa de Ciências Sociais, Lisboa

Bonifácio MF (2009) Uma história de violência politica. Portugal de 1834 a 1851. Tribuna da História, Lisboa

Bonifácio V, Malaquias I, Fernandes J (2007) Solar photography in the nineteenth century: the case of the Infante D. Luiz Observatory in Lisbon (1871–1880). Journal of Astronomical History and Heritage 10:101–113

Cabral MV (1976) O capitalismo em Portugal. A Regra do Jogo, Porto

Canales J (2009) A tenth of a second: a history. The University of Chicago, Chicago

Canas AJDC, Silva PAR (1999) O desenvolvimento das ciências geofísicas em Portugal no século XIX. Contribuição de Brito Capelo. In: Colóquio Vasco da Gama: Os Oceanos e o Futuro, Escola Naval

Carolino LM (2011) O Observatório Astronómico da Escola Politécnica de Lisboa, 1875–1911. In: Lourenço MC, Neto MJ (eds) Património da Universidade de Lisboa-Ciência e Arte. Tinta da China, Lisboa, pp 109–119

Carolino LM (2012) Measuring the Heavens to rule the territory: Filipe Folque and the teaching of astronomy at the Lisbon Polytechnic School and the modernization of the state apparatus in nineteenth century Portugal. Science & Education 21:109–133

Coutinho VHA (1906) Apontamentos para um curso elementar de hydrographia. Tip. do Annuario Commercial, Lisboa

Coutinho VHA, Neves HBB (1920) Relatório da delegação Portuguesa à conferência internacional de hidrografia. Imprensa Nacional, Lisboa

Daston L, Galison P (1992) The image of objectivity. Representations 40:81–128

Diário do Governo no. 91, 1836; no. 13, 1837; no. 69, 1851; no. 93, 1869

Dick SJ (2003) Sky and ocean joined-The US Naval Observatory 1830–2000. Cambridge University, Cambridge

Eça VA (1886) Trabalhos hydrographicos no Ministerio da Marinha e Ultramar. Parecer das Secções reunidas de cartographia e nautica sobre o relatorio e proposta de organisação do socio Francisco Maria Pereira da Silva. Typographia de Adolpho, Modesto e C[a], Lisboa

Eça VA (1892) Nota sobre os estabelecimentos de instrução naval em Portugal principalmente sobre a Escola Naval. Imprensa Nacional, Lisboa

Edney MH (1997) Mapping an empire: the geographical construction of British India, 1765–1843. University of Chicago, Chicago

Escola N (1904) Apontamentos para um Curso Elementar de Hydrographia: Livro Primeiro, 1904–1905. Lithografia da Escola Naval, Lisboa

Ferreira JP (1886) Estudos sobre alguns instrumentos e sobre algumas questões de suma importância em hydrographia. Typographia da Viuva Sousa Neves, Lisboa

Gonzales FJ (1992) El Observatorio de San Fernando (1831–1924). Ministerio de Defensa, Madrid

Guimarães A (1984) Uma corrente do colonialismo Português: a Sociedade de Geografia de Lisboa, 1875–1895. Livros Horizonte, Lisboa

Harley JB (1988) Maps, knowledge and empire. In: Harley JB, Laxton P (eds) (2001) The new nature of maps (reprinted). The Johns Hopkins University, Baltimore, pp 51–81

'Hidrografia', Historical Archive of the Portuguese Navy, Box 241-4

Hoffmann C (2007) Constant differences: Friedrich Wilhelm Bessel, the concept of the observer in early nineteenth century practical astronomy and the history of personal equation. British Journal for the History of Science 40:333–365

Kaiser D (2005) Introduction: moving pedagogy from the periphery to the center. In: Kaiser D (ed) Pedagogy and the practice of science. MIT Press, Cambridge, pp 1–7

Lacerda H (1885, 1886) Pleas to King Luis of Portugal, Historical Archive of the Portuguese Navy, Box 730

Lacerda H (1903a) Apontamentos sobre trabalhos hydrographicos utilisaveis, feitos por aspirantes a guardas-marinhas na vigencia das leis organicas da Escola Naval de 1895 e 1897. Annaes Club Mil Naval 33:91–96

Lacerda H (1903b) Extracto do relatorio da missao hydrographica do anno lectivo de 1901–1902. Annaes do Club Militar Naval 33:97–103, (168–177, 207–214)

Lacerda H (1906) Estado actual do Porto de Lourenço Marques e suas necessidades. Imprensa Lucas, Lisboa

Lacerda H (1907) O Porto de Lourenço Marques. Sociedade de Geografia, Lisboa

Lacerda H (1914) Trabalhos realizados na primeira campanha da missão hidrográfica de Portugal. Anais do Clube Militar Naval 45:472–483

Lacerda H (1926) Questões relativas à vantagem de se unificar os serviços hidrográficos de Portugal e suas colónias. Anais do Clube Militar Naval 57:267–271

Lacerda H (1927) Questões relativas à vantagem de se unificar os serviços hidrográficos de Portugal e suas colónias. Anais do Clube Militar Naval 58:18–25

Latour B (1987) Science in action: how to follow scientists and engineers through society. Harvard University, Cambridge

Lei Orgânica do Real Observatório Astronómico de Lisboa (1878) Imprensa Nacional, Lisboa

Leitão AES (1970) Reflexões sobre a moderna hidrografia portuguesa. Tipografia da Editorial Minerva, Lisboa

Lucas MM (1993) Organização do Império. In: Mattoso J (ed) Torgal LR, Roque J (vol eds) História de Portugal. O Liberalismo (1807–1890), vol V. Editorial Estampa, Lisboa, pp 246–266

Macedo M (2012) Projectar e Construir a Nação. Engenheiros, ciência e território em Portugal no século XIX. Imprensa de Ciências Sociais, Lisboa

Manning T (2004) US Coast Survey vs. Naval Hydrographic Office: A 19th-century rivalry in science and politics. University of Alabama Press, Tuscaloosa

Matos P (1932) Algumas considerações sobre a organização dos serviços hidrográficos portugueses. Anais do Clube Militar Naval LXII:159–175

Mattoso J, Torgal LR, Roque J (eds) (1993) História de Portugal. O Liberalismo (1807–1890), vol. V. Editorial Estampa, Lisboa

Medeiros CA (2004) Os exploradores africanos do ultimo quartel do século XIX. In: João Medina (ed) História de Portugal, vol 11. Ediclube, Lisboa, pp 431–449

Ministério da Marinha (1915) Missão Hidrográfica da Costa de Portugal. Relatorio dos trabalhos efectuados durante a campanha do aviso "5 de Outubro", em 1913. Do Rio Minho a Espinho. Imprensa Nacional, Lisboa

Mónica MF (2009) Fontes Pereira de Melo-uma biografia. Alêtheia Editores, Lisboa

Morais T (1945) Esboço histórico. In: Os primeiros cem anos da Escola Naval, 23-IV-1845 a 23-IV-1945. Escola Naval, Lisboa, pp 15–57

Oom FA (1875) Considerações acerca da organização do Real Observatorio Astronómico de Lisboa. Imprensa Nacional, Lisboa

Proença MC (2010) A questão colonial. In: Rosas F, Rolo MF (eds) História da primeira República portuguesa. Tinta da China, Lisboa, pp 205–228

Pyenson L (1996) On the military and the exact sciences in France. In: Forman P, Sanchez-Ron JM (eds) National Military establishments and the advancement of science and technology. Dordrecht, Boston, pp 135–152

Ramos R (2006) D. Carlos. Círculo de Leitores, Lisboa

Raposo PMP (2008) Charming tools of a demanding trade: the heritage of nineteenth-century astrometry at the Astronomical Observatory of Lisbon. Rittenhous—The Journal of the American Scientific Instrument Enterprise 22:25–46

Raposo PMP (2010) Polity, precision and the stellar heavens: the Royal Astronomical Observatory of Lisbon (1875–1910). Dissertation, University of Oxford

Raposo PMP (2012) O império e o tempo: Notas para uma abordagem à história dos observatórios coloniais Portugueses. In: Diogo MP, Amaral IM (eds) A outra face do império: ciência, tecnologia e medicina (sécs. XIX–XX). Edições Colibri, Lisboa, pp 17–29

Reis AE (2009) Observatório Real da Marinha (1798–1874). CTT, Lisboa
Rozwadovski HM (1996) Small world: forging a scientific maritime culture for oceanography. Isis 87:409–429
Schaffer S (1988) Astronomers mark time. Discipline and the personal equation. Science in Context 2:115–145
Schiavon M (2010) Geodesy and mapmaking in France and Algeria: between army officers and observatory scientists. In: The Heavens on Earth: observatories and astronomy in nineteenth-century science and culture. Duke University, Durham, pp 199–224
Silva AB (1893) Estudo historico e hydrographico sobre a barra e o porto de Lisboa. Imprensa Nacional, Lisboa
Silva FMP (1886) Trabalhos hydrographicos no Ministerio da Marinha e Ultramar. Elementos para um relatorio e proposta de organização. Typ. e Lyt. de Adolpho, Modesto e C^{a}, Lisboa
Silva C (1945) 8ª Cadeira: topografia e hidrografia-breve resumo histórico da sua evolução e actuais condições de funcionamento. In: Os primeiros cem anos da Escola Naval, 23-IV-1845 a 23-IV-1945. Escola Naval, Lisboa, pp 185–190
Simões A et al (2013) Uma história da Faculdade de Ciências da Universidade de Lisboa (1911–1974). Faculdade de Ciências da Universidade de Lisboa
Smith RW (1991) A national observatory transformed: Greenwich in the nineteenth-century. Journal for the History of Astronomy xxiii:5–20
Slotten HR (1993) The dillemas of science in the United States: Alexander Dallas Bache and the US Coast Survey. Isis 84:26–49
Sousa LM (1886) O serviço hydrographico. Annaes do Club Militar Naval XVI:221–226, (280–284, 303–307)
Tavares C (2009) Albert I do Mónaco, Afonso Chaves e a meteorologia nos Açores. Episódios oitocentistas da construção científica do mundo atlântico. Sociedade Afonso Chaves/Centro Inter-Universitário de História da Ciência e da Tecnologia, Lisboa
Teixeira NS (1990) O Ultimatum inglês: política externa e política interna. Alfa, Lisboa
Telo AJ (2004) O modelo político e económico da Regeneração e do Fontismo (1851–1890). In: Medina J (ed) História de Portugal-dos tempos pré-históricos aos nossos dias, vol XI. Ediclube, Amadora, pp 115–156
Vicedo M (2012) The secret lives of textbooks. Isis 103:83–87
Widmalm S (2010) Astronomy as military science: the Case of Sweden. In: Aubin D, Bigg C, Sibbum HO (eds) The Heavens on Earth: observatories and astronomy in nineteenth-century science and culture. Duke University, Durham, pp 174–198

Combinatorial Games and Machines

Lisa Rougetet

Abstract Mathematics and technology interact with each other for their respective benefit. This article will try to illustrate this unavoidable fact by studying the machines built to play games—especially combinatorial games (no chance moves)—and their impact on the development of mathematical ideas. We will see how all began with the simple Nim game and the first machines built to play it (and to win!). Then we will focus on the game of Chess to demonstrate that some mathematical ideas can originate from technologies and that these technologies can enable their concrete achievement.

Keywords Mathematics · Combinatorial game theory · Nim game · Chess · Technological advances in programming · Artificial intelligence · Minimax algorithm · Alpha–Beta pruning · Backward induction · Chess automaton · Game analysis · Complexity · Selectivity · Data storage

1 Introduction

The emergence of artificial intelligence cannot be precisely dated. Indeed, this theme of reflexion is quite old and can be found regularly through history with myths about artificial creatures, construction of automatons, or first attempts to formalize human thoughts. The term "Artificial Intelligence" (AI), coined in 1956 by John McCarthy (1927–2011), can be defined as the construction of computer programs that try to solve tasks that are, at the moment, better fulfilled by humans as they require high-level mental processes. These programs can be created for reasoning, for the understanding of natural languages, for perception, or for example, for games and mathematical practice. We will now focus on these last two points, especially through combinatorial games and through the game of Chess which was the starting point of many reflections about human thinking processes.

L. Rougetet (✉)
Laboratoire d'Informatique Fondamentale de Lille, Lille 1 University Science and Technology, Villeneuve d'Ascq, France
e-mail: lisa.rougetet@gmail.com

R. Pisano (ed.), *A Bridge between Conceptual Frameworks,* History of Mechanism and Machine Science 27, DOI 10.1007/978-94-017-9645-3_24

2 First of All: What is a Combinatorial Game?

In this article we are dealing with special games called combinatorial games. They are defined by properties which are the following (Berlekamp et al. 2001, p 14):

1. There are just two players, often called Left and Right.
2. There are several, usually finitely many, positions, and often a particular starting position.
3. There are clearly-defined rules that specify the moves that either player can make from a given position to its options.
4. Left and Right move alternately, in the game as a whole.
5. Both players know what is going on, *i.e.* there is complete information.
6. There are no chance moves such as rolling dice or shuffling cards.
7. In the normal play convention a player unable to move loses.
8. The rules are such that play will always come to an end because some player will be unable to move. This is called the ending position. So there are no games that are drawn by repetition of moves.

These properties give a strict definition of a combinatorial game. However a more tolerant conception of combinatorial game theory does not always respect the seventh and the eighth points (note for example, that Chess does not fulfil the seventh condition as a match can end up in a draw). The global aim of the combinatorial game theory is to study the nature of the several positions of the game (winning, losing or draw) in order to build a strategy that will lead to a win. Let us now focus on the Nim game, a combinatorial game that enabled great improvements in the mathematical theory of games and for which a machine was specially built to play it.

3 The Starting Point: The Nim Game

The Nim game is one of the most famous combinatorial games. It was introduced for the first time under this name[1] by a mathematician from Harvard, Charles Leonard Bouton (1901) (1869–1922) in an article of the famous journal *Annals of Mathematics*, published in 1901 (Bouton 1901). This article is considered as the starting point of the relatively recent (twentieth century) mathematical theory of combinatorial games. Indeed its mathematical content is at the base of the resolution of the more general class of impartial games[2].

[1] Nim comes from the imperative form of the German verb nehmen, which means to take.

[2] An impartial game is a combinatorial game in which the available moves are similar for both players (which is not the case in Chess for example). The main theorem about the resolution of impartial games was independently found by Roland Parcival Sprague (1894–1967) in 1935 and Patrick Mickael Grundy (1917–1959) in 1939, and states that "every impartial game is just a bogus Nim-heap" (Berlekamp et al. 2001, p 56). This means that, thanks to some modifications

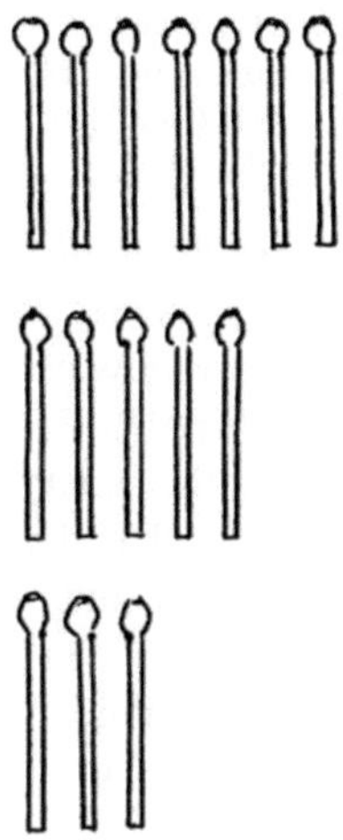

Fig. 1 A possible starting position at Nim. (Eiss 1988, p 188)

Because its resolution is based on the binary system, it was not very difficult to elaborate a program able to play it. This is what we will see just after presenting the rules of the Nim game.

3.1 Presentation of the Nim Game (Bouton 1901)

Bouton justifies his interest to the Nim game on account of its seeming complexity despite its extremely simple and complete mathematical theory. Then he describes the game as followed: upon a table opposing two players, A and B, three piles of objects of any kind are placed, let us say matches. The number of matches in each pile is arbitrary, except that it is well to agree that no two piles shall be equal at the beginning (see Fig. 1 for a possible starting position). Bouton requires this latter condition but actually it is not necessary to respect, for it does not change anything in the resolution of the game.

Alternatively, players select one of the piles, and take from it as many matches as they want: one, two,..., or the whole pile. The first one who takes the last match or matches from the table wins the game.

The entire theory of the Nim game is based on the notion of "safe combinations" (Bouton 1901, p 35). It refers to special positions that allow[3] the player who reaches one of them to win the game at the end (under the condition of playing "without mistake" (Bouton 1901, p 35)). Safe combinations present the following properties (Bouton 1901, p 36):

---

and transformations, we can always identify an impartial game to a determined Nim game (or a succession of Nim games).

[3] It is not necessary for the player who reaches a safe combination during the play to win the entire game. To do that, he must reach a safe combination at every move he makes.

> THEOREM I. If A leaves a safe combination on the table, B cannot leave a safe combination on the table at his next move. [...]
> THEOREM II. If A leaves a safe combination on the table, and B diminishes one of the piles, A can always leave a safe combination.

These theorems and their proof provide the explanation of a possible win when a player reaches a safe combination.

Now, how can we check if a position is a safe combination or not? First, Bouton (1901, pp 35–36) explains that the numbers of matches of the piles have to be written in the binary scale. Then these binary numbers are placed in three horizontal lines so that the units are in the same vertical column. Then if the sum of each column is congruent to 0 mod. 2, the set of numbers on the table forms a safe combination. It is called the Nim-sum. In our example, 7, 5, 3 is not a safe combination and to change this, we would have to remove a single match from one of the three piles[4]. It is up to you now!

3.2 *The Nimatron*

In the Spring of 1940, an electromechanical Nim player machine (see Fig. 2) called *The Nimatron*, invented by two members of the staff of the Westinghouse Electric Company during their lunch break (Condon 1942, p 331), was built and exhibited at the Westinghouse Building of the New York World's fair, where it played more than 100,000 games (and won 90,000 of them). The *Nimatron* could play a game made up of four piles containing at most seven counters.

Condon underlines that the *Nimatron* is not like the other mathematical machines and that it serves no other useful purpose than entertainment, "unless it be to illustrate how a set of electrical relays can be made to make "a decision" in accordance with a fairly simple mathematical procedure". (Condon 1942, p 330) (Fig. 3).

3.3 *Redheffer's Machine*

Despite the fact that the *Nimatron* had no other aim than to entertain the visitors, the construction of a much more improved Nim-playing machine started in 1941. It was designed by Raymond Moos Redheffer (1948, p 343) (1921–2005), an assistant professor of mathematics at the University of California at Los Angeles (Gardner 1959, p 156) who stated "this theory is of such a nature that the computations required can be carried out by simple electrical circuits" (Redheffer 1948, p 343). Redheffer's machine proposed the same arrangement as the *Nimatron*, the equivalent of four piles containing seven counters at most, but weighed only 2.3 kg (5 pounds) against a ton for the *Nimatron*!

[4] The exercise is let to the reader!

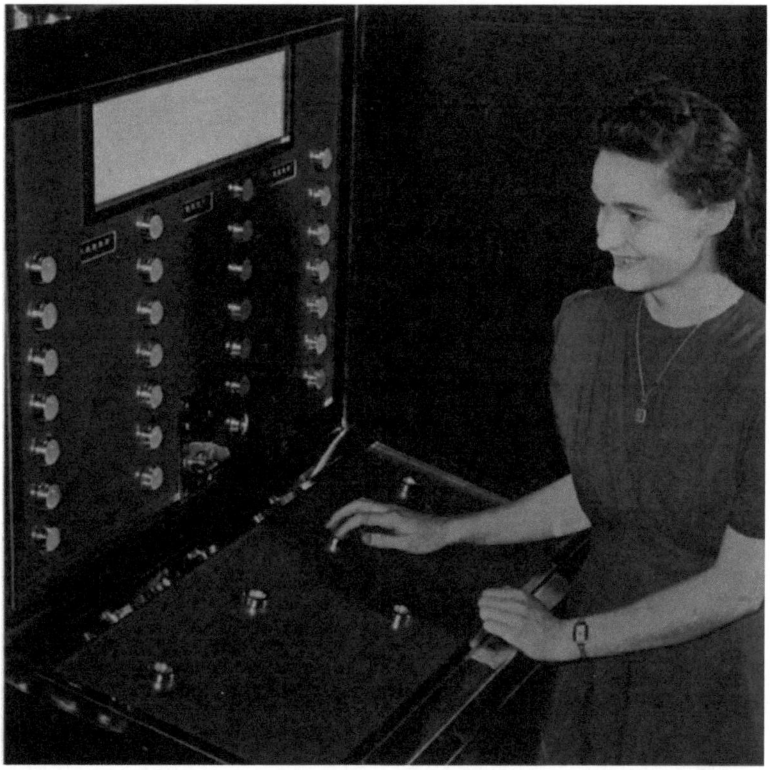

Fig. 2 The *Nimatron* is a machine which is very skilful at playing the game of Nim (Condon 1942, p 330)

Redheffer extended his results in a B.S. Thesis in Mathematics to a more general game where players can remove objects from *k* piles and not only from one[5].

The first task of Redheffer's machine is to convert the number of objects of each pile in the binary system. Each of these numbers is represented by a switch supplied with as many layers of contacts ("pies") as there are digits in the binary number required for representing the maximum number of objects in the pile (see Fig. 4).

Then, the next step is to find the sum of the converted numbers. Redheffer simplifies the problem to an arrangement of connected switches (see Fig. 5): "It follows that *E* and *G* will be connected, and *F* and *H* will be connected, whenever an even number of switches are in the down position" (Redheffer 1948, p 347).

As the machine was planned for a maximum of seven objects in four piles, there are four switches, each having eight positions. Since any numbers between 0 and 7 can be written with three digits in the binary system, there are three indicator lights for each pile. "They are grouped behind a translucent screen covering a large round hole above each switch.

[5] This version of the Nim game is actually due to an American mathematician, Eliakim Hastings Moore (1862–1932), who extended in 1910 the work of Bouton and gave a generalisation of the Nim game when the player can remove matches from k piles. (Moore 1910).

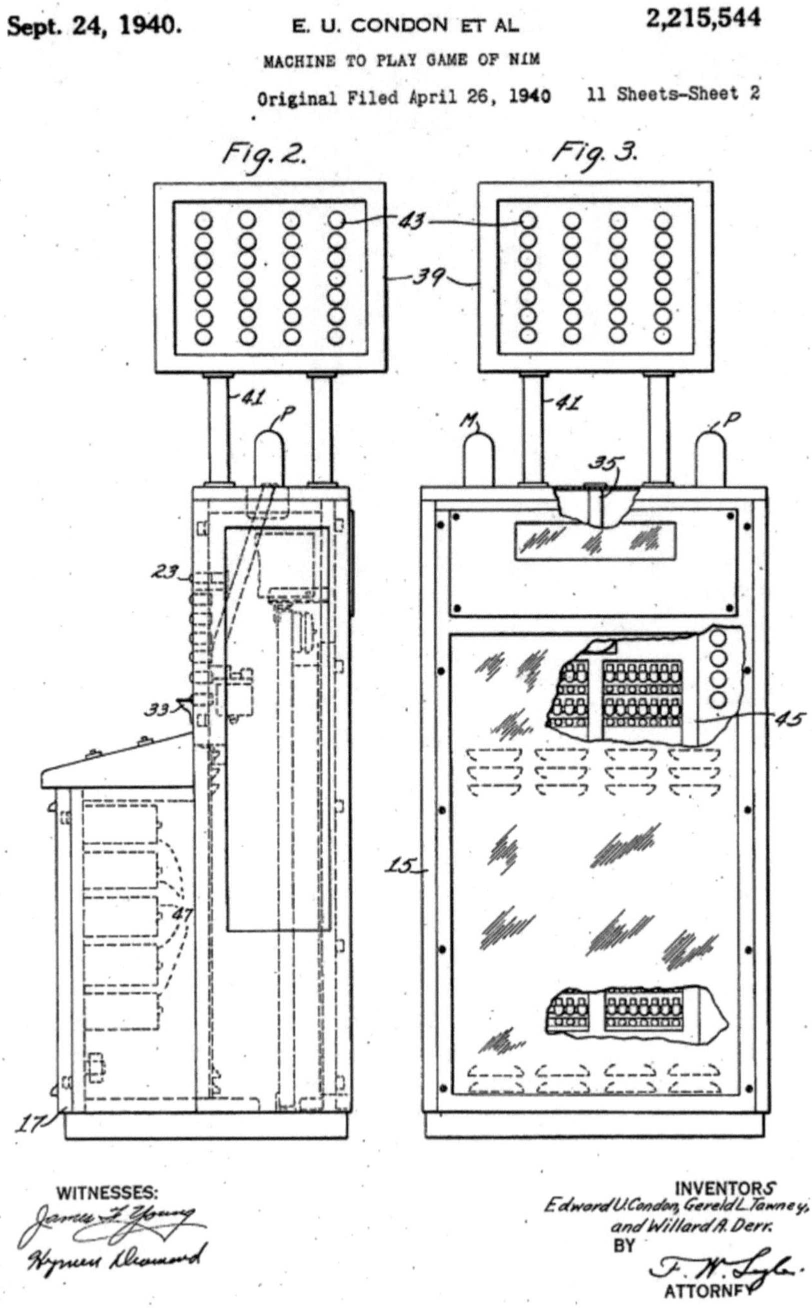

Fig. 3 Description of the patent of the *Nimatron* (US Patent 1940, p 2)

Fig. 4 This figure shows the connections for the pies representing the unit's, two's, and four's digits. (Redheffer 1948, p 345)

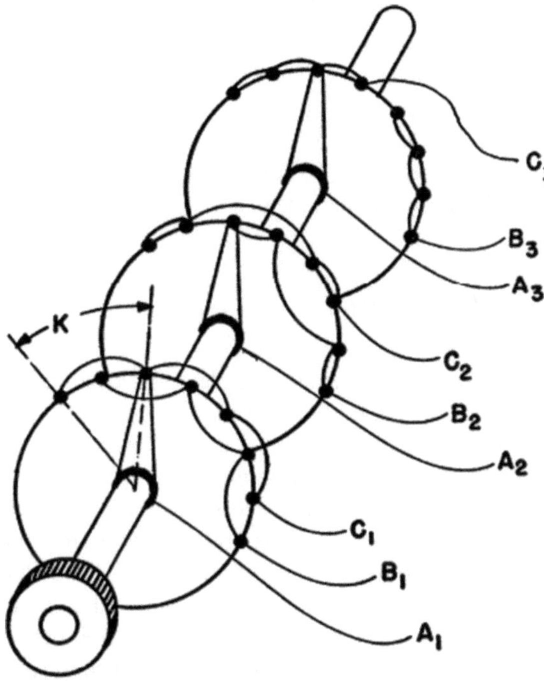

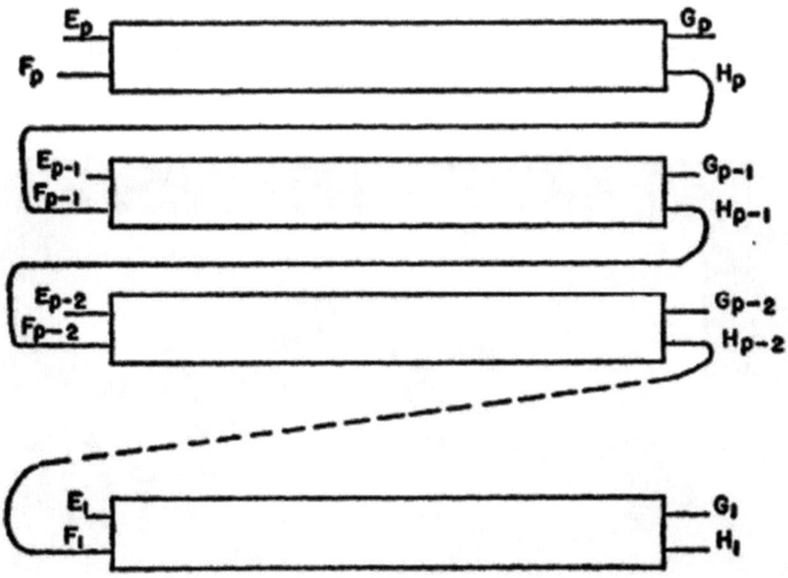

Fig. 5 "[…] but the connections will be interchanged whenever an odd number of switches are in the down position." (Redheffer 1948, p 347)

Neon bulbs are used partly because of their low current requirement and partly because they will not light if two are in series. The input is 110 V." (Redheffer 1948, pp. 348–349). As far as we know, Redheffer's machine was not exhibited in public, so no results about the games it played are available.

Redheffer's machine is a good example that shows the technological evolution (less or new material, lighter components) of a mechanical device initially designed around a mathematical idea.

3.4 *The Nimrod*

A few years later, Ferranti, the electrical engineering and defence electronics equipment firm, designed the first digital computer exclusively dedicated to play Nim, *The Nimrod.* It was exhibited at the Festival of Britain (Exhibition of Science) in May 1951 and afterwards at the Berlin Trade Fair (Industrial Show) in October. Its exhibitions were a great success and few witnesses relate that the most impressive thing about the *Nimrod* was not to play against the machine but to look at all the flashing lights which were supposed to reflect its thinking activity! (See Fig. 6) Some even say that none of the persons who came to play against *Nimrod* noticed the British bar next door, which offered free beverages… (Gardner 1959, p. 156). This particular display was built on purpose to illustrate the algorithm and the programming principles involved.

After the *Nimrod*, it seems that no other machines were built to play Nim. The simplicity of the game and of its solution may have led the scientists to turn to more complicated games both in their rules and in their programing, and whose achievement would require a more developed technology.

Fig. 6 *The Nimrod* at the Berlin Industrial Show on 6 October 1951. It is a 9 by 12 by 5 feet machine that contains 480 vacuum tubes and executes its program almost independently. (http://www.heise.de/newsticker/meldung/Vor-50-Jahren-fing-alles-an-das-erste-Elektronenhirn-in-Deutschland-51722.html)

As we will see in the next section through several examples, there exists a link between games and their underlying mathematical concepts, and their possible creation through a machine or a computer.

4 The Game of Chess

The game of Chess is considered as the King of all games: International players and Grand Masters are seen as the most intelligent people in the world. Thus, it is not surprising if the first researches led in artificial intelligence started to study the game of Chess (but not only, of course) in order to build a program that could play at a relatively good level (and maybe defeat one day the World Champion). But a few decades before these first programs, an electro-mechanical machine able to play a particular endgame in Chess was designed: *El Ajedrecista.*

4.1 El Ajedresista of Torres y Quevedo

El Ajedrecista (the Chess Player in English, see Fig. 7) was the first serious[6] automaton built in 1911 by the Spanish engineer Leonardo Torres y Quevedo (1915) (1852–1936)[7]. Torres worked on the construction of several devices, which can be divided into two groups: automatons and algebraic machines (Vigneron 1914). An automaton refers to a machine that imitates the appearance and the movements of a man or an animal. Torres explains (Vigneron 1914) that the mechanism has to bear its own source of energy which makes it work (a spring for example) and makes it accomplish some gestures, always the same, without any external influence. Automatons must be capable of discernment and adapt themselves to their environment and to the impressions they receive. Through his Chess Player, "merveille d'ingéniosité" (Vigneron 1914, p 59), Torres proved well that it was possible to create a mechanical machine (in fact, electro-mechanical in this case), which could play Chess.

El Ajedrecista was first exhibited to the public in Paris; it was able to play the special endgame configuration of the white king and rook from any position (held by the machine) against the human black king from any position. *El Ajedrecista*

[6] In 1769, the Hungarian Johann Wolfgang von Kempelen (1734–1804) constructed an automaton Chess player, the Mechanical Turk, able to play at a high level against human opponents. In fact, the Turk was a complete stunt as a human chess master could hide inside the machine and operate the moves [...]. Ajeeb, created by Charles Hooper, succeeded the Turk in 1868, but used the same trick with a hidden player.

[7] "Born in Santa Cruz in the province of Santander in Spain in 1852 and educated as a civil engineer, Torres became director of a major laboratory, president of the Academy of Sciences of Madrid, a member of the French Academy of Sciences, and famous as a prolific and successful inventor. Some of his earliest inventions took the form of mechanical analog calculating devices of impressive originality." (Randell 1982, p 331).

Fig. 7 Front view of the Chess Player (Vigneron 1914, p 57)

runs on a rather simple electro-mechanical system: to every position of the system (*i.e.* of the game) corresponds an electro-magnet. This electro-magnet is activated with electrical connexions when switches are arranged in a given position. Since the switches can be placed in different positions, every move on the chessboard can be obtained by a specific combination of the switches. For example, on Fig. 8, the switch M can take two different positions (A or B) that can activate switches N or N'. If N is activated, it has once again three different positions available (E, F and G), which will activate new switches, and so on until all switches have their own position and form a particular combination.

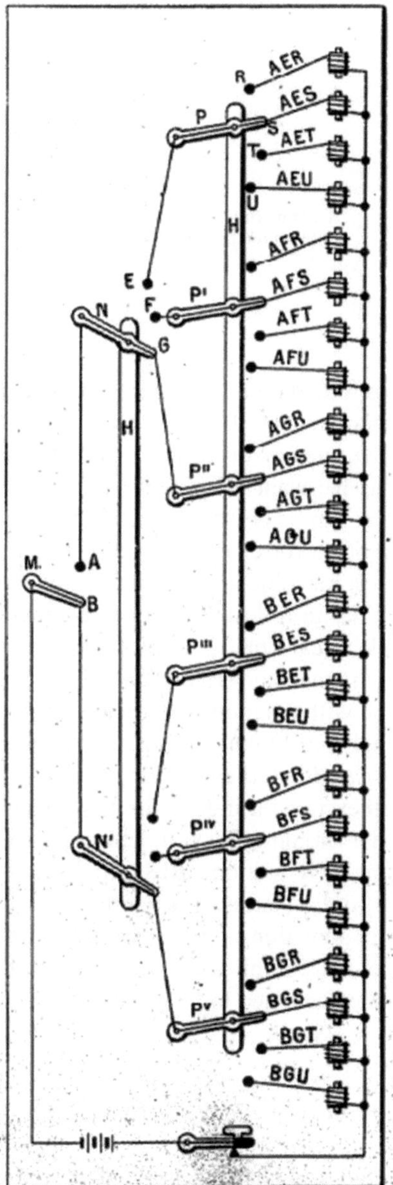

Fig. 3. — Schéma montrant comment on peut déterminer 24 opérations différentes.

Fig. 8 Drawing to show how 24 different operations can be determined in Torres' machine (Vigneron 1914, p 58)

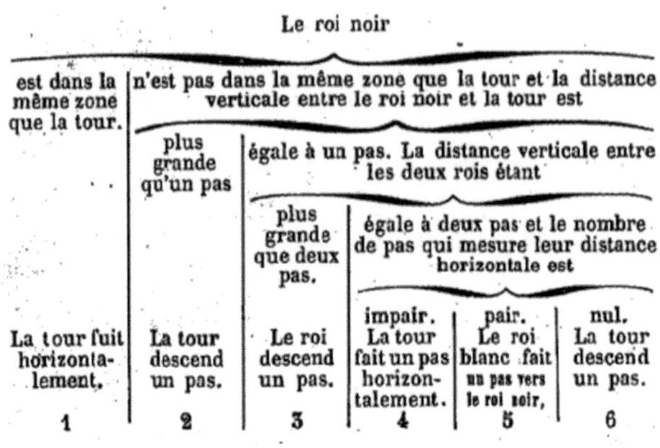

Fig. 9 Set of rules that the automaton has to follow before connecting the switch. (Vigneron 1914, p 60)

Combinations are determined by a set of rules that the automaton has to follow before connecting the switches (Fig. 9).

Torres explains that the number of switches and their associated positions can be increased "as much as we want" (Vigneron 1914, p 59). This means that the number of particular cases can be endlessly increased, which makes the actions of the automaton more complex. Torres considered that there are no essential differences between the simplest machine and a more complicated automaton, since we can provide it the rules it has to follow to execute its moves. The very first version of the automaton used electrical sensing of pieces on the board and a mechanical arm moved the pieces of the machine (Randell 1982, p 332). Some years later, Torres made a second version with magnets underneath the board to move the pieces.

Two paradoxical thoughts arise concerning machines such as *The Nimrod*, *The Nimatron* and *El Ajedrecista*. First, their development and their exhibition did lead to major public events; they remain indeed the first machines ever built to reproduce a situation of a particular game. Even if their popularity was mostly due to the curiosity of the public, their fabrication still symbolises the first achievement of a combinatorial reasoning. And secondly, it is worth noticing that the initial idea was to provide a game and not something more serious. Retrospectively, the reasoning behind these machines is not complicated at all, and it is mainly the advertising made at that moment that contributed to their fame.

4.2 The Beginnings of Artificial Intelligence

Nevertheless these machines did not prevent the development of more serious programs in the 1950s. Some mathematical ideas originate from technologies and these technologies enable their concrete achievement. Creating a program or a machine able to act as the human brain is the main problem of artificial intelligence and the first researches (among others) started with the most noble game of all games: Chess. In 1950, Claude Elwood Shannon (1950) (1916–2001), an electronic engineer and cryptographer, was concerned with the problem of constructing a program for a computer that would enable a computer to play Chess (Shannon 1950). This led him to lay the foundations of computer Chess programming thanks to the Minimax algorithm, a principle found in every game analysis (not just combinatorial games). This idea was also developed by Alan Mathison Turing (1912–1954) in 1953 as part of his research on artificial intelligence. Here is an explanation of the Minimax's principle.

The Minimax algorithm consists in *mini*mizing the maximum loss in a given position. This principle provides the advantage to evaluate the different positions of the game and to choose the most beneficial one, considering the opponent's moves. The Minimax is a heuristic principle that takes into account the hypothesis that our opponent's aim is to maximise his benefit. In the particular case of combinatorial games, the goal of the two players is clearly opposite: when A wants to maximise his profit, B wants to minimise it (to maximise his own profit).

To apply this algorithm, we first need to represent the game in the form of a tree. Every node of the tree corresponds to a possible position of the game and its branches lead to the positions that can be reached from that node. To evaluate the initial position (called the *root* of the tree, level 0), we need to generate the set of positions reachable from that initial position, we obtain the level 1. We do so as many times as necessary to generate levels 2, 3, …, *n*. Then we assign a value to each final position, which gives the quality of the position for one of the two players. For example, in Fig. 10 player A (represented by the root of the tree) is about to play and wants to evaluate his position to minimise his loss.

The first step of the Minimax algorithm is to minimise the set of the final positions for every branch. This gives a value to the three parent nodes of the terminal nodes. We find the following minima (Fig. 11).

As player A wants to maximise his profit, the next step is to find the maximum between the three nodes of the level 1 that will correspond to the value of the position and then give the right move to play (Fig. 12).

The Minimax algorithm is based on simple recursive calculations that alternatively minimise or maximise nodes at a given level, actually a simple mathematical process, which is used in every program. If we apply the Minimax algorithm until the terminal positions, *i.e.* if we apply it for the entire game tree, therefore it is equivalent to apply a *backward induction* reasoning.

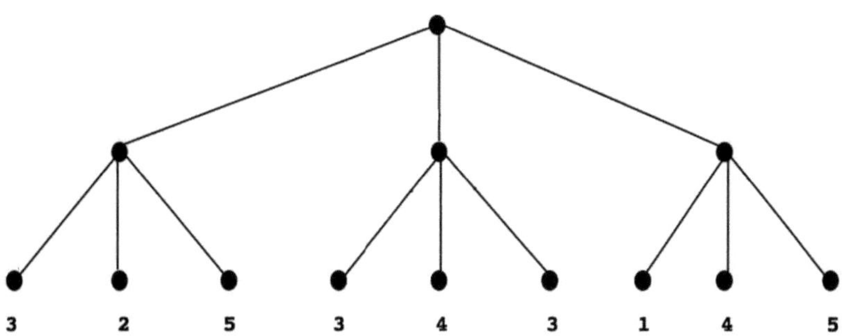

Fig. 10 An evaluation function has evaluated the quality of the final positions from player A's point of view (Alliot and Schiex 1994, p 274)

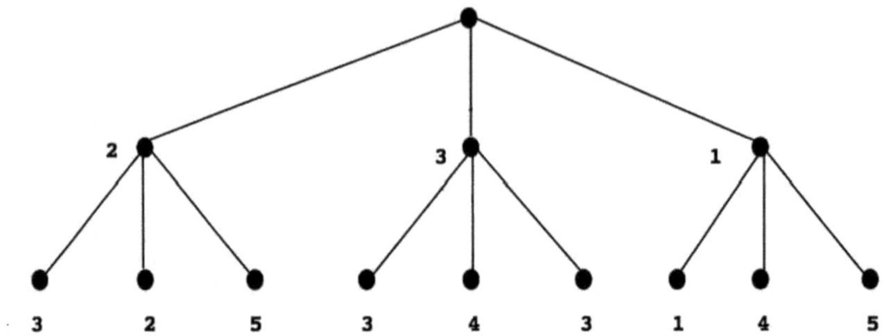

Fig. 11 The minimising step gives the value of the parent nodes of the terminal nodes. They represent player B's strategy that consists of maximising his profit and therefore minimising A's profit. (Alliot and Schiex 1994, p. 274)

Backward induction is used in the resolution of every combinatorial game whenever it is possible. For example, as Chess presents 10^{120} different games, it is impossible to apply backward induction on its game tree, which is also impossible to represent!

Both Shannon and Turing wrote their program based on the Minimax—even if the name did not exist yet—and for machines that did not even exist at that time! It seems (Newell et al. 1958) that there were two other hand simulations between 1951 and 1956 by Frederick Mosteller (1916–2006) and a Russian program but not enough information is available on any of them. Consequently they will not be taken into account in this study. But, as we will see further, the technological advances in computer industry soon gave the opportunity to implement these programs.

In 1956, *Los Alamos Chess* was the first program to run on a computer, MANIAC I, built in 1952 and based on John von Neumann (1903–1957) architecture. *Los Alamos* program is a good example of the system Shannon (1950) described; all alternatives were considered, all continuations were explored to a depth of two moves for each player, the values were determined by a Minimax procedure, and the best alternative was chosen for the move.

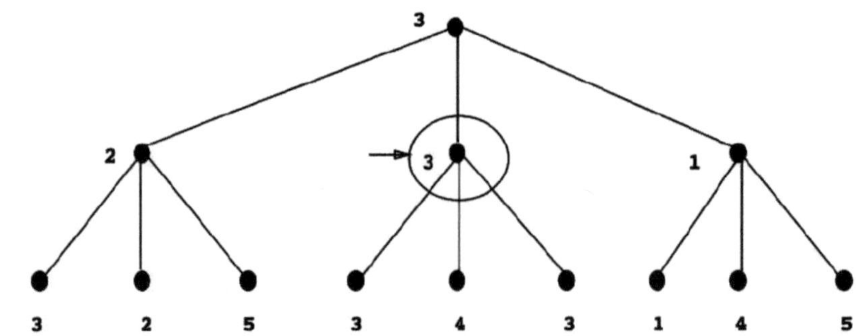

Fig. 12 The value of the initial position is 3 so to maximise his profit, player A will have to choose the second branch (Alliot and Schiex 1994, p. 274)

But the game *Los Alamos* played was a simplified version of Chess on a 6×6 board without Bishops and eliminated special moves such as castling, two-square Pawn moves in the opening and *en passant* captures. These reduced board and rules were chosen in order to carry out the computation within reasonable time limits. In a normal 8×8 game, looking two moves ahead brings about 30^4 that is to say 800,000 continuations to be considered. In the reduced 6×6 game, "only" 160,000 continuations were taken into account. The *Los Alamos Chess* program was developed by Paul Stein and Mark Wells in Los Alamos laboratory. It was able to make a move in about twelve minutes on average and played three games: one against itself, a second against a strong player (it lost), and a third against a beginner (it won).

Two years later a mathematician in the Programming Research Department of IBM, Alex Bernstein, wrote the first program that could play a full game of Chess for the IBM 704, introduced in 1954. The IBM 704 was a very rapid large-scale electronic digital computer, which had performed as many as one billion calculations in a single day in computing the orbit of an artificial satellite (Bernstein and Roberts 1958). Physically, it consisted of several units all under the constant electronic control of the central unit. The principal advanced feature of the 704 was its high-speed magnetic core storage or memory. It replaced the electrostatic or cathode ray tube storage used in the earlier machine systems. A magnetic core was about the size of a pinhead and shaped like a doughnut. Thousands of core were strung on a complex of wires in such a way that several wires passed through the centre of each core. Combinations of electrical pulses on these wires altered the magnetic state of the cores, and a line of cores, some altered, some unaltered, stood for a certain word or number. A word or number stored in the magnetic core memory was available for calculation in 12 millionths of a second. The Type 704 multiplied or divided in 240 μs, or approximately 40,000 operations per second (IBM Archives 1955). Bernstein's program was also in the Shannon tradition (Newell et al. 1958); it could play either Black or White, was capable of playing a complete game of Chess, including moves such as castling, promoting and capture *en passant* and was divided into five parts (1) Input-output, (2) Table generation, (3) Evaluation, (4) Division, and (5) Tree (Bernstein et al. 1958). But an important step was made in the direction of greater sophistication: only a fraction of the legal alternatives and continuations were considered. A set of decision routines was written, which selected a small number (not greater than seven) of strategically good moves. In this way, the program examined 2800 different positions and arrived at the move which, in its estimation, would leave the opponent with the worst possible position. It minimaxed and chose the alternative with the greatest effective value. The time required to the program to make a move was on the average of eight minutes and the machine printed out its move on a sheet of paper. After a mating move or a resignation, the machine printed the score of the game and a paper intended to its opponent with this sentence: "THANK YOU FOR THIS INTERESTING GAME"! (Bernstein and Roberts 1958).

Bernstein's program was the first to give information about radical selectivity, in move generation and analysis.

TABLE 1 Comparison of Current Chess Programs

	Turing	Kister, Stein, Ulam, Walden, Wells (Los Alamos)	Bernstein, Roberts, Arbuckle, Belsky (Bernstein)	Newell, Shaw, Simon (NSS)
Vital statistics				
Date	1951	1956	1957	1958
Board	8 × 8	6 × 6	8 × 8	8 × 8
Computer	Hand simulation	MANIAC-I 11,000 ops./sec	IBM 704 42,000 ops./sec	RAND JOHNNIAC 20,000 ops./sec
Chess program				
Alternatives	All moves	All moves	7 plausible moves Sequence of move generators	Variable Sequence of move generators
Depth of analysis	Until dead (exchanges only)	All moves 2 moves deep	7 plausible moves 2 moves deep	Until dead Each goal generates moves
Static evaluation	Numerical Many factors	Numerical Material, mobility	Numerical Material, mobility Area control King defense	Nonnumerical Vector of values Acceptance by goals
Integration of values	Minimax	Minimax (modified)	Minimax	Minimax
Final choice	Material dominates Otherwise, best value	Best value	Best value	1. First acceptable 2. Double function
Programming				
Language		Machine code	Machine code	IPL-IV, interpretive
Data scheme		Single board No records	Single board Centralized tables Recompute	Single board Decentralized List structure Recompute
Time	Minutes	12 min/move	8 min/move	1-10 hr/move (est.)
Space		600 words	7000 words	Now 6000 words, est. 16,000
Results				
Experience	1 game	3 games (no longer exists)	2 games	0 games Some hand simulation
Description	Loses to weak player Aimless Subtleties of evaluation lost	Beats weak player Equivalent to human with 20 games experience	Passable amateur Blind spots Positional	Good in spots (opening) No aggressive goals yet

Fig. 13 Table of comparison of the different Chess programs in 1958. (Newell et al. 1958, p. 45)

But every increase in sophistication of performance involves an increase in the complexity of the program (Newell et al. 1958). This implies both more program and more computing time per position than with the *Los Alamos* program. Figure 13 compares Chess programs available in 1958 according to their *vital statistics*, their *programming language* or their *results* (Newell et al. 1958, p. 45).

We can observe that Bernstein's program takes 7000 words, the *Los Alamos* program only 600: a factor of about 10. But for time per position, both programs take about the same time to produce the move, 8 and 12 min, respectively. The increase in problem size of the 8 × 8 board over the 6 × 6 board is about five to one; it is approximately cancelled by the increase in speed of the IBM 704 over the MANIAC (also about five to one, counting the increased power of the 704 order code (Newell et al. 1958)). So it can be said that both programs would produce moves in the same 8 × 8 game in the same time. Hence the increase in amount of processing per move in Bernstein's program approximately cancels the gain of 300 to 1 (2800 positions investigated against 800,000 for *Los Alamos*) in selectivity. Newell et al. (1958) deplores this kind of equality between the two programs because selectivity is a very powerful device and speed a very weak one for improving the performance of complex programs. He gives for example the case of both programs that could explore three moves deep instead of two. Then the *Los Alamos* program would take about 1000 times as long as it does to make a move, whereas Bernstein's program would take about 50 times as long.

This aspect emphasises the fact that the time needed for a machine to determine the right move does not only depend on the number of the positions evaluated. The increase of the complexity in the programs also plays an important part in the time needed. But we can allow programs to become more complex because machines that support them become also more powerful and faster for calculation. And as the performance of the machines increases, researchers become more challenging on the programs... This is how improvements are obtained in Chess programming.

With Bernstein's work we understand that selection in programs is an essential device to increase their level of play. It allows to reduce calculations and to overcome the low speed and power of the machines. A major progress in that direction was made with the development of the Alpha–Beta pruning, a very effective selective cut-off of the Minimax algorithm without loss of information, which is still used in nowadays programs. Here is the principle: on Fig. 11's example, we consider that Minimax algorithm has already analysed the first two branches of the tree and is going to deal the third one. Since the root of the tree is a maximising one (to get its value we maximise values among its children nodes), and one of its children nodes has the value three, we already know that the root will have a value at least equal to three. When analysing the third branch, the first leaf gives the value one. But the parent of this leaf is a minimising one so its value will be less or equal to one whatever the values of the other unexplored leaves are. The latter are therefore uninteresting and will not be analysed by Alpha–Beta. The name of 'Alpha–Beta' and the description we gave are found for the first time in an article of 1963 (Edwards and Hart 1963). The writers underline the importance of ordering the branches of the tree to optimise the Alpha–Beta efficiency: if the level is a maximising one, moves likely to generate high-valued positions must be analysed first and reciprocally if the level is a minimising one. "It turns out at best, that is, in the case of perfect ordering the α-β heuristic can cut a tree's exponential growth rate in half, thus allowing almost twice the search depth for the same effort" (Edwards and Hart 1963, p. 3).

4.3 The Achievement of Technology: Deep Blue's Victory

We saw that most of the problems that arose were connected to the speed of calculation of machines. Therefore researches to reduce the analysis of the game tree progressed, and the Alpha–Beta pruning soon became the idea that prevailed in Chess programming. Since the late 1970s, it had been established that Chess computers became stronger as their hardware speed increased. By 1985 engineers thought that a thousand fold increase in hardware speed might be sufficient to produce a World Champion-class Chess machine (Hsu 2002). But they were soon confronted with problems of material such as the size and the number of the components used in Chess machines. Here is an example through transistors and chips described by Feng-hsiung Hsu (b. 1959) (Hsu 2002). Hsu was specialised on the hardware part of the *Deep Blue* project and before he joined IBM in 1989, he worked on the conception of smaller and faster chips that would permit more calculation.

One of the Chess program's main components is the *move generator*. The move generator generates the Chess moves that the program examines. But to work properly, the move generator needs also to generate unexamined, or not yet searched, moves by the program (Hsu 2002). In the *Belle*[8] design, this second task, performed by the *disable-stack*, required a 64-bit wide memory, one bit for each square of the chessboard. Generally, the program is allowed to look up to 128 plies[9] ahead (so 256 words deep to handle). If we assume that six transistors are needed for every bit of memory, then the number of transistors necessary for the disable-stack alone would be at least 1500 for each square of the chessboard, or about 100,000 transistors for the whole board. And the problem was that in 1985, it was nearly impossible to fit the *Belle* move generator into a single chip; consequently the circuit size was too big. One question engineers had to deal with was the possibility to create a new single chip (instead of the actual 64) to complete *Belle* Chess generator. It led to a re-thinking of the necessity of the disable-stack and after a redefining of its function, only ten transistors on average were used for each square. A 150 to 1 reduction! The same problem was faced with the *evaluation function*[10]: "Could the evaluation function be fitted onto a single chip as well? [...] If it is barely possible to fit the move generator onto a single chip, what was the chance of doing the same for a good evaluation function? It doesn't look good, does it?" (Hsu 2002, p. 30). The basic idea to solve this problem was "trade space for time. Chess evaluation functions have spatially repetitive components, and it is possible to use a smaller circuit multiple times to do the same computation" (Hsu 2002, p. 30). Finally, it had been possible to build a single chip Chess move generator and a single chip evaluation function.

This illustrates the important fact that mathematics is a good opportunity for technology to expand and that both cannot be completely detached from each other. In 1996, the year the program *Deep Blue* won one of the six matches against the World Chess Champion Garry Kasparov (born in 1963) in the first ever traditional Chess tournament between man and computer, IBM set a new world record in magnetic data storage density—five billion bits of data per square inch—the equivalent of 312,500 double-spaced typewritten pages in one square inch of disk surface. *Deep Blue* is a combination of special purpose hardware and software with an IBM RISC System/6000, a system capable of examining 200 million moves per second, or 50 billions positions, in the three minutes allocated for a single move in Chess (IBM Archives 2004). Nevertheless, it is a misconception to think that if the computer Chess wins against a human player it is only because it computes faster. The win of the IBM Chess-machine *Deep Blue* against Garry Kasparov in May 1997

[8] Belle is a special purpose Chess machine that was built in the early 1980s by Ken Thompson (b. 1943) and Joe Condon (1935–2012) from the Bell Laboratories. Belle became the first Chess program to play at the US National Master level in 1982 (Hsu 2002).

[9] In computer Chess jargon, a ply designates a move made by one of the two player. In Chess literature a move refers to a move played by White and the answer of Black (or vice-versa), so two plies make a move.

[10] The evaluation function assesses the quality of the positions reached when the Chess machine looks ahead (Hsu 2002).

was the result of many years of researches for the "human" team who had worked on the project. And it is a parallel development between complexity of the programs and efficiency of the machines that permitted this success. Of course, behind all the *Deep Blue* story is hidden the everlasting human desire to create a being at his image. "Writing the technical report and doing the presentation (about the built of a single ship), however, made me realize that I had the basic blueprint to build the Mother of all Chess Machines, a machine that could defeat the World Champion. In other words, I had a chance to pursue one of the oldest holy grails in computer science, and possibly make history" (Hsu 2002, p. 32).

5 Conclusion

The game of Chess is regarded as "one of the most sophisticated of human activities" (Bernstein and Roberts 1958, p. 96). Therefore, it is not surprising if the first attempts to approach human thinking through a machine (problems in simulation of human thinking) were turned toward Chess. It is worth noticing that researches were usually led by mathematicians (Torres, Redheffer, Shannon, Turing, Bernstein, and later the Russian International Grandmaster and computer scientist Mikhail Botvinnik (1911–1995)) and that the first programs were based on basic mathematical concepts (such as the Minimax principle using minimising and maximising functions, the Alpha–Beta pruning using properties of these functions). The evolution of technologies quickly allowed implementation of these programs on machines able to compute faster than humans. And soon, these Chess computer scientists realised that "the level of its chess playing could be considerably improved were this program to be adapted to a bigger and faster machine" (Bernstein 1958, p. 208). So the race for fewer and simpler components to improve the level of the machines started: more powerful hardware would allow more complex search strategy (Marsland and Björnsson 1997, p. 6).

As more and more powerful computers became available, the full pruning capabilities of Alpha–Beta became better known, and so programs permitted more and more calculations. Therefore, improvements of the algorithms are as important as brute-force in the success of Chess programing. Various techniques to improve the move ordering and to make the search more efficient were developed, such as iterative-deepening, use of transposition-tables, and forward pruning (Marsland and Björnsson 1997, p. 8). These various techniques, as well as the establishment of a mathematical model to represent the game through a tree with branches and leaves, are directly connected to the algorithmic aspect of Chess programing. They are quite elementary compare to the deep mathematical theory that arose from Sprague-Grundy theorem, which is actually the mathematical theory of the Nim game and impartial games. Other combinatorial games such as Go or Sprouts use also complex mathematical theory, especially for the endgames. But such a theory for Chess has not emerged; Chess computer scientists successfully performed to represent Chess in an abstract way but, despite the improvement of technologies and computers, no real mathematical theory of Chess permitting, not only a better implementation of the programs, but also to help a player by giving him another perspective of the game, has been found… yet!

References

Alliot JM, Schiex T (1994) Intelligence artificielle et informatique théorique. Cépaduès, Toulouse

Berlekamp ER, Conway JH, Guy R (2001) Winning ways for your mathematical plays, vol 1, 2nd edn. AK Peters, Wellesley

Bernstein A (1958) A chess playing program for the IBM 704. Chess Review 26(7):208–209

Bernstein A, Roberts MV (1958) Computer V. Chess player. Scientific American 198:96–105

Bernstein A, Roberts MV, Arbuckle T, Belsky MA (1958) A chess playing program for the IBM 704. Proceedings of the 1958 western joint computer conference. Los Angeles, California, pp 157–159

Bouton CL (1901) Nim: a game with a complete mathematical theory. The Annals of Mathematics 2nd Series 3(1/4):35–39

Condon EU (1942) The Nimatron. The American Mathematical Monthly 49(5):330–332

Edwards DJ, Hart TP (1963) The α–β Heuristic. Artificial intelligence project RLE and MIT Computation Centre. Memo 30, 28 Oct 1963, revised version of The Tree Prune (TP) Algorithm, AI, Memo 30, 4 Dec 1961

Eiss HE (1988) Dictionary of mathematical games, puzzles, and amusements. Greenwood, New York

Gardner M (1959) Hexaflegons and other mathematical diversions The first scientific American book of puzzles & games. The University of Chicago Press, Chicago

Hsu F (2002) Behind deep blue: building the computer that defeated the world chess champion. Princeton University Press, Princeton

IBM Archives (1955) 704 data processing system. Manual of IBM electronic data processing machines. http://www-03.ibm.com/ibm/history/exhibits/mainframe/mainframe_PP704.html. Accessed 21 Dec 2014

IBM Archives (2004) IBM highlights, 1996–1999. http://www-03.ibm.com/ibm/history/documents/pdf/1996–1999.pdf. Accessed 21 Dec 2014

Marsland TA, Björnsson Y (1997) From Minimax to Manhattan. In: Deep Blue versus Kasparov: The significance for artificial intelligence. AAAI Workshop, 1997, pp 5–18

Moore EH (1910) On the foundations of mathematics. Bulletin of the American Mathematical Society 9(8):402–424

Newell A, Shaw JC, Simon HA (1958) Chess-playing programs and the problem of complexity. IBM Journal Research and Development 2:320–355

Randell B (1982) From analytical engine to electronic digital computer: the contributions of Ludgate, Torres, and Bush. Annals of the History of Computing 4(4):327–341

Redheffer RM (1948) A machine for playing the game Nim. The American Mathematical Monthly 55(6):343–349

Shannon CE (1950) Programming a computer for playing chess. Philosophical Magazine Series 7 41(314):256–275

Torres L (1915) Essais sur l'automatique. Revue générale des sciences pures et appliquées. Tome 26:601–611

US Patent (1940) Condon et al., Machine to play game of Nim. US Patent 2,215,544 24 September 1940, p 2

Vigneron H (1914) Les automates. Nature 2142:56–61

Development of New Steamships and History of the Shipping Industry in the Kingdom of the two Sicilies (1816–1861)

Maria Sirago

Abstract This paper aims at explaining the development of new Neapolitan steamships that started in 1816 by a French merchant, Pierre Andriel. At the same time it outlines the development of the Neapolitan fleet and of the shipyards of Naples and Castellammare, which played an important role in the overall social and political reorganization of Southern Italy. In 1840 the factory of Pietrarsa was built. And in 1851, the steam frigate *Ettore Fieramosca*, with a steam engine from Pietrarsa, was finally assembled in Castellammare: at last, the kingdom started to build its own ships entirely in situ.

Keywords Neapolitan steamship · The royal shipyards of Castellammare and Naples · Southern Italy · Kingdom of Two Sicilies · The royal factory of Pietrarsa · The factories of Henry and Guppy

1 The Development of the Royal Fleet and the Merchant Navy after the Restoration

After the Restoration (1815) king Ferdinand IV decided to rearrange the sea power of the Two Sicilies: the vessels of the Royal fleet, built in the French Period (1806–1815) and used also to protect the merchant navy, had become obsolete. Between 1815 and 1816 rules were established to reorganize the Royal fleet; a General Ordinance of the Royal Navy, published in 1818, set the statutes of the Naval Academy founded in 1735, where naval officers were educated. Vessels with 80 guns, like the French ones in the Napoleonic era (Sirago 2004a, p 54 ss), were first built starting from 1822 in the shipyard of Castellammare, appropriately rearranged for that purpose (Sirago 2009). In the same period, the "Collegi nautici" (naval schools) of Naples and Sorrento, destined to produce many future pilots, enriched their curriculum, with much thanks to the expansion of mathematical and astronomical studies (Sirago 2010). Indeed, they had to prepare their students to face increasing

M. Sirago (✉)
Institute Jacopo Sannazaro, Naples, Italy
e-mail: maria.sirago@gmail.com

R. Pisano (ed.), *A Bridge between Conceptual Frameworks,* History of Mechanism and Machine Science 27, DOI 10.1007/978-94-017-9645-3_25

navigation both to the Black and the Baltic Sea and to the Americas (Giura 1967): it was for this reason that the merchant navy was provided with a new kind of boat, the brigs. Their construction was promoted by royal decree n. 415, promulgated on July 15th, 1816. A subsequent royal decree (n. 837, November 3rd, 1823), included in the navigation law of 1826[1], provided for further facilities for the construction of cargo vessels, particularly those suitable for oceanic lines. As a consequence, in 1825, after 10 years of such a policy, the merchant ships (about 5000, considering large and small ones) had more than doubled. After a long struggle, in 1826, Neapolitan ships finally won equivalent rights to those of the French, English and Spanish, all of whom had obtained specific tax relief in 1818. As a consequence, trade also increased abroad, and ships from the kingdom of the Two Sicilies were employed for oceanic lines (Sirago 2004a, p 64).

2 The Adventure of Steam Engines: Pierre Andriel's First Attempt (1817–1820)

The "adventure" of steam engines in the Kingdom of the Two Sicilies had begun with the French trader Pierre Andriel, from Montpellier. He thought that, thanks to the facilities granted by the kingdom of Naples to foreign investors, steam navigation could have been used also for passenger transport and mail delivery (Sirago 2006). After all, the facts showed evidence of a rapid increase in trade exchanges, due to the developing shipbuilding industry in Castellammare (Sirago 2009). Actually, Andriel wasn't new to this kind of experiments. Following Robert Fulton's example—who had built one of the first steamships in 1807, called *Clermont,* which took 32 hours to go from New York to Albany—about 240 km—(Campaignac 1842 p. X; Armstrong and Williams 2011) whose system was later improved in England (Kennedy 1903), Andriel had decided to organize a similar voyage between London and Paris. After sending a paper about steamships to the French Minister for Navigation, on April 13, 1815, he had been allowed to test this new kind of navigation offshore instead of, as was customary, sailing up and down the river. The company he had created, "Andriel & Pajol", had bought the *Mercey*, a steamship built in Dumbarton, England, in 1814, and renamed it *Eloise* (Fig. 1).

Andriel had left from London at the beginning of March 1816 and had arrived at Le Havre Port on March 18th, where an astonished population had given him a warm welcome (Busson 1970). Then he had sailed up the river until he had reached Paris, on the 29th of the same month (Stratt 1970), proving the superiority of steam navigation to sailing, which was recently being used also for oceanic lines between America and Europe (Andriel 1817, pp 30–36). The following year, Andriel took advantage of his friendship with Fortunato Adolfo Wolf, a merchant from Lorraine who had been living in Naples for some years, and reached him in Naples (Perfetto 1923, p 8). Here, the king granted him a 15-year patent-right ("privativa")

[1] Legge di navigazione e di commercio 1826 dalla Stamperia Reale Napoli.

Fig. 1 The steamship Eloise. (Serristori 1817 picture V)

to promote the building of steamships in the Kingdom of the Two Sicilies, and another one to mine coal to operate steam engines (royal decree, January 17, 1817, n. 616)[2]. Andriel also wrote about all the advantages of steam navigation, first of all the increase of trade: the report was read at the "Istituto di Incoraggiamento" (an Institute created in the French era to promote scientific studies and industrial development) on February 6, 1817 (Andriel 1817). Finally, on April 10, he founded with Wolf a limited partnership for the introduction of the steamships in the Kingdom of the Two Sicilies, called "P. Andriel & C." (Perfetto 1923, p 20). Many famous members belonged to it: among them, the Finance Minister Luigi de' Medici (Vanga 2009), who had always supported him in this business, and Carlo Filangieri (son of the jurist Gaetano), a skilled engineer who had personally tested the new technology (De Lorenzo 1997). Luigi de' Medici promoted the publication of Serristori's essay in Naples (it had been printed in Florence in 1816), because he thought it could help in circulating useful knowledge about the system of navigation: the book included the drawing of *Eloise* (Fig. 1.), Andriel's first steamboat. As soon as he finished collecting funds, Andriel ordered the building of the first of the four steamboats which were destined to the trade routes from Naples to Marseille. Stanislao Filosa. The chief shipbuilder, who was working in the royal shipyard of Castellammare, won the contract, but the steamboat was assembled in "Marine of Vigliena" (near Naples), where it was rapidly built and launched in June 24. It was called "*Ferdinando I*" and had an English engine, which had been personally bought by Andriel in England (Perfetto 1923, pp 21–22). The drawing kept in the Archivio di Stato di Napoli (State Archive of Naples)[3], which is similar to that of the *Elise*, was probably meant to be of the *Ferdinando I*, even though the second steamboat (built 1 year later) had the same name, which was a little bit confusing. Anyway, on September 27th, the steamship, which had 16 first-class cabins and a standard one, sailed for its first voyage, so marking the beginning of steam navigation in the Mediterranean sea (Giura 1976). Meanwhile, Neapolitan monarchy, which was worried about the technical aspects, gave the order to select the captains of the steamships among the officers of the royal navy. As a consequence, the Minister for Navigation allowed

[2] Archivio di Stato, Napoli (State Archives, Naples) Ministero dell'Interno, II inventario, f. 582.

[3] Archivio di Stato, Napoli (State Archives of Naples) fs. 70/bis, cartella B-8, first steamship S.Ferdinando.

1832. Among the participants there were many famous people, apart from Giorgio Sicard himself. The tour included the ports of Messina, Catania, Malta, Ionian Islands, Patrasso, Nauplion, Atene, Smirne, Costantinopoli and the Bosporus, the mouth of the Black Sea, the Asian coasts Smirne and Zante, then returning to Malta, Palermo, Messina and Naples. One of the participants, Marchebeus (1839), was so excited about the cruise— the first "tour" in the East—that he made a quite detailed report of it. In 1834 the patent-right expired, but the king did not want to renew it to the "Società Sicard" (Bianco 1836), which had requested it. The king's intention was to initiate a royal postal service by steamboat (which indeed worked for about 3 years). However, Sicard was able to turn his society into a joint-stock company (a corporation) whose only benefit was to employ sailors from the royal navy, bearing himself the cost of it. He was under the obligation of training them in the new kind of navigation and placing them at the king's disposal. In the same year his son Leopoldo went to Glasgow to buy the third steamboat, called *Maria Cristina*, which was launched the following year. When Giorgio Sicard died, in 1835, the company turned into the limited partnership of the "Amministrazione della navigazione a vapore nel Regno delle Due Sicilie" (Administration of the steamship navigation in the kingdom of the Two Sicilies). Unfortunately, Leopoldo also died, in 1839, when he was only twenty-seven; therefore, the partnership turned into a joint-stock company (a corporation) named Leopoldo Sicard & co, owner of the steamboats *Ferdinando I* (laid up in 1838) and *Maria Cristina* (Perfetto 1923, p. 30 ss).

Meanwhile, the question about the granting of patent-rights to the steamships sparked off a heated discussion. They only lasted for a short term from their release, so the king was asked to intervene in order to promote development of the new navigation system (Lanza 1836). It was for this reason that in 1839 the king decided to give a prize to all the people who were able to build steamships in the kingdom and granted them a free coastal navigation to every ship. Meanwhile, the "Amministrazione per la navigazione a vapore" (Administration of the steamship navigation in the kingdom of the Two Sicilies) was suffering from the loss of the patent-right, the foreign competition and the excessive customs duties imposed by the port of Marseille in France (Millenet 1837), which had formed the first steamship navigation society (Gille 1970). The Administration then decided to buy three steamships to give new life to its activity and increase competitive offers for passengers. A smaller one, called *Furia,* was destined to navigation in the gulf of Naples, whereas the *Mongibello* and the *Ercolano*, made in England with Maudslay engines, were used for transatlantic lanes. The *Ercolano* arrived in Naples on May 15th, 1841, the *Mongibello* by the beginning of June: both were under Ferdinando Cafiero's command. In order to promote knowledge of the operating aspects of the ships, a "tour" of the Gulf was organized on May 23rd for the members of the company[5]. However, the *Mongibello*, usually employed for the route to Marseille, collided with a steamship of the company Rubattino from the kingdom of Sardinia (Perfetto 1923, pp. 62–63). It happened near the port of Genoa during its first voyage back and the company took legal action against the *Mongibello* in order to establish its possible liability of the disaster and claim damages (AAVV 1843; Borrelli 1845).

[5] Giornale delle Due Sicilie (1841) June 2.

In England, in the same period, there had been a shortage of wood supply caused by the great number of ships built in the previous years. Therefore, a new kind of ship was being improved: it had an iron hull, it was lighter and steam propelled. In addition, the paddle wheel had been replaced by the screw, which increased the speed of the ships and gave a real stimulus to steam navigation. As a consequence, in 1846, the "Administration" first bought two steamships in England: the *Vesuvio* and the *Capri*. They had an iron hull like the new models, but still kept paddle wheels. Then, between 1853 and 1854, two more steamships were bought: this time, they had both the iron hull and the screw. Their names were the *Sorrento* and the *Amalfi,* each weighing 300 t: they could carry a great number of passengers and marked the beginning of a new course between Marseille and Trieste (Radogna 1982, p. 56). In the same years, the *Maria Cristina* was being laid up and its boilers were replaced with similar ones built in Naples by the shop Zino & Henry. However, the "Administration" went through a severe crisis and almost became bankrupt also because of the Sicilian risings (1848–1849) and the Crimean war. The Administration gradually recovered from the crash and changed its name in 1858: "Compagnia di navigazione a vapore delle Due Sicilie" (Steam navigation Company of the Two Sicilies). It issued 400 shares, whose greatest part was purchased by members of the noblest families and famous business men (Rothschild 53, Sicad–Radice & C. 20): 60 shares were also bought back by the Company itself. After the reunification (1861), the Company became the most important of Italy. Indeed, it owned six steamers totaling 1801 t, whereas the tonnage of the six steamers belonging to the Rubattino company only reached 1329 tons; the *Florio* company had five steamers. Later, the Italian government entrusted the state postal service to the companies Rubattino and Florio; as a consequence, the Neapolitan Company stopped working in 1865 because of the keen competition (De Matteo 2002).

4 A Survey of the Other Steamship Societies

4.1 The Florio Steamship Society

Since May 1839 the king had pledged his support to steam navigation by granting free coastal navigation and, in addition, giving a 3-year prize to all the people who were able to build steamships in the kingdom or purchase them abroad. Vincenzo Florio availed himself of this opportunity: he was a Sicilian business man who had traded in many business sectors, which had allowed him to make a fortune. He was one of Beniamino Ingham's friends, a rich English trader who had settled in Sicily during the Napoleonic period, when the king had moved to the island with the help of the British navy to escape the French army invading the kingdom of Naples. In 1840 Florio formed the "Società dei battelli a vapore siciliani" (The Sicilian steamer Company) together with Ingham: they purchased the *Palermo* in Greenwich, and by 1841 it sailed from Naples to Palermo, with an intermediate call in Messina. Florio also owned a foundry with 500–800 workers who were employed to fix the new engines. Once he was granted the postal service to Sicily, in 1847 he formed himself the "Impresa I. E V. Florio per la navigazione a vapore dei piroscafi siciliani" (Enterprise I. E V. Florio for the steam navigation of the Sicilian steamers). He bought a wooden steamship, the *Indépendant,* which was named *Diligente* after the risings of 1848–1849: it sailed from Sicily to Malta. In Glasgow he built another steamship, called the "Corriere Siciliano", whose courses were basically two: one concerning the Sicilian cities of Palermo Messina Catania and Siracusa, the other including Palermo, Naples, Civitavecchia, Livorno, Genoa, Marseille. In 1856 the Bourbon government decided to privatize the postal service on Florio's request, who was entrusted with it. The Sicilian trader purchased another steamship in England: it was the *Etna.* In 1858 he also undertook the weekly postal service between Naples and Sicily on contract, the same which the "Compagnia di Navigazione a vapore delle Due Sicilie, had been granted in 1856. By the reunification of Italy, Florio's fleet was composed by five steamships: the *Diligente,* the *Corriere Siciliano,* the *Etna, the Archimede,* and the *Eclettico.* They linked up Palermo–Naples–Marseille–Palermo–Trapani–Agrigento, monthly, and Palermo and Messina to Naples, every week. However, when Garibaldi arrived, the service stopped working for 1 year. In 1861 the *Etna* sank during Gaeta's siege, but the tonnage of Florio's fleet was superior to Rubattino's, even though it had fewer boats. Therefore, when the Italian government decided to let the postal service out on contract, it chose Florio and Rubattino, also considering that the Neapolitan company was suspected to be pro-Bourbon (Cancila 2008, p. 120 ss.).

4.2 The Calabrian–Sicilian Steamship Society

After the liberalization of the steam navigation in 1839, the king published a report where he stated reasons about the advisability of widening this kind of navigation to link up Naples to the ports of the Calabrie (Anonymous 1839a, b). Andrea de Martino, who had piloted the steamships *Ferdinando I* and *Francesco I,* agreed to the king's proposal. In January 1840 he formed the "Società di navigazione per traffico de' battelli a vapore nel Mediterraneo" (Company for steam navigation in the Mediterranean sea) in Naples. He invested 200,000 ducats divided into 400 shares, and bought the steamer *Vesuvio* (built in 1832) in England: the ship reached Naples on December 12th under Raffaele Cafiero's command. It was employed for the following 2 years to sail to Tropea, Messina and Palermo (Radogna 1982). When de Martino died, in May 1842, a new partnership was formed: it was the Società Vicisvinci & C., later named "Società Calabro–Sicula per la navigazione a vapore" (the Calabrian–Sicilan company for steam navigation). Giusepe Vicesvinci was in charge of it in its office in Naples. Two steamships were bought: the *Vesuvio* (restored and renamed, first *Faro,* then *Polifemo*), and the *Duca di Calabria,* which sailed from Naples to Messina with an intermediate call in Calabria. In 1846 a new steamship was ordered in England, the first example of helical iron steamship in Italy: its name was the *Giglio delle Onde* (Fig. 3).

Fig. 3 A steamship in the Neapolitan Gulf. Bayard M H (1832) gravure private collection. (Formicola and Romano 1994, p. 24)

During the Sicilian risings of 1848 the steamships were commandeered until February 1849 and used to carry troops; then, starting from the month of June, the courses to Calabria and Sicily were reactivated. In 1854 the company purchased another steamship in France, the *Calabrese*, which was first piloted by Raffaele Cafiero, then by Salvatore Pampinella. Later, the four steamships were alternatively under the command of Agostino and Antonio Cafiero, and Michele Mancino. Finally, in 1856, the *Polifemo* was laid up, his engines were replaced and it was renamed *Ercole*. After the reunification of Italy, some of the steamships of the company, still managed by Vicesvinci, kept on sailing to the Calabrie (Perfetto 1923, pp. 46–67).

4.3 Other Steamship Societies

During the month of August 1850, as a consequence of the spreading of the screw steamers, a new partnership was formed: it was the "Società anonima per la navigazione dei piroscafi con elica" (the joint-stock company for the screw steamer navigation), which got a royal licence in February 1851. However, the partnership was soon dissolved because it was impossible to collect an adequate amount of shares. The following year the plan was carried through by the Neapolitan shipping agent Giuseppe Cianelli, who formed the partnership "Giuseppe Cianelli & C." in 1853. He bought three screw steamers in England, *Elba, Partenope* and *Newa,* with iron hulls and English engines. Two years later the "Giuseppe Cianelli & C." became a limited partnership renamed "Giuseppe Cianelli & C. Vapori ad elica" ("Giuseppe Cianelli & C. screw steamers)". It only took over the *Elba* and the *Partenope*, which will be used until 1860 to sail weekly to Calabria and Sicily, and, exceptionally, to Ischia and Casamicciola during the "stagione de' bagni" (bathing season). In 1860 old Cianelli retired and the company was bought by one of the partners, Mr. de la Tour, who changed its name into "Società dei vapori ad elica napolitani del conte Francesco de la Tour" (Count Francesco de la Tour's company of Neapolitan screw steamers). During the Thousand's expedition, the two steamers were employed to carry the troops to Sicily; they were reactivated in 1861, but the partnership was wound up in 1864 (Radogna 1982, pp. 99–101).

Finally, it is worth mentioning a first attempt to set up steam navigation service between the kingdom of the Two Sicilies and the United States of America. Indeed, two Neapolitan business men, Domenico Bellini and Enrico Quadri, got a government patent in 1842 which allowed them to sail to the USA with a 400–500 HP steamship. However, the plan was not carried through until 1852, thanks to the two shipowners from Palermo, Luigi e Salvatore de Pace, who already owned some deep-sea sailing ships. They formed a partnership called "Sicula Transatlantica", which had to voyage from Palermo to New York. The helical iron steamer *Sicilia*, built in Glasgow, sailed for the first time in June 1854 and was under Ferdinando Cafiero's command, a captain from Meta di Sorrento. It reached New York in 26 days. However, the enterprise was not really profit-making, therefore the company did not plan any more voyages to America (Cancila 2008, p. 141).

5 The Royal Fleet and the Royal Factory of Pietrarsa

In 1836, under the reign of King Ferdinand II, the government purchased three steamships for postal service and passengers, thus forming the "Delegazione Reale dei pacchetti a Vapore" (Steamboat Royal Delegation). The Delegation purchased in England two wooden ships with paddle wheels for the Royal navy: the *Nettuno* and the *Ferdinando II*. In addition, the English steam schooner *Santa Wenefreda* was also bought for the Royal fleet, after being restored in the shipyard of Castellammare because of the damage done by a fire (Radogna 1982, p. 75 ss). After 3 years, the "Delegazione Reale dei pacchetti a Vapore" was dissolved for lean profits (Anonymous 1839a, b) (Fig. 4).

In 1840 king Ferdinand II supported the purchase of three steamers in England, the *Nettuno,* the *Lilibeo* and the *Peloro,* which were used to carry mail, passengers and goods. This task was assigned to the "Amministrazione generale delle Poste e dei Procacci" (General Post Office), which started its activity in 1841. The following year the king ordered other five steamers: *Rondine*, *Antilope* and *Argonauta*, from England, and *Palinuro* and *Misero,* from France. This event marked the beginning of the state postal service (Radogna 1982, p. 75 ss).

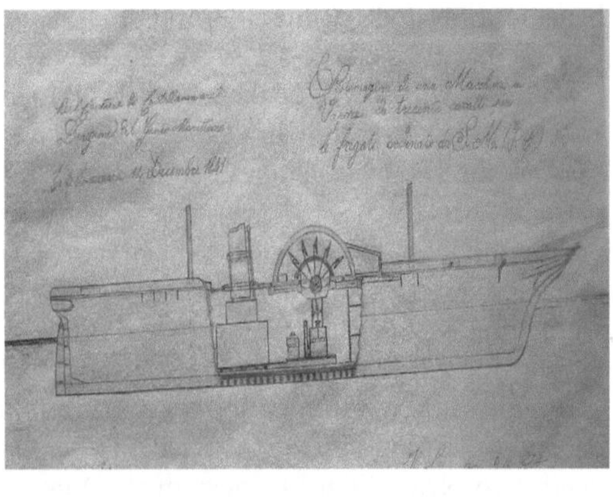

Fig. 4 A picture of a steam engine, Castellammare Royal shipyard (1841), private collection. (Formicola and Romano 1994, p. 124)

Fig. 5 A picture of Castellammare Royal shipyard (1860) private collection. (Formicola and Romano 1994 p. 121)

In 1841 a ship was made on *S. Wenereda*'s model in the basin of Naples: its engines came from England[6]. Starting from 1835, some little steamships (6 HP), with English engines, were also built for the Royal Fleet: they were called "cavafondi" (like the *Vulcano)*, and were used to clean the harbours (Sirago 2004a, p. 55 ss). Their little machines were built in a laboratory founded in 1830 in Torre Annunziata and managed by William Robinson, a Scottish captain who worked for the Bourbon fleet (Garofalo 1997 p 11). Later on, between 1840 and 1849, the shipyard of Castellammare was renovated with the most advanced machinery made in England (Fig. 5).

In 1837 a machine had been already assembled to lift the vessels. Then, by the beginning of 1838, Ferdinand II had formed a technical committee headed by the "expert supervisor" Giovan Battista Staiti, the director of the Naval Engineers, and Mugnai, a hydraulic engineer. They had the task of planning the expansion of the shipyard of Castellammare to adapt it to the new kinds of boats. The area belonging to the merchant shipyard was included in the plan, so that it had to be rebuilt in another place. The works started in 1839, after the king's visit, with the restoration of the roofs which protected the timber used to build the vessels (the shipyard was completed in 1845). At the end of 1843 there had been the successful launch of the 300 HP steam frigate called *Hercules*. From that moment on, it was possible to start building other steamships, which were made with wood and iron from Baltic, both from Sweden and Russia (Sirago 2012). In 1841, the hull of the first steam frigate could be built, following the English models (Tredgold 1840–1841), which had been previously applied for. Later on, two more steam frigates were built (Formicola and Romano 1994) (Fig. 6).

In the same period, the Neapolitan shipyard kept on being used for the refit and the maintenance of the Bourbon sailing fleet, in addition to the assembly of the light surface craft. In order to meet the technical requirements of the new kind of ships, a dry dock ("bacino di raddobbo") was built in 1852, following the plans of the main European ports, which had been working since the first half of the eighteenth century. The new construction consisted in a big masonry tank, which allowed to refit the ships on the spot, instead of doing it with the ship aground, as was customary (Formicola and Romano 1994, p. 141 ss).

[6] Giornale delle Due Sicilie (1841) June 2.

Fig. 6 Naples inauguration of the dry dock (1852), Fergola Salvatore painter, Naples Maritime Military Command. (Formicola and Romano 1994, p. 147)

Since pilots, ships and machinery had always been imported from England and France, the king decided to shake off the foreign influence. In 1837, after Robinson's death, he decided to move the laboratory to the Royal Palace, entrusting Luigi Corsi with the management of it because he had been working for Robinson for a long time. Then, in 1840, the king ordered Carlo Filangieri to look for a suitable place to built a new factory to make domestic steam engines. Filangieri was a naval officer who had been experimenting the most recent innovations about steam machines for a long time. He chose Pietrarsa, a deserted place in Portici, near Naples, where the "Royal Factory" was finally built. The factory was under Robertson's technical direction, the most expert engineer working in the kingdom: he was helped both by Corsi, who became general manager, and Filippo Pinto, an expert "chief artificer" from Naples who had worked in one of the most important factories. In 1841 most of the Royal Factory had been built and 200 workers were producing the most varied steam engines. In the meantime Filangieri decided to found a school for steamship pilots for 20 pupils, which got the royal licence on February 6th. According to his plans, this kind of school had to teach all the useful scientific knowledge to form "perfect engineers" and expert machine inventors and shipbuilders. Similar studies were also taught in the Royal Naval Academy of Naples (Radogna 1978, pp .108–117). In 1844 the "Corpo dei Piloti" had a new staff formed by 100 units; the following year rules were set for the engineers who worked on the steamboats. In 1847 the "Scuola prattica" was rearranged to meet the requirements for the education of sea pilots belonging both to the Navy and to the merchant navy[7]. The factory, which was called "Officina Reale Pirotecnica" (Royal Pyrotechnic Factory), now "Museo delle ferrovie" (Italian National Railway Museum of Pietrarsa), was finished in 10 years The first building was a "Big Hall for Constructions", which was first used to collect pieces of the steam engines bought abroad, and then to refit the ships of the Royal navy (Anonymous 1842; De Rosa 1968, p. 8 ss) (Fig. 7).

However, it was soon possible to make steam engines for the ships and for the railway which was being built; in 1843, a new hoist was made for the shipyard of Castellammare (Formicola and Romano 1994, p. 94 ss).

[7] Biblioteca Napoletana di Storia Patria Naples ms XXIX A 14 ff 238t–257t; Filangieri C Autobiografia manoscritta (1862).

Fig. 7 The Pietrarsa Royal Factory under construction 1840–1841. National Railway Museum of Pietrarsa

In the meanwhile, a new 300 HP steamship was being built: it was like a corvette, or a “brick”, with 26 guns, more expensive than the other boats but surely more functional. At the end of 1844 the steam frigate *Archimede* was successfully launched. By the end of 1848 three steam frigates, called *Ettore*, *Carlo III* and *Sannita*, were built with British engines. In 1851, the steam frigate *Ettore Fieramosca,* built in Castellammare between 1849 and 1850, was finally assembled with a steam engine from Pietrarsa: at last, the kingdom started to build in situ its own ships entirely—both the outer hulls and the engines. These events marked the beginning of the history of southern shipping industry, the only profit-making line of business supported by the king (Davis 1979, p. 131 ss). Since the fifties, the screw navigation system was also experimented because it was easier than the other with the heavy paddled wheels. Therefore, the Royal navy promoted studies to build a couple of screw frigates. Once again, the model had to be found abroad: it was the English frigate *Shannon.* The ship was finished in the summer of 1860 and it was called *Borbone.* In 1861, when the Piedmontese arrived, two more ships were in the shipyard of Castellammare: the frigate *Farnese, Borbone*’s twin ship with an English engine too, and the steam corvette *Etna* (Formicola and Romano 1994, p. 73 ss).

6 A Survey of other Neapolitan Engineering Industries

In 1834 Lorenzo Zino founded an iron foundry in Capodimonte: it was called “Zino & Henry” because he went into business with Francesco Henry. The factory soon expanded, so that it was moved to Ponte della Maddalena (by the Granili area) in 1836: here, small steamers of about 6 up to 10 HP were built. Its production won the gold medal at the industrial exhibition held in Naples in 1837 (Fazzini 1836–1837). Later on, the factory became the ship chandler of the Bourbon navy (De Crescenzo

2002), providing a boiler for the steamer *Maria Cristina* in 1842 (Radogna 1982, p. 120). In 1850, thanks to the contribution of Gregorio Macry's large capital, a Calabrian business man who bought out Zino, the firm expanded and changed its name into Macry, Henry & Co. The orders for the production of metal structures for warships increased (De Rosa 1968, pp. 3–4). By the way, the most important factory was the "Guppy & Co", founded in 1852 by the naval engineer Thomas Richard Guppy. He was an English manufacturer from Bristol who had moved to Naples in 1849 (De Rosa 1968, p. 4 ss). Here he had formed a partnership with John Pattison, an English qualified engineer who had come to Naples in 1842 to supervise the repair shop of the Bayard railway and to open a new one[8]. The "Guppy" soon expanded because its production was considered the best in the kingdom. Since 1855 it also became the ship chandler of "minuterie" ("nuts and bolts"—the raw materials for ships which had to be built or refitted) for the Royal navy. A shipyard dealing with single pieces of the ships was set up by the factory located at the Maddalena. In 1859 Guppy was granted a patent-right for the "Ammigliorazioni delle caldaie a vapore" (the Improvement of the steam boilers), which was concerned with their steam generating capacity[9]. However, the partnership dissolved in 1862, after the reunification of Italy, and Pattison founded another factory with his sons Cristofaro and Thomas Taylo (Mancarello 1999–2000).

7 Steam Engines Experimentation in Naples

Studies on steam engines and its applications had been carried out in the kingdom since the thirties. In that period Bernardino de Angelis had published a booklet dedicated to Ferdinando II, King of the Two Sicilies, where he introduced an innovative machinery. It was destined to be used to refloat steam ships from sandy soundings, which were very common in many southern ports and needed to be frequently cleaned. The drawing of a steam ship was enclosed in the booklet (De Angelis 1830). In 1832 abbot Giuseppe Conti made a request for a patent-right for the invention of a medium pressure steam engine ("macchina a vapore di media pressione"). It should have been smaller than the ones still in use, easier and more suitable for the transport of a greater number of passengers and some goods. This plan was approved by the Neapolitan "Istituto di Incoraggiamento", founded by king Ferdinando in 1778 (Balletta 2007), also because the goods made in Naples would have employed 1000 workers[10]. In 1849 Enrico Buckmaster, an English "mechanical engineer" and an "engineer of the Neapolitan Royal Navy", had been granted a 5-year patent-right for making changes in the steam engines in use, in

[8] Archivio di Stato (State Archive) Naples Ministero Agricoltura Industria Commercio 277/26 February 16 and April 7 1851; July 22 1852.

[9] Archivio di Stato (State Archive) Naples Ministero Agricoltura Industria Commercio 286/21 May 2 1859, with the "Ammigliorazioni" (improvements) of Guppy.

[10] Archivio di Stato (State Archives) Naples Ministero dell'Interno II inventario f 588 fs. 4420, April 4 July 8 August 15 1832; ibid Ministero Industria Agricoltura Commercio 279/13.

order to reduce their size, thus favouring the increase of the loading capacity of the ships[11]. The patent had been granted on condition that all the experiments and the production from the public factory of Pietrarsa were free[12]. Experimentation was being carried out in the factories "Zino & Henry" and "Guppy", but above all in Pietrarsa, where, since the 40ies, sophisticated steam engines had been built for steamers and for the rising railway. In addition, a lot of different machines were assembled, first of all those used in the shipyard of Castellammare. Experimentations kept on being carried out even after the reunification of Italy: indeed, in 1861, Luigi Cognard asked for a patent-right for a new propeller to use on ships[13].

8 Conclusion

Steam navigation had begun as if it were a sort of "adventure" pioneered by foreign business men who had trusted the competence of Neapolitan pilots—particularly, those from Sorrento, heirs to an ancient mariner tradition—and the ability of the naval engineers who were working in the shipyards of Castellammare e Sorrento (Maresca and Passaro 2011). Then, Giorgio Wilding, prince of Butera, had gone through this "adventure" (started by Andriel) thanks to the farsightedness of the Finance Minister Luigi de' Medici and Carlo Filangieri. In a short time, the new kind of navigation quickly spread, also thanks to Giorgio Sicard's patronage, an enterprising foreign business man, and his son Leopoldo, who had become an expert on steam navigation. In 1839, when everybody had the chance to form steam navigation companies, the review "Poliorama Pittoresco" had enthusiastically described this new means of transport which had become widespread so quickly—in only 20 years (Anonymous 1838a, b)—that in 1841 there were six steamships in the Neapolitan merchant navy[14]. However, brigs and big sailing ships were still built and used for the transatlantic lanes, because they could carry large quantities of goods (Clemente 2013). In Sicily, Vincenzo Florio had formed a company by purchasing some ships. This system had also been introduced in the Royal fleet, first with small dredges, then with bigger ships. In only 20 years a notable increase had occurred: in 1856 there were 12 frigates, 4 corvettes, 4 brigantines, 70 gunboats, whose maintenance was in charge of the Officina di Pietrarsa (shop of Pietrarsa) and Henry and Guppy's factories. In 1851 an engine built in Naples had been assembled on the *Ettore Fieramosca,* a steam frigate from the shipyard of Castellammare; in the same period screw steamers were being assembled.

---

[11] Archivio di Stato (State Archives) Naples Ministero Industria Agricoltura Commercio 274 (1847) July 13 (1849) June 16.

[12] Archivio di Stato (State Archives) Naples Ministero Industria Agricoltura Commercio 280 (1849) n° 217 April 22.

[13] Archivio di Stato (State Archives) Naples Ministero Industria Agricoltura Commercio 286/92 (1861) July 11.

[14] Annali Civili del Regno delle Due Sicilie (1841) Tables of merchant vessels.

References

AAVV (1838) Cenno sull'amministrazione della real delegazione de' pacchetti a vapore napoletani, Napoli

AAVV (1838) Il pacchetto a vapore postale. Poliorama Pittoresco, pp 153–154

AAVV (1839a) Leopoldo Sicard, Necrologio. Poliorama Pittoresco, pp 57–59

AAVV (1839b) Sulla navigazione a vapore nelle Calabrie dalle Officine Tipografiche Naples

AAVV (1842) Reale opificio pirotecnico di Pietr'arsa. Poliorama pittoresco, pp 302–304

AAVV (2004) Casa di Re. Un secolo di storia alla Reggia di Caserta 1752–1860. Skira ed Milan

Aliberti G (1976) La vita economica a Napoli nella prima metã dell'Ottocento. Aa Vv Storia di Napoli ESI Naples, vol V, pp 97–359

Andriel P (1817) Coup d'oeil historique sur l'utilité des batiments—a—vapeur dans le Royaume des Deux—Siciles De l'Imprimerie du Ministere de la Secrétairerie d'État Naples

Anonymous (1838a) Cenno sull'amministrazione della real delegazione de' pacchetti a vapore napoletani, Napoli

Anonymous (1838b) Il pacchetto a vapore postale. Poliorama Pittoresco, pp 153–154

Anonymous (1839a) Leopoldo Sicard, Necrologio. Poliorama Pittoresco, pp 57–59

Anonymous (1839b) Sulla navigazione a vapore nelle Calabrie dalle Officine Tipografiche Naples

Anonymous (1842) Reale opificio pirotecnico di Pietr'arsa. Poliorama Pittoresco, pp 302–304

Anonymous (1843) Consultazioni a favore del signor Ferdinando Cafiero capitano e amministratore della navigazione a vapore del Regno delle Due Sicilie (col) bastimento il Mongibello contro il signor Carlo Cazzuolo Capitano del bastimento Polluce Tipografia Bonducciana Florence

Armstrong J, Williams DM (2011) The impact of thecnological change: the early steamship in Britain. Research in Maritime History 47

Balletta F (2007) La Biblioteca e l'Archivio documentale dell'Istituto d'Incoraggiamento di Napoli. In: Di Leo A (ed) L'ingegneria e la sua storia. Ruoli, istituzioni, contesti culturali XIX e XX secolo Marlin Cava de'Tirreni, pp 48–58

Bianco D (1836) Per la societã Sicard intorno alla libera navigazione a vapore del Reame delle Due Sicilie Stamperia del Fibreno Naples

Biondi S (1971) Sul "Real Ferdinando" da Napoli a Ischia il 21 novembre 1824 Naples

Borrelli P (1845) Consultazione per la societã per la navigazione a vapore del Regno delle Due Sicilie contro l'impresa De Luchi, Rubattino & C di Genova Tipografia G Sardi Livorno

Busson JP (1970) Quelques aspects des débuts de la construction navale et de la navigation ã vapeur particulierment ã Rouen et au Havre de 1816 ã 1818. In: Mollat M (ed) Les origins de la navigation ã vapeur. Presses Universitaires de France, Paris, pp 27–40

Campaignac AB (1842) De l'etat actuel de la navigation par la vapeur. L Mathias, Paris, p X

Cancila O (2008) I Florio. Storia di una dinastia imprenditoriale. Bompiani, Milan

Clemente A (2013) Le comunicazioni via mare. Malanima P, Ostuni N (eds) Il Mezzogiorno prima dell'Unitã Fonti. Dati, storiografia Rubattino ed. Soveria Mannelli, pp 96–128

Davis J (1979) Societã e imprenditori nel regno borbonico. Laterza, Bari

De Angelis B (1830) Il bastimento assicurato dagli arrenamenti e la carrozza garantita dall'indocilitã de'cavalli con nuovi metodi per agevolare la navigazione a vapore Naples. In: De Crescenzo G (2002) Le industrie del Regno di Napoli. Grimaldi & C. Editori, Naples

De Lorenzo R (1997) Carlo Filangieri Dizionario Biografico degli italiani 47. www.treccani.it

De Matteo L (2002) "Noi della meridionale Italia". Imprese e imprenditori del Mezzogiorno nella crisi dell'unificazione ESI Naples chapter IV La compagnia di navigazione a vapore delle Due Sicilie

De Rosa L (1968) Iniziativa e capitale straniero nell'industria metalmeccanica del Mezzogiorno 1840–1904. Giannini, Naples

Fazzini (1836–1837) A Fonderie di ferro e costruzioni di macchine per le arti industriali. Poliorama Pittoresco, pp 195–196

Formicola A, Romano C (1994) L'industria navale di Ferdinando II di Borbone. Giannini, Naples

Formicola A, Romano C (2010) Storia della marina da guerra dei Borbone di Napoli II vol part II 1815–1830. Ufficio Storico della Marina Militare, Roma
Garofalo C (1997) Pietrarsa da opificio a museo. Altrastampa, Naples
Gille B (1970) Les premiéres compagnies de navigation ã vapeur avant 1848. In: Mollat M (ed) Les origins de la navigation ã vapeur. Presses Universitaires de France, Paris, pp 97–101
Giura V (1967) Russia, Stati Uniti e Regno di Napoli nell'etã del Risorgimento ESI Napoli
Giura V (1976) Contributo alla storia della navigazione a vapore nel Regno delle Due Sicilie AA VV Fatti e idee di storia economica nei secoli XII-XIX. Studi dedicati a Franco Borlandi Il Mulino Bologna, pp 709–732
Kennedy J (1903) History of steam navigation. C Bircham, Liverpool
Lanza P (1836) Principe di Scordia. I vapori Tip Eveg Lao e Roberti Palermo
Malanima P, Ostuni N (eds) (2013) Il Mezzogiorno prima dell'Unitã Fonti. Dati, storiografia Rubettino ed Soveria Mannelli
Mancarello M (1999–2000) Le attivitã industriali e la cantieristica navale nel Napoletano dal 1800 al 1860. Thesis, Universitty of Lecce, Faculty of Economics and Business
Marchebeus (1839) Voyage de Paris ã Costaninopole par bateau ã vapeur chez. Artus Bertrand, Paris
Maresca M, Passaro B (2011) La Marineria della Penisola Sorrentina Shipowners, shipping and wooden shipbuilding inthe Sorrento Peninsula Confine ed Sorrento
Millenet J (1837) Reflexions sur la question touchant les bateaux ã vapeur de la Poste De l'imprimerie et papeterie du. Fibreno, Naples
Perfetto C (1923) Vicende della Marina Mercantile a vapore nel Reame delle Sicilie dal 1818 al 1860 Stabilimento tipografico G Barca, Napoli
Radogna L (1978) Storia della marina militare delle Due Sicilie (1734–1860). Mursia, Milano
Radogna L (1982) Storia della marina mercantile delle Due Sicilie (1734–1860). Mursia, Milano
Serristori L (1817) Sopra le macchine a vapore Stamperia di Giovanni de Bonis Naples
Sirago M (2004a) Le cittã e il mare. Economia, politica portuale, identitã culturale dei centri costieri del mezzogiorno moderno. ESI, Napoli
Sirago M (2004b) Firts steamship S. Ferdinando tab. 6.25 picture p 273 AAVV Casa di Re. Un secolo di storia alla reggia di Caserta 1752–1860. Skira ed Milan, pp 319–320
Sirago M (2006) Nuove tecnologie: dal vascello alla nave a vapore. In: Buccaro A, Fabbricatore G, Papa LM (eds) Storia dell'Ingegneria Acts of the 1st national conference, Naples 8–9 March 2006. Cuzzzolin ed Naples 2, vol II, pp 671–679
Sirago M (2007) La riorganizzazione del sistema portuale del Regno meridionale. Dal decennio francese all'Unità. Di Leo A ed L'ingegneria e la sua storia. Ruoli, istituzioni, contesti culturali XIX e XX secolo Marlin Cava de'Tirreni, pp 94–107
Sirago M (2009) Il cantiere di Castellammare dal 1784 ai primi del '900. In: Capobianco L (ed) Le donne di Castellammare ricordano e raccontano. Meridione Nord e Sud nel Mondo
Sirago M (2010) Gli studi nautici nel Regno di Napoli (1770–1861). In: Mafrici M, Vassallo C (eds) Sguardi mediterranei tra Italia e Levante (XVII-XIX secolo). Mediterranean Maritime History Network (MMHN) 3rd MMHN onference Izmir 4–7 May 2010. University Press, Malta, pp 19–35
Sirago M (2012) The shipbuilding industry and trade exchanges between the kingdom of the Two Sicilies and the countries of the Baltic and the Black Sea (1734–1861). Journal of Mediterranean Area Studies 6(2 December):81–107
Sirago M (2013) La scoperta del mare. La nascita e lo sviluppo della balneazione a Napoli e nel suo Golfo tra '800 e '900 Intra Moenia ed Naples
Stratt HP (1970) L'evolution de la propulsion marine ã vapeur (1820–1870). In: Mollat M (ed) Les origins de la navigation ã vapeur. Presses Universitaires de France, Paris, pp 73–87
Tredgold T (1840–1841) On the steam engine and on the steam navigation. J Weale London Appendix A and B
Vanga M (2009) Luigi de' Medici Dizionario Biografico degli italiani 73. www.treccani.it

Unrealized Models of Tesla's Table Fountains From 1917

Bratislav Stojiljković and Svetislav Lj. Marković

Abstract This paper presents the results of Tesla's work and research in the field of fountains that was accomplished and completed in 1917. In this period the inventor designed and developed five different models of table fountains. The design of one table fountain was further modified and developed resulting in four completely distinguishable versions of the same device. By exploring the original archival documents, the authors attempted to present the inventor's research in this less known field and the unknown details of his rich achievements. The authors also present CAD models based on original table fountain drawings, which are archived as a part of Tesla's legacy in the Nikola Tesla Museum in Belgrade.

Keywords Fountain · Nikola Tesla · Tesla's pump · Tesla's legacy · Nikola Tesla Museum

1 Introduction

Nikola Tesla (1856–1943), scientist, engineer and inventor, lived and worked in the period which includes the last two decades of the nineteenth century and the first half of the twentieth century. He belongs to that class of rare inventors whose inventions for more than a 100 years have not ceased to be actual and recognizable. His opus constantly arouses a wide interest of various researchers: from science historians, electrical and machine engineers, experts in aviation, telecommunication and military affairs to ecologists, doctors, psychologists and philosophers. This great interest unequivocally indicates that Tesla's creative and scientific work has set some fundamental milestones up on the road towards modern scientific and technological civilization.

B. Stojiljković (✉)
Nikola Tesla Museum, Belgrade, Serbia
e-mail: bratislav.stojiljkovic@tesla-museum.org

S. L. Marković
Technical College, Čačak, Serbia
e-mail: svetom@open.telekom.rs

R. Pisano (ed.), *A Bridge between Conceptual Frameworks,* History of Mechanism and Machine Science 27, DOI 10.1007/978-94-017-9645-3_26

Fig. 1 Nikola Tesla in his forties. (Nikola Tesla's personal archives, MNT, VI/V,20)

Nikola Tesla, American citizen of Serbian origin, was born on July 10th, 1856, in Smiljan, in the Military Border zone of Austrian Empire, now in the Republic of Croatia. He was educated in Smiljan, Gospic, Karlovac, Graz, and Prague. In the European part of his life he worked in Maribor, Budapest, Paris, and Strasbourg. In June 1884 he moved to the USA, expecting to realize his inventive ideas in the "New World" more easily than in Europe. He lived and worked in New York, where he created his universal scientific work that changed the world and accelerated the wheels of technological development of human civilization for centuries to come (Fig. 1).

He made numerous inventions and discoveries in the field of electricity and mechanical engineering. Thanks to his inspired discovery of rotating magnetic field in 1882, the induction motor was designed and a new technology of energy transfer to great distances was introduced, based on the application of polyphase alternate currents. Tesla's contribution in the field of high voltage, wireless transmission and remote control resulted from the study of properties of high-voltage and high-frequency currents. His research in high-frequency currents started in 1890, marking at the same time the beginning of the second stage in developing the method and apparatus for wireless transmission of information and energy. In the next year he invented a device known as the Tesla's oscillator. In 1898, he presented to the public for the first time the application of radio waves for the remote transmission of commands, by remote control of a small vessel. He gave a significant contribution to machine engineering with his original solutions for bladeless turbines and pumps applying his novel principle of exploiting the energy from fluids through friction. He patented solutions in the field of speedometers, worked on the construction of various types of fountains, and one of the inventions confirming he was always way ahead of his time was the patent for a flying machine with vertical take-off (Marincic 1994, 2006).

2 Tesla's Research in the Field of Fountains

Intensive industrial development at the beginning of the last century—resulting directly from Tesla's ingenious inventions in the field of alternate current generators and the transmission of electrical energy—imposed an increased interest for machines which convert energy into useful work on the one hand, and for machines which transmit energy to a fluid by the use of work (pumps, blowers, compressors) on the other hand. Accordingly, Tesla began his investigations in the construction of machines just in the field of turbines. The significance of his contribution to the development of new-type turbines lies in the original design of a machine for the exchange of energy between the fluid and the bladeless discs, based on friction, i.e. so-called frictional turbine. He invented an essentially new type, based on his ideas, definite physical laws and investigations performed (Jovanovic 2001).

In parallel with his examinations of different turbine designs, Tesla searched for ways and possibilities to define, realize and exploit commercially his other ideas and inventions. He did not use the well-established principles of his engineering profession, but always tried to find his own way towards original solutions. One of these less known solutions is his patent for a fountain. This invention differs from usual solutions applied to fountains and aquariums of the time, where the fluid was sprayed or jetted with appropriate devices for decorative purposes (Tesla III Millenium 1996).

Tesla began his work in this field in May 1913, immediately after his first letters patents for turbines were issued. The oldest preserved document of Tesla's work in this field clearly indicates his initial considerations on designs and construction of new fountain models. At the drawing made on July 10th, 1913, he presented a design of a two-stage cascade fountain with decorative plants arranged in the region of overflow cascades and on the top of elevated conical overflow as well (Fig. 2).

The Tesla fountain, a completely new type for that time, introduced a new type of beauty, which relied principally "on a fascinating spectacle of the large mass of fluid in motion and the display of seemingly great power". Water or other fluid would be lifted from the fountain receptacle to the horizontal flared-out top of conduit, to overflow it in the form of a unique circular cascade, making a splendid sight for an observer's eye. By applying only 1/25 of a horsepower (~29.5 W) to the shaft and assuming a lift of 18 in (457.2 mm), more than one hundred gallons (378.5 l) per minute could be propelled to flow over the flared top of conduit, 1 ft (305 mm) in diameter with the depth of the overflowing fluid of approximately 1-half inch (12.7 mm). As the circulation would be extremely rapid, the total quantity of liquid required would be comparatively small. The new type of fountain would allow for "realization of beautiful and striking views through additional colourful illumination and the disposition of voluminous cascades", with relatively low consumption of energy (Marinkovic et al. 2012).

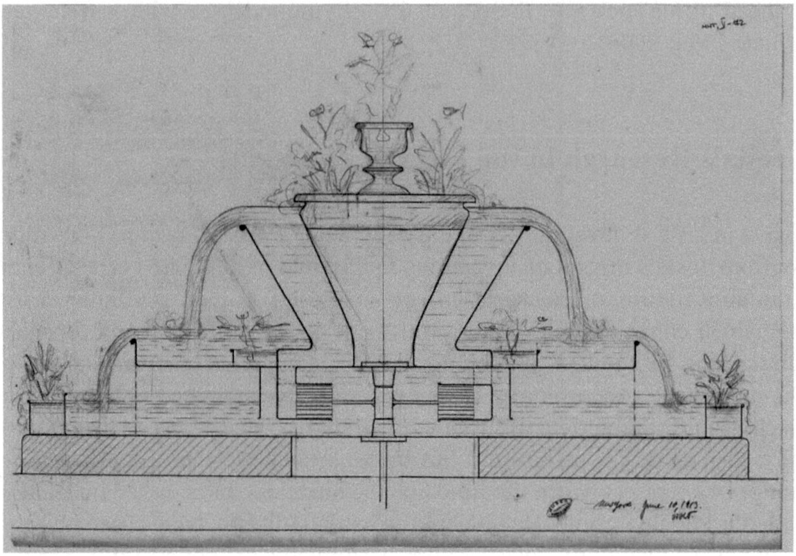

Fig. 2 Vertical section of a two-stage cascade fountain, New York, June 10th, 1913. (Nikola Tesla's personal archives, MNT, V,I-4,78)

Tesla's application for his model of fountain was filed at the United States Patent Office on October 28th, 1913, and the letters patent No. 1.113.716 was issued 1 year later, on October 13th, 1914. After submitting his patent application for "certain new and useful improvements in fountains" Tesla started searching for the ways and options to realize his invention. He offered the model of his cascade fountain, based on the new principle, to Louis Tiffany, inventor of a special stained glass technique, famous designer and the owner of Tiffany Studios. Delay of the US Patent Office in issuing the letters patent for Tesla's fountain, as well as his extremely bad financial situation and impossibility to finance independently the construction of a prototype cascade fountain, on the one hand, as well as the transfer of patent rights to the Tiffany Studio, together with their feasibility estimate of the joint enterprise on the other hand, led to a subsequent divergences of opinions and break of cooperation.

In the period from January to June 1917, Tesla dedicated himself to investigating and realizing new designs of table fountains. In order to make the operation of fountains more beautiful and attractive, he used various shapes and forms for overflow cascades and coloured tempered glass, horizontal and vertical decorative style ornaments, appealing art composition with stylized figures, and decorative arrangement of plant and many-coloured electric light systems.

After several unsuccessful attempts to realize his table fountains, Nikola Tesla made the last one at the end of the second decade of the twentieth century. He worked on the design and construction of three new, very similar models.

The abundance of documents in the archives preserved at the Nikola Tesla Museum in Belgrade facilitates a comprehensive survey of the extent of Tesla's research in this field. They also afford a more distinct perception of time and circumstances in which this great scientist and inventor lived and created. During more than 8 years—with shorter or longer interruptions—Tesla developed numerous models of his fountains, but without any commercial success (Stojiljkovic and Vujovic 2007).

3 Models of Tesla's Table Fountains From 1917

Impossibility of technical production and realization of Tesla fountains (Tesla-Tiffany cascade fountain, Tesla-Tiffany water fountain) resulted in the termination of co-operation between Nikola Tesla and Louis K. Tiffany and his Studio. Therefore, till the end of 1916, Tesla dedicated himself to investigations in other fields of mechanical engineering: speedometers, flow-meters, valvular conduits etc.

However, at the beginning of the next year, Tesla reverts to his investigations in the field of fountains and continues his work, now on the design and construction of new types of this device. From January 18 to January 25, 1917 he defined and designed a new model of the table fountain. Preserved drawings in pencil show four different solutions of the construction of his new fountain (Stojiljkovic 2012).

3.1 A Model of Table Fountain with Arrangements of Decorative Plants

These new solutions resulted from the application of Tesla's constructive improvements and deliberated modifications of the existing fountain models from the previous period. Placing interesting and plentiful decorative plant arrangements was foreseen in all the elaborated variants of his new table fountain, in a separate compartment above the elevated conical overflow. It may be supposed that with the application of such a concept Tesla endeavoured not only to make his new model even more attractive and interesting to the observer's eye, but also to realize his old concept of using attractive plant arrangements in fountain decoration. The oldest preserved document of Tesla's activity in this field unambiguously indicates his initial deliberations on design and construction of such fountains.

At the construction of the new table fountain Tesla applied for the first time his patented disc pump to propel the required quantity of a fluid. By the action of his pump the fluid would be elevated from the receptacle to the level of horizontal, broadened top of the cone, and by its overflow a uniform circular water cascade would be formed. The circular cascade, with its interesting and attractive appearance, would afford an observer a splendid sight of an immense and unconstrained power of water in perpetual motion. The four developed designs of the new fountain differed from one another in the following constructive details:

- the shape and height of the fluid receptacle,
- the position of propelling motor,
- the position and fitting of internal reinforcement of the fluid receptacle, and
- the position and way of embedding the additional electrical lighting.

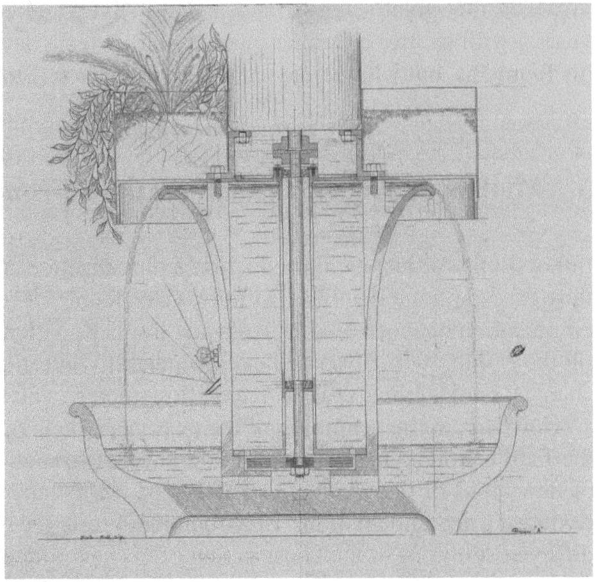

Fig. 3 Drawing "A". The first variant of the new design of Tesla's table fountain. New York, January 18th, 1917. (Nikola Tesla's personal archives, MNT, V,I-4,90)

The technical drawing, marked by Tesla as "Drawing A", shows the vertical section of the first variant of the new fountain model (Fig. 3). It was the first one of a total of three technical drawings from that period, on which Tesla elaborated four different constructions of his new table fountain.

The shape and height of the fluid receptacle in the first fountain variant, and in all other variants of this model as well, have been defined by the way of fitting and embedding some constructive parts, and by the quantity of fluid required for undisturbed operation of the complete system as well. The electric motor is placed on the upper side of the elevated overflow, in its central part, and separated from the remaining space intended for arrangements of decorative plants. The axis of the motor shaft follows the direction of the vertical axis of the fountain. Tesla's pump, with four rotating discs manufactured in a special process, has been attached at the end of the shaft extension on the lower side of the electric motor. The extended shaft of the motor protrudes the internal space of a unique vertical construction; it is fixed to it at places of defined elevation by two independent ball bearings. According to the design, the upper part of this vertical construction is at the same time the receptacle for plant arrangements. Both in its upper and lower part, the unique vertical construction is fastened with a certain number of screws to the inner side of the elevated conical pipe. Interior of the vertical construction, and the space under the propelling electric motor have to be protected and secured from a possible breakthrough of water (Fig. 4).

The part of the table fountain, designed for arrangements of decorative plants, is situated on the upper side of the elevated conical overflow; in its base it forms a circular ring.

Fig. 4 The first variant of Tesla's table fountain and its propelling system with an electric motor (CAD models)

It may be assumed that its construction is made of profiled sheet iron of different thickness, and that its inner surface is protected by special process from harmful effects of water and soil fertilizers. The lateral height of this vessel, together with the height of plants, keep the electric motor hidden from the observer's eye. Electric motor props together with the base of the vessel for decorative plants are fastened by four screws into reinforced recesses on the internal side of the vertical conical conduit. The way of fastening of vertical conical conduit together with the fluid receptacle has not been clearly defined and marked on the technical drawing of this table fountain.

In order to make the operation of his new fountain more attractive and interesting, Tesla designed the installation of electric lighting. As it can be seen on the technical drawing, the system of electric lighting consists most probably of three independent sets, each one of a coloured light-bulb together with its reflecting surface. The sets are arranged at the angle of 120° and fastened horizontally on the outer side of the vertical conduit, approximately at one half of its total height. The light-bulbs are not protected from the effects of water. The reflecting surfaces additionally direct the bulblight upwards, to the lower part of the vessel with decorative plants.

It may be assumed that with such an arrangement of coloured light-bulbs Tesla intended to make a play of coloured light refraction resulting from the light passing through the circular water curtain, and flowing continuously from the horizontal, broadened top of the cone.

Another technical drawing, marked as "Drawing B", illustrates the vertical section of the second variant of Tesla's new table fountain (Fig. 5). According to its shape and height, the fluid receptacle differs from the fluid receptacle in the first variant.

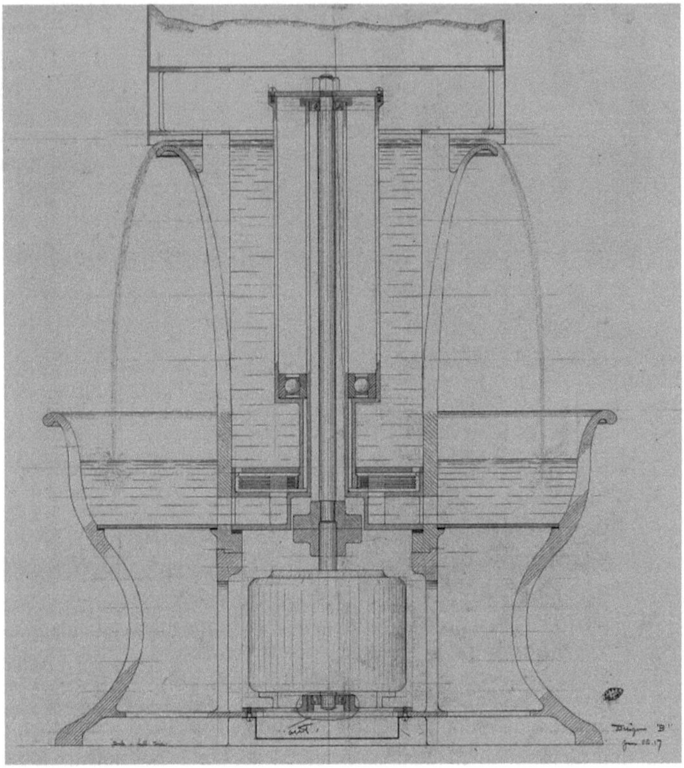

Fig. 5 Drawing "B". Another variant of a new model of Tesla's table fountain. New York, January 22nd, 1917. (Nikola Tesla's personal archives, MNT, V,I-4,91)

The propelling electric motor is placed vertically in the interior of the fluid receptacle, below the elevated conical overflow. The way of transmitting power from the propelling electric motor to the pump discs is based on the partially modified solution described in the patent of the fountain, granted to Tesla in 1914. The working part of the pump consists of four discs. The extension of the motor shaft is fitted into the central nave; the shaft itself is supported by a ball bearing. The cylindrical casing with pump discs is located on the upper side of the extended shaft. The performance of the pump has to enable the elevation of the required quantity of water to the level of horizontal, broadened top of the cone. The moving cylindrical casing is fixed by a ball-bearing to the central nave, in order to hinder its asymmetrical rotation during the fountain operation. The construction of the central nave, together with the construction of the vertical conical conduit, is fixed to the inner vertical reinforcement of the fluid receptacle.

The vertical reinforcement of the receptacle is circular in form, and its outer diameter is equal to the outer diameter of the base of the vertical conical conduit. The propelling electric motor of the fountain is placed in its interior. Both the interior of the central nave and the interior of the vertical circular reinforcement of the fluid receptacle have to be especially secured from the breakthrough of fluid into them. This has been achieved by placing special rubber sealings at the places of their connection with other constructive elements of the fountain. The construction of the electric motor supports is fixed with not less than four screws for the base of the fluid receptacle.

The part of the table fountain, designed for arrangements of decorative plants, is situated on the upper side of the elevated conical overflow, covering it completely with its circular base. Its construction is to be manufactured of profiled sheet iron of different thickness and fastened by four screws into reinforced recesses on the internal side of the vertical conical conduit. The interior of the construction would be protected by particular layers resistant to the influence of water and harmful substances. On the technical drawing of this table fountain the possibility of installing the additional electric lighting has not been indicated. From the analysis of Tesla's constructive solutions, applied on the models of fountains from the previous period, it may be concluded that a certain system of electric lighting could also be realized in the interior of the fluid receptacle. In such a case the electric lighting would consist of two or more light-bulbs of different power and colour, placed horizontally or vertically under the glass bottom of the fountain receptacle. The light-bulbs installed in this way and automatically controlled would by their multi-coloured light enlighten the agitated fluid in the fountain receptacle (Fig. 6).

The third and fourth variant of Tesla's new table fountain are shown in partial vertical section, both of them on the same technical drawing made on January 24th and 25th, 1917 (Fig. 7). Technical solutions applied to these new models result from Tesla's modifications made on the previous designs of the first and second variants of his table fountain from this period. The design of the table fountain, marked as "Drawing C", by its technical solutions resembles the second variant of the same Tesla's model, shown on the drawing of January 22nd, 1917. The shape and height of the fluid receptacle in both variants are very similar and defined for each model according to the mode of installation of their constructive parts. In the third variant of the table fountain, the electric motor is placed vertically in the interior of the fluid receptacle, under the elevated conical overflow. Tesla's pump, with discs for propelling the fluid up to the level of horizontal, broadened conical overflow, is attached to the extension of the shaft of the electric motor approximately at half of its height. The working part of the pump consists of four discs. The extended part of the shaft is supported by two ball-bearings, one of them at the base, and the other one at the top. The ball-bearing at the shaft extension base is fixed to the additional horizontal reinforcement of the fluid receptacle. The other one is fixed to the inner side of the central nave's upper part. The construction of the central nave in its lower part is fixed to the construction of the vertical conical conduit.

Its height is lower than the projected height of the central nave in the second variant of this model of fountain. The space above the horizontal reinforcement in the fluid receptacle had to be specially protected from the breakthrough of the liquid, but this is not clearly defined on the drawing of the fountain. The protection of the opening at the base of the elevated conical overflow, protruded by the extension of the electric motor shaft, is solved by fitting the appropriate rubber sealing and a special screw-cap for its fastening.

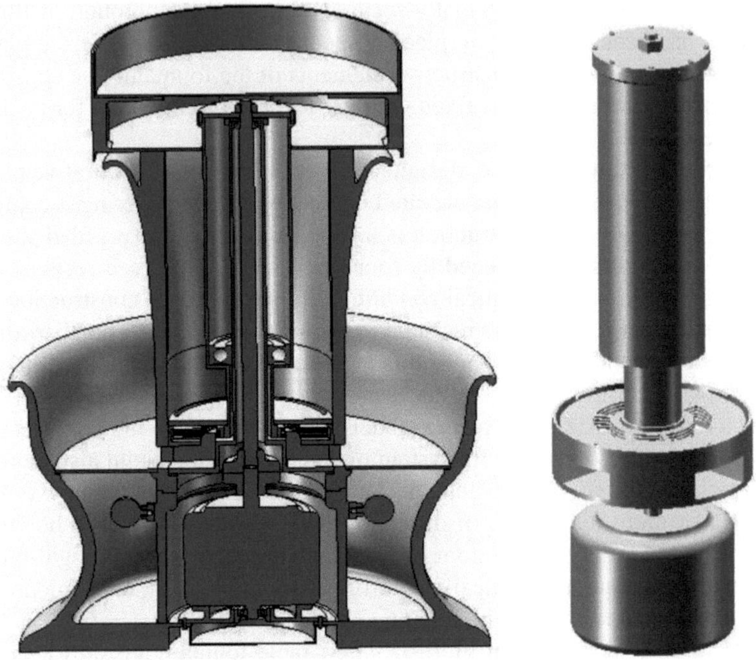

Fig. 6 Vertical section of the second variant of Tesla's table fountain and its propelling system with an electric motor (CAD models)

The electric motor is fixed with screws to the base of the fluid receptacle. Arrangements of decorative plants would be placed in a circular vessel, placed on the upper side of the elevated conical overflow. According to the design, the base of the vessel completely covers the conical overflow. It would have probably been made of profiled sheet iron of different thickness, fastened with four screws to especially reinforced recesses on the inner side of the vertical conical overflow.

In the third variant of the table fountain the installation of the electric lighting has been planned. The lighting system would consist of two or three light-bulbs of different power and colour, placed horizontally under the glass bottom of the fountain receptacle. By their multi-coloured light the installed light-bulbs would enlighten the agitated fluid in the fountain receptacle, making the operation of Tesla's table fountain even more picturesque and attractive (Fig. 8).

In the fourth variant of the table fountain, marked as "Drawing D", the propelling assembly is placed vertically along the central axis in the interior of the elevated conical conduit (Fig. 7). In order to secure the operation of the electric motor, it is installed in a separate cylindrical metal construction of variable cross section.

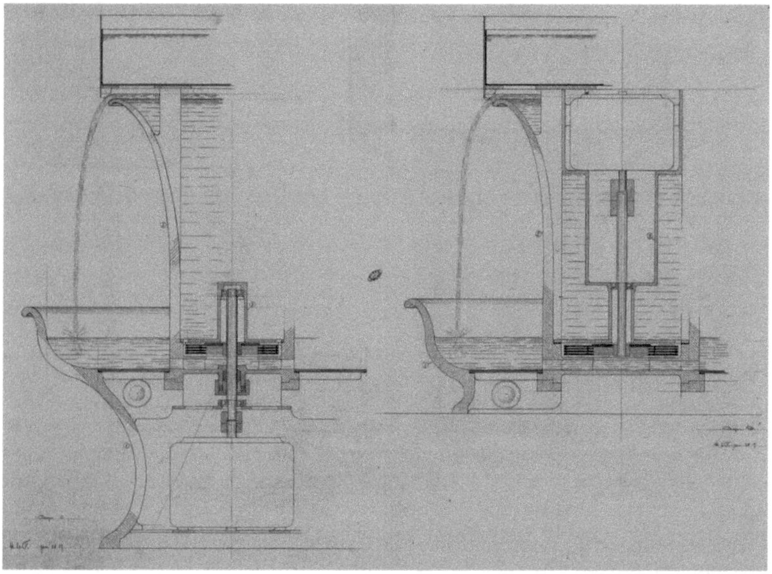

Fig. 7 Drawings "C" and "D", illustrating the third and the fourth variant of Tesla's table fountain; New York, January 24th and 25th, 1917. (Nikola Tesla's personal archives, MNT, V,I-4,92)

Fig. 8 The third variant of Tesla's table fountain and its propelling system with an electric motor (CAD models)

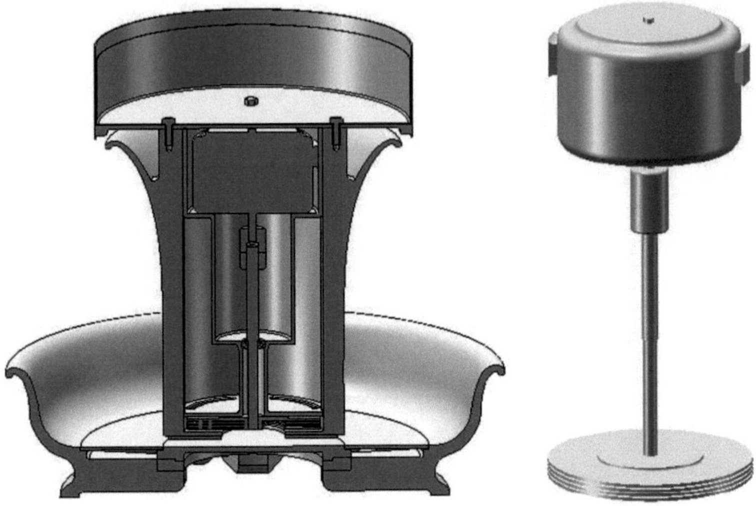

Fig. 9 Vertical section of the fourth variant of Tesla's table fountain and its propelling system with an electric motor (CAD models)

Though on the preserved drawing the spots and the way of direct fixing of some fountain parts are not clearly indicated, it may be assumed that the electric motor is fixed laterally to the inner side of the protective cylindrical construction, this construction being fixed both in its lower part and on the top to the inner side of the elevated conical conduit. Interconnection and fixation of the vertical conical conduit and the fluid receptacle is not clearly indicated neither. The lower part of the conical conduit leans on the profiled ring; this ring is at the same time the horizontal reinforcement of the fountain receptacle and their interconnecting element. The outer diameters of the profiled ring and the base of the vertical conical conduit are equal (Fig. 9).

Tesla's pump with discs is installed at the lower part of the electric motor, at the end of extended part of its shaft. The performance of the pump supplies the required quantity of liquid for the proper operation of the fountain. According to its design, the working part of the pump consists of four discs. The extended shaft of the electric motor leans on a ball-bearing, situated approximately at the half of it, in order to hinder its asymmetrical rotation during the fountain operation. The ball-bearing is fixed to the inner side of the protective cylindrical construction at the top of its second narrowing. The breakthrough of liquid into the protective construction is prevented by special rubber seals at the interconnections with other parts of the fountain.

By its shape and height the fluid receptacle of this model mostly resembles the receptacle illustrated in the first variant of the same table fountain. The vessel for arrangements of decorative plants, by its design, set-up, materials used and the way of protection, is similar to the vessels designed in the second and third variants of the same fountain model. The electric lighting system would consist of two or three light-bulbs of different power and colour, placed horizontally under the glass bottom of the fountain in the interior of the fluid receptacle.

By this alternating and multi-coloured light the installed light-bulbs would enlighten the agitated fluid in the fountain receptacle. The automatically controlled electric lighting would additionally increase the visual effects of this fountain, making it even more attractive and interesting for the observer's eye.

3.2 *Tesla's Disc Pump Feasibility*

At the end of January 1917, Tesla directed his investigations to the search of best possibilities for applying his patented disc pump in the construction of the new fountain models. In the period from January 29th to January 31st, 1917, he worked on the definition and design of complete propelling assemblies and on the conditions of their protection against water. He also considered the use of adequate piping in order to achieve a more uniform directing of the propelled liquid jet, resulting from the operation of his disc pump.

On the back of the technical drawing dated January 22nd, 1917, his new solutions and design improvements considered in that period are drawn in pencil (Fig. 10). The drawing presents:

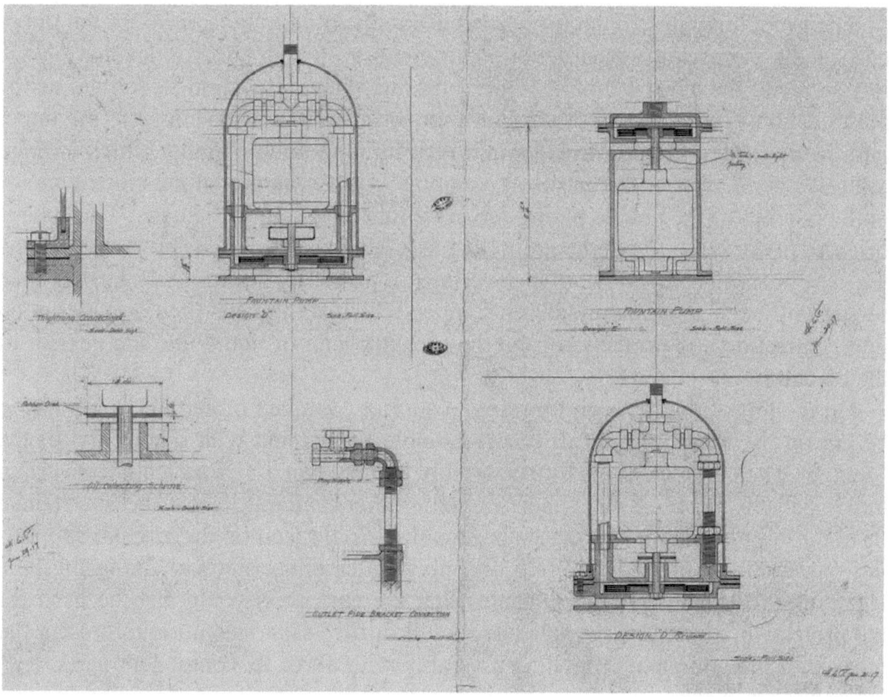

Fig. 10 Variants of installing Tesla's disc pump and solutions for its protection from the breakthrough of water; New York, January 29–31, 1917. (Nikola Tesla's personal archives, MNT, V,I-4,91B)

- two installation variants of the sub-assembly of the propelling electric motor with Tesla's pump;
- three general solutions for protecting the propelling modules against the break-through of liquid into their interior;
- details of piping and their interconnections.

The drawing has been divided by one horizontal and one vertical line into three unequal parts. In this way the content has been clearly defined according to the date of its creation.

The first of the three general solutions for the complete propelling subassembly, together with its protective covering, has been presented on the left side of the drawing. The first solution was drawn on January 29th, 1917 and titled "Fountain pump—Drawing D". The propelling assembly is presented in the vertical section, with three details showing the interconnection of its parts and elements. The supporting circular construction of the propelling assembly is made of two parts, each one to be shaped on the lathe. The upper part of the supporting construction has the purpose of directing the liquid jet into the piping construction, of fixing the propelling electric motor and piping, and of reliably securing the interior of the propelling assembly from the breakthrough of liquid. The lower part has been designed as the casing for Tesla's pump, with openings for directing the required quantity of liquid from the fountain receptacle to its rotating discs.

The propelling electric motor has been fixed vertically by screws for the upper part of the supporting construction. The pump with four discs is located on the lower side of the electric motor, at the end of its shaft. The liquid jet formed by the pump action is displaced first through two vertical circular openings in the upper part of the supporting construction and then into the double piping which is fixed to these openings as an independent assembly. Above the propelling electric motor both branches of the double piping form one tube of a certain diameter. This tube is oriented towards the top of the propelling assembly, protruding its protective covering. The opening on the protective covering, protruded by the vertical tube, has been secured by a specially manufactured stopper, which at the same time provides good interconnection and protection from the breakthrough of liquid into the interior of the assembly.

On the top of the tube, on its external surface, a thread of certain length has to be cut on the lathe in order to secure reliable connection with other parts of the fountain. The complete propelling assembly is surrounded by a protective covering, most probably made of steel sheet or some other material; it may be ascertained from the way of manufacturing its base which would lean on the circular supporting construction of the propelling assembly. Interconnection and fixing of some elements of the construction indicates that the cast glass could also be used for the protective cover. The use of glass would fulfill Tesla's intention to present the interior of the complete propelling assembly with all of its component parts to an interested observer.

The technical drawing marked as "Fountain pump—Design E" illustrates the vertical section of the second variant of the complete propelling assembly, shown on the upper part of the right side. The drawing represents Tesla's work of January

30th, 1917. In this variant, the disc pump is placed above the vertically placed propelling motor. As a consequence of their different position, the parts of the supporting construction had to be adapted to the new design. Now the upper part of the supporting construction is used to displace and direct the liquid jet to the central opening on its top. At the inner side of the central opening, along its total height, a thread has to be cut on the lathe, to enable its connection and reliable fixing to other parts of the fountain. The lower part of the supporting construction has been changed, and the electric motor has been placed into a separate compartment, rendering its protection simpler and more efficient. The propelling electric motor has to be fixed by a circular metallic support for the construction base. A segment of the lower part of the construction has been defined as the casing of the pump, with particular openings to direct the required quantity of liquid toward the four rotating discs of Tesla's pump.

The third general solution of the fountain's propelling assembly results from the application of changes and modifications developed while designing details of the first variant of the system. The propelling assembly together with its protective cover, with changes in construction, has been shown by its vertical section on the part of the drawing marked "Design D—Revised", on the lower part of the right side. The drawing is dated January 31st, 1917.

With the design and development of the third variant of the propelling assembly the first period of Tesla's work and engagement in the field of fountains in 1917 was completed. After a delay of several months he will continue his activity in this field of mechanical engineering and get on with his temporarily ceased investigations.

3.3 *Unique Designs of Table Fountain Models*

In the period from May 11th to June 1st, 1917, Tesla planned and designed new models of table fountains. Each of the four preserved drawings illustrates one model in the full view and in the vertical section. The drawings have been made in pencil on paper. These new models of table fountains are of unique design, with nice and attractive decorations on their receptacles, cascades, elevated overflows and other construction elements. All the models have the same disc pump, placed in the lower part of the fluid receptacle. The propelling assembly is located vertically along the central axis of the fountain. Installation of the electric lighting is planned for almost all the models, in order to make the sight of their operation impressive, fascinating and inspirative.

The preserved drawing dated May 11th, 1917, illustrates the first of the four new table fountains from the period (Fig. 11). The shape and height of the fluid receptacle, for these models as well, are defined by the way of installation of some constructive parts and by the quantity of fluid required for undisturbed operation of all the systems. The fluid receptacle is supported by four independent glass balls, at the angle of 90° between them for the purpose of uniform leaning.

The installed assembly of the electric motor and the pump is a result of new modifications, made by Tesla on the existing construction of January 31st, 1917.

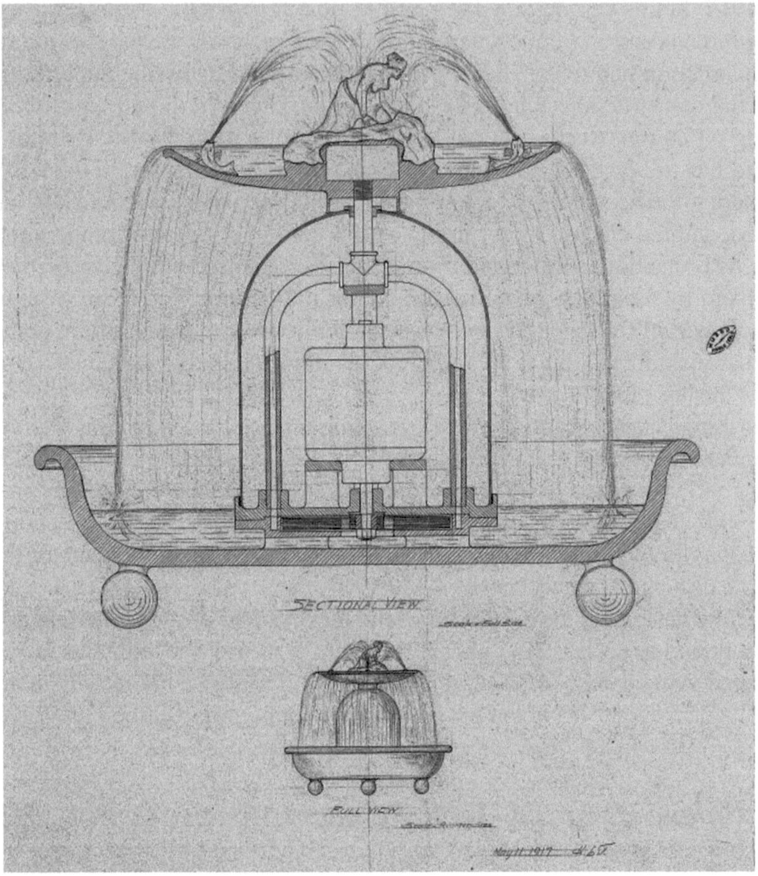

Fig. 11 Vertical section and the full view of the first model of Tesla's table fountain, New York, May 11th, 1917. (Nikola Tesla's personal archives, MNT, V,I-4,94B)

The upper part of the old supporting construction has been modified and adapted to the new requirements due to the fixing of the double piping, protective cover and propelling electric motor. The double piping elements are manufactured by bending a single pipe of a certain diameter. The place where the protective cover has to be fixed to the base of circular supporting construction has been modified, and the interconnection is rendered by sticking their lateral sides together. A rubber ring is placed between contiguous surfaces to secure more reliable sealing of the connection.

The tube which unites both branches of the doubled piping system leans on the covering of the electric motor. The complete propelling assembly, together with the protective covering, is placed in the middle of the fountain receptacle, leaning on its bottom. Its fixing has not been clearly defined on the preserved drawing. The protective covering of the propelling assembly would be made of steel sheet, as the overflow cascade has to be fixed to it. The outer surface of the protective covering would be decoratively painted and in this way protected from harmful effects of water or other liquids (Fig. 12).

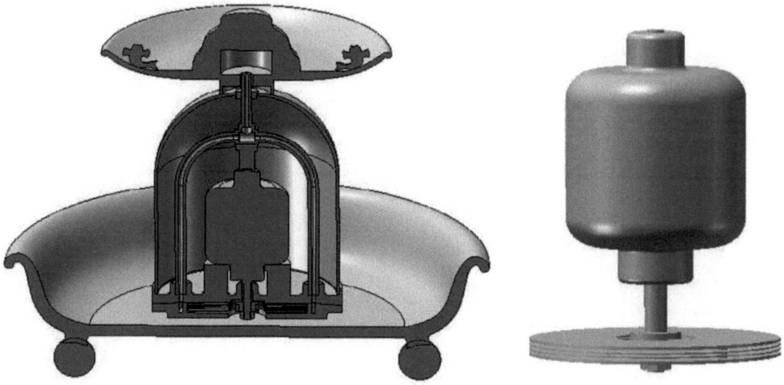

Fig. 12 Vertical section of the first design of Tesla's table fountain and its propelling system with an electric motor (CAD models)

The overflow cascade has been designed as an attractive artistic composition made of figures of a girl and several stylized fishes. In the centre of the cascade there is a rock with a figure of a girl. The girl is slightly bent, with hands leaning on the elevated part of the rock. The jets of fluid outflowing vividly from stylized fishes' mouths are sprayed at her figure. It may be assumed that the overflow cascade is formed out of four stylized fishes, situated at the angle of 90° between them. The liquid displacement from the top of the uniting tube has not been defined i.e. drawn in the vertical section of the fountain. The overflow cascade vessel is concave and filled with water or some other liquid to its edge. Water sprayed from vivid jets will elevate the level in the vessel, causing its overflow and forming a uniform circular water curtain. Electric lighting installation has not been designed for this model of the table fountain.

Another technical drawing, dated May 11th, 1917, illustrates Tesla's table fountain with two overflow cascades (Fig. 13). This fountain is distinguished both by its elegant and attractive construction and its decorations. The fluid receptacle differs in its height and shape from the receptacle of the first fountain from this period. Its outer surface is amply decorated with horizontal and vertical stylized elements, expressing it in an artistic form.

The power transmission from the electric motor to the discs of Tesla's pump is based on a partially altered solution, illustrated on the drawing titled "Fountain pump—Drawing E", dated January 30th, 1917. The propelling electric motor is placed in the interior of the fluid receptacle, under the vertical, elevated overflow. According to the drawing, the working part of the pump consists of six discs of particular design. The fountain has two overflow cascades: the upper one, on the vertical, elevated overflow, and the lower one shaped as a spherical sector. The elevated overflow has a constant diameter along its height, thus differing from conical overflows of earlier designs. The other cascade, i.e. the spherical sector is situated under the vertical overflow, at half of its height, forming with it a unique construction of the fountain. It consists of 12 equal parts filled with tempered glass with lower surface in relief.

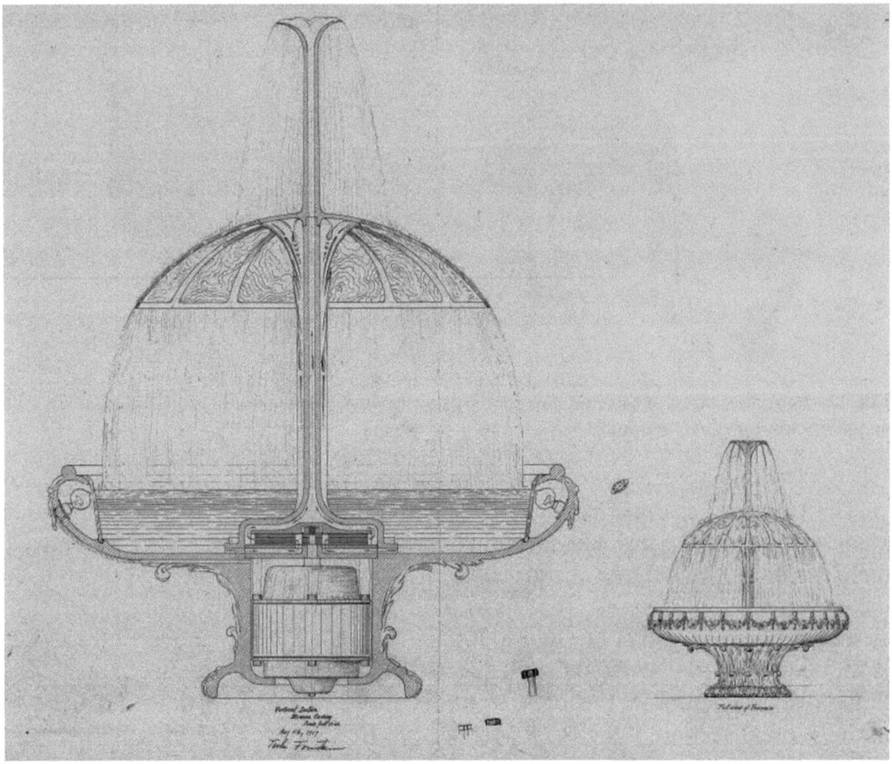

Fig. 13 Vertical section and full view of the second model of Tesla's table fountain; New York, May 11th, 1917 (Nikola Tesla's personal archives, MNT, V,I-4,93)

The base of the jointed construction of the overflow cascades is intentionally made as a casing for the propelling pump, with separated openings for directing the liquid from the fountain receptacle to the rotating discs. The pump would propel the liquid jet to the top of the vertical overflow, from where it would form the first circular water curtain. The water curtain formed in this way would fall to the spherical sector, and flowing along its glass surface would form another circular water curtain at its edge.

Such cascades of circular water curtains could hardly be realized. When the liquid of the first water curtain falls to the second cascade it would cause whirling and turbulence. The planned length of the spherical sector along which the whirling liquid flows downwards is not sufficient to calm the turbulence and render its laminar flowing. Accordingly, the formation of a new regular circular water curtain cannot be expected at the edge of the second cascade.

Installation of the automatically controlled electric lighting has been planned for this model of the table fountain. The lighting system would probably consist of several light-bulbs of different colour, placed horizontally on the inner side of the fluid receptacle. A glass partition between the installed light-bulbs and the fluid in the receptacle protects this part of the receptacle, top to bottom, from the breakthrough of the fluid.

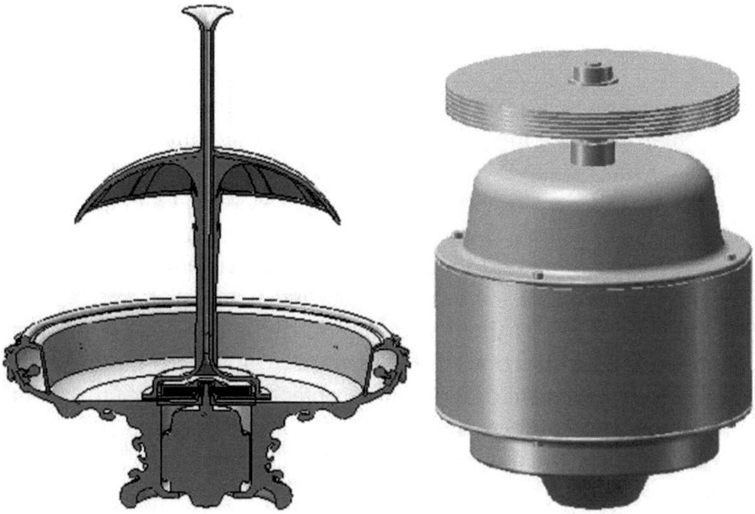

Fig. 14 Vertical section and the propelling system of the second variant of Tesla's table fountain (CAD models)

From the arrangement of the drawn light-bulbs it might be concluded that an even number was planned, but it has not been indicated exactly (Fig. 14).

The table fountain design shown on the technical drawing dated May 21st, 1917 (Fig. 15) unifies several solutions used by Tesla in his previous models.

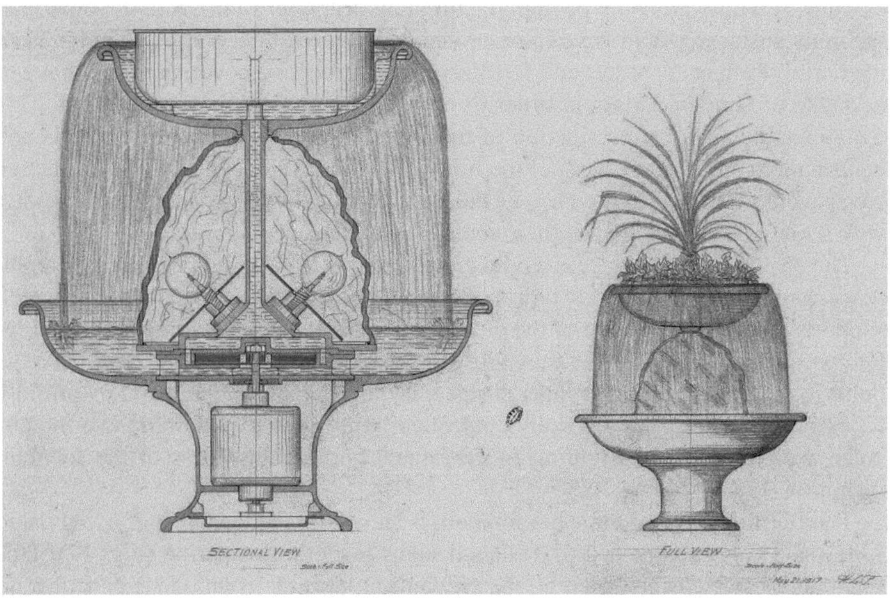

Fig. 15 Full view and vertical section of the third model of Tesla's table fountain; New York, May 21st, 1917 (Nikola Tesla's personal archives, MNT, V,I-4,94)

The fluid receptacle of the third fountain resembles mostly the same element of the second model, designed 10 days earlier. The outer surface of the vessel is not amply decorated with stylized elements, but the constructive details are characterized by their decoratively profiled edges. The propelling assembly of the electric motor and pump depicts a modification of Tesla's design illustrated on the drawing dated January 30th, 1917.

The upper part of the supporting construction is altered and adapted to the requirements of the new table fountain. The design alterations are made because of the way of directing the liquid jet to the top of the overflow, inserting the protective covering and fixing of the vessel for arrangements of decorative plants. The propelling electric motor is fitted vertically in the interior of the fluid receptacle under the overflow construction. It is supported by and fixed for the additional inner horizontal reinforcement of the fluid receptacle of the fountain. The propelling pump rotor consists of an increased number of discs. The pump casing is designed as an independent constructive element, fixed to the base of the fluid receptacle. Its fixing has not been clearly defined on the preserved drawing. A rubber ring is inserted between the contiguous surfaces of the pump casing and the receptacle base, for the purpose of sealing the connection and preventing the water breakthrough towards the propelling electric motor. The construction of the vertical overflow with protective cover is fixed to the upper side of the pump casing. The vertical, elevated overflow has a constant inner diameter along its whole height.

The displaced liquid jet fills the overflow cascade to the top of its edge, from where it flows over making a circular water curtain. The overflow cascade is concave and in its centre there is a vessel for the arrangement of decorative plants. The vessel is circular and its height slightly exceeds the level of liquid in the overflow cascade. It would probably be manufactured by deep extrusion. Both the outer and the inner surfaces would have to be protected by special decorative coatings from the harmful effects of water and fertilizers. The protective covering would have to be made of tempered glass in order to obtain a transparent surface in relief. This would enable additional refraction of multi-coloured light in many directions and contribute to the attractiveness of the full view of the table fountain. The protective cover would be fixed to the base of the vertical overflow by a steel ring, together with a rubber sealing between their contiguous surfaces.

The electric lighting system would consist of two or four light-bulbs of different power and colour. The electric light-bulbs would be installed on separately designed supports on the construction of the elevated overflow. Thin metallic plates with reflective surfaces placed under the bulbs would additionally direct the light from the bulbs to the protective cover inner surface. By their multi-coloured and manifoldly refracted light these automatically controlled light-bulbs would illuminate the circular water curtain, contributing to the beauty and attractiveness of the fountain operation (Fig. 16).

The fourth variant of the table fountain is shown on the technical drawing dated June 1st, 1917. It is the last of Tesla's designs made in this period (Fig. 17). This fountain is distinguished from all the previous models by its unique technical solution of forming a circular water curtain. The water curtain is not formed by the flow

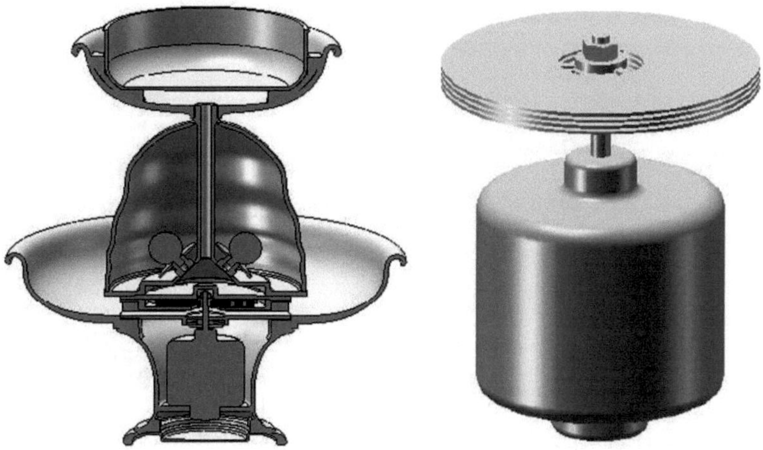

Fig. 16 Vertical section and the propelling system of the third variant of Tesla's table fountain (CAD models)

of liquid over the edge of an elevated vessel, but by displacement of a constant fluid jet through a narrow groove cut in the circular tube. This tube is placed on the inner side of the fountain receptacle along its full circumference, just under the top. The circular water curtain would spray the outer surface of the glass spherical sector and flow down to the fluid receptacle. The interconnection of the pump casing outlet with the circular tube for directing the displaced liquid has not been indicated on the drawing.

The constructive details on the outer surface of the fluid receptacle are clearly defined by decoratively profiled edges.

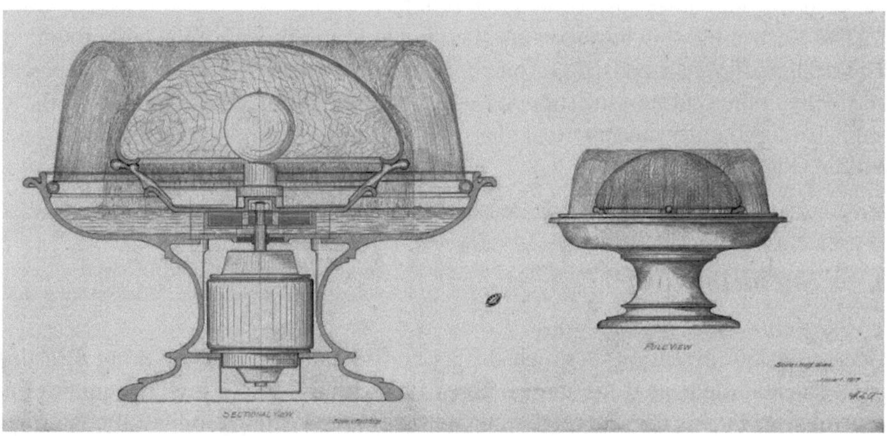

Fig. 17 Full view and vertical section of the fourth model of Tesla's table fountain; New York, June 1st, 1917. (Nikola Tesla's personal archives, MNT, V,I-4,95)

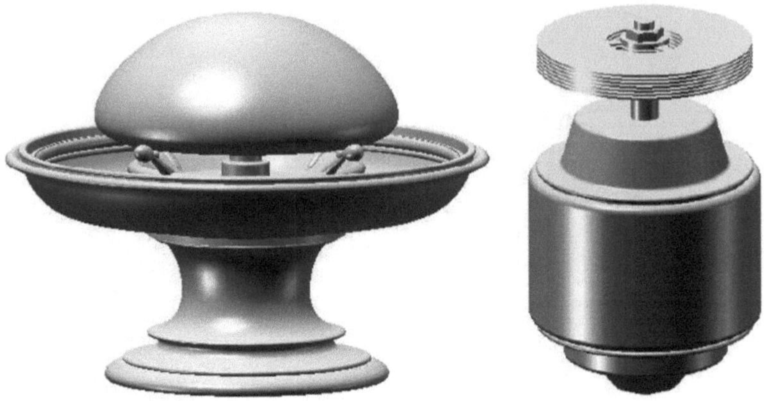

Fig. 18 Full view and the propelling system of the fourth variant of Tesla's table fountain (CAD models)

The propelling electric motor is placed vertically in the interior of the fluid receptacle, under the support of the protective cover and the casing of Tesla's pump. The motor is supported by the additional reinforcement of the fountain receptacle and fixed to it. The pump rotor consists of several thin discs. The pump casing is fixed to the base of the fountain receptacle. On the upper side of the casing there is a nicely shaped cast support, with protective cover and electric lighting of the fountain. The cast support is shaped like a circular vessel which is by its form, style and dimensions in accordance with the design of the fluid receptacle. The support vessel is reinforced on the outer side by four vertical decorative elements, disposed at an angle of 90°. These reinforcements are designed as delicate decorative elements which support the protective cover, emphasizing moderately independent form of its design. The protective cover shaped as a spherical sector would be made of tempered glass, with inner surface in a low relief.

The electric lighting has been installed in the fourth model of the table fountain. The designed system consists of one coloured light-bulb of greater power, placed vertically in the centre of the support vessel, under the protective glass cover. By its intensive light under the spherical glass sector it would enlighten the circular water curtain (Fig. 18).

4 Conclusion

One of Tesla's inventions—which differs considerably from everything else this great scientist dedicated the largest part of his creativity to—was a fountain. For several years during the second decade of the twentieth century, with some interruptions, Tesla worked in this field and developed several fountain models.

His engagement in the table fountain design during 1917 resulted in development and presentation of new and interesting constructions of these devices and the use of his patented disc pump for displacement of the required liquid quantities and uniform operation of the new models. However, it is regrettable that the designed table fountains have not been put into practice and their author could not make any commercial success on the market with them.

After several unsuccessful attempts to achieve and technically realize one of his imagined and designed fountains, Nikola Tesla tried to do it one last time at the end of the second decade of the twentieth century. In the period from 1919 to 1921 he designed three new table fountain models, very similar to one another. He managed to realize one of them and to test its working model in laboratory conditions. The making and testing of this working model is confirmed by five surviving photographs, recording individual segments of its operation. The other two table fountains were not realized, although construction parts of one of them served as a basis for building a working model. The expected commercialization of the conceived projects, as well as the possibility of achieving a certain financial success, failed to happen one more time.

Numerous documents in the archives preserved in the Nikola Tesla Museum demonstrate Tesla's great engagement in this—at first glance less significant—segment of his creative activity. Numerous sketches and technical drawings, a long list of calculations regarding fountain elements dimensions, fluid flow parameters and motor characteristics, as well as precisely defined construction modifications and improvements, provide enough evidence that Tesla dedicated all his inventive energy and passion to this segment of his opus.

At the beginning of 2003, the former Director of Nikola Tesla Museum, Marija Šešić, created an expert team in the field of machine engineering, in order to perform the practical evaluation of selected solutions. After several years of research, when the part of the material related to the water pump was systematized, and with generous support of the curator of the Nikola Tesla Museum, Mr. Bratislav Stojiljković, it was decided that the prototype to be realized should be based on the pump model from 1912. The Company CPS-CAD Professional Systems from Belgrade, applying the latest CAD/CAM solutions, developed and partly modified the pump construction, and created a fully functional prototype. Under the supervision of Professor Miroslav Benišek, the pump has been tested at the Faculty of Machine Engineering in Belgrade, and it was established that the measured parameters corresponded to the results that Nikola Tesla published in his patent documentation.

Ninety-four years after Tesla applied for a patent of his fountain, in Belgrade, on a green space across the Nikola Tesla Museum, Tesla's fountain with two overflow cascades was realized, based on a model designed in 1917. The implemented model symbolizes the extent of Tesla's dedication to inventions in the field of machine engineering from the beginning of the twentieth century to his death. However, a careful analysis of the finished fountain operation cannot disregard the fact that if the heart of the fountain consists of Tesla's inventions in machine engineering, then its brain also bears the signature of Tesla's patents—frequency regulator for the number of revolutions of the electromotor, regulation relay, polyphase motor etc.

References

Jovanovic B (2001) Tesla—duh, delo, vizija. Freemental, Belgrade
Marincic A (1994) The Life of Nikola Tesla. Serbian Academy of Science and Arts, Nikola Tesla Museum and Museum of Science and Technology, Belgrade
Marincic A (2006) The Works of a Genius. Serbian Academy of Science and Arts and Committee for celebration of the 150th anniversary of Nikola Tesla's birth, Belgrade
Marinkovic A, Zrnic N, Stojiljkovic B, Vulicevic M (2012) Tesla's Research in the Field of Mechanical Engineering Focused on Fountains Design. In: Explorations in the history of machines and mechanisms. Proceedings of HMM2012, Springer, Dordrecht, pp 395–420
Nikola Tesla's personal archives, Nikola Tesla Museum, Belgrade
Stojiljkovic B (2012) Rad i stvaralastvo Nikole Tesle u oblasti fontana—Mogucnost izrade 3D animacija i interaktivnih modela. Habilitation thesis for the rank of senior curator. Nikola Tesla Museum, Belgrade
Stojiljkovic B, Vujovic M (2007) Tesla's Fountain—Creating an Object of Beauty. Nikola Tesla Museum, Belgrade
Tesla III Millenium (1996) Proceedings of the Fifth International Conference, 15–18 October 1996. Institut Nikola Tesla, Belgrade

The Design of Timber Trusses in Italy: From Empiricism to Structural Analysis

Emanuele Zamperini

Abstract Until the mid-nineteenth century timber trusses were empirically designed thanks to centuries-old experience. Since the early nineteenth century the theory of statically-determinate trusses had been elaborated for the analysis of metal trusses, but this could not be used for timber trusses, because loads were not applied in joints, a hypothesis required to ignore static indeterminacy. In the late 1850s a specific analysis method was developed for timber trusses: it considered rafters as continuous beams on rigid supports, thus it did not satisfy constitutive and strain-displacement equations. However, theoretical developments of the 1870s allowed the exact calculation of statically-indeterminate structures, thus engineers often continued to resort to empiricism, or simplified methods.

Keywords Timber trusses · Theory of structures · Statically indeterminate structures

1 Introduction

Roof structures have always been one of the areas of greatest technical effort in the construction sector, and therefore one of the most important fields of experimentation. Historiographic studies that have analyzed wooden trusses have focused mainly on the long period of time ranging from their origins to the birth and diffusion of metal structures, ignoring or neglecting their subsequent evolution; indeed historians of construction technology often considered that in 1850–1950 European culture of wooden constructions had been a backward culture without any progress. Moreover the contributions that wooden construction technique has brought to the birth of metal construction have been poorly analyzed.

Therefore, focusing on the Italian case, this paper tries to bridge this historiographical gap by deeply analysing this neglected chapter of history of construction, and by highlighting the complex system of relationships that developed between the cultures of timber and metal constructions.

E. Zamperini (✉)
Department of Civil Engineering and Architecture, University of Pavia, Pavia, Italy
e-mail: emanzamp@yahoo.com

R. Pisano (ed.), *A Bridge between Conceptual Frameworks*, History of Mechanism and Machine Science 27, DOI 10.1007/978-94-017-9645-3_27

2 Italian Traditional Timber Trusses Before the Nineteenth Century

Until the eighteenth century, in most buildings roof coverings were supported by wooden structures, whose complexity depended on the depth of the buildings and the presence or absence of internal walls.

Before the beginning of the nineteenth century, timber trusses were designed by architects and carpenters with redundant members that were not always suitable from a structural point of view; the well-established centuries-old tradition in this kind of constructions, and their standardisation with regard to the span and the distance to cover, allowed designs to be made on an empirical and intuitive basis (Rondelet 1810). Therefore they often have static schemes that are difficult to be analysed with the modern theory of structures.

Italian tradition differed much from the central and northern European one. Italian architecture was characterised by lower sloped roofs, and presented peculiar constructive solutions. Moreover in Italy a plethora of local traditions was present with regard to the arrangement of the members of the trusses, to the details of the carpentry joints, to the iron elements used for their reinforcement, and—above all—to the relation between trusses and other structural elements of the roof.

Two common solutions were employed in buildings without internal walls: the so-called Lombard roof (more diffused), and the Piedmontese roof. Both of them had triangular trusses lying directly on walls or supported by interposed corbels, and their extremities were enclosed in masonry. In Lombard roofs purlins—which bore joists and laths to support the covering—laid directly on trusses, and usually charge the rafters with loads placed at close range, so that they should be better considered as a span load. In Piedmontese roofs the trusses supported a ridge beam; in the span between two trusses one or more couples of rafters were borne by the ridge beam and in their turn they supported joists and laths.

Furthermore in Italian trusses, rafters were usually made of a single continuous squared log from the joint with the tie-beam to the ridge, and were supported at an intermediate point by secondary members (struts, etc.).

On the contrary, central and northern Europe roofs had often quite complex timber frames strongly connected to the supporting walls; secondary elements (purlins or rafters) were linked to the principal trusses with carpentry joints (halved or dovetail joints, tenon-mortise, etc.) so as to form a fully interconnected whole with longitudinal and transverse bracings.

The diffusion of Italian Renaissance and Neoclassical architecture in the rest of Europe had—in recurring phases—a great influence also on construction techniques. Central and northern Europe traditional roof structures were unsuitable for wider spans and lower sloped roofs typical of classical architecture; therefore "Italian style" timber trusses spread over Europe through treatises and through the work of Italian architects. Although the Italian structural conception of a roof necessarily spread in the rest of Europe, nevertheless local craftsmen kept using carpentry details in their tradition, thus creating original and hybrid constructive solutions (Yeomans 1986, 1992; Wünnerman et al. 2012).

3 The Technological Context and the Evolution of Trusses

In Europe the transition from the eighteenth to the nineteenth century marked a fundamental change in the manufacturing, building, and scientific context. In sequential stages during this period, many factors changed the mind-set and practical attitude of carpenters, builders, architects and engineers.

On the one hand, since the beginning of the century the classifying attitude typical of Enlightenment encyclopaedism and engineering positivism led—at least in treatises—to a rationalisation of the various types of traditional trusses and to a standardisation of the link between type and span, yet this brought a significant loss in the variety of carpentry culture. On the other hand, many issues were so complex that a noteworthy evolution of timber trusses took place in Italian constructions. The most important of these issues were: changes in metallurgical industry, evolution of systems for processing timber and metal fastenings (nails, screws, bolts, straps, spiked plates) from manufacturing to machine working, and spread of the theories of strength of materials and structural analysis.

4 The Developments of Metallurgical Industry

Since the late eighteenth century, metallurgical industry advancements—chiefly developed in England (Pedrocco 2004a)—brought profound changes in production and processing of building materials. With respect to trusses, this led to simplification of wooden connections, thanks to a wider use of wrought iron components for reinforcement of joints; moreover it allowed the employment of entirely metallic members instead of those made of timber.

In the first mixed timber–iron trusses the only change was that wrought iron rods were substituted for timber tie-beams. This technical choice could be possible only in roofs that used "Italian style" trusses; in fact, unlike in most of the northern European roofs—if we omit self weight—the tie-beam of these trusses are quite often subject only to tensile forces and not to bending. Later more articulated mixed trusses appeared: the most widely diffused is the Polonceau truss, also called Fink truss in English and German-speaking countries. The French engineer Camille Polonceau (1840) (1813–1859) conceived this truss in 1840 combining timber, cast iron and wrought iron, and using each material to the best of its peculiar strength characteristics: timber is employed for members subject to compression and bending, cast iron for the simply compressed ones, wrought iron for the stretched rods (Fig. 1). Furthermore, since the end of the eighteenth century the availability of

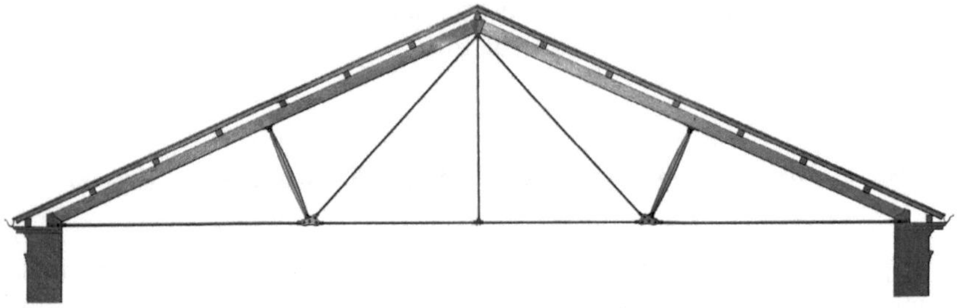

Fig. 1 A 21 m span Polonceau truss designed for the roof of a military riding school. (Castellazzi 1863)

increasing amounts of both cast and wrought iron, and their progressive decrease in price brought also the realisation of entirely metallic structures, at first mainly for the construction of bridges, afterwards also for large span roofs.

5 Structural Analysis

Up to the end of the eighteenth century the emerging theories of structure and of strength of materials had limited practical application and engineers still preferred to use empirical formulas (Timoshenko 1953, p. 66). However in the first decades of the nineteenth century the application of theoretical investigation to the solution of practical problems grew, particularly among French engineers, thanks to the higher technical education provided by the *École Polytechnique* and by the *École nationale des ponts et chaussées*. In addition, the studies (Navier 1826) by Claude-Louis Navier (1785–1836) played a key role in the spread of the theory of strength of materials in practical design of structures.

Even if the first applications of theory of structures to the design of timber trusses goes back at least to the 1820s (Cavalieri San Bertolo 1826), they were based on very simplified methods and limited to exceptional cases. Instead, in the absence of specific building traditions on which to base the empirical design, the spread of metal structures—especially in bridge construction—required employment of existing theories and the development of new appropriate structural analysis methods.

6 A Parallel Path: Theories for Trussed Bridges

After an initial phase in which cast iron arch bridges and suspension bridges prevailed, reticular structures became considerably widespread; this phenomenon was favoured by a series of outward and return cultural exchanges between Europe and North America, regions in which timber trussed bridges diffused since the eighteenth century in imitation of Swiss and Palladian bridges (Guardini 2004, pp. 29–41; Bennett 2004). The progressive diffusion of railways made it necessary to build a great number of bridges, prompting builders to realise more and more daring works. Usually they were statically indeterminate and oversized structures, but the randomness of dimensioning often caused disastrous failures.

Since reticular structures were very difficult to be empirically designed, the spread of these structures required the elaboration of specific structural theories. Thanks to the slenderness of members and to the ease of connections these structures were almost always realised with a rational sequence of triangles arranged so that applied loads acted only at joints. In this period we can identify three main and quite distinct analysis methods.

The first approach decomposes the truss in joints and verifies the equilibrium of each of them. This approach was systematically proposed by Squire Whipple (1847) (1804–1888) who analysed trussed bridges with parallel chords or with upper arched profile (Whipple 1847). The trusses he studied were statically indeterminate due to redundant members; he simply solved the problem of hyperstaticity by neglecting some of the members and then studying the resulting structure with equilibrium equations of joints.

The second approach was conceived by Johann Wilhelm Schwedler (1851) (1823–1894): he analyses trussed beams as a whole (i.e. as if they were a common beam), using bending moment and shear to find axial forces in the members of the truss (Schwedler 1851).

The third approach (also known as the method of moments) was elaborated by August Ritter (1863) (1826–1908) who simplified the calculations of forces in members by making a cross section across the truss and verifying the rotational equilibrium of one of the two parts of the truss under the effect of external loads and internal forces of the cut members (Ritter 1863).

As a result, thanks to these and many other contributors the classic theory of statically-determinate trusses was elaborated approximately by 1865 (Benvenuto 1981; Capecchi and Ruta 2010).

6.1 *The Analysis of Trusses Used in Roofs*

The requirement to apply structural analysis to metallic constructions and the growing use of iron members in timber trusses continued to create in engineers the need to overcome the empirical approach to design of timber trusses, also to train the structural sensitivity of young engineers, and to optimise the use of materials (Sachero 1864, pp. 193–194). Since the 1850s all the handbooks of applied mechanics had a chapter on the design of trusses.

Many attempts to apply the theory of statically determinate trusses to timber ones were carried out (e.g. Morin 1853; Fig. 2) but they conflicted with the peculiarity of these structures, and they lead to heavy simplifications (i.e. considering the rafters discontinuous at the supports given by the secondary members of the truss, or the loads applied only in the joints).

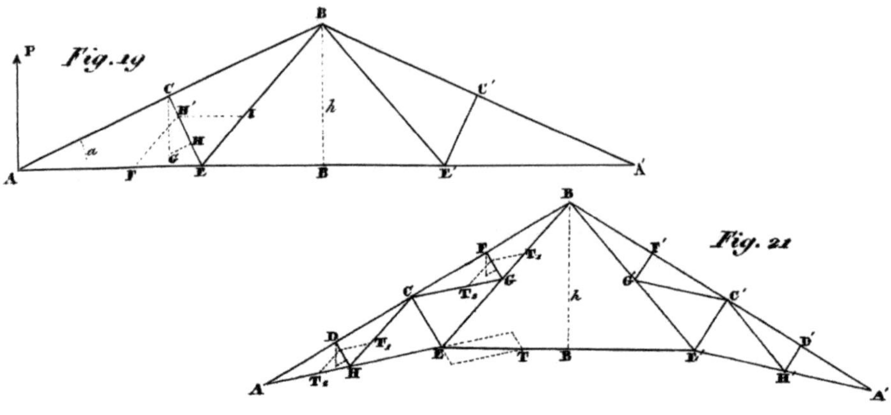

Fig. 2 Polonceau trusses analysed with the method of joints equilibrium. (Morin 1853)

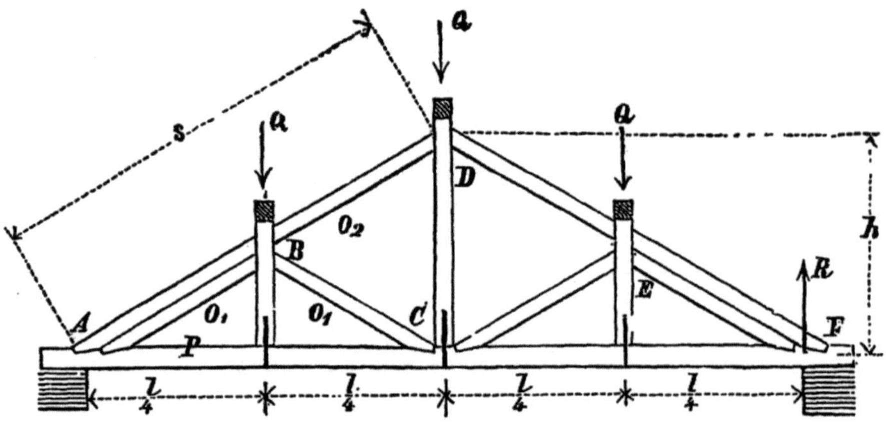

Fig. 3 German truss loaded only at joints. (Breymann 1887)

Indeed the theory of statically-determinate trusses could not be rightly employed with timber trusses because loads weren't applied in the joints, a hypothesis required to ignore the static indeterminacy of the structure. In Germany the use of this theory led to simplification of the schemes of timber trusses and to conformation of their design to the adopted hypothesis (Breymann 1887; Fig. 3).

6.2 *The Theory of Continuous Beams*

An important role in preparing the groundwork for further studies on "Italian-style" statically indeterminate trusses was played by the theory of continuous beams. A general solution for statically indeterminate beams was already elaborated in the 1820s by Navier (1826): in the analysis of continuous beams he assumed as statically indeterminate quantities the reactions of intermediate supports.

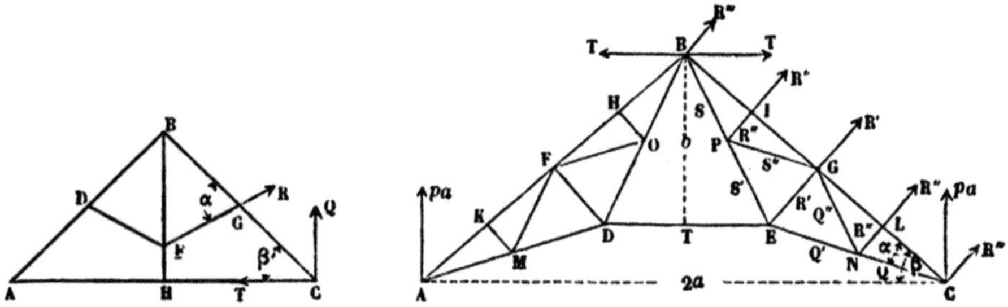

Fig. 4 Examples of trusses from Bresse (1859)

Unfortunately, with this approach to the problem, each necessary equation included all the unknowns, making this approach unsuitable for beams with more than two spans. Benoît Clapeyron (1799–1874) elaborated a further method in 1849: unlike Navier, he chose as hyperstatic unknowns the support bending moments and elaborated the three moments equation that bears his name[1]. The employment of this method made it possible to study continuous beams in quite a simple way.

6.3 *Bresse's Method for Truss Structural Analysis*

In 1859 Jacques Bresse (1822–1883) published his *Cours de mécanique appliquée*, in which he exposed a new method for the analysis of "Italian style" static indeterminate timber trusses with span loads on the rafters. Rafters were studied as continuous beams on the rigid supports given by struts or other secondary reinforcing members; the analysis of the rafters could be easily conducted thanks to the three moments equation. In this way bending moments and subdivision of span loads on joints are obtained, afterwards internal forces in members are calculated verifying analytically the equilibrium of the joints. In his handbook Bresse showed the application of his method to timber trusses with struts and to mixed Polonceau trusses (Bresse 1859, pp. 415–25; Fig. 4).

6.4 *The Spread of the Bresse Method in Italy*

While in more industrialised countries steelworks became the main choice for medium- and long-span roofs, in Italy timber and mixed iron-timber trusses continued

[1] Clapeyron's new approach to the analysis of continuous beams was elaborated in connection to his project for the railway bridge over the Seine in Asnières near Paris built in 1849, but it was first published only in 1855 by Henri Bertot. A detailed description of the events that led to the formulation and publishing of the method of the three moments can be found in Timoshenko (1953, pp. 144–146).

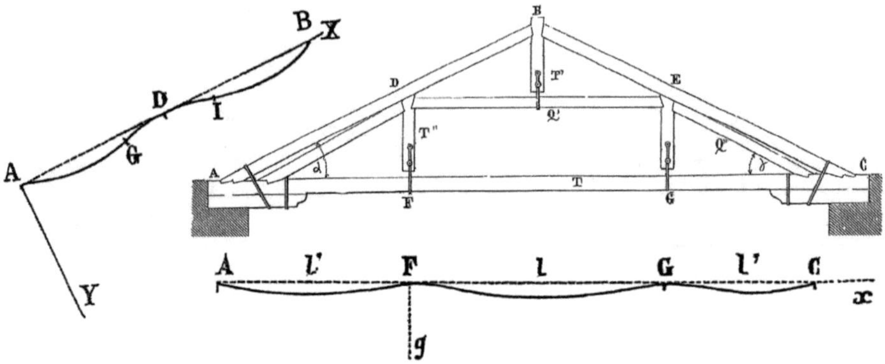

Fig. 5 Analysis of a Palladio truss by Sachero (1864)

to be widely used up to the end of the nineteenth century[2]. Therefore structural analysis of timber trusses continued to be a widely studied topic in Italian journals, and many scholars tried to systematise the subject in technical papers or handbooks. A few years after the publication of Bresse's manual, Celestino Sachero (1821–1908) wrote an essay on the strength of roof structures in which he extended the application of Bresse's method to all timber and mixed trusses most common in Italy[3] (Sachero 1864; Fig. 5).

Sachero was an esteemed scholar and the director of the Italian School of Application of Artillery and Military Engineering, the most important military institution of higher technical education (Turri et al. 2009a); his authoritativeness among the officers of the Corps of Engineers favoured the diffusion of this method in the design of military buildings (Caveglia 1878; Fig. 6), while handbooks used by civil engineers still proposed more simplified methods that schematised timber trusses as statically determinate structures.

The formalisation and subsequent spread of the graphic statics methods (Maxwell 1864; Cullmann 1866; Cremona 1872) favoured in the 1870s an easier employment both of the theory of statically determinate trusses and of Bresse's analysis technique, replacing the laborious equilibrium equations of joints with the Cremona diagram (Caveglia 1876; Fig. 7).

Cesare Ceradini (1879) (1844–1935) made a second attempt to diffuse Bresse's method in Italy. He published in the journal *Il Politecnico* an essay in which—as well as Sachero already did—he provided schemes and calculation drafts for all the most common trusses (Ceradini 1879).

[2] The reasons for this must be sought in the severe weakness of the Italian iron industry. In 1863 the production of cast-iron in most industrialised countries was much higher than in Italy: English production was 123 times higher, and French production was 35 times higher than the Italian (Degli Esposti 2004, p. 324).

[3] Although in his essay Sachero directly referred to Bresse's handbook, in the analysis of a continuous beam he resorted to the most laborious differential equation of the elastic line rather than to the three moments equation.

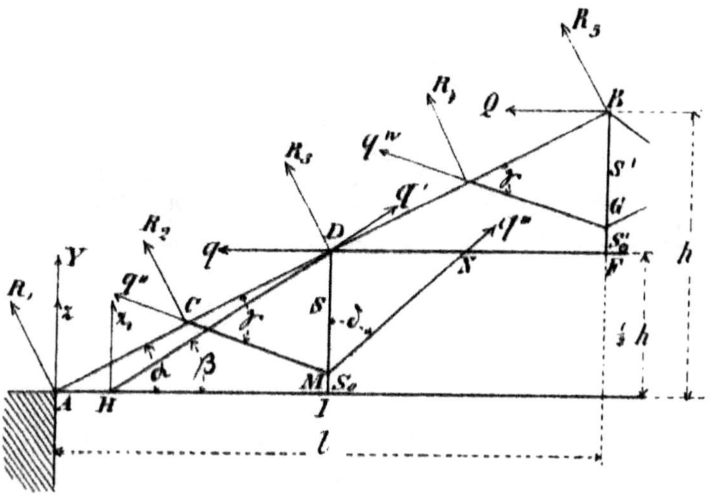

Fig. 6 Scheme for the analysis of a Palladio truss elaborated by Caveglia (1878) on the basis of Sachero's (1864) essay: the rafter is considered as a continuous beam on five supports

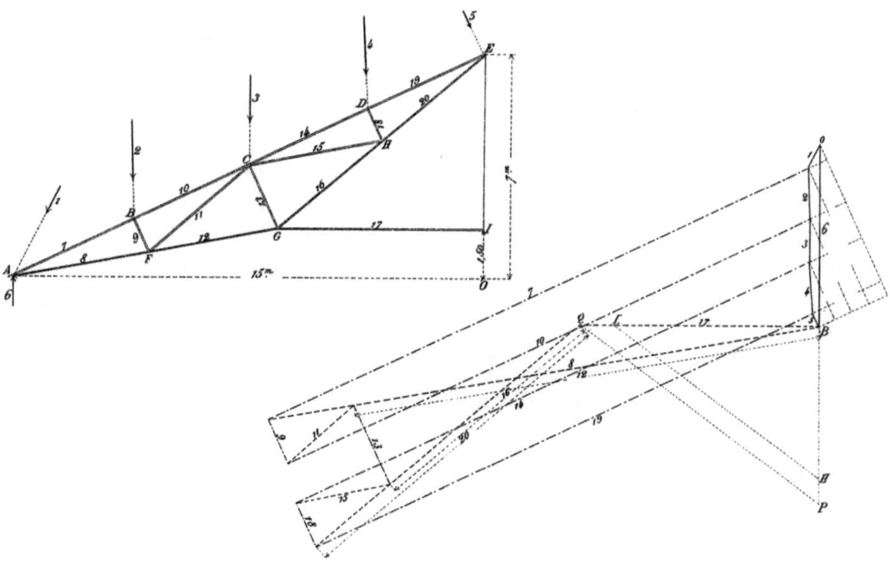

Fig. 7 Graphic analysis of a Polonceau truss by Caveglia (1876)

The fact that Ceradini—at a distance of more than 15 years—repeated the same operation of divulgation already made by Sachero, support the thesis that Bresse's method was at that time still employed almost exclusively by military engineers.

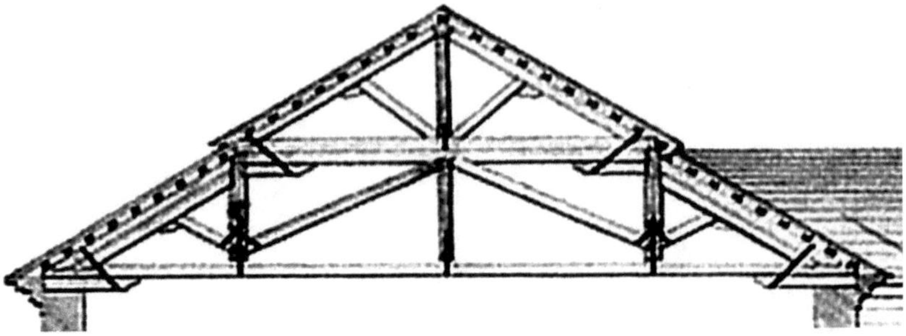

Fig. 8 Complex timber truss designed by Luigi Federico Menabrea in 1840 for the roof of a military riding school. (Fara 2011)

6.5 Effects of Structural Analysis Evolution on Italian Trusses

As clearly as the spread of the theory of statically determined trusses influenced German typologies of trusses, the application of Bresse's method conditioned the evolution of Italian trusses. The necessity of having a clear static scheme compelled designers to reduce the number of members with respect to trusses used before (e.g. the one in Fig. 8) and to modify some constructive details; the need to ignore intermediate supports deformation led to the use of thicker secondary members; in addition the spatial conception of wide and complex roofing as a whole was often disregarded, and these structures were designed as a sequence of unconnected planar trusses (Laner 1998).

New generations of engineers acquired familiarity with structural analysis, which became an integral part of their training and radically changed their mindset, diffusing a primarily analytical mind among them. On the one hand this led to a broader comprehension of the structural behaviour of timber trusses, as it is clearly expressed in this passage written by Crescentino Caveglia (1844–1922):

> …examining the parts that are added to the simplest type of truss in order to increase its span, it is easy to recognise their purpose, that is to transform the bending moments in rafters and tie-beam into longitudinal forces in the same members or in the added ones[4].

On the other hand, as it is affirmed by Edoardo Benvenuto:

> The new scientific theories on trusses have gradually eliminated the differences between one type and the other, making their reading uniform and impoverishing their lexicon: the framework simply becomes a generic composition of tension and compression members, of tie-rods and struts, and the choice between one scheme and another one or the rigorous

[4] Caveglia 1878, p. 120.

evaluation of their convenience are devolved upon the designer's inventiveness. We could say that scientific theory has grown at the expense of technical tradition[5].

6.6 *The Energetic Analysis of Structures*

Another innovation in theory of structures had important consequences on the analysis of trusses: the introduction of energetic methods.

Since 1858 Federico Menabrea (1809–1896) enunciated the so-called principle of least work to which he gave general validity:

> When an elastic system is put in equilibrium under the action of external forces, the work developed by the effect of tension or compression of the links between the various points of the system is a minimum[6].

In his essay Menabrea schematised the continuum body as a set of joints connected by a system of straight elastic members to form a statically indeterminate spatial truss with surplus rods. Therefore, despite the general validity attributed to the principle, Menabrea had reticular structures as his ideal model and he probably elaborated its principle in relation to the studies he had done for the design of trusses to be built in military riding schools (Fara 2011, pp. 20–29).

No direct practical applications of Menabrea's principle over the subsequent years are known. However, in his graduation thesis, Alberto Castigliano (1873) (1847–1884) revised the mentioned principle and expanded its field of application to many kinds of structures, among which were statically indeterminate structures with bent members; he also showed a series of applications to hyperstatic timber and mixed trusses (Castigliano 1873).

Thereafter many engineers published technical papers in which they applied this theory to various types of wooden trusses (Candellero 1890; Chiarle 1891). The complexity of the calculations needed to carry out the exact analysis of a generic truss led Candellero to simplify the various types of structure by fixing some of the geometrical parameters (e.g. the slope of rafters, the angle between struts and rafters, etc.; Fig. 9).

Nevertheless, computations were still quite complicated and this made Italian architects, civil engineers, builders and carpenters employ this method very rarely for timber and steel–timber trusses. In most cases until the mid-twentieth century they continued to design empirically with the reproduction of well-established models or with the aid of tables, to use simplified formulas (Fig. 10), or more rarely Bresse's method.

[5] Benvenuto 1981, pp. 781–782.

[6] Menabrea 1858, p. 1056.

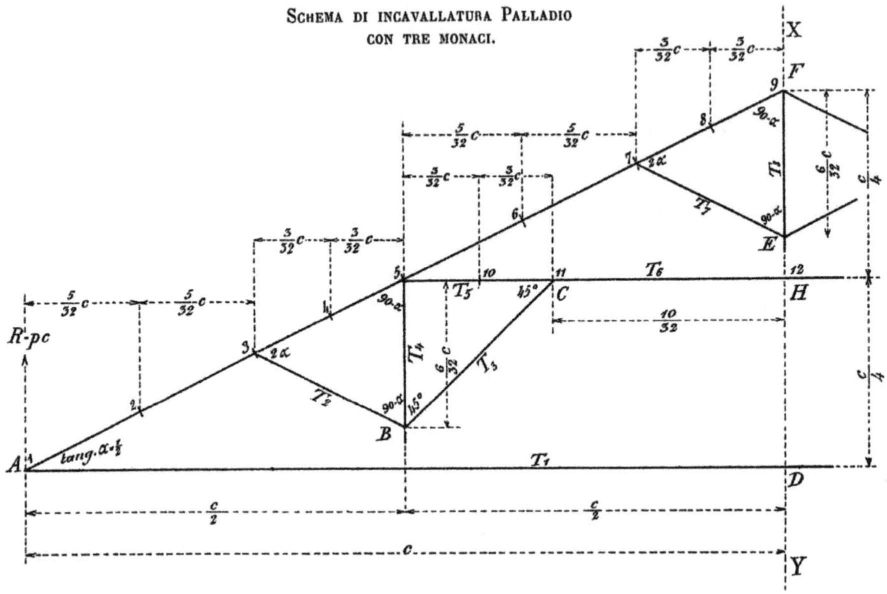

Fig. 9 A Palladio truss with struts. To simplify the formulas and make them practically usable, Candellero (1890) fixes almost all the geometrical parameters of the truss

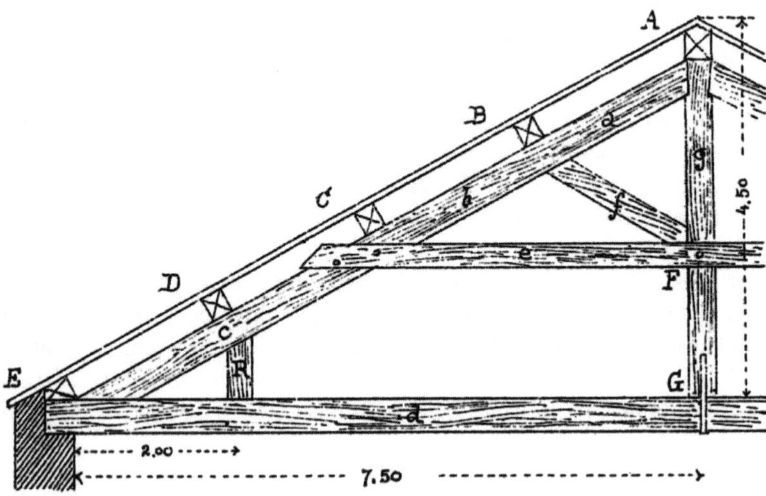

Fig. 10 Truss characterised by a very complex scheme. For its analysis it was necessary to make many simplifications: the small post on the left had to be ignored and the rafter was thus analysed as it was subdivided in three parts of different length (*A–B*, *B–C* and *C–E*), loaded only at joints, neglecting the bending moment due to the load of purlin *D*. (Anonymous 1889)

7 Further Developments in Steel Industry and Marginalisation of Timber Trusses

Since the 1870s the use of new systems for the production of melted steel gradually spread causing a progressive drop in the price of steel (Pedrocco 2004b, p. 290): converters could produce steel from cast-iron (Bessemer converter was conceived in 1855 and Thomas converter in 1878); Martin-Siemens furnaces (devised in 1865) could obtain steel from scraps. In the last decade of the nineteenth century a further step in this direction was taken by the invention of electric furnaces (e.g. the Stassano electric furnace conceived in 1896) that could promote Italian energy independence from imported fuels thanks to the developments of hydroelectric power stations (Turri and Zamperini 2013). Italian steel production grew enormously rising from 3600 t in 1879 to 1,269,500 in 1916 (Degli Esposti 2004, p. 336). This increased the cost effectiveness of steel trusses for minor spans, progressively reducing the field of application of timber and timber–steel trusses. Therefore engineers gradually decreased their interest in timber trusses, concentrating their technical researches at first on steel and then on reinforced concrete structures.

8 Applications of Reinforced Concrete to Roof Trusses

Reinforced concrete started to significantly spread in Italy in the last years of the nineteenth century and at first was used almost exclusively in floors (Zamperini 2013) and later also in foundation slabs or piles and in frame structures (Turri et al. 2009b). Long after the use of reinforced concrete was well-established, trusses made of this material were slow to spread due to the irrationality in using such a material in members subject to tension; nevertheless since the 1920s reinforced concrete trusses have been quite widespread in medium span roofs (15 ÷ 20 m).

Being made with rigid, moment-resistant joints, these trusses are statically indeterminate and have many hyperstatic unknown bending moments (three for each triangle forming the truss, Ciappi (1940, p. 170)), so they were very difficult to be analysed with exact methods.

Therefore engineers refocused their attention on many simplified methods (e.g. the ones originally elaborated by Heinrich Manderla and Emil Winkler in 1879–1882, Heinrich Müller-Breslau, and Otto Mohr in 1892–1893)[7] to calculate secondary stresses (i.e. the ones due to the stiffness of joints, eccentricity of joints with respect to the axes of rods, span loads applied on truss members, etc.) originally elaborated for rigidly connected steel structures, but very rarely employed. These methods required first of all the calculation of primary stresses as if the truss was statically determinate, then the evaluation of stresses due to span loads on members and to rotational stiffness of the joints (Santarella 1934; Ciappi 1940, pp. 169–338).

[7] An early discussion on the theories for the calculation of secondary stresses was done by Grimm (1908). Kurrer (2008, pp. 587–589) also explains how these theories finally led to formulation of the displacement method.

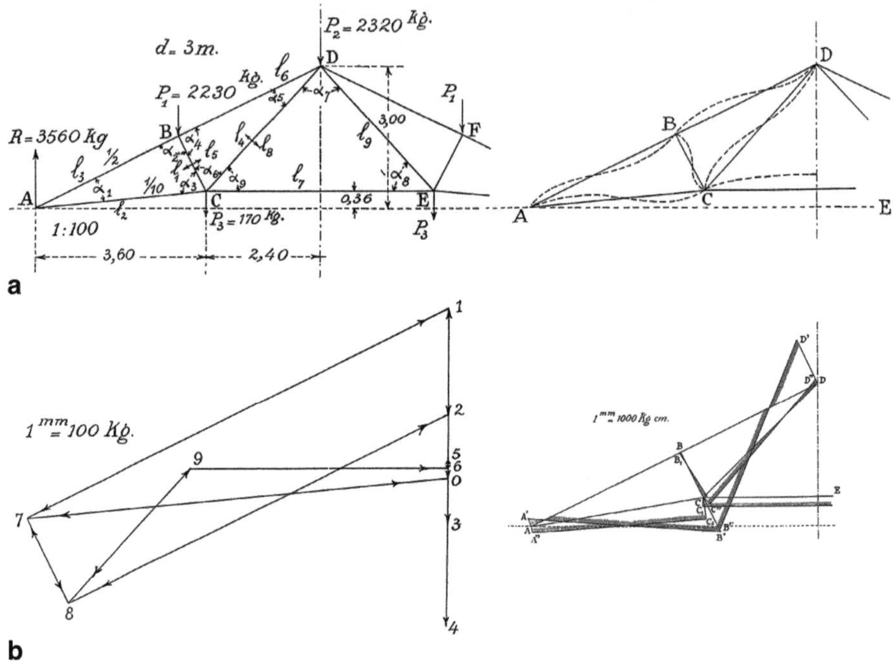

Fig. 11 Schemes for the analysis of principal and secondary stresses in a reinforced concrete Polonceau truss. (Ciappi 1940)

Even if they are quite simplified, these methods were still difficult and laborious to be applied, and their use was never extended to the design of timber structures. Indeed in the case of wooden structures engineers believed that the cost of the complex structural calculations required for an exact analysis were higher than the savings resulting from them. This was due to two factors: the hypothetical material saving due to a better analysis did not lead to large cost savings because of the small cost of the material itself; wooden beams needed for the composition of trusses were usually marketed with established size (typically differing about two centimetres between each other), therefore timber members had always to be oversized by choosing the larger beam (Fig. 11).

9 Timber Processing and Production of Metal Fasteners

During the nineteenth century a series of innovations in timber processing and production of metal fasteners laid the foundations for a radical innovation in the timber structures to be achieved at the beginning of the twentieth century.

Although Venetian sawmills for timber driven by waterwheels date back at least to the fifteenth century, their placing was bound to the presence of a hydraulic head, and their usage was limited to restricted areas (Zamperini 2012, pp. 731–732).

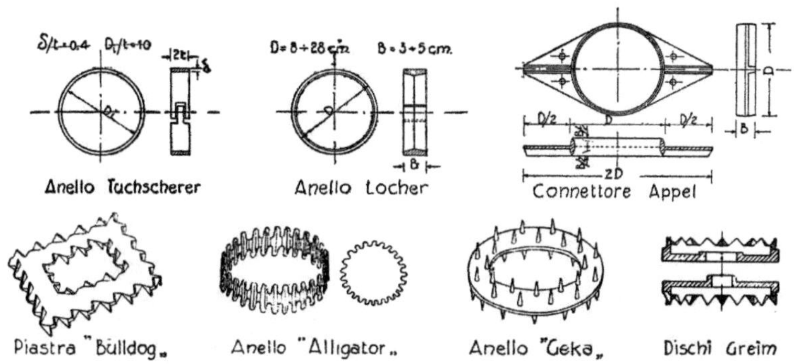

Fig. 12 New types of steel connectors for timber. (Arcangeli 1949)

Only in the late eighteenth century did the invention of steam engine make a huge source of mechanical energy available also in places where natural sources of kinetic energy were absent. This made it possible to locate manufacturing industries even in the large urban centres where goods were conveyed, commercialised, and used; therefore the invention of new types of mechanically powered saws was stimulated. Between the end of the eighteenth and the beginning of the nineteenth century, the circular and the band saw were invented, but their diffusion in Italy was very slow and their application to the woodworking of timber to be employed in carpentry dates back only to the end of the nineteenth century, thanks to a general industrial progress and to the diffusion of electricity generated by hydroelectric plants (Zamperini 2012, pp. 734–737, 739).

Many improvements also occurred in wrought iron and steel processing. The development of the modern rolling techniques, the enhancement of the methods of iron wires and rods drawing, the invention and spread of the screw-cutting lathe, and many other innovations favoured the spread of inexpensive nail with cylindrical shank (Frascio 1929; Wells 1998), of screws (Ferrero 1928; Dickinson 1941) and bolts, and allowed the invention of spiked plates and rings to be used in timber connections (Giordano 1947, pp. 220–237; Arcangeli 1949, pp. 53–56; Fig. 12).

10 New Types of Trusses

After the First World War and during the Fascist autarkic period, the lack of steel imports made its price grow and, consequently, its use decrease. Therefore, while traditional timber trusses continued to be used, engineers began to study and design also a new type of timber structures made mainly of boards and with nailed (Fig. 14), screwed or bolted joints (Fig. 13) whose realisation was possible and economically advantageous due to economic and technological changes described in the previous paragraph.

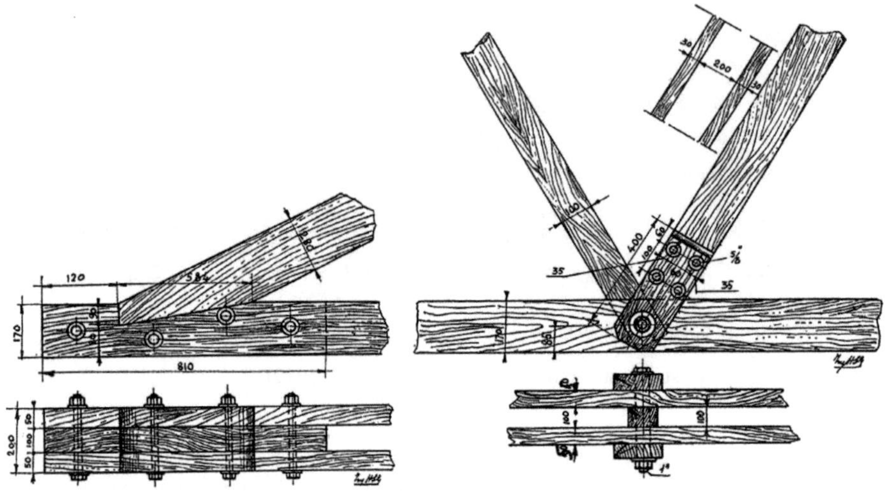

Fig. 13 Details of bolted connections for timber Polonceau truss. (Harley di San Giorgio 1921)

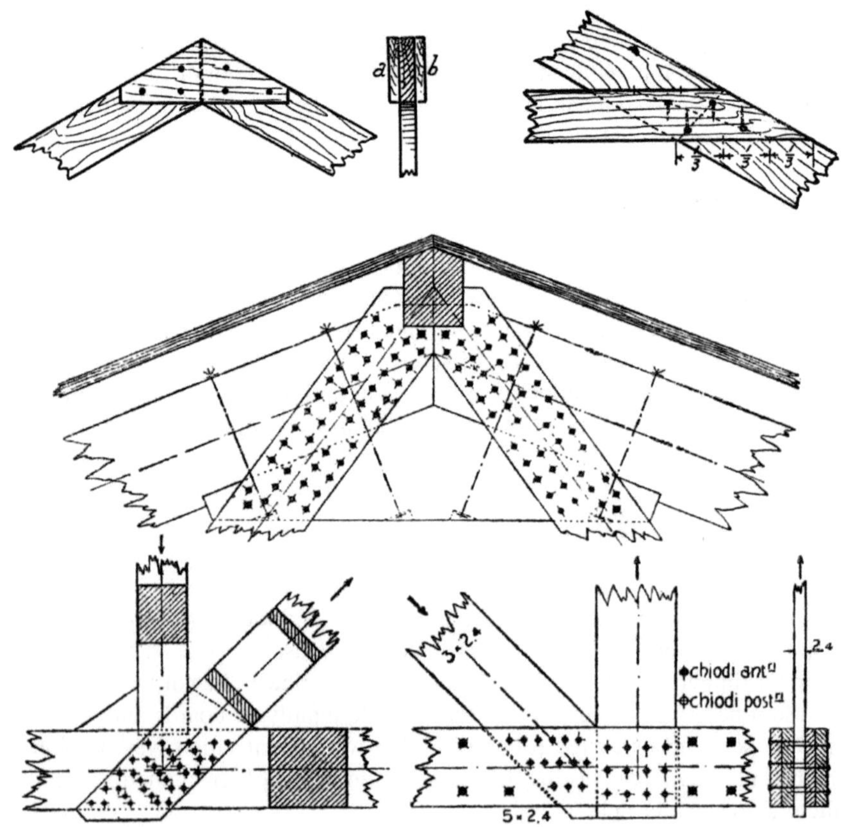

Fig. 14 Details of nailed connection for the joints of timber trusses. On the *top*: truss made only of boards (Rosci 1934). On the *bottom*: truss made of squared logs and boards. (Arcangeli 1949)

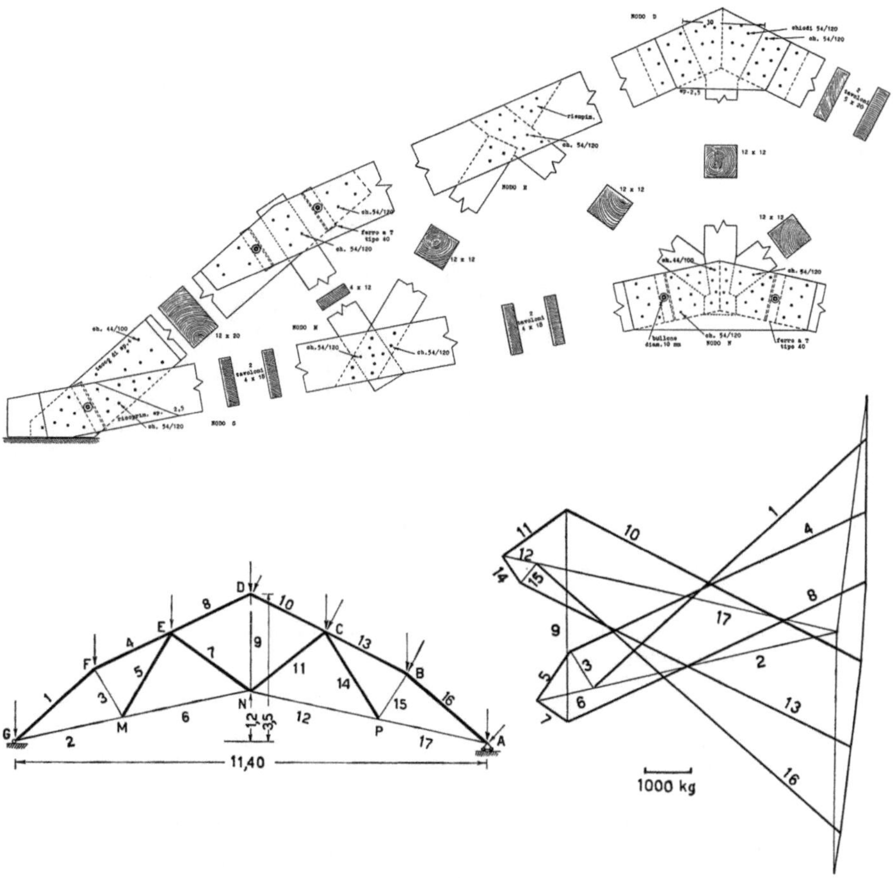

Fig. 15 Timber truss made of squared logs and nailed boards loaded only at joints; structural scheme and analysis made with the Cremona diagram. (Giordano 1964)

Time was then ripe to face up to a radical reversal: timber trusses whose empirical tradition provided an inspiration for the first metallic structures more than one century before, were culturally colonised by the analytical mind of the new generation of engineers, educated in the mindsets of the structural mechanics specifically elaborated for steel structures.

Engineers thus abandoned traditional carpentry schemes to seek solutions easier to be analysed. Thanks to slenderness of members and to the simplicity of nailed, screwed or bolted joints, they could design these new structures according to the schemes of steel constructions, that required less skilled craftsmanship for carpentry joints (Arcangeli 1949, pp. 72). Many timber structures were realised to cover great exhibition halls, or industrial warehouses and plants reproducing metallic three-hinged arches schemes (Gradi 2008, pp. 12–14). In roofs new type of timber trusses were used following the models of steel ones with loads concentrated in joints, thus allowing a proper use of the theory of statically-determinate trusses (Giordano 1947; Arcangeli 1949; Figs. 15 and 16)

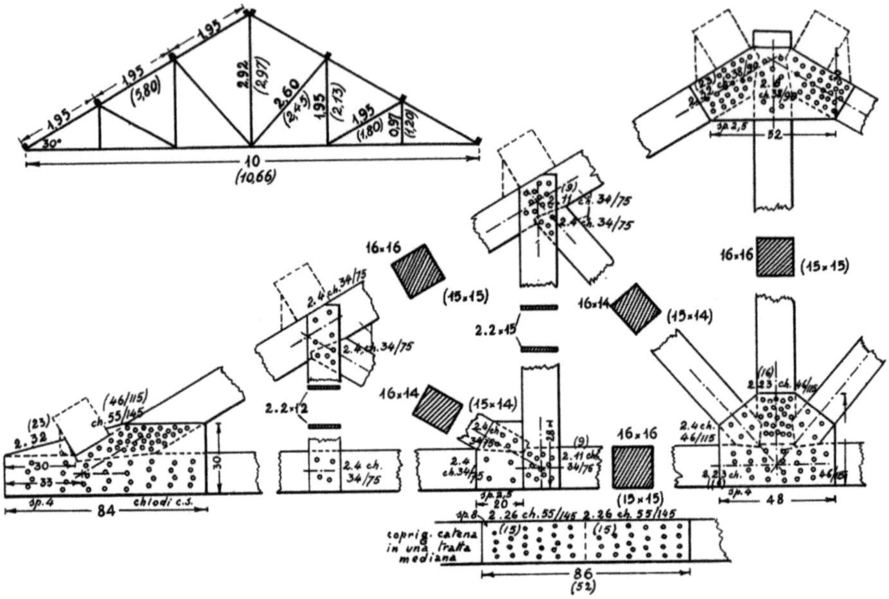

Fig. 16 Timber truss that reproduces the structural schemes of Howe truss, originally used for steel constructions. (Giordano 1947)

11 Conclusion

The events here discussed show an important and almost forgotten chapter of the history of structural mechanics applied to constructions: after the Second World War, the growing analytical character of structural mechanics and the approximate nature of the methods used for assessment of timber trusses made them progressively fall into oblivion, and even the most recent historical essays in this field completely neglect this topic. The rediscovery of these events is therefore a key step in the historiographic interpretation of that branch of engineering which is halfway between theoretical and applied structural mechanics.

Furthermore, through particular events, the technical and cultural path of the Italian timber trusses discussed in this paper outlines and makes it easy to understand a wider and more important chapter of the history of building techniques that linger on in the balance between the traditions of empirical craftsmanship and architecture, and the new positivist method of the emerging engineering.

Although structural analysis was initially required to schematise the complex and traditional reality of timber trusses, at the end of the examined period it eventually came to renew trusses and to shape them according to its own capability: the technological and cultural nature of timber trusses radically changed in the course of only one century and a half.

When studying existing buildings in order to take care of them for preservation purposes, we must be aware that the full comprehension of these cultural and technological changes plays a fundamental role in understanding of the historical and documentary importance of consistency in building materials, and thus in the definition of the most appropriate preservation strategies.

However, even if the history analysed here has a great practical importance, we should not run the risk of reducing the understanding of the real nature of an existing structure to reply to a few basic questions:

- Has it been empirically built by a carpenter or designed by an engineer?
- Is it one of the last backward examples of a disappearing technological culture or a pioneering construction?
- …

Indeed we must never lose sight of the fact that none of our evaluations can ever be uncritically neutral, but they always contain a value judgment; however, the judgment of historical value should not become a tool to discern between what is to be preserved and what can be destroyed.

Moreover, the supposed knowledge of historical events that belong to the early construction of a given structure should not give rise to the illusion of being able to turn back time. The deletion of changes that time and man have made on a structure threatens to deprive future researchers and practitioners of the chance to repeat and—if necessary—correct our interpretation.

Finally we can say that our studies, judgements, and interpretations are necessary and inescapable. Yet the results we will draw from them must be proposed to posterity with cultural action and not imposed by an intervention directly amending the building, an operation that must instead have preservation as its main purpose.

References

Anonymous (1889) Consultazioni tecniche. L'architettura pratica 2

Arcangeli A (1949) Tecnica delle costruzioni. Vol. 2: Le costruzioni in legno. Hoepli, Milano

Bennett L (2004) From craft to science: American Timber Bridges, 1790–1840. APT Bulletin 4:13–19

Benvenuto E (1981) La scienza delle costruzioni e il suo sviluppo storico. Sansoni, Firenze

Bresse JAC (1859) Cours de mécanique appliquée. Première partie. Résistence des matériaux et stabilité des costructions. Mallet–Bachelier, Paris

Breymann GA (1887) Trattato generale di costruzioni civili. Costruzioni in legno. Vallardi, Milano

Candellero C (1890) Travi armate e incavallature considerate come sistemi elastici. Camilla e Bertolero, Torino

Capecchi D, Ruta G (2010) La scienza delle costruzioni in Italia nell'Ottocento. Springer, Milano

Castellazzi G (1863) Relazione a corredo di progetti di massima per cavallerizze coperte. Giornale del Genio Militare 1:71–86

Castigliano A (1873) Intorno ai sistemi elastici. Bona, Torino

Cavalieri San Bertolo N (1826) Istituzioni di architettura statica e idraulica. Cardinali e Frulli, Bologna

Caveglia C (1876) Elementi di statica grafica e loro applicazione allo studio della stabilità delle costruzioni. UTET, Torino

Caveglia C (1878) Corso di costruzioni civili e militari, vol 3. UTET, Torino

Ceradini C (1879) Dell'equilibrio e della stabilità delle capriate. Il Politecnico-Giornale dell'ingegnere architetto civile ed industriale 4–7:8–12
Chiarle A (1891) Le nuove formole per il calcolo esatto delle armature dedotte dalla teoria matematica dell'elasticità. Rivista di artiglieria e genio 2:28–76
Ciappi A (1940) Corso di scienza delle costruzioni. Parte quarta—sezione terza. La grossa armatura dei tetti. Cremonese, Roma
Cremona L (1872) Le figure reciproche nella statica grafica. Bernardoni, Milano
Cullmann C (1866) Die Graphische Statik. Meyer & Zeller, Zürich
Degli Esposti F (2004) Le vicende della siderurgia italiana tra Unità e Seconda Guerra Mondiale. In: Nicodemi W (ed) La civiltà del ferro dalla preistoria al III millennio. Olivares, Milano, pp 323–352
Dickinson HW (1941) Origin and manufacture of wood screws. Trans Newcom Soc 80:79–89
Fara A (2011) Luigi Federico Menabrea (1809–1896). Scienza, ingegneria e architettura militare dal Regno di Sardegna al Regno d'Italia. Olschki, Firenze
Ferrero M (1928) Viti—Dadi. Utensili a filettare. Lattes, Torino
Frascio A (1929) La lavorazione del filo metallico e la fabbricazione di tutti i prodotti che ne derivano. Hoepli, Milano
Giordano G (1947) La moderna tecnica delle costruzioni in legno. Hoepli, Milano
Giordano G (1964) La moderna tecnica delle costruzioni in legno, 3rd edn. Hoepli, Milano
Gradi I (2008) Architettura lignea 1920–1940 in Italia e in Germania (parts 1–2). Bollettino Ingegneri 1–2:2–24
Grimm CR (1908) Secondary stresses in bridge trusses. Wiley, New York
Guardini M (2004) L'intreccio nella teoria delle reticolari. Tecnologos, Cavriana
Harley di San Giorgio O (1921) Manuale di carpenteria. Lattes, Torino
Kurrer KE (2008) The history of the theory of structures: from arch analysis to computational mechanics. Ernst & Sohn, Berlin
Laner F (1998) La concezione delle strutture lignee nel sec. XIX. In: Ventrice P (ed) Tecnica e tecnologia nell'architettura dell'Ottocento. Istituto Veneto di Scienza Lettere ed Arti, Venezia, pp 273–292
Maxwell JC (1864) On reciprocal figures and diagrams of forces. Philosophical Magazine 27:250–261
Menabrea LF (1858) Nouveau principe sur la distribution des tensions dans les systèmes élastiques. Comptes rendus hebdomadaires des séances de l'Académie des sciences 46:1056–1058
Morin A (1853) Leçons de mécanique pratique. Resistence des materiaux. Hachette, Paris
Navier CLMH (1826) Résumé des leçons données à l'École des Ponts et Chaussées sur l'application de la mécanique à l'établissement des constructions et des machines. Firmin Didot, Paris
Pedrocco G (2004a) Siderurgia e Rivoluzione Industriale nell'Inghilterra del XVIII secolo. In: Nicodemi W (ed) La civiltà del ferro dalla preistoria al III millennio. Olivares, Milano, pp 255–270
Pedrocco G (2004b) La siderurgia francese nell'Ottocento tra tradizione e innovazione. In: Nicodemi W (ed) La civiltà del ferro dalla preistoria al III millennio. Olivares, Milano, pp 271–298
Polonceau C (1840) Notice sur nouveau système de charpente en bois et fer. Revue Générale de l'Architecture et des Travaux Publics 1:27–32
Ritter A (1863) Elementare Theorie und Berechnung eiserner Dach- und Brücken-Constructionen. Rümpler, Hannover
Rondelet JB (1810) Traité théorique et pratique de l'art de bâtir. Paris
Rosci L (1934) Calcolo dell'armatura dei tetti con incavallature di tavole. Lavagnolo, Torino
Sachero C (1864) Stabilità delle armature dei tetti. Giornale del Genio Militare 9:193–278
Santarella L (1934) Gli sforzi secondari nelle travi a traliccio in cemento armato. Hoepli, Milano
Schwedler JW (1851) Theorie der Bruckenbalkensysteme. Zeitschrift fur Bauwesen 1:114, 162, 265
Timoshenko SP (1953) History of strength of materials. McGraw-Hill, New York
Turri F, Zamperini E (2013) Sustainable valorisation of the network of hydropower plants of Italian Alpine valleys. In: Hàjek P et al (eds) CESB2013. Sustainable building and refurbishment for next generations. Grada Publishing for CTU, Praha

Turri F, Zamperini E, Cappelletti V (2009a) Military contribution to building technical evolution in Italy (1860–1940). In: Kurrer K et al (eds) Third international congress on construction history, vol 3, pp 1427–1434

Turri F, Zamperini E, Cappelletti V (2009b) The evolution of construction techniques in Italian barracks after national unification. In: Kurrer K et al (eds) Third international congress on construction history, vol 3, pp 1435–1442

Wells T (1998) Nail chronology: the use of technologically derived features. Historical Archaeology 2:78–99

Whipple S (1847) A work on bridge building. Curtiss, Utica

Wünnerman A, Holzer SM, Voigts C (2012) Wooden "Italian" wide-span roofs of German 19th century theatre buildings. In: Carvais R et al (eds) Nuts & bolts of construction history. Culture, technology and society, vol 3. Picard, Paris, pp 309–316

Yeomans D (1986) Inigo Jones's roof structures. Architectural History 29:85–101

Yeomans DT (1992) The trussed roof: its history and development. Scolar Press, Aldershot

Zamperini E (2012) Evoluzione delle tecniche di taglio e segagione dei legnami da opera tra inizio Ottocento e metà Novecento In: D'Agostino S (ed) Storia dell'Ingegneria (4 Convegno Nazionale), pp 729–740

Zamperini E (2013) At the origins of modern construction: early applications of reinforced concrete in floors designed by Italian corps of engineers. In: Avramidou N et al (eds) Importance of place. CICOPBH, Sarajevo, pp 327–340

A Point of View: Sciences, Societies, Cultures and Their Evolutions

Bertrand Bocquet

Abstract Scientific research is now a large international activity and has become very professional with many connections among researchers, but also has formed an alliance with the world's economic sphere. The benefits of research are mainly oriented towards economic competition. More recently, a local to global movement has been born with a goal to promote a *bottom-up* approach to research questions. Participative researches could be the new paradigm of the *knowledge society* promoted by the European Community and the main part of the *Science in society* scientific domain. Many initiatives have been developed at different levels of interaction with the public and they are in a position to continue making significant contributions to the democratization of sciences.

Keywords Science in Society · Participative research · Collective research · Citizen scientist · Commons · Responsibility in research and innovation

1 An Introduction

The relationships between sciences and society are, today, subject to some serious contradictions. A strong rejection of sciences, extending to blind technophilia, has been observed, and the voices of civil society actors have become audible. Some recent technoscientific overflowing, combined with the increasing need for participative democracy, are the main reasons (Sclove 1995). It is urgent that we question the linkage of the sciences, not only between themselves but also with such diverse entities as *profane* knowledge and *Civil Society Organizations* (CSOs). The difficulty in building cooperative support for a wide variety of scientific researches, at the same time interdisciplinary and open to other highly influential partners, is characterized by numerous gaps in understanding and perceptions of differences in goals. Among them are: very few reflections on the choice of disciplines to be involved for global research, failure of methodology for interdisciplinary dialogue,

B. Bocquet (✉)
Sciences, Societies, Cultures in their evolutions, Lille 1 University Science and Technology, Villeneuve d'Ascq Cedex, France
e-mail: Bertrand.Bocquet@univ-lille1.fr

R. Pisano (ed.), *A Bridge between Conceptual Frameworks,* History of Mechanism and Machine Science 27, DOI 10.1007/978-94-017-9645-3_28

little convincing relevance of evaluations making difficult the capitalization of best practices, lack of tools to codify and share knowledge, low recognition of experimental facilities, absence of ethical guidelines, etc. The disciplinary approach to the sciences has, from the beginning of scientific awareness, allowed a considerable inflation of acquired knowledge and, in modern times, the technologies of information and communication have made accessibility an ever more powerful reality. However, it seems important to continue to better synchronize scientific research with the societal and environmental stakes inherent in this acquired and accessible knowledge. A question arises: *How can we imagine a link between the field of the sciences, which are disciplinary, and that of our multiple-faceted society where problems are strongly transdisciplinary*? A scientific mediation appears fundamental to answering the societal stakes.

2 What Possible Perspectives Exist for the Sciences in Society?

In order to re-open debate on scientific and technological questions, the *knowledge society*, as defined by the *European Union,* could open a bottom–up perspective on the sciences in society, replacing the top–down approach known commonly as the *knowledge economy* (Wynne 1991). This potential extension of the field of scientific research answers the question of connections with the sciences of information and communication. Technologies are at a very high performance level today but real communication between them and the public at large remains weak, especially considering that mediation of scientific knowledge is central. *What about scientific mediation for whom and by whom?* The institutionalized and recognized scientific mediations are the ones of the scientific and technological cultural centers and scientific journalism. They constitute the emerged part from an iceberg in the making. The recent phenomenon of *participative researches*, which renews the need for scientific mediation of knowledge, still remains too little known and too parsimoniously funded. Of course, it is necessary to know what sort of participative researches are in the lead. The answer to the implied question is thorough analyses that will lead to a real understanding of research activity and the best practices of such researches. Beyond these analyses, a theoretical approach will be able to develop promotion of participative researches while avoiding exploitation of the CSOs. For example, some research groups use citizens simply as labor for gathering data. The group then does not associate the results to the hypotheses, concepts or methodologies underlying the data. It is the same whether the results are tightly held or are published. Moreover, with the development of data mining, some researches are conducted without the knowledge of the participants, e.g., gathered from their intelligent phones. (Irwin 1995; Stilgoe 2009). Other projects, in the case of more traditional researches by companies, aim to determine the acceptability of selected technologies. Researchers in humanities or social sciences are more and more involved in such programs for questioning people on their using or testing some products. These practices are new but they do not give automatically a scientific caution for complex researches, generally controversial. In these cases, we speak about embedded scientists.

3 A New Scientific Dynamism

A research challenge today is to analyze participative researches, which have different significances. Recently in France, the association for a "Fondation Sciences Citoyennes" has recorded 200 experiences of scientific works realized within civil society Storup (2013). Perhaps, in a short analysis, we can distinguish three different realities, which can be evaluated by public engagement: participative, cooperative or citizen researches.

We use the term *participative research* for the well-known phenomenon of the amateur scientist that has existed for a long time (Gall et al. 2009; Joss and Durant 1995). Typically, astronomy or biodiversity studies benefit from the data collected by citizens interested in these scientific questions. They are trained by a scientist to properly collect thousands of data. There has been a deep mutation of this process by the increasing use of information and communication technologies. Many people are involved in these researches throughout many small associations. I think that we can speak of a truly active scientific mediation. Academic researchers are the main users and managers of these kind of studies.

More recently, a new type of interaction between scientist and citizens emerged. We call this *cooperative research*. In this case, the research program is co-constructed between a CSO and an academic research laboratory. These programs exist in France at a regional level. They have increased slowly since 2005. We have today three different regions on twenty two where these well funded programs exist. The first of them, created in 2005 in the Ile de France region, is the *Institution Citizen Program for Research and Innovation* (PICRI). It was followed in 2009 by the *Social appropriation of sciences* (ASOSc) in the Bretagne region and in 2011 by the *Scientists-Citizens Program* in the Nord-Pas de Calais region. The research subjects are very different, varying from humanities to more technological realizations. It is interesting to note that a number of such submissions are made each year, showing a real interest and a social demand for research. In the same way, the ministry of sustainable development has created a national program called *Exchange and Projects Network on the Piloting of Research and Expertise* funded in 2009.

The last type is the *citizen research* where the research promoters are the CSOs. We have now non-governmental organizations, which have sophisticated measurement means. They, for example, realize counter-measurements for evaluating the quality of an expert's reports but also, they can build a research program in collaboration with an academic department.

At the international level, we observe the same dynamism. The United Nations Educational, Scientific and Cultural Organization (UNESCO) has created a chair at the University of Victoria (Canada) in 2012 entitled Community-based Research and Social Responsibility in Higher Education. This chair is co-animated with Participation Research In Asia (PRIA) located in India.

Special mention must be made of the establishment of science shops. I consider these shops to be tools for developing a scientific *mediation in situation* between actors of civil society and the world of research, with transdisciplinary questions on societal issues, health, environment, energy, etc. Science shops are organizations that provide access to academic knowledge and research at a low cost (Hellemans 2001; Mulder et al. 2006). This structure recognizes the social demand for research, evaluates the research level required (bibliographic study up to experiments) and establishes connections with research units. Three conditions are required for accepting demands. Firstly, the customer organization has no monetary object and no financial means. Secondly, the research results must be published in peer-review or open access journals to insure common availability. Thirdly, the customers must have the capacity to use the results for accomplishing their social objective. There are many different models around the world. In the Netherlands, most shops are found inside universities. They can be attached to departments and known as the physics shop or the biology shop. In Germany, most of these structures are outside universities and are funded directly by foundations or by the government and the European Union. The *Science in Society* area, a part of the European Research Area ERA), is actively promoting a European network of science shops. Note that there already exists an international network named *Living Knowledge*. It can re-evaluate what we expect from science, and in the future, this structured network could influence science itself.

A large part of society challenges knowledge as an *ivory tower*, and consequently also takes little notice of the condition of research realizations. This is true for pure academic researches but also for research projects associated with companies. Citizens in general claim their interest in opening of research practices to their needs. Everybody knows today that scientific research is a key to a better understanding of the global world, but under some conditions. Participative researches underline the contradiction between the dominant questions on which researchers work and the true questions asked by the people. They contribute to the empowerment of CSOs about scientific and technological questions. The logical follow-up is that they can question the starting hypotheses, the methodology employed during the researches and more globally the epistemology. However, do not be afraid. Researchers could be interested by such approaches. The hyper specialization of scientific disciplines no longer allows narrow interests to determine research directions. Moreover, the obligation to get results (publish or perish) and a feel of a *taylorization* of scientific work could make participative researches attractive for researchers.

4 On the Horizon 2020 Initiative *Science in Society*

As I have mentioned before, the program *Science in Society* belongs to the European Research Area. It has existed since 2001 and some evolutions have occurred. A new instrument called *Sciences with and for society* has been defined for *Horizon 2020*. The goal is to develop new original ways for increasing connections between sciences and society and for demonstrating their innovative character as well as their economic, social and environmental sustainability. The principal interests for this instrument are how to fashion a more attractive appearance of science or how to raise the appetite of society for innovation. It hopes to work together with all societal actors during the whole research and innovation process. This approach is called Responsible Research and Innovation. These topics seem to me the true challenges. The technoscientific solutions today are very complex, with many socio-economic, political, ethical and environmental externalities. In this way, we must also consider that the CSOs and stakeholders could participate in the co-constructions of research programs, and not only on their technological acceptability. The true innovative process for research and innovation in this domain will be mixed academic researches with associative members. Four calls are proposed for 2014–2015, but the more interesting call is entitled *Integrating society in science and innovation.* Clearly, the European Union wishes to improve the integration of society in science and innovation.

5 What Conditions are Necessary for Democratization of Sciences?

All these issues put the question of relationships among all disciplinary researches, which have today multiple subdivisions. The specialization of researchers, their professionalization, but also the complexity of their experiments and measurement tools, aim at autonomous sub-disciplines. If this way, defined by Descartes and called reductionism, has led to an extraordinary sum of knowledge, this one today is dramatically fragmented. If disciplinarization remains necessary, the inverse way, defined also by Descartes (understanding the whole by understanding each part) must be seriously investigated today. *What should be meant by seriously?* It signifies that more means must be affected to transdisciplinarity by creating new types of laboratories in each university where researchers come from different scientific horizons. These new laboratories must be connected with the civil society. Each transdisciplinary laboratory will have its own methodology that depends on the social and natural environment. A first consequence is that we are going to reconnect science with society, but also it will increase the *biodiversity* of research subjects (Farkas 2002).

Better national and European supports could make more fertile the hillside of the participative researches. They could gather a large number of different initiatives: collaboration with amateur associations for the collection of data; cooperation for building joint research projects; eventual citizen researches where some associations employ researchers (doctors, engineers) or can fund academic groups; etc. Nevertheless, these experiments, however rich, are still only experimental and leave the majority of profane actors outside of the research system. The emergence from the CSOs of their enhanced knowledge is a chance for the academic world, which should welcome such a chance, not to make it an ultimate research subject, but to find a balanced relationship with the citizens. This opening way could allow us to be more synchronous with reality and to more fully apprehend the stakes of a planet that may in the near future be occupied by about 10 billion human beings.

6 Conclusion

We have seen that participative researches are a real dynamical movement coming from society. They define the relationship between the two spheres: scientific research and society. We are about to know and/or to understand, to verify and/or to answer, but also to put in relation the projects of society and required knowledge. It is quite different with research judged by one's peers or associated with companies. We have seen also that these researches are more and more supported by institutional organizations at regional, national and international levels. In consequence, it is fundamental that academic institutions where research is done (universities, colleges, engineering schools), take a part in this movement. Be sure that real research efforts in that complex lead to forging new forms of knowledge and new methods to get into the sciences in society. For better or for worse, science is now being shared with more and more actors from civil society organizations.

Is it necessary to be afraid to democratize the research?

References

Farkas NE (2002) Bread, cheese, and expertise: Dutch science shops and democratic institutions. Ph.D. Dissertation, Rensselaer Polytechnic Institute. Deprtament of Science and Technology Studies, NY. http://opac.lib.rpi.edu/search/X?SEARCH=Farkas&m=t&SORT=D&searchscope=6 Accessed 22 Dec 2014

Gall E, Millot G, Neubauer C (2009) Participation of civil society organisations in research. Civil organisations, actor in the European system of research and innovation. STACS. http://www.livingknowledge.org/livingknowledge/wp-content/uploads/2011/12/STACS_Final_Report-Partic.research.pdf Accessed 22 Dec 2014

Hellemans A (2001) Science shops provide non-profit alternative. Nature 412(6842):4–5

Irwin A (1995) Citizen science—a study of people, expertise and sustainable development. Routledge, New-York

Joss S, Durant J (eds) (1995) Public participation in science: the role of consensus conferences in Europe. Science Museum Press, London

Mulder HA, Jorgensen MS, Pricope L, Steinhaus N, Valentin A (2006) Science shops as science-society interfaces. Interfaces between science and society 1(48):278–296

Sclove R (1995) Democracy and technology. Guilford Press, New York

Stilgoe J (2009) Citizen scientists: reconnecting science with civil society. Demos, London

Storup B (2013) La recherche participative comme mode de production de savoirs. Un état des lieux des pratiques en France. http://sciencescitoyennes.org/recherche-participative-ou-en-est-on/, in French language. Accessed 22 Dec 2014

Wynne B (1991) Knowledges in context. Science, Technology & Human Values 16(1):111–121

Index

R. Pisano (ed.), *A Bridge between Conceptual Frameworks*, History of Mechanism and Machine Science 27, DOI 10.1007/978-94-017-9645-3

G

M

U